T0306079

Numerical Relativity
Solving Einstein's Equations on the Computer

Aimed at students and researchers entering the field, this pedagogical introduction to numerical relativity will also interest scientists seeking a broad survey of its challenges and achievements. Assuming only a basic knowledge of classical general relativity, this textbook develops the mathematical formalism from first principles, then highlights some of the pioneering simulations involving black holes and neutron stars, gravitational collapse and gravitational waves.

The book contains 300 exercises to help readers master new material as it is presented. Numerous illustrations, many in color, assist in visualizing new geometric concepts and highlighting the results of computer simulations. Summary boxes encapsulate some of the most important results for quick reference. Applications covered include calculations of coalescing binary black holes and binary neutron stars, rotating stars, colliding star clusters, gravitational and magnetorotational collapse, critical phenomena, the generation of gravitational waves, and other topics of current physical and astrophysical significance.

Thomas W. Baumgarte is a Professor of Physics at Bowdoin College and an Adjunct Professor of Physics at the University of Illinois at Urbana-Champaign. He received his Diploma (1993) and Doctorate (1995) from Ludwig-Maximilians-Universität, München, and held postdoctoral positions at Cornell University and the University of Illinois before joining the faculty at Bowdoin College. He is a recipient of a John Simon Guggenheim Memorial Foundation Fellowship. He has written over 70 research articles on a variety of topics in general relativity and relativistic astrophysics, including black holes and neutron stars, gravitational collapse, and more formal mathematical issues.

Stuart L. Shapiro is a Professor of Physics and Astronomy at the University of Illinois at Urbana-Champaign. He received his A.B from Harvard (1969) and his Ph.D. from Princeton (1973). He has published over 340 research articles spanning many topics in general relativity and theoretical astrophysics and coauthored the widely used textbook *Black Holes, White Dwarfs and Neutron Stars: The Physics of Compact Objects* (John Wiley, 1983). In addition to numerical relativity, Shapiro has worked on the physics and astrophysics of black holes and neutron stars, relativistic hydrodynamics, magnetohydrodynamics and stellar dynamics, and the generation of gravitational waves. He is a recipient of an IBM Supercomputing Award, a Forefronts of Large-Scale Computation Award, an Alfred P. Sloan Research Fellowship, a John Simon Guggenheim Memorial Foundation Fellowship, and several teaching citations. He has served on the editorial boards of *The Astrophysical Journal Letters* and *Classical and Quantum Gravity*. He was elected Fellow of both the American Physical Society and Institute of Physics (UK).

Numerical Relativity

Solving Einstein's Equations on the Computer

THOMAS W. BAUMGARTE

Bowdoin College

AND

STUART L. SHAPIRO

University of Illinois at Urbana-Champaign

CAMBRIDGE
UNIVERSITY PRESS

University Printing House, Cambridge CB2 8BS, United Kingdom

One Liberty Plaza, 20th Floor, New York, NY 10006, USA

477 Williamstown Road, Port Melbourne, VIC 3207, Australia

314-321, 3rd Floor, Plot 3, Splendor Forum, Jasola District Centre, New Delhi - 110025, India

103 Penang Road, #05-06/07, Visioncrest Commercial, Singapore 238467

Cambridge University Press is part of the University of Cambridge.

It furthers the University's mission by disseminating knowledge in the pursuit of education, learning and research at the highest international levels of excellence.

www.cambridge.org
Information on this title: www.cambridge.org/9780521514071

First published 2010
Reprinted 2020

A catalogue record for this publication is available from the British Library

ISBN 978-0-521-51407-1 Hardback

Contents

Preface

What is numerical relativity?

General relativity – Einstein's theory of relativistic gravitation – is the cornerstone of modern cosmology, the physics of neutron stars and black holes, the generation of gravitational radiation, and countless other cosmic phenomena in which strong-field gravitation plays a dominant role. Yet the theory remains largely untested, except in the weak-field, slow-velocity regime. Moreover, solutions to Einstein's equations, except for a few idealized cases characterized by high degrees of symmetry, have not been obtained as yet for many of the important dynamical scenarios thought to occur in nature. With the advent of supercomputers, it is now possible to tackle these complicated equations numerically and explore these scenarios in detail. That is the main goal of numerical relativity, the art and science of developing computer algorithms to solve Einstein's equations for astrophysically realistic, high-velocity, strong-field systems.

Numerical relativity has become one of the most powerful probes of relativistic spacetimes. It is the tool that allows us to recreate cataclysmic cosmic phenomena that are otherwise inaccessible in the conventional laboratory – like gravitational collapse to black holes and neutron stars, the inspiral and coalescence of binary black holes and neutron stars, and the generation and propagation of gravitational waves, to name a few. Numerical relativity picks up where post-Newtonian theory and general relativistic perturbation theory leave off. It enables us to follow the full nonlinear growth of relativistic instabilities and determine the final fate of unstable systems. Numerical relativity can also be used to address fundamental properties of general relativity, like critical behavior and cosmic censorship, where analytic methods alone are not adequate. In fact, critical behavior in gravitational collapse is an example of a previously unknown phenomenon that was first discovered in numerical experiments, triggering a large number of analytical studies.

Building a numerical spacetime on the computer means solving equations. The equations that arise in numerical relativity are typically multidimensional, nonlinear, coupled partial differential equations in space and time. They have in common with other areas of computational physics, like fluid dynamics, magnetohydrodynamics, and aerodynamics, all of the usual problems associated with solving such nontrivial systems of equations. However, solving Einstein's equations poses some additional complications that are unique to general relativity. The first complication concerns the choice of *coordinates*. In general relativity, coordinates are merely labels that distinguish points in spacetime; by themselves coordinate intervals have no physical significance. To use coordinate intervals to determine physically measurable proper distances and proper times requires the spacetime metric, but the metric is known only *after* Einstein's equations have been solved. Moreover, as the numerical integrations that determine the metric proceed, the original, arbitrary choice of coordinates often turns out to be bad, because, for example, *singularities* appear in the equations. Encountering such singularities, be they physical or coordinate, results in some of the terms in Einstein's equations becoming infinite, potentially causing overflows in the computer output and premature termination of the numerical integration. It is not always easy to exploit successfully the gauge freedom inherent in general relativity – the ability to choose coordinates in an arbitrary way – and avoid these singularities in a numerical routine.

Treating *black holes* is one of the main goals of numerical relativity, but this poses another complication. The reason is that black holes contain physical spacetime singularities – regions where the gravitational tidal field, the matter density and the spacetime curvature all become infinite. Thus, when dealing with black holes, it is crucial to choose a computational technique that avoids encountering their interior spacetime singularities in the course of the simulation.

Another complication arises in the context of one of the most pressing goals of numerical relativity – the calculation of waveforms from promising astrophysical sources of *gravitational radiation*. Accomplishing this task is necessary in order to provide theoretical waveform templates both for ground-based and space-borne laser interferometers now being designed, constructed and placed into operation world-wide. These theoretical templates are essential for the identification and physical interpretation of gravitational wave sources. However, the gravitational wave components of the spacetime metric usually constitute small fractions of the smooth background metric. Moreover, to extract the waves from the background in a simulation requires that one probe the numerical spacetime in the far-field, or radiation, zone, which is typically at large distance from the strong-field central source. Yet it is the strong-field region that usually consumes most of the computational resources (e.g., spatial resolution) to guarantee accuracy. Furthermore, waiting for the wave to propagate to the far-field region usually takes nonnegligible integration time. Overcoming these difficulties to reliably measure the wave content thus requires that a numerical scheme successfully cope with the problem of vast *dynamic range* – the presence of disparate length and time scales – inherent in a numerical relativity simulation.

These are just some of the subtleties that must be confronted when doing numerical relativity. The payoff is the ability to build a spacetime on the computer that simulates the unfolding of some of the most exciting and exotic dynamical phenomena believed to occur in the physical Universe. Generating such a spacetime – "spacetime engineering" – then allows for an intimate probing of events and physical regimes that cannot be reproduced on Earth and may even be difficult to observe with telescopes. For those that can be detected, numerical relativity is a tool that can be called upon to interpret the observed features.

About this book

The purpose of this book is to provide a basic introduction to numerical relativity for nonexperts. It is a summary of the fundamental concepts as well as a broad survey of some of its most important applications. The book was conceived and written as a guide for readers who want to acquire a working knowledge of the subject, so that on mastery of the material, they can read and critique the scientific literature and begin active research in the field. Our book was born out of necessity: we needed a comprehensive guide to train our own students who want to pursue research with us in numerical relativity. Since we were unable to identify a suitable text to provide such an overview, we decided to write a book ourselves and fill the void.[1] As constructed, the book should also serve as a useful reference for researchers in the field of numerical relativity, as well as a primer for scientists in other areas desiring to get acquainted with our discipline and some of its most significant achievements.

Readers of our book are assumed to have a solid background in the basic theory of general relativity. There are several excellent textbooks that provide such a background. We are most familiar with *Gravitation* by C. W. Misner, K. S. Thorne and J. A. Wheeler (MTW) and will occasionally refer readers to this book for background material. We assume that our readers already have mathematical familiarity with tensors and differential geometry at the level of MTW, or a comparable graduate-level textbook on general relativity, and that they already have surveyed most of the physical applications covered in that book. This prerequisite roughly translates into a basic understanding of the geometric concepts and objects that enter the Einstein field equations, as well as the equations of motion for geodesics and relativistic fluids, the equations of hydrostatic equilibrium for spherical relativistic stars, the geometric and physical properties of black holes, the nature of gravitational radiation, and the concept of gravitational collapse. Beyond these standard topics, which we briefly review in Chapter 1, our book is essentially self-contained.

The question arises as to whether readers either with little or no acquaintance with general relativity can learn something about numerical relativity by reading this book. The question might be especially relevant for experts in other disciplines with related skills, such as computational physicists and astrophysicists, computer scientists, or mathematicians. The answer is that we don't know the answer, but we are eager to find out! It is a fact that when expressed in numerical terms, many of the equations arising in numerical relativity have a form similar to equations found in many other computational disciplines (e.g., fluid dynamics). It is also a fact that advances in the field of numerical relativity have benefited enormously from developments in other fields of computational physics and computer science. We thus hope that colleagues in these and other areas continue to venture into

---

[1] Apparently we were not alone in recognizing this void; well into our own writing another book on numerical relativity appeared, *Introduction to 3+1 Numerical Relativity*, by Alcubierre (2008b).

numerical relativity, and we look forward to learning from them to what extent our book can be of assistance.

To be useful as a textbook, our book contains 300 exercises scattered throughout the text. These exercises vary in scope and difficulty. They are included to assist students and instructors alike in calibrating the degree to which the material has been assimilated. The exercises comprise integral components of the main discussion in the book, so that is why they are inserted throughout the main body of text and not at the end of each chapter. The results of the exercises, and the equations derived therein, are often referred to in the book. We thus urge even casual readers who may not be interested in working through the exercises to peruse the problems and to make a mental note of what is being proven.

The book is designed as a general survey and a practical guide for learning how to use numerical relativity as a powerful tool for tackling diverse physical and astrophysical applications. Not surprisingly, the flavor of the book reflects our own backgrounds and interests. The mathematical presentation is not formal, but it is sound. We believe our overall approach is adequate for the main task of training students who seek to work in the field.

The organization of the book follows a systematic development. We begin in Chapter 1 with a *very* brief review (more of a reminder) of some elementary results in general relativity. In Chapter 2 we recast the equations of general relativity into a form suitable for solving an initial value problem in general relativity, i.e., a problem whereby we determine the future evolution of a spacetime, given a set of well-posed initial conditions at some initial instant of time. Specifically, we recast the familiar covariant, 4-dimensional form of the Einstein gravitational field equations into the equivalent $3 + 1$-dimensional Arnowitt–Deser–Misner (ADM) set of equations. This ADM decomposition effectively slices 4-dimensional spacetime into a continuous stack of 3-dimensional, space-like hypersurfaces that pile up along a 1-dimensional time axis. Two distinct types of equations emerge for the gravitational field in the course of this decomposition: "constraint" equations, which specify the field on a given spatial hypersurface (or "time slice"), and "evolution" equations, which describe how the field changes in time in advancing from one time slice to the next. In Chapter 3 we discuss approaches for solving the constraint equations for the construction of suitable initial data, and we provide some simple examples. In Chapter 4 we summarize a few different coordinate choices (gauge conditions) that have proven useful in numerical evolution calculations. Chapter 5 deals with the right-hand side of Einstein's equations, cataloging some different relativistic stress-energy sources that arise in realistic astrophysical applications, together with their equations of motion. Hydrodynamic and magnetohydrodynamic fluids, collisionless gases, electromagnetic radiation, and scalar fields are all represented here.

This is not a book on numerical methods *per se*. Rather, our emphasis is on deriving and interpreting geometrically various formulations of Einstein's equations that have proven useful for numerical implementation and then illustrating their utility by showing results of numerical simulations that employ them. We do not, for example, present finite difference or other discrete forms of the continuum equations, nor do we provide numerical code. But in Chapter 6 we do review some of the basic numerical techniques used to integrate standard elliptic, hyperbolic and parabolic partial differential equations, and we discuss some methods that help calibrate the accuracy of numerical solutions. These basic techniques comprise the building blocks on which all numerical implementations of the continuum formulations of Einstein's equations are based.

No object is more central to numerical relativity than the black hole. Black holes are featured throughout the book. Chapter 7 discusses some of the quantities (i.e., horizons) that help us locate and diagnose the properties of black holes residing in a numerical spacetime.

As we turn toward physical applications, our discussion proceeds in order of decreasing spacetime symmetry and increasing computational challenge. Some of the spacetimes we build involve vacuum black holes, others contain relativistic matter in various forms. Many of the examples address topical issues in relativistic gravitation or relativistic astrophysics. A substantial fraction are drawn from our own work, a choice triggered by our familiarity with this material and its accessibility, including illustrations. We hope that our colleagues will understand, and forgive us, if we seem to have overrepresented our own work as a result of this choice.

Chapter 8 constructs numerical spacetimes in spherical symmetry, which provides useful insight into gravitational collapse and black hole formation with minimal resources, but is devoid of gravitational waves. To treat gravitational waves we need to abandon spherical symmetry (Birkhoff's theorem!). To set the stage, Chapter 9 reviews some of the basic properties, plausible astrophysical sources, and current and future detectors of gravitational waves, as well as standard extraction techniques for gravitational waves in numerical spacetimes. Chapter 10 then begins our discussion of nonspherical, radiating spacetimes by featuring the collapse of collisionless clusters in axisymmetry.

To maintain long-term numerical stability during simulations in $3 + 1$ dimensions, it proves necessary to modify the ADM system of equations. Chapter 11 shows why this is true and provides alternative formulations in common use that are stable and robust.

Chapters 12 and 13 focus on the inspiral and coalescence of binary black holes, one of the most important applications of numerical relativity and a promising source of detectable gravitational radiation. These chapters treat the two-body problem in classical general relativity theory, and its solution represents one of the major triumphs of numerical relativity. Chapter 12 generates initial data for two black holes in quasistationary circular orbit, the astrophysically most realistic prelude to coalescence. Chapter 13 discusses dynamical simulations of the plunge, merger and ringdown of the two black holes and the associated waveforms. Chapter 14 treats rotating relativistic fluid stars, including numerical equilibrium models and simulations dealing with secular and dynamical instabilities and catastrophic collapse to black holes and neutron stars. Chapters 15 and 16 are the analogs of Chapters 12 and 13 for binary neutron stars. The inspiral and merger of binary neutron stars is not only a promising source of gravitational waves, but also a plausible candidate for at least one class of gamma-ray burst sources. So are black hole-neutron star binaries, which we take up in Chapter 17.

Our book could not have been written without the encouragement and insights provided by our colleagues and collaborators in numerical relativity and related areas. The individuals whose expertise we have drawn on over the years are far too numerous to list here, but we would be totally remiss if we did not thank G. B. Cook, M. W. Choptuik, C. F. Gammie, T. Nakamura, F. A. Rasio, M. Shibata, L. L. Smarr, S. A. Teukolsky, K. S. Thorne, and J. W. York, Jr. for their mentoring. We are very grateful to A. M. Abrahams, M. D. Duez, Y. T. Liu and H. J. Yo for furnishing invaluable notes and to our research groups for material that has found its way into this volume. We thank A. R. Lewis, R. Z. Gabry and A. H. Currier for helping us generate the 3-dimensional geometric illustrations in our book, to P. Spyridis for producing several line plots, and to Z. B. Etienne for providing indispensable technical assistance throughout the writing process. This project would not have been initiated without the support of G. A. Baym, D. K. Campbell, F. K. Lamb, F. K. Y. Lo, B. G. Schmidt and P. R. Shapiro, to whom we are indebted. We gratefully acknowledge the National Science Foundation, the National Aeronautics and Space Administration, and the John Simon Guggenheim Memorial Foundation for funding our research. Finally, we thank our families, to whom we dedicate this volume, for their devotion, encouragement and patience.

As the numerical algorithms continue to be refined and incorporate more physics, and as computer technology continues to advance, we anticipate that numerical relativity will accelerate in importance and use in the future. We can already foreshadow the day when youngsters are routinely downloading simulations of black hole binary coalescence on their iPods, or playing video games involving colliding neutron stars on their video cell phones, or on some new device that we cannot yet imagine! It is our fervent hope that some of the more curious will be motivated to dial into our book and learn something about the physics and mathematics underlying these remarkable simulations, so that they, in turn, may be inspired to produce the next generation of simulations that can go further toward unraveling the mysteries of nature.

Thomas W. Baumgarte
Stuart L. Shapiro

February 4, 2010

Suggestions for using this book

Our book is intended both as a general reference for researchers and as a textbook for use in a formal course that treats numerical relativity. We envision that there are at least two different ways in which the book can be used in the classroom: as the main text for a one-semester course on numerical relativity for students who have already taken an introductory course in general relativity, or as supplementary reading in numerical relativity at the end of an introductory course in general relativity. There *may* be more material in the book than can be covered comfortably in a single semester devoted entirely to numerical relativity. There *certainly* is more material than can be integrated into a supplementary unit on numerical relativity in an introductory course on general relativity. The latter may be true even when such a course is taught as a two-semester sequence, if the course is already broad and comprehensive without numerical relativity.

There are several ways to design a shortened presentation of the material in our book without sacrificing the core concepts or interfering with the logical flow. The amount of material that must be cut out from any course depends, naturally, on the amount of time that is available to devote to the subject. One means of reducing the content while retaining the fundamental ideas in a self-contained format is to restrict the discussion to pure *vacuum* spacetimes, i.e., spacetimes with no matter sources. Such spacetimes can contain gravitational waves and black holes, including binary black holes, but nothing else. Since the solution of the binary black hole problem in vacuum constitutes one of the main triumphs of numerical relativity, and since binary black hole inspiral and merger constitutes one of the most promising sources of detectable gravitational waves, one can still explore a seminal and timely topic in its entirety, even with the restriction to vacuum spacetimes. Of course, all astrophysical applications involving either hydrodynamic or magnetohydrodynamic matter, collisionless matter, or scalar fields, and whole classes of relativistic objects, like neutron stars, supernovae, collapsars, supermassive stars, collisionless clusters, etc. must then be omitted.

We provide a "roadmap" through our book in Table 1 for instructors who wish to restrict their discussion to vacuum spacetimes. The chapters and sections earmarked for inclusion constitute a respectable and self-contained "minicourse" on numerical relativity. Pointers to the relevant appendices are found in these chapters at the appropriate places. In all the sections designated in the table, all matter source terms that are retained in the gravitational field equations can be set to zero. Instructors who have time to cover more ground, but not the entire book, can then augment their discussion by adding material in the

Table 1 Vacuum spacetime "minicourse".

Chapter	Sections
1	1.1, 1.2
2	all
3	all
4	all
5	omit
6	all
7	all
8	8.1
9	all, but black holes only in 9.2
10	omit
11	all
12	all
13	all
14	omit
15	omit
16	omit
17	omit

book involving matter sources on a selective basis. For example, scalar field collapse and critical phenomena are developed in Chapters 5.4 and 8.4. Collisionless matter evolution and cluster collapse and collisions are discussed in Chapters 1.4, 5.3, 8.2, 10, and 14.1.3. Hydrodynamic and magnetohydrodynamic matter evolution, stellar collapse and stellar collisions are treated in Chapters 1.3, 1.4, 5.2, 8.3, 9.2, and 14–17. Each of these topics is developed independently of the others in the book, to first approximation, but they do rely on material covered in earlier chapters of the "minicourse".

There are, of course, other ways to parse and select from the material in the book to fit into a given course schedule. We shall leave it to individual instructors to arrange an alternative program that best suits their aims and the needs of their students.

1 General relativity preliminaries

In this chapter we assemble some of the elements of Einstein's theory of general relativity that we will be working with in later chapters. We assume that the geometric objects and equations that we list, as well as their interpretation, are already very familiar to readers.[1] The discussion below should serve simply as a checklist of a few of the basics that we need to pack with us before embarking on our voyage into numerical spacetime.

Throughout this book we adopt the $(-+++)$ metric signature together with all the sign conventions of Misner *et al.* (1973). Following that book, but in this chapter only, we will display a tensor in spacetime by a symbol in boldface when emphasizing its coordinate-free character, or by its components when the tensor has been expanded in a particular set of basis tensors. However, unlike that book, we will use Latin indices $a, b, \ldots$ instead of Greek letters to denote the spacetime indices of the tensor components, with the values of the indices running from 0 to 3. This choice anticipates a switch we will make to *abstract index notation* in all subsequent chapters of this book. We will introduce this switch in Section 2.1. We adopt the usual Einstein convention of summing over repeated indices. Finally, here and throughout we will use geometrized units in which both the gravitational constant and the speed of light are assigned the values of one, $G = c = 1$.

1.1 Einstein's equations in 4-dimensional spacetime

Cast of characters

The metric tensor of 4-dimensional spacetime (i.e., the 4-metric) is denoted by g_{ab} and determines the invariant interval (distance) between two nearby events in spacetime according to

$$ds^2 = g_{ab}dx^a dx^b. \tag{1.1}$$

Here dx^a are the differences in the coordinates x^a that label events, or points, in spacetime. For a flat spacetime, g_{ab} becomes the Minkowski metric η_{ab}. In Cartesian coordinates with $x^0 = t, x^1 = x, x^2 = y$ and $x^3 = z$, the Minkowski metric components are

$$\eta_{ab} = \mathrm{diag}(-1, 1, 1, 1), \tag{1.2}$$

representing a global inertial or Lorentz frame.

---

[1] They are treated in depth in introductory textbooks on general relativity, such as Misner *et al.* (1973), Weinberg (1972), Wald (1984) and Carroll (2004), to name a few.

In general, the components of the metric tensor are given by the scalar dot products between the four basis vectors $\mathbf{e}_a$ that span the vector space tangent to the spacetime manifold,[2]

$$g_{ab} = \mathbf{e}_a \cdot \mathbf{e}_b. \tag{1.3}$$

In a coordinate basis, the basis vectors are tangent vectors to coordinate lines and may be written as $\mathbf{e}_a = \partial/\partial x^a \equiv \partial_a$. Clearly coordinate basis vectors commute. It is sometimes useful to set up orthonormal basis vectors at a point (an orthonormal tetrad) for which

$$\mathbf{e}_{\hat{a}} \cdot \mathbf{e}_{\hat{b}} = \eta_{\hat{a}\hat{b}}. \tag{1.4}$$

We denote an orthonormal tetrad by carets. In general, orthonormal basis vectors do not form a coordinate basis and do not commute. However, in flat spacetime it is always possible to transform to coordinates which are everywhere orthonormal or globally inertial, whereby the metric is given by equation (1.2) everywhere. For a general spacetime, this is not possible. But we can always choose any particular event in spacetime to be the origin of a local inertial coordinate frame, where $g_{ab} = \eta_{ab}$ at that point and where, in addition, the first derivatives of the metric tensor at that point vanish, i.e., $\partial_a g_{bc} = 0$. An observer in such a coordinate frame is called a local inertial or local Lorentz observer and can use a coordinate basis that forms a local orthonornal tetrad to make measurements as in special relativity. In fact, such an observer will find that all the (nongravitational) laws of physics in this frame are the same as in special relativity ("Principle of Equivalence").

For any set of basis vectors, a 4-vector $\mathbf{A}$ can be expanded in contravariant components,

$$\mathbf{A} = A^a \mathbf{e}_a. \tag{1.5}$$

The scalar product of two 4-vectors $\mathbf{A}$ and $\mathbf{B}$ is

$$\mathbf{A} \cdot \mathbf{B} = (A^a \mathbf{e}_a) \cdot (B^b \mathbf{e}_b) = g_{ab} A^a B^b. \tag{1.6}$$

Now introduce a set of basis 1-forms $\tilde{\omega}^a$ dual to the basis vectors $\mathbf{e}_a$. An arbitrary 1-form $\tilde{\mathbf{B}}$ can be expanded in its covariant components according to

$$\tilde{\mathbf{B}} = B_a \tilde{\omega}^a. \tag{1.7}$$

The scalar product of two 1-forms $\tilde{\mathbf{A}}$ and $\tilde{\mathbf{B}}$ is

$$\tilde{\mathbf{A}} \cdot \tilde{\mathbf{B}} = (A_a \tilde{\omega}^a) \cdot (B_b \tilde{\omega}^b) = g^{ab} A_a B_b, \tag{1.8}$$

where $g^{ab} = \tilde{\omega}^a \cdot \tilde{\omega}^b$ is the inverse of g_{ab}. A basis of 1-forms dual to the basis $\mathbf{e}_a$ always satisfies

$$\tilde{\omega}^a \cdot \mathbf{e}_b = \delta^a{}_b. \tag{1.9}$$

---

[2] Recall that the subscript a in $\mathbf{e}_a$ denotes the ath basis vector, and not the a-component of a basis vector. In 4-dimensional spacetime, there are four independent basis vectors.

Accordingly, the scalar product of a vector with a 1-form does not involve the metric, but only a summation over an index:

$$\mathbf{A} \cdot \tilde{\mathbf{B}} = (A^a \mathbf{e}_a) \cdot (B_b \tilde{\omega}^b) = A^a \delta_a{}^b B_b = A^a B_a. \tag{1.10}$$

The vector $\mathbf{A}$ carries the same information as the corresponding 1-form $\tilde{\mathbf{A}}$, and we often will not make a distinction between them. Their components are related by

$$A_a = g_{ab} A^b, \tag{1.11}$$

or

$$A^a = g^{ab} A_b. \tag{1.12}$$

A coordinate basis of 1-forms may be written $\tilde{\omega}^a = \widetilde{\mathbf{d}x}^a$; geometrically, the basis form $\widetilde{\mathbf{d}x}^a$ may be thought of as surfaces of constant coordinate x^a. An orthonormal basis $\tilde{\omega}^{\hat{a}}$ is denoted by a caret and satisfies the relation

$$\tilde{\omega}^{\hat{a}} \cdot \tilde{\omega}^{\hat{b}} = \eta^{\hat{a}\hat{b}}. \tag{1.13}$$

A particularly useful one-form is $\widetilde{\mathbf{d}f}$, the gradient of an arbitrary scalar function f. In a coordinate basis, it may be expanded according to $\widetilde{\mathbf{d}f} = \partial_a f \widetilde{\mathbf{d}x}^a$, whereby its components are ordinary partial derivatives. The scalar product between an arbitrary vector $\mathbf{v}$ and the 1-form $\widetilde{\mathbf{d}f}$ gives the directional derivative of f along $\mathbf{v}$

$$\mathbf{v} \cdot \widetilde{\mathbf{d}f} = (v^a \mathbf{e}_a) \cdot (\partial_b f \widetilde{\mathbf{d}x}^b) = v^a \partial_a f. \tag{1.14}$$

A change of basis is always allowed, whereby $\mathbf{e}_{a'} = \mathbf{e}_b M^b{}_{a'}$, $\tilde{\omega}^{a'} = M^{a'}{}_b \tilde{\omega}^b$. Here $||M^b{}_{a'}||$ is an arbitrary, nonsingular matrix; its inverse is $||M^{a'}{}_b|| = ||M^{b'}_a||^{-1}$. Under such a change, components of vectors and 1-forms transform according to

$$A^{a'} = M^{a'}{}_b A^b, \quad B_{a'} = B_b M^b{}_{a'}. \tag{1.15}$$

When both of the bases are coordinate bases, then $M^b{}_{a'} = \partial_{a'} x^b$.

The generalization of the above concepts to tensors of arbitrary rank is straightforward. A 4-vector $\mathbf{A}$ and 1-form $\tilde{\mathbf{B}}$ are both tensors of rank 1. An arbitrary tensor can be expanded in its components, given a set of basis vectors and corresponding basis 1-forms. As an example, a mixed rank-2 tensor $\mathbf{T}$ can be expanded in components according to $\mathbf{T} = T^a{}_b \mathbf{e}_a \tilde{\omega}^b$. Here $\mathbf{e}_a \tilde{\omega}^b$ is a direct, or outer, tensor product. The componets of $\mathbf{T}$ transform according to

$$T^{a'}{}_{b'} = M^{a'}{}_c T^c{}_d M^d{}_{b'}. \tag{1.16}$$

The covariant derivative of an arbitrary tensor $\mathbf{T}$ is also a tensor and it measures the change of $\mathbf{T}$ with respect to parallel transport. For the above example of a mixed rank-2 tensor with components $T^a{}_b$, the covariant derivative is a tensor of rank 3 and its components are[3]

$$\nabla_c T^a{}_b = \partial_c T^a{}_b +^{(4)} \Gamma^a{}_{dc} T^d{}_b -^{(4)} \Gamma^d{}_{bc} T^a{}_d, \tag{1.17}$$

where the quantities $^{(4)}\Gamma^a{}_{bc}$ are connection coefficients or, in the special case of coordinate bases, Christoffel symbols, associated with the spacetime metric g_{ab}. The connection coefficients measure the change in the basis vectors and 1-forms with respect to parallel transport. In a coordinate basis they are related to partial derivatives of the metric by[4]

$$^{(4)}\Gamma^a{}_{bc} = g^{ad} \, ^{(4)}\Gamma_{dbc} = \frac{1}{2} g^{ad} (\partial_c g_{db} + \partial_b g_{dc} - \partial_d g_{bc}), \tag{1.18}$$

where the above relation defines $^{(4)}\Gamma_{dbc}$. In a local Lorentz frame the Christoffel symbols vanish. The covariant derivative of a scalar function f is the gradient 1-form; in components, $\nabla_a f = \partial_a f$. The corresponding vector $\nabla^a f$ is normal to the hypersurface $f = $ constant.

Curvature is the true measure of the gravitational field. The Riemann curvature tensor is given by

$$^{(4)}R^a{}_{bcd} = \partial_c \, ^{(4)}\Gamma^a{}_{bd} - \partial_d \, ^{(4)}\Gamma^a{}_{bc} +^{(4)} \Gamma^a{}_{ec} \, ^{(4)}\Gamma^e{}_{bd} -^{(4)} \Gamma^a{}_{ed} \, ^{(4)}\Gamma^e{}_{bc} \tag{1.19}$$

in a coordinate basis.[5] Curvature vanishes if and only if the spacetime is flat. Second covariant derivatives of tensor fields do not commute in general and their difference is related to the Riemann tensor, e.g., for any vector v^a

$$\nabla_a \nabla_b v_c - \nabla_b \nabla_a v_c = v_d \, ^{(4)}R^d{}_{cab}. \tag{1.20}$$

The Riemann tensor obeys a number of symmetries and identities, such as

$$^{(4)}R_{abcd} = -^{(4)} R_{bacd}, \qquad ^{(4)}R_{abcd} = -^{(4)} R_{abdc}, \qquad ^{(4)}R_{abcd} = ^{(4)} R_{cdab} \tag{1.21}$$

as well as the cyclic identity

$$^{(4)}R_{abcd} +^{(4)} R_{adbc} +^{(4)} R_{acdb} = 0 \tag{1.22}$$

and the Bianchi identities

$$\nabla_e \, ^{(4)}R_{abcd} + \nabla_d \, ^{(4)}R_{abec} + \nabla_c \, ^{(4)}R_{abde} = 0. \tag{1.23}$$

The symmetric Ricci tensor and Ricci scalar are formed from the Riemann tensor:

$$^{(4)}R_{ab} = ^{(4)} R^c{}_{acb} \tag{1.24}$$

$$^{(4)}R = ^{(4)} R^a{}_a. \tag{1.25}$$

---

[3] Sometimes the components of the covariant derivative of a tensor are written with a semicolon as $T^a{}_{b;c} \equiv \nabla_c T^a{}_b$.

[4] The expression for a noncoordinate basis involves additional commutation coefficient terms; see, e.g., Misner *et al.* (1973), equation (8.24b).

[5] See Misner *et al.* (1973), equation (11.3), for the components in a noncoordinate basis.

The Ricci tensor $^{(4)}R_{ab}$ is thus the trace of the Riemann tensor. The "trace-free part" is called the Weyl conformal tensor $^{(4)}C_{abcd}$ and, in four dimensions, is given by

$$^{(4)}C_{abcd} = {}^{(4)}R_{abcd} - \frac{1}{2}(g_{ac}{}^{(4)}R_{bd} - g_{ad}{}^{(4)}R_{bc} - g_{bc}{}^{(4)}R_{ad} + g_{bd}{}^{(4)}R_{ac})$$

$$+ \frac{1}{6}(g_{ac}g_{bd} - g_{ad}g_{bc})^{(4)}R. \tag{1.26}$$

It is invariant under conformal transformations and vanishes if and only if the metric is conformally flat (i.e., can be transformed to Minkowski spacetime by a conformal transformation). For manifolds with dimensions ≤ 3, the Weyl tensor is identically zero and the Ricci tensor completely determines the Riemann tensor. In vacuum spacetimes, the Weyl tensor and the Riemann tensor are identical (by virtue of Einstein's equations (1.32) below).

Geodesics

Freely-falling test particles move along geodesic curves in spacetime. The tangent vector u^a of a geodesic curve is parallel propagated, $u^b \nabla_b u^a = 0$. If we introduce coordinates to construct the trajectories and set $u^a = dx^a/d\lambda$, then the geodesic equation becomes

$$0 = u^b \nabla_b u^a = \frac{d^2 x^a}{d\lambda^2} + \Gamma^a{}_{bc}\frac{dx^b}{d\lambda}\frac{dx^c}{d\lambda}, \tag{1.27}$$

where λ is an affine parameter along the curve. For timelike particles with finite rest-mass, we can identify u^a with the particle 4-velocity and λ with proper time. In this case the quantity $a^a = u^b \nabla_b u^a$ is the 4-acceleration of the particle and is zero for geodesic motion. To accommodate null particles with zero rest-mass, we can always define an affine parameter by setting $p^a = dx^a/d\lambda$, where p^a is the particle 4-momentum. In terms of p^a the geodesic equation can be written as

$$0 = p^b \nabla_b p^a = \frac{dp^a}{d\lambda} + \Gamma^a{}_{bc} p^b p^c = 0, \tag{1.28}$$

and may be expressed exactly as in the right-hand side of equation (1.27) in a coordinate representation.

The function

$$L = \frac{1}{2}g_{ab}\dot{x}^a \dot{x}^b, \tag{1.29}$$

where $\dot{x}^a \equiv dx^a/d\lambda$, provides a useful Lagrangian for geodesics. That is, the Euler–Lagrange equations derived from $L = L(x^a, \dot{x}^a)$ yield equations (1.27). The canonically conjugate momentum to the coordinate x^a is defined by

$$p_a \equiv \frac{\partial L}{\partial \dot{x}^a}, \tag{1.30}$$

and is just a covariant component of the 4-momentum of a particle. If the metric is independent of any coordinate x^a, then L is independent of the coordinate and p_a is a constant of the motion. In this case we say that $\mathbf{e}_a = \partial_a$ is a Killing vector of the spacetime, in which case the component $p_a = \mathbf{P} \cdot \mathbf{e}_a$ is conserved, where $\mathbf{P}$ is the particle 4-momentum vector.

The importance of Riemann curvature is reflected in the behavior of two nearby, freely-falling particles moving along two nearby geodesics with nearly equal affine parameters. If $u^a = dx^a/d\lambda$ is the tangent vector to one of the geodesics and n^a is the differential vector connecting the particles at equal values of affine parameter, then n^a satisfies the equation of geodesic deviation,

$$u^c \nabla_c (u^d \nabla_d n^a) = -^{(4)} R^a_{cbd} n^b u^c u^d. \tag{1.31}$$

The quantity on the left measures the relative acceleration of the two particles and it will be zero if and only if the tidal gravitational field, measured by Riemann curvature, is zero.

The Einstein field equations

In general relativity, the gravitational field is measured by the curvature of spacetime, and curvature is generated by the presence of matter, or, more properly, mass-energy. The energy, momentum and stress of matter are represented by the symmetric energy-momentum, or stress-energy, tensor T^{ab}. All nongravitational sources of energy and momentum in the Universe contribute to T^{ab} – all particles, fluids, fields, etc. For pure vacuum spacetimes we have $T^{ab} = 0$.

Einstein's field equations of general relativity relate the geometry of spacetime to the local matter content in the Universe according to

$$G_{ab} = 8\pi T_{ab}, \tag{1.32}$$

where G_{ab} is the symmetric Einstein tensor defined by

$$G_{ab} = {}^{(4)}R_{ab} - \frac{1}{2} g_{ab} {}^{(4)}R. \tag{1.33}$$

As a consequence of the Bianchi identities (1.23), the covariant divergence of G_{ab} vanishes, $\nabla_b G^{ab} = 0$, so equation (1.32) automatically guarantees that

$$\nabla_b T^{ab} = 0. \tag{1.34}$$

Equation (1.34) is the equation of motion governing the flow of energy and momentum for the matter. This equation is the statement that the total energy-momentum of the Universe is conserved. Solving equation (1.32) completely determines the spacetime metric, up to coordinate (gauge) transformations.

Astute readers will notice that a cosmological constant term has been omitted from equation (1.32). This omission has occurred in spite of cosmological evidence[6] that there exists such a term, as Einstein originally proposed, and that the actual field equations are in fact

$$G_{ab} + \Lambda g_{ab} = {}^{(4)}R_{ab} - \frac{1}{2}g_{ab}{}^{(4)}R. \tag{1.35}$$

However, the tiny magnitude inferred for the cosmological constant Λ makes this term completely unimportant for determining the dynamical behavior of relativistic stars, black holes, and most of the applications we treat in this book. Only when considering problems on cosmological scales, like the expansion of the Universe (which certainly affects the the propagation of electromagnetic and gravitational waves produced by local sources at large redshift), or the growth of primordial fluctuations and large-scale structure in the early Universe, is the presence of the Λ term important. For the applications we discuss in this book, and unless specifically stated otherwise, the cosmological constant will be taken to be zero and we will assume that our sources are immersed in an asymptotically flat vacuum spacetime.[7]

Gravitational radiation

Gravitational waves are ripples in the curvature of spacetime that propagate at the speed of light. Once the waves move away from their source in the near zone, their wavelengths are generally much smaller than the radius of curvature of the background spacetime through which they propagate. The waves usually can be described by linearized theory in this far zone region. Introducing Minkowski coordinates, one has

$$g_{ab} = \eta_{ab} + h_{ab}, \quad |h_{ab}| \ll 1, \tag{1.36}$$

where we assume Cartesian coordinates and, ignoring any quasistatic contributions to the perturbations h_{ab} from weak-field sources, consider only the wave contributions. Defining the trace-reversed wave perturbation $\bar{h}_{ab}$ according to

$$\bar{h}_{ab} \equiv h_{ab} - \frac{1}{2}h^c{}_c \eta_{ab}, \tag{1.37}$$

the key equation governing the propagation of a linear wave in vacuum is

$$\Box \bar{h}_{ab} \equiv \nabla^c \nabla_c \bar{h}_{ab} = 0, \tag{1.38}$$

---

[6] Measurements from the Wilkinson Microwave Anisotropy Probe (WMAP) combined with the Hubble Space Telescope yield a value for the cosmological constant of $\Lambda = 3.73 \times 10^{-56}$ cm^{-2}, corresponding to $\Omega_\Lambda \equiv \Lambda/(3H_0)^2 = 0.721 \pm 0.015$, where $H_0 = 70.1 \pm 1.3$ km/s/Mpc is Hubble's constant; Freedman *et al.* (2001); Spergel *et al.* (2007); Hinshaw *et al.* (2009).

[7] It is also possible to restore the cosmological constant, or a slowly-varying term that mimics its effects, by incorporating an appropriate matter source term on the right hand side of equation (1.32). Such a "dark energy" contribution might arise from the stress-energy associated with the residual vacuum energy density (Zel'dovich 1967), or from an as yet unknown cosmic field, like a dynamical scalar field, sometimes referred to as "quintessence" (see, e.g., Peebles and Ratra 1988; Caldwell *et al.* 1998; see Chapter 5.4 for a discussion of dynamical scalar fields).

assuming it satisfies the Lorentz gauge condition

$$\nabla_b \bar{h}^{ab} = 0. \tag{1.39}$$

The Lorentz gauge condition does not yet define the gauge uniquely. Using the remaining gauge freedom we can introduce the transverse-traceless or "TT" gauge, defined by

$$h^{TT}_{a0} = 0, \quad h^{TTa}{}_a = 0, \tag{1.40}$$

which is particularly useful for describing gravitational waves. Gravitational waves are completely specified by two dimensionless amplitudes, h_+ and $h_\times$, representing the two possible polarization states of a gravitational wave. In terms of the polarization tensors e^+_{ab} and $e^\times_{ab}$ we may write a general gravitational wave as

$$h^{TT}_{jk} = h_+ e^+_{ij} + h_\times e^\times_{ij}, \tag{1.41}$$

where the letters $i, j, k, \ldots$ run over spatial indices only. For example, for a linear plane wave propagating in the z-direction, the amplitudes h_+ and $h_\times$ are functions of $t - z$ only and the only nonvanishing components of the polarization tensors are

$$e^+_{xx} = -e^+_{yy} = 1, \quad e^\times_{xy} = e^\times_{yx} = 1. \tag{1.42}$$

A passing gravitational wave drives the relative acceleration of two nearby test particles at a spatial separation ξ_i,

$$\ddot{\xi}_j = \frac{1}{2} \ddot{h}^{TT}_{jk} \xi^k. \tag{1.43}$$

According to equation (1.43), the wave amplitude measures the relative strain between the particles, $\delta \xi / \xi \sim h$. Equation (1.43) is the basis of most gravitational wave detectors.

Gravitational waves carry energy and momentum. The effective stress-energy tensor for gravitational waves is

$$T^{GW}_{ab} = \frac{1}{32\pi} \left\langle \partial_a h^{TT}_{jk} \partial_b h^{TT}_{jk} \right\rangle, \tag{1.44}$$

where $\langle \ \rangle$ denotes an average over several wavelengths and where repeated indices are summed. The power generated in the form of gravitational waves by a weak-field, slow-motion ($v \ll 1$) source is given to leading order by the quadrupole formula,

$$L_{GW} = -\frac{dE}{dt} = \frac{1}{5} \left\langle \mathcal{I}^{(3)}_{ij} \mathcal{I}^{(3)}_{ij} \right\rangle, \tag{1.45}$$

where $\mathcal{I}$ is the "reduced quadrupole moment tensor" of the emitting source, given by

$$\mathcal{I}_{ij} \equiv \int \rho \left(x_i x_j - \frac{1}{3} \delta_{ij} r^2 \right) d^3 x. \tag{1.46}$$

Here $\langle \ \rangle$ denotes an average over several periods of the source, and $r = (x^2 + y^2 + z^2)^{1/2}$. The superscript (3) in the above formula indicates the third time derivative, E is the energy of the source, and, once again, repeated indices are summed. The angular momentum of the source is also being carried off by gravitational waves at a rate

$$\frac{dJ_i}{dt} = -\frac{2}{5}\epsilon_{ijk}\left\langle \mathcal{I}_{jm}^{(2)}\mathcal{I}_{km}^{(3)}\right\rangle. \tag{1.47}$$

Note, however, that no angular momentum is carried off if the source is axisymmetric, a result that is quite general. In the slow-velocity, weak-field approximation, the gravitational wave perturbation as measured by a distant observer is given by

$$h_{jk}^{TT}(t, x_j) = \frac{2}{r}\mathcal{I}_{jk}^{TT(2)}(t - r). \tag{1.48}$$

Here the "TT" part of the reduced mass quadrupole moment is evaluated at retarded time $t' = t - r$ and is found from

$$\mathcal{I}_{jk}^{TT} \equiv P_{jl}P_{km}\mathcal{I}_{lm} - \frac{1}{2}P_{jk}(P_{lm}\mathcal{I}_{lm}), \tag{1.49}$$

where $P_{jk} \equiv \delta_{jk} - n_j n_k$ is the projection tensor that projects out the "TT" components and $n_j = x_j/r$ is a unit vector along the direction of propagation. In the same limit, one can add a radiation-reaction potential Φ^{react}, given by

$$\Phi^{\text{react}} = \frac{1}{5}\mathcal{I}_{jk}^{(5)}x^j x^k, \tag{1.50}$$

to the Newtonian potential in the equations of motion of the source.[8] Such a radiation-reaction potential correctly drains the source of energy and angular momentum at just the rate at which gravitational waves carry off these quantities, but otherwise does not properly account for the post-Newtonian motion of the source.

A self-consistent treatment of gravitational waves that correctly describes their generation in a strong gravitational field to all orders, their evolution in the near-zone and their ultimate emergence and propagation in the far-zone, requires the full machinery of numerical relativity. The same machinery automatically accounts for the back-reaction of the radiation on the source. Forging such machinery is one of the goals of this book.

1.2 Black holes

A black hole is a region of spacetime that cannot communicate with the outside Universe. The boundary of this region is a 3-dimensional hypersurface in spacetime (a spatial 2-surface propagating in time) called the surface of the black hole or the *event horizon*. Nothing can escape from the interior of a black hole, not even light. Spacetime singularities inevitably form inside black holes. Provided the singularity is enclosed by the event horizon, it is "causally disconnected" from the exterior Universe and cannot influence it. Einstein's equations continue to describe the outside Universe, but they break down inside the black hole due to the singularity.

---

[8] Burke (1971).

The most general stationary black hole solution to Einstein's equations is the analytically known Kerr–Newman metric.[9] It is uniquely specified by just three paramters: the mass M, angular momentum J and the charge Q of the black hole. Special cases are the Kerr metric ($Q = 0$), the Reissner–Nordstrom metric ($J = 0$) and the Schwarzschild metric ($J = 0$, $Q = 0$).

Schwarzschild black holes

The Schwarzschild solution[10] for a vacuum spherical spacetime may be written as

$$ds^2 = -\left(1 - \frac{2M}{r}\right) dt^2 + \left(1 - \frac{2M}{r}\right)^{-1} dr^2 + r^2 d\theta^2 + r^2 \sin^2\theta d\phi^2. \qquad (1.51)$$

Written in this form, the radial coordinate r is called the *areal* radius since it is related to the area $\mathcal{A}$ of a spherical surface at r centered on the black hole according to the Euclidean expression $r = (\mathcal{A}/4\pi)^{1/2}$. The Schwarzschild solution holds in the vacuum region of any spherical spacetime, including a spacetime containing matter; it thus applies to the vacuum exterior of a static or collapsing star (Birkhoff's theorem). The mass of this spacetime, as measured by a distant static observer in the vacuum exterior, is M. When the vacuum extends down to $r = 2M$, the exterior spacetime corresponds to a vacuum black hole of mass M. The black hole event horizon is located at $r = 2M$ and is sometimes called the *Schwarzschild radius*. It is also referred to as the "static limit", because static observers cannot exist inside $r = 2M$, and the "surface of infinite redshift", because photons emitted by a static source just outside $r = 2M$ will have infinite wavelength when measured by a static observer at infinity.

Schwarzschild geometry admits the two Killing vectors, $\mathbf{e}_t = \partial_t$ and $\mathbf{e}_\phi = \partial_\phi$. Freely-falling test particles in Schwarzschild geometry thus conserve their energy $E = -p_t$ and orbital angular momentum $l = p_\phi$. Circular orbits of test particles exist down to $r = 3M$. The energy and angular momentum of a particle of rest-mass μ in circular orbit are given by

$$(E/\mu)^2 = \frac{(r - 2M)^2}{r(r - 3M)}, \qquad (1.52)$$

$$(l/\mu)^2 = \frac{Mr^2}{r - 3M}. \qquad (1.53)$$

The circular orbit at $r = 3M$ corresponds to a photon orbit ($E/\mu \to \infty$). Circular Schwarzschild orbits are stable if $r > 6M$, unstable if $r < 6M$.

The singularity in the metric at $r = 2M$ is a coordinate singularity, removable by coordinate transformation, while the singularity at $r = 0$ is a physical spacetime singularity. In fact, the curvature invariant

$$I \equiv {}^{(4)}R_{abcd}{}^{(4)}R^{abcd} = 48M^2/r^6 \qquad (1.54)$$

---

[9] Kerr (1963); Newman *et al.* (1965)
[10] Schwarzschild (1916).

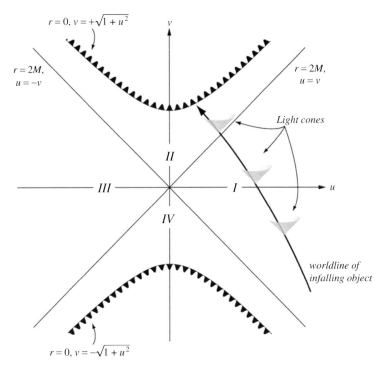

Figure 1.1 A Kruskal–Szekeres diagram. [After Shapiro and Teukolsky (1983).]

clearly blows up at the origin, showing that the tidal gravitational field becomes infinite at the center of the black hole.

One alternative coordinate choice that removes the coordinate singularity at $r = 2M$ is the *Kruskal–Szekeres* coordinate system.[11] In these coordinates, the metric (1.51) takes the form

$$ds^2 = \frac{32M^3}{r}e^{-r/2M}\left(-dv^2 + du^2\right) + r^2 d\theta^2 + r^2\sin^2\theta d\phi^2. \tag{1.55}$$

The original Schwarzschild coordinate system covers only half of the spacetime manifold, while Kruskal–Szekeres coordinates cover the entire manifold. This situation is revealed in the Kruskal–Szekeres diagram shown in Figure 1.1. In this spacetime diagram the timelike coordinate v is plotted vertically and the spacelike coordinate u is plotted horizontally. Region I corresponds to the original region $r > 2M$, "our Universe". Region II is the region $r < 2M$, the "black hole interior". Regions III and IV represent the "other Universe": region III has $r > 2M$ and is asymptotically flat, while region IV has $r < 2M$ and can describe a "white hole". The relationship between Kruskal–Szekeres coordinates u and v and Schwarzschild coordinates t and r depends on the quadrant in the u–v plane.

[11] Kruskal (1960); Szekeres (1960).

We have

$$u = \pm (r/2M - 1)^{1/2}\, e^{r/4M} \cosh(t/4M) \left.\right\} \atop v = \pm (r/2M - 1)^{1/2}\, e^{r/4M} \sinh(t/4M)} \quad r \geq 2M, \qquad (1.56)$$

where the upper sign refers to region I and the lower to region III, while

$$u = \pm (r/2M - 1)^{1/2}\, e^{r/4M} \sinh(t/4M) \left.\right\} \atop v = \pm (r/2M - 1)^{1/2}\, e^{r/4M} \cosh(t/4M)} \quad r \leq 2M, \qquad (1.57)$$

where the upper sign refers to region II, and the lower sign to region IV. The inverse transformations are

$$(r/2M - 1)e^{r/2M} = u^2 - v^2 \ \text{ in I, II, III, IV ;} \qquad (1.58)$$

and

$$t = \begin{cases} 4M \tanh^{-1}(v/u) & \text{in regions I and III,} \\ 4M \tanh^{-1}(u/v) & \text{in regions II and IV.} \end{cases} \qquad (1.59)$$

In the Kruskal–Szekeres diagram, curves of constant r are hyperbolae with asymptotes $u = \pm v$, while curves of constant t are straight lines through the origin.[12] From equation (1.58) we see that the singularity at $r = 0$ is at $v = \pm(1 + u^2)^{1/2}$ and is shown in Figure 1.1 as a saw-toothed curve. Note that radial light rays propagate along $45°$ lines in the Kruskal–Szekeres diagram. Thus timelike worldlines propagating in region II cannot escape the black hole interior and must hit the singularity at $r = 0$.

Other forms of the Schwarzschild metric are useful, particularly in numerical computations. For example, the Schwarzschild metric in *isotropic* radial coordinates is

$$ds^2 = -\left(\frac{1 - M/2\bar{r}}{1 + M/2\bar{r}}\right)^2 dt^2 + \left(1 + \frac{M}{2\bar{r}}\right)^4 \left[d\bar{r}^2 + \bar{r}^2\left(d\theta^2 + \sin^2\theta d\phi^2\right)\right], \qquad (1.60)$$

where the transformation between areal and isotropic coordinates is

$$r = \bar{r}\left(1 + M/2\bar{r}\right)^2. \qquad (1.61)$$

The inverse transformation is

$$\bar{r} = \frac{1}{2}\left[r - M \pm ((r - 2M))^{1/2}\right], \qquad (1.62)$$

and is double valued. Note that the isotropic coordinate $\bar{r}$ describes only the region of Schwarzschild geometry with $r \geq 2M$. The black hole event horizon is located at $\bar{r} = M/2$ in these coordinates. We will have occasion to use this and other coordinate systems for analyzing Schwarzschild black holes in later chapters.

---

[12] See also Figure 8.1 for a more detailed plot of region I; there r_s denotes areal radius.

Kerr black holes

The solution to Einstein's equations describing a stationary, rotating, uncharged black hole of mass M and angular momentum J in vacuum may be expressed in Boyer–Lindquist coordinates[13] in the form

$$ds^2 = -\left(1 - \frac{2Mr}{\Sigma}\right)dt^2 - \frac{4aMr\sin^2\theta}{\Sigma}dtd\phi + \frac{\Sigma}{\Delta}dr^2 + \Sigma d\theta^2$$
$$+ \left(r^2 + a^2 + \frac{2a^2Mr\sin^2\theta}{\Sigma}\right)\sin^2\theta d\phi^2, \tag{1.63}$$

where

$$a \equiv J/M, \quad \Delta \equiv r^2 - 2Mr + a^2, \quad \Sigma \equiv r^2 + a^2\cos^2\theta, \tag{1.64}$$

and where the black hole is rotating in the $+\phi$ direction. Note that when the angular momentum parameter a is zero, the solution reduces to the Schwarzschild metric (1.51). The spin is restricted to the range $0 \leq a/M \leq 1$. The rotating black is stationary and axisymmetric, hence the spacetime possesses two Killing vectors ∂_t and ∂_ϕ. Thus, test particles moving in the field of a rotating black hole conserve their energy $E = -p_t$ and axial component of angular momentum $l = p_\phi$.[14] Circular orbit parameters for particles moving in the equatorial plane of a rotating black hole are analytic and discussed in many references.[15]

The horizon of the black hole is located at r_+, the largest root of the equation $\Delta = 0$,

$$r_+ = M + \left(M^2 - a^2\right)^{1/2}. \tag{1.65}$$

The static limit is the surface within which no static observers exist; it resides at r_0, the largest root of $g_{tt} = 0$:

$$r_0 = M + \left(M^2 - a^2\cos^2\theta\right)^{1/2}. \tag{1.66}$$

The region between the horizon and static limit is called the ergosphere; in this region all timelike observers are dragged around the hole with angular velocity $\Omega > 0$.

Global theorems

A number of extraordinary theorems have been proven over the years that address very general, global properties of black hole spacetimes. We defer to other textbooks for detailed derivation and discussion of these elegant results.[16] Some of these results are encapsulated

[13] Boyer and Lindquist (1967).
[14] There is an additional constant of the motion, called Carter's "fourth constant" that is related to total angular momentum; Carter (1968).
[15] Bardeen *et al.* (1972); Shapiro and Teukolsky (1983).
[16] Hawking and Ellis (1973); Misner *et al.* (1973); Wald (1984); Poisson (2004), and references therein.

in the "four laws of black hole mechanics", which are remarkably similar to the laws of thermodynamics and demonstrate that black holes act like thermodynamic systems. As an example, consider the second law of black hole dynamics, proven by Hawking:[17] *In an isolated system, the sum of the surface areas of all black holes can never decrease.* Consider the implication of this area theorem for a Kerr black hole. The surface area is the area $\mathcal{A}$ of the horizon at some instant of time. Setting $r = r_+$ and $t =$ constant and using equation (1.63) gives the 2-metric on the horizon,

$$^{(2)}ds^2 = \left(r_+^2 + a^2\cos^2\theta\right) d\theta^2 + \frac{(2Mr_+)^2}{r_+^2 + a^2\cos^2\theta}\sin^2\theta d\phi^2, \tag{1.67}$$

from which we may derive $\mathcal{A}$ according to

$$\mathcal{A} = \int \int \sqrt{^{(2)}g}d\theta d\phi, \tag{1.68}$$

where g is the determinant of the 2-metric. Evaluating equation (1.68) yields

$$\mathcal{A} = 8\pi M \left[M + (M^2 - a^2)^{1/2}\right], \tag{1.69}$$

which reduces to $\mathcal{A} = 4\pi(2M)^2$ when $a = 0$. If we define an *irreducible mass* M_{irr} according to

$$\mathcal{A} \equiv 16\pi M_{\text{irr}}^2, \tag{1.70}$$

then we may write (1.69) as

$$M^2 = M_{\text{irr}}^2 + \frac{J^2}{4M_{\text{irr}}^2}. \tag{1.71}$$

Equation (1.71) states that the mass of a Kerr black hole is composed of an irreducible contribution plus a rotational kinetic energy contribution.[18] According to the area theorem, only the rotational energy contribution can be tapped as a source of energy by an external system interacting with the hole, since the irreducible mass can never decrease. For a system of black holes, the sum of the squares of the irreducible masses of all black holes can never decrease, at least classically.

Taking quantum mechanics into account, a black hole is characterized by a well-defined temperature T, emits thermal Hawking radiation, and has an entropy S proportional to its area according to

$$S = \frac{kc^3}{G\hbar}\frac{\mathcal{A}}{4}, \tag{1.72}$$

where k is Boltzmann's constant, $\hbar$ is Planck's constant, and where we have temporarily restored G and c.[19] When black hole evaporation via Hawking radiation is taken into

---

[17] Hawking (1971, 1972, 1973).
[18] Christodoulou (1970); Christodoulou and Ruffini (1971).
[19] Bekenstein (1973, 1975); Hawking (1974, 1975).

account, the *generalized* second law of black hole thermodynamics states that the *total* entropy, the sum of black hole and radiation entropies, never decreases.[20]

1.3 Oppenheimer–Volkoff spherical equilibrium stars

The metric describing the gravitational field of a spherical star may be written in the form

$$ds^2 = -e^{2\Phi}dt^2 + e^{2\lambda}dr^2 + r^2 d\Omega^2, \tag{1.73}$$

where Φ and λ are functions of t and r in general, but functions of areal radius r alone in the case of static equilibrium, and $d\Omega^2 = d\theta^2 + \sin^2\theta d\phi^2$. Suppose that the stellar matter can be described as a perfect fluid and that the equation of state can be written in the form $\rho = \rho(n_b, s)$, where ρ is the total mass-energy density, n_b is the baryon density and s is the specific entropy. The first law of thermodynamics then yields the pressure,

$$P = n_b^2 \frac{\partial(\rho/n_b)}{\partial n_b} = P(n_b, s). \tag{1.74}$$

In many applications the equation of state reduces to a one-parameter equation of state of the form

$$P = P(\rho). \tag{1.75}$$

Such is the case, for example, when the matter is isentropic, as in the case of cold nuclear matter ($s = 0$) or matter in a supermassive star ($s = $ constant).

The equations of stellar structure for spherical equilibrium stars in general relativity are coupled, first-order, ordinary differential equations. Defining a new metric function $m(r)$ by

$$e^\lambda \equiv \left(1 - \frac{2m}{r}\right)^{-1}, \tag{1.76}$$

Einstein's equations give

$$\frac{dm}{dr} = 4\pi r^2 \rho, \tag{1.77}$$

$$\frac{dP}{dr} = -\frac{\rho m}{r^2}\left(1 + \frac{P}{\rho}\right)\left(1 + \frac{4\pi P r^3}{m}\right)\left(1 - \frac{2m}{r}\right)^{-1}, \tag{1.78}$$

$$\frac{d\Phi}{dr} = -\frac{1}{\rho}\frac{dP}{dr}\left(1 + \frac{P}{\rho}\right)^{-1}. \tag{1.79}$$

---

[20] Our treatment throughout focuses on classical general relativity, since it provides an excellent description for the astrophysical systems that we shall consider. Only for "mini" black holes of mass $M \lesssim 10^{15}$ g is the evaporation time scale shorter than the age of the Universe.

The above set of equations is sometimes called the Oppenheimer–Volkoff or OV equations, and sometimes the Tolman–Oppenheimer–Volkoff or TOV equations, of spherical equilibrium.[21] The Newtonian limit is recovered by choosing $P \ll \rho$ and $m \ll r$.

The quantity $m(r)$ can be interpreted as the "mass interior to radius r". The total mass of the star is given by equation (1.77),

$$M = \int_0^R 4\pi r^2 \rho \, dr, \qquad (1.80)$$

where R is the stellar radius (the point where $P = \rho = 0$). Note that the quantity $m(R)$ must equal M so that the interior metric coefficient (1.76) will match smoothly onto the exterior vacuum Schwarzschild metric (1.51). The total rest-mass M_0 is determined from

$$M_0 = \int_0^R 4\pi r^2 \rho_0 (1 - 2m/r)^{-1/2} dr, \qquad (1.81)$$

where ρ_0 is the rest-mass density. The quantities ρ and ρ_0 are related by

$$\rho = \rho_0 (1 + \epsilon), \qquad (1.82)$$

where ϵ is the internal energy density per unit rest mass. For baryonic matter, $\rho_0 = n_b m_b$, where m_b is the mean baryonic rest mass. For bound configurations we have $M < M_0$: the total mass-energy M includes negative gravitational potential energy in addition to the rest-mass-energy M_0 and internal energy.

Equations (1.77)–(1.79) are straightforward to integrate numerically to construct a stellar model: First choose a central density, ρ_c, for which the equation of state (1.75) gives the central pressure P_c. The central boundary conditions

$$m = 0 \quad \text{and} \quad P = P_c \quad \text{at } r = 0, \qquad (1.83)$$

allow one to integrate equations (1.77) and (1.78) beginning at the origin, to get $m(r)$ and $P(r)$, hence $\rho(r)$, for all $0 \leq r \leq R$. It is useful to integrate equation (1.79) simultaneously with the other two equations, choosing an arbitrary value for $\Phi(r = 0)$. Since equation (1.79) is linear in Φ, one can then add a constant value to Φ everywhere so that it matches smoothly onto the Schwarzschild solution at the surface:

$$\Phi(R) = \frac{1}{2} \ln \left(1 - \frac{2M}{R} \right). \qquad (1.84)$$

Integrating the OV equations analytically for a uniform density, incompressible star shows that equilibrium is possible only if

$$\frac{2M}{R} < \frac{8}{9}. \qquad (1.85)$$

In fact, the above limit for the maximum "compaction" M/R of a uniform density sphere applies to spheres of arbitrary density profile, provided the density does not increase outwards.[22]

---

[21] Oppenheimer and Volkoff (1939); Tolman (1939).

[22] See, e.g., Weinberg (1972), Section 11.6. Note that this limit is sometimes referred to as the "Buchdahl limit"; Buchdahl (1959).

Polytropes

Some of the simplest and most useful families of equilibrium models are constructed from an isentropic equation of state of the form,

$$P = K\rho_0{}^\Gamma \quad (K, \ \Gamma \text{ constants}), \tag{1.86}$$

where ρ_0 is the rest-mass density. The constant K is the *polytropic gas constant* and the quantity n defined by $\Gamma \equiv 1 + 1/n$ is called the *polytropic index*. Stellar models constructed from such an equation of state are called *polytropes*. For an equation of state give by equation (1.86), we find from the first law of thermodynamics (or from equation 1.74) that $\rho_0\epsilon = P/(\Gamma - 1)$ and $\rho = \rho_0 + P/(\Gamma - 1)$.

There are a number of physically interesting stars that can be modeled as polytropes in a first approximation. For example, stars supported against collapse by the pressure of noninteracting, nonrelativistic, degenerate fermions can be modeled as $n = 3/2$ polytropes, while stars supported by noninteracting, ultrarelativistic, degenerate fermions can be modeled as $n = 3$ polytropes. In such cases, lower-mass objects are constructed from nonrelativistic fermions, while higher-mass objects are constructed from highly relativistic fermions. White dwarfs, which are supported by the pressure of degenerate electrons, and neutron stars, which are supported by degenerate neutrons, are members of this class of models.[23] When nuclear interactions are included, high-mass neutron stars are better represented by a "stiffer" equation of state, but the resulting models are often crudely modeled as $n = 1$ relativistic polytropes. Another example is a star supported by thermal radiation pressure at constant specific entropy, which can be modeled as an $n = 3$ polytrope.[24]

When using a polytropic equation of state to construct stellar equilibrium models, it is always possible to scale out the constant K. In gravitational units $K^{n/2}$ has units of length, so that we can introduce a new set of nondimensional quantities, often denoted by a bar:

$$\bar{r} \equiv K^{-n/2}r, \quad \bar{\rho}_0 \equiv K^n \rho_0, \qquad \bar{\rho} \equiv K^n \rho,$$

$$\bar{P} \equiv K^n P, \quad \bar{M} \equiv K^{-n/2} M, \quad \bar{M}_0 \equiv K^{-n/2} M_0. \tag{1.87}$$

One can thus set $K = 1$ in numerical integrations and either use the above relations to scale the results to more physical values of K, or express answers in terms of nondimensional ratios that are independent of K (e.g., R/M, $M^2\rho_0$, etc.).

In Figure 1.2 we plot the equilibrium sequence for $n = 1$ polytropes as an example. The turning point along a curve of equilibrium mass *vs.* central density, like the ones plotted here, identifies the maximum mass configuration. It also marks the onset of radial dynamical instability along the sequence. In particular, configurations to the left of the turning point, where $dM/d\rho_c > 0$, are dynamically stable to small radial perturbations and will

---

[23] For ultrarelativistic degenerate fermions there is a maximum mass limit, which for a white dwarf is called the *Chandrasekhar limit* and is about $1.4M_\odot$. S. Chandrasekhar received the Nobel prize in 1983, in part for identifying this important limit (Chandrasekhar 1931).

[24] For a thorough discussion of polytropes and more detailed models of compact objects like white dwarfs, neutron stars and supermassive stars and their stability properties, see Shapiro and Teukolsky (1983) and references therein.

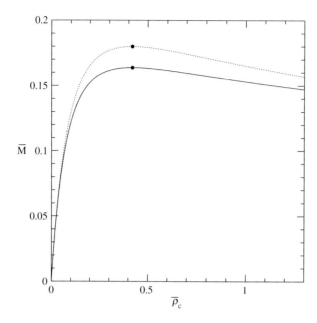

Figure 1.2 Equilibrium sequence for $n = 1$ spherical polytropes. The total mass-energy $\bar{M}$ (solid line) and the rest-mass $\bar{M}_0$ (dotted line) are plotted as functions of the central mass-energy density $\bar{\rho}_c$ along the sequence. The dots indicate the turning points, or the location of the maximum mass configuration, on each curve. The turning points occur at the same density along the two curves.

undergo small amplitude radial oscillations when subjected to such perturbations, while configurations to the right, where $dM/d\rho_c < 0$, are unstable and can undergo catastrophic collapse. For the case of an $n = 1$ polytrope, the turning point occurs at $\bar{\rho}_c = 0.420$ where $\bar{M} = 0.164$ and $\bar{M}_0 = 0.180$.[25]

1.4 Oppenheimer–Snyder spherical dust collapse

Among the most useful analytical solutions of the Einstein equations is the solution of Oppenheimer and Snyder (1939) describing the collapse of a spherical star with uniform density and zero pressure to a Schwarzschild black hole. Though it treats a highly idealized collapse scenario, the analysis is exact and fully nonlinear. The Oppenheimer–Snyder, or OS, solution illustrates many generic features of gravitational collapse and black hole formation. Since the solution is analytic, it is simple to work with and is often used to test and calibrate numerical codes designed to deal with more complicated cases, as we shall see later. Because of the important role that it plays in numerical relativity, we present this classic solution here.

---

[25] We recommend that students newly aquainted with computational physics integrate the OV equations numerically for an $n = 1$ polytrope and reproduce Figure 1.2, together with the quoted values at the turning points, before moving on to some of the more difficult computational challenges that lie ahead.

In the OS solution, each fluid element in the star of mass M follows a radial geodesic, as there is no pressure. The interior metric is given by the familiar (closed Friedmann) line element

$$ds^2 = -d\tau^2 + a^2(d\chi^2 + \sin^2 \chi \, d\Omega^2). \tag{1.88}$$

Here τ is the time coordinate, measured from the onset of collapse, χ is a Lagrangian or comoving radial coordinate and a is related to τ implicitly through the conformal time parameter η,

$$a = \frac{1}{2}a_m(1 + \cos \eta), \tag{1.89}$$

$$\tau = \frac{1}{2}a_m(\eta + \sin \eta). \tag{1.90}$$

The parameter η varies between 0 and π. The spatial coordinates of a fluid element are comoving, with χ, θ and ϕ remaining fixed during the collapse, and the time coordinate τ measures the proper time of a fluid element. This choice of coordinates is called synchronous, Gaussian normal or geodesic. The surface of the star is located at some fixed radial coordinate $\chi = \chi_0$.

The exterior metric is given by the Schwarzschild line element,

$$ds^2 = -\left(1 - \frac{2M}{r_s}\right) dt^2 + \left(1 - \frac{2M}{r_s}\right)^{-1} dr_s^2 + r_s^2 d\Omega^2. \tag{1.91}$$

The surface of the star in these coordinates is at $r_s = R(\tau)$ and follows a radial geodesic according to

$$R = \frac{1}{2}R_0(1 + \cos \eta), \tag{1.92}$$

$$\tau = \left(\frac{R_0^3}{8M}\right)^{1/2} (\eta + \sin \eta), \tag{1.93}$$

where the subscript '0' denotes the value of the radius at $t = 0$. Matching the interior and exterior solutions at the surface yields

$$a_m = \left(\frac{R_0^3}{2M}\right)^{1/2}, \tag{1.94}$$

$$\sin \chi_0 = \left(\frac{2M}{R_0}\right)^{1/2}. \tag{1.95}$$

According to the above equation, χ_0 must lie in the range $0 \le \chi_0 \le \pi/2$.

The fluid 4-velocity $u^a = \partial_\tau$ satisfies the geodesic equations (1.27). In these coordinates, the rest-mass density ρ_0 (which equals the total mass-energy density ρ, since, in the absence

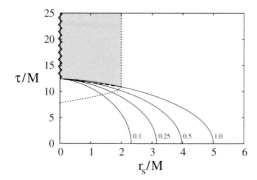

Figure 1.3 Spacetime diagram for Oppenheimer–Snyder spherical collapse to a Schwarzschild black hole. The initial stellar (areal) radius is $R/M = 5$. Worldlines of spherical fluid shells are shown as solid lines and labeled by the interior mass fraction. The event horizon is indicated by the dotted line. The shaded area denotes the region of trapped surfaces and its outer boundary is the apparent horizon. The inner boundary of the region of trapped surfaces is denoted by the dashed line. The spacetime singularity that forms at the center is indicated by the zig-zag line.

of pressure, there is no internal energy either) is a function of proper time alone,

$$\frac{\rho_0(\tau)}{\rho_0(0)} = Q^{-3}(\tau), \tag{1.96}$$

where

$$Q(\tau) = \frac{a}{a_m} = \frac{1}{2}(1 + \cos \eta). \tag{1.97}$$

In our synchronous coordinate system the star thus remains homogeneous throughout the collapse. The proper time for the star to undergo complete collapse is $\tau_{\text{coll}} = \pi(R_0^3/8M)^{1/2}$, as is evident from equations (1.92) and (1.93). At this time a central singularity forms at the center of the star.

It is both instructive and straightforward to probe the spacetime geometry of OS collapse. The spacetime diagram in Figure 1.3 shows the worldlines of infalling Lagrangian fluid elements as well as the location of the black hole event horizon. The event horizon first forms at the center and grows monotonically outward to encompass the entire star. Determining the event horizon requires that the *global* spacetime be known. Since it is known analytically in this example, the location of the event horizon can be determined quite easily: Outgoing null rays in the interior satisfy $ds^2 = 0$ or $d\tau = a(\tau)d\chi$ from equation (1.88). Using equations (1.89) and (1.90) this yields $d\chi/d\eta = 1$. Thus an outgoing ray emitted at $\eta = \eta_e$, $\chi = \chi_e$ follows the trajectory

$$\chi = \chi_e + (\eta - \eta_e). \tag{1.98}$$

The event horizon is the trajectory of an outgoing null ray that originates at the stellar center and intersets the surface of the star just when the surface crosses $R = 2M$. This trajectory traces the worldline of the the last ray that manages to escape to infinity from any point in the interior. According to equation (1.92) the stellar surface crosses $R = 2M$ when $\eta \equiv \eta_{AH} = 2\cos^{-1}(2M/R_0)^{1/2}$ – the reason for calling this value η_{AH} will become "apparent" shortly. This yields for the event horizon trajectory inside the star

$$\chi = \chi_0 + (\eta - \eta_{AH}), \qquad (1.99)$$

which holds from $\eta_{EH} \leq \eta \leq \eta_{AH}$, where $\eta_{EH} = \eta_{AH} - \chi_0$ is the value of η at which the event horizon first forms at the origin. For $\eta \geq \eta_{AH}$, and for all $\tau \geq \tau(\eta_{AH})$, the entire star is inside the event horizon, which remains frozen at $r_s = 2M$, the areal radius of the event horizon for a static Schwarzschild black hole. As is evident in the spacetime diagram, the spacetime singularity which forms at the center is "clothed" by the black hole event horizon, and therefore the black hole exterior is causally disconnected from the singularity. This result is in accord with Penrose's Cosmic Censorship Conjecture,[26] which states that gravitational collapse from well-behaved initial conditions never gives rise to a "naked" singularity (i.e., a singularity not clothed by an event horizon). We shall return to this conjecture later on.

A region of trapped surfaces is also shown in the spacetime diagram. It has the property that an outgoing bundle of null rays emitted at any point within such a region converges, i.e., its cross-sectional area instantaneously decreases. The apparent horizon is the outer boundary of the region of trapped surfaces. The significance of an apparent horizon is that it can be identified by knowing the *local* spacetime geometry and, when it exists, it always resides inside the event horizon. As we will see later on, the appearance of an apparent horizon is gauge dependent; for some time coordinates, an apparent horizon does not appear even when a black hole forms. But the converse is not true, hence whenever an apparent horizon can be identified, it signifies that a black hole has been formed. For stationary spacetimes, the apparent horizon always coincides with the event horizon. Therefore, once the spacetime settles down to a stationary state, finding the apparent horizon is an easy way to locate the event horizon.

For spherical collapse it is particularly easy to locate a region of trapped surfaces, if it exists. Consider a spherical flash of light emitted at $\chi = \chi_e$ in the matter interior. As it propagates outward, the areal radius of the flash satisfies

$$r_s(\eta) = a\sin\chi = \frac{1}{2}a_m(1 + \cos\eta)\sin(\chi_e + \eta - \eta_e), \qquad (1.100)$$

where we have used equation (1.98). If the bundle lies in a region of trapped surfaces then its area must decrease,

$$\left.\frac{d(4\pi r_s^2)}{d\eta}\right|_{\eta=\eta_e} \leq 0, \qquad (1.101)$$

[26] Penrose (1969).

or equivalently

$$\left.\frac{dr_s}{d\eta}\right|_{\eta=\eta_e} \leq 0, \tag{1.102}$$

where equality identifies the boundary of the region of trapped surfaces. Using equation (1.100), equation (1.102) may be evaluated to give

$$\eta_e \geq \pi - 2\chi_e. \tag{1.103}$$

For a flash emitted inside the matter, we have $\chi_e \leq \chi_0$, in which case the earliest that equation (1.103) is satisfied is at

$$\eta_e = \pi - 2\chi_0 = 2\cos^{-1}\left(\frac{2M}{R_0}\right)^{1/2} = \eta_{AH}, \tag{1.104}$$

where we have used equation (1.95). The apparent horizon first appears at the value of η at which the matter surface crosses $r_s = 2M$. For $\tau \geq \tau(\eta_{AH})$ the apparent horizon remains fixed at $r_s = 2M$, coinciding with the event horizon. The inner and outer boundaries of the region of trapped surfaces coincide when the surface is at $R = 2M$. Thereafter, the inner boundary moves inside the matter. According to equation (1.103), it is located at $\chi = \pi/2 - \eta/2$ for $\eta \geq \eta_{AH}$, hence its areal radius is given by

$$r_I = a\sin\chi = \frac{1}{2}a_m(1+\cos\eta)\sin(\pi/2 - \eta/2) = \left(\frac{R_0^3}{2M}\right)^{1/2}\cos^3(\eta/2). \tag{1.105}$$

2 The 3+1 decompostion of Einstein's equations

The major purpose of this book is to describe how to determine the dynamical evolution of a physical system governed by Einstein's equations of general relativity. For all but the simplest systems, analytic solutions for the evolution of such systems do not exist. Hence the task of solving Einstein's equations must be performed numerically on a computer. To construct algorithms to do this we first have to recast Einstein's 4-dimensional field equations (1.32) into a form that is suitable for numerical integration. In this chapter we present such a formulation.

The problem of evolving the gravitational field in general relativity can be posed in terms of a traditional initial value problem or "Cauchy" problem. This is a fundamental problem arising in the mathematical theory of partial differential equations. In classical dynamics, the evolution of a system is uniquely determined by the initial positions and velocities of its constituents. By analogy, the evolution of a general relativistic gravitational field is determined by specifying the metric quantities g_{ab} and $\partial_t g_{ab}$ at a given (initial) instant of time t. In particular, we need to specify the metric field components and their first time derivatives everywhere on some 3-dimensional spacelike hypersurface labeled by coordinate $x^0 = t = constant$. The different points on this surface are distinguished by their spatial coordinates x^i. Now these metric quantities can be integrated forward in time provided we can obtain from the Einstein field equations expressions for $\partial_t^2 g_{ab}$ at all points on the hypersurface. That way we can integrate these expressions to compute g_{ab} and $\partial_t g_{ab}$ on a new spacelike hypersurface at some new time $t + \delta t$, and then, by repeating the process, obtain g_{ab} for all other points x^0 and x^i in the (future) spacetime.[1]

Obtaining the appropriate expressions for $\partial_t^2 g_{ab}$ for such an integration is not so trivial. We require 10 second derivatives and, at first sight, there appear to be 10 field equations, $G_{ab} = 8\pi T_{ab}$, that might furnish them. But note that the Bianchi identities $\nabla_b G^{ab} = 0$ give

$$\partial_t G^{a0} = -\partial_i G^{ai} - G^{bc}\Gamma_{bc}^a - G^{ab}\Gamma_{bc}^c. \qquad (2.1)$$

Since no term on the right hand side of equation (2.1) contains third time derivatives or higher, the four quantities G^{a0} cannot contain second time derivatives. Hence the four

---

[1] Here we are assuming, of course, that suitable boundary conditions and initial data are chosen so that these solutions do indeed exist.

equations

$$G^{a0} = 8\pi T^{a0} \tag{2.2}$$

do not furnish any of the information required for the dynamical evolution of the fields. Rather, they supply four *constraints* on the initial data, i.e., four relations between g_{ab} and $\partial_t g_{ab}$ on the initial hypersurface at $x^0 = t$. The only truly dynamical equations must be provided by the six remaining relations

$$G^{ij} = 8\pi T^{ij}. \tag{2.3}$$

It is not surprising that there is a mismatch between the required number (10) of second time derivatives $\partial_t^2 g_{ab}$ and the available number (6) of dynamical field equations. After all, there is always a fourfold ambiguity associated with the freedom to choose four different coordinates to label points in spacetime. So, for example, we could always choose Gaussian normal coordinates and set $g_{00} = -1$ and $g_{0i} = 0$. That way we have six metric variables g_{ij} to evolve, six dynamical equations (2.3) to provide the required quantities $\partial_t^2 g_{ij}$, and four constraint equations (2.2) that relate g_{ij} and $\partial_t g_{ij}$ on the initial hypersurface. The initial value problem thus appears to be solved, at least in principle.[2]

> **Exercise 2.1** Demonstrate that the constraint equations (2.2), if satisfied initially, are automatically satisfied at later times when the gravitational field is evolved by using the dynamical equations (2.3). Equivalently, show that the relation $\partial_t(G^{a0} - 8\pi T^{a0}) = 0$ will be satisfied at the initial time $x^0 = t$, hence conclude that equation (2.2) will be satisfied at $x^0 = t + \delta t$, etc.
> **Hint:** Use the Bianchi identities together with the equations of energy-momentum conservation to evaluate $\nabla_b(G^{ab} - 8\pi T^{ab})$ at $x^0 = t$.

The above discussion reveals that formulating the Cauchy problem in general relativity logically involves a decomposition of 4-dimensional spacetime into 3-dimensional space and one-dimensional time. In this chapter we will explore how this split induces a natural "3 + 1" decomposition of Einstein's equations and leads to the standard "3 + 1" equations of general relativity. The 3 + 1 equations are entirely equivalent to the usual field equations (1.32) but they focus on the evolution of 12 purely spatial quantities closely related to g_{ij} and $\partial_t g_{ij}$ and the constraints that they must satisfy on spatial hypersurfaces. Once these spatial field quantities are specified on some initial "time slice" (i.e., spatial hypersurface) consistent with the 3 + 1 constraint equations, the 3 + 1 evolution equations can then be

---

[2] Only four of the 12 functions g_{ij} and $\partial_t g_{ij}$ represent true dynamical degrees of freedom that can be independently specified on the initial hypersurface. The reason is as follows: In addition to the four constraint equations, one can choose three arbitrary functions to induce coordinate transformations on the hypersurface without changing its geometry. Plus there exists the freedom to choose the initial hypersurface in the embedding spacetime, which can be accomplished by specifying one other arbitrary function. The remaining $12 - 4 - 3 - 1 = 4$ freely specifiable quantities can be identified with two sets of the pair of metric functions $(g_{ij}, \partial_t g_{ij})$, i.e., the 3-metric and its "velocity". These four functions specify the two dynamical degrees of freedom characterizing a gravitational field in general relativity (e.g., the two polarization states of a gravitational wave). For further discussion, see Chapter 3 below and Wald (1984), Chapter 10.2.

integrated, together with evolution equations for the matter sources, to determine these field quantities at all later times.

The $3 + 1$ formalism has some advantages over the usual 4-dimensional spacetime viewpoint for treating the Cauchy problem. It provides a nice geometric interpretation of the "foliation" of spacetime, i.e., the way in which successive time slices are chosen to fill spacetime. It furnishes (i) four constraint equations that contain no time derivatives but provide relations between the spatial field quantities and their matter sources that must be satisfied on any time slice, and (ii) a convenient set of 12 coupled, first-order, time-evolution equations for the spatial field variables in terms of field and source quantities residing on the slice. The $3 + 1$ formalism also identifies four freely specifiable functions appearing in the metric that are directly associated with the fourfold freedom to choose time and space coordinates arbitrarily. Understanding the geometric role that these four "gauge" functions play in choosing both the foliation of spacetime and the labeling of points on spatial hypersurfaces facilitates our making convenient choices for their values as a numerical evolution unfolds.

The origin of the $3 + 1$ decomposition of Einstein's equations has a long and rich history. Much of the original work was related to the study of the Cauchy problem and the solution of the initial value equations.[3] Other early work was directed toward a Hamiltonian formulation of general relativity in $3 + 1$ dimensions, with an eye toward building a theory of quantum gravity. The work of Arnowitt *et al.* (1962), often referred to as "ADM", has been the most frequently cited study in this category. ADM construct a Hamiltonian density and use it to formulate an action principle to derive a set of evolution equations for the metric functions and their "geometrodynamic conjugate momenta", which are quantities containing first-order time derivatives of the metric. Because of the wide influence of this paper, the standard $3 + 1$ equations that we will derive in this chapter are sometimes referred to as the "ADM equations", which, though equivalent, are not identical to the ones obtained in Arnowitt *et al.* (1962).

Most of the modern focus on the $3 + 1$ approach has been triggered by the necessity of solving Einstein's equations numerically on computers to obtain solutions to physically realistic dynamical systems and to probe fundamental aspects of the theory of general relativity that analytic techniques have been unable to resolve. These are the motivations underlying the treatment presented here. We shall see in later chapters that the goal of achieving numerically *stable* computer solutions, especially when the absence of spatial symmetries requires us to work in all three spatial dimensions, has led to alternative formulations and to crucial modifications of the standard $3 + 1$ equations. But before we describe these modifications, we will derive the standard set of $3 + 1$ equations in this chapter.

To introduce the subject, we shall begin by discussing the initial value problem in electrodynamics, a simpler field theory than general relativity, but often a good place to

---

[3] Darmois (1927); Lichnerowicz (1944); Fourès-Bruhat (1956); see also York, Jr. (1979) and references therein.

gain intuition. Specifically, we will cast Maxwell's equations into $3 + 1$ form in Minkowski spacetime. We will then return to general relativity, introduce a foliation of spacetime, and define the "intrinsic" and "extrinsic" curvature of spacelike hypersurfaces. Next we will relate the 3-dimensional curvature intrinsic to these hypersurfaces to the 4-dimensional curvature of spacetime, and this will give rise to the equations of Gauss, Codazzi and Ricci. Finally, we will use these equations to rewrite Einstein's field equations (1.32) in terms of the 3-dimensional curvatures. The end result will be the complete set of $3 + 1$ equations in standard form, summarized in Box 2.1, and a roadmap for building dynamical spacetimes.

2.1 Notation and conventions

Throughout the remainder of the book we shall, for the most part, adopt *abstract index notation*[4] to represent tensors, as is commonly done in numerical relativity. Specifically, we will use the convention that a variable with Latin indices does not represent a tensor component, but instead represents the abstract, coordinate-free tensor itself. For example, the symbol T_{ab} no longer stands for the covariant "ab" component in a particular basis of the tensor heretofore referred to as **T**. Instead, T_{ab} represents the second-rank, coordinate-free tensor **T** itself. Likewise, the equation $G_{ab} = 8\pi T_{ab}$ is no longer a relation between tensor components, but is instead a coordinate-independent tensor equation.[5] In fact, many equations in Chapter 1, including Einstein's equations (1.32), may be interpreted as tensor equations in abstract index notation, rather than relations equating tensor components.

   In light of our switch in notation, it is useful to revisit some of the other objects and a few representative equations that we have encountered in Chapter 1. In abstract index notation, we denote a basis vector $\mathbf{e}_a$ as $e_{(a)}^b$, for example, where the superscript b indicates that this object is a vector, and the subscript a in parenthesis means that this is the ath basis vector. In the abstract index notation of Wald (1984) components of tensors in a particular basis are distinguished from the abstract tensor itself by displaying the components with Greek indices. For example, the β-component of the ath basis vector, when expanded in its own basis, is $e_{(a)}^{\beta} = \delta_a{}^{\beta}$. Only rarely might we have need to borrow this notation; hence, our references to components of tensors in a specific basis will appear with Latin indices, but the meaning should be clear from the context. For example, the dot product between two vectors can be written as $A^a B_a$ or $g_{ab} A^a B^b$. In a few, very rare instances in the following chapters, we may slip back to boldface for clarity or emphasis when representing a particular tensor.

   We shall also adopt the standard convention whereby the letters $a - h$ and $o - z$ are used for 4-dimensional spacetime indices that run from 0 to 3, while the letters $i - n$ are reserved for 3-dimensional spatial indices that run from 1 to 3.

---

[4] See, e.g., Wald (1984).
[5] Recall, though, that an equality that holds between components of tensors in one frame holds in all frames and in this sense also constitutes a tensor equation.

We denote the 4-dimensional spacetime metric by g_{ab}, the 3-dimensional spatial metric by γ_{ij}, and its conformally related metric by $\bar{\gamma}_{ij}$. All of these are objects that we will encounter in later chapters. Four-dimensional objects associated with g_{ab} are denoted with a superscript $^{(4)}$ in front of the symbol, objects associated with $\bar{\gamma}_{ij}$ carry a bar, and objects related to γ_{ij} carry no decorations. For example, Γ^i_{jk} is associated with γ_{ij}, $\bar{\Gamma}^i_{jk}$ with $\bar{\gamma}_{ij}$, and $^{(4)}\Gamma^i_{jk}$ with g_{ab}. The covariant derivative operator is denoted with D_i and $\bar{D}_i$ when associated with the spatial metric and the conformally related metric, respectively, but with the nabla symbol ∇_a when associated with the 4-dimensional metric g_{ab}. We occasionally use the symbol Δ^{flat} for the flat scalar Laplace operator.

We denote the symmetric and antisymmetric parts of a tensor with brackets ( ) and [ ] around indices in the usual way. For example

$$T_{(ab)} \equiv \frac{1}{2}(T_{ab} + T_{ba}) \quad \text{and} \quad T_{[ab]} \equiv \frac{1}{2}(T_{ab} - T_{ba}) \tag{2.4}$$

represent the symmetrized and antisymmetrized tensors constructed from T_{ab}. We write a flat 4-dimensional spacetime metric as η_{ab} (Minkowski spacetime) and a flat 3-dimensional spatial metric as η_{ij}; these symbols are meant to apply in *any* coordinate system. Only when specifically stated will η_{ab} denote the Minkowski metric in Cartesian (inertial) coordinates with components $\text{diag}(-1, 1, 1, 1)$. Finally, we refer to a 4-dimensional line interval in spacetime as ds^2 and a 3-dimensional line interval on a spatial hypersurface as dl^2.

2.2 Maxwell's equations in Minkowski spacetime

Many of the concepts that we will encounter in this chapter are more transparent in the simpler framework of electromagnetism in special relativity as described by Maxwell's equations. In several places throughout this book we will return to electromagnetism to illustrate various features of Einstein's equations.

Maxwell's equations naturally split into two groups. The first group can be written as

$$\mathcal{C}_E \equiv D_i E^i - 4\pi\rho = 0 \tag{2.5}$$

$$\mathcal{C}_B \equiv D_i B^i = 0, \tag{2.6}$$

where E^i and B^i are the electric and the magnetic fields and ρ is the charge density. Here and throughout, D_i denotes a spatial, covariant derivative with respect to the coordinate x^i. In flat space and Cartesian coordinates, it reduces to an ordinary partial derivative.

The above equations involve only spatial derivatives of the electric and magnetic fields and hold at each instant of time independently of the prior or subsequent evolution of the fields. They therefore constrain any possible configurations of the fields, and are correspondingly called the *constraint equations*.

The second group of Maxwell equations is

$$\partial_t E_i = \epsilon_{ijk} D^j B^k - 4\pi j_i \tag{2.7}$$

$$\partial_t B_i = -\epsilon_{ijk} D^j E^k, \tag{2.8}$$

where j^i is the charge 3-current. These equations describe how the fields evolve forward in time, and are therefore called the *evolution equations*. To completely determine the time evolution of the electromagnetic fields we also have to specify how the sources ρ and j^i evolve according to the net force acting on them. Their motion depends on what forces are acting on them, but the motion of the sources is less relevant for our discussion here. We do note, however, that the total charge is conserved, as can be seen by taking the spatial divergence of equation (2.7) and substituting the constraint (2.5) to get the continuity equation,

$$\frac{\partial \rho}{\partial t} + D_i j^i = 0. \tag{2.9}$$

It is possible to bring Maxwell's equations into a form that is closer to the 3+1 form of Einstein's equations that we will derive in this chapter. To do so, we introduce the vector potential $A^a = (\Phi, A^i)$ and write B^i as

$$B_i = \epsilon_{ijk} D^j A^k. \tag{2.10}$$

By construction, B^i automatically satisfies the constraint (2.6). The two evolution equations (2.7) and (2.8) can be rewritten in terms of E_i and A_i:

$$\partial_t A_i = -E_i - D_i \Phi \tag{2.11}$$

$$\partial_t E_i = D_i D^j A_j - D^j D_j A_i - 4\pi j_i. \tag{2.12}$$

Exercise 2.2 Show that the evolution equations (2.11) and (2.12) preserve the constraint (2.5); i.e., show that

$$\frac{\partial}{\partial t} C_E = 0. \tag{2.13}$$

With the vector potential A_i we have introduced a *gauge* freedom into electrodynamics which is expressed in the freely specifiable gauge variable Φ.

Exercise 2.3 Show that a transformation to a new "tilded" gauge according to

$$\tilde{\Phi} = \Phi - \frac{\partial \Lambda}{\partial t} \tag{2.14}$$

$$\tilde{A}_i = A_i + D_i \Lambda \tag{2.15}$$

leaves the physical fields E^i and B^i unchanged.

The initial value problem in electrodynamics can now be solved in two steps. In the first step, *initial data* (A_i, E_i), together with the sources (ρ, j^i), are specified that satisfy the constraint (2.5). In the second step, these fields are evolved according to the evolution equations (2.11) and (2.12). Before the evolution equations can be solved, a suitable gauge condition has to be chosen.

Exercise 2.4 In the so-called *radiation*, *Coulomb* or *transverse* gauge, the divergence (or longitudinal part) of A_i is chosen to vanish

$$D_i A^i = 0, \qquad (2.16)$$

so that A_i is purely transverse. Show that in this gauge Φ plays the role of a Coulomb potential,

$$D^i D_i \Phi = -4\pi\rho, \qquad (2.17)$$

and that the vector potential A_i satisfies a simple inhomogeneous wave equation

$$\Box A_i \equiv -\partial_t^2 A_i + D^j D_j A_i = -4\pi j_i + D_i(\partial_t \Phi). \qquad (2.18)$$

As we will see, the initial value problem in general relativity shares many features with that in electrodynamics. In the remainder of this chapter we will show how Einstein's equations can be split into a set of constraint and evolution equations. We will also see how the coordinate freedom inherent in Einstein's equations manifests itself as a gauge freedom that is very similar to that associated with the vector potential A_i. Later chapters will deal with how the constraint equations can be solved, how suitable coordinate conditions can be defined, and how this formalism can be used to construct interesting solutions to Einstein's equations.

2.3 Foliations of spacetime

We will assume that the spacetime manifold M we aim to model is 4-dimensional, and will denote the metric in this spacetime by g_{ab} (see Section 2.1 for an explanation of our abstract index convention). Casting Einstein's equations in a 3+1 form amounts to carving this spacetime M into a stack of spatial slices and expressing the 4-dimensional spacetime curvature quantities in terms of 3-dimensional curvature quantities related to the spatial slices.

More formally, we assume that the spacetime (M, g_{ab}) can be foliated into a family of nonintersecting spacelike 3-surfaces Σ, which arise, at least locally, as the level surfaces of a scalar function t that can be interpreted as a global time function (see Figure 2.1 for an illustration). From t we can define the 1-form

$$\Omega_a = \nabla_a t, \qquad (2.19)$$

which is closed by construction,[6]

$$\nabla_{[a}\Omega_{b]} = \nabla_{[a}\nabla_{b]}t = 0. \qquad (2.20)$$

---

[6] Equivalently, we may define the vector field $\Omega^a = \nabla^a t$, which is everywhere normal to the $t = constant$ hypersurface Σ. Like any vector formed from the gradient of a scalar function, the curl of Ω^a must vanish. In the language of differential forms and the exterior calculus, we have $\widetilde{\Omega} = \mathbf{d}t$, from which equation (2.20) follows automatically from the general rules of exterior differentiation: $\mathbf{d}\widetilde{\Omega} = \mathbf{dd}t = 0$; see, e.g., Lightman *et al.* (1975), Problem 8.5.

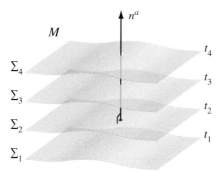

Figure 2.1 A foliation of the spacetime M. The hypersurfaces Σ are level surfaces of the coordinate time t, $\Omega_a = \nabla_a t$. The normal vector n^a is orthogonal to these $t = constant$ spatial hypersurfaces.

The 4-metric g_{ab} allows us to compute the norm of Ω_a, which we call $-\alpha^{-2}$,

$$\|\Omega\|^2 = g^{ab}\nabla_a t \nabla_b t \equiv -\frac{1}{\alpha^2}. \tag{2.21}$$

As we will see more clearly later, α measures how much proper time elapses between neighboring time slices along the normal vector Ω^a to the slice, and is therefore called the *lapse* function. We assume that $\alpha > 0$, so that Ω^a is timelike and the hypersurface Σ is spacelike everywhere.

Exercise 2.5   Show that the normalized 1-form

$$\omega_a \equiv \alpha\Omega_a \tag{2.22}$$

is rotation-free

$$\omega_{[a}\nabla_b\omega_{c]} = 0. \tag{2.23}$$

We can now define the unit normal to the slices as

$$n^a \equiv -g^{ab}\omega_b. \tag{2.24}$$

Here the negative sign has been chosen so that n^a points in the direction of increasing t,

$$n^a\omega_a = -g^{ab}\omega_a\omega_b = 1. \tag{2.25}$$

By construction, n^a is normalized and timelike,

$$n^a n_a = g^{ab}\omega_a\omega_b = -1, \tag{2.26}$$

and may therefore be thought of as the 4-velocity of a "normal" observer whose worldline is always normal to the spatial slices Σ.

With the normal vector we can now construct the spatial metric γ_{ab} that is induced by g_{ab} on the 3-dimensional hypersurfaces Σ,

$$\boxed{\gamma_{ab} = g_{ab} + n_a n_b.} \tag{2.27}$$

Thus γ_{ab} is a projection tensor that projects out all geometric objects lying along n^a. This metric allows us to compute distances within a slice Σ. To see that γ_{ab} is purely spatial, i.e., resides entirely in Σ with no piece along n^a, we contract it with the normal n^a,

$$n^a \gamma_{ab} = n^a g_{ab} + n^a n_a n_b = n_b - n_b = 0. \tag{2.28}$$

Intuitively, γ_{ab} calculates the spacetime distance with g_{ab} and then kills off the timelike contribution (normal to the spatial surface) with $n_a n_b$. The inverse spatial metric can be found by raising the indices of γ_{ab} with g^{ab},

$$\gamma^{ab} = g^{ac} g^{bd} \gamma_{cd} = g^{ab} + n^a n^b. \tag{2.29}$$

Next we break up 4-dimensional tensors by decomposing them into a purely spatial part, which lies in the hypersurfaces Σ, and a timelike part, which is normal to the spatial surface. To do so, we need two projection operators. The first one, which projects a 4-dimensional tensor into a spatial slice, can be found by raising only one index of the spatial metric γ_{ab}

$$\gamma^a_{\ b} = g^a_{\ b} + n^a n_b = \delta^a_{\ b} + n^a n_b. \tag{2.30}$$

Exercise 2.6 Show that $\gamma^a_{\ b} v^b$, where v^a is an arbitrary spacetime vector, is purely spatial.

To project higher rank tensors into the spatial surface, each free index has to be contracted with a projection operator. It is sometimes convenient to denote this projection with a symbol $\perp$, e.g.,

$$\perp T_{ab} = \gamma_a^{\ c} \gamma_b^{\ d} T_{cd}. \tag{2.31}$$

Similarly, we may define the normal projection operator as

$$N^a_{\ b} \equiv -n^a n_b = \delta^a_{\ b} - \gamma^a_{\ b}, \tag{2.32}$$

even though in most cases it is just as easy to write out the normal vectors $n^a n_b$. We can now use these two projection operators to decompose any tensor into its spatial and timelike parts. For example, we can write an arbitrary vector v^a as

$$v^a = \delta^a_{\ b} v^b = (\gamma^a_{\ b} + N^a_{\ b}) v^b = \perp v^a - n^a n_b v^b. \tag{2.33}$$

Exercise 2.7 Show that for the second rank tensor T_{ab} we have

$$T_{ab} = \perp T_{ab} - n_a n^c \perp T_{cb} - n_b n^c \perp T_{ac} + n_a n_b n^c n^d T_{cd}. \tag{2.34}$$

Exercise 2.7 illustrates that the $\perp$ symbol has to be used with some care, since it applies only to the free indices of the tensor that it operates on. To avoid confusion, we will usually write out the projection operators explicitly.

It may be useful to illustrate the above concepts for a familiar example. Consider a Schwarzschild spacetime in isotropic spherical polar coordinates[7]

$$ds^2 = -\left(\frac{1 - M/(2r)}{1 + M/(2r)}\right)^2 dt^2 + \left(1 + \frac{M}{2r}\right)^4 (dr^2 + r^2 d\theta^2 + r^2 \sin^2\theta d\phi^2) \qquad (2.35)$$

and identify the spatial slices Σ with hypersurfaces of constant coordinate time t. Then the components of the 1-form Ω_a in a coordinate basis are simply

$$\Omega_a = (1, 0, 0, 0) \qquad (2.36)$$

and from its normalization (2.21) we find the lapse

$$\alpha = \frac{1 - M/(2r)}{1 + M/(2r)}. \qquad (2.37)$$

The normal vector n^a is then

$$n^a = -g^{ab}\omega_b = \frac{1 + M/(2r)}{1 - M/(2r)}(1, 0, 0, 0), \qquad (2.38)$$

and the spatial metric (2.27) becomes

$$\gamma_{ab} = \left(1 + \frac{M}{2r}\right)^4 \text{diag}\left(0, 1, r^2, r^2 \sin^2\theta\right). \qquad (2.39)$$

It is evident that this metric eliminates any t-components.

Returning to our formal derivation of the $3 + 1$ decomposition we will also need a 3-dimensional covariant derivative that maps spatial tensors into spatial tensors. It is uniquely defined by requiring that it be compatible with the 3-dimensional metric γ_{ab}. We can construct this derivative by projecting all indices present in a 4-dimensional covariant derivative into Σ. For a scalar f, for example, we define

$$D_a f \equiv \gamma_a{}^b \nabla_b f, \qquad (2.40)$$

and for a rank $\binom{1}{1}$ tensor T^a_b

$$D_a T^b_c \equiv \gamma_a{}^d \gamma_e{}^b \gamma_c{}^f \nabla_d T^e_f. \qquad (2.41)$$

The extension to other type tensors is obvious.

> **Exercise 2.8** Show that the 3-dimensional covariant derivative is compatible with the spatial metric γ_{ab}, that is, show that
>
> $$D_a \gamma_{bc} = 0. \qquad (2.42)$$

> **Exercise 2.9** Show that for a scalar product $v^a w_a$, the Leibnitz rule
>
> $$D_a(v^b w_b) = v^b D_a w_b + w_b D_a v^b \qquad (2.43)$$
>
> holds only if v^a and w_a are purely spatial.

---

[7] See, e.g., Misner *et al.* (1973), equation (31.22).

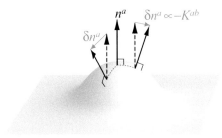

Figure 2.2 The *extrinsic curvature* of a hypersurface in an enveloping spacetime measures how much normal vectors to the hypersurface differ at neighboring points. It therefore measures the rate at which the hypersurface warps as it is carried forward along a normal vector.

The 3-dimensional covariant derivative can be expressed in terms of 3-dimensional connection coefficients, which, in a coordinate basis, are given by

$$\Gamma^a_{bc} = \frac{1}{2}\gamma^{ad}(\partial_c\gamma_{db} + \partial_b\gamma_{dc} - \partial_d\gamma_{bc}).$$
(2.44)

The 3-dimensional Riemann tensor associated with γ_{ij} is defined by requiring that[8]

$$2D_{[a}D_{b]}w_c = R^d_{\ cba}w_d \qquad R^d_{\ cba}n_d = 0$$
(2.45)

for any spatial vector w_d. In a coordinate basis, the components of the Riemann tensor can be computed from

$$R_{abc}^{\quad d} = \partial_b\Gamma^d_{ac} - \partial_a\Gamma^d_{bc} + \Gamma^e_{ac}\Gamma^d_{eb} - \Gamma^e_{bc}\Gamma^d_{ea}.$$
(2.46)

Contracting the Riemann tensor yields the 3-dimensional Ricci tensor $R_{ab} = R^c_{\ acb}$ and the 3-dimensional Ricci scalar $R = R^a_{\ a}$.

Einstein's equations (1.32) relate contractions of the 4-dimensional Riemann tensor $^{(4)}R^a_{\ bcd}$ to the stress–energy tensor. Since we want to rewrite these equations in terms of 3-dimensional objects, we decompose $^{(4)}R^a_{\ bcd}$ into spatial tensors. Not surprisingly, this decomposition involves its 3-dimensional cousin $R^a_{\ bcd}$, but obviously this cannot contain all the information needed. $R^d_{\ abc}$ is a purely spatial object and can be computed from spatial derivatives of the spatial metric alone, while $^{(4)}R^d_{\ abc}$ is a spacetime creature which also contains time derivatives of the 4-dimensional metric. Stated differently, the 3-dimensional curvature $R^a_{\ bcd}$ only contains information about the curvature *intrinsic* to a slice Σ, but it gives no information about what shape this slice takes in the spacetime M in which it is embedded. This information is contained in a tensor called the *extrinsic* curvature.

2.4 The extrinsic curvature

The extrinsic curvature K_{ab} can be found by projecting gradients of the normal vector into the slice Σ (see Figure 2.2). We will also see that the extrinsic curvature is related to

---

[8] See equation (1.20) for the 4-dimensional analog of this expression.

the first time derivative of the spatial metric γ_{ab}. The metric and the extrinsic curvature (γ_{ab}, K_{ab}) can therefore be considered as the equivalent of positions and velocities in classical mechanics – they measure the "instantaneous" state of the gravitational field, and form the fundamental variables in our initial value formulation. Mathematicians often refer to the metric as the first, and the extrinsic curvature the second, fundamental form.

The projection of the gradient of the normal vector $\gamma_a{}^c\gamma_b{}^d\nabla_c n_d$ can be split into a symmetric part, also known as the expansion tensor

$$\theta_{ab} = \gamma_a{}^c\gamma_b{}^d\nabla_{(c}n_{d)}, \tag{2.47}$$

and an antisymmetric part, also known as the rotation 2-form or twist,

$$\omega_{ab} = \gamma_a{}^c\gamma_b{}^d\nabla_{[c}n_{d]}. \tag{2.48}$$

Exercise 2.10 Show that the twist ω_{ab} has to vanish as a consequence of n^a being rotation-free (see exercise 2.5).

We now define the extrinsic curvature, K_{ab}, as the negative expansion

$$\boxed{K_{ab} \equiv -\gamma_a{}^c\gamma_b{}^d\nabla_{(c}n_{d)} = -\gamma_a{}^c\gamma_b{}^d\nabla_c n_d.} \tag{2.49}$$

By definition, the extrinsic curvature is symmetric and purely spatial. It measures the gradient of the normal vectors n^a. Since the latter are normalized, they can only differ in the direction in which they are pointing, and the extrinsic curvature therefore provides information on how much this direction changes from point to point across a spatial hypersurface, as illustrated in Figure 2.2. As a consequence, the extrinsic curvature measures the rate at which the hypersurface deforms as it is carried forward along a normal.

Exercise 2.11 Show that the extrinsic curvature of $t = constant$ hypersurfaces of the Schwarzschild metric (2.35) vanishes.

Alternatively, we can express the extrinsic curvature in terms of the acceleration of the unit normal vector field

$$a_a \equiv n^b\nabla_b n_a. \tag{2.50}$$

Exercise 2.12 Show that the acceleration a_a is purely spatial, $n^a a_a = 0$.

Exercise 2.13 Show that the acceleration a_a is related to the lapse α according to

$$a_a = D_a \ln\alpha. \tag{2.51}$$

Exercise 2.14 Find the acceleration a_a for the normal observer (2.38) in a Schwarzschild spacetime.

Expanding the right hand side of (2.49) and using the identity $n^d\nabla_c n_d = 0$ together with the definition of a_b we find

$$K_{ab} = -\gamma_a{}^c\gamma_b{}^d\nabla_c n_d = -(\delta_a^c + n_a n^c)(\delta_b^d + n_b n^d)\nabla_c n_d$$
$$= -(\delta_a^c + n_a n^c)\delta_b^d\nabla_c n_d = -\nabla_a n_b - n_a a_b. \tag{2.52}$$

Finally, we can write the extrinsic curvature as

$$K_{ab} = -\frac{1}{2}\mathcal{L}_{\mathbf{n}}\gamma_{ab}, \tag{2.53}$$

where $\mathcal{L}_{\mathbf{n}}$ denotes the *Lie derivative* along n^a. The concept and properties of the Lie derivative are sketched in Apppendix A. Here we simply note that the Lie derivative along a vector field X^a measures by how much the changes in a tensor field along X^a differ from a mere infinitesimal coordinate transformation generated by X^a. For a scalar f, the Lie derivative reduces to the partial derivative

$$\mathcal{L}_{\mathbf{X}}f = X^b D_b f = X^b \partial_b f; \tag{2.54}$$

for a vector field v^a the Lie derivative is the commutator, so that in a coordinate basis,

$$\mathcal{L}_{\mathbf{X}}v^a = X^b \partial_b v^a - v^b \partial_b X^a = [X, v]^a, \tag{2.55}$$

and for a 1-form ω_a the Lie derivative is given by

$$\mathcal{L}_{\mathbf{X}}\omega_a = X^b \partial_b \omega_a + \omega_b \partial_a X^b. \tag{2.56}$$

It then follows that for a tensor T^a_b of rank $\binom{1}{1}$ the Lie derivative is

$$\mathcal{L}_{\mathbf{X}}T^a_b = X^c \partial_c T^a_b - T^c_b \partial_c X^a + T^a_c \partial_b X^c. \tag{2.57}$$

Generalization to tensors of arbitrary rank follows naturally. Lie differentiation satisfies the chain rule and the usual addition properties obeyed by covariant differentiation. Also, in all of the above expressions for the Lie derivative one may replace the partial derivatives with covariant derivatives.

Since n^a is a timelike vector, equation (2.53) illustrates the intuitive interpretation of the extrinsic curvature as a geometric generalization of the "time derivative" of the spatial metric γ_{ab}. Obviously, the spatial metric γ_{ab} on two different slices Σ may differ by virtue of a coordinate transformation. Equation (2.53) states that, in addition to a mere coordinate transformation (which by itself would yield $\mathcal{L}_{\mathbf{n}}\gamma_{ab} = 0$, see Appendix A), γ_{ab} changes proportionally to K_{ab}.

To derive equation (2.53), we write γ_{ab} in terms of g_{ab} and n_a and use (A.13) and (A.19):

$$\mathcal{L}_{\mathbf{n}}\gamma_{ab} = \mathcal{L}_{\mathbf{n}}(g_{ab} + n_a n_b) = 2\nabla_{(a}n_{b)} + n_a\mathcal{L}_{\mathbf{n}}n_b + n_b\mathcal{L}_{\mathbf{n}}n_a$$
$$= 2(\nabla_{(a}n_{b)} + n_{(a}a_{b)}) = -2K_{ab}. \tag{2.58}$$

The last equality results from equation (2.52). The extrinsic curvature is often defined by equation (2.53), from which our definition (2.49) as well as (2.52) can be derived. Obviously, this logical development is completely equivalent and the choice is merely a matter of taste.

The trace of the extrinsic curvature, often called the *mean curvature*,

$$K = g^{ab}K_{ab} = \gamma^{ab}K_{ab}, \tag{2.59}$$

also has a nice geometrical interpretation. To see this, we take the trace of (2.53) to find

$$K = \gamma^{ab} K_{ab} = -\frac{1}{2}\gamma^{ab}\mathcal{L}_{\mathbf{n}}\gamma_{ab} = -\frac{1}{2\gamma}\mathcal{L}_{\mathbf{n}}\gamma = -\frac{1}{\gamma^{1/2}}\mathcal{L}_{\mathbf{n}}\gamma^{1/2} = -\mathcal{L}_{\mathbf{n}}\ln\gamma^{1/2}. \quad (2.60)$$

Since $\gamma^{1/2}d^3x$ is the proper volume element in the spatial slice Σ, the negative of the mean curvature measures the fractional change in the proper 3-volume along n^a.[9]

2.5 The equations of Gauss, Codazzi and Ricci

The metric γ_{ab} and the extrinsic curvature K_{ab} cannot be chosen arbitrarily. Instead, they have to satisfy certain constraints, so that the spatial slices "fit" into the spacetime M. In order to find these relations, we have to relate the 3-dimensional Riemann tensor $R^a{}_{bcd}$ of the hypersurfaces Σ to the 4-dimensional Riemann tensor ${}^{(4)}R^a{}_{bcd}$ of M. To do so, we first take a completely spatial projection of ${}^{(4)}R^a{}_{bcd}$, then a projection with one index projected in the normal direction, and finally a projection with two indices projected in the normal direction. All other projections vanish identically because of the symmetries of the Riemann tensor. A decomposition of ${}^{(4)}R^a{}_{bcd}$ into spatial and normal pieces therefore involves these three different types of projections.

> **Exercise 2.15** Following the example of exercise 2.7, show that the 4-dimensional Riemann tensor ${}^{(4)}R_{abcd}$ can be written as
>
> $$\begin{aligned}{}^{(4)}R_{abcd} &= \gamma_a{}^p\gamma_b{}^q\gamma_c{}^r\gamma_d{}^s\,{}^{(4)}R_{pqrs} - 2\gamma_a{}^p\gamma_b{}^q\gamma_{[c}{}^r n_{d]}n^s\,{}^{(4)}R_{pqrs} \\ &\quad - 2\gamma_c{}^p\gamma_d{}^q\gamma_{[a}{}^r n_{b]}n^s\,{}^{(4)}R_{pqrs} + 2\gamma_a{}^p\gamma_{[c}{}^r n_{d]}n_b n^q n^s\,{}^{(4)}R_{pqrs} \\ &\quad - 2\gamma_b{}^p\gamma_{[c}{}^r n_{d]}n_a n^q n^s\,{}^{(4)}R_{pqrs}. \qquad (2.61)\end{aligned}$$

The above projections give rise to the equations of Gauss, Codazzi and Ricci, which we will derive below. Given that ${}^{(4)}R^a{}_{bcd}$ involves up to second time derivatives of the metric, while $R^a{}_{bcd}$ only contains space derivatives, we may already anticipate that these relations will involve the extrinsic curvature and its time derivative.

The Riemann tensor is defined in terms of second covariant derivatives of a vector. To relate the 4-dimensional Riemann tensor to its 3-dimensional counterpart, it is therefore natural to start by relating the corresponding covariant derivatives to each other. We first expand the definition of the spatial gradient of a spatial vector V^b as

$$\begin{aligned}D_a V^b &= \gamma_a{}^p\gamma_q{}^b\nabla_p V^q = \gamma_a{}^p(g_q{}^b + n_q n^b)\nabla_p V^q = \gamma_a{}^p\nabla_p V^b - \gamma_a{}^p n^b V^q\nabla_p n_q \\ &= \gamma_a{}^p\nabla_p V^b - n^b V^e\gamma_a{}^p\gamma_e{}^q\nabla_p n_q = \gamma_a{}^p\nabla_p V^b + n^b V^e K_{ae}, \qquad (2.62)\end{aligned}$$

where we have used $n_q V^q = 0$, and hence $n_q\nabla_p V^q = -V^q\nabla_p n_q$, as well the definition of the extrinsic curvature (2.49).

---

[9] See also Poisson (2004), Section 2.3.8. Alternatively, from equation (2.49) or (4.7) we have that $K = -\nabla_a n^a = -(1/V)dV/d\tau$, hence K measures the expansion of normal observers, or the fractional rate of change, with respect to proper time τ, of the proper volume V of a bundle of normal observers; see, e.g., Misner *et al.* (1973), equation (22.2).

Exercise 2.16 Show that

$$\nabla_a V^a = \frac{1}{\alpha} D_a(\alpha V^a) \tag{2.63}$$

for any spatial vector V^a.

Hint: One possible derivation uses equations (2.51) and (2.62); a more elegant approach starts with the identity (A.44).

Exercise 2.17 Show that

$$D_a D_b V^c = \gamma^p_a \gamma^q_b \gamma^c_r \nabla_p \nabla_q V^r - K_{ab} \gamma^c_r n^p \nabla_p V^r - K_a{}^c K_{bp} V^p. \tag{2.64}$$

We can now use equation (2.64) to relate the 3- and 4-dimensional Riemann tensors to each other. Writing the definition of the 3-dimensional Riemann tensor (2.45) as

$$R^{dc}{}_{ba} V_d = 2 D_{[a} D_{b]} V^c \tag{2.65}$$

we can insert the second derivative (2.64) to find

$$R^{dc}{}_{ba} V_d = 2\gamma^p_a \gamma^q_b \gamma^c_r \nabla_{[p} \nabla_{q]} V^r - 2K_{[ab]} \gamma^c_r n^p \nabla_p V^r - 2K_{[a}{}^c K_{b]p} V^p. \tag{2.66}$$

The second term on the right hand side vanishes because K_{ab} is symmetric, and the first term can be rewritten in terms of the 4-dimensional Riemann tensor, which yields

$$R_{dcba} V^d = \gamma^p_a \gamma^q_b \gamma^r_c{}^{(4)} R_{drqp} V^d - 2K_{c[a} K_{b]d} V^d \tag{2.67}$$

after relabeling some indices and lowering the index c. Since this relation has to hold for any arbitrary spatial vector V^d, we have

$$\boxed{R_{abcd} + K_{ac} K_{bd} - K_{ad} K_{cb} = \gamma^p_a \gamma^q_b \gamma^r_c \gamma^s_d{}^{(4)} R_{pqrs}.} \tag{2.68}$$

This equation is called *Gauss' equation*. It relates the full spatial projection of $^{(4)}R^a{}_{bcd}$ to the 3-dimensional $R^a{}_{bcd}$ and terms quadradic in the extrinsic curvature.

Next, we want to consider projections of $^{(4)}R^a{}_{bcd}$ with one index projected in the normal direction. This will involve a spatial derivative of the extrinsic curvature

$$D_a K_{bc} = \gamma^p_a \gamma^q_b \gamma^r_c \nabla_p K_{qr} = -\gamma^p_a \gamma^q_b \gamma^r_c (\nabla_p \nabla_q n_r + \nabla_p(n_q a_r)). \tag{2.69}$$

Since $\gamma^q_b n_q = 0$, only the gradient of n_q will give a nonzero contribution in the second term, namely

$$\gamma^p_a \gamma^q_b \gamma^r_c a_r \nabla_p n_q = -a_c K_{ab}. \tag{2.70}$$

We therefore have

$$D_a K_{bc} = -\gamma^p_a \gamma^q_b \gamma^r_c \nabla_p \nabla_q n_r + a_c K_{ab}. \tag{2.71}$$

Since K_{ab} is symmetric, the last term disappears when antisymmetrizing to give

$$D_{[a} K_{b]c} = -\gamma^p_a \gamma^q_b \gamma^r_c \nabla_{[p} \nabla_{q]} n_r. \tag{2.72}$$

By the definition of the Riemann tensor, this can be rewritten as

$$\boxed{D_b K_{ac} - D_a K_{bc} = \gamma^p_a \gamma^q_b \gamma^r_c n^s \, {}^{(4)}R_{pqrs}.}$$

(2.73)

This equation is known as the *Codazzi equation*. Note that Gauss' equation (2.68) and the Codazzi equation (2.73) depend only on the spatial metric, the extrinsic curvature and their spatial derivatives. They can be thought of as the integrability conditions allowing the embedding of a 3-dimensional slice Σ with data (γ_{ab}, K_{ab}) inside a 4-dimensional manifold M with g_{ab}. As we will see in the next section, these two equations give rise to the "constraint" equations.

However, before deriving the constraint equations in the next section, we first consider the last remaining projection of ${}^{(4)}R^a_{bcd}$, namely with two indices projected in the normal direction. This will involve a "time" derivative of K_{ab}, and therefore we first compute

$$\mathcal{L}_{\mathbf{n}} K_{ab} = n^c \nabla_c K_{ab} + 2 K_{c(a} \nabla_{b)} n^c$$

$$= -n^c \nabla_c \nabla_a n_b - n^c \nabla_c (n_a a_b) - 2 K_{c(a} K^c_{b)} - 2 K_{c(a} n_{b)} a^c.$$

(2.74)

Here we have used equation (2.52) to expand both terms. We can now insert

$${}^{(4)}R_{dbac} n^d = 2 \nabla_{[c} \nabla_{a]} n_b,$$

(2.75)

which yields

$$\mathcal{L}_{\mathbf{n}} K_{ab} = -n^d n^c \, {}^{(4)}R_{dbac} - n^c \nabla_a \nabla_c n_b - n^c a_b \nabla_c n_a -$$
$$n^c n_a \nabla_c a_b - 2 K^c_{(a} K_{b)c} - 2 K_{c(a} n_{b)} a^c.$$

(2.76)

Using the definition of $a_b = n^c \nabla_c n_b$ and the relation

$$n^c \nabla_a \nabla_c n_b = \nabla_a a_b - (\nabla_a n^c)(\nabla_c n_b) = \nabla_a a_b - K^c_a K_{cb} - n_a a^c K_{cb}$$

(2.77)

several terms cancel and we find

$$\mathcal{L}_{\mathbf{n}} K_{ab} = -n^d n^c \, {}^{(4)}R_{dbac} - \nabla_a a_b - n^c n_a \nabla_c a_b - a_a a_b - K^c_b K_{ac} - K_{ca} n_b a^c.$$

(2.78)

Exercise 2.18 Show that $\mathcal{L}_{\mathbf{n}} K_{ab}$ is purely spatial,

$$n^a \mathcal{L}_{\mathbf{n}} K_{ab} = 0.$$

(2.79)

Since $\mathcal{L}_{\mathbf{n}} K_{ab}$ is purely spatial, projecting the two free indices in (2.78) leaves the left hand side unchanged and results in

$$\mathcal{L}_{\mathbf{n}} K_{ab} = -n^d n^c \gamma^q_a \gamma^r_b \, {}^{(4)}R_{drqc} - \gamma^q_a \gamma^r_b \nabla_q a_r - a_a a_b - K^c_b K_{ac}.$$

(2.80)

Exercise 2.19 Show that

$$D_a a_b = -a_a a_b + \frac{1}{\alpha} D_a D_b \alpha.$$

(2.81)

Finally, we simplify (2.80) with the help of equation (2.81), and find

$$\boxed{\mathcal{L}_{\mathbf{n}} K_{ab} = n^d n^c \gamma^q_a \gamma^r_b {}^{(4)}R_{drcq} - \frac{1}{\alpha} D_a D_b \alpha - K^c_b K_{ac}.}$$ (2.82)

Equation (2.82) is Ricci's equation.[10] It relates the "time" derivative of K_{ab} to a projection of the 4-dimensional Rieman tensor with two indices projected in the "time" direction.

2.6 The constraint and evolution equations

Now we have assembled all the necessary tools, and can rewrite Einstein's field equations in a $3 + 1$ form. Basically, we just need to take the equations of Gauss, Codazzi and Ricci and eliminate the 4-dimensional Rieman tensor using Einstein's equations

$$G_{ab} \equiv {}^{(4)}R_{ab} - \frac{1}{2} {}^{(4)}Rg_{ab} = 8\pi T_{ab}.$$ (2.83)

The last few sections dealt with purely geometrical objects; we will now invoke Einstein's equations to link these geometrical objects to physical properties of spacetimes. We will first derive the constraint equations from Gauss' equation (2.68) and the Codazzi equation (2.73), and will then derive the evolution equations from (2.53) and the Ricci equation (2.82).

Contracting Gauss' equation (2.68) once, we find

$$\gamma^{pr} \gamma^q_b \gamma^s_d {}^{(4)}R_{pqrs} = R_{bd} + K K_{bd} - K^c_d K_{cb},$$ (2.84)

where K is the trace of the extrinsic curvature, $K = K^a_a$. A further contraction yields

$$\gamma^{pr} \gamma^{qs} {}^{(4)}R_{pqrs} = R + K^2 - K_{ab} K^{ab}.$$ (2.85)

The left hand side can be expanded into

$$\gamma^{pr} \gamma^{qs} {}^{(4)}R_{pqrs} = (g^{pr} + n^p n^r)(g^{qs} + n^q n^s) {}^{(4)}R_{pqrs} = {}^{(4)}R + 2n^p n^r {}^{(4)}R_{pr}.$$ (2.86)

Note that the term $n^p n^r n^q n^s {}^{(4)}R_{pqrs}$ vanishes identically because of the symmetry properties of the Riemann tensor. We also have

$$2n^p n^r G_{pr} = 2n^p n^r {}^{(4)}R_{pr} - n^p n^r g_{pr} {}^{(4)}R = 2n^p n^r {}^{(4)}R_{pr} - n^p n^r (\gamma_{pr} - n_p n_r) {}^{(4)}R$$

$$= 2n^p n^r {}^{(4)}R_{pr} + {}^{(4)}R = \gamma^{pr} \gamma^{qs} {}^{(4)}R_{pqrs},$$ (2.87)

---

[10] In the general relativity literature, equations (2.68) and (2.73) are often jointly referred to as the *Gauss–Codazzi equations*. In the differential geometry literature, however, (2.68) is known as the *Gauss equation* (which represents a generalization of Gauss's *Theorema Egregium*), while (2.73) is called the *Codazzi* or *Codazzi–Mainardi relation*. The name *Ricci equation* for (2.82) is also more commonly used in differential geometry books than in general relativity books.

where we have used equation (2.86) in the last equality. Inserting this into the contracted Gauss' equation (2.85) yields

$$2n^p n^r G_{pr} = R + K^2 - K_{ab}K^{ab}. \tag{2.88}$$

We now define the energy density ρ to be the total energy density as measured by a normal observer n^a,

$$\rho \equiv n_a n_b T^{ab}. \tag{2.89}$$

Using Einstein's equation (2.83) together with equations (2.88) and (2.89), we obtain

$$\boxed{R + K^2 - K_{ab}K^{ab} = 16\pi\rho.} \tag{2.90}$$

Equation (2.90) is the *Hamiltonian constraint*.

Contracting the Codazzi equation (2.73) once yields

$$D_b K_a^{\ b} - D_a K = \gamma_a^p \gamma^{qr} n^{s\ (4)} R_{pqrs}. \tag{2.91}$$

The right hand side is

$$\gamma_a^p \gamma^{qr} n^{s\ (4)} R_{pqrs} = -\gamma_a^p (g^{qr} + n^q n^r) n^{s\ (4)} R_{qprs} = -\gamma_a^p n^{s\ (4)} R_{ps} - \gamma_a^p n^q n^r n^{s\ (4)} R_{qprs}. \tag{2.92}$$

The last term vanishes again because of the symmetries of $^{(4)}R_{efgd}$, while the first term on the right hand side can be rewritten using

$$\gamma_a^q n^s G_{qs} = \gamma_a^q n^{s\ (4)} R_{qs} - \frac{1}{2}\gamma_a^q n^s g_{qs} \,^{(4)}R = \gamma_a^q n^{s\ (4)} R_{qs}. \tag{2.93}$$

Here the last equality holds because $\gamma_a^q n^s g_{qs} = \gamma_{as} n^s = 0$. Collecting terms and inserting into equation (2.91) we obtain

$$D_b K_a^{\ b} - D_a K = -\gamma_a^q n^s G_{qs}. \tag{2.94}$$

We now define S_a to be the momentum density as measured by a normal observer n^a,

$$S_a \equiv -\gamma_a^b n^c T_{bc}, \tag{2.95}$$

and find

$$\boxed{D_b K_a^{\ b} - D_a K = 8\pi S_a.} \tag{2.96}$$

Equation (2.96) is the *momentum constraint*.

> **Exercise 2.20** Consider a swarm of particles of rest-mass m and proper (comoving) number density n, all moving with the same 4-velocity u^a. The stress-energy tensor for such a swarm is $T^{ab} = mnu^a u^b$. Determine the energy density ρ and momentum density S_a for the swarm and provide a simple physical interpretation for the terms in your expressions.

The Hamiltonian constraint (2.90) and the momentum constraint (2.96) are the direct equivalent of the constraints (2.5) and (2.6) in electrodynamics. They involve only the spatial metric, the extrinsic curvature, and their spatial derivatives. They are the conditions that allow a 3-dimensional slice Σ with data (γ_{ab}, K_{ab}) to be embedded in a 4-dimensional manifold M with data (g_{ab}). Field data (γ_{ab}, K_{ab}) that are being imposed on a timeslice Σ have to satisfy the two constraint equations. We will discuss strategies for solving the constraint equations and finding initial data that represent a snapshot of the gravitational fields at a certain instant of time in Chapter 3.

The evolution equations that evolve the data (γ_{ab}, K_{ab}) forward in time can be found from (2.53), which can be considered as the definition of the extrinsic curvature, and the Ricci equation (2.82). However, the Lie derivative along n^a, $\mathcal{L}_{\mathbf{n}}$, is not a natural time derivative since n^a is not dual to the surface 1-form Ω_a, i.e., their dot product is not unity but rather

$$n^a \Omega_a = -\alpha g^{ab} \nabla_a t \nabla_b t = \alpha^{-1}. \tag{2.97}$$

Instead, consider the vector

$$t^a = \alpha n^a + \beta^a, \tag{2.98}$$

which is dual to Ω_a for any spatial *shift vector* β^a,

$$t^a \Omega_a = \alpha n^a \Omega_a + \beta^a \Omega_a = 1. \tag{2.99}$$

It will prove useful to choose t^a to be the congruence along which we propagate the spatial coordinate grid from one time slice to the next slice. In other words, t^a will connect points with the same spatial coordinates on neighboring time slices. Then the shift vector β^a will measure the amount by which the spatial coordinates are shifted within a slice with respect to the normal vector, as illustrated in Figure 2.4. As we have noted before, the lapse function α measures how much proper time elapses between neighboring time slices along the normal vector. The lapse and the shift therefore determine how the coordinates evolve in time. The choice of α and β^a is quite arbitrary, and we will postpone a discussion of some common choices to Chapter 4. The freedom to choose these four gauge functions α and β^a completely arbitrarily embodies the four-fold coordinate degrees of freedom inherent in general relativity.[11] Specifically, the lapse function reflects the freedom to choose the sequence of time slices, pushing them forward by different amounts of proper time at different spatial points on a slice and thus exploiting "the many-fingered nature of time".[12] The shift vector reflects the freedom to relabel spatial coordinates on each slice in an arbitrary way. Observers who are "at rest" relative to the slices follow the normal congruence n^a and are called either normal or Eulerian observers, while observers

[11] Recall that β^a is spatial and therefore subject to the constraint that $n^a \beta_a = 0$, hence only three of its components may be freely specified.
[12] See, e.g., Misner *et al.* (1973), p. 527.

following the congruence t^a are called coordinate observers. If matter is present it moves entirely independently of the coordinates with 4-velocity u^a.

The duality $t^a \nabla_a t = 1$ implies that the integral curves of t^a are naturally parametrized by t. As a consequence, all (infinitesimal) vectors $t^a dt$ (and hence $\alpha n^a dt$) originating on one spatial slice Σ_t will end on the same spatial slice Σ_{t+dt} (unlike the corresponding vectors $n^a dt$, which generally would end on different slices).[13] This also implies that the Lie derivative of any spatial tensor along t^a is again spatial (see also exercise 2.18).

> **Exercise 2.21** (a) Show that the Lie derivative of the projection operator along αn^a vanishes:
>
> $$\mathcal{L}_{\alpha\mathbf{n}} \gamma^a{}_b = 0. \tag{2.100}$$
>
> (b) Show that the Lie derivative of any spatial tensor along αn^a is again spatial.

Consider now the Lie derivative of K_{ab} along t^a,

$$\mathcal{L}_t K_{ab} = \mathcal{L}_{\alpha\mathbf{n}+\beta} K_{ab} = \alpha \mathcal{L}_{\mathbf{n}} K_{ab} + \mathcal{L}_\beta K_{ab}, \tag{2.101}$$

which follows from the definition of the Lie derivative. Here we can insert the Ricci equation (2.82) to eliminate $\mathcal{L}_{\mathbf{n}} K_{ab}$.

Before we do so, we first rewrite the projection of $^{(4)}R_{abcd}$ that appears in equation (2.82) as

$$n^d n^c \gamma^q{}_a \gamma^r{}_b \, ^{(4)}R_{drcq} = \gamma^{cd} \gamma^q{}_a \gamma^r{}_b \, ^{(4)}R_{drcq} - \gamma^q{}_a \gamma^r{}_b \, ^{(4)}R_{rq}. \tag{2.102}$$

Next we can replace the first term on the right hand side above by substituting Gauss' equation (2.84) and the second term by substituting Einstein's equations:

$$n^d n^c \gamma^q{}_a \gamma^r{}_b \, ^{(4)}R_{drcq} = R_{ab} + K K_{ab} - K_{ac} K^c{}_b - 8\pi \gamma^q{}_a \gamma^r{}_b \left(T_{rq} - \frac{1}{2} g_{rq} T \right), \tag{2.103}$$

where $T = T_{ab} g^{ab}$. We now define the spatial stress and its trace according to

$$S_{ab} \equiv \gamma^c{}_a \gamma^d{}_b T_{cd} \qquad S \equiv S^a{}_a. \tag{2.104}$$

We can then evaluate the last term in equation (2.103) as

$$\gamma^q{}_a \gamma^r{}_b g_{rq} g^{ef} T_{ef} = \gamma_{ab} (\gamma^{ef} - n^e n^f) T_{ef} = \gamma_{ab} (S - \rho). \tag{2.105}$$

Inserting these expressions into (2.82) and (2.101), we find

$$\boxed{\mathcal{L}_t K_{ab} = -D_a D_b \alpha + \alpha (R_{ab} - 2 K_{ac} K^c{}_b + K K_{ab}) - 8\pi \alpha \left(S_{ab} - \frac{1}{2} \gamma_{ab} (S - \rho) \right) + \mathcal{L}_\beta K_{ab}.}$$

$$\tag{2.106}$$

---

[13] Simply stated, the change in t along the vector t^a is $dt = t^a \nabla_a t = 1$ and is thus the same value at all points on the hypersurface Σ_t. Hence the vector congruence t^a connects the hypersurface $t = constant$ (Σ_t) to the hypersurface $t + dt = constant$ (Σ_{t+dt}). In the language of differential forms, the spatial slice Σ_t represented by the 1-form $\widetilde{dt}$ is pierced by the vector t^a by the same amount everywhere on the hypersurface surface, $\langle \widetilde{dt}, t^a \rangle = t^a \nabla_a t = 1$, implying that t^a connects the two neighboring hypersurfaces.

This is the *evolution equation* for the extrinsic curvature. Note that all differential operators and the Ricci tensor R_{ab} are associated with the spatial metric γ_{ab}.

Exercise 2.22 Show that raising an index in equation (2.106) yields

$$\mathcal{L}_t K^a_{\ b} = -D^a D_b \alpha + \alpha(R^a_{\ b} + K K^a_{\ b}) - 8\pi\alpha(S^a_{\ b} - \frac{1}{2}\gamma^a_{\ b}(S - \rho))$$
$$+ \mathcal{L}_\beta K^a_{\ b}. \tag{2.107}$$

The evolution equation for the spatial metric γ_{ab}, the last missing piece, can be found directly from equation (2.53), again using equation (2.98),

$$\boxed{\mathcal{L}_t \gamma_{ab} = -2\alpha K_{ab} + \mathcal{L}_\beta \gamma_{ab}.} \tag{2.108}$$

The coupled evolution equations (2.106) and (2.108) determine the evolution of the gravitational field data (γ_{ab}, K_{ab}). Together with the constraint equations (2.90) and (2.96) they are completely equivalent to Einstein's equations (2.83). Note we have succeeded in recasting Einstein's equations, which are second order in time in their original form, as a coupled set of partial differential equations that are now first order in time. As in electrodynamics, *the evolution equations conserve the constraint equations*, i.e., if the field data (γ_{ab}, K_{ab}) satisfy the constraints at some time t and are evolved with the evolution equations, then the data will also satisfy the constraint equations at all later times (see exercises 2.1 and 2.2).

2.7 Choosing basis vectors: the ADM equations

So far, we have expressed our equations in a covariant, coordinate-independent manner, i.e., the basis vectors $e^a_{(b)}$ have been completely arbitrary and have no particular relationship to the 1-form Ω_a or to the congruence defined by t^a. It is quite intuitive, though, that things will simplify if we adopt a coordinate system that reflects our $3+1$ split of spacetime in a natural way. We will see that the Lie derivative in the evolution equations (2.106) and (2.108) then reduces to a partial derivative with respect to coordinate time and, as an additional benefit, we will be able to ignore all timelike components of spatial tensors.

To do so, we first introduce a basis of three spatial vectors $e^a_{(i)}$ (the subscript $i = 1, 2, 3$ distinguishes the vectors, not the components; we again refer the reader to Section 2.1 for a summary of our notation) that reside in a particular time slice Σ,[14]

$$\Omega_a e^a_{(i)} = 0. \tag{2.109}$$

We extend our spatial vectors to other slices Σ by Lie dragging along t^a,

$$\mathcal{L}_t e^a_{(i)} = 0, \tag{2.110}$$

as illustrated in Figure 2.3.

---

[14] In the language of differential forms, the spatial vectors $e^a_{(i)}$ do not pierce the spatial hypersurface Σ: $\langle \widetilde{\Omega}, e_{(i)} \rangle = \Omega_a e^a_{(i)} = 0$.

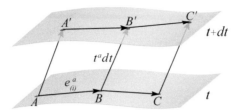

Figure 2.3 Spatial basis vectors $e_{(i)}^a$ are Lie dragged from one spacelike slice to the next along the coordinate congruence t^a. Consequently, these basis vectors connect points with the same spatial coordinates on neighboring slices (e.g., points A, B and C on slice t have the same spatial coordinates as points A', B' and C', respectively, on slice $t + dt$).

> **Exercise 2.23** Show that a spatial vector $e_{(i)}^a$ that is Lie dragged along t^a remains spatial, i.e., show that
>
> $$\mathcal{L}_t(\Omega_a e_{(i)}^a) = 0. \tag{2.111}$$

As the fourth basis vector we pick $e_{(0)}^a = t^a$. Recall that we want to set the vector congruence t^a to be tangent to the coordinate line congruence and therefore connect points with the same spatial coordinates on neighboring time slices. The duality condition (2.99) then implies that $e_{(0)}^a$ has the components[15]

$$t^a = e_{(0)}^a = (1, 0, 0, 0). \tag{2.112}$$

This means that the Lie derivative along t^a reduces to a partial derivative with respect to t : $\mathcal{L}_t = \partial_t$ (see equation A.10).

From equation (2.109) we now find

$$\Omega_a e_{(i)}^a = -\frac{1}{\alpha} n_a e_{(i)}^a = 0, \tag{2.113}$$

which, since the $e_{(i)}^a$ span Σ, implies that the covariant spatial components of the normal vector have to vanish,

$$n_i = 0. \tag{2.114}$$

Since spatial tensors vanish when contracted with the normal vector, this also means that all components of spatial tensors with a contravariant index equal to zero must vanish. For the shift vector, for example, this implies $n_a \beta^a = n_0 \beta^0 = 0$ and hence

$$\beta^a = (0, \beta^i). \tag{2.115}$$

Solving equation (2.98) for n^a then yields the contravariant components

$$n^a = (\alpha^{-1}, -\alpha^{-1}\beta^i), \tag{2.116}$$

---

[15] Strictly, equation (2.112) is not a tensor equation, but rather a specification of components in the adopted basis. In the abstract index notation of Wald (1984), the index a appearing on the left hand side would be denoted with a Greek letter. Since the meaning is clear from the context, we do not make this distinction in the few places that it arises in this book.

and from the normalization condition $n_a n^a = -1$ we find

$$n_a = (-\alpha, 0, 0, 0). \tag{2.117}$$

From the definition of the spatial metric (2.27) we have

$$\gamma_{ij} = g_{ij}, \tag{2.118}$$

meaning that the metric on Σ is just the spatial part of the 4-metric. Since zeroth components of spatial contravariant tensors have to vanish, we also have $\gamma^{a0} = 0$. The inverse metric can therefore be expressed as

$$g^{ab} = \gamma^{ab} - n^a n^b = \begin{pmatrix} -\alpha^{-2} & \alpha^{-2}\beta^i \\ \alpha^{-2}\beta^j & \gamma^{ij} - \alpha^{-2}\beta^i\beta^j \end{pmatrix}. \tag{2.119}$$

Exercise 2.24 Show that

$$\gamma^{ik}\gamma_{kj} = \delta^i{}_j. \tag{2.120}$$

Equation (2.120) implies that γ^{ij} and γ_{ij} are 3-dimensional inverses, and can hence be used to raise and lower spatial indices of spatial tensors. For example, the covariant form of the shift vector is

$$\beta_i = \gamma_{ij}\beta^j. \tag{2.121}$$

We can now invert (2.119) and find the components of the 4-dimensional metric

$$g_{ab} = \begin{pmatrix} -\alpha^2 + \beta_l\beta^l & \beta_i \\ \beta_j & \gamma_{ij} \end{pmatrix}. \tag{2.122}$$

Equivalently, the line element may be decomposed as

$$ds^2 = -\alpha^2 dt^2 + \gamma_{ij}(dx^i + \beta^i dt)(dx^j + \beta^j dt), \tag{2.123}$$

which is often refered to as the metric in 3 + 1 form. We may interpret this line element as the Pythagorean theorem for a 4-dimensional spacetime, $ds^2 = -$ (proper time between neighboring spatial hypersurfaces)^2 + (proper distance within the spatial hypersurface)^2. This equation thus determines the invariant interval between neighboring points A and B, as illustrated in Figure 2.4.

Exercise 2.25 Show that the determinant $g = \det(g_{ab})$ of the spacetime metric g_{ab} can be written as

$$\sqrt{-g} = \alpha\sqrt{\gamma}, \tag{2.124}$$

where $\gamma = \det(\gamma_{ij})$ is the determinant of the spatial metric γ_{ij}.
Hint: Recall that for any square matrix A_{ij} the following is true: $(A^{-1})_{ij} = $ cofactor of $A_{ji}/\det A$.

Exercise 2.26 Use equation (2.123) directly to determine the proper time $d\tau$ measured by a clock carried by a normal observer n^a in a coordinate time interval dt.

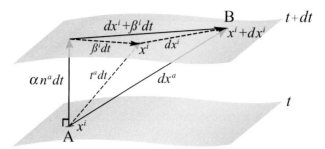

Figure 2.4 Pythagorean theorem in $3 + 1$ dimensional spacetime. The normal vector αn^a and the time vector t^a connect points on two neighboring spatial slices. The shift vector β^i resides in the slice and measures their difference. The infinitesimal displacement vector dx^a connects two nearby, but otherwise arbitrary, points on neighboring slices (e.g., the point A at x^i on slice t and the point B at $x^i + dx^i$ on slice $t + dt$). The total displacement vector $dx^a = t^a dt + dx^i$, where dx^i is the spatial vector drawn in the figure, may be decomposed alternatively into two vectors that form the legs of a right-triangle, $dx^a = (\alpha n^a dt) + (dx^i + \beta^i dt)$, as shown. Using this decomposition to evaluate the invariant interval $ds^2 = dx^a dx_a$, commonly expanded as in equation (1.1), yields the Pythagorean theorem, equation (2.123).

The entire content of any spatial tensor is available from their spatial components. This is obviously true for contravariant components, since their zeroth component vanishes, but also holds covariant components. Therefore, the entire content of the decomposed Einstein equations is contained in their spatial components alone, and we can rewrite the Hamiltonian constraint (2.90),

$$R + K^2 - K_{ij}K^{ij} = 16\pi\rho, \qquad (2.125)$$

the momentum constraint (2.96),

$$D_j(K^{ij} - \gamma^{ij}K) = 8\pi S^i, \qquad (2.126)$$

the evolution equation for the extrinsic curvature (2.106),

$$\begin{aligned} \partial_t K_{ij} = -D_i D_j \alpha + \alpha(R_{ij} - 2K_{ik}K^k_{\ j} + K K_{ij}) - 8\pi\alpha(S_{ij} - \tfrac{1}{2}\gamma_{ij}(S - \rho)) \\ + \beta^k D_k K_{ij} + K_{ik}D_j\beta^k + K_{kj}D_i\beta^k, \end{aligned} \qquad (2.127)$$

and the evolution equation for the spatial metric (2.108),

$$\partial_t \gamma_{ij} = -2\alpha K_{ij} + D_i\beta_j + D_j\beta_i. \qquad (2.128)$$

Equations (2.125)–(2.128) are equivalent to Einstein's equations (2.83) and comprise the "standard" $3 + 1$ equations. Sometimes they are referred to as the "ADM" equations after Arnowitt, Deser and Misner,[16] even though these equations had been derived earlier,[17] and even though Arnowitt *et al.* (1962) derived the equations in a different form.[18]

---

[16] Arnowitt *et al.* (1962).
[17] E.g., Darmois (1927); Lichnerowicz (1944); Fourès-Bruhat (1956).
[18] See, e.g., Anderson and York, Jr. (1998) for a discussion.

The shift terms in (2.127) and (2.128) arise from the Lie derivatives $\mathcal{L}_\beta \gamma_{ij}$ and $\mathcal{L}_\beta K_{ij}$. In (2.128) it is convenient to express the Lie derivative in terms of the covariant derivative D_i to eliminate the term $\beta^k D_k \gamma_{ij}$, but in (2.127) the covariant derivatives in the shift terms may be replaced with the partial derivatives ∂_i (see Appendix A).

Exercise 2.27 Show that the determinant $\gamma = \det(\gamma_{ij})$ of the spatial metric and the trace $K = K^i_{\ i}$ of the extrinsic curvature satisfy the equations

$$\partial_t \ln \gamma^{1/2} = -\alpha K + D_i \beta^i \tag{2.129}$$

and

$$\partial_t K = -D^2 \alpha + \alpha \left(K_{ij} K^{ij} + 4\pi(\rho + S) \right) + \beta^i D_i K, \tag{2.130}$$

where $D^2 = \gamma^{ij} D_i D_j$ is the Laplace operator associated with γ_{ij}.

Box 2.1   The standard $3 + 1$ or ADM equations

In the standard $3 + 1$ decomposition, the metric is written as

$$ds^2 = -\alpha^2 dt^2 + \gamma_{ij}(dx^i + \beta^i dt)(dx^j + \beta^j dt). \tag{2.131}$$

Einstein's equations are then decomposed into the *Hamiltonian constraint*,

$$R + K^2 - K_{ij} K^{ij} = 16\pi\rho, \tag{2.132}$$

the *momentum constraint*,

$$D_j(K^{ij} - \gamma^{ij} K) = 8\pi S^i, \tag{2.133}$$

the *evolution equation for the spatial metric*,

$$\partial_t \gamma_{ij} = -2\alpha K_{ij} + D_i \beta_j + D_j \beta_i \tag{2.134}$$

(which is really a definition of the extrinsic curvature), and the *evolution equation for the extrinsic curvature*,

$$\partial_t K_{ij} = \alpha(R_{ij} - 2K_{ik} K^k_{\ j} + K K_{ij}) - D_i D_j \alpha - 8\pi\alpha(S_{ij} - \tfrac{1}{2}\gamma_{ij}(S - \rho))$$
$$+ \beta^k \partial_k K_{ij} + K_{ik} \partial_j \beta^k + K_{kj} \partial_i \beta^k. \tag{2.135}$$

Useful contractions of the two evolution equations are

$$\partial_t \ln \gamma^{1/2} = -\alpha K + D_i \beta^i \tag{2.136}$$

and

$$\partial_t K = -D^2 \alpha + \alpha \left(K_{ij} K^{ij} + 4\pi(\rho + S) \right) + \beta^i D_i K, \tag{2.137}$$

where we have used the Hamiltonian constraint (2.132) in deriving (2.137). The matter source terms appearing in the above equations are defined by

$$\rho = n_a n_b T^{ab}, \qquad S^i = -\gamma^{ij} n^a T_{aj}, \qquad S_{ij} = \gamma_{ia} \gamma_{jb} T^{ab}, \qquad S = \gamma^{ij} S_{ij}. \tag{2.138}$$

Exercise 2.28   Argue that when treating the weak-field, slow-velocity (Newtonian) limit of general relativity in nearly inertial coordinates, the lapse function is given by

$$\alpha = 1 + \Phi, \tag{2.139}$$

where Φ is the Newtonian gravitational potential (which we assume to vanish at spatial infinity).

It is instructive to compare the standard $3 + 1$ gravitational field equations (2.125)–(2.128) with Maxwell's equations of electrodynamics as summarized in Section 2.2. Evidently, the two sets of equations have a very similar structure. The two $3 + 1$ evolution equations (2.127) and (2.128) are quite similar to Maxwell's evolution equations (2.11) and (2.12) if we identify the vector potential A_i with the spatial metric γ_{ij} and the electric field E_i with the extrinsic curvature K_{ij}. The right hand sides of both (2.128) and (2.11) involve a field variable and a spatial derivative of a gauge variable, while the right hand sides of both (2.127) and (2.12) involve matter source terms as well as second spatial derivatives of the second field variable (which in 2.127 are hidden in the Ricci tensor). The constraint (2.5) constrains the divergence of the electric field and can therefore be identified with the momentum constraint (2.126), which constrains the divergence of the extrinsic curvature. The most important differences between the two theories are also obvious: electromagnetism is a linear, vector field theory, while general relativity is a nonlinear, tensor field theory.

Equations (2.125)–(2.128) represent the spatial components of the corresponding equations (2.90), (2.96), (2.106) and (2.108). As a further simplification, it turns out that in evaluating the right hand sides of equations (2.125)–(2.128), which involves expanding derivatives or evaluating the Ricci tensor R_{ij}, we can disregard all components with indices equal to 0. Expanding the term $D_a D_b \alpha$ in (2.106), for example, yields $\partial_a \partial_b \alpha - \partial_c \alpha \, \Gamma^c_{ba}$. Here, the index c runs from 0 to 3, even if we restrict a and b to be spatial. From their definition (2.44) with $\gamma^{a0} = 0$, it is clear that

$$\Gamma^0_{ba} = 0, \tag{2.140}$$

so that we can ignore the component with $c = 0$, and the connection coefficients can be computed from spatial components alone

$$\Gamma^i_{jk} = \frac{1}{2} \gamma^{il} (\partial_k \gamma_{lj} + \partial_j \gamma_{lk} - \partial_l \gamma_{jk}). \tag{2.141}$$

Similarily, the Ricci tensor R_{ij} can be found from the spatial components γ_{ij} alone,

$$R_{ij} = \partial_k \Gamma^k_{ij} - \partial_j \Gamma^k_{ik} + \Gamma^k_{ij} \Gamma^l_{kl} - \Gamma^k_{il} \Gamma^l_{jk}, \tag{2.142}$$

or, in terms of second derivatives of the metric,

$$R_{ij} = \frac{1}{2} \gamma^{kl} \left(\partial_i \partial_l \gamma_{kj} + \partial_k \partial_j \gamma_{il} - \partial_i \partial_j \gamma_{kl} - \partial_k \partial_l \gamma_{ij} \right) + \gamma^{kl} \left(\Gamma^m_{il} \Gamma_{mkj} - \Gamma^m_{ij} \Gamma_{mkl} \right). \tag{2.143}$$

Exercise 2.29 Find the connection coefficients for a spherically symmetric spatial metric of the form

$$\gamma_{ij} = \psi^4 \, \eta_{ij} = \psi^4 \, \text{diag}\left(1, r^2, r^2 \sin^2 \theta\right), \qquad (2.144)$$

where the function ψ, which we will refer to as the *conformal factor* in Chapter 3, depends on r alone.
Hint: Problem 7.6 in Lightman *et al.* (1975).

Exercise 2.30 Show that for any spatial vector V^i, $\nabla_i V_j = D_i V_j$, but that, in general, $\nabla_i V^j \neq D_i V^j$ and $\nabla^i V_j \neq D^i V_j$.
Hint: Use equation (2.62).

As a consistency check, it is useful to return to the Schwarzschild spacetime in isotropic spherical polar coordinates that we considered earlier in Section 2.3. Comparing the spacetime metric (2.35) with the $3 + 1$ metric (2.123) we can identify the lapse as

$$\alpha = \frac{1 - M/(2r)}{1 + M/(2r)} \qquad (2.145)$$

(as we have already noted), the shift

$$\beta^i = 0 \qquad (2.146)$$

and the spatial metric

$$\gamma_{ij} = \left(1 + \frac{M}{2r}\right)^4 \text{diag}\left(1, r^2, r^2 \sin^2 \theta\right). \qquad (2.147)$$

Since γ_{ij} is independent of time and $\beta^i = 0$, the evolution equation (2.128) immediately yields

$$K_{ij} = 0, \qquad (2.148)$$

as we have previously discovered in exercise 2.11. The connection coefficients can be found from exercise 2.29 with $\psi = 1 + M/(2r)$. The nonvanishing components of the 3-dimensional Ricci tensor are

$$R_{rr} = -\frac{8r\,M}{(2r^2 + Mr)^2}, \qquad R_{\theta\theta} = \frac{4r^3\,M}{(2r^2 + Mr)^2}, \qquad R_{\phi\phi} = \sin^2 \theta\, R_{\theta\theta}. \qquad (2.149)$$

Recall that in vacuum the 4-dimensional Ricci tensor vanishes, of course, $^{(4)}R_{ij} = 0$. Here, however, we are computing the 3-dimensional Ricci tensor R_{ij}, which is nonzero.

We can now convince ourselves that this solution indeed satisfies the constraint and evolution equations (2.125)–(2.128). Taking the trace of the Ricci tensor (2.149) yields $R = 0$, as it must to satisfy the Hamiltonian constraint (2.125) with $K_{ij} = \rho = 0$. In vacuum the momentum constraint (2.126) only involves the extrinsic curvature, which is zero, so that this equation is satisfied trivially. We have used the evolution equation for the metric (2.128) to compute the extrinsic curvature, so it is obviously satisfied. The only

Table 2.1 The lapse α, shift β^i, spatial metric γ_{ij}, extrinsic curvature K_{ij} and mean curvature K for a Schwarzschild spacetime in different coordinate systems. Here $\eta_{ij} = \delta_{ij}$ and $l^i = l_i = x^i/r$ with $r^2 = x^2 + y^2 + z^2$ in Cartesian coordinates, or $\eta_{ij} = \text{diag}(1, r^2, r^2 \sin^2 \theta)$ and $l^i = l_i = (1, 0, 0)$ in spherical polar coordinates.

	Schwarzschild	Isotropic	Painlevé–Gullstrand	Kerr–Schild
α	$\left(1 - \frac{2M}{r}\right)^{1/2}$	$\frac{1 - M/(2r)}{1 + M/(2r)}$	1	$\left(1 + \frac{2M}{r}\right)^{-1/2}$
β^i	0	0	$\left(\frac{2M}{r}\right)^{1/2} l^i$	$\frac{2M}{r} \alpha^2 l^i$
γ_{ij}	$\text{diag}\left(\left(1 - \frac{2M}{r}\right)^{-1}, r^2, r^2 \sin^2 \theta\right)$	$\left(1 + \frac{M}{2r}\right)^4 \eta_{ij}$	η_{ij}	$\eta_{ij} + \frac{2M}{r} l_i l_j$
K_{ij}	0	0	$\left(\frac{2M}{r^3}\right)^{1/2} \left(\eta_{ij} - \frac{3}{2} l_i l_j\right)$	$\frac{2M\alpha}{r^2} \left(\eta_{ij} - \left(2 + \frac{M}{r}\right) l_i l_j\right)$
K	0	0	$\frac{3}{2} \left(\frac{2M}{r^3}\right)^{1/2}$	$\frac{2M\alpha^3}{r^2} \left(1 + \frac{3M}{r}\right)$

nontrivial equation is the evolution equation for the extrinsic curvature (2.127). In addition to the Ricci tensor (2.149) it involves the second derivatives of the lapse, $D_i D_j \alpha$.

> **Exercise 2.31** Compute the second derivatives $D_i D_j \alpha$ for the Schwarzschild solution in isotropic coordinates.

Exercise 2.31 demonstrates that the two terms cancel exactly so that $\partial_t K_{ij} = 0$. This is not a surprising result, of course, but it is reassuring to see that the formalism works as we expect.

Identifying the lapse, shift and spatial metric from a given spacetime metric g_{ab} is straightforward whenever the shift vanishes, since the lapse and spatial metric can be read off immediately from (2.122). For nonzero shift the spatial metric and the covariant components of the shift β_i can be identified from (2.122). By inverting the spatial metric the contravariant components of the shift β^i can be determined, after which the lapse can be found from g_{00}, again using equation (2.122).

In the following exercises we encourage the reader to work through this formalism for several well-known solutions to Einstein's equations. We summarize some of the results for a Schwarzschild spacetime, expressed in a number of different coordinate systems, in Table 2.1.[19]

> **Exercise 2.32** In spherical polar *Painlevé–Gullstrand* coordinates the line element of a Schwarzschild spacetime is
>
> $$ds^2 = -dt^2 + \left(dr + \left(\frac{2M}{r}\right)^{1/2} dt\right)^2 + r^2(d\theta^2 + \sin^2 \theta d\phi^2). \quad (2.150)$$

---

[19] In Table 2.1, the column "isotropic" refers to spatial slices of constant Schwarzschild time t represented in an isotropic spatial coordinate system. Other slices of the Schwarzschild spacetime may also be expressed in isotropic coordinates; see exercise H.1.

(a) Identify the lapse α, the shift β^i and note that the spatial metric γ_{ij} is *flat*, which is a remarkable property.
(b) Compute the extrinsic curvature K_{ij} and the Ricci tensor R_{ij}.
(c) Verify that this solution satisfies the constraint and evolution equations (2.125)–(2.128) with $\partial_t K_{ij} = 0$.

Exercise 2.33 In *Kerr–Schild* coordinates, or ingoing *Eddington–Finkelstein* coordinates, the line element of a Schwarzschild spacetime is

$$ds^2 = (\eta_{ab} + 2Hl_al_b)\,dx^a dx^b. \tag{2.151}$$

Here the 4-vector l^a is null with respect to both g_{ab} and η_{ab}, $\eta_{ab}l^al^b = g_{ab}l^al^b = 0$, and, in Cartesian coordinates, can be written

$$l_t = -l^t = 1, \qquad l_i = l^i = \frac{x^i}{r} \tag{2.152}$$

with $r^2 = x^2 + y^2 + z^2$. We have also defined

$$H = \frac{M}{r}. \tag{2.153}$$

Recall that in Cartesian coordinates $\eta_{ab} = \mathrm{diag}(-1, 1, 1, 1)$. Identify the lapse α, the shift β^i and the spatial metric γ_{ij}, and show that the extrinsic curvature is

$$K_{ij} = \frac{2H\alpha}{r}(\eta_{ij} - (2 + H)l_il_j). \tag{2.154}$$

Exercise 2.34 In *Boyer–Lindquist* coordinates, the spacetime metric of a rotating *Kerr* black hole is

$$ds^2 = -\left(1 - \frac{2Mr}{\rho^2}\right)dt^2 - 4\frac{Mar\sin^2\theta}{\rho^2}dt d\phi$$
$$+ \frac{\rho^2}{\Delta}dr^2 + \rho^2 d\theta^2 + \frac{\sin^2\theta}{\rho^2}\left((r^2 + a^2)^2 - a^2\Delta\sin^2\theta\right)d\phi^2 \tag{2.155}$$

where

$$\Delta = r^2 - 2Mr + a^2 \tag{2.156}$$

and

$$\rho^2 = r^2 + a^2\cos^2\theta, \tag{2.157}$$

and where $a = J/M$ is the angular momentum per unit mass.
(a) Identify the lapse α, the shift β^i and the spatial metric γ_{ij}. Evaluate $t^a t_a$ and determine its sign both inside and outside the ergosphere.
(b) Show that in the asymptotic region where $r \gg M$ and $r \gg a$ the only nonvanishing component of the extrinsic curvature is given by

$$K_{r\phi} = \frac{3Ma}{r^2}\sin^2\theta. \tag{2.158}$$

Another useful set of coordinates for rotating Kerr black holes is the *Kerr–Schild* coordinate system employed in exercise 2.33 for a Schwarzschild black hole. In fact, the metric, lapse and shift still take the same form as in equation (2.151) of exercise 2.33, except that the vectors l^a and the function H now depend on the angular momentum parameter $a = J/M$. Specifically, l^a takes the form

$$l_a = \left(1, \frac{rx + ay}{r^2 + a^2}, \frac{ry - ax}{r^2 + a^2}, \frac{z}{r}\right), \tag{2.159}$$

where we assume that the axis of rotation is aligned with the z-axis, and the function H is

$$H = \frac{Mr}{r^2 + a^2 \cos^2 \theta}, \tag{2.160}$$

where $\cos \theta = z/r$. From $\eta_{ab} l^a l^b = 0$ we find that r must be related to the Cartesian coordinates x, y and z by

$$\frac{x^2 + y^2}{r^2 + a^2} + \frac{z^2}{r^2} = 1. \tag{2.161}$$

All these functions reduce to their nonrotating counterparts in exercise 2.33 when $a = 0$.

We close this chapter with two final comments about the role of the lapse α and the shift β^i. The constraint equations (2.125) and (2.126) are independent of these functions. This is what we should expect, since the lapse and the shift determine how the coordinates evolve from one time slice Σ to the next, whereas the constraint equations represent integrability conditions which have to be satisfied *within* each slice. Therefore, the constraints have to be independent of how the coordinates evolve, and the lapse and the shift can enter only the evolution equations.

We also point out that the decomposed Einstein equations (2.125)–(2.128) do not provide any equations for α and β^i. Again, this is not surprising, since these functions represent the coordinate freedom of general relativity. The lapse and shift are therefore arbitrary, and must be determined by imposing gauge conditions. Clearly, different gauge conditions will lead to different functions for the spatial metric γ_{ij} and the extrinsic curvature K_{ij} when they are used in the evolution equations. Even for a stationary spacetime, like Schwarzschild, most choices for the lapse and shift will lead to a time-dependent spatial metric function. Only for special choices of the lapse and the shift will the spatial metric of a stationary spacetime remain time-independent. This will be the case if our time-vector t^a, defined in equation (2.98), is aligned with a Killing vector of the spacetime,

$$\xi^a = t^a = \alpha n^a + \beta^a. \tag{2.162}$$

We refer to a lapse and shift that generate a Killing vector according to equation (2.162) as a *Killing lapse* and *Killing shift*. The lapses and shifts that we listed in Table 2.1 are examples of such Killing lapses and Killing shifts. In that context we identified them directly from the stationary spacetime metric. If we used those lapses and shifts in a dynamical evolution, the metric would indeed remain time-independent.

In Chapter 4 we will discuss and compare a few canonical gauge conditions that form the basis of choices frequently adopted in dynamical simulations. We will further highlight some specific choices when we summarize some of these simulations in subsequent chapters.

3 Constructing initial data

As we have seen in Chapter 2, the spatial metric γ_{ij}, the extrinsic curvature K_{ij} and any matter fields have to satisfy the Hamiltonian constraint (2.132),

$$R + K^2 - K_{ij}K^{ij} = 16\pi\rho, \tag{3.1}$$

and the momentum constraint (2.133),

$$D_j(K^{ij} - \gamma^{ij}K) = 8\pi S^i \tag{3.2}$$

on every spacelike hypersurface Σ. Before we can evolve the fields to obtain a spacetime that satisfies Einstein's equations, we have to specify gravitational fields (γ_{ij}, K_{ij}) on some initial spatial slice Σ that are compatible with the constraint equations. These fields can then be used as "starting values" for a dynamical evolution obtained by integrating the evolution equations (2.135) and (2.134).[1]

Clearly, the four constraint equations (3.1) and (3.2) cannot determine all of the gravitational fields (γ_{ij}, K_{ij}). Since both γ_{ij} and K_{ij} are symmetric, 3-dimensional tensors, they together have 12 independent components. The four constraint equations can only determine four of these, leaving eight undetermined. Four of these eight undetermined functions are related to coordinate choices: three specify the spatial coordinates within the slice Σ, and these coordinates can be chosen arbitrarily without changing the physical properties of the slice. One function, associated with the time coordinate, can be used to specify the choice of the hypersurface Σ. For any given spacetime solution to Einstein's equations, choosing different hypersurfaces on which to obtain initial data, and then propagating these data, regenerates physically equivalent spacetimes. This leaves four undetermined functions that represent the two dynamical degrees of freedom characterizing a gravitational field in general relativity, e.g., two independent sets of values for the conjugate pair (γ_{ij}, K_{ij}). These two dynamical degrees of freedom correspond to the two polarization modes of a gravitational wave in general relativity.[2]

It is quite intuitive that the state of a dynamical field, like a gravitational wave, cannot be determined from constraint equations. Waves satisfy hyperbolic equations, and their state

---

[1] If there is any matter present, then the matter fields must be specified on the initial hypersurface as well, and the matter evolution equations, $\nabla_b T^{ab} = 0$, must be integrated simultaneously with the field evolution equations to build a spacetime. Some typical matter sources and their evolution equations are discussed in Chapter 5.

[2] This is the same number of degrees of freedom in the linearized theory of general relativity appropriate for a weak gravitational field; the field equations in this limit reduce to propagation equations for a spin-2 linear field in Minkowski spacetime.

at any time depends on their past history. Constraint equations, on the other hand, tend to be elliptic in nature and constrain the fields in space at one instant of time, independently of their past history. It is therefore natural that the constraint equations serve to constrain only the "longitudinal" parts of the fields, while the "transverse" parts, related to the dynamical degrees of freedom, remain freely specifiable.

Ideally one would like to separate unambiguously the longitudinal from the transverse parts of the fields at some initial time, freely specifying the latter and then solving the constraints for the former. Given the nonlinear nature of general relativity such a rigorous separation is not possible; instead, all these fields are entangled in the spatial metric and the extrinsic curvature. We can nevertheless introduce decompositions of γ_{ij} and K_{ij} that allow for a convenient split of constrained from freely specifiable variables. These decompositions often amount to an approximate split of transverse from longitudinal pieces of the field and, in any case, serve to simplify the solution of the resulting constraint equations.

Typically, then, the solution of Einstein's initial value equations proceeds along the following lines. We first decide which field variables we want to determine by solving constraint equations. This amounts to choosing a particular *decomposition* of the constraint equations. We then have to make choices for the remaining, freely specifiable variables. These choices should reflect the physical or astrophysical situation at hand, but may also be guided by any resulting simplifications that they induce in the constraint equations. Lastly, we must solve these equations for the constrained field variables.

We point out that this situation is similar to what we encounter in electrodynamics. As we have seen in Chapter 2.2, Maxwell's equations also split into constraint and evolution equations. The constraint equations (2.5) and (2.6) have to be satisfied by any electric and magnetic field at each instant of time, but they are not sufficient to completely determine these fields. Consider the equation for the electric field E^i,

$$D_i E^i = 4\pi\rho. \tag{3.3}$$

Given an electrical charge density ρ, we can solve this equation for one of the components of E^i, but not all three of them. For example, we could make certain choices for E^x and E^y, and then solve (3.3) for E^z, even though we might be troubled by the asymmetry in singling out one particular component in this approach. Alternatively, we may prefer to write E^i as some "background" field $\bar{E}^i$ times some overall scaling factor, say ψ^4,

$$E^i = \psi^4 \bar{E}^i. \tag{3.4}$$

We could now insert (3.4) into (3.3), make certain choices for all three components of the background field $\bar{E}^i$, and then solve (3.3) for the scaling factor ψ^4. Though it might not be so useful for treating Maxwell's equations, such an approach leads to a very convenient and tractable system for Einstein's equations, as we shall discuss in the following sections.

3.1 Conformal transformations

3.1.1 Conformal transformation of the spatial metric

By analogy with our electromagnetic example (3.4) we begin by writing the spatial metric γ_{ij} as a product of some power of a positive scaling factor ψ and a *background* metric $\bar{\gamma}_{ij}$,

$$\gamma_{ij} = \psi^4 \bar{\gamma}_{ij}. \tag{3.5}$$

This identification is a *conformal transformation* of the spatial metric.[3] We call ψ the *conformal factor*, and $\bar{\gamma}_{ij}$ the *conformally related* metric. Taking ψ to the fourth power turns out to be convenient, but is otherwise arbitrary. In 3 dimensions it is natural to use

$$\bar{\gamma}_{ij} = \gamma^{-1/3}\gamma_{ij}, \tag{3.6}$$

where γ is the determinant of γ_{ij} and $\gamma = \psi^{12}$. This particular choice results in $\bar{\gamma} = 1$, but we can instead choose any normalization for $\bar{\gamma}$. Loosely speaking, the conformal factor absorbs the overall scale of the metric, which leaves five degrees of freedom in the conformally related metric.

To avoid any misunderstandings, we would like to emphasize that in this book we only consider conformal transformations of the spatial metric. In a different context it may also be useful to study conformal transformations of the spacetime metric. While many of the general properties remain the same, some of the specific results that we derive below hold only in three (spatial) dimensions.[4]

Superficially, the conformal transformation (3.5) is just a mathematical trick, namely, rewriting one unknown as a product of two unknowns in order to make solving some equations easier. At a deeper level, however, the conformal transformation serves to define an equivalence class of manifolds and metrics.[5]

Inserting the transformation law (3.5) into (2.141) we find that, in three dimensions, the connection coefficients must transform according to

$$\Gamma^i_{jk} = \bar{\Gamma}^i_{jk} + 2(\delta^i{}_j \bar{D}_k \ln \psi + \delta^i{}_k \bar{D}_j \ln \psi - \bar{\gamma}_{jk} \bar{\gamma}^{il} \bar{D}_l \ln \psi). \tag{3.7}$$

Here we have used

$$\gamma^{ij} = \psi^{-4} \bar{\gamma}^{ij}, \tag{3.8}$$

where $\bar{\gamma}^{ij}$ is the inverse of $\bar{\gamma}_{ij}$. From now on we will denote all objects associated with the conformal metric $\bar{\gamma}_{ij}$ with a bar. In the above equations ψ must be treated as a scalar function in covariant derivatives (as opposed to a scalar density; *cf.* Appendix A.3).

---

[3] See Lichnerowicz (1944); York, Jr. (1971).

[4] Note that the Weyl, or conformal, tensor (1.26), is invariant only under conformal transformations of the spacetime metric.

[5] Equivalence classes of conformally related manifolds share some geometric properties. For conformally related spacetimes, for example, the Weyl tensor (1.26) is identical up to a coordinate transformation. For 3-dimensional spaces that are conformally related the Bach tensor (3.15) below will be the same, again up to a coordinate transformation. One class of manifolds that we will encounter repeatedly are conformally flat, for which the Bach tensor (3.15) vanishes.

Exercise 3.1   Verify the transformation law (3.7) from (2.141).

Exercise 3.2   Show that the covariant derivative associated with the connection (3.7) is compatible with the conformally related metric,

$$\bar{D}_i \bar{\gamma}_{jk} = 0. \tag{3.9}$$

For the Ricci tensor we similarly find

$$R_{ij} = \bar{R}_{ij} - 2\left(\bar{D}_i \bar{D}_j \ln\psi + \bar{\gamma}_{ij}\bar{\gamma}^{lm}\bar{D}_l\bar{D}_m \ln\psi\right)$$
$$+ 4\left((\bar{D}_i \ln\psi)(\bar{D}_j \ln\psi) - \bar{\gamma}_{ij}\bar{\gamma}^{lm}(\bar{D}_l \ln\psi)(\bar{D}_m \ln\psi)\right), \tag{3.10}$$

and for the scalar curvature

$$R = \psi^{-4}\bar{R} - 8\psi^{-5}\bar{D}^2\psi. \tag{3.11}$$

Here $\bar{D}^2 = \bar{\gamma}^{ij}\bar{D}_i\bar{D}_j$ is the covariant Laplace operator associated with $\bar{\gamma}_{ij}$.

Exercise 3.3   Verify equations (3.10) and (3.11).

Inserting the scalar curvature (3.11) into the Hamiltonian constraint (3.1) yields

$$8\bar{D}^2\psi - \psi\bar{R} - \psi^5 K^2 + \psi^5 K_{ij}K^{ij} = -16\pi\psi^5\rho, \tag{3.12}$$

which, for a given choice of the conformally related metric $\bar{\gamma}_{ij}$, we may interpret as an equation for the conformal factor ψ. The extrinsic curvature K_{ij} has to satisfy the momentum constraint (3.2), and it will be useful to rescale K_{ij} conformally as well. We will do that in Section 3.1.3, after discussing in Section 3.1.2 some elementary solutions to (3.12) for which $K_{ij} = 0$.

3.1.2 Elementary black hole solutions

At this point it is instructive to consider some simple, but physically interesting, solutions to the constraint equation. Consider vacuum solutions for which the matter source terms vanish ($\rho = 0 = S^i$, etc.) and focus on a "moment of time symmetry". At a moment of time symmetry, all time derivatives of γ_{ij} are zero and the 4-dimensional line interval has to be invariant under time reversal, $t \to -t$. The latter condition implies that the shift must satisfy $\beta^i = 0$ and, hence, by equation (2.134), the extrinsic curvature also has to vanish everywhere on the slice, i.e., $K_{ij} = 0 = K$.[6] On such a time slice the momentum constraints (3.2) are satisfied trivially. The Hamiltonian constraint (3.12) reduces to

$$\bar{D}^2\psi = \frac{1}{8}\psi\bar{R}. \tag{3.13}$$

---

[6] A 4-geometry is said to be time-symmetric if there exists such a time slice.

Let us further choose the conformally related metric to be flat,

$$\bar{\gamma}_{ij} = \eta_{ij}. \tag{3.14}$$

Whenever this is the case, we call the physical spatial metric γ_{ij} *conformally flat*.

We again emphasize that we only consider conformal transformations of the spatial metric in this text. Accordingly, "conformal flatness" refers, for our purposes, to the spatial metric and not the spacetime metric. In four or any higher dimensions, we can evaluate the Weyl tensor to examine whether any given metric is conformally flat. This is a consequence of the fact that the Weyl tensor (1.26) is invariant under conformal transformations of the spacetime metric – this explains why it is often called the *conformal tensor*.[7] Since it vanishes for a flat metric, it must also vanish for all geometries that are conformally related to the flat metric. In three dimensions, however, the Weyl tensor vanishes identically, so that it no longer provides a useful diagnostic. For a spatial metric, we may instead evaluate the *Bach* or *Cotton–York* tensor

$$B^{ij} = \gamma^{1/3} [ikl] D_k \left(R_l{}^j - \frac{1}{4} \delta_l{}^j R \right), \tag{3.15}$$

which vanishes if and only if the spatial geometry is conformally flat.[8] In equation (3.15), $[ikl]$ is the completely antisymmetric, or permutation, symbol, with $[123] = 1$.

As an aside, we note that any spherically symmetric spatial metric is always conformally flat, meaning that we can always write such a metric as $\gamma_{ij} = \psi^4 \eta_{ij}$. For any spherically symmetric space, we may hence assume conformal flatness without loss of generality.

Assuming conformal flatness dramatically simplifies all calculations, since $\bar{D}_i$ reduces to the flat covariant derivative (and in particular to partial derivatives in Cartesian coordinates). Moreover, the Ricci tensor and scalar curvature associated with the conformally related metric must now vanish, $\bar{R}_{ij} = \bar{R} = 0$. Under this assumption, the Hamiltonian constraint becomes the remarkably simple Laplace equation

$$\bar{D}^2 \psi = 0, \tag{3.16}$$

where $\bar{D}^2$ is now the flat Laplace operator. We will be interested in *asymptotically flat* solutions that satisfy

$$\psi \to 1 + \mathcal{O}(r^{-1}) \quad \text{for} \quad r \to \infty \tag{3.17}$$

where r is the coordinate radius. Spherically symmetric solutions are

$$\psi = 1 + \frac{\mathcal{M}}{2r}. \tag{3.18}$$

We will show in exercise 3.20 below that, in this particular case, the constant $\mathcal{M}$ is in fact the black hole mass M. It shouldn't come as a great surprise that this is just the

---

[7] See, e.g., page 130 in Carroll (2004).
[8] See Eisenhart (1926); York, Jr. (1971); see also Problem 21.22 in Misner *et al.* (1973). Note that the Bach tensor is a tensor density rather than a tensor; see Appendix A.3.

Schwarzschild solution in isotropic coordinates, which we have seen before in Chapter 2. Inspection of the metric (2.147),

$$dl^2 = \gamma_{ij}dx^i dx^j = \left(1 + \frac{M}{2r}\right)^4 \eta_{ij}dx^i dx^j = \left(1 + \frac{M}{2r}\right)^4 \left(dr^2 + r^2(d\theta^2 + \sin^2\theta d\phi^2)\right),$$

(3.19)

shows that it is explicitly written in the form (3.5), where the conformally related metric is the flat metric in spherical polar coordinates. This solution forms the basis of the so-called *puncture* methods for black holes, which we will discuss in much greater detail in Chapters 12.2.2 and 13.1.3.

While the formalism of conformal decomposition may appear unnecessarily technical and perhaps confusing initially, this example demonstrates that it provides an extremely powerful tool for constructing solutions to Einstein's equations. In fact, once the formalism has been developed, it is much easier to derive the Schwarzschild solution by going through the above steps then deriving it from Einstein's equations directly. Perhaps even more impressively, we will see below that we can trivially generalize this method to construct multiple black hole initial data. Before we do that, though, it is useful to discuss the single black hole solution in more detail.

The solution (3.19) is singular at $r = 0$. However, we can show that this singularity is only a coordinate singularity by considering the coordinate transformation

$$r = \left(\frac{M}{2}\right)^2 \frac{1}{\hat{r}},$$

(3.20)

under which the isotropic Schwarzschild metric (3.19) becomes

$$dl^2 = \left(1 + \frac{M}{2\hat{r}}\right)^4 \left(d\hat{r}^2 + \hat{r}^2(d\theta^2 + \sin^2\theta d\phi^2)\right).$$

(3.21)

The geometry described by metric (3.21) evaluated at a radius $\hat{r} = a$ is identical to that of the metric (3.19) evaluated at $r = a$. The mapping (3.20) therefore maps the metric into itself, and is hence an *isometry*. In particular, this demonstrates that the origin $r = 0$ is isomorphic to spatial infinity, which is perfectly regular. The same conclusion can be reached by considering a coordinate transformation between the isotropic radius r to a Schwarzschild areal radius R. This demonstrates that the isotropic radius r covers only the black hole exterior, and that each Schwarzschild R corresponds to two values of the isotropic radius r.

> **Exercise 3.4** Find the coordinate transformation between the isotropic radius r and Schwarzschild radius R that brings the isotropic Schwarzschild metric (3.19) into the Schwarzschild form
>
> $$dl^2 = \left(1 - \frac{2M}{R}\right)^{-1} dR^2 + R^2(d\theta^2 + \sin^2\theta d\phi^2).$$
>
> (3.22)

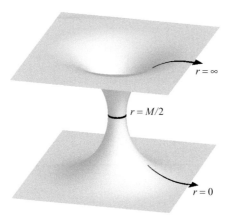

Figure 3.1 Schematic embedding diagram of Schwarzschild geometry at a moment of time symmetry with one degree of rotational freedom suppressed ($\theta = \pi/2$). The spatial metric is given by equation (3.19). Here a spatial slice of Schwarzschild in the equatorial plane is embedded as a 2-dimensional surface (paraboloid) in a Euclidean 3-space.

The isotropic radius r corresponding to the smallest areal (or circumferential) radius R is $r = M/2$, which we refer to as the black hole *throat*. The throat is located in the horizontal symmetry plane on the circle of smallest circumference in the embedding diagram[9] in Figure 3.1. In Chapter 7.2 (equation 7.26) we will see that, for a single Schwarzschild black hole, the throat coincides with both the apparent and event horizons.

The isometry (3.20) maps points on the throat into themselves, and applying the isometry twice yields the identity transformation. We can therefore think of (3.20) as a reflection in the throat. In the embedding diagram in Figure 3.1 this is a reflection across the horizontal symmetry plane. The geometry close to the origin ($r \to 0$) is identical to the geometry near infinity ($r \to \infty$). We can therefore think of the geometry described by the solution (3.19) as two separate, identical universes, which are connected by a throat, or a so-called *Einstein–Rosen* bridge. Equivalently, a time-symmetric slice of Schwarzschild as depicted in Figure 3.1 corresponds to the $v = 0$ ($t = 0$) hypersurface in the Kruskal–Szekeres diagram in Figure 1.1. On this diagram the throat at areal radius $R = 2M$ connecting the two asymptotically flat universes is located at the origin, $(u, v) = (0, 0)$.

So far we have only rediscovered the vacuum Schwarzschild solution, and that alone would hardly justify the effort of having developed all this decomposition formalism. The formalism is very powerful, however, and allows for the construction of much more general solutions. In Chapters 12 and 15, for example, we will use this approach to construct binary black hole and neutron star initial data. To catch a glimpse of how useful this formalism is, we point out that it is almost trivial to generalize our *one* black hole solution (3.18) to an arbitrary number of black holes at a moment of time symmetry.[10] Since (3.16) is linear,

---

[9] See, e.g., Misner *et al.* (1973), Chapters 23.8 and 31.6 for the construction of embedding diagrams for Schwarzschild geometry.

[10] E.g., Brill and Lindquist (1963).

Figure 3.2 Schematic embedding diagram of the geometry described by metric (3.23) for two black holes at a moment of time symmetry. This is the three-sheeted topology, which does not satisfy an isometry across each throat.

we obtain the solution simply by adding the individual contribution of each black hole according to

$$\psi = 1 + \sum_\alpha \frac{\mathcal{M}_\alpha}{2r_\alpha}. \tag{3.23}$$

Here $r_\alpha = |x^i - C_\alpha^i|$ is the (coordinate) separation from the center C_α^i of the αth black hole. The total mass of the spacetime is the sum of the coefficients $\mathcal{M}_\alpha$. However, since the total mass will also include contributions from the black hole interactions, $\mathcal{M}_\alpha$ can be identified with the mass of the αth black hole only in the limit of large separations. Particularly interesting astrophysically and for the generation of gravitational waves is the case of binary black holes, in which case (3.23) reduces to

$$\psi = 1 + \frac{\mathcal{M}_1}{2r_1} + \frac{\mathcal{M}_2}{2r_2}. \tag{3.24}$$

This simple solution to the constraint equations for two black holes instantaneously at rest at a moment of time symmetry can be used as initial data for head-on collisions of black holes (see Chapter 13.2).

We can now define mappings equivalent to (3.20), which represent reflections through the αth throat. In general, the existence of other black holes destroys the symmetry that we found for a single black hole. Each Einstein–Rosen bridge therefore connects to its own asymptotically flat Universe. Drawing an embedding diagram for such a geometry yields several different "sheets", where each sheet corresponds to one Universe. A geometry containing N black holes may contain up to $N + 1$ different asymptotically flat universes (see Figure 3.2).

If desired, however, the isometry across the throats can be restored as follows. Recall that equation (3.16) is equivalent to the Laplace equation in electrostatics, so that we

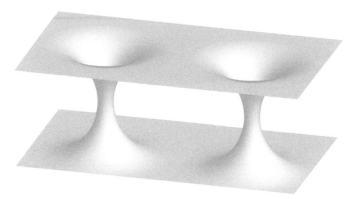

Figure 3.3 Schematic embedding diagram of a "symmetrized" two black hole solution. This is a two-sheeted topology, in which two Einstein–Rosen bridges connect two identical, asymptotically flat universes.

Figure 3.4 Illustration of a wormhole black hole solution.

can borrow the method of *spherical inversion images*[11] to analyze it. For each throat in (3.23) we can add terms inside that throat that correspond to images of the other black holes. Doing so, the solution (3.23) becomes "symmetrized" so that the reflection through each throat is again an isometry. In other words, each Einstein–Rosen bridge connects to the *same* asymptotically flat Universe, and the geometry consists of only two asymptotically flat universes, which are connected by several Einstein–Rosen bridges (see Figure 3.3).

   For two equal-mass black holes we may also interpret this solution as a *wormhole* black hole solution. To see this, consider the solution illustrated in Figure 3.3 for two throats of equal mass. Cut off the bottom Universe at the two throats, which leaves two "open-ended" throats hanging down from the top Universe. We can now identify these two open ends with each other, effectively gluing them together. As illustrated in Figure 3.4, the two throats now form a "wormhole" that connects to a single, asymptotically flat (but multiply connected) Universe. Given the original isometry conditions across the throats, and given

---

[11] See Misner (1963); Lindquist (1963).

that they have the same mass, the resulting metric is smooth across the throat and a valid solution to the Hamiltonian constraint.[12] In cylindrical coordinates the metric becomes

$$dl^2 = \psi^4(d\rho^2 + dz^2 + \rho^2 d\phi^2),\tag{3.25}$$

where the corresponding conformal factor is given by[13]

$$\psi = 1 + \sum_{n=1}^{\infty} \frac{1}{\sinh(n\mu)}\left(\frac{1}{\sqrt{\rho^2 + (z + z_n)^2}} + \frac{1}{\sqrt{\rho^2 + (z - z_n)^2}}\right).\tag{3.26}$$

Here $z_n = \coth(n\mu)$, and μ is a free parameter. In exercise 3.21 we will see that the total mass of this system, which we will identify with the "ADM mass" in Section 3.5, is

$$M_{\text{ADM}} = 4\sum_{n=1}^{\infty}\frac{1}{\sinh(n\mu)}.\tag{3.27}$$

The proper distance L along the spacelike geodesic connecting the throats, or equivalently the proper length of a geodesic loop through the wormhole, is

$$L = 2\left(1 + 2\mu\sum_{n=1}^{\infty}\frac{n}{\sinh(n\mu)}\right).\tag{3.28}$$

The parameter μ is seen to parametrize both the mass and separation of the two holes. Since the solution can be rescaled to arbitrary physical mass, μ effectively determines the dimensionless ratio L/M_{ADM}, the parameter that, apart from mass, distinguishes one binary from another in this class of initial data.

As we have seen, the solution to the Hamiltonian constraint equation for a system containing more than one vacuum black hole at a moment of time symmetry is by no means unique. The different solutions satisfy different inversion properties on the throats, and represent solutions to the Hamiltonian constraint in different *topologies*. If viewed from only one "Universe", the different solutions satisfy different boundary conditions on the throats. This difference leads to a different initial gravitational wave content in the sense that the dynamical evolution of these initial data would lead to different gravitational wave signals, at least for the initial burst. We will discuss initial data for multiple black holes in much more detail in Chapter 12, and will postpone a further discussion of these issues until then.

Exercise 3.5 *Brill waves* are defined as nonlinear, axisymmetric gravitational waves in vacuum spacetimes that admit a moment of time symmetry.[14] Consider a spatial metric in cylindrical coordinates

$$dl^2 = \psi^4\left(e^q(d\rho^2 + dz^2) + \rho^2 d\phi^2\right),\tag{3.29}$$

[12] Misner (1960) originally derived this solution by starting with a 3-dimensional "donut" solution. Part of this donut ultimately forms the tube of the wormhole, while one point on the original donut is pulled apart towards infinity to form the asymptotically flat Universe.
[13] See, e.g., Anninos *et al.* (1994).
[14] Brill (1959). See also Eppley (1977).

where $q(\rho, z)$ is an arbitrary, axisymmetric function that introduces a deviation from conformal flatness and that can be considered a measure of the gravitational wave amplitude. Show that at a moment of time symmetry the conformal factor ψ satisfies

$$\nabla^2 \psi = -\frac{\psi}{8} \left(\frac{\partial^2 q}{\partial \rho^2} + \frac{\partial^2 q}{\partial z^2} \right), \qquad (3.30)$$

where ∇^2 is the flat space Laplace operator in three dimensions. Solving this nonlinear elliptic equation provides a surprisingly simple way of constructing nonlinear gravitational wave initial data.

3.1.3 Conformal transformation of the extrinsic curvature

Return now to the development of the conformal decomposition of the constraint equations. We have conformally transformed the spatial metric, but before we proceed we also have to decompose the extrinsic curvature. It is convenient to split K_{ij} into its trace K and a traceless part A_{ij} according to

$$K_{ij} = A_{ij} + \frac{1}{3} \gamma_{ij} K, \qquad (3.31)$$

and to conformally transform K and A_{ij} separately. A priori it is not clear how to transform K and A_{ij}, and our only guidance for inventing rules is that the transformation should bring the constraint equations into a simple and solvable form. Consider the transformations

$$A^{ij} = \psi^\alpha \bar{A}^{ij} \qquad (3.32)$$

$$K = \psi^\beta \bar{K}, \qquad (3.33)$$

where α and β are two so far undetermined exponents.

Exercise 3.6 Show that the divergence of any symmetric, traceless tensor A^{ij} which transforms according to (3.32) satisfies

$$D_j A^{ij} = \psi^{-10} \bar{D}_j (\psi^{10+\alpha} \bar{A}^{ij}). \qquad (3.34)$$

Exercise 3.6 immediately suggests the choice $\alpha = -10$, i.e.,

$$A^{ij} = \psi^{-10} \bar{A}^{ij}, \qquad (3.35)$$

which implies $A_{ij} = \psi^{-2} \bar{A}_{ij}$. With this choice a symmetric traceless tensor A^{ij} has zero divergence if and only if $\bar{A}^{ij}$ does. This is not the only possible choice for the exponent α, though, and we will use a different scaling in Chapter 11.

Inserting the above expressions into the momentum constraint (3.2) yields

$$\psi^{-10} \bar{D}_j \bar{A}^{ij} - \frac{2}{3} \psi^{\beta-4} \bar{\gamma}^{ij} \bar{D}_j \bar{K} - \frac{2}{3} \beta \psi^{\beta-5} \bar{K} \bar{\gamma}^{ij} \bar{D}_j \psi = 8\pi S^i. \qquad (3.36)$$

Our desire to simplify equations motivates the choice $\beta = 0$, so that we treat K as a conformal invariant, $K = \bar{K}$. With these choices, the Hamiltonian constraint now becomes

$$8\bar{D}^2\psi - \psi\bar{R} - \frac{2}{3}\psi^5 K^2 + \psi^{-7}\bar{A}_{ij}\bar{A}^{ij} = -16\pi\psi^5\rho, \qquad (3.37)$$

and the momentum constraint is

$$\bar{D}_j\bar{A}^{ij} - \frac{2}{3}\psi^6\bar{\gamma}^{ij}\bar{D}_j K = 8\pi\psi^{10}S^i. \qquad (3.38)$$

> **Exercise 3.7** Consider the weak-field limit of equation (3.37) under the same assumptions as in Exercise 2.28. Then compare with the Poisson equation for the Newtonian gravitational potential Φ to show that
>
> $$\psi = 1 - \frac{1}{2}\Phi \qquad (3.39)$$
>
> in the weak-field limit, assuming suitable boundary conditions for ψ and Φ.

In addition to the spatial metric and extrinsic curvature, it may also be necessary to transform the matter sources ρ and S^i in (3.37) and (3.38) to insure uniqueness of solutions.[15] We will largely ignore this issue in this chapter, but it is nevertheless instructive to discuss its origin in passing.

We start by considering the linear equation

$$\nabla^2 u = fu \qquad (3.40)$$

on some domain Ω. Here f is some given function, and we will assume $u = 0$ on the boundary $\partial\Omega$. If f is nonnegative everywhere, we can apply the *maximum principle* to show that $u = 0$ everywhere. The point is that if u were nonzero somewhere in Ω, say positive, then it must have a maximum somewhere. At the maximum the left hand side of (3.40) must be negative, but the right hand side is nonnegative if $f \geq 0$, which is a contradiction. Clearly, the argument works the same way if u is negative somewhere, implying that $u = 0$ everywhere if $f \geq 0$.

Now consider the nonlinear equation

$$\nabla^2 u = fu^n, \qquad (3.41)$$

and assume there exist two positive solutions u_1 and $u_2 \geq u_1$ that are identical, $u_1 = u_2$, on the boundary $\partial\Omega$. The difference $\Delta u = u_2 - u_1$ must then satisfy an equation

$$\nabla^2\Delta u = nf\tilde{u}^{n-1}\Delta u, \qquad (3.42)$$

where $\tilde{u}$ is some positive function satisfying $u_1 \leq \tilde{u} \leq u_2$. Applying the above argument to Δu, we see that the maximum principle implies $\Delta u = 0$ and hence uniqueness of solutions

---

[15] See, e.g., York, Jr. (1979).

if and only if $nf \geq 0$, i.e., if the coefficient and exponent in the source term of (3.41) have the same sign.

Inspecting the Hamiltonian constraint (3.37) we see that the matter term $-16\pi\psi^5\rho$ features the "wrong signs": it has a negative coefficient (assuming a positive matter density ρ), but a positive exponent for ψ. Therefore the maximum principle cannot be applied, and the uniqueness of solutions cannot be established. Exercise 3.8 explores this issue for an analytical example.

> **Exercise 3.8**   Consider the Hamiltonian constraint (3.37) at a moment of time symmetry, $K_{ij} = 0$, and under the assumption of conformal flatness and spherical symmetry. Also assume boundary conditions $\partial_r\psi = 0$ at $r = 0$, and $\psi \to 1$ for $r \to \infty$. Now consider a constant density star with
>
> $$\rho(r) = \begin{cases} \rho_0, & r < r_0 \\ 0, & r \geq r_0. \end{cases} \tag{3.43}$$
>
> In the following we will consider r_0 as given, and will study the solutions as a function of the density ρ_0.
>
> (a) Show that the *Sobolev functions*
>
> $$u_\nu(r) \equiv \frac{(\nu r_0)^{1/2}}{\left(r^2 + (\nu r_0)^2\right)^{1/2}} \tag{3.44}$$
>
> satisfy the equation
>
> $$\bar{D}^2 u_\nu = \frac{1}{r^2}\frac{\partial}{\partial r}\left(r^2\frac{\partial u_\nu}{\partial r}\right) = -3u_\nu^5 \tag{3.45}$$
>
> for any constant ν. Conclude that for $r < r_0$ the solution for ψ is given by $\psi_{\text{int}} = Cu_\nu$, and find the value of C.
>
> (b) For $r \geq r_0$ the solution is given by $\psi_{\text{ex}} = 1 + \mu/r$, where μ is another yet undetermined constant. The interior and exterior solutions already satisfy the differential equation and boundary conditions individually, but to obtain a global solution we still need to enforce that their function values and first derivatives match at $r = r_0$. These two conditions fix the constants μ and ν for a given background density ρ_0. Show that the conditions can be combined to yield an equation for ν,
>
> $$\rho_0 r_0^2 = \frac{3}{2\pi}f^2(\nu), \tag{3.46}$$
>
> where $f(\nu) \equiv \nu^5/(1 + \nu^2)^3$.
>
> (c) Exploring the properties of the function $f(\nu)$ show that no solutions exist if
>
> $$\rho_0 > \rho_{\text{crit}} = \frac{3}{2\pi r_0^2}\frac{5^5}{6^6}. \tag{3.47}$$
>
> Further show that for any $\rho_0 < \rho_{\text{crit}}$, there are *two* solutions ν, and hence two distinct solutions ψ, that satisfy the Hamiltonian constraint for the matter distribution (3.43). Clearly, the solutions are not unique.[16]

---

[16] See Baumgarte *et al.* (2007) for a further exploration of this solution as well as related issues. Note that for homogeneous fluid stars in hydrostatic *equilibrium*, there is only one physically relevant solution to the Hamiltonian constraint

Uniqueness of solutions can be restored, however, by introducing a conformal rescaling of the density. With $\rho = \psi^\delta \bar{\rho}$, where $\delta \leq -5$ and where $\bar{\rho}$ is now considered a given function, the matter term carries the "right signs", and the maximum principle can be applied to establish the uniqueness of solutions. Furthermore, in the example of exercise 3.8 the solutions are unique locally even for unscaled density sources – at least for matter densities smaller than the critical density – and there is some evidence that this property is generic.[17] If so, a numerical algorithm can still iterate towards the desired solution, given suitable background data, as long as the iteration starts with a sufficiently "close" initial guess.

Most decompositions use the conformal rescaling of the spatial metric and the extrinsic curvature as introduced above. Different decompositions then proceed by decomposing $\bar{A}_{ij}$ in different ways. In the following sections we will discuss the transverse-traceless and the conformal thin-sandwich decompositions.

3.2 Conformal transverse-traceless decomposition

Any symmetric, traceless tensor can be split into a transverse-traceless part that is divergenceless and a longitudinal part that can be written as a symmetric, traceless gradient of a vector. We can therefore decompose $\bar{A}^{ij}$ as

$$\bar{A}^{ij} = \bar{A}^{ij}_{TT} + \bar{A}^{ij}_L, \tag{3.48}$$

where the transverse part is divergenceless,

$$\bar{D}_j \bar{A}^{ij}_{TT} = 0, \tag{3.49}$$

and where the longitudinal part satisfies

$$\bar{A}^{ij}_L = \bar{D}^i W^j + \bar{D}^j W^i - \frac{2}{3} \bar{\gamma}^{ij} \bar{D}_k W^k \equiv (\bar{L} W)^{ij}. \tag{3.50}$$

Here W^i is a vector potential, and it is easy to see that the *longitudinal operator* or *vector gradient* $\bar{L}$ produces a symmetric, traceless tensor.[18] We can now write the divergence of $\bar{A}^{ij}$ as

$$\bar{D}_j \bar{A}^{ij} = \bar{D}_j \bar{A}^{ij}_L = \bar{D}_j (\bar{L} W)^{ij} = \bar{D}^2 W^i + \frac{1}{3} \bar{D}^i (\bar{D}_j W^j) + \bar{R}^i{}_j W^j \equiv (\bar{\Delta}_L W)^i, \tag{3.51}$$

where $\bar{\Delta}_L$ is the *vector Laplacian*.

equation. Moreover, the critical density at which the central pressure becomes infinite in an equilibrium star, which occurs when $M/R = 4/9$, is smaller than the critical density given by equation (3.47). Hence the latter density is never reached along an equilibrium sequence of stars of constant mass M but increasing density and compaction.

[17] See Baumgarte *et al.* (2007); Walsh (2007).

[18] Vectors ξ^i satisfying $(\bar{L}\xi)^{ij} = 0$ are called *conformal Killing vectors* (see exercise A.7 in Appendix A), which suggests why $\bar{L}$ is also called the *conformal Killing operator*.

Exercise 3.9 Consider a flat conformally related metric $\bar{\gamma}_{ij} = \eta_{ij}$ in spherical polar coordinates, and assume that the only nonvanishing component of the vector W^i is the radial component W^r. Then show that the only nonzero component of the vector Laplacian is

$$(\bar{\Delta}_L W)^r = \frac{4}{3} \frac{\partial}{\partial r} \left(\frac{1}{r^2} \frac{\partial}{\partial r} \left(r^2 W^r \right) \right). \tag{3.52}$$

Note that $\bar{A}_{TT}^{ij}$ and $\bar{A}_L^{ij}$ are transverse and longitudinal with respect to the conformal metric $\bar{\gamma}_{ij}$, which is why this decomposition is called the *conformal* transverse-traceless decomposition. Alternatively, one can adopt a *physical* transverse-traceless decomposition, where the corresponding tensors are transverse and longitudinal with respect to the physical metric γ_{ij}.

Inserting the conformally related quantities into the momentum constraint (3.38) yields

$$\boxed{(\bar{\Delta}_L W)^i - \frac{2}{3} \psi^6 \bar{\gamma}^{ij} \bar{D}_j K = 8\pi \psi^{10} S^i.} \tag{3.53}$$

The Hamiltonian constraint remains in its form (3.37). We now see that we can freely choose the conformally related metric $\bar{\gamma}_{ij}$, the mean curvature K and the transverse-traceless part of the conformally related extrinsic curvature, $\bar{A}_{TT}^{ij}$. Given these choices, we can then solve the Hamiltonian constraint (3.37) for the conformal factor ψ and the momentum constraint (3.53) for the vector potential W^i. Knowing these quantities, we can construct finally the physical solutions γ_{ij} and K_{ij}. We summarize this *conformal transverse-traceless*, or "CTT", decomposition in Box 3.1.

Box 3.1   The conformal transverse-traceless (CTT) decomposition

Freely specifiable variables are $\bar{\gamma}_{ij}$, K and $\bar{A}_{TT}^{ij}$. Given these, the momentum constraint

$$(\bar{\Delta}_L W)^i - \frac{2}{3} \psi^6 \bar{\gamma}^{ij} \bar{D}_j K = 8\pi \psi^{10} S^i \tag{3.54}$$

is solved for W^i, and the Hamiltonian constraint

$$8 \bar{D}^2 \psi - \psi \bar{R} - \frac{2}{3} \psi^5 K^2 + \psi^{-7} \bar{A}_{ij} \bar{A}^{ij} = -16\pi \psi^5 \rho, \tag{3.55}$$

where

$$\bar{A}^{ij} = \bar{A}_{TT}^{ij} + \bar{A}_L^{ij} = \bar{A}_{TT}^{ij} + (\bar{L} W)^{ij}, \tag{3.56}$$

is solved for ψ. The physical solution is then constructed from

$$\gamma_{ij} = \psi^4 \bar{\gamma}_{ij} \tag{3.57}$$

and

$$K_{ij} = A_{ij} + \frac{1}{3} \gamma_{ij} K = \psi^{-2} \bar{A}_{ij} + \frac{1}{3} \gamma_{ij} K. \tag{3.58}$$

Before discussing some simple solutions to equation (3.53), it is useful to count degrees of freedom again, as we did at the beginning of this chapter. We started out with six independent variables in both the spatial metric γ_{ij} and the extrinsic curvature K_{ij}. Splitting off the conformal factor ψ left five degrees of freedom in the conformally related metric $\bar{\gamma}_{ij}$ (once we have specified its determinant $\bar{\gamma}$). Of the six independent variables in K^{ij} we moved one into its trace K, two into $\bar{A}_{TT}^{ij}$ (which is symmetric, traceless, and divergenceless), and three into $\bar{A}_L^{ij}$ (which is reflected in its representation by a vector). Of the 12 original degrees of freedom, the constraint equations determine only four, namely the conformal factor ψ (Hamiltonian constraint) and the longitudinal part of the traceless extrinsic curvature $\bar{A}_L^{ij}$ (momentum constraint). Four of the remaining eight degrees of freedom are associated with the coordinate freedom – three spatial coordinates hidden in the spatial metric and a time coordinate that is associated with K. This leaves four physical degrees of freedom undetermined – two in the conformally related metric $\bar{\gamma}_{ij}$, and two in the transverse part of the traceless extrinsic curvature $\bar{A}_{TT}^{ij}$. These freely specifiable degrees of freedom carry the dynamical degrees of freedom of the gravitational fields. All others are either fixed by the constraint equations or represent coordinate freedom.

We have reduced the Hamiltonian and momentum constraints to equations for the conformal factor ψ and the vector potential W^i, from which the longitudinal part of the extrinsic curvature is constructed. These quantities can be solved for only after choices have been made for the remaining quantities in the equations, namely the conformally related metric $\bar{\gamma}_{ij}$, the transverse-traceless part of the extrinsic curvature $\bar{A}_{TT}^{ij}$, the trace of the extrinsic curvature K, and, if present, any matter sources. The choice of these *background data* has to be made in accordance with the physical or astrophysical situation that one wants to represent. Physically, the choice affects the gravitational wave content present in the initial data, in the sense that a dynamical evolution of data constructed with different background data leads to different amounts of emitted gravitational radiation. It is often not clear how a suitable background can be constructed precisely, and we will return to this issue on several occasions. Given its loose association with the transverse parts of the gravitational fields, one often sets $\bar{A}_{TT}^{ij}$ equal to zero in an attempt to minimize the gravitational wave content in the initial data.

The freedom in choosing the background data can also be used to simplify the equations. Focus again on vacuum solutions, so that $\rho = S^i = 0$. We will now assume *maximal slicing* $K = 0$ (see Chapter 4.2), which amounts to assuming that the initial slice Σ has a certain shape in the spacetime M – namely one that maximizes its volume. In this case the momentum constraint (3.53) decouples from the Hamiltonian constraint

$$(\bar{\Delta}_L W)^i = 0 \tag{3.59}$$

and can therefore be solved independently. If we further assume conformal flatness, $\bar{\gamma}_{ij} = \eta_{ij}$, the vector Laplacian simplies and, in Cartesian coordinates, reduces to

$$\partial^j \partial_j W^i + \frac{1}{3}\partial^i \partial_j W^j = 0. \tag{3.60}$$

Solutions to this equation are often called *Bowen–York solutions*.[19] We will encounter the above operator on several occasions, and two different approaches for solving it in the presence of nonzero right-hand sides are discussed in Appendix B. As examples of these two approaches we will derive next two well-known Bowen–York solutions to the momentum constraints (3.60), one describing a spinning black hole and one describing a boosted black hole. These solutions form the basis for many numerical solutions to the constraint equations, for example the "puncture" initial data for binary black holes that we shall discuss in Chapter 12.2.

A spinning black hole

In one approach, the vector W^i is decomposed as

$$W_i = V_i + \partial_i U, \tag{3.61}$$

in which case equation (3.60) reduces to the coupled set of Possion equations

$$\partial^j \partial_j U = -\frac{1}{4} \partial_j V^j \tag{3.62}$$

$$\partial^j \partial_j V_i = 0 \tag{3.63}$$

(see Appendix B). We can derive a simple solution by assuming $V_i = 0$. The general spherically symmetric solution for U is then given by $U = a - b/r$, where a and b are arbitrary constants and where $r^2 = x^2 + y^2 + z^2$. Inserting this into the decomposition (3.61) we find

$$W^i = \eta^{ij} \partial_j U = b \frac{x^i}{r^3} = b \frac{l^i}{r^2} = b X^i, \tag{3.64}$$

where we have used the normal vector $l^i = x^i/r$ and defined $X^i = l^i/r^2$ for convenience. In spherical polar coordinates the only nonvanishing component of W^i is $W^r = b/r^2$ and, given that this solution is spherically symmetric, we can use equation (3.52) to verify immediately that this solution satisfies the momentum constraint (3.59). We can also use X^i to generate another solution that is not spherically symmetric, as shown in exercise 3.10.

Exercise 3.10 Show that

$$W^i = \bar{\epsilon}^{ijk} X_j J_k \tag{3.65}$$

is a solution to equation (3.59), assuming conformal flatness and that J^i is some vector satisfying $\bar{D}_i J_j = 0$. Here $\bar{\epsilon}^{ijk}$ is the three-dimensional Levi-Civita tensor associated with the conformally related metric $\bar{\gamma}_{ij}$, so that $\bar{D}_i \bar{\epsilon}^{jkl} = 0$.

---

[19] Bowen and York, Jr. (1980).

Inserting the solution (3.65) into equation (3.50) we now find

$$\bar{A}_L^{ij} = (\bar{L} W)^{ij} = \frac{6}{r^3} l^{(i} \bar{\epsilon}^{j)kl} J_k l_l. \qquad (3.66)$$

Since J_i satisfies $\bar{D}_i J_j = 0$, it must be a vector with constant coefficients when expressed in Cartesian coordinates. Exercise 3.11 shows that for large r the extrinsic curvature (3.66) agrees with that of a Kerr black hole (see exercise 2.34), suggesting that our solution describes a rotating black hole with angular momentum J_i. We will confirm this identification in exercise 3.29.

Exercise 3.11 Show that in spherical polar coordinates the only nonvanishing components of W^i in equation (3.65) and the extrinsic curvature $\bar{A}_L^{ij}$ in equation (3.66) are

$$W^\phi = -\frac{J}{r^3} \qquad (3.67)$$

and

$$\bar{A}_{r\phi}^L = \frac{3J}{r^2} \sin^2 \theta. \qquad (3.68)$$

Here J is the magnitude of the vector J^i aligned with the polar axis. Further show that, with suitable choices for $\bar{A}_{ij}^{TT}$ and ψ, this solution agrees asymptotically with the Kerr solution of exercise 2.34.

In order to construct a complete solution to the constraint equations we would have to insert the extrinsic curvature (3.66) back into the Hamiltonian constraint (3.37) and solve for the conformal factor ψ. This equation is nonlinear and in general can only be solved numerically. We will return to this problem in several places throughout this book; in Chapter 12.2.2, for example, we will discuss two approaches for solving this equation in the context of binary black holes. In the meantime, we can find an approximate, analytical solution by expanding around the nonrotating case $J = 0$, as outlined in exercise 3.12.

Exercise 3.12 Consider the solution (3.68) for a rotating black hole and assume $\bar{A}_{TT}^{ij} = 0$ to show that the only nonvanishing source term in the Hamiltonian constraint (3.37) arises from

$$\bar{A}_{ij} \bar{A}^{ij} = \frac{18 J^2}{r^6} \sin^2 \theta. \qquad (3.69)$$

For $J = 0$ we recover the solution (3.18), which we will denote here as $\psi^{(0)}$. The leading-order correction to this solution must therefore scale with J^2, which suggests the ansatz

$$\psi = \psi^{(0)} + \frac{J^2}{\mathcal{M}^4} \psi^{(2)} + \mathcal{O}(J^4) = 1 + \frac{\mathcal{M}}{2r} + \frac{J^2}{\mathcal{M}^4} \psi^{(2)} + \mathcal{O}(J^4). \qquad (3.70)$$

Show that inserting this ansatz into the Hamiltonian constraint (3.37) yields

$$\bar{D}^2 \psi^{(2)} = -\frac{9}{4} \frac{\mathcal{M}^4 r}{(r + \mathcal{M}/2)^7} \sin^2 \theta. \qquad (3.71)$$

Now express the angular dependence $\sin^2 \theta$ in terms of the Legendre polynomials $P_0(\cos \theta) = 1$ and $P_2(\cos \theta) = (3 \cos^2 \theta - 1)/2$, split $\psi^{(2)}$ into

$$\psi^{(2)} = \psi_0^{(2)}(r) P_0(\cos \theta) + \psi_2^{(2)}(r) P_2(\cos \theta), \tag{3.72}$$

and show that the resulting equations are solved by[20]

$$\psi_0^{(2)}(r) = -\left(1 + \frac{M}{2r}\right)^{-5} \frac{M}{5r} \left(5\left(\frac{M}{2r}\right)^3 + 4\left(\frac{M}{2r}\right)^4 + \left(\frac{M}{2r}\right)^5\right) \tag{3.73}$$

and

$$\psi_2^{(2)}(r) = -\frac{1}{10}\left(1 + \frac{M}{2r}\right)^{-5}\left(\frac{M}{r}\right)^3. \tag{3.74}$$

With the conformal factor ψ we now have a complete solution to the constraint equations describing a rotating black hole. But we already know that the solution for a stationary, rotating black hole is given by the Kerr solution of exercise 2.34. This raises an interesting question, namely whether the above Bowen–York rotating black hole is identical to the Kerr black hole. The answer is no, even though it is not as easy to see this as one might think.

One complication is that the Kerr solution describes a complete spacetime solution, whereas the Bowen–York solution describes initial data only. These initial data represent the solution on a certain time slice Σ, but a priori it is not clear on which slice. For example, there is no reason to expect that this slice corresponds to a slice of constant Boyer–Lindquist time t (see Chapter 1.2 and exercise 2.34), making any direct comparison very difficult. We could evolve the Bowen–York initial data dynamically, thereby constructing a spacetime solution. Again, we would have no reason to expect that this spacetime solution would be represented in Boyer–Lindquist or any other coordinate system in which the Kerr solution is known, and it would be hard to decide whether or not the two spacetimes are identical, i.e., whether or not there exists a coordinate transformation that relates the two. One way to distinguish the resulting spacetimes is to compare gauge-invariant quantities, and one such example is the emitted gravitational radiation as measured by a distant observer. The Kerr solution is stationary and does not emit any gravitational radiation, but a dynamical evolution of the Bowen–York initial data *does* lead to a burst of gravitational radiation before it settles down into a stationary solution.[21] This demonstrates that the Bowen–York data do not represent a spatial slice of the Kerr solution. Instead, it may be considered a Kerr solution plus an initial perturbation.

This observation raises the next question: how does this perturbation arise in the CTT decomposition? All evidence points towards our assumption of conformal flatness, $\bar{\gamma}_{ij} = \eta_{ij}$. Given this assumption, the resulting data can represent a Kerr black hole only if the Kerr spacetime admits a slicing on which the spatial metric is conformally flat. Whether or not the Kerr spacetime admits such a slicing is again difficult to decide, since we have

---

[20] Gleiser *et al.* (1998).
[21] See Gleiser *et al.* (1998) as well as the discussion in Burko *et al.* (2006).

no a priori knowledge of which slicing this might be. We can nevertheless consider certain families of slices Σ and evaluate the Bach tensor (3.15). Slices of constant Boyer–Lindquist time t, for example, are not conformally flat, nor are axisymmetric foliations that smoothly reduce to slices of constant Schwarzschild time in the Schwarzschild limit.[22] This suggests very strongly (but does not prove) that, for nonzero angular momentum, the Kerr spacetime does not admit spatial slices that are conformally flat.

A boosted black hole

In an alternative approach to solving (3.60) we decompose W^i as

$$W_i = \frac{7}{8}V_i - \frac{1}{8}\left(\partial_i U + x^k \partial_i V_k\right) \tag{3.75}$$

(see equation B.7 in Appendix B), in which case (3.60) becomes equivalent to the set

$$\partial^j \partial_j U = 0 \tag{3.76}$$

$$\partial^j \partial_j V_i = 0. \tag{3.77}$$

Note that both equations are now homogeneous. We can construct a simple solution by assuming $U = 0$ and writing the solution for V_i as

$$V_i = -\frac{2P_i}{r}, \tag{3.78}$$

where P_i is an arbitrary vector with constant coefficients when expressed in Cartesian coordinates (the general solution also allows for an arbitrary constant term, which would drop out, however, when we compute $\bar{A}_L^{ij}$ below). Inserting V_i into the decomposition (3.75) yields

$$W^i = -\frac{1}{4r}(7P^i + l^i l_j P^j), \tag{3.79}$$

where again $l^i = x^i/r$, and, from equation (3.50),

$$\bar{A}_L^{ij} = (\bar{L}W)^{ij} = \frac{3}{2r^2}\left(P^i l^j + P^j l^i - (\eta^{ij} - l^i l^j)l_k P^k\right). \tag{3.80}$$

In exercise 3.32 we will see that the linear momentum of this solution is P^i. By virtue of the linearity of the momentum constraint (3.60) we can add several terms of this form to obtain a solution describing multiple, boosted black holes. If we would like these black holes to spin, we could further add terms of the form (3.66). In fact, these solutions form the basis of the binary black hole initial data that we discuss in Chapter 12.2. To complete these data, of course, we still need to solve the Hamiltonian constraint (3.37), as described in Chapter 12.2.2. As for the spinning black holes, this can in general only

---

[22] See Garat and Price (2000).

be done numerically, but we can again compute approximate solutions as an expansion around the known Schwarzschild solution for $P^i = 0$.

> **Exercise 3.13** Follow the approach outlined in exercise 3.12 to construct an approximate, analytical solution to the Hamiltonian constraint (3.37) for the extrinsic curvature (3.80). First show that
>
> $$\bar{A}_{ij}\bar{A}^{ij} = \frac{9P^2}{2r^4}\left(1 + 2\cos^2\theta\right), \qquad (3.81)$$
>
> where $P^2 = P_i P^i$ is the square of the momentum's magnitude. Then make an ansatz
>
> $$\psi = \psi^{(0)} + \frac{P^2}{\mathcal{M}^2}\psi^{(2)} + \mathcal{O}(P^4) = 1 + \frac{\mathcal{M}}{2r} + \frac{P^2}{\mathcal{M}^2}\psi^{(2)} + \mathcal{O}(P^4) \qquad (3.82)$$
>
> and show that the Hamilton constraint is solved, to first order in P^2, by
>
> $$\psi^{(2)} = \psi_0^{(2)}(r)P_0(\cos\theta) + \psi_2^{(2)}(r)P_2(\cos\theta) \qquad (3.83)$$
>
> where[23]
>
> $$\psi_0^{(2)} = \left(1 + \frac{\mathcal{M}}{2r}\right)^{-5}\frac{\mathcal{M}}{16r}\left[\left(\frac{\mathcal{M}}{2r}\right)^4 + 5\left(\frac{\mathcal{M}}{2r}\right)^3 + 10\left(\frac{\mathcal{M}}{2r}\right)^2 + 10\left(\frac{\mathcal{M}}{2r}\right) + 5\right] \qquad (3.84)$$
>
> and
>
> $$\psi_2^{(2)} = \frac{1}{20}\left(1 + \frac{\mathcal{M}}{2r}\right)^{-5}\left(\frac{\mathcal{M}}{2r}\right)^2\left[84\left(\frac{\mathcal{M}}{2r}\right)^5 + 378\left(\frac{\mathcal{M}}{2r}\right)^4 + 658\left(\frac{\mathcal{M}}{2r}\right)^3\right.$$
> $$\left. + 539\left(\frac{\mathcal{M}}{2r}\right)^2 + 192\left(\frac{\mathcal{M}}{2r}\right) + 15\right] + \frac{21}{5}\left(\frac{\mathcal{M}}{2r}\right)^3\ln\left(\frac{\mathcal{M}/(2r)}{1 + \mathcal{M}/(2r)}\right). \qquad (3.85)$$

Before closing this section it may be useful to discuss one more technical aspect.[24] We have seen that the longitudinal part $\bar{A}_L^{ij}$ is constructed from the vector potential W^i. According to the decomposition (3.48) we can add to this longitudinal part a transverse part $\bar{A}_{TT}^{ij}$ to construct a general solution to the momentum constraint. According to equation (3.49) the transverse part $\bar{A}_{TT}^{ij}$ has to be divergence-free, and we want to discuss briefly how a divergence-free tensor can be constructed from a symmetric tracefree tensor $\bar{M}^{ij}$. Consider a solution Y^i to the equation

$$\bar{\Delta}_L Y^i = \bar{D}_j \bar{M}^{ij}. \qquad (3.86)$$

If we define

$$\bar{A}_{TT}^{ij} = \bar{M}^{ij} - (\bar{L}Y)^{ij}, \qquad (3.87)$$

[23] Gleiser *et al.* (2002).
[24] Cook (2000).

then $\bar{A}_{TT}^{ij}$ is clearly divergence-free,

$$\bar{D}_j \bar{A}_{TT}^{ij} = \bar{D}_j \bar{M}^{ij} - \bar{D}_j (\bar{L}Y)^{ij} = \bar{D}_j \bar{M}^{ij} - \bar{\Delta}_L Y^i = 0. \tag{3.88}$$

In fact, this construction can be conveniently embedded in the above solution of the constraint equations. Given that $\bar{L}$ is linear, equation (3.48) can be written

$$\bar{A}^{ij} = \bar{A}_{TT}^{ij} + \bar{A}_L^{ij} = \bar{M}^{ij} - (\bar{L}Y)^{ij} + (\bar{L}W)^{ij} = \bar{M}^{ij} + (\bar{L}V)^{ij}, \tag{3.89}$$

where we have defined

$$V^i = W^i - Y^i. \tag{3.90}$$

Given that the vector Laplacian is linear, we also have

$$\bar{\Delta}_L V^i = \bar{\Delta}_L W^i - \bar{\Delta}_L Y^i = \frac{2}{3} \psi^6 \bar{\gamma}^{ij} \bar{D}_j K - \bar{D}_j \bar{M}^{ij} + 8\pi \psi^{10} S^i. \tag{3.91}$$

Solving this equation for V^i instead of equation (3.53) for W^i allows us to construct a general solution $\bar{A}^{ij}$ from an arbitrary symmetric tracefree tensor $\bar{M}^{ij}$ (equation 3.89) directly, rather than having to first find a transverse part $\bar{A}_{TT}^{ij}$.

3.3 Conformal thin-sandwich decomposition

Solving the conformal transverse-traceless decomposition yields data γ_{ij} and K_{ij} intrinsic to one spatial slice Σ, but this solution does not tell us anything about how it will evolve in time away from Σ, nor does the formalism allow us to determine any such time evolution. In some circumstances, for example when we are interested in constructing equilibrium or quasiequilibrium solutions, we would like to construct data in such a way that they do have a certain time evolution. The *conformal thin-sandwich*, or "CTS" approach[25] offers an alternative to the transverse-traceless decomposition that does allow us to determine the evolution of the spatial metric. Instead of providing data for γ_{ij} and K_{ij} on one time slice, it provides data for γ_{ij} on two time slices, or, in the limit of infinitesimal separation of the two slices, data for γ_{ij} and its time derivative.

We start by defining u_{ij} as the traceless part of the time derivative of the spatial metric,

$$u_{ij} \equiv \gamma^{1/3} \partial_t (\gamma^{-1/3} \gamma_{ij}), \tag{3.92}$$

in terms of which the evolution equation (2.134) becomes

$$u^{ij} = -2\alpha A^{ij} + (L\beta)^{ij}. \tag{3.93}$$

Here L is the vector gradient defined in equation (3.50), except that the "unbarred" L is defined in terms of the physical metric γ_{ij}. We also define

$$\bar{u}_{ij} \equiv \partial_t \bar{\gamma}_{ij} \tag{3.94}$$

[25] See York, Jr. (1999), as well as Isenberg (1978) and Wilson and Mathews (1995) for related earlier approaches.

and

$$\bar{\gamma}^{ij}\bar{u}_{ij} \equiv 0. \tag{3.95}$$

It may seem odd that we need both of these definitions to specify $\bar{u}_{ij}$. The reason for this is the arbitrariness in choosing the determinant of the conformal metric $\bar{\gamma}$. So far we have only chosen $\bar{\gamma}_{ij}$ and hence $\bar{\gamma}$ on the slice Σ, but we have not yet specified a normalization $\bar{\gamma}$ away from Σ. If defined from equation (3.94) alone, this would leave $\bar{u}_{ij}$ defined only up to some overall factor. This factor is fixed with the help of equation (3.95), which in effect determines the time derivative of $\bar{\gamma}$:

$$0 = \bar{\gamma}^{ij}\bar{u}_{ij} = \bar{\gamma}^{ij}\partial_t\bar{\gamma}_{ij} = \partial_t \ln \bar{\gamma}. \tag{3.96}$$

With this result we can now show that

$$u_{ij} = \psi^4 \bar{u}_{ij}. \tag{3.97}$$

Exercise 3.14 Derive equation (3.97).

Exercise 3.15 Show that

$$(L\beta)^{ij} = \psi^{-4}(\bar{L}\beta)^{ij}. \tag{3.98}$$

From equations (3.97) and (3.98), together with the scaling relation (3.35), we can conformally transform equation (3.93) to obtain

$$\bar{A}^{ij} = \frac{\psi^6}{2\alpha}\left((\bar{L}\beta)^{ij} - \bar{u}^{ij}\right). \tag{3.99}$$

At this point it seems natural to introduce a conformally rescaled or "densitized" lapse,

$$\alpha = \psi^6\bar{\alpha}, \tag{3.100}$$

in terms of which (3.99) becomes

$$\bar{A}^{ij} = \frac{1}{2\bar{\alpha}}\left((\bar{L}\beta)^{ij} - \bar{u}^{ij}\right). \tag{3.101}$$

This equation relates $\bar{A}^{ij}$ to the shift vector β^i. Inserting this equation into the momentum constraint (3.38), yields an equation for the shift,

$$(\bar{\Delta}_L\beta)^i - (\bar{L}\beta)^{ij}\bar{D}_j \ln(\bar{\alpha}) = \bar{\alpha}\bar{D}_j(\bar{\alpha}^{-1}\bar{u}^{ij}) + \frac{4}{3}\bar{\alpha}\psi^6\bar{D}^iK + 16\pi\bar{\alpha}\psi^{10}S^i. \tag{3.102}$$

We can now construct a solution of the thin-sandwich formulation as summarized in Box 3.2. We first choose the background metric $\bar{\gamma}_{ij}$ as well as its time derivative $\bar{u}_{ij}$. Given choices for the densitized lapse $\bar{\alpha}$ and the trace of the extrinsic curvature K, we can then solve the Hamiltonian constraint (3.37) and the momentum constraint (3.102) for the conformal factor ψ and the shift β^i. With these solutions, we can then construct $\bar{A}^{ij}$ from equation (3.101) and finally the physical quantities γ_{ij} and K_{ij}. This version of the conformal thin-sandwich formalism is sometimes called the "original" version, in contrast to the "extended" version that we will discuss below.

Box 3.2   The original conformal thin-sandwich (CTS) decomposition

Freely specifiable variables are $\bar{\gamma}_{ij}$, $\bar{u}_{ij}$, K and $\bar{\alpha}$. Given these, the momentum constraint

$$(\bar{\Delta}_L \beta)^i - (\bar{L}\beta)^{ij}\,\bar{D}_j \ln(\bar{\alpha}) = \bar{\alpha}\,\bar{D}_j(\bar{\alpha}^{-1}\bar{u}^{ij}) + \frac{4}{3}\bar{\alpha}\psi^6\bar{D}^i K + 16\pi\bar{\alpha}\psi^{10}S^i \qquad (3.103)$$

is solved for β^i, and the Hamiltonian constraint

$$\bar{D}^2\psi - \frac{1}{8}\psi\bar{R} - \frac{1}{12}\psi^5 K^2 + \frac{1}{8}\psi^{-7}\bar{A}_{ij}\bar{A}^{ij} = -2\pi\psi^5\rho, \qquad (3.104)$$

where

$$\bar{A}^{ij} = \frac{1}{2\bar{\alpha}}\left((\bar{L}\beta)^{ij} - \bar{u}^{ij}\right), \qquad (3.105)$$

is solved for ψ. The physical solution is then constructed from

$$\gamma_{ij} = \psi^4\bar{\gamma}_{ij}$$

$$K_{ij} = A_{ij} + \frac{1}{3}\gamma_{ij}K = \psi^{-2}\bar{A}_{ij} + \frac{1}{3}\gamma_{ij}K \qquad (3.106)$$

$$\alpha = \psi^6\bar{\alpha}.$$

It is again instructive to count the degrees of freedom, and to compare with the transverse-traceless decomposition of Section 3.2. There, we found that of the 12 independent variables in γ_{ij} and K_{ij}, four were determined by the constraint equations, four were related to the coordinate freedom, and four represented the dynamical degrees of freedom of general relativity. The latter eight conditions can be chosen freely. In the thin-sandwich formalism we count a total of 16 independent variables, of which we can freely choose 12: five each in $\bar{\gamma}_{ij}$ and $\bar{u}_{ij}$, and one each for $\bar{\alpha}$ and K. The four remaining variables, ψ and β^i, are then determined by the constraint equations. The four new independent variables are accounted for by the lapse $\bar{\alpha}$ and the shift β^i, which are absent in the transverse-traceless decomposition. The CTT approach only deals with quantities intrinsic to one spatial slice Σ, and hence only requires coordinates on Σ. The thin-sandwich approach, on the other hand, also takes into account the evolution of the metric off the slice, and therefore requires coordinates in a neighborhood of Σ. As a consequence, the lapse α and the shift β^i, which describe the evolution of the coordinates away from Σ, appear in the CTS approach, but not in the CTT decomposition. The four new degrees of freedom hence reflect the time derivatives of the coordinates.

Instead of fixing the densitized lapse $\bar{\alpha}$, we can specify alternatively the time derivative of the mean curvature $\partial_t K$ together with K. From equation (2.137) we can then solve

$$D^2\alpha = -\partial_t K + \alpha\left(K_{ij}K^{ij} + 4\pi(\rho + S)\right) + \beta^i D_i K \qquad (3.107)$$

for the lapse together with equations (3.37) and (3.102). Condition (3.107) involves the physical Laplace operator D^2, but in this context the conformal Laplace operator $\bar{D}^2$ would be more handy to evaluate. As it turns out, we can express equation (3.107) in terms of $\bar{D}^2$ by combining it with the Hamiltonian constraint (3.37) to yield

$$\bar{D}^2(\alpha\psi) = \alpha\psi\left(\frac{7}{8}\psi^{-8}\bar{A}_{ij}\bar{A}^{ij} + \frac{5}{12}\psi^4 K^2 + \frac{1}{8}\bar{R} + 2\pi\psi^4(\rho + 2S)\right)$$
$$- \psi^5\partial_t K + \psi^5\beta^i\bar{D}_i K. \tag{3.108}$$

The freely specifiable quantities are now the mean curvature K and the conformal metric $\bar{\gamma}_{ij}$, together with their time derivatives $\partial_t K$ and $\bar{u}_{ij}$. We can then solve the Hamiltonian constraint (3.37) for the conformal factor ψ, the momentum constraint (3.102) for the shift β^i, and equation (3.108) for the lapse α. We summarize this "extended" version of the conformal thin-sandwich decomposition in Box 3.3.[26]

Box 3.3   The extended conformal thin-sandwich (CTS) decomposition

Freely specifiable variables are $\bar{\gamma}_{ij}$, $\bar{u}_{ij}$, K and $\partial_t K$. Given these, the momentum constraint

$$(\bar{\Delta}_L\beta)^i - (\bar{L}\beta)^{ij}\bar{D}_j \ln(\bar{\alpha}) = \bar{\alpha}\bar{D}_j(\bar{\alpha}^{-1}\bar{u}^{ij}) + \frac{4}{3}\bar{\alpha}\psi^6\bar{D}^i K + 16\pi\bar{\alpha}\psi^{10}S^i \tag{3.109}$$

is solved for β^i, the Hamiltonian constraint

$$\bar{D}^2\psi - \frac{1}{8}\psi\bar{R} - \frac{1}{12}\psi^5 K^2 + \frac{1}{8}\psi^{-7}\bar{A}_{ij}\bar{A}^{ij} = -2\pi\psi^5\rho \tag{3.110}$$

is solved for ψ, and a combination of the Hamiltonian constraint and the trace of the evolution equation for K_{ij}

$$\bar{D}^2(\alpha\psi) = \alpha\psi\left(\frac{7}{8}\psi^{-8}\bar{A}_{ij}\bar{A}^{ij} + \frac{5}{12}\psi^4 K^2 + \frac{1}{8}\bar{R} + 2\pi\psi^4(\rho + 2S)\right) - \psi^5\partial_t K + \psi^5\beta^i\bar{D}_i K \tag{3.111}$$

is solved for the product $\alpha\psi = \bar{\alpha}\psi^7$. In the above equations $\bar{A}^{ij}$ is given by

$$\bar{A}^{ij} = \frac{1}{2\bar{\alpha}}\left((\bar{L}\beta)^{ij} - \bar{u}^{ij}\right). \tag{3.112}$$

The physical solution is then constructed from

$$\gamma_{ij} = \psi^4\bar{\gamma}_{ij}$$

$$K_{ij} = A_{ij} + \frac{1}{3}\gamma_{ij}K = \psi^{-2}\bar{A}_{ij} + \frac{1}{3}\gamma_{ij}K. \tag{3.113}$$

Exercise 3.16   Derive equation (3.108) from equations (3.107) and (3.37).

[26] See Pfeiffer and York, Jr. (2003a). For some results on the uniqueness and nonuniqueness of solutions to these equations see Pfeiffer and York, Jr. (2005); Baumgarte *et al.* (2007); Walsh (2007); and references therein.

The extended version of the thin-sandwich formalism seems particularly useful for the construction of equilibrium or quasiequilibrium data, since it allows us to set the time derivatives of the conformal metric and the mean curvature to zero,

$$\bar{u}_{ij} = 0 \quad \text{and} \quad \partial_t K = 0. \tag{3.114}$$

In this case equation (3.99) reduces to

$$\bar{A}^{ij} = \frac{\psi^6}{2\alpha}(\bar{L}\beta)^{ij}, \tag{3.115}$$

which looks very similar to expression (3.50) for $\bar{A}_L^{ij}$ in the conformal transverse-traceless decomposition. Here, however, $\bar{A}^{ij}$ is not longitudinal because of the extra factor of ψ^6/α. Further assuming maximal time-slicing, $K = 0$, equations (3.37), (3.102) and (3.108) now form the coupled system

$$\bar{D}^2\psi = \frac{1}{8}\psi\bar{R} - \frac{1}{8}\psi^{-7}\bar{A}_{ij}\bar{A}^{ij} - 2\pi\psi^5\rho, \tag{3.116}$$

$$(\bar{\Delta}_L\beta)^i = 2\bar{A}^{ij}\bar{D}_j(\alpha\psi^{-6}) + 16\pi\alpha\psi^4 S^i, \tag{3.117}$$

$$\bar{D}^2(\alpha\psi) = \alpha\psi\left(\frac{7}{8}\psi^{-8}\bar{A}_{ij}\bar{A}^{ij} + \frac{1}{8}\bar{R} + 2\pi\psi^4(\rho + 2S)\right), \tag{3.118}$$

for ψ, β^i and α. With $\bar{u}_{ij} = 0$, $K = 0$ and $\partial_t K = 0$ the only remaining freely specifiable quantity is the conformal metric $\bar{\gamma}_{ij}$. The equations further simplify under the assumption of conformal flatness $\bar{\gamma}_{ij} = \eta_{ij}$, in which case $\bar{R} = 0$ and the differential operators become much easier to invert.

Exercise 3.17 Show that under the additional assumption of conformal flatness, the CTS equations (3.116)–(3.118) reduce to

$$\partial^i\partial_i\psi = -\frac{1}{8}\psi^{-7}\bar{A}_{ij}\bar{A}^{ij} - 2\pi\psi^5\rho, \tag{3.119}$$

$$\partial^j\partial_j\beta^i + \frac{1}{3}\partial^i\partial_j\beta^j = 2\bar{A}^{ij}\partial_j(\alpha\psi^{-6}) + 16\pi\alpha\psi^4 S^i, \tag{3.120}$$

$$\partial^i\partial_i(\alpha\psi) = \alpha\psi\left(\frac{7}{8}\psi^{-8}\bar{A}_{ij}\bar{A}^{ij} + 2\pi\psi^4(\rho + 2S)\right), \tag{3.121}$$

in Cartesian coordinates.

The equations for the shift (3.117) and (3.120) again involve the vector Laplacian, which we previously encountered in the conformal transverse-traceless decomposition of Section 3.2.[27] We discuss strategies for solving this operator in Appendix B. Interestingly, we will rediscover the shift condition (3.117) in Chapter 4.5, where we will find that it is identical to the "minimal distortion" condition. The conformal thin-sandwich formalism therefore

---

[27] Recall equation (3.60) in Cartesian coordinates and the discussion of Bowen–York solutions that follow.

reduces to the Hamiltonian constraint for the conformal factor, the minimal distortion condition for the shift, and the maximal slicing condition for the lapse.

If initial data for a time evolution calculation are constructed from the CTT decomposition, then the lapse and shift have to be chosen independently of the construction of initial data. The CTS formalism, on the other hand, provides a lapse and a shift together with the initial data γ_{ij} and K_{ij}. Obviously, once the initial data are determined, the lapse and shift can always be chosen freely in performing subsequent evolution calculations. However, the original relation between the time derivative of γ_{ij} and $\bar{u}_{ij}$ only applies when the lapse and shift as obtained from the CTS solution are employed in the dynamical simulation.

Before closing this section it may be of interest to consider the circumstances under which the CTS formalism can reproduce a complete spacetime solution, in the sense that a dynamical evolution of the initial data, using the lapse and the shift obtained from the CTS solution, would lead to a time-independent solution. We would expect this to be possible only if the spacetime is stationary, i.e., possesses a timelike Killing vector field ξ^a. For the metric coefficients to be independent of time during an evolution, our time-vector t^a, defined in (2.98), then has to be be aligned with ξ^a. This is the case if the lapse and the shift are the Killing lapse and Killing shift, as we discussed in the context of equation (2.162).

To obtain a Killing lapse and shift we need to choose $\bar{u}_{ij} = 0$ and $\partial_t K = 0$ in the extended CTS formalism, but these conditions are not sufficient. The point is that we also have to choose the conformally related metric $\bar{\gamma}_{ij}$, as well as K, and these choices may or may not represent the stationary slices of the spacetime that we are seeking. For simple spherical spacetimes, and suitable boundary conditions, the choice $K = 0$ always gives rise to a static solution of Einstein's equations. For example, in vacuum with $\bar{\gamma}_{ij}$ chosen to be flat and $K = 0$, there is a particular choice of boundary conditions that yields the familiar isotropic form of the Schwarzschild metric (2.35); other choices of boundary conditions yield different static solutions, also with maximal slicing.[28] In the presence of matter, static solutions to the equations of motion $\nabla_b T^{ab}$ must be solved simultaneously to obtain static spacetimes.

By contrast, as we discussed in Section 3.2, there exists strong evidence that rotating Kerr black holes do not admit spatial slices that are conformally flat. Adopting conformal flatness, then, would preclude our obtaining a spatial slice of a rotating Kerr black hole. The dynamical evolution of any CTS initial data that we might construct for a rotating black hole in vacuum thus must display some time dependence that we can interpret as a gravitational wave perturbation of our rotating hole. This issue serves as a motivation for the "waveless" approximation, yet another decomposition of the constraint equations, which we discuss briefly in the following section. Before discussing this approach, we note that numerical CTS solutions for a rotating black hole, adopting

---

[28] See equations (4.23)–(4.25). Given that these solutions are spherically symmetric they are automatically conformally flat. A coordinate transformation from the areal radius R in equations (4.23)–(4.25) to an isotropic coordinate r (see exercise H.1) brings the spatial metric into the form $\psi^4 \eta_{ij}$.

conformal flatness and maximal slicing, appear to give a maximum dimensionless spin of $J_{\rm ADM}/M_{\rm ADM}^2 \lesssim 0.94$ and not unity, as we would obtain for a slice of Kerr.[29] The ADM measure of the total angular momentum ($J_{\rm ADM}$) and mass ($M_{\rm ADM}$) quoted here are defined in Section 3.5.[30]

3.4 A step further: the "waveless" approximation

The conformal thin-sandwich formalism of Section 3.3 seems more suitable for the construction of equilibrium or quasiequilibrium initial data than the transverse-traceless formalism of Section 3.2 because we can freely determine some time derivatives of quantities rather than the quantities themselves. Basically, in the CTT formalism the freely specifiable variables are the conformally related metric and parts of the extrinsic curvature, and in the CTS formalism the latter are replaced with the time derivative of the conformally related metric. For equilibrium data it is much easier to make a well-motivated choice for the time derivative of a quantity – typically zero – than for a quantity itself.

In the CTS formalism, we therefore have a natural choice for the time derivative of the conformally related metric when constructing equilibrium data, but there still is very little guidance as to how to choose the conformally related metric itself. Many applications adopt conformal flatness, so that the conformally related metric is simply a flat metric. This choice simplifies the equations quite dramatically, but a priori it is not clear whether or not it leads to a good approximation for quasiequilibrium, for example, for binary black holes or neutron stars in nearly circular orbit (see Chapters 12 and 15).

To address this problem, the "waveless" approximation[31] goes one step further and replaces the conformally related metric as a freely specifiable variable with the time derivative of the extrinsic curvature.[32] Here we will discuss this approximation only qualitatively and refer the reader to the literature for details.

Recall that to replace pieces of the extrinsic curvature with the time derivative of the conformally related metric as freely specifiable variables, the CTS formalism employs the evolution equation for the metric, e.g., equations (2.134) or (3.93). The waveless approximation is an extension of this proceedure. The key idea is to replace the conformally related metric with the time derivative of the extrinsic curvature as freely specifiable variables, now employing the evolution equation for the extrinsic curvature, equation (2.135). In addition, one can choose the time derivatives of both $\bar{\gamma}_{ij}$ and $\bar{A}_{ij}$ to satisfy a helical symmetry in a near-zone, and to set them to zero in the far zone.[33] The asymptotic behavior of these

[29] Lovelace *et al.* (2008).

[30] Measured in terms of the *quasilocal* spin J_S and mass M_S of the black hole (see Chapter 7.4 for definitions), the maximum dimensionless spin for such a CTS solution is $J_S/M_S^2 \lesssim 0.99$; Lovelace *et al.* (2008).

[31] Shibata *et al.* (2004).

[32] Several other approaches have been suggested for the construction of equilibrium models; see, e.g., Blackburn and Detweiler (1992); Andrade *et al.* (2004); Friedman and Uryū (2006).

[33] A helical symmetry makes the system appear stationary in the rotating frame of a binary; compare Chapters 12.3.1 and 15.1. Imposing a (nontrivial) helical symmetry globally is generally not compatible with asymptotic flatness.

quantities is so chosen to eliminate any standing waves in the far zone, which explains the name of this approximation.

In the context of Chapter 2 we have considered equation (2.135) as an equation that determines the time derivative of the extrinsic curvature for a given spatial metric. Instead, we now consider the time derivative of the extrinsic curvature to be given, and we would like to solve for the spatial metric, or at least its conformally related part. To do so we focus on the Ricci tensor R_{ij} that appears on the right-hand side of equation (2.135). Using equation (3.10) we can write this in terms of the Ricci tensor $\bar{R}_{ij}$ associated with the conformally related metric $\bar{\gamma}_{ij}$. When expressed in the form (2.143) this tensor contains a combination of second-order spatial derivatives of the conformally related metric, making it a rather complicated differential operator. Fortunately we still have a gauge freedom that we can use to our advantage. As it turns out, we can make a certain gauge choice under which these second derivatives simplify dramatically, leaving only the last second-order term in (2.143).[34] This last second-order term forms an elliptic operator that we can invert to find the conformally related metric.[35]

In summary, the "waveless" approximation uses the equations of the CTS formalism, as in Box 3.3, together with equation (2.135). Given a suitable gauge choice, the Ricci tensor in equation (2.135) turns into an elliptic operator on $\bar{\gamma}_{ij}$. The freely specifiable quantities are now K and its time derivative, as well as the time derivatives of both $\bar{\gamma}_{ij}$ and $\bar{A}_{ij}$. Instead of making an ad hoc choice for $\bar{\gamma}_{ij}$ we can now choose the time derivative of $\bar{A}_{ij}$ and obtain $\bar{\gamma}_{ij}$ as a solution of equation (2.135). Presumably, this approach provides a further advantage over the CTS formalism of Section 3.3 for the construction of equilibrium and quasiequilibrium initial data, since we can freely set more time derivatives of quantities equal to zero, rather than specifying (i.e., guessing) the quantities themselves.

As a concrete example we may return to the example of a rotating black hole. As we discussed in both Section 3.2 and at the end of Section 3.3, assuming conformal flatness $\bar{\gamma}_{ij} = \eta_{ij}$ in either the CTT or CTS decomposition never leads to a spatial slice of a vacuum, stationary, rotating Kerr black hole solution, but instead to a solution that may be interpreted as a rotating black hole plus some gravitational radiation. In the "waveless" approximation, by contrast, we would choose the time derivatives of $\bar{\gamma}_{ij}$, $\bar{A}_{ij}$ and K to vanish, and would then obtain the conformally related metric $\bar{\gamma}_{ij}$ as a result of the calculation. In this approach we could indeed find a slice of the Kerr solution, without any gravitational wave perturbation.[36]

More generally, the "waveless" approximation allows us to construct stationary slices of stationary spacetimes exactly, independently of any choice of the conformally related

---

[34] We will use similar gauges leading to this simplification of the Ricci tensors in several places in later chapters; see Chapters 4.3, 11.3 and 11.5.

[35] One subtlety arises from the fact that the differential operator acting on $\bar{\gamma}_{ij}$ involves $\bar{\gamma}_{ij}$ itself. To avoid this problem it is possible to express this operator – or, in fact, all differential operators in the problem – in terms of a flat reference metric.

[36] Assuming, of course, that we have imposed suitable boundary conditions.

metric. In many parts of this book we will be more interested in solutions that do not possess exact Killing vectors, for example black hole or neutron star binaries. We will see, however, that these spacetimes do admit approximate helical Killing vectors. For these situations it is possible that the "waveless" approximation produces initial data that represent this symmetry more accurately than initial data that require an ad hoc choice for the conformally related metric. We will discuss some results for binary neutron stars obtained with the "waveless" approximation in Chapter 15.3.

3.5 Mass, momentum and angular momentum

There are several important global conserved quantities that characterize an isolated system, such as its total mass and angular momentum. Associated with these global parameters, which are well-defined only for asymptotically flat spacetimes, are conservation laws that state that the rate of loss of these quantities from an isolated system is equal to the rate at which matter, fields and gravitational waves carry them away.[37] Once we have solved the constraint equations and constructed a complete set of initial data for a system, we can then determine the values of the global conserved parameters associated with the system. During a numerical evolution, monitoring the degree to which these parameters are conserved provides a very useful check on the accuracy of the numerical integration. Here we assemble a few useful formulae for evaluating some of these parameters.

Suppose the system contains matter. Then we can derive an expression for its conserved *rest mass* M_0 (sometimes called the *baryon* mass, if the matter is composed of baryons) from the continuity equation,

$$\nabla_a(\rho_0 u^a) = 0, \tag{3.122}$$

where ρ_0 is the rest-mass density. Integrating this expression over a 4-dimensional region of spacetime Ω yields

$$\int_\Omega d^4x \sqrt{-g} \nabla_a(\rho_0 u^a) = 0. \tag{3.123}$$

Using Gauss's theorem we can relate the divergence of $\rho_0 u^a$ inside the region Ω to the value of $\rho_0 u^a$ on the region's 3-dimensional boundary $\partial\Omega$,

$$\int_\Omega d^4x \sqrt{-g} \nabla_a(\rho_0 u^a) = \int_{\partial\Omega} d^3\Sigma_a \rho_0 u^a, \tag{3.124}$$

where $d^3\Sigma_a = \epsilon \mathcal{N}_a \sqrt{\gamma} d^3x$ and $\mathcal{N}^a$ is the *outward*-pointing unit normal vector on $\partial\Omega$. When $\partial\Omega$ is spacelike, the factor $\epsilon = -1$ and when $\partial\Omega$ is timelike, $\epsilon = +1$.

Now imagine a "pill-box"-shaped spacetime region that is bounded by two spatial slices Σ_1 and Σ_2 as well as a timelike hypersurface residing entirely outside the source, as illustrated in Figure 3.5. In this case only the spatial surfaces contribute to the surface

[37] See, e.g., Misner *et al.* (1973), Chapters 19 and 20.

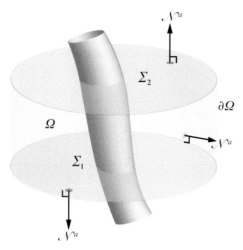

Figure 3.5 A "pill-box"-shaped spacetime region Ω is bounded by a 3-dimensional surface $\partial\Omega$ consisting of two spatial slices Σ_1 and Σ_2, as well as a timelike hypersurface that lies in the exterior vacuum. An isolated, gravitating source in the interior is confined within the shaded worldtube. The outward-pointing normal vectors $\mathcal{N}_a$ to $\delta\Omega$ are also shown.

integral. On Σ_2 the normal vector $\mathcal{N}^a$ points toward the future, and therefore coincides with the normal vector n^a of the spacetime foliation introduced in Section 2.3. From equation (2.117) we then have

$$\mathcal{N}_a u^a = n_a u^a = -\alpha u^t. \tag{3.125}$$

On Σ_1, $\mathcal{N}^a$ points toward the past, while n^a points toward the future, which introduces a negative sign between them. The conservation law (3.124) can therefore be written as

$$\int_{\Sigma_1} d^3x \sqrt{\gamma}\, \alpha u^t \rho_0 - \int_{\Sigma_2} d^3x \sqrt{\gamma}\, \alpha u^t \rho_0 = 0. \tag{3.126}$$

Equation (3.126) immediately implies that the rest mass, defined as

$$\boxed{M_0 = \int_{\Sigma} d^3x \sqrt{\gamma}\, \alpha u^t \rho_0,} \tag{3.127}$$

is conserved. The rest mass is thus determined by a 3-dimensional volume integral over a spatial region spanning the matter source.

Defining the total mass-energy of the system is more subtle, since it cannot be defined locally in general relativity. One useful measure of mass-energy is provided by the *ADM mass*, M_{ADM}, named after Arnowitt, Deser and Misner.[38] The ADM mass measures the total mass-energy of an isolated gravitating system at any instant of time measured within a spatial surface enclosing the system at infinity. Formally, M_{ADM} is defined by an integral

---

[38] See Arnowitt *et al.* (1962).

over the 2-dimensional surface at infinity, $\partial \Sigma_\infty$, of a spatial slice Σ according to[39]

$$M_{\mathrm{ADM}} = \frac{1}{16\pi} \int_{\partial \Sigma_\infty} \sqrt{\gamma} \gamma^{jn} \gamma^{im} (\partial_j \gamma_{mn} - \partial_m \gamma_{jn}) dS_i. \qquad (3.128)$$

Here $dS_i = \sigma_i \sqrt{\gamma^{\partial \Sigma_\infty}} d^2z$ is the outward-oriented surface element, where the z^i's are coordinates on $\partial \Sigma_\infty$, $\gamma_{ij}^{\partial \Sigma_\infty}$ is the induced metric on $\partial \Sigma_\infty$, and σ^i is the unit normal ($\sigma^i \sigma_i = 1$) to $\partial \Sigma_\infty$.[40] This definition requires the spacetime to be asymptotically flat and the spacetime metric to approach the Minkowski metric sufficiently quickly with increasing distance from the source,

$$g_{ab} - \eta_{ab} = \mathcal{O}(r^{-1}). \qquad (3.129)$$

The form of the integrand appearing in equation (3.128) is clearly not covariant and gives the correct answer only when it is evaluated in asymptotically Cartesian coordinates (in which case $\sqrt{\gamma} = 1$ on $\partial \Sigma_\infty$). We will provide a covariant expression shortly (see equation 3.145).[41] The existence of such a surface integral to determine the mass-energy of an isolated system is consistent with the principle that the total mass-energy of such a system can always by determined by a measurement performed by a distant observer (e.g., using Kepler's third law).

By definition, the spatial surface $\partial \Sigma_\infty$ must be taken out to infinity, in which case M_{ADM} is rigorously conserved. In numerical applications, the integral is often evaluated on a large surface at a distant, but finite, radius from the gravitating source, in the asymptotically flat region of spacetime. In this case, M_{ADM} will change in time whenever there is a flux of matter or gravitational radiation passing across the surface. However, the rate of change of M_{ADM} will exactly reflect the rate at which mass-energy is carried across the surface by these fluxes.

Using the identity

$$\partial_j \gamma_{mn} - \partial_m \gamma_{jn} = \Gamma_{mnj} - \Gamma_{jmn}, \qquad (3.130)$$

the integral (3.128) can be rewritten as

$$M_{\mathrm{ADM}} = \frac{1}{16\pi} \int_{\partial \Sigma_\infty} \gamma^{jn} \gamma^{im} (\Gamma_{mnj} - \Gamma_{jmn}) dS_i = \frac{1}{16\pi} \int_{\partial \Sigma_\infty} (\Gamma^i - \gamma^{im} \Gamma_{mn}^n) dS_i \qquad (3.131)$$

where $\Gamma^i = \gamma^{nj} \Gamma_{nj}^i$.

---

[39] Ó Murchadha and York (1974), equation (7). See also Misner *et al.* (1973), equation (20.7), or Wald (1984), equation (11.2.14), for equivalent expressions.

[40] The surface element dS_i can be written alternatively as $dS_i = \frac{1}{2}\epsilon_{ilm}\widetilde{dx}^l \wedge \widetilde{dx}^m$ ($= \epsilon_{ilm} dx^l dx^m$ in an integral), where $\epsilon_{ilm} = \sqrt{\gamma}[ilm]$ is the 3-dimensional Levi-Civita symbol and $[ilm]$ is the total antisymmetrization symbol. See, e.g., Lightman *et al.* (1975), Problem 8.10; see also Appendix C and Poisson (2004), Section 3.2.1 for discussion of alternative expressions for surface elements and Section 3.3 for a derivation of Gauss's theorem.

[41] A "gauge-invariant" form of equation (3.128) for asymptotic Minkowski spacetimes is provided by equation (85) in York, Jr. (1979): $M_{\mathrm{ADM}} = \frac{1}{16\pi} \int_{\partial \Sigma_\infty} D^j(h_j^i - \delta_j^i \operatorname{tr} h) dS_i$, where $\gamma_{ij} = \eta_{ij} + h_{ij}$, and where $h_i^k = \mathcal{O}(r^{-1})$.

To get comfortable with the above formulation of M_{ADM}, we will evaluate this mass for a Schwarzschild spacetime as described by the Kerr–Schild metric (see exercise 2.33 and Table 2.1). Taking derivatives of the spatial metric

$$\gamma_{ij} = \eta_{ij} + 2Hl_i l_j, \qquad (3.132)$$

where $H = M/r$ and $l^i = x^i/r$ with $r^2 = x^2 + y^2 + z^2$, we find

$$\Gamma^i = \frac{\alpha^4 H}{r}(3 + 8H)l^i \quad \text{and} \quad \gamma^{im}\Gamma^n_{mn} = -\frac{\alpha^4 H}{r}l^i. \qquad (3.133)$$

The metric evaluated on a 2-sphere S^2 at large radius $r \gg M$ is

$$\gamma^{\partial\Sigma_\infty}_{ij} dz^i dz^j = r^2(d\theta^2 + \sin^2\theta d\phi^2), \qquad (3.134)$$

so that the oriented surface element becomes

$$dS_i = \sigma_i \sqrt{\gamma^{\partial\Sigma_\infty}} d^2 z = l_i r^2 \sin\theta d\theta d\phi. \qquad (3.135)$$

Inserting these expressions into equation (3.131) we obtain

$$M_{\text{ADM}} = \lim_{r\to\infty} \frac{1}{16\pi} \int_{S^2} \frac{\alpha^4 H}{r}(4 + 8H)l^i l_i r^2 \sin\theta d\theta d\phi = M. \qquad (3.136)$$

While this result is not surprising, the identification of M_{ADM} with the parameter M appearing in the Kerr–Schild metric is reassuring.

To bring the expression for M_{ADM} into a more familiar and useful form for numerical applications, we employ the conformal transformation of the spatial metric according to equation (3.5). With equation (3.7) the integrand in equation (3.131) can then be expressed as

$$\Gamma^i - \gamma^{im}\Gamma^n_{mn} = \psi^{-4}(\bar{\Gamma}^i - \bar{\gamma}^{im}\bar{\Gamma}^n_{mn}) - 8\psi^{-5}\bar{D}^i\psi. \qquad (3.137)$$

Substituting this relation into the integrand in equation (3.131) and assuming that the asymptotic behavior of the conformal factor satisfies

$$\psi = 1 + \mathcal{O}(r^{-1}), \qquad (3.138)$$

we find

$$M_{\text{ADM}} = \frac{1}{16\pi} \int_{\partial\Sigma_\infty} (\bar{\Gamma}^i - \bar{\gamma}^{im}\bar{\Gamma}^n_{mn}) d\bar{S}_i - \frac{1}{2\pi} \int_{\partial\Sigma_\infty} \bar{D}^i\psi d\bar{S}_i. \qquad (3.139)$$

Note that in writing the above expression we used the fact that dS_i is conformally invariant on $\partial\Sigma_\infty$, i.e., $d\bar{S}_i = dS_i$.

> **Exercise 3.18**   Show that in Cartesian coordinates, when the conformal metric satisfies $\bar{\gamma} = 1$, then $\bar{\Gamma}^n_{mn} = 0$, so that equation (3.139) simplies to
>
> $$\boxed{M_{\text{ADM}} = \frac{1}{16\pi} \int_{\partial\Sigma_\infty} \bar{\Gamma}^i d\bar{S}_i - \frac{1}{2\pi} \int_{\partial\Sigma_\infty} \bar{D}^i\psi d\bar{S}_i,} \qquad (3.140)$$
>
> where $\bar{\Gamma}^i = -\partial_j\bar{\gamma}^{ij}$.

Exercise 3.19 Consider again a Schwarzschild spacetime described in Kerr–Schild coordinates (see exercise 2.33 and Table 2.1). Now assume a conformal decomposition of the spatial metric as in equation (3.6), so that the conformal factor becomes

$$\psi = \gamma^{1/12} = \left(1 + \frac{2M}{r}\right)^{1/12}. \tag{3.141}$$

(a) Verify that the conformally related metric satisfies

$$\bar{\gamma}_{ij} = \left(1 + \frac{2M}{r}\right)^{-1/3}\left(\delta_{ij} + \frac{2M}{r}l_i l_j\right) = \delta_{ij} + \left(\frac{2M}{r}l_i l_j - \frac{2M}{3r}\delta_{ij}\right) + O\left(\frac{1}{r^2}\right), \tag{3.142}$$

and that its inverse satisfies

$$\bar{\gamma}^{ij} = \delta^{ij} - \left(\frac{2M}{r}n^i n^j - \frac{2M}{3r}\delta^{ij}\right) + O\left(\frac{1}{r^2}\right). \tag{3.143}$$

(b) Apply equation (3.140) to determine M_{ADM}. In particular, show that the first integral in that expression contributes $2M/3$ and the second contributes $M/3$ to M_{ADM}.

If the conformal metric falls off sufficiently fast,

$$\bar{\gamma}_{ij} = \eta_{ij} + \mathcal{O}(r^{-1+a}), \tag{3.144}$$

with $a > 0$, then the first integral vanishes and the ADM mass reduces to

$$M_{\mathrm{ADM}} = -\frac{1}{2\pi}\int_{\partial\Sigma_\infty} \bar{D}^i \psi \, d\bar{S}_i = -\frac{1}{2\pi}\int_{\partial\Sigma_\infty} D^i \psi \, dS_i. \tag{3.145}$$

In the writing the last expression we used the fact that we are evaluating the surface integral at infinity, whereby we can drop the bar over the variables in the integrand. Similarly, we can replace the operator D by its flat-space counterpart. We note, however, that not every conformal metric satisfies the fall-off condition (3.144); we have already seen a counterexample in exercise 3.19.

Exercise 3.20 Show that the constant M appearing in the isotropic form of the metric for a Schwarzschild spacetime, equation (2.35), is the ADM mass M_{ADM}.

Exercise 3.21 Show that the ADM mass of the wormhole solution (3.26) is given by equation (3.27).

With the help of Gauss's theorem the 2-dimensional surface integral (3.145) can be converted into a 3-dimensional volume integral,[42]

$$M_{\mathrm{ADM}} = -\frac{1}{2\pi}\int_{\partial\Sigma_\infty} \bar{D}^i \psi \, d\bar{S}_i = -\frac{1}{2\pi}\int_\Sigma d^3x \sqrt{\bar{\gamma}}\,\bar{D}^2\psi. \tag{3.146}$$

---

[42] An identical integral can be written in terms of the unbarred (physical) metric: $M_{\mathrm{ADM}} = -\frac{1}{2\pi}\int_\Sigma d^3x \sqrt{\gamma}D^2\psi$. In fact, the interior metric can be replaced by a flat spatial metric, in which case $M_{\mathrm{ADM}} = -\frac{1}{2\pi}\int_\Sigma d^3x \nabla^2\psi$, where ∇^2 is the flat-space Laplacian operator. The choice is usually made depending on which form of the integrand is particularly useful numerically.

Substituting the Hamiltonian constraint (3.37) then yields[43]

$$M_{\text{ADM}} = \frac{1}{16\pi} \int_\Sigma d^3x \sqrt{\bar\gamma} \left(16\pi \psi^5 \rho + \psi^{-7} \bar A_{ij} \bar A^{ij} - \psi \bar R - \frac{2}{3} \psi^5 K^2 \right). \quad (3.147)$$

Of course, equations (3.146) and (3.147) only apply if the integrand is regular everywhere inside the volume Σ. They therefore do not apply, for example, if Σ contains a singularity arising from a black hole. In this case equation (3.146) must be modified so that Σ does not contain the singularity:

$$M_{\text{ADM}} = -\frac{1}{2\pi} \int_{\partial\Sigma_\infty} \bar D^i \psi d\bar S_i = -\frac{1}{2\pi} \int_\Sigma d^3x \sqrt{\bar\gamma} \bar D^2 \psi - \frac{1}{2\pi} \int_{\partial\Sigma_{\text{ex}}} \bar D^i \psi d\bar S_i, \quad (3.148)$$

where $\partial\Sigma_{\text{ex}}$ is some suitable inner surface of Σ that surrounds the black hole singularity, *excising* it from the interior of Σ. Generalization of equation (3.148) for multiple black holes is obvious, as is the substitution of the Hamiltonian constraint (3.37) for $\bar D^2 \psi$ in the integrand of the first term on the right hand side.

> **Exercise 3.22** Use equation (3.140) with $\bar\gamma = 1$, together with Hamiltonian constraint (3.37), to derive the following general expression for M_{ADM}, which allows for a black hole singularity inside $\partial\Sigma_{\text{ex}}$:
>
> $$\boxed{\begin{aligned} M_{\text{ADM}} &= \frac{1}{16\pi} \int_\Sigma d^3x \left(16\pi \psi^5 \rho + \psi^{-7} \bar A_{ij} \bar A^{ij} - \psi \bar R - \frac{2}{3} \psi^5 K^2 + \partial_i \bar\Gamma^i \right) \\ &\quad + \frac{1}{16\pi} \int_{\Sigma_{\text{ex}}} (\bar\Gamma^i - 8 \bar D^i \psi) d\bar S_i. \end{aligned}}$$
>
> $$(3.149)$$
>
> This formula for evaluating M_{ADM} has proven very useful as a diagnostic in numerical simulations involving black holes.[44]
>
> **Exercise 3.23** Assume a perfect gas, so that the stress–energy tensor is given by
>
> $$T^{ab} = (\rho_0 + \epsilon\rho_0 + P)u^a u^b + P g^{ab} \quad (3.150)$$
>
> where ϵ is the specific internal energy density and P the pressure (see Section 5.2.1). Show that under the assumptions of exercise 2.28 the difference between the ADM mass M_{ADM} and the rest-mass M_0 is given by
>
> $$M_{\text{ADM}} - M_0 = T + W + U \quad (3.151)$$
>
> in the Newtonian limit, where T is the kinetic energy,
>
> $$T = \frac{1}{2} \int \rho_0 v^2 d^3x, \quad (3.152)$$

---

[43] Note that in Chapter 11.5 we adopt a different convention for the conformal rescaling of A_{ij} via equation (11.34), whereby $\psi = e^\phi$, the second term in equation (3.147) becomes $\psi^5 \tilde A_{ij} \tilde A^{ij}$, and $\bar\gamma = 1$.

[44] Cao *et al.* (2008); see footnote 43.

W the gravitational potential energy,

$$W = \frac{1}{2} \int \rho_0 \Phi d^3 x, \tag{3.153}$$

and U is the internal energy,

$$U = \int \rho_0 \epsilon d^3 x. \tag{3.154}$$

Exercise 3.23 reinforces our interpretation of the ADM mass as the total mass-energy. Given this interpretation, the ADM mass for the boosted black hole solution of exercise 3.13 may seem quite confusing at first sight.

Exercise 3.24 Show that the ADM mass of the boosted black hole solution of exercise 3.13 is

$$M_{\text{ADM}} = \mathcal{M} + \frac{5}{8} \frac{P^2}{\mathcal{M}} + \mathcal{O}(P^4). \tag{3.155}$$

This result seems to suggest that the kinetic energy of a black hole is $5P^2/(8\mathcal{M})$, and not $P^2/(2\mathcal{M})$, as we might have expected. However, the blame for this result does not go to the ADM mass, but instead to our incorrect interpretation of the parameter $\mathcal{M}$ as the black hole's mass. In Chapter 7.1 we will find that the *irreducible mass* M_{irr}, based on the black hole's horizon area (see equations 1.70 and 7.2), is a suitable definition for the black hole's quasilocal mass. This mass differs from $\mathcal{M}$ for nonzero momentum P. In exercise 7.20 we will find that if the ADM mass (3.155) is expressed in terms of M_{irr} instead of $\mathcal{M}$, we indeed find the expected Newtonian expression $P^2/(2M_{\text{irr}})$ for the kinetic energy.

The boosted black hole solution treated in exercises 3.13 and 3.24 is constructed by specifying initial data on a conformally flat, $t = constant$ slice for a nonspinning black hole endowed with some linear momentum P_i . As we discussed in Section 3.2, the analogous approach for constructing a rotating black hole did *not* lead to a slice of a rotating Kerr black hole (recall that this approach generates a solution that has a perturbation containing gravitational radiation). Likewise, there is no reason to expect that the same approach to constructing nonspinning "boosted" black hole initial data will lead to a time slice of a Schwarzschild black hole spacetime that has been "boosted" by applying a conventional Lorentz boost to the static Schwarzschild solution (it does not). We follow the later approach to constructing a boosted black hole in exercise 3.25 and show that it leads to the expected result for the ADM mass on $t = constant$ slices in a boosted coordinate system. Specifically, if the black hole mass is M in a static frame, then we find $M_{\text{ADM}} = \gamma M$ in an asymptotic frame in which the black hole is observed to be moving with velocity v, where $\gamma = 1/\sqrt{1 - v^2}$. This result is reassuring.

Exercise 3.25 Consider the Schwarzschild metric (2.35), expressed in Cartesian isotropic coordinates, in the limit of $r = (x^2 + y^2 + z^2)^{1/2} \gg M$,

$$ds^2 = - \left(1 - \frac{2M}{r} \right) dt^2 + \left(1 + \frac{2M}{r} \right) (dx^2 + dy^2 + dz^2). \tag{3.156}$$

Now boost this metric in the z-direction by applying a Lorentz transformation

$$x = \bar{x}, \qquad y = \bar{y}, \qquad z = \gamma(\bar{z} - v\bar{t}), \qquad t = \gamma(\bar{t} - v\bar{z}) \qquad (3.157)$$

where $\gamma = (1 - v^2)^{-1/2}$.

(a) Find the new spacetime metric in the boosted (barred) coordinate system and show that the spatial metric on $\bar{t} = constant$ slices is given asymptotically by

$$\gamma_{\bar{i}\bar{j}} = \left(1 + \frac{2M}{r}\right) \eta_{\bar{i}\bar{j}} + 2\gamma^2 v^2 \frac{2M}{r} z_{\bar{i}\bar{j}}, \qquad (3.158)$$

where r is now given in terms of the boosted coordinates as $r = (\bar{x}^2 + \bar{y}^2 + \gamma^2(\bar{z} - v\bar{t})^2)^{1/2}$, and where we have defined $z_{\bar{i}\bar{j}} = \mathrm{diag}(0, 0, 1)$.

(b) The easiest approach for evaluating the ADM mass may be to use expression (3.128), keeping only leading-order terms when evaluating the integral on a sphere of constant $\bar{r} = (\bar{x}^2 + \bar{y}^2 + \bar{z}^2)^{1/2}$ with $\bar{r} \gg M$ and $\bar{r} \gg v\bar{t}$. Without loss of generality, we may set $\bar{t} = 0$. Show that

$$\partial_{\bar{j}} \gamma_{\bar{m}\bar{n}} = -\frac{2M}{r^2} m_{\bar{j}}(\eta_{\bar{m}\bar{n}} + 2\gamma^2 v^2 z_{\bar{m}\bar{n}}), \qquad (3.159)$$

where $m_{\bar{j}} = \partial r / \partial x^{\bar{j}} = (\bar{x}, \bar{y}, \gamma^2 \bar{z})/r$, and note that the surface element is $dS_{\bar{i}} = l_{\bar{i}}\bar{r}^2 \sin\theta d\theta d\phi$, where $l_{\bar{i}} = x_{\bar{i}}/\bar{r} = (\bar{x}, \bar{y}, \bar{z})/\bar{r}$.

(c) Now show that $\bar{r}/r = (1 + \gamma^2 v^2 \mu^2)^{-1/2}$, where $\mu = \bar{z}/\bar{r} = \cos\theta$, insert all the above expressions into (3.128), carry out the trivial integration over ϕ, and show that the integral reduces to

$$M_{\mathrm{ADM}} = \gamma^2 \frac{M}{2} \int_{-1}^{1} \frac{d\mu}{(1 + \gamma^2 v^2 \mu^2)^{3/2}}. \qquad (3.160)$$

(d) Finally, evaluate the integral (3.160) to find

$$M_{\mathrm{ADM}} = \gamma M. \qquad (3.161)$$

Exercise 3.26   Consider the Brill wave defined in exercise 3.5, subject only to the following regularity and asymptotic restrictions on q:

$$q = 0 \quad \text{for} \quad \rho = 0, \qquad (3.162)$$

$$\partial_\rho q = 0 \quad \text{for} \quad \rho = 0, \qquad (3.163)$$

$$\partial_z q = 0 \quad \text{for} \quad z = 0, \qquad (3.164)$$

$$q \sim r^{-a}, \ a \geq 2. \qquad (3.165)$$

(a) Show that M_{ADM} can be evaluated as

$$M_{\mathrm{ADM}} = \frac{1}{2\pi} \int \left(\frac{D_i \psi}{\psi}\right)^2 d^3x, \qquad (3.166)$$

and thereby prove that the mass-energy is positive definite for any $q \neq 0$.[45] In the context of a gravitational wave at a moment of time symmetry, the mass-energy is sometimes referred to as the *Brill mass*.

---

[45] Brill (1959); Eppley (1977).

(b) Consider the following form for q,

$$q = \frac{A\rho^2}{1 + (r/\lambda)^n},$$ (3.167)

where A and λ are constants, $r^2 = \rho^2 + z^2$, and $n \geq 4$. Choose $n = 5$ and $\lambda = 1$, integrate either equation (3.146) or equation (3.166) numerically for different A between $10^{-2} \leq A \leq 16$ and make a plot of M_{ADM} vs. A. Check your integrations by showing that for small amplitude A, the mass is proportional to A^2 and satisfies $M_{\mathrm{ADM}}/A^2 \approx 1.4 \times 10^{-2}$.

On further probing, things can sometimes get a little confusing, as when evaluating the ADM mass for Schwarzschild spacetime in Painlevé–Gullstrand coordinates (see exercise 2.32 or Table 2.1). All the above expressions yield an ADM mass of zero in these coordinates. The reason for this is that the shift in Painlevé–Gullstrand coordinates does not fall off sufficiently fast, violating the condition (3.129). If nothing else, this result provides a warning to us that we must be careful to check that the metric satisfies the correct asymptotic conditions in the adopted coordinates when applying the above formulae to calculate the mass. It also motivates a search for other mass definitions.

For example, another possibility for determining the mass is the *Bondi–Sachs mass*. For stationary spacetimes, the ADM and Bondi–Sachs mass are identical. But they are different for dynamical spacetimes emitting radiation (gravitational or otherwise). The ADM mass, evaluated at spatial infinity, remains strictly conserved, while the Bondi–Sachs mass, evaluated at null infinity, decreases in response to the outgoing radiation. We refer the reader to other references for a detailed discussion and formulae for the Bondi–Sachs mass.[46]

One other measure of mass that has proven particularly useful in recent numerical applications is the *Komar mass*,[47] M_{K}, which can be defined for a spacetime that admits a timelike Killing vector, say $\xi_{(t)}^a$. Contracting this Killing vector with the (4-dimentional) Ricci tensor we can define a current

$$J_{(t)}^a = \xi_{(t)}^b \, {}^{(4)}R^a_{\ b}.$$ (3.168)

The Ricci tensor is related to the stress-energy tensor through Einstein's equations, suggesting that this current may lead to a reasonable mass definition. As it turns out, the current (3.168) is conserved, meaning that its divergence vanishes. To see this, we compute

$$\nabla_a J_{(t)}^a = (\nabla_a \xi_{(t)}^b) \, {}^{(4)}R^a_{\ b} + \xi_{(t)}^b \nabla_a \, {}^{(4)}R^a_{\ b}.$$ (3.169)

Here the first term vanishes because of Killing's equation and the second one can be rewritten using the contracted Bianchi identity

$$\nabla_a J_{(t)}^a = \frac{1}{2} \xi_{(t)}^b \nabla_b \, {}^{(4)}R = 0,$$ (3.170)

[46] See, e.g., Poisson (2004), Sections 4.3, and Section 11.2, and references therein.
[47] See Komar (1959). Our discussion is adapted from Carroll (2004), Section 6.4.

where the second term vanishes because the directional derivative of $^{(4)}R$ along a Killing vector has to be zero. Just as the existence of the conserved matter current $\rho_0 u^a$ gave rise to a conserved rest-mass (3.127) in the beginning of this section, the conserved current (3.168) gives rise to the conserved Komar mass,

$$M_{\mathrm{K}} = \frac{1}{4\pi} \int_{\Sigma} d^3 x \sqrt{\gamma} n_a J^a_{(t)}. \tag{3.171}$$

Assuming that the Killing field $\xi^a_{(t)}$ is properly normalized in the asymptotically flat region of spacetime ($\xi^a_{(t)} \xi^{(t)}_a = -1$), the above definition of the Komar mass leads to the expected result for the value of the mass in familiar cases, as we shall see below. On the other hand, in a vacuum spacetime we have $^{(4)}R_{ab} = 0$, so that $J^a_{(t)} = 0$, so we might conclude from equation (3.171) that the Komar mass vanishes for a black hole. This conclusion is incorrect, however, since for a black hole the volume integral (3.171) contains a singularity at which $^{(4)}R_{ab}$ is not defined.[48] However, the current (3.168) can be written as a total derivative. As a consequence we can convert the volume integral into a surface integral at infinity that can be evaluated even for black holes.

To show that $J^a_{(t)}$ can be written as a total derivative we start with the definition of the Riemann tensor,

$$\nabla_a \nabla_b \xi^c_{(t)} - \nabla_b \nabla_a \xi^c_{(t)} = {}^{(4)}R^c_{dab} \xi^d_{(t)}. \tag{3.172}$$

Contracting this equation and using the relation $\nabla_a \xi^a_{(t)} = 0$, which holds for any Killing vector, we find

$$\nabla_b \nabla_a \xi^b_{(t)} = {}^{(4)}R_{ab} \xi^b_{(t)}, \tag{3.173}$$

so that, by equation (3.168), we have

$$J^a_{(t)} = \nabla_b \nabla^a \xi^b_{(t)}. \tag{3.174}$$

Inserting equation (3.174) into equation (3.171) we then have

$$M_{\mathrm{K}} = \frac{1}{4\pi} \int_{\Sigma} d^3 x \sqrt{\gamma} n_a \nabla_b \nabla^a \xi^b_{(t)}. \tag{3.175}$$

Since $\nabla^a \xi^b_{(t)}$ is antisymmetric, we can use Gauss's theorem to convert this volume integral into the surface integral at infinity,

$$\boxed{M_{\mathrm{K}} = \frac{1}{4\pi} \int_{\partial \Sigma_{\infty}} dS_b n_a \nabla^a \xi^b_{(t)}.} \tag{3.176}$$

Equation (3.176), which can be used even in the case of black holes, could have been the starting point for our derivation of the Komar mass.

---

[48] We already faced a similar issue above when we discussed volume integrals for M_{ADM} in the case of spacetimes containing singularities.

We now bring this expression into a form that is easier to evaluate in terms of standard $3 + 1$ variables. We assume that the coordinates are chosen to reflect the symmetry imposed by the Killing vector $\xi_{(t)}^a$, meaning that our lapse and shift are the Killing lapse and shift as defined in equation (2.162), and thus satisfy

$$\xi_{(t)}^a = t^a = \alpha n^a + \beta^a. \tag{3.177}$$

Recalling that $dS_b = d^2 z \sqrt{\gamma^{\partial\Sigma_\infty}} \sigma_b$, we can rewrite the integrand in equation (3.176) as

$$
\begin{aligned}
n_a \sigma_b \nabla^a \xi_{(t)}^b &= -n_a \sigma_b \nabla^b \xi_{(t)}^a = -n_a \sigma_b \nabla^b (\alpha n^a + \beta^a) \\
&= -n_a \sigma_b (n^a \nabla^b \alpha + \alpha \nabla^b n^a + \nabla^b \beta^a) = \sigma_b \nabla^b \alpha + \sigma_b \beta^a \nabla^b n_a \\
&= \sigma_b \nabla^b \alpha - \sigma_b \beta^a K^b_{\ a} = \sigma_i D^i \alpha - \sigma_i \beta^j K^i_{\ j},
\end{aligned}
\tag{3.178}
$$

where we have used Killing's equation, and equation (2.52), as well as the relations $n_a n^a = -1$, $n_a \beta^a = 0$ and $n^a \sigma_a = 0$. Inserting this into equation (3.176) finally yields

$$\boxed{M_{\mathrm{K}} = \frac{1}{4\pi} \int_{\partial\Sigma_\infty} dS_i (D^i \alpha - \beta^j K^i_{\ j}).} \tag{3.179}$$

In many cases, the term $\beta^j K^i_{\ j}$ falls off rapidly enough so that it can be neglected in the integral (3.179). A counterexample is the Schwarzschild spacetime in Painlevé–Gullstrand coordinates, for which the shift term carries the entire contribution to the Komar mass.

Exercise 3.27 Evaluate the Komar mass M_{K} for a Schwarzschild black hole in (1) Schwarzschild coordinates and (2) Painlevé–Gullstrand coordinates.

Evidently, exercise 3.27 shows that the Komar mass can handle Painlevé–Gullstrand coordinates while the ADM mass cannot. On the other hand, the definition of the Komar mass assumes the existence of a timelike Killing vector, which the ADM mass does not. This property of the Komar mass in fact provides a powerful diagnostic tool for numerical spacetimes: by searching for solutions for which the ADM mass and Komar mass are both well-defined and identical, we can identify spacetimes that admit a timelike Killing vector and are hence in stationary equilibrium. For example, we shall apply this diagnostic in Chapter 12.3.3 to construct (quasi-)stationary spacetimes containing binary black holes in circular equilibrium.

Finally, we can turn the surface integral (3.179) back into the volume integral

$$M_{\mathrm{K}} = \frac{1}{4\pi} \int_\Sigma d^3 x \sqrt{\gamma} (D^2 \alpha - \beta^j D^i K_{ij} - K_{ij} D^i \beta^j), \tag{3.180}$$

provided the integrand contains no singularities. Otherwise, we can excise the singularity and restrict the volume integral to the region outside the excision surface just as we did in deriving equation (3.148). With the help of equation (2.137), the momentum constraint (2.133) and the evolution equation (2.134) this integral can be converted into

$$M_K = \int_\Sigma d^3x \sqrt{\gamma} \left(\alpha(\rho + S) - 2\beta^i S_i \right), \tag{3.181}$$

where we have assumed again that $\xi^a_{(t)} = t^a$ is a Killing vector, so that we have $\partial_t \gamma_{ij} = \partial_t K = 0$. For vacuum spacetimes this volume integral vanishes, which is not surprising in the light of the discussion following equation (3.171). In practice, this realization means that the Komar mass in equation (3.179) can be evaluated on any surface $\partial \Sigma$ enclosing all matter sources and singularities, and does not need to be evaluated at infinity on $\partial \Sigma_\infty$ (in contrast to the ADM mass).

> **Exercise 3.28** Show that, under the same weak-field, slow-velocity assumptions as in exercises 2.28, 3.7 and 3.23, the difference between the Komar mass M_K and the rest-mass M_0 becomes
>
> $$M_K - M_0 = 3T + 2W + U + 3\Pi, \tag{3.182}$$
>
> in the Newtonian limit. Here we have defined the quantity
>
> $$\Pi = \int P d^3x. \tag{3.183}$$

Equation (3.182) may seem strange at first. Recall, however, that the Komar mass is defined only in the presence of a timelike Killing vector, that is, for stationary spacetimes. Such spacetimes are in dynamical equilibrium, so that they obey the virial theorem given by

$$2T + W + 3\Pi = 0 \tag{3.184}$$

in the Newtonian limit.[49] Combining equation (3.184) with (3.182) shows that, just like the ADM mass, the Komar mass may be interpreted as the total mass-energy (*cf.* exercise 3.23). As we have discussed above, this argument can be turned around to identify systems in dynamical equilibrium by searching for spacetimes for which the ADM and Komar mass are equal. In fact, a relativistic formulation of the virial theorem can be derived by imposing the equality of the ADM and Komar mass.[50]

We now turn to the *angular momentum* of an isolated system. First we shall treat the case of an axisymmetric spacetime, for which there exists is a global rotational Killing vector, $\xi^a_{(\phi)}$. Retracing the steps that lead to the Komar mass integral (3.176), we can construct a surface integral for a conserved angular momentum according to

$$\boxed{J_K = -\frac{1}{8\pi} \int_{\partial\Sigma_\infty} dS_b n_a \nabla^a \xi^b_{(\phi)}.} \tag{3.185}$$

---

[49] See, e.g., Shapiro and Teukolsky (1983), equation (7.1.21).
[50] See Gourgoulhon and Bonazzola (1994).

Equation (3.185) is the *Komar angular momentum* expression for an asymptotically flat, axisymmetric spacetime. This quantity measures the angular momentum about the symmetry axis and gives the expected magnitude in familiar examples, as we will see in exercise 3.29 below. The fact that the surface can be chosen arbitrarily can be used to prove that in an axisymmetric spacetime gravitational radiation carries no angular momentum.[51] The integrand can be rewritten as

$$\sigma_b n_a \nabla^a \xi^b_{(\phi)} = -\sigma_b n_a \nabla^b \xi^a_{(\phi)} = \sigma_b \xi^a_{(\phi)} \nabla^b n_a = -\sigma_j \xi^i_{(\phi)} K_i{}^j, \qquad (3.186)$$

where we have used Killing's equation, the relation $\xi^a_{(\phi)} n_a = 0$, equation (2.52) and, in the last step, the fact that both $\xi^a_{(\phi)}$ and σ_b are spatial. Inserting the above expression into equation (3.185) yields

$$J_K = \frac{1}{8\pi} \int_{\partial\Sigma_\infty} dS_j \xi^i_{(\phi)} K_i{}^j. \qquad (3.187)$$

Adopting Cartesian coordinates, we can assign the symmetry axis to be along x^i, in which case the spatial rotational Killing vector about this axis becomes[52]

$$\xi^{(\phi)}_k = \xi^{(i)}_k = \epsilon_{kml} e^m_{(i)} x^l, \qquad (3.188)$$

where $e^m_{(i)} = \delta_i{}^m$ is the basis vector along x^i and ϵ_{kml} is the 3-dimensional Levi-Civita tensor, both evaluated in the asymptotically flat region. The angular momentum about the x^i axis may then be calculated from

$$\boxed{J^K_i = \frac{\epsilon_{ijk}}{8\pi} \int_{\partial\Sigma_\infty} dS_l x^j K^{kl}.} \qquad (3.189)$$

If desired, this integral can also be converted into a volume integral.

> **Exercise 3.29** Show that the angular momentum of solution (3.66) is J_i (which, by virtue of exercises 3.11 and 2.34 also shows that the angular momentum of a Kerr black hole is $J = aM$).

In the absence of axisymmetry, it is often useful to evaluate the *ADM angular momentum*, which is defined as

$$J^{ADM}_i = \frac{1}{8\pi} \int_{\partial\Sigma_\infty} dS_m (K^m{}_n - \delta^m{}_n K) \xi^n_{(i)}, \qquad (3.190)$$

or, substituting equation (3.188),

$$\boxed{J^{ADM}_i = \frac{1}{8\pi} \epsilon_{ijn} \int_{\partial\Sigma_\infty} dS_m x^j (K^{mn} - \delta^{mn} K).} \qquad (3.191)$$

---

[51] Lightman *et al.* (1975), Problem 18.9
[52] See, e.g., Lightman *et al.* (1975), Problem 10.9.

This expression requires some stronger asymptotic gauge conditions beyond asymptotic flatness, but these conditions are met by a conformally flat metric, $\bar{\gamma}_{ij} = \eta_{ij}$.[53]

Exercise 3.30 Prove that J_i^{ADM} reduces to J_i^{K} in axisymmetry.
Hint: First choose the surface $\partial\Sigma_\infty$ to be a 2-sphere and show that the piece of the integral involving K in the expressions for J_i^{ADM} vanishes. Now generalize to arbitrary surfaces.

Exercise 3.31 Show that in a stationary, axisymmetric and nonsingular spacetime the mass-energy can be calculated from

$$M_{\text{ADM}} = M_{\text{K}} = \int_\Sigma d^3x \sqrt{\gamma}\, n_a(2T^a{}_b - \delta^a{}_b T)\xi_{(t)}^b, \qquad (3.192)$$

and the angular momentum can be calculated from

$$J^{\text{ADM}} = J^{\text{K}} = -\int_\Sigma d^3x \sqrt{\gamma}\, n_a T^a{}_b \xi_{(\phi)}^b. \qquad (3.193)$$

Here we assume that the interior volume Σ covers the entire matter distribution and contains no singularities.

Finally, if we replace the rotational Killing vector in equation (3.190) by the spatial translational Killing vector, we obtain an expression for the *ADM linear momentum*. A spatial translational Killing vector along x^i in the asymptotically flat region takes the form

$$\xi_{(i)}^n = e_{(i)}^n. \qquad (3.194)$$

Inserting this relation into (3.190) yields the linear momentum along x^i:

$$\boxed{P_i^{\text{ADM}} = \frac{1}{8\pi} \int_{\partial\Sigma_\infty} dS_m(K^m{}_i - \delta^m{}_i K).} \qquad (3.195)$$

Once again, the surface integrals given above for J_i^{ADM} and P_i^{ADM} can be converted into volume integrals, if desired.

Exercise 3.32 Show that the linear momentum of solution (3.80) is P_i^{ADM}. (Recall that this solution assumes $K = 0$.)

Exercise 3.33 Show that the linear momentum of the boosted black hole in exercise 3.25 is given by $P_i^{\text{ADM}} = \gamma M v_i$, where $\gamma = 1/\sqrt{1 - v^2}$.

The importance of monitoring the conserved quantities assembled here when performing a numerical simulation of an evolving system cannot be overstated. As we mentioned

---

[53] York, Jr. (1979); Gourgoulhon *et al.* (2002).

earlier, the degree to which they are conserved helps calibrate the accuracy of the simulation. Often these quantities are evaluated on a large, but finite, surface in the asymptotically flat region near the outer boundary of the computational grid. In this case, any changes in the calculated mass, angular momentum or linear momentum of the system must be accounted for by the integrated flux of these quantities carried off by any matter or gravitational waves crossing the surface. We will see how well this principle works in action in some representative simulations described in later chapters.

4 | Choosing coordinates: the lapse and shift

In Chapter 2 we performed a $3 + 1$ decomposition of Einstein's field equations and have seen that these can be split into two distinct sets: constraint equations and evolution equations. The constraint equations contain no time derivatives and relate field quantities on a given $t = constant$ spacelike hypersurface. The evolution equations contain first-order time derivatives that tell us how the field quantities change from one hypersurface to the next. In Chapter 3 we have brought the constraint equations into a form that is suitable for numerical implementation, that is, we cast the equations in terms of spatial differential operators that can be inverted with standard numerical techniques. We will provide a brief introduction to some common numerical algorithms for solving these (elliptic) equations in Chapter 6. The $3 + 1$ evolution equations that we derived, e.g., equation (2.134) for γ_{ij}, and equation (2.135) for K_{ij}, are not quite ready for numerical integration. For one thing, we have yet to impose *coordinate conditions* by specifying the *lapse* function α and the *shift* vector β^i that appear in these equations. The lapse and shift are freely specifiable gauge variables that need to be chosen in order to advance the field data from one time slice to the next. As it turns out, finding kinematical conditions for the coordinates that allow for a well-behaved, long time evolution is nontrivial in general. However, geometric insight and numerical experimentation can be combined to produce good gauge choices for treating many of the most important physical and astrophysical problems requiring numerical relativity for solution, as we shall see.

What constitutes a "good" coordinate system? Clearly, the adopted coordinates must not allow the appearance of any singularities, which could have dire consequences for a numerical simulation. Such a singularity, which is often associated with a black hole, could be either a coordinate singularity or a physical singularity. Recall, for example, the case of a Schwarzschild black hole in Schwarzschild coordinates. Singularities associated with the Schwarzschild radius $r_s = 2M$ (the horizon) are coordinate in origin and are removable by a coordinate transformation. Singularities at $r_s = 0$, however, are physical and invariant, resulting from the infinite curvature at the origin, and are not removable by a coordinate transformation. We know that encountering such a physical singularity during a space voyage would be disastrous for any traveler; be assured that encountering one in a numerical simulation would be equally disastrous for a numerical relativist![1] Such a singularity will result in one or more of the field variables blowing up to infinity,

---

[1] Unless one's code is specifically designed to explore the nature of singularities; see, e.g., Berger (2002).

leading to underflows and overflows in the output and eventually causing the code to crash.

To avoid coordinate singularities associated with horizons, like the one at $r_s = 2M$ that we discussed above, black hole simulations have sometimes been carried out using "horizon penetrating" coordinates in which light cones do not pinch-off at the horizon as they do in Schwarzschild coordinates. Kerr–Schild coordinates provide one example of a "horizon penetrating" coordinate system that is well-behaved at $r_s = 2M$. Many simulations also have relied on "singularity-avoiding" gauge conditions to prevent or postpone the appearance of physical singularities in the computational domain. These gauges have been especially important in treating stellar collapse to black holes, where physical singularities are not present in the initial spacetime, but inevitably arise following the formation of a black hole. More recently, black hole "excision" techniques to prevent the appearance of singularities have been developed whereby the black hole interior and its curvature singularity are excised from the computational domain altogether. In codes that solve the partial differential equations on a discrete spacetime coordinate lattice, it has proven adequate in some cases to retain the black hole interior and its singularity within the computational domain by simply avoiding placing the singularity on any lattice point on which the variables are evaluated. This is the trick commonly used to evolve "puncture" black holes, which contain interior coordinate singularities, rather than physical singularities, as we discussed in Chapter 3.1.2.[2] In these simulations it is again the particular choice of coordinates that prevents numerical lattice points from reaching the physical spacetime singularity.[3]

In fact, it has even been possible in some simulations to replace the black hole interior and its spacetime singularity with smooth, but otherwise arbitrary, initial data in order to evolve the spacetime numerically. The interior data in such cases is "junk" in that it typically fails to satisfy the constraint equations. However, these "junk-filling" initial data are completely adequate to permit a reliable evolution of the exterior field, as long as suitable gauge conditions are implemented to insure that the computational scheme allows no information to leak out from the black hole interior to the exterior during the evolution. Given the causal nature of the black hole event horizon, it is not surprising that such a scheme can be implemented. Examples of all of these approaches will be discussed later on.

But now we are jumping ahead ourselves. First we must step back and note that the problem of picking an appropriate coordinate system typically is split into two parts: choosing a *time slicing* (i.e., a time coordinate), and picking a *spatial gauge* (i.e., spatial coordinates). The time slicing determines what shape the spatial slices Σ take in the enveloping spacetime. The lapse α determines how the shape of the slices Σ changes in time, since it relates the advance of proper time to coordinate time along the normal vector n^a connecting one spatial slice to the next, as illustrated in Figure 2.4. Picking a time

---

[2] Recall the discussion following equation (3.19).
[3] See Section 4.5 and Chapter 13.1.3.

slicing or a time coordinate therefore amounts to making a choice for the lapse function. Letting the lapse vary with position across the spatial slice takes advantage of the freedom that proper time can advance at different rates at different points on a given slice ("the many-fingered nature of time"). The shift β^i, on the other hand, determines how spatial points at rest with respect to a normal observer n^a are relabeled on neighboring slices. The spatial gauge is therefore imposed by a choice for the shift vector.

In the rest of this chapter we will discuss a few different gauge choices that are commonly used in numerical simulations. We will focus here on some of the simpler conditions that lend themselves to straightforward geometric interpretation and provide us with valuable intuition. In later chapters we will observe these and other choices in action when we study actual simulations that construct numerical spacetimes. There we will see how well different gauge conditions perform in different physical situations. A brief summary of a few common gauge choices is provided in Box 4.1 for the reader eager and willing to order from a limited, but representative, menu.

4.1 Geodesic slicing

Since the lapse α and the shift β^i can be chosen freely, let us first consider the simplest possible choice,

$$\boxed{\alpha = 1, \quad \beta^i = 0.}$$

(4.1)

In the context of numerical relativity this gauge choice is often called *geodesic slicing*; the resulting coordinates are also known as *Gaussian-normal coordinates*.[4]

Recall that coordinate observers move with 4-velocities $u^a = t^a = e^a_{(0)}$ (i.e., spatial velocities $u^i = 0$). Thus with $\beta^i = 0$, coordinate observers coincide with normal observers ($u^a = n^a$). With $\alpha = 1$, the proper time intervals that they measure agree with coordinate time intervals. Their acceleration is given by equation (2.51),

$$a_b = D_b \ln \alpha = 0.$$

(4.2)

Evidently, since their acceleration vanishes, normal observers are freely-falling and therefore follow geodesics, hence the name of this slicing condition. Clearly, the evolution equations (2.134) and (2.135) simplify significantly when geodesic slicing is adopted.

> **Exercise 4.1** Consider the Robertson–Walker metric for a flat Friedmann cosmology,
>
> $$ds^2 = -dt^2 + a^2(t)\,\eta_{ij}dx^i dx^j,$$
>
> (4.3)
>
> where the expansion factor $a(t)$ is a function of time only. We can immediately read off the lapse and shift to be $\alpha = 1$ and $\beta^i = 0$, showing that this spacetime is geodesically sliced.
>
> (a) Identify the spatial metric γ_{ij} and find the extrinsic curvature K_{ij}.

---

[4] Darmois (1927) used these coordinates in his very early development of the 3+1 decomposition.

(b) Assume that the matter stress-energy tensor can be described by a homogeneous and isotropic perfect fluid comoving with coordinate observers,

$$T_{ab} = (\rho^* + P)u_a u_b + P g_{ab}, \tag{4.4}$$

whereby the fluid 4-velocity satisfies $u^a = n^a$. Here ρ^* is the total mass-energy density and P is the pressure of the fluid. Derive the familiar Friedmann equations

$$\frac{\ddot{a}}{a} + 2\left(\frac{\dot{a}}{a}\right)^2 - 4\pi(\rho^* - P) = 0$$

$$3\frac{\ddot{a}}{a} + 4\pi(\rho^* + 3P) = 0 \tag{4.5}$$

from the constraint and evolution equations.

Despite its simplicity, geodesic slicing tends to form coordinate singularities very quickly during an evolution. This result is not surprising, since geodesics tend to focus in the presence of gravitating sources. Coordinate observers therefore approach each other, collide, and thereby form a coordinate singularity. This can be seen quite easily from the evolution equation (2.137) for the trace of the extrinsic curvature which, for geodesic slicing and a comoving, perfect fluid, reduces to

$$\partial_t K = K_{ij}K^{ij} + 4\pi(\rho + 3P) \geq 0. \tag{4.6}$$

The above inequality holds since $K_{ij}K^{ij}$ is nonnegative, and so is $\rho + 3P$, provided that the strong energy condition holds.[5] This means that K grows monotonically in time, and that the expansion of normal observers,

$$\nabla_a n^a = g^{ab}\nabla_a n_b = (g^{ab} + n^a n^b)\nabla_a n_b = \gamma^{ab}\nabla_a n_b = -K, \tag{4.7}$$

decreases monotonically in time. In geodesic slicing, equation (2.136) becomes

$$\partial_t \ln \gamma^{1/2} = -K, \tag{4.8}$$

which shows that the coordinate volume element of the normal observers goes to zero when K grows without bound.[6] The geometric situation is depicted in Figure 2.2. This behavior results in a coordinate singularity.

As an example, consider a weak gravitational wave that is initially centered on the origin of an otherwise flat vacuum spacetime. After a brief interaction the wave disperses and leaves behind flat space. Also consider a set of coordinate observers that are at rest with respect to each other initially. The gravitational wave packet carries energy and hence attracts the observers gravitationally, who, initially, start moving toward the origin of the spacetime. Once the gravitational wave has dispersed, the observers are no longer attracted gravitationally to the center, but they continue to coast toward each other until they form

[5] See Hawking and Ellis (1973), Section 4.3 for a discussion of energy conditions.
[6] See the discussion following equation (2.60) and the related footnote.

a coordinate singularity. As the following exercise demonstrates, we can even estimate the time at which this singularity will form.

> **Exercise 4.2** Consider a weak gravitational wave packet centered on the origin in a vacuum spacetime at a moment of time symmetry ($K_{ij} = 0$.)
> (a) Argue that after some finite time t_0, the trace of the extrinsic curvature at the origin will acquire a positive-definite value, $K_0 > 0$.
> (b) Split K_{ij} into its trace and its traceless part, and integrate equation (4.6) to find a lower bound for K as a function of time.
> (c) Find an upper limit for the time at which a coordinate singularity will develop, as $K \to \infty$. Express your answer in terms of t_0 and K_0.

For a Friedmann cosmology represented by a Robertson–Walker metric, geodesic slicing presents no such difficulty. For the case described in exericse 4.1, for example, K certainly increases monotonically, but it starts out at negative infinity when $a = 0$ at $t = 0$ and approaches zero as $a \to \infty$ and $t \to \infty$. For this special case of a homogeneous, expanding Universe, K thus does not grow without bound, and, away from the initial curvature singularity (the "big bang"), the coordinates remain regular. Small density enhancements or other perturbations on the homogeneous background, however, can cause a local focusing of geodesic observers, and then geodesic slicing might again develop coordinate singularities.

> **Exercise 4.3** Consider evolving the vacuum spacetime for a Schwarzschild black hole of mass M. Refer to the Kruskal–Szekeres diagram depicted in Figure 1.1 and take the initial time slice to be the $v = 0$ hypersurface at a moment of time symmetry. Suppose we adopt geodesic slicing to perform the evolution. Show that the coordinate time at which the geodesic slices will hit the physical curvature singularity at the origin, and thereby bring the evolution to a screeching halt, is $t = \pi M$.
> **Hint:** Follow the motion of the normal observer located at the point $u = 0$ on the initial slice.

The situation explored in exercise 4.3 reveals the difficulty of evolving a vacuum space-time containing a black hole with geodesic slicing. The problem is further elucidated in the left-hand panel in Figure 4.1. There it is shown how geodesic slices, beginning at a moment of time symmetry, fail to cover a substantial portion of the black hole exterior by the time they hit the central singularity. We cannot evolve the spacetime into the future once we encounter the singularity, because of the breakdown of the equations. Were we to use the same slicing condition to treat a more realistic spacetime containing, say, by some additional matter or radiation in the exterior of the black hole, any exploration of poten-tially interesting physical phenomena associated with these exterior sources would have to terminate once the slices hit the singularity. We thus might never learn what observable features the presence of a black hole might imprint on these exterior sources. Such an unfortunate situation motivates our search for a better gauge choice.

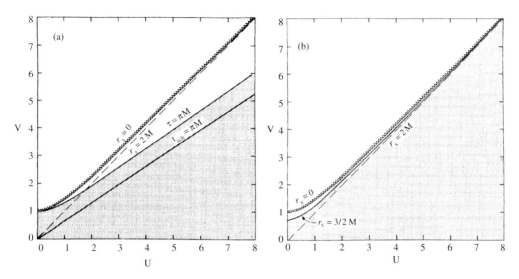

Figure 4.1 The shaded regions on the Kruskal–Szekeres diagrams represent the development of the initial $v = 0$ time slice in a Schwarzschild spacetime for *geodesic* slicing in the left panel (a) and *maximal* slicing in the right panel (b). The geodesic slices hit the singularity at (areal) radius $r_s = 0$ after a coordinate time $t = \pi M$ (= proper time for any normal observer), during which only a small fraction of the black hole exterior is covered. The maximal slices, on the other hand, cover the entire black hole exterior and approach a limiting surface at $r_s = (3/2)M$ in the interior as $t \to \infty$. [After Smarr and York (1978a).]

4.2 Maximal slicing and singularity avoidance

From equation (4.7) it is evident that the divergence of normal observers can be controlled by imposing a suitable condition on the mean curvature K. If K is specified as a function of both space and time, equation (2.137) becomes an elliptic equation for the lapse α,

$$D^2\alpha = -\partial_t K + \alpha \left(K_{ij} K^{ij} + 4\pi(\rho + S) \right) + \beta^i D_i K. \tag{4.9}$$

Evidently this condition simplifies when K is chosen to be a constant in both space and time. A common choice is the *maximal slicing* condition

$$\boxed{K = 0,} \tag{4.10}$$

which we have already encountered in the context of initial data.[7] If we assume maximal slicing not only on one slice, but at all times, then the time derivative of K must vanish as well,

$$K = 0 = \partial_t K. \tag{4.11}$$

---

[7] For some applications it may be advantageous to choose K to be a constant value different from zero. It can be shown that this choice leads to time slices that are asymptotically null in asymptotically flat spacetimes.

With this choice equation (4.9) reduces to

$$D^2\alpha = \alpha\left(K_{ij}K^{ij} + 4\pi(\rho + S)\right),$$ (4.12)

which we can solve for the lapse α independently of the shift β^i.

The maximal slicing condition (4.12) can be expressed in various ways. Combining with the Hamiltonian constraint (2.132) we obtain

$$D^2\alpha = \alpha\left(R - 4\pi(3\rho - S)\right).$$ (4.13)

Alternatively, we have already found in Chapter 3.3 that a combination with the conformally decomposed Hamiltonian constraint (3.37) yields

$$\bar{D}^2(\alpha\psi) = \alpha\psi\left(\frac{7}{8}\psi^{-8}\bar{A}_{ij}\bar{A}^{ij} + \frac{1}{8}\bar{R} + 2\pi\psi^4(\rho + 2S)\right)$$ (4.14)

(see equation 3.108).

> **Exercise 4.4** Prove that choosing $K = 0$ extremizes the volume of spatial slices.
> **Hint:** Define the volume of a bounded portion S of the slice Σ according to $\text{vol}(S) = \int_S d^3x \sqrt{\gamma}$. Deform Σ infinitesimally along a vector $d^a = cn^a + b^a$, where c is a (small) arbitrary constant, $n^a b_a = 0$, and $c = 0 = b^a$ on the boundary δS of S. Show that $\delta\text{vol}(S) = \int_S d^3x \sqrt{\gamma}(-cK)$ and conclude.[8]

Maximal slicing is not an entirely unfamiliar concept. Consider a soap film spanning a fixed closed wire loop, which then forms a two-dimensional hypersurface in a three-dimensional Euclidian space. If gravity can be neglected, the potential energy of the soap film is due only to surface tension and hence is proportional to its surface area. The film therefore assumes a shape of minimal area. With an argument similar to the one in exercise 4.4 it can be shown[9] that this shape satisfies $K = 0$. In a Euclidean geometry, such extremal surfaces with $K = 0$ are minimal, while in a pseudo-Riemannian geometry they are maximal.

By construction, maximal slicing prevents the focusing of normal observers that we have found for geodesic slicing. Equation (4.7) implies that in maximal slicing the divergence of normal observers vanishes or, equivalently, that the normal congruence is expansion free. This means that maximal slices are "volume preserving" along the normal congruence n^a, which can also be seen from equation (2.60) and the discussion which immediately follows it. In exercise 2.5 we have also established that the normal congruence is irrotational. Together this means that normal observers in maximal slicing move like irrotational and incompressible fluid elements. The incompressible property prevents the focusing of normal observers.

---

[8] See York, Jr. (1979), Section 8.5 for futher discussion.
[9] See, e.g., Lightman *et al.* (1975), Problem 9.31.

In exercise 2.11 we found that the extrinsic curvature of $t = constant$ slices of a Schwarz-schild black hole in isotropic coordinates, equation (1.60), vanishes. We now realize that in this time coordinate the foliation of Schwarzschild spacetime is maximally sliced. It is straightforward to verify that the lapse function characterizing this metric satisfies the maximal slicing condition, equation (4.12). The same statements must apply to a Schwarzschild black hole in Schwarzschild coordinates, equation (1.51), since the $t = constant$ time slices are the same as in isotropic coordinates, as only the radial coordinate is different. Slices of constant coordinate time in Kerr–Schild and Painlevé–Gullstrand coordinates, on the other hand, are not maximally sliced.

> **Exercise 4.5** Verify explicitly that the lapse function for a Schwarzschild black hole in isotropic or Schwarzschild coordinates in Table 2.1 satisfies the maximal slicing condition (4.12), but that the lapse functions in Kerr–Schild and Painlevé–Gullstrand coordinates do not.

We point out that the $t = constant$ slices of Schwarzschild for the isotropic and Schwarz-schild metrics in Table 2.1 are by no means the only maximal slices of Schwarzschild spacetime. In fact, we can derive an entire family of time-independent, maximal slices of Schwarzschild as follows.[10] Start with the Schwarzschild solution in Schwarzschild coordinates

$$ds^2 = -f_0 dt^2 + f_0^{-1} dr_s^2 + r_s^2 d\Omega^2, \tag{4.15}$$

where we have introduced the function $f_0(r_s) = 1 - 2M/r_s$ for ease of notation. Now consider a new time coordinate $\bar{t}$ that is related to the old time coordinate t according to

$$\bar{t} = t + h(r_s), \tag{4.16}$$

where $h(r_s)$ is the so-called "height" function. By allowing $h(r_s)$ to depend on r_s alone we ensure that the resulting metric will again be time-independent.[11] In a spacetime diagram, $h(r_s)$ measures how far $\bar{t} = constant$ surfaces "lift off" the familiar $t = constant$ surfaces. Using $dt = d\bar{t} - h' dr_s$, where $h' \equiv dh/dr_s$, we can now transform the metric (4.15) to find

$$ds^2 = -f_0 d\bar{t}^2 + 2 f_0 h' d\bar{t} dr_s + (f_0^{-1} - f_0 h'^2) dr_s^2 + r_s^2 d\Omega^2. \tag{4.17}$$

Comparing (4.17) with the line element in the $3 + 1$ form (2.123) we can identify the spatial metric, shift and lapse associated with $\bar{t} = constant$ surfaces as

$$\gamma_{ij} = \text{diag}\left((1 - f_0^2 h'^2)/f_0, r_s^2, r_s^2 \sin^2\theta\right), \qquad \beta^{r_s} = \frac{f_0^2 h'}{1 - f_0^2 h'^2}, \qquad \alpha^2 = \frac{f_0}{1 - f_0^2 h'^2}. \tag{4.18}$$

---

[10] See Reinhart (1973); Estabrook *et al.* (1973); Beig and Ó Murchadha (1998); see also our discussion in Chapter 8.1.
[11] We could relax this assumption, but will instead postpone a derivation of time-dependent maximal slices to Chapter 8.1.

Not surprisingly, we recover the familiar $t = constant$ slices for $h = constant$. From equation (2.116) we then construct the normal vector $n^a = \alpha^{-1}(1, -\beta^i)$. For the $\bar{t} = constant$ slices to be maximal we must have

$$K = -\nabla_a n^a = -|g|^{-1/2}\partial_a(|g|^{1/2}n^a) = 0. \tag{4.19}$$

Using $|g|^{1/2} = \alpha\gamma^{1/2} = r_s^2 \sin\theta$ (see exercise 2.25), and noting that all time derivatives must vanish, equation (4.19) reduces to

$$\frac{d}{dr_s}\left(r_s^2\left(\frac{f_0}{1 - f_0^2 h'^2}\right)^{1/2}f_0 h'\right) = 0. \tag{4.20}$$

As we might expect, equation (4.20) forms a second-order differential equation for h. We can immediately obtain a first integral,

$$r_s^2\left(\frac{f_0}{1 - f_0^2 h'^2}\right)^{1/2}f_0 h' = C, \tag{4.21}$$

where C is some constant of integration, or

$$f_0^2 h'^2 = \frac{C^2}{f_0 r_s^4 + C^2}. \tag{4.22}$$

Inserting (4.22) into the expressions (4.18) we obtain the spatial metric

$$dl^2 = f^{-2}(r_s; C)dr_s^2 + r_s^2(d^2\theta + \sin^2\theta d^2\phi), \tag{4.23}$$

the lapse

$$\alpha = f(r_s; C) \tag{4.24}$$

and the shift

$$\beta^{r_s} = \frac{Cf(r_s; C)}{r_s^2}, \tag{4.25}$$

where the function $f(r_s; C)$ is given by

$$f(r_s; C) = \left(1 - \frac{2M}{r_s} + \frac{C^2}{r_s^4}\right)^{1/2}. \tag{4.26}$$

The constant C parametrizes each member of the family. Evidently, the familiar $t = constant$ slices of a Schwarzschild spacetime in Schwarzschild coordinates are recovered for $C = 0$.

> **Exercise 4.6** Show that in the spherical polar coordinates of the metric (4.23) the extrinsic curvature is given by
>
> $$K^i{}_j = \frac{C}{r_s^3}\text{diag}(-2, 1, 1), \tag{4.27}$$

which confirms immediately that the slices given by equations (4.23)–(4.25) are indeed maximal for all values of C. More adventurous readers may also verify that these slices satisfy the $3 + 1$ constraint and evolution equations.

Since the maximal slicing expressions (4.12) through (4.14) constitute spatial, second-order, partial differential equations for the lapse, two boundary conditions are required to specify a unique solution. For asymptotically flat spacetimes it is natural to require $\alpha \to 1$ for the outer boundary at $r_s \to \infty$. The second boundary condition depends on the physical situation, the location of the inner boundary, the adopted spatial coordinates (e.g., Cartesian *vs.* spherical polar coordinates), etc. For example, for spherically symmetric spacetimes without singularities, one might adopt spherical polar coordinates and impose regularity at the origin, whereby $\partial_{r_s} \alpha = 0$. In other cases there may be some freedom associated with the choice of an inner boundary condition, as we will now illustrate. Consider a Schwarzschild spacetime and again take the initial time slice to be at a moment of time symmetry, e.g., the $v = 0$ hypersurface in the Kruskal–Szekeres diagram, Figure 1.1. Focus on the upper right-hand quadrant in the diagram (i.e., $u > 0$) and take the black hole throat at $u = 0$ to be the inner boundary. Now consider the lapse function appearing in the Schwarzschild or isotropic metric in Table 2.1; the two functions represent the same lapse, but in different radial coordinates. Note that this lapse satisfies $\alpha = 0$ at our inner boundary, where the isotropic radius is $r = M/2$ and the areal radius $r_s = 2M$. Adopting $\alpha = 0$ as our inner boundary condition we find that the resulting maximal slices will be the hypersurfaces of constant Schwarzschild time t appearing as straight lines through the origin in the Kruskal–Szekeres diagram. The situation is illustrated in Figure 8.1. The lapse function obtained by solving equation (4.12) will be the same function that we have been looking at in Table 2.1. If we combine this lapse with a vanishing shift we obtain a Killing lapse and shift, meaning we can construct the Killing vector t^a from equation (2.98). As a consequence, no metric coefficients change in time for this gauge choice (making this a "static slicing" of Schwarzschild) and they are given by the familiar static metric coefficients in standard Schwarzschild or isotropic coordinates. As seen in the figure, the slices terminate at Schwarzschild time $t = \infty$ and never penetrate the black hole interior.

Suppose instead we choose the inner boundary condition on the lapse to be symmetric across the throat by setting $\partial_{r_s} \alpha = 0$ there. Again take the initial time slice to be the moment of time symmetry on the $v = 0$ axis for $u > 0$, so that the solution of equation (4.12) at $t = 0$ is $\alpha = 1$.[12] As shown in Figure 8.3, the resulting foliation is now quite different from the previous foliation, although both are maximal. We postpone a detailed discussion of this particular slicing until Chapter 8.1, where we shall derive the complete solution for the spacetime analytically. For now it suffices to note that the resulting metric on successive time slices changes with time (this is therefore an example of a "dynamical slicing" of Schwarzschild). More significantly, the time slices manage to penetrate the

---

[12] The shift is zero initially (moment of time symmetry), but we shall allow for a nonzero shift as the evolution proceeds in order to require the radial coordinate to be the areal radius.

black hole interior, but they never encounter the central singularity. Instead, the slices asymptote to a limiting surface at areal radius $r_s = 3M/2$ in the black hole hole interior. This property makes maximal slicing an example of a "singularity avoiding" slicing condition. The situation is summarized in the right-hand panel of Figure 4.1, where we see that these maximal slices succeed in covering the entire black hole exterior by $t = \infty$. Able to penetrate into the interior, this dynamic slicing solution yields a lapse and other metric coefficients that do not exhibit the familiar coordinate singularities at $r_s = 2M$ that plague the previous nonpenetrating, static solution. By avoiding the central singularity, this particular dynamic slicing solution[13] is able to cover the entire black hole exterior and thereby follow future evolution in the exterior "forever", at least in principle. This capability is in stark contrast to geodesic slicing and is the main reason why dynamic maximal slicing is considered a "good" gauge choice for building numerical spacetimes containing black holes.[14]

It may be helpful to clarify the relation between these two maximal slicings of Schwarz-schild with the family of time-independent slicings in equations (4.23)–(4.25). In the latter, the lapse and shift are the Killing lapse and shift, which makes this solution time-independent. This Killing lapse satisfies the boundary conditions that we described above only for two special cases. For $C = 0$, we have $\alpha = 0$ at $r_s = 2M$, and we recover "static slicing", yielding hypersurfaces of constant Schwarzschild time. For $C = 3\sqrt{3}M^2/4$, we have $\partial_{r_s}\alpha = 0$ on the limiting surface at $r_s = 3M/2$; this is the slice to which the "dynamical slicing" of above asymptotes as $t \to \infty$.[15] We will return to this discussion in Chapters 8.1 and 13.1.3.

How do these maximal slices manage to avoid hitting the central singularity after penetrating the black hole interior? After all, the maximum proper time it takes *any* timelike observer in the black interior ($r_s < 2M$) to reach the central singularity is πM, the value approached by a freely-falling observer who starts from rest just outside the horizon (see exercise 4.3). Since the advance of proper time of a normal observer is given by $d\tau = \alpha dt$, it is necessary that the lapse function plummet to zero at late times in the black hole interior in order to prevent the observer from reaching the singularity as $t \to \infty$. This behavior is sometimes referred to as the "collapse of the lapse", which we will now explore.

The collapse of the lapse can be illustrated in the following simple model problem.[16] In vaccum, the maximal slicing condition (4.13) is

$$D^2\alpha - \alpha R = 0. \tag{4.28}$$

Obviously, both the Laplace operator D^2 and the Ricci scalar R depend on the spatial metric γ_{ij}. Our simplification now lies in taking γ_{ij} to be flat, and the Ricci scalar R

---

[13] Sometimes it is referred to as the "extended maximal foliation" of Schwarzschild.
[14] "Grid stretching" near the throat and other complications that can arise to make late-time black hole evolution numerically inaccurate when adopting dynamic maximal slicing will be discussed in Chapter 8.
[15] See also Hannam *et al.* (2008) for a more detailed discussion.
[16] Smarr and York (1978a); York, Jr. (1979).

to be some positive constant R_0 inside some radius r_0, and zero outside. The curvature and metric are no longer consistent, but it turns out that this model has a qualitative behavior not unlike the dynamical maximal slicing solution described above for a vacuum Schwarzschild spacetime.

In spherical symmetry, general solutions to (4.28) can be found quite easily both inside and outside r_0. Imposing the boundary conditions $\alpha = 1$ at $r = \infty$ and $d\alpha/dr = 0$ at $r = 0$, and matching both α and its first derivative at r_0, yields the solution

$$
\alpha = \begin{cases} \dfrac{1}{\cosh x_0} \dfrac{\sinh x}{x}, & x < x_0 \\[2mm] 1 + \dfrac{\tanh x_0 - x_0}{x}, & x \geq x_0, \end{cases} \tag{4.29}
$$

where

$$
x = r \sqrt{R_0}, \quad \text{and} \quad x_0 = r_0 \sqrt{R_0}. \tag{4.30}
$$

The solutions are naturally parametrized by the dimensionless parameter x_0, which is a measure of the strength of the scalar curvature R_0.

Exercise 4.7 Verify that equation (4.29) is the desired solution to the model problem (4.28).

The minimum value for α occurs at the origin and takes the value

$$
\alpha_{\min} = \frac{1}{\cosh x_0}. \tag{4.31}
$$

As x_0 increases, $\alpha_{\min}$ approaches zero; see Figure 4.2. For strong fields, and hence large x_0, we find the asymptotic behavior

$$
\alpha_{\min} \sim \exp(-x_0). \tag{4.32}
$$

This relation is responsible for the collapse of the lapse, which we can see as follows: Suppose we guess that as the maximal foliation proceeds, the interior field-strength parameter x_0 increases linearly with time t according to

$$
x_0 = t/\tau_e + \text{constant}, \tag{4.33}
$$

where τ_e is some constant. We then would expect that at late times the minimum lapse will decay as

$$
\alpha_{\min} \sim \exp(-t/\tau_e). \tag{4.34}
$$

We might also guess that the e-folding time constant τ_e would be comparable to M, the mass of the source. In fact, such late-time exponential decay of the lapse has been found in numerous numerical simulations employing maximal slicing.[17] For the dynamic

---

[17] See, e.g., Smarr and York (1978a); Evans (1984); Petrich *et al.* (1985).

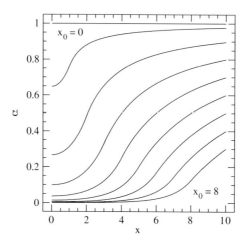

Figure 4.2 The lapse α as a function of dimensionless radius x for select values of the parameter x_0, plotted according to the analytic model. The parameter x_0 varies between $x_0 = 0$ (top curve, $\alpha = 1$) and $x_0 = 8$ (bottom curve) in increments of unity. The "collapse of the lapse" is evident in the strong-field region. [After Smarr and York (1978a).]

maximally-sliced Schwarzschild spacetime discussed above and derived in Chapter 8.1, it can be shown analytically[18] that the e-folding time is given by $\tau_e = 3\sqrt{6}/4M \sim 1.837M$, which is very close to the value 1.82 inferred from some of the earliest numerical calculations.[19] Profiles of the lapse for the exact solution are shown in Figure 8.4.

While many of its geometric properties are very desirable, maximal slicing also has some computational disadvantages. The conditions (4.12)–(4.14) are spatial elliptic equations for the lapse function α. Even though they are linear, such equations are often costly to invert numerically in two and three dimensions. Since maximal slicing is only a coordinate gauge condition, truly physical results extracted from a simulation should not be affected at all if the maximal slicing condition is modified, or if the condition is satisfied only approximately. This realization suggests that, instead of solving an elliptic equation, one could convert that equation into a parabolic equation, which is much faster to solve numerically.[20] In fact, one way of solving an elliptic equation, called "relaxation", is to introduce a time variable, recast the equation as a parabolic equation, and look for steady-state solutions.[21]

An evolution that adopts an "approximate" maximal slicing condition presumably will generate data that satisfy condition (4.10) only approximately. Suppose that during the course of the evolution the data give rise to nonzero values of the mean curvature, K, in violation of condition (4.10). Maintaining the condition $\partial_t K = 0$ after this occurs would allow K to remain nonzero, which would not be consistent with maximal slicing. It would

[18] Petrich *et al.* (1985); Beig and Ó Murchadha (1998).
[19] Smarr and York (1978a).
[20] Balakrishna *et al.* (1996); Shibata (1999a).
[21] See Chapter 6, where we discuss the classification of partial differential equations, as well as computational strategies for solving them.

be better to impose an alternative condition that drives K back to zero. We can accomplish this task by employing the condition

$$\partial_t K = -cK, \tag{4.35}$$

where c is some positive constant. Note that the above condition is completely consistent with (4.11). Inserting the above relation into equation (4.9) we now find

$$D^2\alpha = \alpha\left(K_{ij}K^{ij} + 4\pi(\rho + S)\right) + \beta^i D_i K + cK, \tag{4.36}$$

which is still an elliptic equation for the lapse α.

We now convert equation (4.36) into a parabolic equation by adding a "time" derivative of the lapse,

$$\partial_\lambda\alpha = D^2\alpha - \alpha\left(K_{ij}K^{ij} + 4\pi(\rho + S)\right) - \beta^i D_i K - cK, \tag{4.37}$$

where λ is some time parameter. Setting $\lambda = \epsilon t$, we can convert this expression into

$$\partial_t\alpha = \epsilon D^2\alpha - \epsilon\alpha\left(K_{ij}K^{ij} + 4\pi(\rho + S)\right) - \epsilon\beta^i D_i K - \epsilon cK, \tag{4.38}$$

where the parameter ϵ now acts as an effective diffusion constant. We note that we can write the new lapse condition (4.38) more compactly as

$$\partial_t\alpha = -\epsilon(\partial_t K + cK). \tag{4.39}$$

This slicing condition is often referred to as a *K-driver* condition.

Equation (4.35), corresponding to the exponential decay of K to zero as time increases, is the solution to equation (4.39) as $\epsilon \to \infty$. However, setting ϵ too large may produce a numerical instability, since some numerical prescriptions for solving parabolic equations must satisfy a Courant condition that limits the size of the allowed time step (see Chapter 6.2.4). For sufficiently large ϵ, this time step may have to be chosen significantly smaller than the one used to evolve the gravitational field equations. In this case equation (4.38) or (4.39) can be solved by breaking up each evolution time step into several substeps. This approach is equivalent to solving equation (4.36) by relaxation, except that the relaxation process is not necessarily carried out to convergence, since it suffices to impose maximal slicing only approximately.[22]

4.3 Harmonic coordinates and variations

Consider a contraction of the 4-dimensional connection coefficients

$$^{(4)}\Gamma^a \equiv g^{bc}\,^{(4)}\Gamma^a_{bc} = -\frac{1}{|g|^{1/2}}\partial_b\left(|g|^{1/2}g^{ab}\right). \tag{4.40}$$

---

[22] See also Duez *et al.* (2003), who typically use five substeps per evolution time step, with $\epsilon \approx 0.6$ and $c \approx 0.1$.

One way to impose a gauge condition is to set these quantities equal to some predetermined *gauge source functions* H^a,

$$^{(4)}\Gamma^a = H^a. \tag{4.41}$$

In particular, we may choose these gauge source functions to vanish, which defines *harmonic coordinates*

$$\boxed{^{(4)}\Gamma^a = 0.} \tag{4.42}$$

Exercise 4.8 explains why these coordinates are called "harmonic".

> **Exercise 4.8** Show that in harmonic coordinates, the coordinates x^a themselves are harmonic functions,
>
> $$\nabla^2 x^a \equiv \nabla^b \nabla_b x^a = 0. \tag{4.43}$$

Inserting the metric (2.119) into equation (4.40) shows that in harmonic coordinates the lapse and shift satisfy the coupled set of hyperbolic equations[23]

$$(\partial_t - \beta^j \partial_j)\alpha = -\alpha^2 K \tag{4.44}$$

$$(\partial_t - \beta^j \partial_j)\beta^i = -\alpha^2 \left(\gamma^{ij} \partial_j \ln \alpha + \gamma^{jk} \Gamma^i_{jk} \right). \tag{4.45}$$

> **Exercise 4.9** Derive equations (4.44) and (4.45).

Harmonic coordinates have played an important role in the mathematical development of general relativity, since they bring the 4-dimensional Ricci tensor $^{(4)}R_{ab}$ into a particularly simple form.

> **Exercise 4.10** Show that in terms of $^{(4)}\Gamma^a$, the Ricci tensor $^{(4)}R_{ab}$ can be written
>
> $$^{(4)}R_{ab} = -\frac{1}{2}g^{cd}\partial_d \partial_c g_{ab} + g_{c(a}\partial_{b)}\,^{(4)}\Gamma^c + \,^{(4)}\Gamma^c\,^{(4)}\Gamma_{(ab)c}$$
>
> $$+ 2g^{ed}\,^{(4)}\Gamma^c_{e(a}\,^{(4)}\Gamma_{b)cd} + g^{cd}\,^{(4)}\Gamma^e_{ad}\,^{(4)}\Gamma_{ecb}. \tag{4.46}$$

With the Ricci tensor expressed in terms of second derivatives of the metric (as in equation 2.143), the second derivatives appear in four different terms with the free indices appearing in all possible different positions. As demonstrated in exercise 4.10, three of these four terms can be absorbed in the derivative of the connection coefficients $^{(4)}\Gamma^a$, leaving only one term that acts as a wave operator. In harmonic coordinates, where $^{(4)}\Gamma^a = 0$, Einstein's equations therefore reduce to a set of nonlinear wave equations,[24] which is why all of the early hyperbolic formulations of Einstein's equations are based on these coordinates.[25] We will return to these considerations in Chapter 11.

---

[23] See Smarr and York (1978b); York, Jr. (1979).
[24] De Donder (1921); Lanczos (1922).
[25] E.g., Choquet-Bruhat (1952); Fischer and Marsden (1972).

Completely harmonic coordinates have been adopted in only a few $3 + 1$ simulations,[26] although a formalism based on "generalized harmonic coordinates", has been employed very successfully.[27] We will discuss this approach in much greater detail in Chapter 11.3. More common in $3 + 1$ calculations is *harmonic slicing*, in which only the time-component $^{(4)}\Gamma^0$ is set to zero. Combining harmonic slicing with a zero shift yields a particularly simple equation for the lapse,

$$\partial_t \alpha = -\alpha^2 K,\qquad(4.47)$$

which, after inserting equations (2.136) for αK, can be integrated to yield

$$\boxed{\alpha = C(x^i)\gamma^{1/2}.}\qquad(4.48)$$

Here $C(x^i)$ is a constant of integration that may depend on the spatial coordinates x^i, but not on time. This condition is identical to keeping the densitized lapse $\bar{\alpha} = \gamma^{-1/2}\alpha$ constant (see equation 3.100).

> **Exercise 4.11**   Show that $t = constant$ slices of the Schwarzschild spacetime in isotropic coordinates (equations 2.145 through 2.148) are harmonic.[28]

The harmonic slicing condition (4.48) is just about as simple as the geodesic slicing condition (4.1), but it provides for a much more stable numerical evolution.[29] It does not focus coordinate observers and in some cases has allowed for long time evolutions.[30] However, there is no guarantee that harmonic slicing will lead to well-behaved coordinates in more general situations[31] and it has been pointed out that the singularity avoidance properties of harmonic slicing are weaker than those, for example, of maximal slicing.[32]

Equation (4.48) is an example of a coordinate condition in which the lapse can be found algebraically, without having to solve complicated and computer-intensive differential equations, as is necessary for maximal slicing. To generalize this condition, we can decorate the right hand side of equation (4.47) with a positive but otherwise arbitrary function $f(\alpha)$,[33]

$$\partial_t \alpha = -\alpha^2 f(\alpha) K.\qquad(4.49)$$

For $f = 1$ this condition obviously reduces to the harmonic slicing condition (4.47) above. For $f = 0$ (and $\alpha = 1$ initially), it reduces to geodesic slicing (Section 4.1). Formally, maximal slicing (Section 4.2) corresponds to $f \to \infty$.[34] For $f = 2/\alpha$, the condition

[26] See Landry and Teukolsky (1999); Garfinkle (2002).
[27] See Pretorius (2005a,b).
[28] Other time-independent harmonic slices of Schwarzschild and Kerr–Newman spacetimes for which the lapse does not vanish on the horizon have been derived by Bona and Massó (1988) and Cook and Scheel (1997).
[29] See, e.g., Shibata and Nakamura (1995); Baumgarte and Shapiro (1998).
[30] See Cook and Scheel (1997) and Baumgarte *et al.* (1999) for some examples.
[31] See, e.g., Alcubierre (1997); Alcubierre and Massó (1998); Khokhlov and Novikov (2002).
[32] Shibata and Nakamura (1995); Garfinkle (2002).
[33] See Bona *et al.* (1995).
[34] See exercise H.4 for an example.

(4.49) can be integrated to yield

$$\alpha = 1 + \ln \gamma, \tag{4.50}$$

where we have used equation (2.136) and have chosen the constant of integration to be unity. This quite popular slicing condition is often called "1+log" slicing. As an algebraic slicing condition it has the virtue of being extremely simple to implement and fast to solve. It has also been found to have stronger singularity avoidance properties than harmonic slicing. The latter can be motivated by the observation that f becomes large when α becomes small, so that it probably behaves more like maximal slicing than harmonic slicing.

In the above derivation we assumed $\beta^i = 0$, which may or may not be a good choice. Allowing for a nonzero shift, the condition (4.49) with $f = 2/\alpha$ may be generalized to include an advective shift term,

$$\boxed{(\partial_t - \beta^j \partial_j)\alpha = -2\alpha K.} \tag{4.51}$$

Equation (4.51) deserves to be boxed, since it has proven to be an extremely successful and robust (hyperbolic) slicing condition. It is currently adopted in in many "moving puncture" binary black hole simulations. We will discuss these simulations, and the role of the slicing condition (4.51), in much greater detail in Chapter 13.1.3.

Before proceeding we point out that the "advective" version of the 1+log condition, equation (4.51), can be written as

$$n^a \nabla_a \alpha = \mathcal{L}_{\mathbf{n}} \alpha = -2K. \tag{4.52}$$

This means that this slicing condition is covariant in the sense that it does not depend on the choice of the shift. The "nonadvective" version (4.49), on the other hand, is not covariant, since the "direction" of the partial derivative $\partial_t \alpha$ does depend on the shift. Stated differently, the nonadvective derivative $\partial_t \alpha$ takes a derivative in the direction of the time vector t^a, which is coordinate dependent, whereas the advective term $-\beta^j \partial_j \alpha$ shifts the direction back along the normal vector n^a, which has a geometric, coordinate-independent meaning.

4.4 Quasi-isotropic and radial gauge

In the previous sections we have focused primarily on time slicing conditions that specify the lapse function α. We now turn to gauge conditions for the spatial coordinates, i.e., conditions that specify the shift vector β^i. As is the case when picking a lapse, an important goal when choosing a shift is to provide for a stable, long-term dynamical evolution. In addition, it is often desirable to bring the spatial metric into a simple form. For asymptotically flat spacetimes, for example, one might like the metric at large distances to be related straightforwardly to the Schwarzschild metric in some familiar coordinate system. One might also like gravitational radiation to be easily identifiable as, for example, the transverse-traceless components of an asymptotically flat metric.

Loosely speaking, two different strategies can be employed when constructing a spatial gauge condition. One strategy is to define a geometric condition on the spatial metric from which a gauge condition can be derived. An example of such a condition is the minimal distortion gauge, which we will discuss in Section 4.5. Alternatively, we can impose an algebraic condition on the spatial metric directly. For example, we can set some of its components to zero in order to simplify the Einstein equations. This latter approach is the basis of the quasi-isotropic and radial gauge conditions, which we will discuss in this section. Both of these gauges have played important roles in the evolution of axisymmetric spacetimes. We shall assume that the spacetimes considered in this section are axisymmetric and we will adopt spherical polar coordinates to treat them.

In general the spatial metric γ_{ij} has six independent components. Using our three degrees of spatial coordinate freedom we can impose three conditions on the metric, and thereby reduce the number of its independent variables to three. In spherical polar coordinates, the *quasi-isotropic gauge* is defined by the three conditions

$$\gamma_{r\theta} = \gamma_{r\phi} = 0 \tag{4.53}$$

and

$$\gamma_{\theta\theta}\gamma_{\phi\phi} - (\gamma_{\theta\phi})^2 = \gamma_{rr}\gamma_{\phi\phi}r^2, \tag{4.54}$$

which reduces the metric to the form[35]

$$dl^2 = A^2(dr^2 + r^2 d\theta^2) + B^2 r^2(\sin\theta d\phi + \xi d\theta)^2. \tag{4.55}$$

Demanding that conditions (4.53) and (4.54) hold at all times, we can insert them into the evolution equation (2.134), and this results in a set of three coupled differential equations for the three components of the shift β^i.

> **Exercise 4.12**   For axisymmetric spacetimes in the absence of net angular momentum we can set $\xi = 0$ and $K^r{}_\phi = 0$. Argue that in this case the quasi-isotropic shift satisfies $\beta^\phi = 0$ and
>
> $$r\partial_r\left(\frac{\beta^r}{r}\right) - \partial_\theta\beta^\theta = \alpha(2K^r{}_r + K^\phi{}_\phi), \tag{4.56}$$
>
> $$r\partial_r\beta^\theta + \partial_\theta\left(\frac{\beta^r}{r}\right) = 2\alpha\frac{K^r{}_\theta}{r}. \tag{4.57}$$

To introduce a somewhat more "natural" set of variables, we can replace A and B by the conformal factor

$$\psi^6 = A^2 B \tag{4.58}$$

and define a new variable,

$$\eta = \ln(A/B), \tag{4.59}$$

---

[35] See, e.g., Smarr (1979b); Bardeen and Piran (1983); Evans (1984); Abrahams and Evans (1988); Shapiro and Teukolsky (1992a); Abrahams *et al.* (1994).

which is a measure of the anisotropy of the spatial slices. The two variables η and ξ are "dynamical" or "radiative" variables that serve to measure gravitational waves at large distance from the gravitating source. We will explore some concrete examples that use these variables in Chapter 10.[36] In terms of ψ, η and ξ, the line element (4.55) now takes the form

$$dl^2 = \psi^4 \left[e^{2\eta/3}(dr^2 + r^2 d\theta^2) + e^{-4\eta/3} r^2 (\sin\theta d\phi + \xi d\theta)^2 \right], \qquad (4.60)$$

In spherical symmetry we can set, without loss of generality, $\eta = \xi = 0$, in which case the spatial metric reduces to the familiar form $\psi^4 \eta_{ij}$. In vacuum we therefore recover the Schwarzschild solution in isotropic coordinates, which explains the name "quasi-isotropic" gauge.

A related coordinate condition is the *radial gauge*, for which condition (4.54) is replaced by the relation

$$\gamma_{\theta\theta}\gamma_{\phi\phi} - (\gamma_{\theta\phi})^2 = r^4 \sin^2\theta, \qquad (4.61)$$

while conditions (4.53) remain unchanged. The spatial metric then takes the form

$$dl^2 = A^2 dr^2 + B^{-2} r^2 d\theta^2 + B^2 r^2 (\sin\theta d\phi + \xi d\theta)^2. \qquad (4.62)$$

Equations for the shift vector can again be derived from equation (2.134) by demanding that conditions (4.53) and (4.61) hold at all times.

In spherical symmetry we can set $B = 1$ and $\xi = 0$ without loss of generality, in which case the metric (4.62) takes the form of the Schwarzschild metric in Schwarzschild coordinates (see equation 3.22). This result again suggests that natural dynamical variables in the radial gauge are ξ and η.[37]

> **Exercise 4.13**   The radial gauge has been used quite often with *polar slicing* in axisymmetry.[38] Polar slicing is defined by
>
> $$K_T \equiv K^\theta_{\ \theta} + K^\phi_{\ \phi} = 0. \qquad (4.63)$$
>
> Show that in spherical symmetry (for which $\beta^\theta = \beta^\phi = 0$) the radial gauge and polar slicing conditions in vacuum imply $\beta^r = 0$.

While both the quasi-isotropic and radial gauges have been used extensively for calculations in spherical symmetry and axisymmetry, they are particularly convenient only in spherical polar coordinates. Such coordinates are natural for spherical and axisymmetric spacetimes, but they suffer from coordinate singularities at $r = 0$ and along the symmetry axis where $\theta = 0$ and $\theta = \pi$. Most calculations in full $3 + 1$ dimensions have adopted Cartesian coordinates, which do not have such singularities. In addition, given that in spherical symmetry the radial gauge reduces to Schwarzschild coordinates, this gauge

[36] See also Appendix F.
[37] Bardeen and Piran (1983) take $\eta = B^2 - 1$ for the dynamical variable in place of equation (4.59).
[38] See, e.g., Bardeen and Piran (1983); Shapiro and Teukolsky (1986).

condition is susceptible quite generally to developing coordinate singularities near the horizon of any black hole that may be present in the spacetime.

4.5 Minimal distortion and variations

In Chapter 3 we found that the conformally related metric $\bar{\gamma}_{ij}$ has five independent functions, two of which correspond to true dynamical degrees of freedom and three to coordinate freedom. For a stable and accurate numerical evolution it is desirable to eliminate purely coordinate-related fluctuations in $\bar{\gamma}_{ij}$. To accomplish this, one may want to construct a gauge condition that minimizes the time rate of change of the conformally related metric. This gauge condition is called *minimal distortion*.[39] A related but recently less popular gauge condition is *minimal strain*, which minimizes the time rate of change of the spatial metric instead of the conformally related metric.

In Chapter 3.3 we introduced u_{ij} as the traceless part of the time derivative of the spatial metric,

$$u_{ij} \equiv \gamma^{1/3} \partial_t (\gamma^{-1/3} \gamma_{ij}),\tag{4.64}$$

(see equation 3.92). Since u_{ij} is traceless, we can decompose it into a transverse-traceless and a longitudinal part

$$u_{ij} = u_{ij}^{TT} + u_{ij}^{L},\tag{4.65}$$

similar to the decomposition of the traceless part of the extrinsic curvature in Chapter 3.2. The divergence of the transverse part vanishes,

$$D^j u_{ij}^{TT} = 0,\tag{4.66}$$

and the longitudinal part can we written as the vector gradient of a vector X^i,

$$u_{ij}^{L} = D_i X_j + D_j X_i - \frac{2}{3} \gamma_{ij} D^k X_k = (LX)^{ij}.\tag{4.67}$$

Since $\bar{\gamma}_{ij} = \gamma^{-1/3} \gamma_{ij}$ is a vector density of weight $-2/3$, the right hand side of (4.67) can be identified with the Lie derivative of $\bar{\gamma}_{ij}$ along the vector X^i,

$$u_{ij}^{L} = \gamma^{1/3} \mathcal{L}_{\mathbf{X}} \bar{\gamma}_{ij}\tag{4.68}$$

(see exercise A.10 in Appendix A). Evidently, the longitudinal part can be interpreted as arising from a change of coordinates, generated by X^i. It represents the coordinate effects in the time development, which can therefore be eliminated by choosing u_{ij}^{L} to vanish. This leaves only the transverse part u_{ij}^{TT}, which implies that the divergence of u_{ij} itself must vanish

$$D^j u_{ij} = 0.\tag{4.69}$$

---

[39] See Smarr and York (1978a,b), whose derivation we will follow closely.

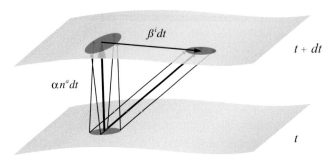

Figure 4.3 Schematic illustration of mimimal distortion, where the shift vector is chosen so that it minimizes coordinate shear in the metric. If a sphere is transported from one time slice at t to a neighboring slice at $t + dt$ along the normal vector, it typically will be sheared into an ellipse. By adopting a minimal distortion shift vector, some of the shear can be eliminated, depending on the spacetime geometry and the choice of time coordinate. [After Smarr and York (1978a).]

Combining this with equation (3.99) yields

$$D^j(L\beta)_{ij} = 2D^j(\alpha A_{ij}) \qquad (4.70)$$

or

$$\boxed{(\Delta_L \beta)^i = 2A^{ij}D_j\alpha + \frac{4}{3}\alpha\gamma^{ij}D_jK + 16\pi\alpha S^i,} \qquad (4.71)$$

where we have replaced the divergence of A^{ij} with the momentum constraint, equation (3.38).[40] Equation (4.71) is the *minimal distortion condition* for the shift vector β^i. The geometric interpretation of minimal distortion is elucidated in exercise 4.14 and Figure 4.3. The quantity u_{ij} can be viewed as a "distortion tensor" that measures the change in the shape (but not the size) of a small spheroid as it evolves from one time slice to a neighboring one. It is analogous to the shear strain instrinsic to a thin shell as it is deformed.[41]

> **Exercise 4.14** Derive the minimal distortion condition (4.71) by varying the action
>
> $$\mathcal{A} \equiv \int u_{ij}u^{ij}\gamma^{1/2}d^3x \qquad (4.72)$$
>
> with respect to β^i, keeping the boundary fixed, and requiring $\delta\mathcal{A} = 0$. Note that $\mathcal{A}$ it is a nonnegative global measure of the magnitude of the time rate of change of the conformal metric. The quantity $u_{ij}u^{ij}$ is analogous to the energy density and the integral $\mathcal{A}$ to the total shear stretching energy of a deformed thin shell. Taking a second variation with respect to β^i shows that equation (4.71) *minimizes* $\mathcal{A}$.
> **Hint:** Substitute equation (3.93) into equation (4.72).

It is also useful to express equation (4.71) in terms of conformally related quantities. Using $(L\beta)^{ij} = \psi^{-4}(\bar{L}\beta)^{ij}$ and $D_jS^{ij} = \psi^{-10}\bar{D}_j(\psi^{10}S^{ij})$ for any symmetric, traceless

---

[40] Equation (4.71) also follows from equation (3.102), using equation (4.69).
[41] York, Jr. (1979).

tensor, we find

$$(\Delta_L \beta)^i = \psi^{-4} \left((\bar{\Delta}_L \beta)^i + (\bar{L}\beta)^{ij} \bar{D}_j \ln \psi^6 \right),$$ (4.73)

and hence

$$\boxed{(\bar{\Delta}_L \beta)^i + (\bar{L}\beta)^{ij} \bar{D}_j \ln \psi^6 = 2\psi^{-6} \tilde{A}^{ij} \bar{D}_j \alpha + \frac{4}{3}\alpha \bar{\gamma}^{ij} D_j K + 16\pi \psi^4 \alpha S^i.}$$ (4.74)

> **Exercise 4.15** Adopt the conformal scaling relation of equation (3.97), $u_{ij} = \psi^4 \bar{u}_{ij}$, and show that condition (4.69) is equivalent to the relation
>
> $$\bar{D}^i(\psi^6 \bar{u}_{ij}) = 0.$$ (4.75)
>
> Then derive equation (4.74) directly from equation (4.75).

The relation between the shift condition that we found in the context of the conformal thin-sandwich formalism (see Chapter 3.3) and the minimal distortion shift is explored in exercise 4.16.

> **Exercise 4.16** Derive the minimal distortion shift condition (4.74) by starting with the momentum constraint as expressed by equation (3.102). Adopt equation (4.75) and use equation (3.101) to eliminate $\bar{u}_{ij}$.

The minimal distortion condition (4.74) is a set of coupled elliptic equations for the components of the shift vector. In higher dimensions, it requires appreciable computational resources to solve such a system numerically. But just as we discussed in Section 4.2 for the case of maximal slicing, it is reasonable to expect that an "approximate minimal distortion" condition may lead to a coordinate system with similar geometric properties. One possible simplification lies in replacing the covariant derivative operators in equation (4.74) with the corresponding flat space operators.[42] The operators then become ordinary partial derivatives in Cartesian coordinates, which simplifies the form of the equations and reduces the computational effort required to solve them.

A related spatial gauge condition is based on the "connection functions"

$$\bar{\Gamma}^i \equiv \bar{\gamma}^{kl} \bar{\Gamma}^i_{kl}$$ (4.76)

that we shall introduce in Chapter 11.5 in connection with the BSSN formulation of the $3 + 1$ equations. The BSSN formulation also assumes that $\bar{\gamma} = \det(\bar{\gamma}_{ij}) = 1$ in Cartesian coordinates, so that

$$\bar{\Gamma}^i = -\partial_j \bar{\gamma}^{ij}.$$ (4.77)

In Section 4.3 we introduced their 4-dimensional counterparts (4.40) which we set to zero to define harmonic coordinates. In complete analogy we could now set the connection

---

[42] Shibata (1999b).

functions $\bar{\Gamma}^i$ to define "conformal three-harmonic" coordinates. Alternatively, consider setting their time derivative to zero[43]

$$\partial_t \bar{\Gamma}^i = 0. \tag{4.78}$$

We can compute the time derivative of the connection functions by combining equation (4.77) with the evolution equation for the spatial metric, equation (2.134). Inserting the result (see equation 11.43) into equation (4.78) then yields the *Gamma-freezing* condition

$$\bar{\gamma}^{lj} \partial_j \partial_l \beta^i + \frac{1}{3} \bar{\gamma}^{li} \beta^j_{,jl} + \beta^j \partial_j \bar{\Gamma}^i - \bar{\Gamma}^j \partial_j \beta^i + \frac{2}{3} \bar{\Gamma}^i \partial_j \beta^j$$
$$= 2\tilde{A}^{ij} \partial_j \alpha - 2\alpha \left(\bar{\Gamma}^i_{jk} \tilde{A}^{kj} - \frac{2}{3} \bar{\gamma}^{ij} \partial_j K - \bar{\gamma}^{ij} S_j + 6\tilde{A}^{ij} \partial_j \ln \psi \right). \tag{4.79}$$

Here we have used the rescaling law

$$\tilde{A}^{ij} = \psi^4 A^{ij}, \tag{4.80}$$

that is used more commonly in the context of the BSSN formulation than the law (3.35) that we employed earlier. To distinguish the two rescaling laws we use a tilde instead of a bar in equations (4.79) and (4.80). The relation to minimal distortion can be seen by combining equations (4.78), (4.77) and (3.94), which yields

$$\partial_j (\bar{u}^{ij}) = 0. \tag{4.81}$$

Exercise 4.17 Derive equation (4.81).

Equation (4.81) can now be compared to the corresponding divergence criterion (4.75) for minimal distortion.

The condition (4.79) forms a complicated set of coupled elliptic equations for the components of the shift vector. In analogy to our conversion of maximal slicing into a parabolic evolution equation via the K-driver condition (4.39), we can convert these elliptic equations for the shift into parabolic evolution equations by approximating the Gamma freezing condition (4.78) with the *Gamma-driver*,[44]

$$\partial_t \beta^i = k(\partial_t \bar{\Gamma}^i + \eta \bar{\Gamma}^i), \tag{4.82}$$

where k and η are positive constants. Inserting equation (11.43) for $\partial_t \bar{\Gamma}^i$ results in a parabolic equation for β^i, in complete analogy to the K-driver condition (4.38) for the lapse. We can go a step further and construct a *hyperbolic Gamma-driver* as follows[45]

$$\boxed{\begin{aligned} \partial_t \beta^i &= \tfrac{3}{4} B^i \\ \partial_t B^i &= \partial_t \bar{\Gamma}^i - \eta B^i. \end{aligned}} \tag{4.83}$$

[43] Alcubierre and Brügmann (2001).

[44] See Alcubierre and Brügmann (2001); Duez *et al.* (2003).

[45] Campanelli *et al.* (2006).

Here the factor of $3/4$ is somewhat arbitrary, but leads to good numerical results, and the parameter η is typically on the order of $1/(2M)$, where M measures the total mass of the spacetime. The condition (4.83) is sometimes called the "nonshifting shift", in contrast to the "shifting shift" condition for which all time derivatives ∂_t in equation (4.83) are replaced with advective derivatives $\partial_t - \beta^j \partial_j$. This replacement has the effect of advecting various unstable "gauge" modes off the computational grid. Various versions of these gauge conditions have been adopted in simulations, and we refer the reader to the literature for comparative studies and analysis.[46] The hyperbolic Gamma-driver condition (4.83), together with the "1+log" slicing condition (4.51), have been extremely successful in dynamical "moving puncture" black hole simulations, which we will discuss in detail in Chapter 13.1.3.

Box 4.1   Lapse and shift conditions: a sampler

Geodesic slicing, or *Gaussian normal coordinates*, are defined by

$$\alpha = 1 \qquad \beta^i = 0. \tag{4.84}$$

Maximal slicing assumes $K = 0 = \partial_t K$, which results in the lapse equation

$$D^2 \alpha = \alpha \left(K_{ij} K^{ij} + 4\pi (\rho + S) \right). \tag{4.85}$$

Sometimes maximal slicing is imposed "approximately" via a *K-driver*, as in equation (4.39). *Harmonic coordinates* satisfy

$$^{(4)}\Gamma^a = 0, \tag{4.86}$$

which results in conditions (4.44) and (4.45) for the lapse and shift. An important variation is *1+log slicing*

$$(\partial_t - \beta^j \partial_j)\alpha = -2\alpha K. \tag{4.87}$$

For the *minimal distortion* gauge condition, the shift satisfies

$$(\Delta_L \beta)^i = 2A^{ij} D_j \alpha + \frac{4}{3}\alpha \gamma^{ij} D_j K + 16\pi \alpha S^i. \tag{4.88}$$

A related condition may be defined in terms of the "conformal connection functions" $\bar{\Gamma}^i$, which leads to the hyperbolic *Gamma-driver* condition for the shift,

$$\partial_t \beta^i = \frac{3}{4} B^i, \qquad \partial_t B^i = \partial_t \bar{\Gamma}^i - \eta B^i, \tag{4.89}$$

where η is a constant of order $1/(2M)$. Replacing the time derivatives ∂_t in equation (4.89) with advective derivatives $\partial_t - \beta^j \partial_j$ is one useful variation.

We note in closing that in simulations of gravitational collapse to black holes, minimal distortion and its close relatives often lead to poor coordinate resolution in the strong-field

---

[46] van Meter *et al.* (2006); see also Gundlach and Martin-Garcia (2006b) for a discussion of the mathematical properties of the resulting evolution system.

central regions. The reason for this behavior can be understood in terms of a simplified model equation[47] for condition (4.74),

$$\bar{D}^2 \beta^i = 16\pi S^i, \tag{4.90}$$

where we shall assume conformal flatness $\bar{\gamma}_{ij} = \eta_{ij}$ so that $\bar{D}^2$ is a flat Laplace operator. If matter is collapsing, the density is increasing, which, according to the continuity equation (e.g., equation 5.12), leads to a negative divergence of the matter flux S^i. Taking the divergence of equation (4.90) we therefore find

$$\bar{D}^2 \bar{D}_i \beta^i = 16\pi \bar{D}_i S^i < 0. \tag{4.91}$$

Adopting boundary conditions to insure that β^i vanishes assymptotically, we find from equation (4.91) that the divergence of the shift is positive, $\bar{D}_i \beta^i > 0$.[48] This means that the shift moves coordinate observers *away* from the collapsing matter, thereby decreasing the coordinate resolution of this matter. This response is unfortunate, since it is precisely the strong field that forms around the high-density, collapsing matter that leads to black hole formation and hence this region is where we desire maximum resolution. To compensate for this coordinate "blow-out" arising in minimal distortion (and related gauge choices), it is possible to add an extra contribution to the shift vector that points inward towards the collapsing region and increases the resolution there.[49]

---

[47] Shibata (1999b).

[48] Viewed as Poisson's equation $\bar{D}^2 \Phi = -4\pi \rho_q$ for the electromagnetic "potential" $\Phi \equiv \bar{D}_i \beta^i$, equation (4.91) has a positive "charge density" $\rho_q \equiv -4\bar{D}_i S^i$ on the right hand side, for which the potential must then be positive.

[49] See Shibata (1999b); Duez *et al.* (2003).

5 Matter sources

In nonvacuum spacetimes, the stress-energy tensor T^{ab} contributes source terms in the $3+1$ field equations. The stress-energy tensor accounts for all sources of energy-momentum in spacetime, excluding gravity. It thus arises from all forms of matter, electromagnetic fields, neutrinos, scalar fields, etc. in the Universe. For brevity, we shall sometimes refer to these sources collectively as the "matter sources" and the terms that they contribute in the $3+1$ equations as the "matter source terms". Matter source terms appear in the Hamiltonian constraint equation (2.132), the momentum constraint equation (2.133), and the evolution equation (2.135). The evolution equations for the "matter" sources are given by $\nabla_b T^{ab} = 0$, which express the conservation of the total 4-momentum in spacetime. Here T^{ab} is the *total* stress-energy tensor of the system. These conservation equations must be solved simultaneously with the $3+1$ evolution equations for the gravitational field to determine the entire foliation of spacetime. Some of the quantities appearing in the stress-energy tensor require auxiliary equations. These auxiliary equations include, for example, the continuity equation and an equation of state in the case of hydrodynamic matter, and Maxwell's equations in the case of an electromagnetic field, and so on.

As shown in Chapter 2, the matter source terms ρ, S_i and S_{ij} appearing in the $3+1$ equations for the gravitational field are projections of the stress-energy tensor into n^a and Σ and are given by

$$
\boxed{
\begin{aligned}
\rho &= n_a n_b T^{ab}, \\
S_i &= -\gamma_{ia} n_b T^{ab}, \\
S_{ij} &= \gamma_{ia} \gamma_{jb} T^{ab}.
\end{aligned}
}
\tag{5.1}
$$

The quantity ρ is the total mass-energy density as measured by a normal observer, S_i is the momentum density and S_{ij} is the stress. Finally, S is defined as the trace of S_{ij},

$$
\boxed{S = \gamma^{ij} S_{ij}.}
\tag{5.2}
$$

In the following sections, we discuss some of the most important "matter" sources that arise in astrophysical applications. These sources include hydrodynamic fluids, magnetohydrodynamic plasmas threaded by magnetic fields, radiation gases (e.g., photon and neutrino), collisionless matter, and scalar fields.

5.1 Vacuum

At the risk of stating the obvious, vacuum spacetimes are characterized by the vacuum stress-energy tensor

$$\boxed{T^{ab} = 0.}$$

(5.3)

Spacetimes containing black holes and (or) gravitational waves, and nothing else, are characterized by such a stress-energy tensor in Einstein's field equations. Vacuum spacetimes are simpler to deal with numerically since they require no additional energy-momentum conservation equations or auxiliary field equations to solve. On the other hand, vacuum black hole spacetimes, which are among the most important astrophysically, have all the numerical complications in the gravitational field sector that arise from the spacetime singularities in the black hole interiors. Handling these complications will be a major issue that we will address in subsequent chapters.

5.2 Hydrodynamics

Relativistic hydrodynamic matter is an important source of stress-energy in many astrophysical applications. Loosely speaking, a hydrodynamic description of matter is appropriate whenever the mean free path of a particle due to collisions with neighboring particles is much shorter than the characteristic size or local scale length of the system. Many of the global properties of relativistic stars, like neutron stars, are described by a hydrodynamic fluid. Also, the very early Universe is filled with hydrodynamic matter (e.g., baryons and electrons) coupled to thermal radiation.

In this section we seek to cast the basic equations of relativistic hydrodynamics and magnetohydrodynamics into forms most suitable for building dynamical spacetimes numerically, i.e., into forms best suited for numerical integration together with the Einstein equations in $3 + 1$ form for the gravitational field. Although our discussion is reasonably self-contained, readers of this section will find it helpful to have been exposed previously to a treatment of thermodynamics, hydrodynamics and electrodynamics in curved spacetime at an elementary level.[1]

5.2.1 Perfect gases

The stress-energy tensor of a perfect gas is given by

$$\boxed{T^{ab} = \rho_0 h u^a u^b + P g^{ab}.}$$

(5.4)

Here u^a is the fluid 4-velocity, ρ_0 is the rest-mass density, P is the pressure, and h is the specific enthalpy

$$h = 1 + \epsilon + P/\rho_0,$$

(5.5)

---

[1] See, e.g., Misner *et al.* (1973), Section 22.1–22.4.

where ϵ is the specific internal energy density. The total mass-energy density as measured by an observer comoving with the fluid is then given by $\rho^* = \rho_0(1 + \epsilon)$.

The equations of motion for the fluid can be derived from the local conservation of energy-momentum,

$$\nabla_b T^{ab} = 0, \tag{5.6}$$

and the conservation of rest-mass,

$$\nabla_a(\rho_0 u^a) = 0. \tag{5.7}$$

Wilson scheme

The resulting equations can be cast in various forms, depending on the fluid variables chosen for numerical integration. One of the most straightforward schemes that readily lends itself to numerical integration was introduced by Wilson.[2] His scheme was designed to mimic the standard Eulerian equations of hydrodynamics in Newtonian theory, for which the large body of expertise that has been acquired to solve the equations numerically can be brought to tackle the relativistic system. To accomplish this goal, one defines a rest-mass density variable

$$D \equiv \rho_0 W, \tag{5.8}$$

an internal energy-density variable

$$E \equiv \rho_0 \epsilon W, \tag{5.9}$$

and a momentum-density variable

$$S_a \equiv \rho_0 h W u_a = (D + E + PW)u_a. \tag{5.10}$$

Here we have defined W to be the Lorentz factor between normal and fluid observers,

$$W \equiv -n_a u^a = \alpha u^t. \tag{5.11}$$

Exercise 5.1 (a) Determine the relation between the density ρ^* and the source term ρ defined in equation (5.1). Interpret your answer physically.
(b) Argue that $\rho^* = \rho$ whenever a normal observer is a comoving observer.
(c) Argue that $\rho^* = \rho$ for a static metric, like the metric (1.73) for an Oppenheimer–Volkoff equilibrium star.
(d) Argue that $\rho^* = \rho = \rho_0$ for the metric (1.88) describing Oppenheimer–Snyder spherical dust collapse.

Exercise 5.2 Show that the spatial vector S_i defined by equation (5.10) is the fluid contribution to the source term S_i appearing in the $3 + 1$ decomposition of Einstein's field equations and defined by equation (5.1).

---

[2] Wilson (1972b, 1979); see also Hawley *et al.* (1984).

In terms of these variables, the equation of continuity $\nabla_a(\rho_0 u^a) = 0$ becomes

$$\partial_t(\gamma^{1/2} D) + \partial_j(\gamma^{1/2} D v^j) = 0. \tag{5.12}$$

Here $v^j = u^j/u^t$ is the fluid 3-velocity with respect to a coordinate observer, and γ is the determinant of the spatial metric γ_{ij}. Contracting equation (5.6) with u^a yields the energy equation,

$$\partial_t(\gamma^{1/2} E) + \partial_j(\gamma^{1/2} E v^j) = -P\left(\partial_t(\gamma^{1/2} W) + \partial_i(\gamma^{1/2} W v^i)\right), \tag{5.13}$$

while the spatial components of (5.6) yield the relativistic Euler equations,

$$\partial_t(\gamma^{1/2} S_i) + \partial_j(\gamma^{1/2} S_i v^j) = -\alpha\gamma^{1/2}\left(\partial_i P + \frac{S_a S_b}{2\alpha S^t}\partial_i g^{ab}\right). \tag{5.14}$$

In deriving equations (5.12), (5.13) and (5.14), we have used the identity $(-g)^{1/2} = \alpha\gamma^{1/2}$ (see exercise 2.25).

Exercise 5.3 Derive equations (5.12), (5.13) and (5.14).

Exercise 5.4 (a) Show that by contracting equation (5.6) with u^a the energy equation may be written in the alternative form

$$\frac{d\rho^*}{d\tau} = \frac{(\rho^* + P)}{\rho_0}\frac{d\rho_0}{d\tau}, \tag{5.15}$$

where $d/d\tau \equiv u^b \nabla_b$ is the rate of change with proper time following a fluid element. (b) Show that by contracting equation (5.6) with the projection tensor $P^{bc} = g^{bc} + u^b u^c$ the Euler equation may be written in the form

$$(\rho^* + P)u^b \nabla_b u_a = \nabla_a P - u_a u^b \nabla_b P. \tag{5.16}$$

(c) Use the first law of thermodynamics and equation (5.15) to show that perfect gas motion is adiabatic, i.e., $ds/d\tau = 0$, where s is the entropy per unit rest-mass. Though entirely equivalent to equations (5.13) and (5.14), equations (5.15) and (5.16) usually are not the most useful form of the equations of relativistic hydrodynamics for numerical integration in an initial-value (evolution) problem.

For many purposes it is useful to employ a simple "Γ-law" equation of state (EOS) of the form

$$P = (\Gamma - 1)\rho_0\epsilon. \tag{5.17}$$

Realistic applications involving relativistic objects are rarely described by EOSs obeying this simple form. However, a Γ-law EOS provides a computationally practical, albeit crude, approximation that can be adapted to mimic the gross behavior of different states of matter in many applications. For example, to model a stiff nuclear EOS in a neutron star, one can adopt a moderately high value of Γ in a Γ-law EOS, e.g., $\Gamma \approx 2$. By contrast, to model a moderately soft, thermal radiation-dominated EOS governing a very massive or supermassive star, one can set $\Gamma = 4/3$. For isentropic flow, a Γ-law EOS is equivalent to the equation of state of a polytrope,

$$P = K\rho_0^{\Gamma}, \quad \Gamma = 1 + 1/n \tag{5.18}$$

where n is the polytropic index and K is the gas constant. However, for nonisentropic flow, which is always the case when encountering a shock, K is no longer constant throughout the fluid, and equations (5.17) and equations (5.18) are no longer equivalent.

> **Exercise 5.5** Use the first law of thermodynamics for isentropic flow to show the equivalence of equations (5.17) and equations (5.18).

When employing a Γ-law EOS the source term on the right-hand side of the energy equation (5.13) can be rearranged to yield

$$\partial_t(\gamma^{1/2} E_*) + \partial_j(\gamma^{1/2} E_* v^j) = 0, \tag{5.19}$$

where we have introduced a new energy variable E_* defined as

$$E_* \equiv (\rho_0 \epsilon)^{1/\Gamma} W. \tag{5.20}$$

This simplification has great computational advantages, since the time derivatives on the right-hand side of (5.13) are difficult to handle in strongly relativistic fluid flow.

> **Exercise 5.6** Use the first law of thermodynamics to show that the left-hand side of equation (5.19) actually describes the evolution of s, the entropy per unit mass, following a fluid element:
>
> $$\partial_t(\gamma^{1/2} E_*) + \partial_j(\gamma^{1/2} E_* v^j) = \frac{\alpha \gamma^{1/2} \rho_0}{\Gamma} \left(\frac{E_*}{W}\right)^{(1-\Gamma)} T \frac{ds}{d\tau}, \tag{5.21}$$
>
> where $d/d\tau = u^a \nabla_a$ is the rate of change with proper time following the fluid. Hence equation (5.19) guarantees that the flow is adiabatic, which is always true for a perfect gas in the absence of shocks.

For given values of u_i, W can be found from the normalization relation $u_a u^a = -1$,

$$W = \alpha u^t = \left(1 + \gamma^{ij} u_i u_j\right)^{1/2}, \tag{5.22}$$

and v^i from

$$v^i = \alpha \gamma^{ij} u_j / W - \beta^i. \tag{5.23}$$

Equations (5.12), (5.14) and (5.19), expressing the conservation of rest-mass, momentum and energy, form a convenient set that can integrated simultaneously to determine the evolution of a perfect gas in the absence of shocks. However, this particular set of equations must be suitably modified to handle the appearance of shock discontinuities. Shocks occur whenever fluid elements collide supersonically, which can occur during stellar collapse, accretion flows, stellar collisions, etc. Mathematically, shocks arise whenever fluid characteristic curves cross, leading to sharp discontinuities in the fluid variables. For realistic fluids containing viscosity, the shock transition region has a finite thickness amounting to several particle collision mean-free-paths, and the fluid variables vary continuously across this finite interval. Two strategies are commonly adopted to treat shocks when the details of the transition region are of no physical consequence and the perfect gas equations are used throughout, whereby the thickness of the transition region is allowed to be arbitrarily thin. The more traditional approach is to add an artificial viscosity term to the equations.[3] This term mimics the effect of physical viscosity, except that it is employed only in the vicinity of a shock and serves to spread the shock transition region over a few spatial grid spacings in finite-difference codes.[4] For this purpose, the artificial viscosity term P_{vis} is nonzero only where the fluid is compressed and is added to the pressure on the right-hand sides of both the energy equation (5.13) and the Euler equation (5.14). Such a term has the approximate form

$$P_{\mathrm{vis}} = \begin{cases} C_{\mathrm{vis}} \rho_0 (\delta v)^2 & \text{for } \delta v < 0, \\ 0 & \text{otherwise,} \end{cases} \tag{5.24}$$

where $\delta v = \partial_k v^k \Delta x$, Δx is the local spatial grid size and C_{vis} is a dimensionless constant of order unity. Adding such a term allows the fluid to satisfy the Rankine–Hugoniot "jump", or junction, conditions across the shock, which we shall derive below. These conditions ensure the continuity of rest-mass, momentum and energy flux across any surface in the fluid, including a shock front. For a fluid element traversing a shock, the junction conditions serve to convert some of the bulk kinetic energy into internal energy and to increase the entropy. Artificial viscosity schemes have the virtue of being quite robust and very easy to implement. For shocks occurring in Newtonian fluids with modest Mach numbers, artificial viscosity enables the fluid to satisfy the Rankine–Hugoniot jump conditions to reasonable accuracy. Artificial viscosity has also been used successfully in many relativistic applications,[5] but it can lead to less satisfactory results for ultrarelativistic flows or high Mach numbers.[6]

> **Exercise 5.7** Show that the addition of P_{vis} to the pressure in the stress-energy tensor modifies equation (5.19) according to
>
> $$\partial_t (\gamma^{1/2} E_*) + \partial_j (\gamma^{1/2} E_* v^j) = - \left(\frac{E_*}{W} \right)^{(1-\Gamma)} \frac{P_{\mathrm{vis}}}{\Gamma} \partial_a (W \gamma^{1/2} v^a), \tag{5.25}$$
>
> where $v^a = u^a / u^t$. Comparing with equation (5.21), we see that the role of P_{vis} is to generate the entropy jump required across a shock discontinuity.

---

[3] von Neumann and Richtmeyer (1950).
[4] Finite-difference techniques for integrating partial differential equations are discussed in Chapter 6.2.
[5] May and White (1966); Wilson (1972b); Shapiro and Teukolsky (1980); Hawley *et al.* (1984); Shibata (1999a); Duez *et al.* (2003).
[6] Winkler and Norman (1986).

High-resolution shock-capturing (HRSC) schemes

An alternative approach to handling shocks in finite-difference algorithms involves recasting the equations in a "flux-conservative form" and adopting a so-called "high-resolution shock-capturing scheme" or HRSC scheme.[7] In such a scheme, one divides up the spatial domain into a discrete set of contiguous cells. At the center of each cell is a grid point where one keeps track of the fluid variables as they evolve in time. One treats all fluid variables as constant in each grid cell. The discontinuous fluid variables at the grid interfaces serve as initial conditions for a local Riemann shock tube problem, which can be treated either exactly or approximately. The general Riemann problem involves the evolution of a gas in which, initially, the fluid variables are everywhere constant on either side of an interface, but discontinuous across the interface. The solution to this idealized problem is known[8] and provides the basis for constructing an HRSC scheme. Allowing for discontinuities, including shocks, lies at the core of such schemes, and does not require any additional artificial viscosity. Constructing Riemann solvers for HRSC schemes requires knowledge of the local characteristic structure of the equations to be solved. This has motived the development of several flux-conservative hydrodynamics schemes for which this characteristic structure can be determined.[9] A key feature of the hydrodynamical equations in flux-conservative form is that they do not contain any derivatives of the fluid variables in the source terms on the right-hand sides, in contrast to equation (5.14), which contains a pressure gradient.

The flux-conservative equations of relativistic hydrodynamics take on the general form

$$\boxed{\partial_t \mathcal{U} + \partial_i \mathcal{F}^i = \mathcal{S}} \tag{5.26}$$

where $\mathcal{U}$ is the state vector of *conserved* variables built out of the so-called *primitive* fluid variables $\mathcal{P} = (\rho_0, v^i, P)$, the $\mathcal{F}^i$ are the flux vectors (one for each spatial dimension, i), and where the source vector $\mathcal{S}$ does not contain any derivatives of the primitive fluid variables. A particularly useful choice which can be written in this way is[10]

$$\mathcal{U} = \begin{pmatrix} \tilde{D} \\ \tilde{S}_j \\ \tilde{\tau} \end{pmatrix} = \begin{pmatrix} \gamma^{1/2} W \rho_0 \\ \gamma^{1/2} \alpha T^0{}_j \\ \alpha^2 \gamma^{1/2} T^{00} - \tilde{D} \end{pmatrix}, \tag{5.27}$$

where $\tilde{D}$ and $\tilde{S}_j$ differ from D and S_j defined in equations (5.8) and (5.10) by a factor of $\gamma^{1/2}$. The flux vectors $\mathcal{F}^i$ are then given by

$$\mathcal{F}^i = \begin{pmatrix} \tilde{D} v^i \\ \alpha \gamma^{1/2} T^i{}_j \\ \alpha^2 \gamma^{1/2} T^{0i} - \tilde{D} v^i \end{pmatrix}, \tag{5.28}$$

---

[7] See the text by Toro (1999), or the review by Martí and Müller (1999), and references therein.
[8] See, e.g., Courant and Friedrichs (1948).
[9] See, e.g., review by Font (2000).
[10] Font *et al.* (2000, 2002).

and the source vector $\mathcal{S}$ by

$$\mathcal{S} = \begin{pmatrix} 0 \\ \frac{1}{2}\alpha\gamma^{1/2}T^{ab}g_{ab,j} \\ \alpha\gamma^{1/2}(T^{a0}\partial_a\alpha - {}^{(4)}\Gamma^0{}_{ab}T^{ab}\alpha) \end{pmatrix}. \tag{5.29}$$

The first row of this equation is just the continuity equation (5.12) expressing rest-mass conservation. The second row arises from the equation of energy-momentum conservation $\nabla_a T^a{}_b = 0$, or

$$\partial_t(\sqrt{-g}\,T^0{}_a) + \partial_i(\sqrt{-g}\,T^i{}_a) = \frac{1}{2}\sqrt{-g}\,T^{bc}\partial_a g_{bc}. \tag{5.30}$$

The second row is simply the $a = j$, or j-momentum, component of equation (5.30). The third row arises from projecting the equations of motion along the normal vector: $n_a\nabla_b T^{ab} = 0$, and subtracting off the continuity equation (5.12) to achieve the desired form. The subtraction removes the conserved rest-mass contribution from the energy density so that the internal energy density, which equals the (typically small) difference between the total and rest-mass energy densities, can be integrated directly without being corrupted by numerical cancelation errors.

> **Exercise 5.8**   On first glance it appears that the source term in the energy equation (i.e., the third row in $\mathcal{S}$), which we will call $s_{\tilde{\tau}}$, contains explicit time derivatives of the lapse and shift, as well as the spatial metric.
>
> (a) Show that by expanding out the source term $s_{\tilde{\tau}}$ and using the definition of the Christoffel symbol, the time derivatives of the lapse and shift miraculously cancel out.
>
> (b) Now let us recast the source term to get a convenient expression free of time derivatives. First use $n_a\nabla_b T^{ab} = 0$ to show that
>
> $$s_{\tilde{\tau}} = -\alpha\gamma^{1/2}T^{ab}\nabla_b n_a = \alpha\gamma^{1/2}T^{ab}(K_{ab} + a_b n_a), \tag{5.31}$$
>
> where K_{ab} is the extrinsic curvature and $a_b = n^a\nabla_a n_b = D_b(\ln\alpha)$ (see exercise 2.13) is the 4-acceleration of the normal observer.
>
> (c) Next show from $n^a D_a\alpha = 0$ that $D_0\alpha = \beta^i D_i\alpha = \beta^i\partial_i\alpha$.
>
> (d) Finally, show that $K_{00} = \beta^i\beta^j K_{ij}$ and $K_{0i} = \beta^j K_{ij}$. Conclude that the source term can be written as
>
> $$s_{\tilde{\tau}} = \alpha\gamma^{1/2}[(T^{00}\beta^i\beta^j + 2T^{0i}\beta^j + T^{ij})K_{ij} - (T^{00}\beta^i + T^{0i})\partial_i\alpha], \tag{5.32}$$
>
> which contains *no* explicit time derivatives.

The basic strategy is to integrate the coupled set of equations (5.26) for the conservative variables $(\tilde{D}, \tilde{S}_j, \tilde{\tau})$ from time t to $t + \Delta t$, and then combine the conservative variables at the new time to solve algebraically for the new set of primitive variables (ρ_0, v^i, P). The procedure is repeated for each subsequent time step. For a self-consistent determination of the background gravitational field, Einstein's equations in $3 + 1$ form must be integrated in time simultaneously to determine the spacetime metric.

Implementing an HRSC scheme to solve equation (5.26) then begins by calculating the primitive variables $\mathcal{P}$ on each cell interface (the "reconstruction step"). At most grid points, the computed value of the primitive variable takes into account the variation of the variable at the nearest points to the interface to high order in the adopted grid spacing. When a discontinuity is identified at the interface (as in the case of a shock) by a "slope-limiter" that looks for changes in the nearby slopes of variables, the order is reduced. Next, the reconstructed data are used as initial data for a local Riemann problem (the "Riemann solver step"). The net flux $\mathcal{F}^i$ at each cell interface is given by the solution to this Riemann problem. Exact Riemann solvers require knowledge of the eigenvectors of the system, while approximate Riemann solvers do not. The later, which provide simpler HRSC schemes and are based on solving dispersion relations for the wave speeds in the fluid, are often adequate.[11] Once the flux is employed to compute the conservative variables $\mathcal{U}$ at the new time step, these values are used to recover the primitive variables $\mathcal{P}$ on the new time level (the "recovery step"). This may not be trivial because, while the functional relations $\mathcal{U}(P)$ are analytic, the inverse relations $\mathcal{P}(U)$ are usually not and must be solved numerically. The whole process is then repeated for the next time step.

We now catalog the source terms ρ, S_i, S_{ij} and S in terms of the primitive fluid variables for a perfect gas. Substituting the fluid stress-energy tensor (5.4) into equations (5.1)–(5.2) yields the fluid contributions to the source terms:

$$\rho_{\text{fluid}} = \rho_0 h W^2 - P, \tag{5.33}$$

$$S_i^{\text{fluid}} = \rho_0 h W u_i, \tag{5.34}$$

$$S_{ij}^{\text{fluid}} = P\gamma_{ij} + \frac{S_i^{\text{fluid}} S_j^{\text{fluid}}}{\rho_0 h W^2}, \tag{5.35}$$

$$S^{\text{fluid}} = 3P + \rho_0 h(W^2 - 1). \tag{5.36}$$

We insert the superscript "fluid" to remind us that the contribution from the perfect fluid must be added to the contributions from all other nongravitational sources of energy-momentum.

Smoothed particle hydrodynamics (SPH) schemes

Smooth particle hydrodynamics, or SPH, provides yet another way of treating a relativistic fluid. The SPH method was originally introduced by Lucy (1977) and Gingold and Monaghan (1977) to handle Newtonian fluids in $3 + 1$ dimensions. Most of the early

---

[11] One of the simplest shock-capturing schemes that does not require knowledge of the eigenvectors is the HLL scheme (Harten *et al.*, 1983), which has been shown to perform with an accuracy comparable to more sophisticated Riemann solvers in shock tube problems when coupled to a high-order reconstruction method like PPM ("piecewise parabolic method"; Colella and Woodward (1984)).

applications involved Newtonian stellar hydrodynamics. SPH was later adapted to treat fully relativistic fluids, particularly in the context of relativistic core collapse, binary neutron star mergers and binary black hole–neutron star mergers.[12] It has also been useful in cosmology simulations of structure formation, where fluid baryonic matter and collisionless "dark" matter must be evolved simultaneously.

SPH is a Lagrangian method that follows the behavior of fluid elements, represented by a large sample of particles. The "forces" that govern the motion of the particles are constructed from the equations of hydrodynamics. At any time, the positions of the particles are assumed to be distributed in proportion to the fluid rest-mass density. Given a finite distribution of particles, the density is determined statistically by introducing a "smoothing kernel" in a Monte Carlo integral over the distribution. Pressure-gradient forces are also calculated by kernel estimation, using the particle positions, rather than by direct evaluation (e.g., finite-differencing) of the hydrodynamic equations.

The Lagrangian formulation employed in SPH introduces a Lagrangian time derivative d/dt in the fundamental hydrodynamic equations (5.12)–(5.14). The Lagrangian time derivative follows changes in the properties of a given fluid element along its worldline and is related to the Eulerian time derivative ∂_t that measures changes at a fixed point in space according to $d/dt = \partial_t + v^j \partial_j$, where the fluid velocity v^j is given by equation (5.23). Substituting this Lagrangian time derivative in equation (5.12) gives the Lagrangian continuity equation,

$$\frac{d\rho_*}{dt} + \rho_* \partial_j v^j = 0, \tag{5.37}$$

where $\rho_* \equiv \gamma^{1/2} D \equiv \alpha u^t \gamma^{1/2} \rho_0$.[13] This conservative form of the continuity equation allows us to define a fixed set of particles, each of which is labeled by subscript "a" and has a constant rest-mass m_a. Each particle has an instantaneous position x_a^j that moves according to $dx_a^j/dt = v_a^j$. For each particle we define a "smoothing length" h_a, which represents the physical size of the particle. Thus, a particle does not have a delta-function density profile, but instead represents a spherically symmetric density distribution of finite radius (with typical radius $= 2h_a$) centered at the particle position. The density at each particle is then determined as a locally weighted average by summing over all the particles residing within this radius:

$$(\rho_*)_a = \sum_b m_b W_{ab}. \tag{5.38}$$

Here W_{ab} is a smoothing (or interpolation) kernel. It can be calculated for a pair of particles as a function of $r_{ab} = (x_{ab}^2 + y_{ab}^2 + z_{ab}^2)^{1/2}$, the coordinate distance from particle a to its neighbor b, and h_a. A second-order differentiable form often used for W was introduced

---

[12] See Chapters 16 and 17 for details and references.
[13] This density variable should not be confused with ρ^* defined below equation (5.5).

by Monaghan and Lattanzio (1985) and is given by

$$W(r, h) = \frac{1}{\pi h^3} \begin{cases} 1 - \frac{3}{2}\left(\frac{r}{h}\right)^2 + \frac{3}{4}\left(\frac{r}{h}\right)^3, & 0 \le \frac{r}{h} < 1, \\ \frac{1}{4}\left[2 - \left(\frac{r}{h}\right)\right]^3, & 1 \le \frac{r}{h} < 2 \\ 0, & \frac{r}{h} \ge 2. \end{cases} \quad (5.39)$$

Note that W is normalized so that when integrated over all space, $\int W(r, h) 4\pi r^2 dr = 1$. Tracking a fixed number of particles in a simulation and evaluating their density according to equation (5.38) is equivalent to solving the continuity equation.

Exercise 5.9 Many of the applications of SPH in general relativity assume *conformally flat* spacetimes, for which the metric is restricted to be of the form

$$ds^2 = -\alpha^2 dt^2 + \psi^4 \eta_{ij}(dx^i + \beta^i dt)(dx^j + \beta^j dt). \quad (5.40)$$

We have already encountered conformal flatness in Chapter 3.1.2, and will discuss this approximation and its domain of validity in the context of dynamical simulations in greater detail in Chapters 16.2 and 17.2.1.

Here we will adopt this metric, together with a Γ-law EOS, and assemble the remaining relativistic Lagrangian hydrodynamic equations used in many SPH applications.

(a) Show that the Euler equation can be written in Lagrangian form as

$$\frac{d\tilde{u}_i}{dt} = -\frac{\alpha\psi^6}{\rho_*}\partial_i P - \alpha h u^0 \partial_i \alpha + \tilde{u}_j \partial_i \beta^j + \frac{2h\alpha(\gamma_n^2 - 1)}{\gamma_n \psi}\partial_i \psi, \quad (5.41)$$

where the specific momentum is defined by

$$\tilde{u}_i \equiv h u_i. \quad (5.42)$$

(b) Show that the energy equation can be written in Lagrangian form as

$$\frac{de_*}{dt} + e_* \partial_i v^i = 0, \quad (5.43)$$

where $e_* \equiv \gamma^{1/2} E \equiv \alpha u^t \psi^6 (\rho_0 \epsilon)^{1/\Gamma}$. Note that in the absence of shocks, the adiabatic energy equation can be replaced by the polytropic equation (5.18).

The pressure gradient appearing the Lagrangian Euler equation (see, e.g., exercise 5.9) is calculated according to

$$\frac{1}{(\rho_*)_a}\partial_i P_a = -\sum_b m_b \left(\frac{P_b}{(\rho_*)_b^2} + \frac{P_a}{(\rho_*)_a^2}\right)\partial_i W_{ab}. \quad (5.44)$$

Other terms depend on the metric functions and their derivatives.[14] Artificial viscosity can be incorporated in a relativistic SPH scheme to handle shocks.[15]

---

[14] Recipes for choosing the smoothing kernels and lengths and for evaluating the full set of relativistic SPH equations in conformally flat spacetimes are given in, e.g., Oechslin *et al.* (2002) and Faber *et al.* (2004), and references therein.

[15] Siegler and Riffert (2000); Oechslin *et al.* (2002).

SPH is perhaps the simplest hydrodynamics scheme to implement for simulating multi-dimensional fluid systems. It is also well suited for tracing the Lagrangian flow of matter, which is particularly convenient when different fluid species are present and mix. However, a large number of particles are required, and their positions must be carefully tracked, in order to minimize numerical errors. Low-resolution SPH calculations tend to be noisy, and this noise can lead to spurious diffusion of SPH particles and spurious viscosity, independent of any real physical mixing and physical viscosity.[16]

Rankine–Hugoniot conditions

Shock waves pose the most serious challenge for any numerical hydrodynamical scheme. The ability to resolve the sharp discontinuities in the fluid parameters associated with a shock is typically what distinguishes one code from the next. This fact motivates a brief discussion here of a few of the key equations that relate the fluid variables across a shock front.[17]

The Rankine–Hugoniot junction conditions across a relativistic shock discontinuity can be derived easily from the fundamental hydrodynamic equations (5.6) and (5.7). Integrating $\nabla_b T^{ab}$ and $\nabla_a(\rho_0 u^a)$ over a "pill box" centered on the shock front, and using Gauss' theorem, gives the junction conditions,

$$[\rho_0 u^a \hat{z}_a] = 0, \tag{5.45}$$

and

$$[T^{ab}\hat{z}_b] = 0, \tag{5.46}$$

where $\hat{z}^a$ is the spacelike normal vector to the front. Here the bracket denotes the difference between quantities on the two sides of the shock front; $[V] \equiv V_+ - V_-$, where "+" labels the front side of the shock (the side upstream) and "−" labels the back side (the side downstream). The first condition can be written as

$$F \equiv \rho_0^+ u_+^a \hat{z}_a = \rho_0^- u_-^a \hat{z}_a, \tag{5.47}$$

where F is the conserved flux of rest-mass across the front. Using definition (5.4) in the second condition allows us to write

$$F(h_+ u_+^a - h_- u_-^a) = \hat{z}^a (P_- - P_+). \tag{5.48}$$

---

[16] For a study of spurious transport effects in Newtonian SPH calculations, and other potential difficulties, see Hernquist (1993) and Lombardi *et al.* (1999), and references therein.

[17] For a detailed treatment of relativistic shocks, see, e.g., Taub (1948); Lichnerowicz (1967); Landau and Lifshitz (1959); Novikov and Thorne (1973). The discussion here is patterned after Evans (1984).

Contracting equation (5.48) alternately with u_a^+ and u_a^-, and taking the difference of the two equations, yields the relativistic Rankine–Hugoniot relation,

$$h_+^2 - h_-^2 = \left(\frac{h_+}{\rho_0^+} + \frac{h_-}{\rho_0^-} \right)(P_+ - P_-). \tag{5.49}$$

Exercise 5.10   Take the nonrelativistic limit of equation (5.49) to get the standard Rankine–Hugoniot relation

$$\epsilon_+ - \epsilon_- = \frac{P_+ + P_-}{2\rho_0^+ \rho_0^-}(\rho_0^+ - \rho_0^-). \tag{5.50}$$

To determine the "jump" in rest-mass density across a shock front it is useful to cast equation (5.49) in nondimensional form,

$$H^2 - 1 = \delta^2(y - 1)\left(\frac{H}{\eta} + 1 \right), \tag{5.51}$$

where we have introduced the nondimensional variables $H = h_+/h_-$, $y = P_+/P_-$, $\eta = \rho_0^+/\rho_0^-$ and $\delta^2 = p_-/(\rho_0^- h_-)$. The ratio H can be eliminated from the above equation by introducing an EOS. Employing our Γ-law EOS (5.17) and defining $q \equiv 1/h_-$ gives a quadratic equation for the density ratio η, parametrized by q, and to be solved as a function of the pressure ratio y:

$$[y(\Gamma - 1) + (\Gamma q + 1)]\,\eta^2 - q\,[y(\Gamma + 1) + (\Gamma - 1)]\,\eta - (1 - q)y(y + \Gamma - 1) = 0. \tag{5.52}$$

The parameter q measures the degree to which the flow is relativistic: $q \to 1$ for nonrelativistic (NR) flow, while $q \to 0$ for extremely relativistic (ER) flow. Taking the nonrelativistic limit of equation (5.52) gives

$$\eta = \frac{y(\Gamma + 1) + (\Gamma - 1)}{y(\Gamma - 1) + (\Gamma + 1)}. \tag{5.53}$$

The parameter y measures the strength of the shock: $y \to \infty$ for strong shocks, while $y \to 1$ for weak shocks (acoustic waves). Taking the strong shock limit of equation (5.53) gives the familiar Newtonian result for the maximum compression behind a strong shock,

$$\eta \to \frac{\Gamma + 1}{\Gamma - 1} \quad \text{(strong NR shock).} \tag{5.54}$$

For an extremely relativistic shock, equation (5.52) gives

$$\eta = \left[\frac{y(y + \Gamma - 1)}{y(\Gamma - 1) + 1} \right]^{1/2}. \tag{5.55}$$

By contrast with the Newtonian result, the maximum compression behind a strong relativistic shock has no upper limit, but increases steadily with y according to

$$\eta \to \left[\frac{y}{\Gamma - 1} \right]^{1/2} \quad \text{(strong ER shock).} \tag{5.56}$$

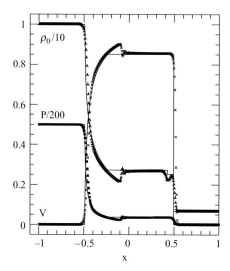

Figure 5.1 A one-dimensional relativistic Riemann shock tube test for an ideal fluid with an adiabatic index $\Gamma = 2$. The plot shows the numerically evolved rest density ρ_0 (triangles), pressure P (squares), and velocity v (crosses) at $t = 0.5$. The analytic values are indicated by solid curves. This calculation used the Wilson scheme with artificial viscosity $P_{\mathrm{vis}} = C_{\mathrm{vis}} A (\delta v)^2$, $A = \gamma^{1/2} \epsilon D$, $\delta v = 2 \partial_k v^k \Delta x$, and $C_{\mathrm{vis}} = 1$. [From Duez *et al.* (2003).]

Not surprisingly, the strongest and most relativistic shocks pose the most stringent test for a numerical hydrodynamics code, as the discontinuities are steepest in this regime.

Tests

Testing a code designed to handle relativistic hydrodynamics is an essential and nontrivial phase of code construction. Convergence tests often serve as the initial step in code validation, providing the first line of defense against coding errors and bugs, including simple typos.[18] To fully calibrate the reliability and robustness of a code it is necessary to run a suite of test-bed problems, the solutions to which are known. One might start by fixing the spacetime to be Minkowski, which allows the fluid sector of the code to be tested in special relativity. Performing a relativistic Riemann shock tube problem is particularly instructive in this case. The problem provides a "pure" hydrodynamics test, since no complications from the numerical treatment of the gravitational field equations can arise. It also provides a rather strenuous test, since a shock wave reflects the full nonlinear character of the hydrodynamic equations.

We illustrate such a test in Figure 5.1 for a simple one-dimensional shock tube. We consider an ideal fluid and adopt a Γ-law EOS with $\Gamma = 2$. At $t = 0$ we set $v^x = 0$ everywhere while for $x < 0$ we set $\rho_0 = 10$, $P = 100$ and for $x > 0$ we set $\rho_0 = 1$,

[18] See Chapter 6.4.

$P = 1$. The solution is evolved with a second-order, finite difference scheme based on the Wilson method using artificial viscosity. Here a grid of 400 points is employed over a domain $x \in (-1, 1)$. The shock is modeled rather well, apart from an "overshoot" in quantities at the rarefaction (i.e., low density expansion) front. It has been shown[19] that in finite difference schemes using artificial viscosity, such an overshoot arises in numerical solutions even when the grid spacing goes to zero. This limitation of artificial viscosity methods is particularly serious when strong shocks are present; HRSC techniques help overcome this difficulty.

> **Exercise 5.11**   Show that the results plotted in Figure 5.1 are consistent with equation (5.55).

In curved spacetime, the set of known, analytic solutions to the equations of relativistic hydrodynamics is not large, particularly for dynamical spacetimes. Holding a stable, static, spherical star constructed from the OV hydrostatic equilibrium equations in stable equilibrium (see Chapter 1.3) provides one simple test. Holding a stable, stationary, rotating star constructed from the stationary equilibrium equations in stable equilibrium (see Chapter 14.1) constitutes a more challenging test, particularly if it is differentially rotating and therefore subject to spurious redistribution of angular momentum by numerical viscosity. The analytic Oppenheimer–Snyder solution[20] for the collapse of a spherical, homogeneous ball of dust from rest at finite radius provides an analytic dynamical spacetime for testing the ability of a code to handle catastrophic collapse of a fluid to a black hole. The interior solution is analytic in geodesic slicing and comoving radial coordinates and must be transformed numerically in order to compare with numerical integrations performed in different time slicings and/or spatial coordinates.[21] Alternatively, the analytic solution, which is easily expressed in closed-Friedmann form in the matter interior and Schwarzschild in the vacuum exterior, can be employed to construct various scalar invariants (e.g., areal radii $R(\tau)$ of Lagrangian fluid elements as functions of proper time τ in the interior, and Riemann curvature invariants in the exterior) that can readily test a dynamical simulation performed in an arbitrary gauge.

During any numerical simulation it is useful to monitor quantities whose values ought to be conserved. For example, the global conserved quantities discussed in Section 3.5 provide useful checks. The total rest mass of the system M_0 must be conserved, provided we account for any rest mass that leaves the computational domain. The ADM mass M_{ADM}, the total linear momentum P^i_{ADM} and angular momentum J^i_{ADM} are also conserved, provided we account for net losses carried off by any matter *and* gravitational radiation that leave the computational domain. Other hydrodynamic quantities are useful to monitor in special cases. For example, the relativistic Bernoulli integral is conserved along flow

---

[19] Norman and Winkler (1983).
[20] See Chapter 1.4.
[21] See Petrich *et al.* (1985, 1986), who construct the Oppenheimer–Snyder solution for both maximal and polar time slicing and isotropic radial coordinates.

lines for adiabatic, stationary flow with 4-velocity u^a in a stationary gravitational field:[22]

$$hu^t = constant \text{ (along flow lines).} \tag{5.57}$$

For a uniformly rotating, stationary, perfect fluid star, we have

$$\frac{h}{u^t} = constant, \tag{5.58}$$

where the constant holds *everywhere* inside the fluid. This result is satisfied whenever the spacetime has two Killing vectors, $\xi^a_{(t)} = (\partial/\partial t)^a$ and $\xi^a_{(\phi)} = (\partial/\partial\phi)^a$, to reflect stationarity and axisymmetry.[23]

The Kelvin–Helmholtz theorem states that the relativistic circulation,

$$\mathcal{C}(c) = \oint_c hu_a \lambda^a d\sigma, \tag{5.59}$$

is conserved for isentropic flow along an arbitrary closed "fluid" contour c.[24] Here σ is a Lagrangian parameter which labels fluid elements on the contour c, and λ^a is the tangent vector to the contour [i.e., $\lambda^a = (\partial/\partial\sigma)^a$]. Conservation of $\mathcal{C}$ can be verified by computing

$$\begin{aligned}
\frac{d}{d\tau}\mathcal{C}(c) &= \oint_c d\sigma\, u^b \nabla_b (hu_a \lambda^a) \\
&= \oint_c d\sigma \left[\lambda^a u^b \nabla_b (hu_a) + (hu_a) u^b \nabla_b \lambda^a\right] \\
&= \oint_c d\sigma [\lambda^a u^b \nabla_b (hu_a) + (hu_a)\lambda^b \nabla_b u^a] \\
&= -\oint_c d\sigma\, \lambda^a \nabla_a h \\
&= 0.
\end{aligned} \tag{5.60}$$

Here, to derive the third line from the second line, we use the fact that $u^a = (\partial/\partial\tau)^a$ and λ^a are coordinate basis vectors, and thus commute according to

$$u^b \nabla_b \lambda^a = \lambda^b \nabla_b u^a. \tag{5.61}$$

We also have used $u_a u^a = -1$ and the Euler equation for isentropic flow in the form

$$u^b \nabla_b (hu_a) = -\nabla_a h \tag{5.62}$$

to obtain the fourth line. Note that it is the derivative of $\mathcal{C}(c)$ with respect to the *proper* time τ that vanishes, so that the circulation has to be evaluated on hypersurfaces of constant proper time as opposed to constant coordinate time.

---

[22] Lightman *et al.* (1975), Problem 14.7, p. 83

[23] *ibid.*, Problem 16.17, p. 95. Actually, the result holds even when $\xi^a_{(t)}$ and $\xi^a_{(\phi)}$ are not Killing vectors separately but combine to form a helical Killing vector $\xi^a_{\text{hel}} = \xi^a_{(t)} + \Omega\xi^a_{(\phi)}$, where Ω is the angular velocity of the fluid; see Chapter 15.2 and equation (15.46) for a proof.

[24] See Landau and Lifshitz (1959); Taub (1959); Evans (1984); Saijo *et al.* (2001).

Exercise 5.12 Show that for general (e.g., nonisentropic) flows, the Euler equation is

$$u^b \nabla_b (h u_a) = -\frac{1}{\rho_0} \nabla_a P = -\nabla_a h + T \nabla_a s, \qquad (5.63)$$

and the circulation changes according to

$$\frac{d}{d\tau} \mathcal{C}(c) = -\oint_c d\sigma \lambda^a \frac{1}{\rho_0} \nabla_a P = \oint_c d\sigma \lambda^a T \nabla_a s. \qquad (5.64)$$

Comment on the implications of this result for a shock.

5.2.2 Imperfect gases

There are many important astrophysical applications that involve *imperfect* gases characterized by viscosity, conductivity and (or) radiation. For example, viscosity can drive nonaxisymmetric instabilities in rotating stars, while radiation can lead to the cooling and contraction of stars. However, most of the numerical work to date in relativistic hydrodynamics has focused on perfect fluid sources.[25] This emphasis is physically reasonable for tracking evolution on *dynamical* (e.g., free-fall) time scales, but is not adequate for following evolution on *secular* (e.g., viscous, conduction or radiative) time scales, which are typically much longer. As the field of numerical relativity matures, exploration of secular behavior over many dynamical timescales is likely to accelerate. For this reason, we shall now briefly summarize additional contributions to the stress-energy tensor arising from some of these nonideal effects.

Viscosity

The contribution of viscosity to the stress-energy tensor is[26]

$$\boxed{T_{\text{visc}}^{ab} = -2\eta \sigma^{ab} - \zeta \theta P^{ab}} \qquad (5.65)$$

where $\eta \geq 0$ is the coefficient of dynamic, or shear, viscosity, $\zeta \geq 0$ is the coefficient of bulk viscosity and $\sigma^{ab}, \theta, P^{ab}$ are the shear, expansion and projection tensor of the fluid:

$$\theta = \nabla_a u^a, \qquad (5.66)$$

$$P^{ab} = g^{ab} + u^a u^b, \qquad (5.67)$$

$$\sigma^{ab} = \tfrac{1}{2} \left(P^{ac} \nabla_c u^b + P^{bc} \nabla_c u^a \right) - \tfrac{1}{3} \theta P^{ab}. \qquad (5.68)$$

Exercise 5.13 Show that in the presence of viscosity, the energy (entropy) equation (5.19) becomes

$$\partial_t (\gamma^{1/2} E_*) + \partial_j (\gamma^{1/2} E_* v^j) = \frac{\alpha \gamma^{1/2}}{\Gamma} \left(\frac{E_*}{W} \right)^{(1-\Gamma)} \left(2\eta \sigma^{ab} \sigma_{ab} + \zeta \theta^2 \right). \quad (5.69)$$

---

[25] But see, e.g., Duez *et al.* (2004) and Chapter 14.2.4 for relativistic hydrodynamical simulations with viscosity.
[26] See Misner *et al.* (1973), Section 22.3

Thus conclude by comparing with equation (5.21) that viscosity generates entropy at a rate

$$\rho_0 T \frac{ds}{d\tau} = \left(2\eta\sigma^{ab}\sigma_{ab} + \zeta\theta^2\right).\tag{5.70}$$

Exercise 5.14   Show that in the presence of viscosity, the momentum equation (5.14) becomes

$$\partial_t(\gamma^{1/2}S_i) + \partial_j(\gamma^{1/2}S_i v^j) =$$

$$-\alpha\gamma^{1/2}\left(\partial_i P + \frac{S_a S_b}{2\alpha S^t}\partial_i g^{ab}\right) + \alpha\gamma^{1/2}\eta\sigma_{ab}\partial_i g^{ab} + 2\partial_a\left(\alpha\gamma^{1/2}\eta\sigma_{ib}g^{ab}\right)$$

$$+\frac{1}{2}\alpha\gamma^{1/2}\zeta\theta P_{ab}\partial_i g^{ab} + \partial_a\left(\alpha\gamma^{1/2}\zeta\theta P_{ib}g^{ab}\right).\tag{5.71}$$

The presence of time derivatives of u_a in σ_{ab} and the time derivative of σ_{ab} on the right hand side of the nonconservative, relativistic Navier–Stokes equations (5.69) and (5.71) require subtle handling. Since viscosity is usually a small perturbation on the dynamical flow, it is often sufficient to split off the viscous terms and integrate them separately ("operator splitting") in a lower-order, but convergent, scheme.[27] When recast in conservative form as in equation (5.26), the dynamical equations for the "conservative" variables $(\tilde{D}, \tilde{S}_j, \tilde{\tau})$ appear unchanged when nonideal gas contributions are added to the stress-energy tensor. However, the character of the equations is altered, as there now are additional derivatives (including time derivatives) of the primitive variables that appear in these variables and in their source terms. Once again, a lower-order scheme to handle these perturbative terms is often adequate.

Heat and radiation diffusion

The contribution of heat flux (i.e., conduction) to the stress-energy tensor is

$$\boxed{T_{\text{heat}}^{ab} = u^a q^b + u^b q^a,}\tag{5.72}$$

where the heat-flux 4-vector q^a is given by

$$q^a = -\lambda_{\text{th}} P^{ab}\left(\nabla_b T + Ta_b\right).\tag{5.73}$$

In equation (5.73), T is the temperature, $a^a = u^b\nabla_b u^a$ is the fluid 4-acceleration, and λ_{th} is the coefficient of thermal conduction.

A useful application of the thermal conduction formalism is heat transport via thermal radiation (e.g., photons or neutrinos) treated in the *diffusion approximation*, for which

$$\lambda_{\text{th}} = \frac{4}{3}\frac{b_R T^3}{\bar{\chi}}.\tag{5.74}$$

---

[27] See Duez *et al.* (2004) and Chapter 14.2.4 for such an approach.

Here b_R is a constant that depends on the type of radiation: for thermal photons it is the usual radiation constant b;[28] for each flavor of nondegenerate thermal neutrino or antineutrino (chemical potential $= 0$) the constant is $(7/16)b$. The quantity $\bar{\chi}$ is the Rosseland mean opacity of the interacting gas particles.[29] Radiation transport may be treated in the diffusion approximation whenever the mean-free path of the energy carriers (e.g., photons) due to interactions with the gas particles is short compared to the system scale size, in which case we say that the gas is "optically thick" to radiation.

5.2.3 Radiation hydrodynamics

In general, a gas is neither optically thick nor optically thin (i.e., transparent) everywhere, nor is the radiation always in thermal equilibrium with the matter. In such cases we cannot treat radiation transport in the diffusion approximation as discussed above. To handle the more general case, the radiation field can be described by a radiation stress-energy tensor,

$$T_{\text{rad}}^{ab} = \int \int d\nu \, d\Omega \, I_\nu \, N^a N^b,$$ (5.75)

where $I_\nu = I(x^a; N^i; \nu)$ is the specific intensity of radiation at x^a with frequency ν and 4-momentum p^a moving in the direction $N^a \equiv p^a/h\nu$, where h is Planck's constant and the integral is over all frequencies and solid angles. Here ν, I_ν and $d\Omega$ are all measured in the local Lorentz frame of a fiducial observer with 4-velocity u^a, whereby $h\nu = -p_a u^a$. The specific intensity is related to the radiation phase-space distribution function f defined by equation (5.191) below according to

$$f = \frac{c^2}{h^4}\left(\frac{I_\nu}{\nu^3}\right).$$ (5.76)

Like f, the ratio I_ν/ν^3 is a Lorentz invariant.

To ascertain the physical meaning of the components of T_{rad}^{ab}, consider a spherically symmetric spacetime. In spherical symmetry, the direction of radiation at an arbitrary point in space (r, θ, ϕ) is specified by polar angle Θ, measured with respect to the outward radial direction. In other words, the intensity at any instant of time is independent of the azimuthal direction Φ at that spatial point, so that $I_\nu = I_\nu(t, r; \mu)$, where $\mu = \cos\Theta$. We use this information in the next exercise to compute the components of T_{rad}^{ab}.

Exercise 5.15 At each point (t, r, θ, ϕ) in a spherical spacetime set up an orthonormal set of basis vectors $\{e_{(\hat{t})}^a, e_{(\hat{r})}^a, e_{(\hat{\theta})}^a, e_{(\hat{\phi})}^a\}$ where $e_{(\hat{r})}^a$ points along the outward radial direction from the center of the spatial coordinate system. Show that in such a frame

---

[28] $b = 8\pi^5 k^4/(15 c^3 h^3) = 7.56464 \times 10^{-15} \text{erg/cm}^3/\text{deg}^4$.

[29] See Shapiro and Teukolsky (1983), equation (I.28), with $\chi_\nu \equiv \kappa_\nu \rho_0$, for the definition of Rosseland mean opacity.

the components of T^{ab}_{rad} are

$$
T^{\hat{a}\hat{b}}_{\text{rad}} = \begin{pmatrix} E & F & 0 & 0 \\ F & P & 0 & 0 \\ 0 & 0 & \frac{1}{2}(E-P) & 0 \\ 0 & 0 & 0 & \frac{1}{2}(E-P) \end{pmatrix}. \tag{5.77}
$$

The quantities appearing in equation (5.77) are the frequency-integrated radiation moments constructed at each point from the frequency-integrated intensity $I = I(t, r; \mu)$, measured in the orthonormal frame. The intensity I is

$$
I = \int d\nu\, I_\nu, \tag{5.78}
$$

the radiation energy density is

$$
E(t, r) = 2\pi \int \int d\nu\, d\mu\, I_\nu, \tag{5.79}
$$

the radiation energy flux is

$$
F(t, r) = 2\pi \int \int d\nu\, d\mu\, \mu\, I_\nu, \tag{5.80}
$$

and the radiation pressure is[30]

$$
P(t, r) = 2\pi \int \int d\nu\, d\mu\, \mu^2 I_\nu. \tag{5.81}
$$

In the above expressions, the frequency ν is integrated from 0 to ∞, while the cosine μ is integrated from -1 to 1.

The dynamical equations for the radiation moments are given by

$$
\nabla_b T^{ab}_{\text{rad}} = -G^a \tag{5.82}
$$

where G^a is the radiation 4-force density expressing the interaction of the matter with the radiation,

$$
G^a = \int \int d\nu\, d\Omega\, [\chi_\nu I_\nu - \eta_\nu]\, N^a. \tag{5.83}
$$

Here χ_ν is the total opacity and η_ν is the emissivity of the gas. The hydrodynamical equations for the fluid then become $\nabla_b T^{ab}_{\text{fluid}} = -\nabla_b T^{ab}_{\text{rad}} = G^a$, which must be solved simultaneously with equations (5.82) for the radiation moments *and* the equation of radiative transport (the "photon Boltzmann equation" or the "radiation kinetic equation") for I_ν. The contribution of the radiation must also be included in the gravitational field source terms, equations (5.1)–(5.2).

Exercise 5.16 Assume that the scattering opacity (superscript "s") is elastic and forward-backward symmetric and that the absorption opacity and thermal emissivity

---

[30] In this section we use the symbol P for *radiation* pressure alone, as defined by equation (5.81). Throughout the rest of the book P is used to denote the *total* pressure.

(superscript "a") are related by Kirchoff's law, $\eta_\nu^a / \chi_\nu^a = B_\nu(T)$, where $B_\nu(T)$ is the Planck function[31] and T is the matter temperature.

(a) Show that in the rest frame of the fluid, the orthonormal components of G^a are given by

$$G^{\hat{0}} = \int\int d\nu d\Omega \chi_\nu^a (I_\nu - B_\nu) = \int d\nu \chi_\nu^a (E_\nu - 4\pi B_\nu), \qquad (5.84)$$

$$G^{\hat{i}} = \int d\nu (\chi_\nu^a + \chi_\nu^s) F_\nu^{\hat{i}} \qquad (5.85)$$

where

$$E_\nu = \int d\Omega I_\nu \qquad (5.86)$$

is the specific radiation energy density and $F_\nu^{\hat{a}}$ is the specific radiation flux 4-vector defined by

$$F_\nu^{\hat{0}} = 0, \quad F_\nu^{\hat{i}} = \int d\Omega I_\nu \, N^{\hat{i}}. \qquad (5.87)$$

Thus the flux satisfies $F^a u_a = 0$, where u^a is the fluid 4-velocity.

(b) Adopt "gray-body" scattering and absorption opacities, whereby $\chi_\nu = \kappa \rho_0$, with κ a constant independent of frequency. Show that

$$G^{\hat{0}} = \kappa^a \rho_0 (E - 4\pi B), \qquad (5.88)$$

$$G^{\hat{i}} = (\kappa^a + \kappa^s) \rho_0 F^{\hat{i}} \qquad (5.89)$$

gives the radiation 4-force density, where $4\pi B = bT^4$,

$$E = \int\int d\nu d\Omega I_\nu = \int d\nu E_\nu \qquad (5.90)$$

is the total integrated radiation energy density, and

$$F^{\hat{0}} = 0, \quad F^{\hat{i}} = \int\int d\nu d\Omega I_\nu N^{\hat{i}} = \int d\nu F_\nu^{\hat{i}} \qquad (5.91)$$

is the total integrated radiation flux 4-vector.

(c) Argue that for this system a frame-independent expression for G^a is

$$G^a = \kappa^a \rho_0 (E - 4\pi B) u^a + (\kappa^a + \kappa^s) \rho_0 F^a. \qquad (5.92)$$

As derived in exercise 5.38, the equation of radiative transfer takes on the general form

$$p^a \frac{\mathcal{D}(I_\nu/\nu^3)}{dx^a} = \left(\frac{\delta(I_\nu/\nu^3)}{\delta\lambda} \right)_{\text{coll}} = (\eta_\nu - \chi_\nu I_\nu)/\nu^2, \qquad (5.93)$$

---

[31] Actually, B_ν is the Planck function for photons, but it is the thermal intensity corresponding to a Fermi–Dirac distribution in the case of neutrinos; we will use the same symbol for both cases. In many problems involving neutrino transport in stars, the neutrino energies are much larger than their masses, in which case the neutrinos can be treated as massless.

where the left-hand side is defined by equation (5.196) below and the right-hand side accounts for the gain or loss of photons due to interactions with the matter: η_ν is the emissivity and χ_ν is the absorption coefficent.[32] Consider the special case of a spherically symmetric medium in comoving coordinates, for which the metric may be cast in the form

$$ds^2 = -e^{2\Psi}dt^2 + e^{2\Lambda}dr^2 + R^2(d\theta^2 + \sin^2\theta d\phi^2), \tag{5.94}$$

where r is a Lagrangian radial coordinate and Ψ, Λ and R are functions of r and t only. In the comoving frame, the equation of radiation transfer is then[33]

$$\mathcal{D}_t(I_\nu/\nu^3) + \mu\mathcal{D}_r(I_\nu/\nu^3) - \nu\left[\mu\mathcal{D}_r\Psi + \mu^2\mathcal{D}_t\Lambda + (1-\mu^2)(U/R)\right](\partial(I_\nu/\nu^3)/\partial\nu)$$

$$+ (1-\mu^2)\{(\Gamma/R) - \mathcal{D}_r\Psi + \mu[(U/R) - \mathcal{D}_t\Lambda]\}(\partial(I_\nu/\nu^3)/\partial\mu) = (\eta_\nu - I_\nu\chi_\nu)/\nu^3, \tag{5.95}$$

where we have introduced two operators

$$\mathcal{D}_t = e^{-\psi}(\partial/\partial t), \quad \mathcal{D}_r = e^{-\Lambda}(\partial/\partial r), \tag{5.96}$$

and two auxiliary variables

$$U = \mathcal{D}_t R, \quad \Gamma = \mathcal{D}_r R. \tag{5.97}$$

Note that for radiation problems it is always convenient to work in the rest frame of the fluid, since that is the frame in which details of the radiation and matter interactions are most easily specified and where the intensity in local thermodynamic equilibrium is close to a Planck function at the local gas temperature. However, it is not always convenient or possible to work in the comoving frame, which is also the Lagrangian fluid frame, when solving the coupled $3+1$ equations for the combined matter–radiation–fluid system, particularly when the flow deviates from spherical symmetry.

> **Exercise 5.17** Consider the collapse of a homogeneous dust sphere from rest at a finite radius (i.e., Oppenheimer–Snyder collapse, as described in Chapter 1.4), for which the interior metric may be written in the familiar closed-Friedmann form
>
> $$ds^2 = -d\tau^2 + a^2(\tau)[d\chi^2 + \sin^2\chi(d\theta^2 + \sin^2\theta d\phi^2)]. \tag{5.98}$$
>
> Show that in this background spacetime the comoving equation of radiation transfer becomes
>
> $$\frac{\partial I_\nu}{\partial\tau} + 3\frac{\dot{a}}{a}I_\nu + \mu\frac{1}{a}\frac{\partial I_\nu}{\partial\chi} - \frac{\dot{a}}{a}\nu\frac{\partial I_\nu}{\partial\nu} + (1-\mu^2)\frac{\cot\chi}{a}\frac{\partial I_\nu}{\partial\mu} = \eta_\nu - \chi_\nu I_\nu, \tag{5.99}$$
>
> where $\dot{a} = da/d\tau$.

The radiation hydrodynamics problem is very difficult to solve in general, since the intensity is a function of 6-dimensional phase space plus time, and the radiation transport

---

[32] See exercises 5.37 and 5.38 and the related discussion and references.
[33] Lindquist (1966); Mihalas and Mihalas (1984), p. 443.

equation (5.93) with complicated radiation-fluid interaction terms (including scattering) has a nontrivial integrodifferential character. Even in spherical symmetry, the intensity varies with frequency, radial position *and* angular direction, plus time, so that even the frequency-integrated transport equation is $2 + 1$ dimensional (*cf.* equation 5.95). Approximation schemes are often used to simplify the problem. They sometimes involve by-passing the full radiation transport equation and only solving coupled, partial differential equations for the lowest *moments* of the transport equation, integrated over frequency and subject to approximate, physically plausible, "closure" relations to truncate the infinite set of moment equations.

Exercise 5.18 Define the "variable Eddington factor" f_{Edd} according to $P = f_{\text{Edd}}E$, where E and P are the radiation moments given by equations (5.79) and (5.81). In general, the variation of f_{Edd} in space and time can only be determined by solving the full transport equation. Show, however, that it is quite simple to calculate f_{Edd} in the extreme opposite limits of (a) isotropic radiation and (b) outward radial beaming of the radiation, in a spherically symmetric spacetime. How is the flux F given by equation (5.80) related to E in the two opposite limits?

Exercise 5.19 Consider a system which is sufficiently optically thick that in the fluid rest frame the radiation field is *nearly* isotropic everywhere. However, allow for a net flux of radiation at each point. This is the physical situation that applies to the interior of a star, either in equilibrium or during collapse. Define the local radiation energy density E as in equation (5.90) and the local radiation flux four-vector F^a as in equation (5.91).
(1) Argue that for an isotropic radiation field, the orthonormal components of the stress-energy tensor in the rest-frame of the fluid are given by

$$T_{\text{rad}}^{\hat{i}\hat{j}} = P\delta^{\hat{i}\hat{j}}, \quad P = E/3. \tag{5.100}$$

(2) Show that for such a system the radiation stress-energy tensor in a general frame may be written

$$\boxed{T_{\text{rad}}^{ab} = Eu^a u^b + F^a u^b + F^b u^a + P(g^{ab} + u^a u^b),} \tag{5.101}$$

where u^a is the fluid 4-velocity. Thus the equation of motion for the radiation field, equation (5.82), together with equation (5.92) for G^a, provides four equations for the four radiation variables E and F^i, with P given by equation (5.100) and F^0 obtained by solving $F^a u_a = 0$.

The truncated moment equations are essentially equivalent to the equations of motion for the radiation field, equations (5.82). The matter profile enters the source terms in the moment equations through the radiation 4-force density G^a, and the moments appear in the equations of motion of the fluid, so the combined radiation-hydrodynamical system must be evolved simultaneously.

Thorne[34] has constructed a "projected, symmetric trace-free tensor" (PSTF) formalism for handling radiative transfer in relativistic systems. It consists of an infinite hierarchy

[34] Thorne (1981).

of partial differential equations corresponding to an infinite number of moments of the radiative transfer equation. The formalism is particularly straightforward to implement for systems with spherical or planar symmetry, in which case the tensor moments reduce to scalar functions analogous to E, F and P defined in exercise 5.15.[35]

The simple collapse scenario described in exercise 5.20 provides one test of a relativistic radiative hydrodynamics code in curved spacetime. It is specifically designed to check how well a numerical scheme that solves the coupled equations of relativistic hydrodynamics, gravitational field evolution and radiative transport handles collapse to a black hole. As shown in the exercise, an analytic solution can be found[36] when the matter and metric follow the Oppenheimer–Snyder (OS) solution (see equation 5.98 and Chapter 1.4) and the radiation can be treated in the thermal diffusion approximation. The resulting scenario may be called "thermal OS collapse".[37]

> **Exercise 5.20** Consider the thermal radiation flux emitted during collapse to a black hole of a spherical, homogeneous, stellar dust-ball initially subjected to a small isothermal temperature fluctuation. Take the mass of the star to be M and the initial radius to be R_0 and assume that the matter density, velocity and the gravitational field are described by the OS solution. Assume further that the internal energy and pressure generated by the temperature perturbation are much smaller than ρ^*, the total mass-energy density, and are dominated by thermal radiation. Treat the radiation in the diffusion approximation.
>
> (a) Use equation (5.72) to write the total stress-energy tensor T^{ab} as
>
> $$T^{ab} = \rho^* u^a u^b + P P^{ab} + q^a u^b + q^b u^a, \qquad (5.102)$$
>
> where q^a is given by equation (5.73) with λ_{th} given by equation (5.74) and $\bar{\chi} \equiv \kappa \rho_0$, where κ is a constant. Evaluate $u_a \nabla_b T^{ab} = 0$ to obtain the law of local energy conservation with radiation,
>
> $$\frac{d\rho^*}{d\tau} = \frac{\rho^* + P}{\rho_0} \frac{d\rho_0}{d\tau} - \nabla_a q^a - a_a q^a. \qquad (5.103)$$
>
> (b) Substitute $3P = E = bT^4$, where E is the thermal radiation energy density in the fluid rest frame and evaluate equation (5.103) for OS collapse to show
>
> $$\partial_{\tilde{\tau}} E_c = \frac{1}{\sin^2(\chi_0 z)} \partial_z \left[\sin^2(\chi_0 z) \partial_z E_c \right], \qquad (5.104)$$
>
> where $E_c(\tilde{\tau}, z) \equiv E Q^4$ is the radiation energy density "corrected for adiabatic collapse", $Q(\tau)$ is given by equation (1.97), $z \equiv \chi/\chi_0, 0 \leq z \leq 1$, is a nondimensional

---

[35] See Thorne *et al.* (1981); Rezzolla and Miller (1994, 1996); Zampieri *et al.* (1996); Balberg *et al.* (2000), and references therein, for applications of the PSTF moment formalism to some astrophysical problems involving relativistic radiative hydrodynamics in spherical symmetry.

[36] Shapiro (1989).

[37] Shapiro (1996) has numerically integrated the relativistic Boltzmann equation coupled to the radiation moment equations to obtain the interior radiation field without approximation for this problem. Optically thin as well as thick cases were considered. The analytic solution obtained in exercise 5.20 is in good agreement with numerical result in the optically thick case.

Lagrangian (comoving) radius, and $\tilde{\tau}$ is a nondimensional time given by

$$\tilde{\tau} = \int_0^\tau d\tau' \frac{Q(\tau')}{3\kappa\rho_0(0)R_0^2} \left[\frac{\sin\chi_0}{\chi_0}\right]^2$$

$$= \frac{1}{4\kappa\rho_0(0)R_0} \left[\frac{R_0}{8M}\right]^{1/2} \left(\eta + \frac{4}{3}\sin\eta + \frac{1}{6}\sin 2\eta\right) \left[\frac{\sin\chi_0}{\chi_0}\right]^2. \quad (5.105)$$

In equation (5.105) η is the OS conformal time parameter. The quantity $\tilde{\tau}$ measures time in units of the radiation diffusion time scale across R_0; it is related to proper time τ through η (see equation 1.93).

(c) Impose the "zero-temperature" boundary condition at the surface and regularity at the origin,

$$E_c(\tilde{\tau}, 1) = 0 = \partial_z E_c(\tilde{\tau}, 0), \quad \tilde{\tau} \geq 0. \quad (5.106)$$

Then solve equation (5.104) subject to the isothermal initial condition

$$E_c(0, z) = E_0, \quad 0 \leq z < 1. \quad (5.107)$$

Hint: Make the substitution

$$E_c(\tilde{\tau}, z) = \frac{\chi_0}{\sin(\chi_0 z)} f(\tilde{\tau}, z), \quad (5.108)$$

in which case equation (5.104) becomes

$$\partial_{\tilde{\tau}} f = \partial_z^2 f + \chi_0^2 f. \quad (5.109)$$

Then expand f in a Fourier sine series consistent with the boundary conditions,

$$f(\tilde{\tau}, z) = \sum_{n=1}^\infty B_n(\tilde{\tau})\sin(n\pi z), \quad (5.110)$$

substitute into equation (5.109) to get each time-dependent coefficient B_n up to a constant factor, and use the initial data to determine the constant. Obtain finally

$$E_c(\tilde{\tau}, z) = 2E_0 \left[\frac{\sin\chi_0}{\sin(\chi_0 z)}\right] e^{\chi_0^2 \tilde{\tau}} \sum_{n=1}^\infty \left[\frac{(-1)^{n+1}}{n\pi} e^{-n^2\pi^2\tilde{\tau}} \sin(n\pi z) \left(\frac{n^2\pi^2}{n^2\pi^2 - \chi_0^2}\right)\right].$$

$$(5.111)$$

(d) Show that the radiation flux $F \equiv (q_a q^a)^{1/2}$ in the fluid rest frame is given by

$$F(\tilde{\tau}, z) = -\frac{1}{3\kappa\rho_0(0)R_0 Q^2} \left[\frac{\sin\chi_0}{\chi_0}\right] \partial_z E_c(\tilde{\tau}, z), \quad (5.112)$$

and evaluate it at the surface to get the emergent flux,

$$F(\tilde{\tau}, 1) = \frac{2}{3} \frac{E_0}{\kappa\rho_0(0)R_0 Q^2} \left[\frac{\sin\chi_0}{\chi_0}\right] e^{\chi_0^2 \tilde{\tau}} \sum_{n=1}^\infty e^{-n^2\pi^2\tilde{\tau}} \left(\frac{n^2\pi^2}{n^2\pi^2 - \chi_0^2}\right). \quad (5.113)$$

Equations (5.82) have been cast in conservative form for the case of a nearly isotropic radiation field obeying the conditions set forth in exercise 5.19.[38] It is then possible to unite the matter and radiation equations in a single HRSC scheme that evolves both the fluid and the radiation field simultaneously. Test simulations with such a scheme involving radiation shocks and nonlinear waves propagating in Minkowski spacetime yield good agreement with analytic results. When used in conjunction with a $3 + 1$ scheme for the gravitational field, the method can reproduce the "thermal OS collapse" solution derived in exercise 5.20 quite well.

Important applications of radiative hydrodynamics in relativistic spacetimes include neutrino transport during stellar core collapse and supernovae explosions, photon emission from gas accretion onto black holes and neutron stars, and photon propagation, decoupling and re-ionization in the Big Bang Universe. In many of these examples, the radiation field can play an important dynamical role in influencing the flow of gas, as well as contributing a source of observable energy flux.

5.2.4 Magnetohydrodynamics

Magnetic fields play a crucial role in determining the evolution of many relativistic objects. In any highly conducting astrophysical plasma, a frozen-in magnetic field can be amplified appreciably by gas compression or shear. Even when an initial seed field is weak, the field can grow in the course of time to significantly influence the gas-dynamical behavior of the system. In problems where the self-gravitation of the magnetized gas can be ignored, calculations can be performed in a fixed, stationary background spacetime. In this case the metric does not have to be evolved numerically. Some important gas accretion problems fall into this category, including accretion onto neutron stars and black holes. In many other problems, the effect of the magnetized gas on the metric cannot be ignored, and the gas, the magnetic fields *and* the metric must be evolved self-consistently. The final fate of many of these astrophysical systems, which often involve compact objects and their distinguishing observational signatures, may hinge on the role that magnetic fields play during the evolution. Some of these systems are promising sources of gravitational radiation for detection by laser interferometers. Others may be responsible for gamma-ray bursts. Examples of astrophysical scenarios involving strong-field, dynamical spacetimes in which magnetic fields may play a decisive role include core collapse in supernovae, magnetorotational collapse of hypermassive neutron stars and supermassive stars, the merger of neutron star–neutron star and black hole–neutron star binaries, and the suppression of r-mode instabilities in rotating neutron stars.[39]

In many astrophysical applications involving magnetic fields, the gas is highly ionized and an excellent conductor of current. The ideal magnetohydrodynamic (MHD)

---

[38] Farris *et al.* (2008).
[39] See, e.g., Baumgarte and Shapiro (2003b) for a brief discussion and references. Many of these applications will be discussed in Chapters 14, 16 and 17.

approximation applies in the limit of infinite conductivity and it is precisely that regime on which we shall focus below.

Electromagnetic field equations

We decompose the Faraday tensor F^{ab} as

$$F^{ab} = n^a E^b - n^b E^a + n_d \epsilon^{dabc} B_c, \tag{5.114}$$

where $\epsilon_{abcd} = \sqrt{-g}\,[abcd]$ is the Levi-Civita tensor, $[abcd]$ is the completely antisymmetric symbol and where E^a and B^a are the electric and magnetic fields observed by a normal observer n^a. Both fields are purely spatial ($E^a n_a = B^a n_a = 0$), and one can easily show that

$$E^a = F^{ab} n_b, \quad B^a = \frac{1}{2} \epsilon^{abcd} n_b F_{dc}. \tag{5.115}$$

The electromagnetic stress-energy tensor

$$4\pi T^{ab}_{\text{em}} = F^{ac} F^b{}_c - \frac{1}{4} g^{ab} F_{cd} F^{cd} \tag{5.116}$$

may be written in terms of E^a and B^a according to

$$4\pi T^{ab}_{\text{em}} = \tfrac{1}{2}(n^a n^b + \gamma^{ab})(E_i E^i + B_i B^i) \\ + 2n^{(a} \epsilon^{b)cd} E_c B_d - (E^a E^b + B^a B^b). \tag{5.117}$$

Here $\epsilon^{abc} = n_d \epsilon^{dabc}$ is the familiar 3-dimensional, spatial Levi-Civita tensor. Along with the electromagnetic field, we shall assume the presence of a perfect fluid, so that the total stress-energy tensor is given by

$$T^{ab} = \rho_0 h u^a u^b + P g^{ab} + T^{ab}_{\text{em}}. \tag{5.118}$$

Exercise 5.21 Insert the electromagnetic stress-energy tensor into equations (5.1)–(5.2) to obtain the electromagnetic contributions to the source terms in the $3+1$ gravitational field equations:

$$4\pi \rho_{\text{em}} = \frac{1}{2}(E_i E^i + B_i B^i) = \frac{1}{2}(\mathbf{E}^2 + \mathbf{B}^2), \tag{5.119}$$

$$4\pi S^{\text{em}}_i = \epsilon_{ijk} E^j B^k = (\mathbf{E} \times \mathbf{B})_i, \tag{5.120}$$

$$4\pi S^{\text{em}}_{ij} = -E_i E_j - B_i B_j + \frac{1}{2}\gamma_{ij}(\mathbf{E}^2 + \mathbf{B}^2), \tag{5.121}$$

$$4\pi S_{\text{em}} = \frac{1}{2}(\mathbf{E}^2 + \mathbf{B}^2). \tag{5.122}$$

The above results are not surprising: expressed in terms of the electromagnetic field components as measured by a normal observer, n^a, i.e., an observer who is at rest with respect to the slices Σ, the $3 + 1$ source terms have the same form as in flat space.[40] Thus $\rho_{\rm em}$ is the familiar energy-density for an electromagnetic field, $S_i^{\rm em}$ is the Poynting vector, etc.

In many astrophysical applications, we can assume perfect conductivity. In the limit of infinite conductivity, Ohm's law yields the ideal MHD condition:

$$F^{ab}u_b = 0. \tag{5.123}$$

To see this, rewrite the Faraday tensor in terms of $E_{(u)}^a$ and $B_{(u)}^a$, the magnetic and electric fields measured by an observer u^a at rest with respect to the fluid, to obtain

$$F^{ab} = u^a E_{(u)}^b - u^b E_{(u)}^a + u_d \epsilon^{dabc} B_c^{(u)}, \tag{5.124}$$

where $B_{(u)}^a u_a = 0 = E_{(u)}^a u_a$. Then the ideal MHD condition (5.123) simply states that the electric field $E_{(u)}^a = F^{ab}u_b$ vanishes in the fluid rest frame, as required for a perfect conductor.

> **Exercise 5.22**   Show that the covariant generalization of Ohm's law is
>
> $$J^a - \rho_e u^a = \sigma F^{ab}u_b, \tag{5.125}$$
>
> where J^a is the electromagnetic current 4-vector, $\rho_e = -J^a u_a$ is the charge density as seen by an observer comoving with the fluid 4-velocity u_a, F^{ab} is the Faraday tensor and σ is the electrical conductivity. Argue that the perfect conductivity condition (5.123) follows immediately from equation (5.125) in the limit of infinite conductivity.

In the ideal MHD case, F^{ab} is completely determined by $B_{(u)}^a$,

$$F^{ab} = \epsilon^{abcd} u_c B_d^{(u)}, \tag{5.126}$$

$$B_{(u)}^a = \frac{1}{2}\epsilon^{abcd} u_b F_{dc}. \tag{5.127}$$

> **Exercise 5.23**   Show that the condition that the electric field vanish in the fluid rest frame may be written $u_a E^a = 0$, or
>
> $$\alpha E_i = -\epsilon_{ijk}(v^j + \beta^j)B^k, \tag{5.128}$$
>
> where $v^i = u^i/u^t$. Note that when evaluated in a Minkowski spacetime, the last equation reduces to the familiar flat spacetime expression $E_i = -\epsilon_{ijk}v^j B^k$ or $\mathbf{E} = -\mathbf{v} \times \mathbf{B}$.

It might appear that the fluid rest-frame components of the magnetic field are the most convenient choice for integrating the evolution equations for the electromagnetic field, due to the vanishing of the electric field in that frame. However, as we shall see below, the magnetic equations are easiest to express in terms of the *normal* components of the magnetic field. Of course, the two sets of components are easily related.

---

[40] *cf.* exercise 5.1 in Misner *et al.* (1973).

Exercise 5.24   Prove the relationship between the components of the electromagnetic fields as measured by a normal observer and by an observer at rest in the fluid:

$$E^a = -\epsilon^{abc} u_b B_c^{(u)}, \tag{5.129}$$

$$B^a = -n_b u^b B_{(u)}^a + n_b B_{(u)}^b u^a. \tag{5.130}$$

To derive the magnetic field evolution equations, introduce the dual of the Faraday tensor according to

$$F^{*ab} = \frac{1}{2}\epsilon^{abcd} F_{cd}. \tag{5.131}$$

Exercise 5.25   Show that

$$B_{(u)}^a = -F^{*ab} u_b. \tag{5.132}$$

Substituting equation (5.126) into equation (5.131) yields

$$F^{*ab} = u^b B_{(u)}^a - u^a B_{(u)}^b. \tag{5.133}$$

Inserting equation (5.133) into the "magnetic" Maxwell equations, $\nabla_a F^{*ac} = 0$, yields

$$\partial_a \left[(-g)^{1/2} \left(u^c B_{(u)}^a - u^a B_{(u)}^c \right) \right] = 0, \tag{5.134}$$

or

$$\partial_a \left[\gamma^{1/2} W \left(v^c b^a - v^a b^c \right) \right] = 0, \tag{5.135}$$

where we have defined

$$b^a \equiv B_{(u)}^a / (4\pi)^{1/2} \tag{5.136}$$

and $v^a = u^a / u^t$. We can now split the above equation into two pieces by introducing the magnetic field variable

$$\mathcal{B}^i = (4\pi)^{1/2} \gamma^{1/2} W (b^i - v^i b^t). \tag{5.137}$$

The variable $\mathcal{B}^i$ is essentially the spatial component of the magnetic field B^i as measured by a normal observer, as shown in the next exercise.

Exercise 5.26   Show the relation

$$\mathcal{B}^i = (\gamma)^{1/2} B^i. \tag{5.138}$$

Setting the index $c = t$ in equation (5.135) yields the magnetic constraint equation,

$$\partial_i \mathcal{B}^i = 0 \tag{5.139}$$

(i.e., $D_i B^i = 0$), the condition for "no magnetic monopoles". Setting $c = i$ yields the magnetic evolution (induction) equation,

$$\partial_t \mathcal{B}^i - \partial_j \left(v^i \mathcal{B}^j - v^j \mathcal{B}^i \right) = 0. \tag{5.140}$$

Integrating this set of equations together with the fluid equations typically forces us to translate between the two sets of magnetic variables, b^a and $\mathcal{B}^i$. The components of b^a are best obtained by solving equation (5.137) together with the orthogonality relation $b^a u_a = 0$.

Analytically, the magnetic induction equation preserves the constraint condition (5.139) if the constraint is satisfied on the initial time slice, but considerable care must be taken to ensure that this happens in a numerical evolution. Defining a vector potential A^i according to $B^i = \epsilon^{ijk} D_j A_k$ and evolving A^i instead of B^i automatically guarantees that the constraint will be satisfied, but the resulting evolution equation for A^i introduces a second-order spatial derivative term that can make the numerical evolution too diffusive. High-order finite difference schemes for evolving the induction equation while maintaining the divergence constraint to roundoff precision are called *constrained transport* schemes.[41] One version, called *flux-interpolated constrained transport*,[42] consists in replacing the induction equation flux computed at each spatial grid point with linear combinations of the flux computed at that point and neighboring points. The combination assures both that second-order accuracy is maintained and that the constraint $D_i B^i = 0$ is strictly enforced. Another means of maintaining this divergence constraint is called *hyperbolic divergence cleaning*[43] and involves adding a new scalar field to the induction equation as well as a new hyperbolic equation containing $D_i B^i$ to describe the evolution of this new field. The role of the new system is to transport divergence errors to the computational domain boundaries with maximal allowable speed, while simultaneously damping them.

The electromagnetic field equations derived in the ideal MHD limit above are considerably simpler to treat than the full set of Maxwell's equations without approximation. However, there are regimes when it becomes necessary to work with Maxwell's equations in full generality. For example, when an electromagnetic field reaches the surface of a star and then propagates out into the exterior vacuum, the fields are no longer frozen into in a highly conducting plasma and the MHD approximation breaks down. The general form of Maxwell's equations can be cast into $3+1$ form, which facilitates their integration in conjunction with $3+1$ equations for the gravitational field. Begin by decomposing the electromagnetic current 4-vector J^a according to

$$J^a = n^a \rho_e + j^a, \tag{5.141}$$

where ρ_e and j^a are the charge density and spatial current as observed by a normal observer n^a ($j^a n_a = 0$). With these definitions, Maxwell's equations,

$$\nabla_b F^{ab} = 4\pi J^a, \tag{5.142}$$

and

$$\nabla_{[a} F_{bc]} = 0, \tag{5.143}$$

---

[41] Evans and Hawley (1988).
[42] Tóth (2000).
[43] See, e.g., Dedner *et al.* (2002) and references therein.

can be brought into $3 + 1$ form as follows:[44]

$$D_i E^i = 4\pi \rho_e \tag{5.144}$$

$$\partial_t E^i = \epsilon^{ijk} D_j(\alpha B_k) - 4\pi \alpha j^i + \alpha K E^i + \mathcal{L}_\beta E^i \tag{5.145}$$

$$D_i B^i = 0 \tag{5.146}$$

$$\partial_t B^i = -\epsilon^{ijk} D_j(\alpha E_k) + \alpha K B^i + \mathcal{L}_\beta B^i. \tag{5.147}$$

Exercise 5.27 Verify the $3 + 1$ form of Maxwell's equations given above.

The charge conservation equation,

$$\nabla_a J^a = 0, \tag{5.148}$$

which is implied by equation (5.142), becomes

$$\partial_t \rho_e = -D_i(\alpha j^i) + \alpha K \rho_e + \mathcal{L}_\beta \rho_e. \tag{5.149}$$

Exercise 5.28 Show that the familiar form of Maxwell's equations in special relativity can be recovered easily by evaluating equations (5.144)–(5.147) for a Minkowski spacetime with $\gamma_{ij} = \eta_{ij}$, where η_{ij} is the flat spatial metric in an arbitrary coordinate system, $\alpha = 1$, $K = 0$ and $\beta^i = 0$.

Exercise 5.29 Derive the MHD induction equation (5.140) from equation (5.147). **Hint:** Take the trace of the $3 + 1$ evolution equation for $\partial_t \gamma_{ij}$ and combine it with the Lie derivative $\mathcal{L}_\beta B^i$ to give

$$\alpha K B^i + \mathcal{L}_\beta B^i = D_j(\beta^j B^i - \beta^i B^j) - B^i \partial_t \ln \gamma^{1/2}, \tag{5.150}$$

where we have used the magnetic constraint (5.146). Substitute (5.150) together with the ideal MHD equation (5.128) into Faraday's law (5.147) to get the result.

When treating an electromagnetic field at the boundary between a highly conducting plasma and a vacuum, such as at the surface of a star, the required electromagnetic field evolution equations switch from being the MHD induction equation for the magnetic field in the matter interior to the full set of B-field *and* E-field evolution equations in the vacuum exterior. Boundary conditions for B^i and E^i just outside the surface are necessary to extend the integrations into the exterior. These boundary values may be obtained by matching fields across the surface using the familiar "junction conditions" of electrodynamics in the rest-frame of the fluid at the surface.[45]

It is computationally simpler, and in some cases it may even be more appropriate, to treat the region outside the surface of a star as an extended atmosphere of very low-density plasma, in which case the MHD approximation may still hold. In the end, the choice depends on the physical situation, as well as the specific questions, being addressed.

---

[44] See, e.g., Thorne and MacDonald (1982).
[45] See, e.g., Misner *et al.* (1973), Section 21.13.

Equations of baryon, energy and momentum conservation

The evolution equations for an MHD plasma are straightforward generalizations of the hydrodynamical equations derived earlier in this chapter for a nonmagnetic gas. In the MHD limit the electromagnetic piece of the total energy-momentum tensor can be written in terms of b^a according to

$$T^{ab}_{\text{em}} = b^2 u^a u^b + \frac{1}{2} b^2 g^{ab} - b^a b^b, \qquad (5.151)$$

where $b^2 = b^a b_a$.

> **Exercise 5.30** Show that in the MHD limit, the electromagnetic source terms appearing in the $3 + 1$ gravitational field equations can be expressed in terms of b^a as follows:
>
> $$\rho_{\text{em}} = b^2 \left(W^2 - \frac{1}{2} \right) - (\alpha b^t)^2, \qquad (5.152)$$
>
> $$S^{\text{em}}_i = b^2 u_i W - \alpha b^t b_i, \qquad (5.153)$$
>
> $$S^{\text{em}}_{ij} = b^2 \left(u_i u_j + \frac{1}{2} \gamma_{ij} \right) - b_i b_j, \qquad (5.154)$$
>
> $$S_{\text{em}} = b^2 \left(\gamma^{ij} u_i u_j + \frac{3}{2} \right) - \gamma^{ij} b_i b_j. \qquad (5.155)$$

Equation (5.12) expressing baryon conservation and equation (5.13) expressing energy conservation for a perfect gas (or equation 5.19 for a Γ-law EOS) remain unchanged. The reason that the energy equation is unchanged is that there is no Joule heating by the electromagnetic field in the MHD limit (see exercise 5.31 below).

> **Exercise 5.31** (a) Use equations (5.116), (5.142) and (5.143) to show that the equations of motion of the electromagnetic field satisfy
>
> $$\nabla_a T^{ab}_{\text{em}} = -F^{bc} J_c. \qquad (5.156)$$
>
> (b) Use this relation to show that, in the presence of an electromagnetic field, equation (5.13) becomes
>
> $$\partial_t (\gamma^{1/2} E) + \partial_j (\gamma^{1/2} E v^j) = $$
> $$-P \left(\partial_t (\gamma^{1/2} W) + \partial_i (\gamma^{1/2} W v^i) \right) - \left(\alpha \gamma^{1/2} \right) u_b F^{bc} J_c, \qquad (5.157)$$
>
> while equation (5.19) becomes
>
> $$\partial_t (\gamma^{1/2} E_*) + \partial_j (\gamma^{1/2} E_* v^j) = -u_b F^{bc} J_c \left(\frac{E_*}{W} \right)^{(1-\Gamma)} \left(\frac{\alpha \gamma^{1/2}}{\Gamma} \right). \qquad (5.158)$$
>
> (c) Argue that, in the case of a perfect conductor, the terms involving J^a in the above equations vanish.

The equation of momentum conservation may be written in terms of the 4-vector S_a^*, a generalization of the momentum-density S_a defined in equation (5.10) in the absence of a magnetic field:

$$S_a^* = (\rho_0 h + b^2)W u_a. \tag{5.159}$$

Exercise 5.32 Show that S_i, the momentum density appearing as a source term in the $3+1$ gravitational field equations and defined in equation (5.1), is given by

$$S_i = (\rho_0 h + b^2)W u_i - \alpha b_i b^t, \tag{5.160}$$

and is thus *not* equal to S_i^* in general.

The equation of momentum conservation, obtained from $\nabla_a T^a{}_b = 0$, is then a generalization of equation (5.14) and may be written in the form[46]

$$\partial_t \left(\gamma^{1/2} \left(S_i^* - \alpha b_i b^t \right) \right) + \partial_j \left(\gamma^{1/2} \left(S_i^* v^j - \alpha b_i b^j \right) \right) =$$
$$- \alpha \gamma^{1/2} \left(\partial_i \left(P + \frac{b^2}{2} \right) + \frac{1}{2} \left(\frac{S_a^* S_c^*}{\alpha S^{*t}} - \alpha b_a b_c \right) \partial_i g^{ac} \right). \tag{5.161}$$

There are alternative ways to cast the basic MHD equations, all of which are equivalent analytically but are quite different when implemented numerically. For example, another way of writing the momentum conservation equation is to use equation (5.156) in the form $\nabla_b T_{\text{fluid}}^{ab} = -\nabla_b T_{\text{em}}^{ab} = F^{ab} J_b$ to get[47]

$$\partial_t(\gamma^{1/2} S_i^{\text{fluid}}) + \partial_j(\gamma^{1/2} S_i^{\text{fluid}} v^j) =$$
$$- \alpha \gamma^{1/2} \left(\partial_i P + \frac{S_a^{\text{fluid}} S_b^{\text{fluid}}}{2\alpha S_{\text{fluid}}^t} \partial_i g^{ab} \right) + \alpha \gamma^{1/2} F_{ia} J^a, \tag{5.162}$$

where S_a^{fluid} is given by equation (5.10). The above expression exhibits the relativistic generalization of the familiar $\mathbf{J} \times \mathbf{B}$ Newtonian force term on the right-hand side.

Exercise 5.33 Show that the electromagnetic term appearing on the right-hand side of equation (5.162) may be expanded in terms of the normal field components E^i and B^i to give

$$\alpha \gamma^{1/2} F_{ia} J^a = \alpha \frac{\gamma^{1/2}}{4\pi} E_i(D_j E^j) \tag{5.163}$$
$$- \frac{\gamma^{1/2}}{4\pi} B^j \left(\epsilon_{jik}(\partial_t E^k - \beta^l \partial_l E^k + E^l \partial_l \beta^k - \alpha K E^k) + \partial_i(\alpha B_j) - \partial_j(\alpha B_i) \right).$$

Note that there is a nasty time derivative of the electric field appearing on the right-hand side of equation (5.163). This is probably the most challenging term in equation (5.162) for numerical implementation, although it is $\mathcal{O}(v^2/c^2)$ times smaller than the last two terms

---

[46] De Villiers and Hawley (2003).
[47] Wilson (1975); Sloan and Smarr (1985); Baumgarte and Shapiro (2003b).

on the right-hand side of equation (5.163) and is likely to be small in most applications. In such cases, it may be useful to invoke "operator splitting" to estimate this term by extrapolating from the two previous timesteps, integrate the resulting system of equations, use the result to improve the estimate of this term, and then iterate. Such an approach, or some other low-order approximation, may be adequate to account for the contribution of this term. Alternatively, it simply might be preferable to use equation (5.161) in lieu of equation (5.162) to avoid the extra time derivative term.

> **Exercise 5.34** Here you will take the Newtonian limit to recover a familiar expression for the Newtonian MHD equation of momentum conservation. Take $g_{00} \rightarrow -(1 + 2\Phi)$, where Φ is the Newtonian potential, to find
>
> $$\frac{1}{2} \frac{S_a^{\text{fluid}} S_b^{\text{fluid}}}{\alpha S_{\text{fluid}}^t} \partial_i g^{ab} \rightarrow -\frac{1}{2} \rho \partial_i g^{00} = \rho \partial_i \Phi. \qquad (5.164)$$
>
> Then evaluate equation (5.162) in the Newtonian limit to obtain in Cartesian coordinates ($\gamma^{1/2} = 1$)
>
> $$\partial_t S_i^{\text{fluid}} + \partial_j (S_i^{\text{fluid}} v^j)$$
>
> $$= -\partial_i P - \rho \partial_i \Phi + \rho_e E_i - \frac{1}{8\pi} \partial_i (B^j B_j) + \frac{1}{4\pi} B^j \partial_j B_i, \qquad (5.165)$$
>
> where $S_{\text{fluid}}^i \rightarrow \rho v^i$, or, equivalently,
>
> $$\rho \frac{dv^i}{dt} = -\partial_i (P + P_{\text{M}}) - \rho \partial_i \Phi + \frac{1}{4\pi} B^j \partial_j B^i + \rho_e E^i. \qquad (5.166)$$
>
> Here we have introduced the Lagrangian time derivative $d/dt = \partial_t + v^j \partial_j$, defined the magnetic pressure
>
> $$P_{\text{M}} \equiv \frac{B^2}{8\pi}, \quad B^2 \equiv B_j B^j, \qquad (5.167)$$
>
> and used Maxwell's constraint equation $\nabla_i E^i = 4\pi \rho_e$ for the electric field. Note that for a neutral plasma $\rho_e = 0$, so that the electric field E^a disappears entirely from the above Newtonian equation.

Neither equation (5.161) nor equation (5.162) is in flux-conservative form. To obtain the equation in conservative form we must evolve the total stress-energy tensor T^{ab} directly and solve algebraically for the primitive variables.[48] The desired system has the identical form as equations (5.26)–(5.29) derived earlier, only now the total stress-energy tensor T^{ab} is given by equations (5.118) and (5.151), and $\tilde{S}_j = \gamma^{1/2} S_j$ employs equation (5.160) for the total momentum density. The goal of the numerical evolution is to integrate the combined system (5.26) and (5.140) for the conservative fluid and magnetic field variables $\mathcal{U} = (\tilde{D}, \tilde{S}_j, \tilde{\tau}, \tilde{B}^i)$, where $\tilde{B}^i \equiv \gamma^{1/2} B^i = \mathcal{B}^i$, and then combine these variables at each

---

[48] Koide *et al.* (1999); Komissarov (1999); Gammie *et al.* (2003); Duez *et al.* (2005b); Shibata and Sekiguchi (2005); Del Zanna *et al.* (2007).

Box 5.1 The relativistic MHD equations

The coupled set of relativistic MHD equations can be written in conservative form as follows:

$$\partial_t \rho_* + \partial_j(\rho_* v^j) = 0, \tag{5.168}$$

$$\partial_t \tilde{S}_i + \partial_j(\alpha\sqrt{\gamma}\, T^j{}_i) = \frac{1}{2}\alpha\sqrt{\gamma}\, T^{ab}g_{ab,i}, \tag{5.169}$$

$$\partial_t \tilde{\tau} + \partial_i(\alpha^2\sqrt{\gamma}\, T^{0i} - \rho_* v^i) = s_{\tilde{\tau}}, \tag{5.170}$$

$$\partial_t \tilde{B}^i + \partial_j(v^j \tilde{B}^i - v^i \tilde{B}^j) = 0, \tag{5.171}$$

where

$$\rho_* = \gamma^{1/2}\alpha u^t \rho_0 = \gamma^{1/2} D, \quad T^{ab} = (\rho_0 h + b^2)u^a u^b + (P + b^2/2)g^{ab} - b^a b^b, \tag{5.172}$$

$\tilde{\tau}$ is given by equation (5.27), $s_{\tilde{\tau}}$ is given by equation (5.32), $\tilde{S}_i = \gamma^{1/2} S_i$ is given by equation (5.160) and where

$$b^a \equiv B^a_{(u)}/(4\pi)^{1/2}, \quad b^2 \equiv b^a b_a. \tag{5.173}$$

To obtain the comoving magnetic field $B^a_{(u)}$, hence b^a, from the evolved normal field B^a, where $\tilde{B}^i \equiv \gamma^{1/2} B^i = \mathcal{B}^i$, introduce the projection operator $P_{ab} = g_{ab} + u_a u_b$ to write

$$B^a_{(u)} = -\frac{P^a{}_b B^b}{n_c u^c}, \tag{5.174}$$

from which one obtains

$$B^0_{(u)} = u_i B^i/\alpha, \quad B^i_{(u)} = \frac{B^i/\alpha + B^0_{(u)} u^i}{u^t}. \tag{5.175}$$

For a nonmagnetic gas, the above MHD equations reduce to the equations of relativistic hydrodynamics in conservative form given by equations (5.26)–(5.29).

---

time step to solve algebraically for the primitive variables $\mathbf{P} = (\rho_0, v^i, P, B^i)$. We collect the coupled set of MHD evolution equations in conservative form[49] in Box 5.1.

As in the nonmagnetic case, highly accurate shock-capturing methods can be applied to solve this set of equations. No artificial viscosity is needed, in contrast to the case for nonconservative schemes. However, recovering $\mathbf{P}$ by inverting the system of algebraic equations $\mathcal{U} = \mathcal{U}(\mathbf{P})$ can be computationally expensive. Finally, Einstein's equations in $3 + 1$ form must be integrated simultaneously to determine the spacetime metric.

Tests

Komissarov[50] has proposed a suite of challenging one-dimensional tests for relativistic MHD codes in Minkowski spacetime. The tests involve the propagation of nonlinear,

---

[49] Duez *et al.* (2005b).
[50] Komissarov (1999).

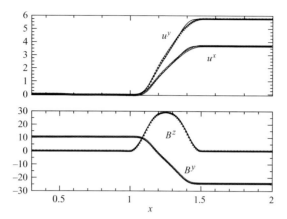

Figure 5.2 Nonlinear Alfvén wave test for an MHD fluid obeying a Γ-law EOS with $\Gamma = 4/3$. Symbols show simulation results with resolution $\Delta x = 0.01$ and solid lines give the exact solution. The profiles are shown at time $t = t_{\text{final}} = 2.0$. The computational domain is $x \in (-2, 2)$. Only the region $0.3 \le x \le 2.0$ is shown in this graph. [From Duez *et al.* 2005b.]

relativistic MHD waves in a gas obeying a Γ-law EOS. Most of the tests start with discontinuous initial data at $x = 0$, with homogeneous profiles in either half-space. Some of the tests involve shocks, others represent rarefaction waves. Exact solutions exist for all the results[51] and can be used for code calibration.

One useful test for a relativistic MHD code in flat spacetime is the propagation of a nonlinear Alfvén wave, which is a transverse hydromagnetic wave. Such a wave travels along magnetic field lines similar to the way in which a wave propagates along an elastic string under tension when it is plucked. The initial data for this test consist of left ($x < -W/2$) and right ($x > W/2$) fluid states separated by a width $W = 0.5$ at $t = 0$. The two states are joined by continuous functions in the region $x \in (-W/2, W/2)$ at $t = 0$.[52] The resulting fluid-field pattern propagates with a constant speed in the x-direction. Figure 5.2 shows the results of a simulation using a HRSC scheme that evolves both the fluid and the radiation field relativistic MHD code.

There are no discontinuities in this problem, so errors should converge to second order in Δx for a code that is second-order accurate.[53] To demonstrate this, consider a grid function g with error $\delta g = g - g^{\text{exact}}$. Calculate the L1 norm of δg (the "average" of δg) by summing over every grid point i:

$$L1(\delta g) \equiv \Delta x \sum_{i=1}^{N} |g_i - g^{\text{exact}}(x_i)|, \tag{5.176}$$

where $N \propto 1/\Delta x$ is the number of grid points. Figure 5.3 shows the L1 norms of the errors in u^x, u^y, B^y and B^z at $t = t_{\text{final}} = 2.0$. From the figure we conclude that the errors in u^x,

---

[51] See, e.g., Komissarov (1999); Cabannes (1970).
[52] See Komissarov (1997) or Duez *et al.* (2005b) for details of the setup of initial data together with an analytic solution.
[53] See Chapter 6.4 for a discussion of code validation and convergence.

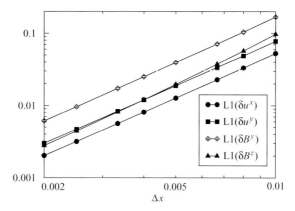

Figure 5.3 L1 norms of the errors in u^x, u^y, B^y and B^z for the nonlinear Alfvén wave test at $t = t_{\text{final}} = 2.0$. This log-log plot shows that the L1 norms of the errors in u^x, u^y and B^y are proportional to $(\Delta x)^2$, and are thus second-order convergent. The error in B^z goes as a slightly higher power of Δx. [From Duez *et al.* 2005b.]

u^y and B^y converge at second order in Δx. The error in B^z converges at slightly better than second order in Δx.

Next, for many applications in numerical relativity it is important to test that a relativistic MHD code can evolve the equations accurately in the strong gravitational field of a black hole. One such test checks that the code can maintain stationary, adiabatic, spherical accretion onto a Schwarzschild black hole, in accord with the relativistic Bondi accretion solution.[54] In particular, the relativistic Bondi solution is unchanged in the presence of a divergence-free radial magnetic field. This solution provides a useful diagnostic both for a numerical scheme that evolves the MHD equations in a fixed background metric as well as for one that evolves the spacetime metric self-consistently.[55]

To test the capability of a code to handle truly dynamical gravitational and MHD fields simultaneously, one can consider a gravitational wave oscillating in an initially homogeneous, uniformly magnetized fluid. The gravitational wave will, in general, induce Alfvén and magnetosonic waves.[56] Duez *et al.* (2005a) have performed a detailed analysis of this problem and provide an analytic solution for the perturbations in a form which is suitable for comparison with numerical results.

This test problem is 1-dimensional. Consider a linear, standing gravitational wave whose amplitude varies in the z-direction:

$$h_+(t, z) = h_{+0} \sin kz \cos kt, \qquad (5.177)$$

$$h_\times(t, z) = h_{\times 0} \sin kz \cos kt, \qquad (5.178)$$

---

[54] Shapiro and Teukolsky (1983), Appendix G.
[55] See De Villiers and Hawley (2003) and Gammie *et al.* (2003) for implementations of the Bondi test employing a fixed background metric, and Duez *et al.* (2005b) for an implementation in which the metric is evolved.
[56] Moortgat and Kuijpers (2003, 2004); Källberg *et al.* (2004).

where k is the wave number, and h_{+0} and $h_{\times 0}$ are constants. Assume that at $t = 0$, the magnetized fluid is unperturbed:

$$P(0, z) = P_0, \qquad \rho_0(0, z) = \rho_0, \tag{5.179}$$

$$v^i(0, z) = 0, \qquad B^i(0, z) = B_0^i. \tag{5.180}$$

Subsequently, the gravitational wave excites the MHD modes of the fluid. The gravitational wave is unaffected by the fluid to linear order, and the metric perturbation, $h_{\mu\nu}(t, z)$, in the transverse-traceless (TT) gauge can be calculated from equations (5.177) and (5.178). The perturbations in pressure $\delta P(t, z)$, velocity $\delta v^i(t, z)$, and magnetic field $\delta B^i(t, z)$ can be computed analytically. The solution[57] remains valid as long as we are in the linear regime. It is a superposition of the three eigenmodes of the homogeneous system (Alfvén, slow magnetosonic and fast magnetosonic waves), plus a particular solution that oscillates at the frequency of the gravitational wave.

The numerical simulation[58] adopts geodesic slicing ($\alpha = 1$, $\beta^i = 0$). The fluid evolves with a Γ-law EOS with $\Gamma = 4/3$. The computational domain is $z \in (-1, 1)$ and spans two wavelengths of the gravitational wave ($k = 2\pi$) and is covered by 200 grid points in the z-direction. At time $t = 0$, the metric is given by $g_{ab}(0, z) = \eta_{ab} + h_{ab}(0, z)$, where $\eta_{ab} = \mathrm{diag}(-1, 1, 1, 1)$ is the Minkowski metric, and the nonzero components of $h_{ab}(0, z)$ are

$$h_{xx}(0, z) = -h_{yy}(0, z) = h_+(0, z), \tag{5.181}$$

$$h_{xy}(0, z) = h_{yx}(0, z) = h_\times(0, z). \tag{5.182}$$

Periodic boundary conditions are enforced on both the matter and gravitational field quantities at the upper and lower boundaries in z.

Figure 5.4 shows a comparison between the analytic solution and numerical simulation for three selected perturbed variables. The simulation employs the same HRSC relativistic MHD code used to generate Figures 5.3 and 5.4 and couples it to a general relativistic $3 + 1$ BSSN scheme to evolve the metric.[59] Good agreement is shown for the MHD variables over many periods of the gravitational wave. Good agreement is also found for the metric perturbations. The pressure perturbation, however, is seen to differ from the analytic solution by a slight secular drift. (In fact, all variables eventually exhibit a drift away from the analytic solution, but the drift is first noticeable in the case of the pressure.)

This secular drift is *not* due to numerical error, but rather is an effect of the nonlinear terms which are neglected in the analytic solution. To demonstrate that it is not a numerical error, simulations at resolutions of 50, 100, and 200 grid points were performed, and convergence was obtained to second order to a solution with nonzero drift. Since the discrepancy is due to nonlinear terms, choosing smaller initial mass-energy density and

---

[57] Duez *et al.* (2005a).
[58] Duez *et al.* (2005b).
[59] See Chapter 11.5 for a description of the BSSN scheme for the evolution of the gravitational field.

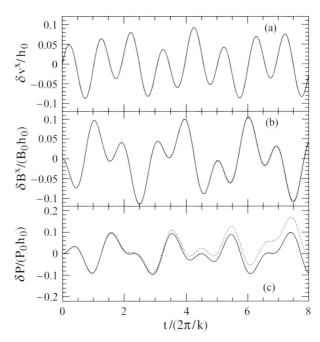

Figure 5.4 Analytic and numerical solutions for the perturbations of a magnetized fluid due to the presence of a linear, standing gravitational wave. The thick solid and thin dotted lines represent, respectively, the analytic and numerical solutions, though the two lines are not readily distinguishable in plots (a) and (b). All quantities are evaluated at $z = 1/8$ and are normalized as indicated. Time is normalized by the gravitational wave period. [From Duez *et al.* (2005b).]

smaller gravitational wave strength leads to a smaller discrepancy with the analytic solution, as was also demonstrated. Finally, the various MHD modes were extracted from the output via velocity projections and the code was shown to correctly represent all three MHD waves, plus the contribution from the particular solution.

It is not trivial to devise a simple test problem to calibrate the ability of a general relativistic code to evolve an electromagnetic field in a strongly-gravitating, *dynamical* spacetime in which a black hole forms. It is harder still to devise for such a spacetime a problem that calibrates a code's performance both in the highly conducting MHD regime *and* the nonconducting vacuum regime. But consider the following scenario for the collapse of a magnetized spherical star to a black hole: Adopt the approximation that the electromagnetic fields are sufficiently weak that the matter and gravitational fields can be described by the unperturbed Oppenheimer–Snyder solution for the collapse of a spherical, homogeneous dust ball to a Schwarzschild black hole (see Chapter 1.4.) Assume the matter to be perfectly conducting and threaded by a nearly uniform interior and a dipole exterior magnetic field at the onset of collapse. Determine the subsequent evolution of the interior and exterior magnetic and electric fields without approximation: calculate the fields analytically in the matter interior, assuming the MHD equations, and numerically in the vacuum exterior. Apply the familiar electromagnetic junction conditions to match the fields across the stellar surface at each instant of time in order to determine the inner boundary conditions for the integration of exterior electromagnetic fields.

Exercise 5.35 Here we seek the relativistic generalization of Newtonian initial data describing a uniformly magnetized dust ball of radius R, matched onto an exterior dipole magnetic field. In orthonormal spherical polar coordinates, the Newtonian interior is characterized by the constant magnetic field

$$B^{\hat{i}} = B(\cos\theta, -\sin\theta, 0), \quad \text{(Newtonian interior)} \quad (5.183)$$

assuming the field is aligned with the z-axis, while the exterior is a pure dipole field given by

$$B^{\hat{i}} = \frac{BR^3}{r^3}(\cos\theta, \frac{1}{2}\sin\theta, 0) \quad \text{(Newtonian exterior)}. \quad (5.184)$$

The junction conditions on the surface of the star require that the normal (i.e., radial) component of the field be continuous across the stellar surface. The tangential discontinuity in B^i implies the presence of a surface current.
(a) Use the OS interior metric (1.88) to show that a relativistic generalization of the uniform interior field may be specified by

$$B^{\hat{\chi}} = B\cos\theta \left(\frac{\chi}{\sin\chi}\right)^2 \left(\frac{\chi_s}{\sin\chi_s}\right)^{-2}, \quad (5.185)$$

$$B^{\hat{\theta}} = -B\sin\theta \left(\frac{\chi}{\sin\chi}\right) \left(\frac{\chi_s}{\sin\chi_s}\right)^{-2}. \quad (5.186)$$

In particular, show that the above field satisfies the constraint $D_i B^i = 0$ and reduces to the uniform interior Newtonian field in stars of low compaction.
(b) Use the Schwarzschild metric (1.51) together with Maxwell equations (5.142) and (5.143) in vacuum to show that the relativistic generalization of the dipole exterior field is given by[60]

$$B_{\hat{r}} = -\frac{6\mu_d \cos\theta}{r_s^3} x_s^2 \left(x_s \ln(1 - x_s^{-1}) + (1 + x_s^{-1}/2)\right), \quad (5.187)$$

$$B_{\hat{\theta}} = \frac{6\mu_d \cos\theta}{r_s^3} x_s^2 \left(x_s(1 - x_s^{-1})^{1/2} \ln(1 - x_s^{-1}) + \frac{1 - x_s^{-1}/2}{(1 - x_s^{-1})^{1/2}}\right). \quad (5.188)$$

Here r_s is the Schwarzschild (areal) radius, $x_s \equiv r_s/(2M)$, and μ_d is the magnetic dipole moment.
Hint: Introduce a vector potential A_a via $F_{ab} = \nabla_b A_a - \nabla_a A_b$ and solve for A_a for an exterior vacuum dipole field in axisymmetry.
(c) Use the junction conditions to show that μ_d is given by

$$\mu_d = -B\frac{R_s^3}{6X_s^2} \left(X_s \ln(1 - X_s^{-1}) + (1 - X_s^{-1}/2)\right)^{-1}, \quad (5.189)$$

---

[60] Wasserman and Shapiro (1983).

where R_s and X_s are the values of r_s and x_s on the stellar surface, and where B is the factor appearing in equations (5.185) and (5.186).

Exercise 5.36 Show that the evolution of the interior magnetic field in exercise 5.35 is given analytically by equations (5.185) and (5.186) all during OS collapse, provided the parameter B varies with proper time τ according to

$$B(\tau) = B_0 \left(\frac{\rho_0(\tau)}{\rho_0(0)} \right)^{2/3}, \qquad (5.190)$$

where $\rho_0(\tau)$ is given by equation (1.96).

The solution to the "magnetized OS collapse" problem formulated above has been determined[61] and this simple scenario has been used to experiment with several alternative time slicing conditions for handling black hole formation and the associated appearance of singularities. The choice of time slicing is important for enabling the evolution of the exterior electromagnetic field to late times. The choices considered ranged from "singularity avoiding" time coordinates, like maximal time slicing, to "horizon penetrating" time coordinates, like Kerr–Schild slicing, accompanied by "black hole excision".[62] The latter choice allows for the integraton of the exterior electromagnetic fields arbitrarily far into the future. At late times the longitudinal magnetic field in the exterior transforms into a transverse electromagnetic wave; part of the electromagnetic radiation is captured by the hole and the rest propagates outward and escapes. The field pattern at various times is shown in Figures 5.5 and 5.6. The electromagnetic field strength at a fixed exterior radius dies out as t^{-4}, which agrees with the $t^{-(2l+2)}$ decay rate expected from perturbation theory[63] and applicable here to an $l=1$ dipole field threading a spherical star undergoing gravitational collapse (see Figure 5.7).

5.3 Collisionless matter

Several important astrophysical systems are made up of particles of *collisionless* matter. In such systems, the mean-free-path for particle-particle interactions is much longer than the scale of the system. Equivalently, the mean time for particle collisions is much longer than the dynamical or "crossing" time scale of the system, i.e., the time it takes for a particle to cross from one side of the system to the other. Systems of particles obeying this condition are in the opposite limit from the hydrodynamical gases we treated above. One example of a collisionless system is a star cluster, a large, self-gravitating, N-body system in which the individual particles – the stars – interact exclusively via gravitation. In the strictly collisionless limit, a cluster of finite total mass is treated as an infinite swarm of point particles, each of infinitesimal mass. The stars then move in the smooth, background gravitational field

[61] Baumgarte and Shapiro (2003a).
[62] For a discussion of black hole excision during stellar collapse, see Chapter 14.2.3.
[63] Price (1972b).

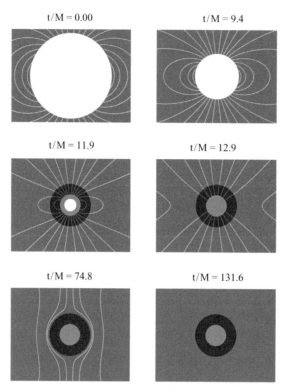

t/M = 0.00   t/M = 9.4

t/M = 11.9   t/M = 12.9

t/M = 74.8   t/M = 131.6

Figure 5.5 Snapshots of the exterior magnetic field lines on select Kerr–Schild time slices for a homogeneous, conducting dust ball of mass M that collapses from rest from an initial areal radius $R_s(0) = 4M$. Points are plotted in a meridional plane in areal radius. The white shaded sphere covers the matter interior; the black shaded area covers the region inside the event horizon; the gray shaded area covers the region inside $r_s = M$ that is excised from the numerical grid once the surface passes inside. The initial growth of the (longitudinal) field is due to flux freezing in the interior and is followed by a burst of (transverse) electromagnetic radiation in the vacuum exterior once the surface approaches the horizon at $r_s = 2M$. In this time coordinate, using excision, the exterior electromagnetic field can be evolved reliably to arbitrary late times. By the end of the integration, all exterior electromagnetic fields in the vicinity of the black hole have been captured or radiated away. [From Baumgarte and Shapiro (2003a).]

established by their cumulative, smooth mass-energy distribution. In reality, the distribution of stars in a cluster is not continuous, but discrete, so that the gravitational field governing their motion is granular. Consequently, the stars undergo stochastic, small-angle deflections from their smooth orbits due the cumulative role of many distant "gravitational encounters" (i.e., small-angle, Coulomb scattering). Very occasionally, stars may even wander close to their neighbors and experience large-angle, gravitational deflections. Moreover, since stars are not point particles and have finite sizes, they can even undergo contact collisions with other stars. Typically, gravitational scattering due to the granularity of the gravitational field is not important in the evolution of a large star cluster over a dynamical (orbital) time scale, but only on a much longer, "relaxation" time scale. Physical collisions

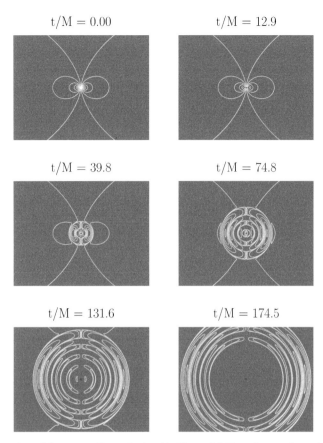

Figure 5.6 Far-zone view of the same collapse depicted in Figure 5.5. Note the transformation of the dipole field from a quasistatic longitudinal field to a transverse electromagnetic wave. [From Baumgarte and Shapiro (2003a).]

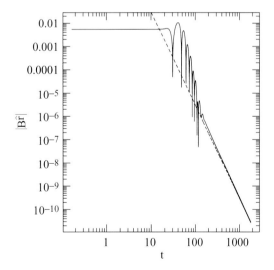

Figure 5.7 The absolute value of the orthonormal component of the radial magnetic field $|B^{\hat{r}}|$ at $r_s = 20M$ as a function of time. The dashed line is a power law varying like t^{-4}, the expected decay rate for an initial static dipole electromagnetic field in a spherical star undergoing collapse to a black hole. [From Baumgarte and Shapiro (2003a).]

resulting from direct impact are also unimportant on dynamical timescales in astrophysical star clusters. For realistic systems, the star cluster may be treated as a perfectly collisionless gas over dynamical time scales, while the effects of gravitational scattering or collisions can be handled as perturbations if the evolution is tracked over longer, secular time scales.[64]

Globular clusters consisting of $N \sim 10^5$ stars, galaxies with $N \sim 10^{12}$ stars and clusters of galaxies consisting of $N \sim 10^3$ galaxies are all familiar examples of nearly collisionless clusters as defined above. Another important example of a collisionless gas includes cosmological "dark matter", which comprises over 90% of the mass in the Big Bang Universe. Dark matter halos, which might be clusters of neutrinos, neutralinos, axions, or some other weakly interacting massive particle, are collisionless clusters. Here we shall be particularly interested in relativistic systems. Astrophysical realizations of collisionless relativistic systems include relativistic star clusters (including large-N clusters of compact stars and (or) black holes) and collisionless dark matter halos consisting of relativistic particles.

Readers may find it useful to be familiar with the elementary principles of relativistic kinetic theory before reading onward,[65] although the discussion below is reasonably self-contained.

In kinetic theory, a large N-body system is described by a dimensionless phase-space distribution function $f(x^a, p^a)$. Assume for the moment that all particles in the system have the same rest mass. Then the function f is is related to the number density of particles in 6-dimensional phase space according to

$$f = \frac{dN}{d^3\hat{x}\,d^3\hat{p}}, \tag{5.191}$$

where we have expressed the volume elements in physical space, $d^3\hat{x}$, and in momentum space, $d^3\hat{p}$, in an orthonormal basis. The product $d^3\hat{x}d^3\hat{p}$, as well as the distribution function f, is a Lorentz invariant. The distribution function f satisfies the relativistic *Boltzmann equation*,

$$\frac{\mathcal{D}f}{d\lambda} \equiv \left(\frac{dx^a}{d\lambda}\right)\frac{\partial f}{\partial x^a} + \left(\frac{dp^a}{d\lambda}\right)\frac{\partial f}{\partial p^a} = \left(\frac{\delta f}{\delta \lambda}\right)_{\text{coll}}. \tag{5.192}$$

Here the derivative is along the particle trajectory in phase space and λ is an affine parameter along that trajectory defined so that the momentum is given by

$$p^a \equiv \frac{dx^a}{d\lambda}. \tag{5.193}$$

For particles of finite rest mass m, we have $\lambda = \tau/m$, where τ is the particle proper time. The Boltzmann equation (5.192) is a continuity equation in phase space, which says that particles are created and destroyed along their trajectories in phase space by virtue of

---

[64] See, e.g., Lightman and Shapiro (1978), Spitzer (1987), and Binney and Tremaine (1987), and references therein, for detailed discussions of star cluster dynamics and evolution, and the role of gravitational encounters and collisions, in Newtonian theory.

[65] See, e.g., Misner *et al.* (1973), Section 22.6.

collisions, the rate of which is given by the term on the right hand side. Excluding all long-range forces other than gravitation, so that the particle paths are geodesics except at occasional points where scattering or collisions occur, we then have

$$\frac{dp^a}{d\lambda} = -\Gamma^a{}_{bc} p^b p^c. \tag{5.194}$$

Exercise 5.37 Show that the Boltzmann equation may be written in the form

$$\frac{\mathcal{D}f}{d\lambda} = p^a \frac{\mathcal{D}f}{dx^a} = \left(\frac{\delta f}{\delta \lambda}\right)_{\text{coll}}, \tag{5.195}$$

where

$$\frac{\mathcal{D}}{dx^a} \equiv \frac{\partial}{\partial x^a} - \Gamma^b{}_{ac} p^c \frac{\partial}{\partial p^b}. \tag{5.196}$$

Exercise 5.38 Derive the relativistic equation of radiation transport (5.93) from the Boltzmann equation (5.195), accounting for photon emission and absorption.[66] To start, use $f \propto I_\nu / \nu^3$ to show that in an arbitrary frame the transfer equation can be cast in the form

$$p^a \frac{\mathcal{D}(I_\nu / \nu^3)}{dx^a} = \left(\frac{\delta(I_\nu / \nu^3)}{\delta \lambda}\right)_{\text{coll}} = e - a(I_\nu / \nu^3), \tag{5.197}$$

where e is an invariant source term due to photon emission and a is an invariant sink term due to photon absorption. Setting $h = 1 = c$, evaluate equation (5.197) in an inertial frame to recover the familiar form of the equation of radiation transfer,

$$\frac{\partial I_\nu}{\partial t} + n^i \frac{\partial I_\nu}{\partial x^i} = \eta_\nu - \chi_\nu I_\nu, \tag{5.198}$$

where we identify e as the invariant emissivity and a as the invariant absorption coefficient (opacity) according to

$$e = \eta_\nu / \nu^2, \qquad a = \nu \chi_\nu. \tag{5.199}$$

Focus now on physical scenarios in which collisions or scatterings are either absent entirely or unimportant on time scales of interest. In that case the right hand side of equation (5.192) may be set equal to zero; the resulting equation is usually referred to as the *collisionless Boltzmann equation*, or the *Vlasov equation*:

$$\frac{\mathcal{D}f}{d\lambda} = 0. \tag{5.200}$$

This equation shows that the distribution function in a collisionless gas is conserved along the orbit of each particle in phase space, which is known as *Liouville's theorem*. Equation (5.200) is sometimes called the *Liouville equation*.[67]

---

[66] See Lindquist (1966), Castor (1972) and Mihalas and Mihalas (1984) for a detailed discussion; see Shapiro and Teukolsky (1983), Appendix I, for a brief introduction.

[67] This result can also hold in a collisional gas, provided the collisions satisfy *detailed balancing*. Detailed balancing always applies, for example, to systems in thermodynamic equilibrium.

Let us cast the Liouville equation in a form most suitable for treating a swarm of collisionless particles with a wide distribution of masses.[68] Consider a small group of particles near a particular event x^a in spacetime, with 4-momentum near a particular value p_a. As seen in its own rest frame, this group occupies a 3-dimensional volume d^3V_x in physical space and a 4-dimensional volume d^4V_p in momentum space. In a general coordinate basis, d^3V_x and d^4V_p are given by

$$d^3V_x = \frac{p^t}{m}\sqrt{-g}dx^1dx^2dx^3 \tag{5.201}$$

$$d^4V_p = \frac{-dp_tdp_1dp_2dp_3}{\sqrt{-g}} = \frac{m}{p^t}\frac{dm\,dp_1dp_2dp_3}{\sqrt{-g}}, \tag{5.202}$$

where $m = -(p_ap^a)^{1/2}$ is the rest-mass of a particle with 4-momentum p_a. In deriving the second equality for d^4V_p, we differentiated this "mass hyperboloid" relation to write $m\,dm = -p^t\,dp_t$, which we then substituted into the first equality to change variables from p_t to m. If there are dN particles in the group, then the number density in phase space, or distribution function, F is defined by

$$F(x^a, p_b) = \frac{dN}{d^3V_xd^4V_p}. \tag{5.203}$$

It is often convenient to specify a point in phase space by the the set of coordinates (x^a, p_a). This set is particularly useful whenever the spacetime possesses a Killing vector ξ^a, in which case the scalar $p_a\xi^a$ is conserved along geodesics. When the Killing vector is a coordinate basis vector, say ∂_K, then the component of p_a along that basis vector, p_K, is a constant of the motion. Sometimes it is useful to employ other coordinate sets, such as (x^a, p_i, m).

For collisionless matter, the particles move along geodesics and the distribution function satisfies the Liouville equation in the form

$$\frac{\mathcal{D}}{d\tau}F = 0, \tag{5.204}$$

where the Liouville operator $\mathcal{D}/d\tau$ represents differentiation with respect to proper time along the trajectory of a particle in phase space:

$$\frac{\mathcal{D}}{d\tau} \equiv \left(\frac{dx^a}{d\tau}\right)\frac{\partial}{\partial x^a} + \left(\frac{dp_a}{d\tau}\right)\frac{\partial}{\partial p_a} = \frac{g^{ab}p_b}{m}\frac{\partial}{\partial x^a} - \frac{1}{2m}\frac{\partial g^{bc}}{\partial x^a}p_bp_c\frac{\partial}{\partial p_a}. \tag{5.205}$$

The smoothed-out stress-energy tensor of a cluster of particles is determined by the distribution function according to

$$\boxed{T^{ab} = \int p^ap^b\left(\frac{F}{m}\right)d^4V_p.} \tag{5.206}$$

---

[68] Our discussion is patterned after Ipser and Thorne (1968).

The smoothed-out gravitational field of the cluster of particles is given by the metric g_{ab}, which, as always, is a solution of Einstein's equations $G_{ab} = 8\pi T_{ab}$. To solve the field equations in $3 + 1$ form we need the matter source terms given by equations (5.1) (5.2), which are computed as quadratures over the distribution function F, using equation (5.206).

Exercise 5.39 Show that the source terms are given by

$$\rho = \int m(W^2 F) d^4 V_p, \tag{5.207}$$

$$S_i = \int m(W u_i F) d^4 V_p, \tag{5.208}$$

$$S_{ij} = \int m(u_i u_j F) d^4 V_p, \tag{5.209}$$

$$S = \rho - \int m F d^4 V_p, \tag{5.210}$$

where W is defined in equation (5.11).

For many calculations, it is adequate to consider systems in which all the particles have the same rest mass, m_0. The reason is that, according to the equivalence principle, all particles follow the same geodesic paths, regardless of their mass, so this is not a physically significant restriction for a collisionless gas. There are several equivalent ways of restricting our general description to particles of the same rest mass. One way is to write

$$F(x^a, p_b) = f(x^a, p_i)\delta(m - m_0). \tag{5.211}$$

The resulting 4-dimensional quadratures over F that determine the matter source terms now reduce to 3-dimensional quadratures over f. For example, inserting the the right hand side in equation (5.202) for $d^4 V_p$ gives

$$\rho = m_0 \int (W^2 f) \left(\frac{dp_1 dp_2 dp_3}{W \gamma^{1/2}}\right), \tag{5.212}$$

and similarly for the other source terms. This is equivalent to demanding that the invariant 4-dimensional volume, $d^4 V_p$, in momentum space obey the constraint $-p^a p_a = m_0^2$, which allows us to generate an invariant, 3-dimensional volume in momentum space,

$$d^3 V_p = \int d^4 V_p \delta[(-p_a p^a)^{1/2} - m_0] = \left(\frac{m_0}{p^t}\right)\frac{dp_1 dp_2 dp_3}{(-g)^{1/2}} = \frac{dp_1 dp_2 dp_3}{W \gamma^{1/2}}, \tag{5.213}$$

where the integral is over dp_t. When we now employ $d^3 V_p$ to integrate over the phase-space distribution function f, defined by

$$f(x^a, p_i) = \frac{dN}{d^3 V_x d^3 V_p}, \tag{5.214}$$

(recall equation 5.191), the stress-energy tensor defined in equation (5.206) becomes

$$T^{ab} = \int p^a p^b \left(\frac{f}{m_0} \right) d^3 V_p,$$

(5.215)

and the quadratures for the matter sources reduce to the form given in equation (5.212).

Exercise 5.40   Show that when the collisionless gas consists of a single mass species, the Liouville equation (5.204) reduces to

$$\frac{\partial f}{\partial t} + \left(\frac{dx^i}{dt} \right) \frac{\partial f}{\partial x^i} + \left(\frac{dp_i}{dt} \right) \frac{\partial f}{\partial p_i} = 0$$

(5.216)

or, explicitly,

$$\frac{\partial f}{\partial t} + \frac{g^{ia} p_a}{p^t} \frac{\partial f}{\partial x^i} - \frac{1}{2 p^t} \frac{\partial g^{ab}}{\partial x^i} p_a p_b \frac{\partial f}{\partial p_i} = 0.$$

(5.217)

For stationary spacetimes, the distribution function of a collisionless gas is independent of time and must be a function of the dynamical constants (or "integrals") of orbital motion. Likewise, any function of the constants of the motion yields a stationary solution of the collisionless Boltzmann equation. This result is often quoted as *Jeans' theorem*. This proposition is not surprising, looking at equation (5.216), since any f that is a function only of the momenta p_a conjugate to the Killing vectors ξ^a in the spacetime clearly provides a stationary solution (recall that $dp_a/dt = 0$). Jeans' theorem is particularly useful in constructing equilibrium solutions. Equilibrium solutions provide initial data for full scale dynamical evolution simulations, which are necessary to determine the dynamical stability of collisionless equilibria and the final fate of unstable systems.

For spherical, static equilibrium systems, the general solution to the Liouville equation depends on just two constants of the motion, the "energy at infinity" E and the total angular momentum J. Thus the distribution function has the form $F = F(E, J, m)$. Adopt a familiar Schwarzschild coordinate system (t, r, θ, ϕ) to describe the static spherical spacetime, whereby the metric may be written as

$$ds^2 = -e^{2\Phi} dt^2 + e^{2\Lambda} dr^2 + r^2 (d\theta^2 + \sin^2 \theta d\phi^2),$$

(5.218)

where Φ and Λ are functions of r and the energy and angular momentum are given by

$$E = -p_t,$$

(5.219)

$$J = \left(p_\theta^2 + \frac{p_\phi^2}{\sin^2 \theta} \right)^{1/2}.$$

(5.220)

Exercise 5.41   Show that E and J defined in equations (5.219) and (5.220) are constants of the motion in a static spherical spacetime.

These are the first steps in the construction of a spherically symmetric equilibrium configuration. The dynamical equation for the matter, the Liouville equation, is automatically solved given $F = F(E, J, m)$. However the stress-energy tensor (5.206), as well as the matter source terms given in equations (5.226)–(5.229) below, cannot be determined from F until the metric is known. The metric, in turn, is determined by solving Einstein's field equations, which requires the stress-energy tensor. The task of constructing an equilibrium configuration therefore amounts to solving a set of coupled integro-differential equations self-consistently. The equations are quite straightforward to solve in the case of spherical symmetry, especially when the distribution of stellar velocities at each point in the cluster is isotropic. In this case the equations can be made to look like the familiar OV equations for fluid stars in spherical equilibrium.[69]

> **Exercise 5.42**   Show that the velocity profile of particles in a spherical cluster in static equilibrium is isotropic as measured by a local observer whenever the distribution function is independent of angular momentum, i.e., $F = F(E, m)$.

The problem is conceptually similar, but quite a bit more complicated, in the case of stationary axisymmetric equilibria. Nonspherical clusters with and without rotation are included in this category. The simplest distribution functions that can give rise to nonspherical, axisymmetric equilibria are of the form $F = F(E, J_z, m)$, where E is the particle energy and J_z is the particle angular momentum along the symmetry axis. By suitable choice of the distribution in J_z, one can construct models that are either prolate or oblate. A depletion in the J_z-distribution produces a prolate configuration, while an enhancement produces an oblate one.[70] We shall return to the construction of collisionless equilibria and their utility as initial data in Chapter 8.2.

Now we wish to discuss how the spacetime for a relativistic cluster of collisionless matter can be determined when the cluster is undergoing dynamical evolution, including catastrophic collapse to a black hole. The basic goal is to solve the Liouville equation for the matter simultaneously with the $3 + 1$ equations for the gravitational field. There are at least two different numerical strategies that have been employed successfully to track the dynamical evolution of relativistic clusters. We describe both of them in the sections immediately below. Applications of these techniques to problems involving collisionless matter in strong gravitational fields are treated in Chapters 8 and 10.

Particle methods

Particle methods, sometimes referred to as "particle-mesh" or "particle-in-cell" methods, are widely used in plasma physics and in Newtonian gravitation for the study of collisionless systems.[71] The method has been extended to general relativity, yielding a robust mean-field, N-body evolution scheme in the context of the $3 + 1$ equations that has been applied to spherical and axisymmetric clusters.[72] In this approach, a statistical representation of

---

[69] See Misner *et al.* (1973), Section 25.7.
[70] See Shapiro and Teukolsky (1993a,b).
[71] See, e.g., Hockney and Eastwood (1981) and Sellwood (1987) for a discussion and review of applications to Newtonian gravitation and plasma physics.
[72] Shapiro and Teukolsky (1985a,b,c, 1986, 1992b) and references therein.

the initial distribution function is constructed by specifying initial positions and velocities for a large but finite number N of discrete particles. The motion of these particles is then calculated by integrating simultaneously N geodesic equations in the mean gravitational field of the system. The source terms of the $3+1$ equations are determined by smearing out each particle over a small spatial volume, and adding up the contributions of all particles.

Working with collisionless matter via particle simulations has several advantages over fluid systems. The collisionless matter equations are ordinary differential equations (the geodesic equations), while the hydrodynamic equations are more challenging partial differential equations. Furthermore, collisionless matter is not subject to shocks or other hydrodynamical discontinuities. These complications require special care or sophisticated handling of the fluid equations for adequate resolution, as we discussed in Section 5.2 of this chapter.

Consider a collisionless gas sampled by a discrete number of particles. Group together into distinct categories those particles at a point in spacetime that have the same mass and 4-velocity. Then the stress-energy tensor is

$$T^{ab} = \sum_A m_A n_A u_A^a u_A^b.$$ (5.221)

Here m_A is the rest-mass of category A particles, u_A^a is their 4-velocity, and n_A is their comoving number density.[73] The quantity n_A is proportional to a δ-function in the limit of point particles, but it can be treated as a continuous, averaged quantity when the number of particles is infinite, the mass of each particle becomes infinitesimal, and the particle distribution is smooth.

The equation of motion of the particles is simply the geodesic equation

$$u^a \nabla_a u^b = 0,$$ (5.222)

where we drop the label A when referring to any one of the particles.

Exercise 5.43 Show that equation (5.222) follows from the conservation of stress-energy, $\nabla_b T^{ab} = 0$, and the conservation of particles, $\nabla_a (n u^a) = 0$.

Exercise 5.44 Show that the equations describing the motion of a particle in phase space can be written in terms of the $3+1$ metric as

$$\frac{du_i}{dt} = -\alpha u^0 \partial_i \alpha + u_k \partial_i \beta^k - \frac{1}{2u^0} u_j u_m \partial_i \gamma^{jm},$$ (5.223)

$$\frac{dx^j}{dt} = \gamma^{jk} \frac{u_k}{u^0} - \beta^j,$$ (5.224)

$$\alpha u^0 = \left(1 + \gamma^{ij} u_i u_j \right)^{1/2}.$$ (5.225)

Exercise 5.45 Derive the Newtonian limit of the equations of motion in exercise 5.44.

---

[73] See, e.g., Misner *et al.* (1973) Section 5.4.

Hint: In the Newtonian limit, $\gamma^{ij} \to \eta^{ij}$, $\alpha \to 1 + \Phi$, where Φ is the gravitational potential, and $\beta^k \to 0$.

Thus, given the metric at any time t, equations (5.223)–(5.225) may be integrated for the new positions and 4-velocities of the particles at $t + \Delta t$. Once the new particle positions and velocities are determined, the new source terms needed in the field equations can be computed as a discrete sum over the particles using equations (5.207)–(5.210):

$$\rho = \sum_A m_A n_A W^2, \tag{5.226}$$

$$S_i = \sum_A m_A n_A W u_i^A \tag{5.227}$$

$$S_{ij} = \sum_A m_A n_A u_i^A u_j^A \tag{5.228}$$

$$S = \rho - \sum_A m_A n_A. \tag{5.229}$$

We regard each particle as being its own category A, so that the sum over A is a sum over particles. Then

$$n_A = \frac{1}{\Delta^3 V_x} = \frac{1}{W \gamma^{1/2} \Delta x^1 \Delta x^2 \Delta x^3}. \tag{5.230}$$

Here $\Delta^3 V_x$ is the invariant 3-volume as measured in a frame comoving with particle A (see equation 5.201). In writing the last equality in equation (5.230) we divided up coordinate space into bins of coordinate volume $\Delta x^1 \Delta x^2 \Delta x^3$ and we treated the particle as smeared out over the entire bin in which it resides. Note that more sophisticated binning and averaging algorithms can be constructed which assign fractions of particles to neighboring bins in order to calculate smoother and statistically more reliable source terms from the discrete particle sample.

Once the source terms have been assembled, the $3 + 1$ equations can be integrated to give the mean gravitational field quantities at the new time. The whole cycle can be repeated to evolve the particles to the next time step in the mean field, and so on.

Phase space methods

Sampling the phase space distribution function f by a discrete set of particles does not determine f, but only moments of f, like the source terms of the field equations.[74] Sometimes it is important to acquire the full knowledge of f to understand the underlying dynamical state of a collisionless system. Many complicated phenomena of great astrophysical interest, such as "violent relaxation", can only be understood by a study of the detailed evolution of f. Violent relaxation, which can lead to the virialization of a cluster

---

[74] The discussion in this section is patterned after Rasio *et al.* (1989b).

that is initially far from dynamical equilibrium, is the result of collisionless damping of perturbations by "phase mixing".[75] Studying violent relaxation requires that the phase space density be tracked in detail, including all distortions due to phase mixing.

A phase space method does not rely on any statistical representation of the system by particles. Instead, a phase space method explicitly constructs the smooth distribution function of matter in phase space. The source terms of the field equations are obtained by direct numerical quadratures of f over the momentum (or velocity) space, as in equation (5.212). This has the great advantage of eliminating random statistical fluctuations in the data while at the same time providing the full distribution function of the system. However, very few phase space methods have so far been developed successfully, even in the much simpler framework of Newtonian gravity. Part of the problem is the extreme complexity of working in phase space instead of physical space. The large number of dimensions in phase space (already three in spherical symmetry where physical space has only one) would already discourage many attempts. In addition, distribution functions often have irregular structures that can be hard to represent accurately on a numerical grid in phase space. Such irregular structures can arise from discontinuities in the initial data, but even in the case of very smooth initial data, phase mixing will usually produce increasingly intricate "fine-grained" structures.

One phase space method developed for general relativistic systems seems to work well, as least for the spherical systems to which it has been applied.[76] The method exploits Liouville's theorem to determine the evolution of the phase space distribution function directly. The method consists of three basic steps to propagate the distribution function f from time t_1 to $t_2 > t_1$: (1) compute the source terms for the field equations (e.g., equation 5.212) by integrating f at t_1 over momentum space; (2) evolve the $3 + 1$ field equations to determine the metric at t_2; (3) compute f at time t_2, assuming that it satisfies the Liouville equation (5.216). The key idea is to use Liouville's theorem for step (3), which can be written as

$$f(t_2, x_2, p_2) = f(t_1, x_1, p_1), \tag{5.231}$$

where (x_1, p_1) represents the position in phase space at time t_1 of a test particle that will actually reach the position (x_2, p_2) at time t_2. By interpolating the metric in the interval (t_1, t_2) one can integrate the equations of motion, equations (5.223)–(5.225), to construct the actual trajectory of such a test particle. Since f is known at all times $t \leq t_2$ one can actually determine $f(t_2, x_2, p_2)$ for all (x_2, p_2).

The original scheme along these lines was developed in Newtonian theory.[77] However, it was soon discovered that in many cases the scheme developed numerical instabilities, rendering the results inaccurate.[78] The reason was that in the original scheme f was

---

[75] Lynden-Bell (1967).
[76] Rasio *et al.* (1989b).
[77] Fujiwara (1981, 1983).
[78] Inagaki *et al.* (1984); Nishida (1986).

constructed on a grid of points in phase space. In employing equation (5.231) to determine f at t_2 by placing a particle at (x_2, p_2) and integrating backward in time from t_2 to t_1, the old position of the particle (x_1, p_1) is not a grid point at t_1 in general. As a result, one must perform a (multidimensional) interpolation on the grid to get $f(t_1, x_1, p_1)$ and thereby the new grid point value $f(t_2, x_2, p_2)$. The error in the interpolation on the grid at t_1 propagates to the new grid at t_2, where it becomes an error on the values of f at the grid points. This leads to an amplification of the error, which is reflected in the failure to conserve rest mass and other conserved quantities, and may even lead to the violation of the positivity of the distribution function. The problem is most severe when the distribution function exhibits discontinuities, or even a mild degree of phase mixing.

One cure adopted in the relativistic scheme that eliminates interpolation error is to extend the value of t_1 in equation (5.231) to $t = 0$, where the distribution function is given by the initial data and is therefore known everywhere to arbitrary accuracy. When f is required at some point in phase space (e.g., in performing the quadratures for the source terms), this point is simply tracked along a dynamical path all the way back to $t = 0$, where f can be accurately evaluated from the initial data. For problems involving discontinuous distribution functions or large degrees of phase mixing, more points can be added in phase space to do the quadratures to the required accuracy. There is never any need to introduce any grid at all in phase space, since intermediate values of f are never needed and therefore need not be stored. The only disadvantage, of course, is that the computational time per time step increases dramatically with time, since longer and longer trajectories have to be constructed to evaluate f at a given point.

5.4 Scalar fields

A classical real scalar field provides one of the simplest sources of stress-energy in Einstein's equations; a classical complex scalar field furnishes another example, only slightly more complicated. Upon quantization, the momentum eigenstates of a scalar field are observable as particles. Scalar fields give rise to particles of spin 0, while vector fields (like the electromagnetic field) give rise to particles of spin 1 (like the photon, in the case of electromagnetism), and tensor fields of rank two or higher give rise to higher-spin particles. A complex scalar field has two degrees of freedom instead of just one, and it can be interpreted as a particle and an antiparticle. Real fields are their own antiparticles. A neutral π-meson is an example of a real scalar field, while the charged π^+- and π^--mesons are described by complex scalar fields.

Interest in scalar fields has been stimulated in recent years by theoretical developments in particle physics and cosmology. Both the standard model of elementary particles as well as their superstring extensions involve scalar fields, although the existence of a fundamental elementary scalar particle has yet to be confirmed by an accelerator experiment. For example, the Higgs boson, which serves to generate masses for the $W^\pm$ and Z^0 gauge vector bosons in the electroweak theory of Weinberg, Salam and Glashow, is a particle of finite mass described by a neutral spin 0 scalar field. Its discovery is a top priority of experimental particle physics. There are many attempts at unifying the standard model with gravitation at the quantum level, like string theory. Typically, these theories give rise to 4-dimensional "effective" models in which the usual spacetime metric g_{ab} of gravity is accompanied by one or more scalar fields. The massive scalar dilaton and the (pseudo-) scalar axion are two examples.

In cosmology, the inflationary phase in the early Universe can be driven by the vacuum energy density provided by the potential of a time-dependent, but slowly varying scalar field $\varphi(t)$ called an "inflaton". It is also possbile that the existence of "dark energy" in the Universe, which manifests itself as a nonzero cosmological constant in the standard Big Bang model, may be represented by the vacuum energy density associated with the potential of yet another scalar field at a much lower characteristic energy ("quintessence").

Candidates for "dark matter" in the Universe include bosonic, as well as fermionic particles. It is an issue of wide speculation whether or not "boson stars" could actually arise during the gravitational condensation of bosonic dark matter in the early Universe. Boson stars are self-gravitating, stationary equilibrium configurations constructed from complex scalar fields in asymptotically flat spacetimes.[79] They are macroscopic quantum states that are supported against gravitational collapse by the Heisenberg uncertainty principle; they can be modeled by classical scalar fields. Like neutron stars, boson stars can have highly relativistic gravitational fields and yet are nonsingular and have no event horizons. However, if the scalar fields have self-interactions, then, unlike neutron stars, boson stars can be very massive. It is therefore not surprising that boson stars are sometimes invoked as alternatives to black holes to model massive, compact stars. By contrast with boson stars, "soliton stars", which are constructed from real scalar fields, are not stationary but periodic, both in the spacetime geometry and the matter field. Nonsingular, self-gravitating stationary solutions do not exist for real, massive scalar fields.

Apart from their possible physical significance, scalar fields serve as very useful tools for probing strong gravitational field phenomena and for learning how to do numerical relativity. The dynamical equation governing a scalar field is the simple, classical Klein–Gordon equation (see discussion below). In contrast to the partial differential equations describing hydrodynamic or magnetohydrodynamic matter, which can exhibit shock waves and other discontinuities that require special handling (see Section 5.2), the Klein–Gordon equation does not tend to develop discontinuities from smooth initial data and is thus straightforward to integrate. As a result, integrating the scalar wave equation is often chosen as the first test of a new numerical evolution scheme, a choice that can prove instructive even when the equation is integrated in flat spacetime. Another consequence of the simple, well-behaved nature of the scalar wave equation is that it is fairly easy to implement advanced numerical techniques such as "adaptive mesh refinement"

---

[79] See, e.g., Kaup (1968); Ruffini and Bonazzola (1969); Colpi *et al.* (1986); Lee and Pang (1992); Seidel and Suen (1990, 1991); Yuan *et al.* (2004); Schunck and Mielke (2003).

(AMR)[80] for high-resolution integrations of the equation, particularly in $1 + 1$ dimensional simulation (e.g., spherical symmetry). The detailed study of gravitational collapse and black hole formation using AMR with a scalar wave source has lead to the discovery of black hole "critical pheonomena".[81] Here the behavior observed in dynamical simulations of configurations at the onset of black formation along a one-parameter family of initial data assumes many of the features of a phase transition in a statistical mechanical system.

Finally, scalar fields generate radiation fields – waves which travel at the speed of light locally, have amplitudes that fall-off with radius like r^{-1} at large distances from a central source, and carry away mass-energy – even in spherical symmetry! By contrast, gravitational waves, which are an important yield of numerical simulations in general relativity, cannot arise in spherical symmetry, as we know from Birkhoff's theorem. So working with scalar fields in spherical symmetry allows one to gain experience tackling some of the computational challenges of wave generation and propagation in dynamical spacetimes without having to integrate in more than one spatial dimension.

The stress-energy tensor of a real scalar field, $\varphi(x^a)$, is given by

$$\boxed{T_{ab} = \nabla_a \varphi \nabla_b \varphi - \frac{1}{2} g_{ab} g^{cd} \nabla_c \nabla_d \varphi - g_{ab} V(\varphi),}$$ (5.232)

where $V(\varphi)$ is the potential. The potential may be decomposed according to

$$V(\varphi) = \frac{1}{2} m^2 \varphi^2 + V_{\text{int}},$$ (5.233)

where m is the mass of the field (i.e., the mass of the momentum eigenstates, or "particles", when the field is quantized), and V_{int} is an interaction potential.[82] When $m = 0$ the field is massless; when $V_{\text{int}} = 0$ the field is noninteracting, apart from gravitation ("minimal coupling"). The equations of motion (i.e., the scalar field equations) follow from $\nabla_b T^{ab} = 0$, which gives the Klein–Gordon equation with a potential term,

$$\nabla_a \nabla^a \varphi - m^2 \varphi - \frac{dV_{\text{int}}}{d\varphi} = 0.$$ (5.234)

> **Exercise 5.46** Calculate the total energy density for a real scalar field measured in the rest frame of an observer in Minkowski spacetime and give a physical interpretation for each of the contributions in your expression.

Consider the case of a massless field in the absence of an interaction potential (i.e., $V = 0$), for which equation (5.234) reduces to the massless Klein–Gordon equation in curved spacetime,

$$\nabla_a \nabla^a \varphi = 0.$$ (5.235)

---

[80] Berger and Oliger (1984); see Chapter 6.2.5.
[81] Choptuik (1993). See Chapter 8 for a discussion.
[82] Here we set $\hbar = 1$ in addition to $c = 1 = G$, so that terms like $(mc/\hbar)^2$ become m^2.

In the special case of a stationary black hole background, the only nonsingular, time-independent solution to equation (5.235) is $\varphi = 0$. If the initial value of φ varies in space or time, it will evolve to this solution eventually. This result was demonstrated by Price[83] and is consistent with black-hole uniqueness ("no-hair") theorems.[84]

> **Exercise 5.47** Consider the propagation of a massless, noninteracting scalar field in a static spherical spacetime,
>
> $$ds^2 = -e^{2\Phi(r)}dt^2 + e^{2\Lambda(r)}dr^2 + r^2(d\theta^2 + \sin^2\theta d\phi^2). \qquad (5.236)$$
>
> Decompose the field into spherical harmonics
>
> $$\varphi(t, r, \theta, \phi) = \frac{\psi}{r} Y_{lm}(\theta, \phi). \qquad (5.237)$$
>
> Show that the scalar field equation reduces to
>
> $$-\partial_t \partial_t \psi + \partial_{r_*} \partial_{r_*} \psi = V_l(r)\psi, \qquad (5.238)$$
>
> where r_* is a "generalized tortoise coordinate" satisfying
>
> $$\frac{dr_*}{dr} = e^{\Lambda - \Phi}, \qquad (5.239)$$
>
> and where $V_l(r)$ is an effective potential given by
>
> $$V_l(r) = e^{2\Phi}\left(\frac{l(l+1)}{r^2} + \frac{e^{-2\Lambda}}{r}\partial_r(\Phi - \Lambda)\right). \qquad (5.240)$$
>
> **Note:** In the case of flat spacetime, the solutions to the above radial wave equation are known analytically for all l.[85] For a static vacuum Schwarzschild spacetime, the numerical solutions exhibit three separate phases: an initial burst, a quasinormal ringing phase, and a power-law tail phase. During the late-time tail phase, all multipoles of the scalar wave decay at finite r_* according to $\psi \sim t^{-(2l+3)}$ as $t \to \infty$ ("Price's theorem" for scalar waves). The same radiative behavior is also observed for scalar wave propagation in the static spacetime of a spherical equilibrium (OV) star.[86]

In exercise (5.48) we construct a conserved energy integral that provides a useful check on numerical integrations of the scalar wave equation in static spacetimes, as derived in exercise (5.47).

> **Exercise 5.48** In a static spacetime $\xi^a = \partial/\partial t$ is a Killing vector and $J^a = T^{ab}\xi_b$ is a conserved current, i.e., $\nabla_a J^a = 0$.
> (a) Prove the conservation law
>
> $$\oint_{\delta\Omega} J^a d^3\Sigma_a = 0, \qquad (5.241)$$
>
> where $\delta\Omega$ is the closed 3-surface enclosing a 4-volume Ω. The surface element $d^3\Sigma_a$ is given by $d^3\Sigma_a = (1/3!)\epsilon_{abcd}dx^b dx^c dx^d$, where ϵ_{abcd} is the Levi-Civita tensor.

[83] Price (1972a,b).
[84] See, e.g., Wald (1984).
[85] Burke (1971).
[86] See, e.g., Kokkotas and Schmidt (1999); Pavlidou *et al.* (2000), and references therein.

(Note: here $\delta\Omega$ is a 3-surface that lies *inside* the worldtube of the matter depicted in Figure 3.5.)

(b) Adopt spherical polar coordinates and choose Ω to be the volume confined by the spherical radii r_1 and r_2 and the times t_1 and t_2. Show that the conservation law (5.241) can be written as

$$\mathcal{E}(t_2) - \mathcal{E}(t_1) = \mathcal{J}(r_2) - \mathcal{J}(r_1), \tag{5.242}$$

where

$$\mathcal{E}(t) \equiv \int_{r=r_1}^{r_2} \int_{\theta=0}^{\pi} \int_{\phi=0}^{2\pi} J^0 \sqrt{-g} \, dr \, d\theta \, d\phi \tag{5.243}$$

is the energy of the matter field contained between r_1 and r_2 at time t, and

$$\mathcal{J}(r) \equiv \int_{t=t_1}^{t_2} \int_{\theta=0}^{\pi} \int_{\phi=0}^{2\pi} J^r \sqrt{-g} \, dt \, d\theta \, d\phi \tag{5.244}$$

is the radial flux across r, integrated over time between t_1 and t_2. Discuss the meaning of equation (5.242).

An interesting mathematical connection exists between the massless Klein–Gordon equation (5.235) and the evolution equation for a perfect, irrotational fluid. As shown in exercise (5.49), the hydrodynamic evolution of a relativistic fluid obeying the special EOS $P = \rho^*$ can actually be determined by integrating the wave equation for a real, massless, noninteracing scalar field!

Exercise 5.49 The relativistic vorticity tensor is defined as

$$\omega_{ab} = P^c{}_a P^d{}_b \left[\nabla_d (h u_c) - \nabla_c (h u_d) \right], \tag{5.245}$$

where u^a is the 4-velocity, $h = (\rho^* + P)/\rho_0$ is the specific enthalpy, $\rho^* = \rho_0(1 + \epsilon)$ is the total mass-energy density, ρ_0 is the rest-mass density, and $P^{ab} = g^{ab} + u^a u^b$ is the projection tensor.

(a) Show that for a perfect fluid, equation (5.62) may be used to recast equation (5.245) as

$$\omega_{ab} = \left[\nabla_b (h u_a) - \nabla_a (h u_b) \right]. \tag{5.246}$$

(b) Argue that if the vorticity is zero, the quantity $h u_a$ can be expressed as the gradient of a potential,

$$h u_a = \nabla_a \varphi. \tag{5.247}$$

What kind of flow does this correspond to physically?

(c) Show that the continuity equation now becomes

$$\nabla_a \left[(\rho_0 / h) \nabla^a \varphi \right] = 0. \tag{5.248}$$

(d) The equation of state relates ρ_0 to h, and h is found from the normalization condition $h = (-\nabla^a \varphi \nabla_a \varphi)^{1/2}$. Relate ρ_0 to h for a polytropic gas where P is given by equation (5.18).

(e) Consider the extreme equation of state $P = \rho^*$. Find Γ and show that for this equation of state, equation (5.248) simplies to yield

$$\nabla_a (\nabla^a \varphi) = 0. \tag{5.249}$$

What is the sound speed in this gas?

To integrate the Klein–Gordon equation (5.235) numerically it is useful to cast it into first-order form. For this purpose, introduce the following new variables:

$$\Pi \equiv \frac{-1}{\alpha} \left(\partial_t \varphi - \beta^i \partial_i \varphi \right), \tag{5.250}$$

$$\psi_i \equiv \partial_i \varphi. \tag{5.251}$$

For a general $3 + 1$ metric, the Klein–Gordon equation (5.235) becomes

$$\partial_t \varphi = \beta^i \partial_i \varphi - \alpha \Pi, \tag{5.252}$$

$$\partial_t \Pi = \beta^i \partial_i \Pi - \alpha g^{ij} \partial_j \psi_i + \alpha g^{ij} \Gamma^k{}_{ij} \psi_k - g^{ij} \psi_j \partial_i \alpha + \alpha K \Pi, \tag{5.253}$$

$$\partial_t \psi_i = \beta^j \partial_j \psi_i + \psi_j \partial_i \beta^j - \alpha \partial_i \Pi - \Pi \partial_i \alpha. \tag{5.254}$$

It is noteworthy that the definition of ψ_i, equation (5.251), turns into a set of three constraints,[87]

$$C_i \equiv \partial_i \varphi - \psi_i = 0, \tag{5.255}$$

that must be satisfied at all times. When solved exactly, the system of equations (5.252)–(5.254) automatically preserves the constraints $C_i = 0$, provided they are satisfied initially. However, in a numerical simulation, truncation errors and boundary errors can cause C_i to wander away from zero. Hence it is useful to monitor the evolution of C_i to help calibrate the accuracy of the simulation. In the same way, monitoring how well the $3 + 1$ constraints are maintained serves to calibrate the accuracy of the gravitational field integrations. The Klein–Gordon system of first-order, symmetric hyperbolic equations thus provides a simple arena for exploring the growth of constraint violations in numerical simulations of hyperbolic systems and devising methods for controlling them.

Evaluating the scalar field equation (5.234) for a Robertson–Walker (RW) metric (see exercise 4.1), assuming the field is everywhere homogeneous ($\nabla_i \varphi = 0$) gives a simple second-order ordinary differential equation for the evolution of $\varphi(t)$,

$$\ddot{\varphi} + 3H\dot{\varphi} + m^2 \varphi + \frac{d V_{\text{int}}}{d\varphi} = 0. \tag{5.256}$$

In equation (5.256) the dot denotes time differentiation, $H(t) = \dot{a}/a$ is Hubble's constant, and $a(t)$ is the expansion parameter appearing in the RW metric. This equation is the usual starting point of most discussions of inflation in the early Universe, in which case

---

[87] Scheel *et al.* (2004).

φ is called an "inflaton" scalar field. The same equation is also invoked as the evolution equation for a simple candidate for a dynamical source of dark energy.[88]

A popular interaction potential used in many studies is a quartic self-interaction, which has the form

$$V_{\text{int}}(\varphi) = \frac{1}{4}\lambda\varphi^4, \tag{5.257}$$

where λ is a dimensionless coupling constant. While this interaction is often employed for illustrative purposes, plausible models arising in particle physics, like the Higgs field in standard electroweak theory, *do* contain φ^4 interactions. The only other possibility for models involving only scalar fields is an interaction of the form φ^3, since theories containing φ^n are not renormalizable for $n > 4$.[89]

Much of the astrophysically relevant work on scalar fields in asymptotically flat spacetimes involves complex scalar fields. Part of the reason is that massive complex fields can form stationary equilibrium configurations (boson stars). The stress-energy tensor for a massive, self-interacting complex field is

$$T_{ab} = \frac{1}{2}\left[(\nabla_a\Phi\nabla_b\Phi^* + \nabla_b\Phi\nabla_a\Phi^*) - g_{ab}(\nabla^a\Phi\nabla_a\Phi^*)\right] - g_{ab}V(\Phi, \Phi^*). \tag{5.258}$$

The corresponding equations of motion are

$$\nabla_a\nabla^a\Phi - m^2\Phi - \lambda|\Phi|^2\Phi = 0 \tag{5.259}$$

and its complex conjugate. Here we have assumed $V(\Phi, \Phi^*) = \frac{1}{2}m^2|\Phi\Phi^*| + \frac{1}{4}\lambda|\Phi\Phi^*|^2$. The complex scalar field can be split into two real scalar fields according to $\Phi = \varphi_1 + i\varphi_2$, where φ_1 and φ_2 are real scalar functions. The two real, coupled, second-order equations that result may then be cast into first-order form by introducing new variables as in equation (5.250) and proceeding in a similar fashion.

As a final twist, we can endow the complex scalar fields with charge e, as well as mass m, and allow them to interact with an electromagnetic field. Express the Faraday tensor for the electromagnetic field, F_{ab}, in terms of the electromagnetic vector potential, A_a,

$$F_{ab} = \nabla_a A_b - \nabla_b A_a. \tag{5.260}$$

Define the operator $\mathcal{D}_a \equiv \nabla_a + ieA_a$. The total stress-energy tensor for this system then becomes

$$T_{ab} = \frac{1}{2}(\mathcal{D}_a^*\Phi^*)(\mathcal{D}_b\Phi) + \frac{1}{2}(\mathcal{D}_a\Phi)(\mathcal{D}_b^*\Phi^*) - \frac{1}{2}g_{ab}(\mathcal{D}_c\Phi)(\mathcal{D}^{c*}\Phi^*) - g_{ab}V(\Phi, \Phi^*) \\ + \frac{1}{4\pi}F_{ac}F_b{}^c - \frac{1}{16\pi}g_{ab}F_{cd}F^{cd}, \tag{5.261}$$

---

[88] See Carroll (2004), Chapter 8.7 and 8.8, for a discussion.
[89] See, e.g., Peskin and Schroeder (1995) for a discussion.

where $V(\Phi, \Phi^*) = \frac{1}{2}m^2|\Phi\Phi^*| + V_{\text{int}}(\Phi, \Phi^*)$. The corresponding equations of motion[90] for the scalar fields are

$$\nabla_a \nabla^a \Phi + i e A^a (2\nabla_a \Phi + i e A_a \Phi) + i e \Phi \nabla_a A^a - 2\frac{\partial V(\Phi, \Phi^*)}{\partial \Phi^*} = 0, \qquad (5.262)$$

and its complex conjugate, while Maxwell's equations for the electromagnetic field are

$$\nabla^b F_{ab} = 4\pi J_a, \qquad (5.263)$$

where the conserved 4-current J_a is given by

$$J_a = i e \left[\Phi \mathcal{D}_a^* \Phi - \Phi^* \mathcal{D}_a \Phi \right]. \qquad (5.264)$$

Recall that the remaining Maxwell equations, $\nabla_{[c} F_{ab]} = 0$, are automatically satisfied by equation (5.260).

The above scalar and electromagnetic field equations must then be solved in conjunction with the $3 + 1$ equations for the gravitational field to determine the complete foliation of spacetime. For this purpose the source terms (5.1)–(5.2) must be computed from T^{ab}. Once again, it is convenient to cast the second-order evolution equations into first-order form to solve them numerically. While the system of equations for a charged, complex scalar field is not trivial to solve numerically, it is simpler than the relativistic MHD system of equations for an ionized gas discussed in Section 5.3. For example, there are no shocks with scalar fields. As a result, a charged scalar field affords the opportunity to explore the behavior of charged "matter" in the presence of an electromagnetic field in curved spacetime, including near black holes, with a somewhat more modest computational effort.

---

[90] i.e., the Euler–Lagrange equations, most easily obtained by varying the total Lagrangian with respect to Φ, Φ^* and A_a independently; see, e.g., Hawking and Ellis (1973), p. 68.

6 Numerical methods

As we have seen, Einstein's field equations in $3+1$ form consist of a set of nonlinear, multidimensional, coupled partial differential equations in space and time. The equations of motion of the matter fields that may be present are typically of a similar nature. Except for very idealized problems with special symmetries, such equations must be solved by numerical means, often on supercomputers. Just as there is no unique analytic formulation of the $3+1$ field equations,[1] there is no unique prescription by which a partial differential equation may be cast into a form suitable for numerical integration. Standard numerical algorithms for treating such equations may be found in many textbooks on numerical methods, as well as in textbooks, monographs and review articles on compuational physics. This branch of applied mathematics is a rich area of ongoing investigation; it progresses with each advance in computer technology. It would take us too far afield to review the subject in any depth here. Instead, we shall present a brief introduction to some of the basic numerical concepts and associated techniques, focusing on those most often employed to solve the partial differential equations that arise in numerical relativity. Although our treatment is rudimentary, we hope that it is sufficient to convey the flavor of the subject, especially to readers unfamiliar with the basic ideas. Throughout our discussion we shall refer the reader to some of the literature where further details and other references can be found.[2]

6.1 Classification of partial differential equations

So far in this book our focus has been on casting Einstein's equations and the equations of motion for any matter sources into a form that can be solved numerically with standard techniques. Most of the resulting equations are second-order partial differential equations and can be classified into three categories: elliptic, parabolic or hyperbolic.

The prototypical example of an *elliptic* equation is Poisson's equation,[3]

$$\boxed{\partial_x^2 \phi + \partial_y^2 \phi = \rho,}$$

(6.1)

---

[1] We will explore alternative formulations in Chapter 11.

[2] See, e.g., the many references cited in Press *et al.* (2007), who provide much more discussion of portions of the material surveyed in this section.

[3] For illustrative purposes in this section it is convenient to focus on problems that have at most two independent variables.

where ρ is a source term that may depend on position, or even on ϕ up to first-order derivatives. For vanishing sources this equation is Laplace's equation. We have encountered elliptic equations, for example, in the Hamiltonian constraint (3.37) and in the maximal slicing condition (4.12).

An example of a *parabolic* equation is the diffusion equation,

$$\partial_t \phi - \partial_x(\kappa \, \partial_x \phi) = \rho, \tag{6.2}$$

where κ is the diffusion coefficient. We have seen a parabolic equation when we converted the maximal slicing condition into the "driver" condition (4.38).

The prototypical example of a hyperbolic equation is the wave equation,

$$\partial_t^2 \phi - c^2 \, \partial_x^2 \phi = \rho, \tag{6.3}$$

where c is the constant wave speed.

Exercise 6.1 Verify that any function

$$\phi = g(x + ct) + h(x - ct) \tag{6.4}$$

satisfies the wave equation (6.3) for $\rho = 0$.

We have encountered hyperbolic equations several times, but often they have been well disguised and have not appeared exactly in this form. To make contact with those examples we introduce the first time derivative of ϕ as a new independent variable, say $-k$, in which case we can rewrite the wave equation (6.3) as the pair of equations

$$\begin{aligned}\partial_t \phi &= -k \\ \partial_t k &= -c^2 \, \partial_x^2 \phi - \rho.\end{aligned} \tag{6.5}$$

Interestingly, this form is very similar to the pair of $3 + 1$ evolution equations (2.134) and (2.135), except not quite. We can identify ϕ with γ_{ij} and k with K_{ij}. The 3-dimensional generalization of the second space derivative $\partial_x^2 \phi$ would the Laplacian of ϕ. A similar term, acting like the Laplacian of γ_{ij}, is hidden in the spatial Ricci tensor on the right-hand side of (2.135) (see the fourth term in 2.143), but the Ricci tensor also contains other, mixed second derivatives. These other terms spoil the hyperbolicity of the evolution equations in the standard $3 + 1$ form (equations 2.134 and 2.135) and motivate the development of alternative formulations of Einstein's equations. We will revisit this issue in much greater detail in Chapter 11.

The form (6.5) is not a particularly elegant representation of a wave equation, since it contains first-order time derivatives but second-order space derivatives. We can fix that quite easily by also introducing the space derivative of ϕ as a new independent variable. With $l \equiv \partial_x \phi$ we now find the system

$$\begin{aligned}\partial_t \phi &= -k \\ \partial_t k + c^2 \partial_x l &= -\rho \\ \partial_t l + \partial_x k &= 0,\end{aligned} \tag{6.6}$$

where the last equation holds because the partial derivatives must commute. In a more compact notation we can write this as

$$\partial_t \mathbf{u} + \mathbf{A} \cdot \partial_x \mathbf{u} = \mathbf{S}, \tag{6.7}$$

where $\mathbf{u} = (\phi, k, l)$ is the solution vector, $\mathbf{S} = (-k, -\rho, 0)$ is the source vector, and where

$$\mathbf{A} = \begin{pmatrix} 0 & 0 & 0 \\ 0 & 0 & c^2 \\ 0 & 1 & 0 \end{pmatrix} \tag{6.8}$$

is the velocity matrix. This is a form of the wave equation that is similar to what we have encountered, for example, in the context of harmonic coordinates (4.44) and (4.45) and hydrodynamics (5.26).

From the solution (6.4) it is evident that part of the solution ϕ, namely g, travels along lines $x + ct = constant$, while the other part h travels along $x - ct = constant$. These lines are called the *characteristic curves*; they are those curves along which partial information about the solution propagates. Even if we cannot derive the general solution analytically, we can find the corresponding *characteristic speeds* dx/dt from the eigenvalues of the velocity matrix $\mathbf{A}$. In our example, these eigenvalues are $\pm c$ and zero, as we would expect.[4]

> **Exercise 6.2** Instead of introducing k and l it may be more elegant to define a pair of *characteristic* variables
>
> $$u = (\partial_t - c\,\partial_x)\phi \quad v = (\partial_t + c\,\partial_x)\phi. \tag{6.9}$$
>
> Relate u and v to the functions g and h in the general solution (6.4). Then define $\mathbf{u} = (\phi, u, v)$ and bring the wave equation into the form (6.7). Show that the velocity matrix $\mathbf{A}$ is now diagonal, and verify that it has the same eigenvalues as before.

To return to our classification of second-order partial differential equations, consider the general equation

$$A\,\partial_\xi^2\phi + 2B\,\partial_\xi\partial_\eta\phi + C\,\partial_\eta^2\phi = \tilde{\rho} \tag{6.10}$$

where the coefficients A, B and C are real, differentiable, and do not vanish simultaneously. Also, the source term $\tilde{\rho}$ may depend on ϕ, but only up to first-order derivatives. Whether this equation is elliptic, parabolic, or hyperbolic then depends on the coefficients A, B and C:[5]

- If $AC - B^2 > 0$, then we can find a coordinate transformation from (ξ, η) to some (x, y) that brings equation (6.10) into the form (6.1); such equations are *elliptic*.
- If $AC - B^2 = 0$, then we can find a coordinate transformation that brings equation (6.10) into the form (6.2); such equations are *parabolic*.

---

[4] The vanishing eigenvalue in (6.7) is associated with the equation $\partial_t\phi = -k$ which "propagates" information along $x = constant$.

[5] A similar classification can be defined in higher dimensions.

- Finally, if $AC - B^2 < 0$, then we can find a coordinate transformation that brings equation (6.10) into the form (6.3); such equations are *hyperbolic*.

Only hyperbolic equations have real (as opposed to imaginary) characteristics. We also point out that hyperbolicity comes in various different flavors, which have slightly different consequences for the properties of the solutions. We briefly discuss these different notions of hyperbolicity in Chapter 11.1, and refer to the literature[6] for a more detailed and rigorous treatment.

Exercise 6.3   Consider the radial wave equation

$$\nabla^a \nabla_a \phi = \frac{1}{\sqrt{-g}} \partial_a (\sqrt{-g} g^{ab} \partial_b \phi) = 0, \tag{6.11}$$

for the evolution of a scalar field $\phi = \phi(t, r)$ in a Schwarzschild spacetime, expressed in Kerr–Schild coordinates (see exercise 2.33). Assume that ϕ falls of with $1/r$ at large radii, and introduce a new variable $\Phi \equiv r\phi$. Show that Φ satisfies the equation

$$-(1 + 2H)\partial_t^2 \Phi + 4H \partial_r \partial_t \Phi + (1 - 2H)\partial_r^2 \Phi - \frac{2H}{r} \partial_t \Phi + \frac{2H}{r} \partial_r \Phi - \frac{2H}{r^2} \Phi = 0$$

$$\tag{6.12}$$

where $H = M/r$. Verify that this equation is hyperbolic both inside and outside the event horizon at $r = 2M$. Then bring this equation into the first-order form (6.7) and show that the two nontrivial characteristic speeds are

$$c_1 = -1, \quad c_2 = \frac{1 - 2H}{1 + 2H}. \tag{6.13}$$

Show that these characteristic speeds correspond to the two radial null geodesics of the Kerr–Schild metric. Integrate the two characteristic speeds to find that the ingoing and outgoing characteristics satisfy

$$t + r = constant \qquad t - r = 4M \ln |r - 2M| + constant, \tag{6.14}$$

and explain why the name "outgoing" is misleading (see Figure 6.1).

The different types of partial differential equations require different kinds of boundary and/or initial conditions.[7] Both parabolic and hyperbolic equations constitute intial value problems, meaning that we have to define initial values of the fields (and possibly their time derivatives) on a $t = constant$ spatial hypersurface. The differential equations then tell us how these initial fields evolve with time. We may also have to impose spatial boundary conditions on the outer boundaries of our computational domain. Elliptic equations, on the other hand, determine a solution on a given spatial hypersurface. No initial data are required, but we must supply boundary values at the outer edge(s) of our computational domain.

Boundary conditions can take various forms. For example, *Dirichlet* conditions specify the values of the solution functions on the boundary, while *Neumann* conditions specify

---

[6] See, e.g., Reula (1998).
[7] See, e.g., Mathews and Walker (1970), Chapter 8, for more detailed discussion.

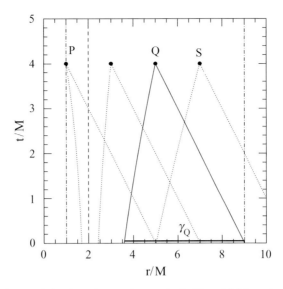

Figure 6.1 Backwards characteristics through several events in the Kerr–Schild metric of a Schwarzschild black hole. The interval γ_Q is the *domain of dependence* of the event Q; the cone defined by γ_Q and Q is the *domain of determinacy* of the interval γ_Q. The dashed line tracks the event horizon, and the dashed-dotted lines mark some hypothetical computational boundaries.

their gradients on the boundary. We will encounter various other boundary conditions in the course of this book.

At this point it is useful to discuss hyperbolic equations in some more detail. The concept of characteristics leads to the notion of the *domain of dependence*. For hyperbolic systems information travels along characteristics. Since no information can travel faster than the fastest characteristics, the solution at a certain event can only be affected by those events that lie inside the *causal past*, or *past cone* defined by the fastest ingoing and outgoing characteristics. In the context of relativity we are quite used to this concept, since causality demands that an event can only be affected by events in its past light cone.

For example, consider the wave equation (6.12) for a scalar wave ϕ that propagates in a Schwarzschild spacetime expressed in Kerr–Schild coordinates. In Figure 6.1 we plot several events in this spacetime together with their backward characteristics. Consider the event Q, whose past characteristics are marked by solid lines. To completely determine $\phi(t, r)$ at Q, we would have to provide initial data $\phi(0, r)$ and $\partial_t \phi(0, r)$ inside the past cone defined by the two backward characteristics, namely on the interval γ_Q. This is the domain of dependence of the point Q.

In reverse, for any interval γ on the r-axis, there is a region of the spacetime in which all events depend only on initial data provided on γ. This region is called the *domain of determinacy*. For γ_Q in Figure 6.1 the domain of determinacy is the cone defined by γ_Q and the event Q, bordered by solid lines.

Suppose we want to obtain a solution to the wave equation (6.12). We will have to provide initial data on an interval γ that extends from a certain radius r_{min} to a radius r_{max} at, say, $t = 0$. If we want to construct the solution only in the domain of determinacy of γ, then the solution is completely determined by the initial data on γ, and no boundary conditions are needed. This situation, however, is rarely the case. It is more typical that we would like to construct the solution in the entire domain between r_{min} and r_{max} for all $t > 0$.

For concreteness, imagine we want to find ϕ in the domain between $r_{min}/M = 1$ and $r_{max}/M = 9$, marked by the dashed-dotted lines in Figure 6.1. The event Q would still be completely determined by the initial data, but the event S, for example, would not. One of its backward characteristics intersects the outer boundary at $r_{max}/M = 9$. The event S is therefore outside the domain of determinacy of γ, and the solution at S depends on more information than is provided by the initial data. This missing information now has to be provided by the boundary conditions. The boundary condition at the outer boundary r_{max} has to specify the information that propagates along the ingoing characteristic that originates on the outer boundary. For example, this could be an *outgoing-wave* boundary condition which, as the name suggests, insures that no wave (i.e., no information) enters the domain through the outer boundary.

The situation is different at the inner boundary r_{min}. Consider, for example, the event P, which lies on the boundary r_{min}. Since we have chosen r_{min} to be inside the event horizon, both characteristics originate from a larger r, and neither one intersects the boundary r_{min}. The event P is therefore completely determined by the initial data (and, had we chosen P at a later time, by the outer boundary condition at r_{max}). There is no need to impose a boundary condition at r_{min}, and in fact it would be inconsistent with the equations. This property will be important when we discuss black hole excision in Chapter 13.1.

Following this general discussion of partial differential equations, we now turn to computational methods that can be used to solve these equations.

6.2 Finite difference methods

Finite difference methods for general applications in computational physics have been treated in great detail in many references.[8] Specific applications to numerical relativity have been discussed in a few review articles[9] and many journal articles. Here we provide a brief introduction that touches on some of the important aspects of the subject.

6.2.1 Representation of functions and derivatives

In a finite difference approximation a function $f(t, x)$ is represented by values at a discrete set of points. At the core of finite difference approximation is therefore a discretization of

---

[8] e.g., Richtmyer and Morton (1967); Roache (1976); Press *et al.* (2007).
[9] e.g., Smarr (1979a); Evans *et al.* (1989).

the spacetime, or a numerical *grid*. Instead of evaluating f at all values of x, for example, we only consider discrete values x_i. As we discuss in more detail in Section 6.2.2, these gridpoints may reside either at the centers or the vertices of gridcells, but this distinction is irrelevant for our purposes in this section. The distance between the gridpoints x_i is called the *grid spacing* Δx, which in principle may depend on x (or rather x_i). For *uniform* grids, for which Δx is constant, we have

$$x_i = x_0 + i \, \Delta x. \tag{6.15}$$

If the solution depends on time we also discretize the time coordinate, for example as

$$t^n = t^0 + n \Delta t, \tag{6.16}$$

where the superscript n denotes the nth time level and should not be confused with an exponent. In more than one space dimension the other dimensions are discretized in the same manner. The result of this is a spacetime lattice, on which all functions can be evaluated. The finite difference representation of the function $f(t, x)$, for example, is

$$f_i^n = f(t^n, x_i) + \text{truncation error}. \tag{6.17}$$

Here (and only here) we have explicitly added the "truncation error" as a reminder that f_i^n only approaches the correct value of f at t^n and x_i as the finite difference solution converges to the correct solution. We will discuss the finite difference error in much more detail below.

Differential equations involve derivatives, so we must next discuss how to represent derivatives in a finite difference representation. Consider a partial derivative of $f(x)$ with respect to x. Assuming that $f(x)$ can be differentiated to sufficiently high order and that it can be represented as a Taylor series, we have

$$f_{i+1} = f(x_i + \Delta x) = f(x_i) + \Delta x (\partial_x f)_{x_i} + \frac{(\Delta x)^2}{2} (\partial_x^2 f)_{x_i} + \mathcal{O}(\Delta x^3). \tag{6.18}$$

Solving for $(\partial_x f)_{x_i} = (\partial_x f)_i$ we find

$$(\partial_x f)_i = \frac{f_{i+1} - f_i}{\Delta x} + \mathcal{O}(\Delta x). \tag{6.19}$$

In the limit $\Delta x \to 0$ equation (6.19) is just the definition of the partial derivative, so this result should not come as a great surprise. The truncation error of this expression is linear in Δx, and it turns out that we can do better. Consider the Taylor expansion to the point x_{i-1},

$$f_{i-1} = f(x_i - \Delta x) = f(x_i) - \Delta x (\partial_x f)_{x_i} + \frac{(\Delta x)^2}{2} (\partial_x^2 f)_{x_i} + \mathcal{O}(\Delta x^3). \tag{6.20}$$

Subtracting (6.20) from (6.18) we now find

$$(\partial_x f)_i = \frac{f_{i+1} - f_{i-1}}{2\Delta x} + \mathcal{O}(\Delta x^2), \tag{6.21}$$

which is second order in Δx, meaning that the truncation error drops by a factor of four when we reduce the grid spacing by a factor of 2.[10] The key point is that we are able to combine the two Taylor expansions in such a way that the leading order error term cancels out, leaving us with a higher order representation of the derivative. This cancellation only works out for uniform grids, when Δx is independent of x. This is one of the reasons why many current numerical relativity applications of finite difference schemes work with uniform grids.

> **Exercise 6.4** The finite difference representation (6.21) being second-order implies that it should be *exact* for any arbitrary polynomial up to second order. Verify that it indeed gives the correct derivative of a polynomial $f(x) = a + bx + cx^2$, independently of x and Δx.

We call equation (6.19) a *one-sided* derivative, since it uses only neighbors on one side of x_i, and (6.21) a *centered* derivative. In general, centered derivatives lead to higher order schemes than one-sided derivatives for the same number of gridpoints. Exercise 6.4 shows that we can also construct one-sided, higher-order difference schemes, but they will involve more than two gridpoints.

> **Exercise 6.5** Construct a second order, one-sided finite difference approximation of $\partial_x f$, i.e., generalize equation (6.19), using only neighbors on one side of x_i, so that the truncation error becomes $\mathcal{O}(\Delta x^2)$. Verify your result by showing that it gives the exact result for an arbitrary polynomial $f(x) = a + bx + cx^2$, as in exercise 6.4.

Higher-order derivatives can be constructed in a similar fashion. Adding the two Taylor expansions (6.18) and (6.20) all terms odd in Δx drop out and we find for the second derivative

$$(\partial_x^2 f)_i = \frac{f_{i+1} - 2f_i + f_{i-1}}{(\Delta x)^2} + \mathcal{O}(\Delta x^2). \qquad (6.22)$$

This expression is second order because the third order terms in equations (6.18) and (6.20) cancel out in the addition. An alternative derivation of the second derivative (6.22) proceeds as follows. First write the first derivative at an indermediate grid point $x_{i+1/2}$ as $(f_{i+1} - f_i)/\Delta x$ and similar at $x_{i-1/2}$. The second derivative at x_i is then the derivative of the derivative, which we can compute from the difference between $(\partial_x f)_{i+1/2}$ and $(\partial_x f)_{i-1/2}$. The result is equation (6.22), and it is second-order accurate because all differences involved in the derivation were properly centered.

Numerical relativity codes often use finite-difference representations that are higher than second order. These can be derived in complete analogy to our above derivation of the second-order stencils. The key idea is that we want to express the derivative of a function at a certain grid point, say $(\partial_x f)_i$, as a linear combination of function values at this and

---

[10] This is correct only in the limit of small Δx since for finite Δx the higher-order error terms also contribute to the truncation error.

neighboring grid points. To do so we can write the function values at these neighboring grid points in terms of a Taylor expansion about the grid point i, and then combine these expressions in such a way that all terms up to a desired order cancel out. For centered, second-order derivatives, we only need the immediate neighbors, i.e., the grid points $i + 1$ and $i - 1$, but for higher-order expressions a larger number of grid points is required. As an example, we ask the reader to work out the first and second derivative of a function to fourth order in the exercise below.

Exercise 6.6 Show that centered, fourth-order finite difference representations of the first and second derivatives of a function f are given by

$$(\partial_x f)_i = \frac{1}{12 \Delta x} (f_{i-2} - 8 f_{i-1} + 8 f_{i+1} - f_{i+2}) \tag{6.23}$$

and

$$(\partial_x^2 f)_i = \frac{1}{12 (\Delta x)^2} (-f_{i-2} + 16 f_{i-1} - 30 f_i + 16 f_{i+1} - f_{i+2}), \tag{6.24}$$

where we have omitted the truncation error, $\mathcal{O}(\Delta x^4)$.

6.2.2 Elliptic equations

As an example of a simple, one-dimensional elliptic equation consider

$$\partial_x^2 f = s. \tag{6.25}$$

For concreteness, let us assume that the solution f is a symmetric function about $x = 0$, in which case we can restrict the analysis to positive x and impose a Neuman condition at the origin,

$$\partial_x f = 0 \qquad \text{at } x = 0. \tag{6.26}$$

(Note that antisymmetry would result in the Dirichlet condition $f = 0$ at $x = 0$.) Let us also assume that f falls off with $1/x$ for large x, which results in the *Robin* boundary condition

$$\partial_x (x f) = 0 \qquad \text{as } x \to \infty. \tag{6.27}$$

We will further assume that the source term s is some known function of x.

We now want to solve the differential equation (6.25) subject to the boundary conditions (6.26) and (6.27) numerically. To do so, we first have to construct a numerical grid that covers an interval between $x_{\min} = 0$ and $x_{\max}$. Unless we compactify[11] the physical interval $[0, \infty]$ with the help of a new coordinate, for example $\xi = x/(1 + x)$, finite computer

---

[11] To compactify is to bring the outer boundary at $x = \infty$ into a finite value $\xi < \infty$ by means of a coordinate transformation from x to ξ.

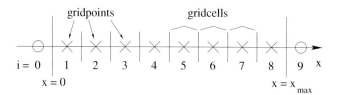

Figure 6.2 A cell-centered grid with eight grid points. The virtual grid points 0 and 9 lie outside of the domain between $x = 0$ and $x = x_{max}$ and do not need to be allocated in the computational grid.

resources will allow a uniform grid to extend to only a finite value x_{max}. We then divide the interval $[x_{min}, x_{max}]$ into N *grid cells*, leading to a grid spacing of

$$\Delta x = \frac{x_{max} - x_{min}}{N}. \tag{6.28}$$

We can choose our grid points to be located either at the center of these cells, which would be referred to as a *cell-centered* grid, or on the vertices, which would be refered to as a *vertex-centered* grid. For a cell-centered grid we have N grid points located at

$$x_i = x_{min} + (i - 1/2)\Delta x, \qquad i = 1, \ldots, N, \tag{6.29}$$

whereas for a vertex centered grid we have $N + 1$ grid points located at

$$x_i = x_{min} + (i - 1)\Delta x, \qquad i = 1, \ldots, N + 1. \tag{6.30}$$

For now, the difference between between cell-centered and vertex-centered grids only affects the implementation of boundary conditions, but not the finite difference representation of the differential equation itself. For concreteness, we will use a cell-centered grid as illustrated in Figure 6.2.

We are now ready to finite difference the differential equation (6.25) together with the boundary conditions (6.26) and (6.27). We define two arrays, f_i and s_i, which represent the functions f and s at the grid points x_i for $i = 1, \ldots, N$.[12] In the interior of our domain we can represent the differential equation (6.25) as

$$f_{i+1} - 2f_i + f_{i-1} = (\Delta x)^2 s_i \qquad i = 2, \ldots, N - 1, \tag{6.31}$$

where we have used the second-order expression (6.22) for the second derivative of f. At the lower boundary point $i = 1$ the neighbor $i - 1$ does not exist in our domain, and, similarly, at the upper boundary point $i = N$ the point $i + 1$ does not exist. At these points we have to implement the boundary conditions (6.26) and (6.27), which can be done in many different ways.

One approach is the following. Consider, at the lower boundary, the *virtual grid point* x_0, positioned at $x = -\Delta x/2$. The two grid points x_0 and x_1 then bracket the boundary point $x_{min} = 0$ symmetrically. Using the centered differencing (6.21) we can then write the

---

[12] In many computing languages, including C++, the first element of an array is refered to as $i = 0$.

boundary condition (6.26) as

$$(\partial_x f)_{1/2} = \frac{f_1 - f_0}{\Delta x} = 0, \tag{6.32}$$

or

$$f_1 = f_0, \tag{6.33}$$

to second order in Δx. For $i = 1$ we can now insert equation (6.33) into equation (6.31), which yields

$$f_{i+1} - f_i = (\Delta x)^2 s_i \qquad i = 1. \tag{6.34}$$

We have used the virtual grid point x_0 to formulate the lower boundary condition, but it does not appear in the final finite difference equation, and therefore does not need to be included in the computational arrays.

We can use a similar strategy at the upper boundary. With the help of a virtual grid point x_{N+1} we can write the boundary condition (6.27) to second order in Δx as

$$f_{N+1} = \frac{x_N}{x_{N+1}} f_N = \frac{x_N}{x_N + \Delta x} f_N. \tag{6.35}$$

We can again insert this into (6.31) for $i = N$ and find

$$\left(\frac{x_i}{x_i + \Delta x} - 2 \right) f_i + f_{i-1} = (\Delta x)^2 s_i \qquad i = N. \tag{6.36}$$

Equations (6.31), (6.34) and (6.36) now form a coupled set of N linear equations for the N elements f_i that we can write as

$$
\begin{pmatrix}
-1 & 1 & 0 & 0 & 0 & 0 & & 0 \\
1 & -2 & 1 & 0 & 0 & 0 & & 0 \\
0 & \ddots & \ddots & \ddots & 0 & 0 & & 0 \\
0 & 0 & 1 & -2 & 1 & 0 & & 0 \\
0 & 0 & 0 & \ddots & \ddots & \ddots & & 0 \\
0 & 0 & 0 & 0 & 1 & -2 & & 1 \\
0 & 0 & 0 & 0 & 0 & 1 & & x_N/(x_N + \Delta x) - 2
\end{pmatrix}
\cdot
\begin{pmatrix}
f_1 \\ f_2 \\ \vdots \\ f_i \\ \vdots \\ f_{N-1} \\ f_N
\end{pmatrix}
= (\Delta x)^2
\begin{pmatrix}
s_1 \\ s_2 \\ \vdots \\ s_i \\ \vdots \\ s_{N-1} \\ s_N
\end{pmatrix}
\tag{6.37}
$$

or, in a more compact form,

$$\mathbf{A} \cdot \mathbf{f} = (\Delta x)^2 \, \mathbf{S}. \tag{6.38}$$

The solution is given by

$$\mathbf{f} = (\Delta x)^2 \, \mathbf{A}^{-1} \cdot \mathbf{S}, \tag{6.39}$$

where $\mathbf{A}^{-1}$ is the inverse of the matrix $\mathbf{A}$, so that we have reduced the problem to inverting an $N \times N$ matrix. All equations entering (6.37) are accurate to second order, so that the solution $\mathbf{f}$ approaches the correct solution with $(\Delta x)^2$ as the leading order error term.

The matrix $\mathbf{A}$ is *tridiagonal*, meaning that the only nonzero entries appear on the diagonal and its immediate neighbors. This helps tremendously, since tridiagonal matrices can be inverted quite easily at small computational cost,[13] even for moderately large N. In particular, matrix inversion of tridiagonal matrices can be implemented in a way that does not require storing all N^2 elements (the vast majority of which vanish anyway), but instead only the diagonal band with its two neighbors.

The situation becomes more complicated in higher dimensions. Consider, for example, Poisson's equation in Cartesian coordinates in flat space,

$$\nabla^2 f = \partial_x^2 f + \partial_y^2 f = s. \tag{6.40}$$

We now construct a 2-dimensional grid in complete analogy to the 1-dimensional grid discussed above. We will denote the N_x grid points along the x-axis by x_i, and the N_y grid points along the y-axis by y_j. The functions $f(x, y)$ and $s(x, y)$ are then represented on a 2-dimenional grid and we write, for example,

$$f_{i,j} = f(x_i, y_j). \tag{6.41}$$

If we choose equal grid spacing in the x and y directions, $\Delta = \Delta x = \Delta y$, we can finite difference equation (6.40) as

$$f_{i+1,j} + f_{i-1,j} + f_{i,j+i} + f_{i,j-1} - 4f_{i,j} = \Delta^2 s_{i,j}. \tag{6.42}$$

As before we can write these coupled, linear equations (together with finite difference representations of the boundary conditions) as one big matrix equation provided we absorb the 2-dimensional arrays $f_{i,j}$ and $s_{i,j}$ into 1-dimensional vector arrays. We can do this, for example, by constructing vector arrays F_I and S_I of length $N_x \times N_y$, where the super-index I, given by

$$I = i + N_x(j - 1), \tag{6.43}$$

runs over the entire 2-dimensional grid (i, j) (see Figure 6.3). Given a super-index I we can reconstruct i and j from

$$j = I \bmod N_x + 1 \tag{6.44}$$

$$i = I - N_x(j - 1). \tag{6.45}$$

Exercise 6.7 Consider a 3-dimensional grid of size $N_x \times N_y \times N_z$. Construct a 1-dimensional index I that runs over the entire grid (i, j, k), and provide expressions that recover i, j and k from I.

---

[13] See, e.g., Press *et al.* (2007).

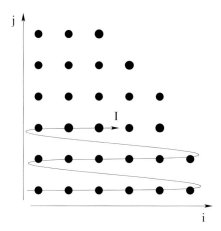

Figure 6.3 An illustration for a super-index I that sweeps through a 2-dimensional grid (i, j).

We can now write (6.42) as

$$F_{I+1} + F_{I-1} + F_{I+N_x} + F_{I-N_x} - 4F_I = \Delta^2 S_I \tag{6.46}$$

and cast the problem as a matrix equation for the vector F_I. For grid points (i, j) in the interior (i.e., away from the boundaries) the equation takes the form

$$
\begin{pmatrix}
\ddots & & \ddots & \ddots & \ddots & & \ddots & \\
\cdots & 1 & \cdots & 1 & -4 & 1 & \cdots & 1 & \cdots \\
& \ddots & & \ddots & \ddots & \ddots & & \ddots
\end{pmatrix}
\cdot
\begin{pmatrix}
\vdots \\
F_{I-N_x} \\
\vdots \\
F_{I-1} \\
F_I \\
F_{I+1} \\
\vdots \\
F_{I+N_x} \\
\vdots
\end{pmatrix}
= \Delta^2
\begin{pmatrix}
\vdots \\
S_{I-N_x} \\
\vdots \\
S_{I-1} \\
S_I \\
S_{I+1} \\
\vdots \\
S_{I+N_x} \\
\vdots
\end{pmatrix}.
\tag{6.47}
$$

Evidently the matrix **A** is no longer tridiagonal. If the boundary conditions cooperate (and they often do), it still is *band diagonal*, however, which means that the only nonvanishing elements reside on bands parallel to the diagonal (this is also called "tridiagonal with fringes"). The size of **A** is now $N_x N_y \times N_x N_y$, but special software exists that takes advantage of the band-diagonal structure, and requires storing only those nonvanishing bands. In this case we need five bands. Exercise 6.7 shows that even for a 2-dimensional covariant Laplace operator, as opposed to the flat Laplace operator in equation (6.40), we are already up to nine nonvanishing bands.

Exercise 6.8 Instead of Poisson's equation in flat space as given by equation (6.40), consider the 2-dimensional covariant version,

$$\nabla^2 f = \nabla^a \nabla_a f = g^{ab} \nabla_a \nabla_b f = g^{ab} \partial_a \partial_b f - \Gamma^a \partial_a f = s. \qquad (6.48)$$

Here the indices a and b only run over x and y, f is a scalar function, and $\Gamma^a = g^{bc} \Gamma^a_{bc}$. Do not assume the metric g_{ab} to be diagonal. Retrace the steps that led from the differential equation (6.40) to the finite difference matrix equation (6.47) and find the matrix $\mathbf{A}$ (away from the boundaries).

Not surprisingly, the problem becomes even more involved in three dimensions. Exercise 6.9 shows that a covariant Laplace operator in three dimensions leads to 19 nonvanishing bands in matrix $\mathbf{A}$.

Exercise 6.9 Consider Poisson's equation in three spatial dimensions in flat space,

$$\partial_x^2 f + \partial_y^2 f + \partial_z^2 f = s. \qquad (6.49)$$

(a) Retracing the steps from equation (6.40) to equation (6.47), determine the structure of the matrix $\mathbf{A}$.
(b) Consider the covariant Laplace equation (6.48) in three dimensions and Cartesian coordinates and find $\mathbf{A}$ for this problem.

The problem becomes so large, especially in three dimensions, that solving the equations by *direct* matrix inversion is either impossible or impractical; it simply becomes too expensive computationally. We therefore have to look for alternative methods of solution.

Sticking with our example in two dimensions, another approach becomes apparent if we rewrite equation (6.42) in the form

$$f_{i,j} = \frac{1}{4}(f_{i+1,j} + f_{i-1,j} + f_{i,j+i} + f_{i,j-1}) - \frac{\Delta^2}{4} s_{i,j}. \qquad (6.50)$$

Evidently, finite differencing the flat-space Laplace operator results in each grid function at every grid point being directly related to the average value of its nearest neighbors. We could then start with an initial guess for a solution and then sweep through the grid and update $f_{i,j}$ at each grid point according to equation (6.50). This approach is an example of a *relaxation* scheme. Many variations are possible. If we only use values from the previous sweep on the right hand side of equation (6.50), this scheme is known as *Jacobi's* method; if instead we use values that have already been updated in the current sweep, it is called the *Gauss–Seidel* method, which converges faster. Unfortunately, neither method converges rapidly enough to be of much practical use, even on modest grids.

An improvement over these methods is most easily explained in terms of the *residual* $R_{i,j}$ at each grid point, defined as

$$\mathcal{R}_{i,j} = f_{i+1,j} + f_{i-1,j} + f_{i,j+i} + f_{i,j-1} - 4f_{i,j} - \Delta^2 s_{i,j}, \qquad (6.51)$$

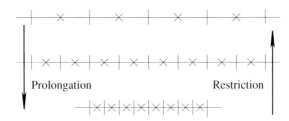

Figure 6.4 An illustration of a multigrid method.

which measures the deviation between the right-hand and left-hand sides of equation (6.42). In terms of the residual we can write Jacobi's method as

$$f_{i,j}^{N+1} = f_{i,j}^{N} + \frac{1}{4}\mathcal{R}_{i,j}^{N}, \tag{6.52}$$

where the superscript N labels the Nth sweep through the grid. This recipe implies that in each sweep we correct each grid point by adding a number that is proportional to the local residual. The idea behind *successive overrelaxation* (SOR) is to overcorrect each grid point, and to modify equation (6.52) according to

$$f_{i,j}^{N+1} = f_{i,j}^{N} + \frac{\omega}{4}\mathcal{R}_{i,j}^{N}, \tag{6.53}$$

where ω is some positive number. We refer to $\omega < 1$ as underrelaxation, and to $1 < \omega < 2$ as overrelaxation; for values of ω larger than two the scheme does not converge. For a good choice of ω, with $1 < \omega < 2$, overrelaxation can improve convergence significantly.

For most applications in three dimensions even SOR is not sufficiently fast, so that alternative techniques are needed. One alternative is a *multigrid* method, which combines the advantages of direct solvers and relaxation.[14]

In multigrid methods, which we illustrate in Figure 6.4, the numerical solution is computed on a hierarchy of computational grids with increasing grid resolution. The finer grids may or may not cover all the physical space that is covered by the coarser grids. The numerical solution is then computed by completing sweeps through the grid hierarchy. The coarse grid is sufficiently small so that we can compute a solution with a direct solver (i.e., direct matrix inversion). This provides the "global" features of the solution, albeit on a coarse grid and hence with a large local truncation error. We then interpolate this approximate solution to the next finer grid. This interpolation from a coarser grid to a finer grid is called a "prolongation", and we point out that the details of this interpolation depend on whether the grid is cell-centered (as illustrated in Figure 6.4) or vertex-centered. On the finer grid we can then apply a relaxation method, for example a Gauss–Seidel sweep. While this method is too slow to solve the problem globally, as we have discussed above, it is very well suited to improve the solution locally. This step is often called a "smoothing

[14] An introduction to multigrid and "full approximation storage" methods, together with references, can be found in Press *et al.* (2007).

sweep". After this smoothing sweep the solution can be prolonged to the next finer grid, where the procedure is repeated. Once we have smoothed the solution on the finest grid, we start ascending back to coarser grids. The interpolation from a finer grid to a coarser grid is called a "restriction". The coarser grids now "learn" from the finer grids by comparing their last solution with the one that comes back from a finer grid. This comparison provides an estimate for the local truncation error, which can be accounted for with the help of an artificial source term. On each grid we again perform smoothing sweeps, again improving the solution because we inherit the smaller truncation error from the finer grids. On the coarsest grid we again perform a direct solve, as on the finer grids with an artificial source term that corrects for the truncation error, to improve the global features of the solution. These sweeps through the grid hierarchy can be repeated until the solution has converged to a predetermined accuracy.

Another popular method is the *conjugate gradient* method, which is also designed for large, sparse matricies. There are discussions in the literature where the method is implemented for applications in numerical relativity.[15]

The good news is that several software packages that provide many of the most efficient and robust elliptic solvers are publically available. These routines are all coded up, including for parallel environments.[16] With the help of these packages the user "only" needs to specify the elements of the matrix $\mathbf{A}$ (preferably in some band-diagonal structure so that only the nonvanishing bands need to be stored) together with the source term $\mathbf{s}$ on the right hand side. The user can then choose between a number of different methods to solve the elliptic equations.

Before closing this section we briefly discuss *nonlinear* elliptic equations. Consider an equation of the form

$$\nabla^2 f = f^n g, \qquad (6.54)$$

where g is a given function and n is some number. Equation (6.54) reduces to the linear equation (6.40) when $n = 0$. The equation is still linear for $n = 1$ and, upon finite differencing, again gives rise to coupled, linear equations that are straighforward to solve by the same matrix techniques discussed above. We note that equation (4.9) for the lapse function α in maximal slicing contains a linear source term of this form, with $n = 1$. But what about the situation for other values of n, resulting in a nonlinear equation for f? For example, the Hamiltonian constraint (3.37) for the conformal factor ψ contains several source terms with $n = 1$, $n = 5$ and $n = -7$.[17] As we shall now sketch, we can solve such a nonlinear

---

[15] See, e.g., Oohara *et al.* (1997).

[16] Examples include *PETSc*, available at `acts.nersc.gov/petsc/`, and *LAPACK*, available at `www.netlib.org/lapack/`. Some of the algorithms discussed above are also implemented as modules (or "thorns") within *Cactus* (available at `www.cactuscode.org`), a parallel interface for performing large-scale computations on multiple platforms.

[17] We remind the reader that, depending on the sign of the exponent n, the solution to equation (6.54) may not be unique; see exercise 3.8 for an example.

equation with very similar matrix techniques, provided we *linearize* the equation first and then iterate to get the nonlinear solution.

Let us denote the solution after N iteration steps as f^N. We could take a very crude approach and insert f^N on the right-hand side of equation (6.54) and solve for the next iteration f^{N+1} on the left-hand side. Defining the correction δf from

$$f^{N+1} = f^N + \delta f \tag{6.55}$$

we can write this scheme as

$$\nabla^2 \delta f = -\nabla^2 f^N + (f^N)^n g. \tag{6.56}$$

The right-hand side is the negative residual of the equation after N steps,

$$\mathcal{R}^N = \nabla^2 f^N - (f^N)^n g, \tag{6.57}$$

so equation (6.56) becomes

$$\nabla^2 \delta f = -\mathcal{R}^N \quad \text{(not recommended)}. \tag{6.58}$$

Evidently this is now a linear equation for the corrections δf, which we can solve with the methods discussed above. The corrections δf also become smaller as the residual decreases, as one would hope. We can do much better, however, and construct an iteration that converges much faster, by inserting equation (6.55) on the right-hand side of equation (6.54) as well and using a Taylor expansion

$$(f^{N+1})^n = (f^N)^n + n(f^N)^{n-1}\delta f + \mathcal{O}(\delta f^2). \tag{6.59}$$

We truncate after the linear term (so that the resulting equation is still linear), insert into (6.54), and find

$$\nabla^2 \delta f - ng(f^N)^{n-1}\delta f = -\mathcal{R}^N \quad \text{(recommended)}. \tag{6.60}$$

We can think of the difference between equation (6.58) and equation (6.60) as having truncated the Taylor expansion after the first term on the right-hand side of equation (6.59), and thereby retaining only the first term on the left-hand side in equation (6.60). Clearly we expect faster convergence for the latter, which usually is the case.

The price that we pay for the faster convergence is the new term involving δf that appears on the left-hand side of (6.60). The good news is that this term can be dealt with quite easily. Equation (6.60) is of the form

$$\nabla^2 f + uf = s, \tag{6.61}$$

where, comparing with the model equation (6.25) that we considered above, the term uf is new. The exact finite difference form of this equations depends on what kind of Laplace operator we are dealing with. For a flat Laplace operator in two dimensions, for example, we could finite difference (6.61) as

$$f_{i+1,j} + f_{i-1,j} + f_{i,j+i} + f_{i,j-1} + (u_{i,j} - 4)f_{i,j} = \Delta^2 s_{i,j}. \tag{6.62}$$

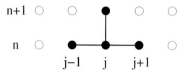

Figure 6.5 The finite-differencing stencil for the "forward-time centered-space" (FTCS) differencing scheme.

Comparing with (6.42) we see that the new term uf only affects the coefficient of the element $f_{i,j}$. In the resulting matrix equation (6.47) we can therefore account for this new term by adding $u_{i,j}$ to the diagonal of the matrix $\mathbf{A}$. To solve equation (6.60), we simply subtract $ng(f^N)^{n-1}$ from the diagonal. This can be done very easily, and leads to a significantly faster iteration than using equation (6.58).

6.2.3 Hyperbolic equations

We now turn to hyperbolic equations. As a model of hyperbolic equations, consider a "scalar" version of equation (6.7)

$$\partial_t u + v \partial_x u = 0. \tag{6.63}$$

Equation (6.63) is sometimes referred to as the *model advective equation,* for obvious reasons. For simplicity it does not contain any source terms, and the wave speed v is constant. The equation is satisfied exactly by any function of the form $u(t, x) = u(x - vt)$. In contrast to the elliptic equations of Section 6.2.2 the equation has a time derivative in addition to the space derivative, and thus requires initial data. A finite difference representation of this time derivative must involve *at least* two neighboring time levels. We will first consider two-level schemes.

As we have seen in Section 6.2.1, we can achieve a second-order differencing scheme for the space derivative in equation (6.63) by using the centered finite difference expression

$$(\partial_x u)_j^n = \frac{u_{j+1}^n - u_{j-1}^n}{2\Delta x} + \mathcal{O}(\Delta x^2). \tag{6.64}$$

Since we have decided to use a two-level scheme, a one-sided and first-order expression

$$(\partial_t u)_j^n = \frac{u_j^{n+1} - u_j^n}{\Delta t} + \mathcal{O}(\Delta t) \tag{6.65}$$

for the time derivative will have to suffice for now. Inserting both finite-difference representations into equation (6.63) we can solve for u_j^{n+1} and find

$$u_j^{n+1} = u_j^n - \frac{v}{2}\frac{\Delta t}{\Delta x}(u_{j+1}^n - u_{j-1}^n). \tag{6.66}$$

For reasons that are quite obvious this differencing scheme is called *forward-time centered-space,* or FTCS (see Figure 6.5). It is an example of an *explicit* scheme, meaning that we can solve for the grid function u_j^{n+1} at the new time level $n + 1$ directly in terms of function values on the old time level n. This is a very convenient feature, because, given all the values at the time level n, we can simply loop through the entire grid at the new time level $n + 1$ and update each grid point independently of all the other grid points; i.e., the u_j^{n+1} are uncoupled.

Unfortunately, however, FTCS is fairly useless. To see this, we perform a *von Neumann stability analysis*, which is basically a linear eigenmode analysis to test a time-dependent numerical scheme for stability. For constant coefficients, as in the case of our model equation (6.63), we can write the solution $u(t, x)$ to our continuum hyperbolic differential equation as a superposition of eigenmodes $e^{i(\omega t + kx)}$. Here k is a spatial wave number, $\omega = \omega(k)$ the wave frequency, and, at the risk of stating the obvious, $i = \sqrt{-1}$ (and *not* a grid index). We can determine the dispersion relation between ω and k by inserting the modal decomposition back into the differential equation. A real ω, for which $e^{i\omega t}$ has a magnitude of unity, yields sinusoidally oscillating modes, while the existence of a complex piece in ω leads to exponentially growing or damping modes. In the case of exponential growth, the magnitude of $e^{i\omega t}$ will exceed unity.

We can perform a similar spectral analysis of the finite difference equation. Write the eigenmode for u_j^n as

$$u_j^n = \xi^n e^{ik(j\Delta x)}. \tag{6.67}$$

Here the quantity ξ plays the role of $e^{i\omega \Delta t}$ and is called the *amplification factor*: $u_j^n = \xi u_j^{n-1} = \xi^2 u_j^{n-2} \ldots = \xi^n u_j^0$. We can find the dependence of ξ on wave number k by inserting equation (6.67) into the finite difference form of the differential equation. For the scheme to be stable, the magnitude ξ must be smaller or equal to unity for *all* k,

$$\boxed{|\xi(k)| \leq 1} \quad \text{von Neumann stability criterion.} \tag{6.68}$$

To perform a von Neumann stability anaylsis of the FTCS scheme we substitute the decomposition (6.67) into (6.66) and find

$$\xi(k) = 1 - i\frac{v\Delta t}{\Delta x}\sin k\Delta x. \tag{6.69}$$

Equation (6.69) shows that the magnitude of ξ is greater than unity for all k, indicating that this scheme is *unstable*. In fact, we have $|\xi| > 1$ independently of our choice for Δx and Δt, which makes this scheme *unconditionally unstable*. That is bad.

Exercise 6.10 Derive equation (6.69).

The good news is that there are several ways of fixing this problem. For example, we could replace the term u_j^n on the right-hand side of equation (6.66) by the spatial average

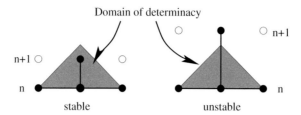

Figure 6.6 The Courant condition for many explicit finite difference implementations of hyperbolic equations requires the new grid point for u_j^{n+1} to lie inside the domain of derterminacy of the finite difference stencil on the previous time level n.

$(u_{j+1}^n + u_{j-1}^n)/2$, in which case equation (6.66) becomes

$$u_j^{n+1} = \frac{1}{2}(u_{j+1}^n + u_{j-1}^n) - \frac{v}{2}\frac{\Delta t}{\Delta x}(u_{j+1}^n - u_{j-1}^n). \qquad (6.70)$$

This differencing scheme is called the *Lax* method.

> **Exercise 6.11**   Show that a von Neumann analysis for the Lax method results in the amplification factor
>
> $$\xi = \cos k\Delta x - i\frac{v\Delta t}{\Delta x}\sin k\Delta x. \qquad (6.71)$$

The von Neumann stability criterion (6.68) then implies that we must have

$$\boxed{\frac{|v|\Delta t}{\Delta x} \le 1} \qquad (6.72)$$

for stability. This is known as the *Courant–Friedrichs–Lewy* condition, or *Courant condition* for short, and it holds for many explicit finite difference schemes for hyperbolic equations. We call the ratio between $|v|\Delta t$ and Δx the *Courant factor*.

Recalling that v represents the speed of a characteristic, we may interpret the Courant condition in terms of the domain of determinacy that we introduced in Section 6.1. Consider the situation sketched in Figure 6.6. The Courant condition (6.72) states that the grid point for u_j^{n+1} at the new time level $n + 1$ has to reside inside the domain of determinacy of the interval spanned by the finite difference stencil at the time level n. This makes intuitive sense: if u_j^{n+1} were outside this domain, its physical specification would require more information about the past than we are providing numerically, which may trigger an instability.

As we have seen, the Lax scheme is stable as long as we have chosen Δt sufficiently small so that it satisfies the Courant conditon (6.72). This makes the Lax scheme *conditionally stable*.

It seems somewhat like a miracle that simply replacing a grid function by a local average manages to change the numerical scheme from unconditionally unstable to conditionally stable. This change can be interpreted in very physical terms, however. Adding and subtracting u_j^n on the right hand side of equation (6.70) allows us to rewrite the equation as

$$u_j^{n+1} = u_j^n + \frac{1}{2}(u_{j+1}^n - 2u_j^n + u_{j-1}^n) - \frac{v}{2}\frac{\Delta t}{\Delta x}(u_{j+1}^n - u_{j-1}^n), \qquad (6.73)$$

or

$$\frac{u_j^{n+1} - u_j^n}{\Delta t} = -v\frac{u_{j+1}^n - u_{j-1}^n}{2\Delta x} + \frac{(\Delta x)^2}{2\Delta t}\frac{u_{j+1}^n - 2u_j^n + u_{j-1}^n}{(\Delta x)^2}. \qquad (6.74)$$

But equation (6.74) is a finite-difference representation of the differential equation

$$\partial_t u + v\partial_x u = D\,\partial_x^2 u, \qquad (6.75)$$

where the term on the right-hand side is essentially a *diffusion* term, with parameter $D = (\Delta x)^2/(2\Delta t)$ serving as a constant coefficient of diffusion. Such a term is identical to the one appearing in the right-hand side of the equations of particle and radiative diffusion and thermal conduction. It is also similar to the dissipation term arising from shear viscosity in the Navier–Stokes equation (see, e.g., the terms proportional to η on the right hand side of 5.71). Clearly, in the context of the model advective equation, this term is purely numerical in nature and disappears, for a constant Courant factor $|v|\Delta t/\Delta x$, in the limit $\Delta x \to 0$. We call this effect *numerical viscosity*; it tends to stabilize numerical schemes, but also introduces numerical errors in the form of anomalous diffusion and dispersion.[18] In particular, a Courant stability analysis for the Lax scheme shows that for stable evolution the magnitude of the amplification factor is always less than unity unless $|v|\Delta t \equiv \Delta x$ (in which case $|\xi| = 1$); this feature implies the amplitude of any wave will decrease spuriously with time as it propagates. A related effect is anomalous dispersion, an additional price we pay for stablity in the Lax scheme and many other finite difference schemes for hyperbolic systems. When the Courant factor is not precisely unity, there are phase errors that arise for modes of different wave numbers k. Anomalous dispersion is most serious for high wave modes $k\Delta x \gtrsim 1$, corresponding to small length scales $\lambda \lesssim \Delta x$. But we should not really expect to be able to probe features on any scale unless we are prepared to resolve that scale with many grid points, in which case anomalous dispersion is usually not a problem.

The *explicit* addition of a diffusive term to the finite-difference representation of a hyperbolic equation can sometimes be exploited to great advantage. If the term is of sufficiently high order (e.g., ∂_x^n where $n \geq 4$) then such a term serves to damp the very high frequency modes that can destabilize the numerical integration of an evolved variable. Moreover, if such a dissipative term is explicilty multiplied by an overall factor (i.e., "diffusion coefficient") that is sufficiently small in magnitude, then the new term will not otherwise distort the numerical solution appreciably. This is the basic idea behind the implementation of *Kreiss–Oliger dissipation*[19] for stabilizing finite-difference integrations

---

[18] The numerical viscosity introduced here should not be confused with the artificial viscosity introduced in Chapter 5.2.1.
[19] Kreiss and Oliger (1973).

of hyperbolic systems. The technique has proven very useful for stabilizing many different numerical integration schemes employed in numerical relativity.

There are a number of other ways of constructing stable finite-difference schemes for the model equation (6.63). A popular alternative to the Lax scheme is *upwind differencing*

$$\frac{u_j^{n+1} - u_j^n}{\Delta t} = -v \begin{cases} \dfrac{u_j^n - u_{j-1}^n}{\Delta x}, & v > 0 \\[2ex] \dfrac{u_{j+1}^n - u_j^n}{\Delta x}, & v > 0. \end{cases} \tag{6.76}$$

This scheme borrows its name from the fact that for a wave with $v > 0$, that travels "to the right", say, the new grid function u_j^{n+1} is affected only by the "upwind" grid points to its left, i.e., that lie in the region through which the wave travels before reaching x_j.

> **Exercise 6.12** Consider the "left-sided" upwind scheme for $v > 0$. Perform a von Neumann stability analysis and show that this scheme is stable as long as the Courant condition (6.72) is satisfied. Also show that the "left-sided" scheme is unstable for $v < 0$.

As written in equation (6.76) the upwind scheme treats even the spatial derivatives only to first order in Δx. This can be improved by replacing the right-hand sides with one-sided, higher-order finite-difference approximations (for example with the stencil found in exercise 6.5).

It would be desirable, however, to have a scheme that is second order in both space *and* time. One way to construct such a code is to abandon two-level schemes, and instead consider a three-level scheme. We can then construct centered derivatives both for the time-derivative,

$$(\partial_t u)_j^n = \frac{u_j^{n+1} - u_j^{n-1}}{2\Delta t} + \mathcal{O}(\Delta t^2), \tag{6.77}$$

and the space derivative (6.64). Putting these two together we have the *leap-frog* scheme,

$$u_j^{n+1} = u_j^{n-1} - v \frac{\Delta t}{\Delta x} (u_j^{n+1} - u_j^{n-1}) \tag{6.78}$$

(see Figure 6.7).

> **Exercise 6.13** Perform a von Neumann stability analysis to show that the leap-frog scheme is stable, as long as the Courant condition (6.72) is satisfied, in which case $|\xi| = 1$ holds exactly. Thus demonstrate that there is no amplitude damping in this scheme.

Some researchers prefer two-level schemes over three-level schemes because three-level schemes require initial data on two different time levels, which can be somewhat awkward. The leap-frog scheme has the additional disadvantage that, if picturing the computational grid as a chess board, it only connects fields of the same color (see Figure 6.7). "Black"

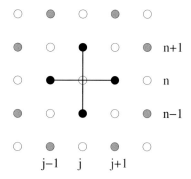

Figure 6.7 Finite difference stencil for the leap-frog scheme.

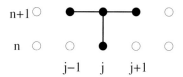

Figure 6.8 Finite difference stencil for an implicit scheme.

grid points can therefore evolve completely independently of "white" grid points, and the two sets of grid points may drift apart as numerical error accumulates differently for the two sets of points. If necessary, this problem can be solved by artificially adding a very small viscous term that links the two sets together. Once these potential issues are resolved, leap-frog is a very simple, accurate and powerful method.

Yet another way of constructing a stable two-level scheme is to use *backward* time differencing instead of forward differencing in evaluating the right-hand side of equation (6.66). We simply evaluate the right-hand side of this equation at the time level $n + 1$ instead of n (which results in a truncation error of the same linear order in Δt). This approach then yields the "backward-time, centered-space" scheme,

$$u_j^{n+1} = u_j^n - \frac{v}{2}\frac{\Delta t}{\Delta x}(u_{j+1}^{n+1} - u_{j-1}^{n+1}) \tag{6.79}$$

(see Figure 6.8). Performing a von Neumann stability analysis we find the amplification factor

$$\xi = \frac{1}{1 + iC\sin k\Delta x}, \tag{6.80}$$

where $C = v\Delta t/\Delta x$ is the Courant factor, and hence we obtain

$$|\xi(k)| \leq 1 \tag{6.81}$$

for *all* values of Δt. This finding means that this scheme is *unconditionally stable*. The size of the step size Δt is no longer restricted by stability, and instead is limited only by accuracy requirements. As we will discuss in Section 6.2.4, this property is even more

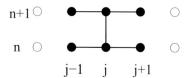

Figure 6.9 Finite difference stencil for the Crank–Nicholson scheme.

important for parabolic equations, but we shall discuss some consequences already in the present context.

The disadvantage of the backward differencing scheme (6.79) is that we can no longer solve for the new grid function u_j^{n+1} at the new time t^{n+1} explicitly in terms of old grid functions at t^n alone. Instead, equation (6.79) now couples u_j^{n+1} with the its closest neighbors u_{j+1}^{n+1} and u_{j-1}^{n+1}. This coupling provides an implicit linear relation between the new grid functions, and is therefore an example of an *implicit* finite-differencing scheme (see Figure 6.8). We can no longer sweep through the grid and update one point at a time; instead we now have to solve for all grid points simultaneously. Writing equation (6.79) at all interior grid points, and then taking into account the boundary conditions, leads to a system of equations quite similar to those we discussed in the context of elliptic equations in Section 6.2.2. Inverting the resulting linear matrix equation directly may be feasable in one spatial dimension (e.g., the matrix is tridiagonal in this example), but it is much harder in higher dimensions. One way to approach this problem is to cast the grid functions into a single vector array and call upon a routine designed to invert sparse matricies. Another way is to deal with only one spatial dimension at a time, updating each dimension in succession, and iterating until convergence. This later method is an example of the *alternating-direction implicit* or ADI method.

The scheme (6.79) is still only first order in time because we are using a one-sided expression for the time derivative. A second-order scheme would be time-centered, meaning that we should estimate the the time derivative at the mid-point between the two time levels n and $n + 1$,

$$(\partial_t u)_j^{n+1/2} = \frac{u_j^{n+1} - u_j^n}{\Delta t} + \mathcal{O}(\Delta t^2). \tag{6.82}$$

Implementing time-centering means that we also have to evaluate the space derivative at this same midpoint $n + 1/2$, which we can do by averaging between the values at n and $n + 1$. This approach yields the *Crank–Nicholson* scheme

$$u_j^{n+1} = u_j^n - \frac{v}{4}\frac{\Delta t}{\Delta x}\left((u_{j+1}^{n+1} - u_{j-1}^{n+1}) + (u_{j+1}^n - u_{j-1}^n)\right), \tag{6.83}$$

illustrated in Figure 6.9. Evidently, we can interpret the Crank–Nicholson scheme as the average between the FTCS scheme (6.66) and the backwards differencing scheme (6.79). Crank–Nicholson is second order in both space and time, and exercise 6.14 shows that it is unconditionally stable.

Exercise 6.14 Perform a von Neumann stability analysis for the Crank–Nicholson scheme (6.83) and show that

$$\xi = \frac{1 + i\beta}{1 - i\beta},\tag{6.84}$$

where $\beta = (v/2)(\Delta t/\Delta x)\sin k\Delta x$, so that $|\xi(k)| = 1$ exactly.

As we found for the fully implicit method (6.79) it is not possible to solve for the new grid functions u_j^{n+1} in the Crank–Nicholson scheme (6.83) explicitly. An alternative to inverting a tridiagonal matrix as in the implicit scheme described above (or the ADI approach in multidimensions) is the *iterative Crank–Nicholson* scheme that uses a *predictor-corrector* approach. In the predictor step we predict the new values u_j^{n+1} by using the fully explicit FTCS scheme (6.66)

$$^{(1)}u_j^{n+1} = u_j^n - \frac{v\Delta t}{2\Delta x}(u_{j+1}^n - u_{j-1}^n),\tag{6.85}$$

which, as we have seen above, would be unconditionally unstable by itself. In a subsequent corrector step we use these predicted values $^{(1)}u_j^{n+1}$ together with the u_j^n to obtain a time-centered approximation for the spatial derivative on the right hand side of equation (6.83). This step yields the corrected values of the grid function,

$$^{(2)}u_j^{n+1} = u_j^n - \frac{v\Delta t}{4\Delta x}\left((^{(1)}u_{j+1}^{n+1} - {}^{(1)}u_{j-1}^{n+1}) + (u_{j+1}^n - u_{j-1}^n)\right).\tag{6.86}$$

The corrector step can be repeated an arbitrary number times N, always using the previous values $^{(N-1)}u_j^{n+1}$ on the right-hand side to find new corrected values $^{(N)}u_j^{n+1}$. In principle this iteration should converge to the Crank–Nicholson scheme (6.83), but the stability analysis in exercise 6.15 shows that one must carry out two corrector steps and not more.[20]

Exercise 6.15 (a) Perform a von Neumann stability analysis for the iterative Crank–Nicholson scheme and find the amplification factor after $N = 0$, 1 and 2 iteration steps.
(b) Now expand the amplification factor (6.84) of Exercise 6.14 for the (noniterative) Crank–Nicholson scheme in powers of β, and show that the first few terms agree with those you found in part (a).
(c) The result of part (b) suggests that the amplification factor ξ of the iterative Crank–Nicholson scheme converges for large N to its non-iterative counterpart found in exercise 6.14 with $|\xi| = 1$, as we would expect. However, it turns out that it does so in an *alternating* pattern. To see this, find the magnitude of ξ for the first few N and show that $|\xi| > 1$ for $N = 0$ and 1, but $|\xi| < 1$ for $N = 2$ and 3, provided the Courant condition $\beta^2 \leq 1$ holds. For $N = 4$ and 5 we have $|\xi| > 1$ again, and so forth. This finding shows that the smallest value of N for which the iterative Crank–Nicholson scheme is stable with $|\xi| < 1$ is $N = 2$. Since this step is already second order accurate, there is no reason to carry out more corrections.

[20] See Teukolsky (2000).

The iterative Crank–Nicholson scheme is an explicit two-level scheme that is second order in both space and time. As we will see in Section 6.2.4 it can also handle second-order space derivatives as they appear in wave equations when they are written in the form (6.5). Since this form is very similar to the $3 + 1$ equations and related formulations (see Chapter 11), the iterative Crank–Nicholson scheme has often been used in numerical relativity simulations.

One restriction of the Crank–Nicholson scheme, as well as its iterative cousin, is that it is only second-order accurate. It is quite straightforward to generalize the space differencing to higher order – as, for example, in exercise 6.6 – but we cannot implement a higher-order time differencing in a two-level scheme. A popular alternative to these complete finite-difference schemes is therefore the *method of lines* (or MOL for short).

The basic idea of the method of lines is to finite difference the space derivatives only. In the above approaches we represented the function $u(t, x)$ on a spacetime lattice, on which the function u takes the values $u_j^n = u(t^n, x_j)$. Now we introduce a spatial grid only, at least for now, so that the function values at these grid point, $u_j(t) = u(t, x_j)$, remain functions of time. As a result, our partial differential equation for $u(t, x)$ becomes a set of ordinary differential equations for the grid values $u_j(t)$.

As a concrete example we will return to our model advection equation (6.63). We will assume $v > 0$, and could choose to use a one-sided, first-order spatial derivative stencil as in the upwind scheme (6.76). This would lead us to the set of *ordinary* differential equations

$$\frac{du_j}{dt} = -v\frac{u_j - u_{j-1}}{\Delta x}. \tag{6.87}$$

Equations (6.87) have to be integrated at all grid points i, except possibly on the boundaries. The boundary conditions may either result in ordinary differential equations also, or algebraic equations, in which case the resulting system is called a "system of differential algebraic equations" (or DAE).

The next question is how to integrate the ordinary differential equations. To start with something very simple we could assume a fixed time step Δt and use forward *Euler differencing* as in equation (6.65),

$$\left(\frac{du_j}{dt}\right)^n = \frac{u_j^{n+1} - u_j^n}{\Delta t} + \mathcal{O}(\Delta t). \tag{6.88}$$

This approach introduces a "time" grid again, labeled by n, and it should not come as a great surprise that the resulting equation is the upwind scheme (6.76). Likewise, it should not be surprising that all our stability considerations and the Courant condition (6.72) on Δt apply exactly as above.

The appealing feature of the method of lines, however, is that we can use *any* method for the integration of the ordinary differential equations that we like. In fact, many such methods, including very efficient, high-order methods, are precoded and readily available. One such algorithm is the ever-popular Runge–Kutta method. To implement, say, a fourth-order scheme for our model equation (6.63), we could adopt the fourth-order differencing stencil of exercise 6.6 to replace the spatial derivative, yielding

$$\frac{du_j}{dt} = -\frac{v}{12\Delta x}(u_{i-2} - 8u_{j-1} + 8u_{j+1} - u_{i+2}),\tag{6.89}$$

and then integrate this set of ordinary differential equations with a fourth-order Runge–Kutta method. Clearly, this approach is much easier than implementing a fourth-order time differencing scheme from scratch. Moreover, it is quite straightforward to generalize this approach to even higher order. Many readily available ordinary differential equation integrators are also equipped with an automated step size control, so that the equations can be integrated in time to a given predetermined accuracy. All of these considerations make the method of lines a very attractive algorithm.

6.2.4 Parabolic equations

Consider now the prototype parabolic equation (6.2), which, if we assume κ to be constant and the source tem ρ to vanish, reduces to

$$\partial_t \phi - \kappa \partial_x^2 \phi = 0.\tag{6.90}$$

An equation of this form arises in particle and radiative diffusion, as well as in heat conduction. To finite difference this equation we can start with the simple explicit FTCS scheme as in Section 6.2.3, now using equation (6.22) to represent the spatial derivative, and find

$$\phi_j^{n+1} = \phi_j^n + \frac{\kappa \Delta t}{(\Delta x)^2}(\phi_{j+1}^n - 2\phi_j^n + \phi_{j-1}^n).\tag{6.91}$$

Exercise 6.16 Perform a von Neumann stability analysis for the finite difference equation (6.91) and show that it is stable as long as

$$\frac{2\kappa \Delta t}{(\Delta x)^2} \leq 1.\tag{6.92}$$

The stability criterion (6.92) is the analogue of the Courant condition (6.72) for parabolic equations. It can be interpreted quite easily in physical terms: condition (6.92) states that the time step Δt must not exceed the time scale required to random walk (diffuse) a distance across one spatial cell, $\Delta t_{\text{ran.walk}} = (\Delta x)^2/2\kappa$.

Exercise 6.17 Consider the two-dimensional diffusion equation

$$\partial_t \phi - \kappa(\partial_x^2 \phi + \partial_y^2 \phi) = 0.\tag{6.93}$$

(a) Finite difference this equation in analogy to the FTCS scheme (6.91), with the same uniform grid spacing $\Delta = \Delta x = \Delta y$ in the x and y direction, and show that this scheme is stable as long as the condition

$$\frac{4\kappa\,\Delta t}{\Delta^2} \leq 1 \qquad\qquad (6.94)$$

is satisfied.

(b) Recall our earlier discussion suggesting that an elliptic equation of the form

$$\partial_x^2\phi + \partial_y^2\phi = 0 \qquad\qquad (6.95)$$

can be solved be integrating the parabolic equation (6.93) until an equilibrium solution with $\partial_t\phi = 0$ has been achieved (see, e.g., our discussion of maximal slicing in Chapter 4.2). Show that using the maximum allowed time step in your FTCS finite difference representation of equation (6.93) yields Jacobi's method (6.52), which, as we have discussed in Section 6.2.2, may prove too slow for most applications in three dimensions.

Unfortunately, the stability criterion (6.92) imposes quite a severe limitation on the size of the time step. Unlike in the Courant condition (6.72), where Δt scales linearly with the grid size Δx, it now has to decrease with the square of Δx as we refine the resolution. For many applications this restriction makes this scheme impractical, since it takes too many timesteps to simulate a physical process for a sufficiently long total time.

As we have seen in Section 6.2.3 we can overcome this problem by constructing an *implicit* differencing scheme. We now evaluate the right hand side of equation (6.91) at the time level $n + 1$ instead of n, which yields

$$\phi_j^{n+1} = \phi_j^n + \frac{\kappa\,\Delta t}{(\Delta x)^2}(\phi_{j+1}^{n+1} - 2\phi_j^{n+1} + \phi_{j-1}^{n+1}) \qquad\qquad (6.96)$$

(see Figure 6.8). Exercise 6.18 shows that this implicit scheme is *unconditionally stable*, meaning it is stable without any condition on the time step Δt. Clearly this is a huge advantage, since the step size is now limited only by accuracy requirements, and no longer by stability constraints.

> **Exercise 6.18** Perform a von Neumann stability analysis and show that the scheme (6.96) is unconditionally stable.

Of course, the implicit nature of the scheme also has disadvantages, as we have discussed in Section 6.2.3. Instead of being able to sweep through the grid and update one grid point at a time, we now have to solve for the grid functions at all grid points simultaneously.

Given that with an implicit method the size of the time step Δt is limited by accuracy requirements only, it is well-worth improving the accuracy of the scheme. The best candidate for improvement is the time derivative, which in equation (6.96) is still implemented only to first order in the truncation error. A second-order scheme would be centered, meaning that we must evaluate the time derivative at the mid-point between the two time-levels n and $n + 1$. Centering is accomplished by averaging the spatial derivatives at the two time levels, and, as in our discussion in Section 6.2.3, such averaging leads to the *Crank–Nicholson* scheme

$$\phi_j^{n+1} = \phi_j^n + \frac{\kappa \Delta t}{2(\Delta x)^2} \left((\phi_{j+1}^{n+1} - 2\phi_j^{n+1} + \phi_{j-1}^{n+1}) + (\phi_{j+1}^n - 2\phi_j^n + \phi_{j-1}^n) \right). \quad (6.97)$$

This scheme is second order in both space and time, and exercise 6.19 shows that it is unconditionally stable.

> **Exercise 6.19**   Perform a von Neumann stability analysis for the Crank–Nicholson scheme (6.97), find the amplification factor and show that the scheme is unconditionally stable.

We close this section by recalling that wave equations can be written as the coupled system (6.5), where, similar to the parabolic equation (6.90), the second equation contains a first time derivative and a second space derivative (but of opposite sign). Not surprisingly, the computational methods discussed in this section are useful for integrating (hyperbolic) wave equations as well. Since the $3 + 1$ equations and other decompositions related to it are quite similar to equation (6.5) in structure, these methods are important for many applications in numerical relativity.

6.2.5 Mesh refinement

As we have discussed in Section 6.2.1, many current numerical relativity codes use a uniform grid spacing to cover the entire spatial domain. Given that computational resources are limited, so that we can afford only a finite number of grid points, such a "unigrid" implementation may pose a problem, especially for a dynamical simulation in three spatial dimensions. Imagine, for concreteness, a simulation of a strong-field gravitational wave source, like a compact binary containing neutron stars or black holes. On the one hand we have to resolve these sources well, so as to minimize truncation error in the strong-field region. On the other hand, the grid must extend into the weak-field region at large distances from the sources, so as to minimize error from the outer boundaries and to enable us to extract the emitted gravitational radiation accurately (see Chapter 9). One possible solution to this classic "dynamic range" problem is to introduce a new coordinate system that serves to cover a larger spatial region for the same number of grid points. For example, replacing a uniform radial grid by one proportional to the logarithm of the radius can extend the computational domain out to larger distances for the same number of grid points. It is also possible to introduce a coordinate system that maps spatial infinity to a finite coordinate value ("compactification").[21] These approaches, however, may also introduce new problems. Consider, for example, the propagation of outgoing gravitational radiation. Its physical wavelength is fixed but in these new coordinate systems that have

---

[21] Recall the discussion following equation (6.27), which suggests why compactification can be particularly useful for imposing asymptotic boundary conditions.

their highest spatial resolution in the strong-field near zone the radiation is resolved by fewer and fewer grid points as it propagates out to larger distances. This effect may thus spoil the quality of wave extraction if the coordinate transformations are too naive.

> **Exercise 6.20**   Show how radial resolution diminishes with increasing radius r for a grid that is uniform in the logarithmic coordinate $y = \ln r$.

A very promising alternative is mesh refinement, which has been widely developed and used in the computational fluid dynamics community and is becoming increasingly popular in numerical relativity. In fact, adaptive mesh refinement was instrumental in the discovery of critical phenomena in general relativity[22] and has played a key role in the simulations of binary black holes.[23]

The basic idea underlying mesh refinement techniques is to perform the simulation not on one numerical grid, but on several, as in the multigrid methods for elliptic equations that we discussed in Section 6.2.2 (see Figure 6.4). A coarse grid covers the entire space, and extends to large physical separations. Wherever finer resolution is needed to resolve small-scale structures, as is the case, for example, of a compact binary emitting gravitational radiation, a finer grid is introduced. Typically, the grid spacing on the finer grid is half that on the next coarser grid, but clearly other refinement factors can be chosen. The hierarchy can be extended, and typical mesh refinement applications employ multiple refinement levels.

While the concept is quite simple, many of the details and the implementation of mesh refinement are fairly subtle. In particular, the boundary conditions imposed on the refined grids have to be posed and implemented with some care, since otherwise waves will reflect off these interfaces, leading to spurious numerical artifacts.[24]

Two versions of mesh refinement can be implemented. In the simpler version, called *fixed mesh refinement* or FMR, it is assumed that the refined grids will be needed only at known locations in space that remain fixed throughout the simulation. The center of a pulsating star, for example, may remain fixed at the origin, so that nested refinements boxes of fixed size and centered at the origin are adequate to refine the computational domain. The situation is more complicated for objects that are moving, as is the case for a coalescing binary star system. In this case we do not know a priori the trajectories of the companion stars, hence do not know which regions need refining. Moreover, these regions will be changing as the system evolves and the stars move. Clearly, we would like to move the refined grids with the stars. Such an approach, whereby the grid is relocated during the simulation to give optimal resolution at each time step, is called *adaptive mesh refinement* or AMR.[25] An example of an AMR implementation in numerical relativity, simulating the

---

[22] See Chapter 8.4.

[23] See Chapter 13.

[24] A discussion of these issues in the context of numerical relativity can be found, for example, in Schnetter *et al.* (2004) and the references therein.

[25] See Berger and Oliger (1984).

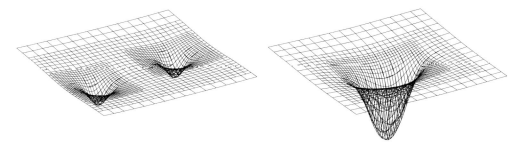

Figure 6.10 A simulation of neutron star coalescence using adaptive mesh refinement. Shown are the lapse function at an early time before merger (left panel) and a late time after merger (right panel) in the inner part of the computational grid, together with a mesh (representing every other grid point used in the simulation) that shows how the refinement levels track the motion of the neutron stars. [From Evans *et al.* (2005).]

head-on collision of two neutron stars, is shown in Figure 6.10. Other examples utilizing FMR and AMR will be cited when we discuss applications in later chapters.

6.3 Spectral methods

A general introduction to spectral methods can be found in several books and monographs on the subject.[26] Applications to numerical relativity have been described in many papers and review articles.[27] Here we shall restrict our own discussion to a brief summary of some of the key features of spectral methods.

6.3.1 Representation of functions and derivatives

Spectral methods approximate the solution to a differential equation, say $u(x)$, as a truncated series in some complete set of basis functions $\phi_k(x)$,

$$u(x) \simeq u^{(N)}(x) = \sum_{k=0}^{N} \tilde{u}_k \phi_k(x). \tag{6.98}$$

Here the coefficients $\tilde{u}_k$, which do not depend on x, are called the spectral coefficients. Evidently we can express derivatives of $u^{(N)}(x)$ *analytically* in terms of derivatives of the known basis functions $\phi_k(x)$, for example

$$\partial_x u^{(N)}(x) = \sum_{k=0}^{N} \tilde{u}_k \partial_x \phi_k(x). \tag{6.99}$$

[26] See, e.g., Gottlieb and Orszag (1977) and Boyd (2001).
[27] See, e.g., Bonazzola *et al.* (1999b), Gourgoulhon (2002) and Pfeiffer *et al.* (2003b). Our treatment draws significantly from the discussions in these references.

Given that we can represent derivative operators exactly, and given that for many applications contributions from the higher order basis functions ϕ_k decrease very quickly with k, the numerical error in spectral methods often drops off exponentially with N. To achieve the same accuracy as finite difference methods (for which the error typically falls of with some small power of the number of gridpoints N), spectral methods often require significantly fewer computational resources, which makes them a very powerful and attractive alternative. On the downside, spectral methods are often more complicated to implement then finite difference methods, and they are less suitable for the modeling of situations in which discontinuities (in either functions or their derivatives) may occur.

The general idea underlying spetral methods is to write all expressions in the differential equation and its boundary conditions in terms of the $N + 1$ basis functions $\phi_k(x)$. We then derive and solve a set of $N + 1$ equations for the $N + 1$ spectral coefficients. The approximate solution $u^{(N)}(x)$ is then given in terms of the spectral coefficients by equation (6.98). Several choices have to be made along the way – in particular we have to pick a set of basis functions, and we have to decide how exactly to construct the conditions on the spectral coefficients imposed by the differential equation – but before discussing these in more general terms it may be useful to work through a simple example problem.

6.3.2 A simple example

Consider the 1-dimensional, linear differential equation[28]

$$\partial_x^2 u - (x^6 + 3x^2)u = 0, \tag{6.100}$$

subject to the boundary conditions

$$u(-1) = u(1) = 1. \tag{6.101}$$

The exact solution to this equation is given by

$$u(x) = \exp\left(\frac{x^4 - 1}{4}\right). \tag{6.102}$$

For basis functions $\phi_k(x)$ we could simply use polynomials x^k, but to satisfy the boundary conditions (6.101) automatically we will instead employ a polynomial expansion in the form[29]

$$u^{(N)} = 1 + (1 - x^2)(\tilde{u}_0 + \tilde{u}_1 x + \tilde{u}_2 x^2 + \cdots + \tilde{u}_N x^N). \tag{6.103}$$

---

[28] This section follows Section 1.2 of Boyd (2001), who provides this example.
[29] This expansion is not exactly in the form of equation (6.98), but nevertheless illustrates very nicely some of the important properties of spectral solutions.

Our goal now is to determine the coefficients $\tilde{u}_k$ in such a way so as to make the residual, defined as

$$R^{(N)} = \partial_x^2 u^{(N)} - (x^6 + 3x^2)u^{(N)}, \qquad (6.104)$$

small. It is not clear, a priori, how to measure the size of this residual. One possible approach is to evaluate the residual at several points x_j.

For concreteness, assume $N = 2$, in which case the residual (6.104) becomes

$$\begin{aligned} R^{(2)} = {} & (2\tilde{u}_2 + 2\tilde{u}_0) - 6\tilde{u}_1 x - (3 + 3\tilde{u}_0 + 12\tilde{u}_2)x^2 - 3\tilde{u}_1 x^3 \\ & + 3(\tilde{u}_0 - \tilde{u}_2)x^4 + 3\tilde{u}_1 x^5 + (-1 - \tilde{u}_0 + 3\tilde{u}_2)x^6 \\ & - \tilde{u}_1 x^7 + (\tilde{u}_0 - \tilde{u}_2)x^8 + \tilde{u}_1 x^9 + 10\tilde{u}_2 x^{10}. \end{aligned} \qquad (6.105)$$

We now need three equations for the three unknown coefficients $\tilde{u}_0$, $\tilde{u}_1$ and $\tilde{u}_2$, so we evaluate this residual at three points, say $x_0 = -1/2$, $x_1 = 0$, and $x_2 = 1/2$. The resulting set of linear equations can be represented as the matrix equation

$$\begin{pmatrix} \dfrac{659}{256} & -\dfrac{1683}{512} & \dfrac{1171}{1024} \\[2mm] -2 & 0 & 2 \\[2mm] \dfrac{659}{256} & -\dfrac{1683}{512} & \dfrac{1171}{1024} \end{pmatrix} \begin{pmatrix} \tilde{u}_0 \\[2mm] \tilde{u}_1 \\[2mm] \tilde{u}_2 \end{pmatrix} = \begin{pmatrix} -\dfrac{49}{64} \\[2mm] 0 \\[2mm] -\dfrac{49}{64} \end{pmatrix}. \qquad (6.106)$$

The solution is given by

$$\tilde{u}_0 = \tilde{u}_2 = -\frac{784}{3807} \qquad \tilde{u}_1 = 0. \qquad (6.107)$$

Given the structure of the differential equation (6.100) and the boundary condition (6.101) we could have guessed that the solution must be symmetric, which implies that all odd coefficients $\tilde{u}_j$, including $\tilde{u}_1$, must be zero. That means that we only had to search for two coefficients rather than three, illustrating the fact that it is often useful to think about the solution before constructing it numerically. This also means that $\tilde{u}_3$ must be zero as well, so that our solution's leading order error term is the fourth order term.

We plot our spectral solution, together with the exact solution and the error, in Figure 6.11. The relative error is on the order of 2%, which is remarkable small given that effectively we constructed the spectral solution with only two free parameters. Clearly, it would not be possible to achieve this accuracy with only two grid points in a finite difference approach. The accuracy of the spectral solution does depend, however, on the choices that we have made along the way, namely the basis functions ϕ_k and the evaluation of the residual. In this simple example problem we have made more or less ad hoc choices, partly motivated by the boundary conditions. In the next section we will discuss "smarter" choices that are known to work well under quite general circumstances.

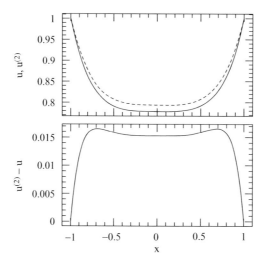

Figure 6.11 The top panel shows the exact solution $u(x)$ (solid line) and the spectral solution $u^{(2)}(x)$ (dashed line) to the differential equation (6.100), subject to boundary conditions (6.101). The bottom panel shows the absolute error $u^{(2)}(x) - u(x)$.

Exercise 6.21 The *Chebychev polynomials* $T_n(x)$ satisfy the differential equation

$$(1 - x^2)^{1/2} \frac{d}{dx}\left((1 - x^2)^{1/2}\frac{dT_n(x)}{dx}\right) = -n^2 T_n(x) \qquad (6.108)$$

on the interval $[-1, 1]$. Solutions $T_n(x)$ exist only for certain values of the eigenvalue n, which have to be determined together with the solution. Show that regular solutions satisfy the boundary conditions

$$\begin{aligned} T_n' &= n^2 T_n, & x &= 1, \\ T_n' &= -n^2 T_n, & x &= -1. \end{aligned} \qquad (6.109)$$

Even if we didn't know already that the solutions are in fact polynomials we might try expanding them in terms of polynomials,

$$T_n(x) \simeq \sum_{k=0}^{N} a_k x^k. \qquad (6.110)$$

For simplicity truncate at $N = 2$ so that $T_n(x) \simeq a_0 + a_1 x + a_2 x^2$, and express the differential equation (6.108) and boundary conditions (6.109) in terms of the coefficients a_0, a_1 and a_2. You need three equations for the three coefficients; two follow from the boundary conditions, and so only one more condition needs to be supplied by the differential equation. To get this last condition you could evaluate the differential equation at one point, e.g., $x = 0$. Instead, leave x undetermined and impose the differential equation for an *arbitrary* value x_0. Form a set of linear equations for the coefficients a_0, a_1 and a_2 and show that you can find nontrivial solutions only for the eigenvalues $n = 0$, 1 and 2. Also find the corresponding eigenvectors, from which you can construct the Chebychev polynomials. Your solutions should be independent of x_0, indicating that they satisfy the differential equation (6.108) at all points, which in turn means that your solutions are not just approximate solutions, but in fact give the correct lowest-order Chebychev polynomials (see equation 6.115 below).

6.3.3 Pseudo-spectral methods with Chebychev polynomials

Consider a differential equation

$$Lu(x) = s(x), \tag{6.111}$$

subject to a suitable set of boundary conditions, where L is some differential operator, and $s(x)$ a source function. Our goal is then to find a set of spectral coefficients $\tilde{u}_k$ associated with the Nth order expansion $u^N(x)$ of $u(x)$ so that the residual

$$R = Lu^{(N)} - s \tag{6.112}$$

becomes small. Different spectral methods differ in how this residual is evaluated.

For example, we could project the residual back into the basis functions $\phi_k(x)$ by performing an overlap integral of equation (6.112) with each of the functions $\phi_k(x)$. If the basis functions $\phi_k(x)$ all satisfy the boundary conditions, this approach is called a *Galerkin* method. If on the other hand the $\phi_k(x)$ do not satisfy the boundary conditions individually and therefore have to be combined in such a way that the combination satisfies the boundary conditions – which results in extra equations for the spectral coefficients – this method is called a *tau* method.

Alternatively we can evaluate the residual (6.112) at a certain set of $N + 1$ points x_i, called the *collocation points*, to derive $N + 1$ equations for the $N + 1$ spectral coefficients $\tilde{u}_k$.[30] This approach, which we used in the simple example presented in Section 6.3.2, is called a *pseudo-spectral* or *collocation* method. It is also the most popular method in numerical relativity, and we will therefore focus on this approach.

We still have to choose a set of basis functions. For periodic problems, a Fourier expansion in sines and cosines is recommended, but for most applications in numerical relativity the most suitable set of basis functions are Chebychev polynomials.[31] The properties of the Chebychev polynomials also suggest a particular choice for the collocation points.

Chebychev polynomials are defined on the interval $[-1, 1]$ as

$$T_n(\cos\theta) = \cos(n\theta) \tag{6.113}$$

---

[30] We may later have to replace some of these equations to account for the boundary conditions.
[31] Boyd (2001) offers the following "Moral Principle 1":

1. When in doubt, use Chebychev polynomials unless the solution is spatially periodic, in which case ordinary Fourier series is better.
2. Unless you're sure another set of basis functions is better, use Chebychev polynomials.
3. Unless you're really, really sure that another set of basis functions is better, use Chebychev polynomials.

and satisfy the singular Sturm–Liouville problem

$$(1 - x^2)^{1/2} \frac{d}{dx} \left((1 - x^2)^{1/2} \frac{dT_n(x)}{dx} \right) = -n^2 T_n(x). \tag{6.114}$$

The first few polynomials are

$$T_0(x) = 1,$$
$$T_1(x) = x,$$
$$T_2(x) = 2x^2 - 1, \tag{6.115}$$
$$T_3(x) = 4x^3 - 3x,$$
$$T_4(x) = 8x^4 - 8x^2 + 1,$$

(*cf.* exercise 6.21). Higher-order polynomials can be constructed from the recurrence relation

$$T_{n+1} = 2x\, T_n(x) - T_{n-1}(x), \tag{6.116}$$

which holds for $n \geq 1$. Chebychev polynomials form a complete set and are orthogonal in the interval $[-1, 1]$ with a weight $w(x) = (1 - x^2)^{-1/2}$,

$$(T_i, T_j) \equiv \frac{2 - \delta_{i0}}{\pi} \int_{-1}^{1} T_i(x) T_j(x) w(x) dx = \delta_{ij}, \tag{6.117}$$

where we have defined (T_i, T_j), the scalar product between T_i and T_j.

The polynomial $T_N(x)$ has N zeros in $[-1, 1]$ located at

$$x_i = \cos \left(\frac{\pi(k + 1/2)}{N} \right), \qquad k = 0, 1, \ldots, N - 1. \tag{6.118}$$

Between these zeros, the Chebychev polynomials take maxima of value 1 or minima of -1 at locations

$$x_i = \cos \left(\frac{\pi k}{N} \right), \qquad k = 0, 1, \ldots, N. \tag{6.119}$$

At the endpoints at $x = \pm 1$ they take values

$$T_n(-1) = (-1)^n, \qquad T_n(1) = 1. \tag{6.120}$$

The facts that the Chebychev polynomials oscillate between $+1$ and -1, and that two successive Chebychev polynomials assume their extrema at locations that are staggered, means that the truncation error due to the neglect of terms of order higher than N is spread more or less "evenly" over the interval $[-1, 1]$. This feature is one of the properties that makes these polynomials so attractive for spectral methods.

Using the orthogonality of the Chebychev polynomials, we can invert equation (6.98) for Chebychev polynomials and compute the coefficients $\tilde{u}_k$ from

$$\tilde{u}_i = (u, T_i) = \frac{2 - \delta_{i0}}{\pi} \int_{-1}^{1} T_i(x) u(x) w(x) dx. \tag{6.121}$$

In reality, however, we cannot evaluate this integral exactly on a computer. A very accurate method for computing this integral is either Gauss integration, in which case the collocation points are the $N+1$ zeros of $T_{N+1}(x)$ (given by equation 6.118), or Gauss–Lobatto integration, for which the collocation points are the $N+1$ extrema of $T_N(x)$ (given by equation 6.119). The latter set includes the boundary at $x = \pm 1$, while the former does not.

In the case of Gauss–Lobatto integration, the integral (6.121) for the scalar product between $u(x)$ and $T_j(x)$ reduces to the discrete expression

$$\tilde{u}_i = \frac{2}{Nc_i} \sum_{k=0}^{N} \frac{1}{c_k} u_k T_i(X_k), \tag{6.122}$$

where we have defined

$$c_i = \begin{cases} 2, & k = 0 \text{ or } k = N, \\ 1, & k = 1, \ldots, N-1, \end{cases} \tag{6.123}$$

and where the collocation points X_i are given by the extrema (6.119). The u_k are the values of the function $u(x)$ at the collocation points,

$$u_i = u^{(N)}(X_i) = \sum_{k=0}^{N} \tilde{u}_k T_k(X_i). \tag{6.124}$$

Evidently, with the help of (6.122) and (6.124) we can transform back and forth between the u_i's or the $\tilde{u}_k$'s, meaning that we can represent the function $u^{(N)}$ by either of the two sets. Realizing that the Chebychev polynomials are just cosines in disguise (see their definition equation 6.113), we can perform these transformations very efficiently with the help of fast Fourier transform (FFT) techniques, which is another reason for the popularity of Chebychev polynomials.

Evaluated at the collocation points, the orthogonality relation (6.117) becomes the discrete expression

$$\frac{2}{Nc_k} \sum_{k=0}^{N} \frac{1}{c_k} T_i(X_k) T_j(X_k) = \delta_{ij}. \tag{6.125}$$

Exercise 6.22 Verify equation (6.125) by using equations (6.122) and (6.124).

6.3.4 Elliptic equations

For simplicity, imagine we want to solve the same 1-dimensional elliptic equation

$$\partial_x^2 f(x) = s(x) \tag{6.126}$$

considered at the beginning of Section 6.2.2. We now write all expressions in this equation in terms of Chebychev polynomials. Before we can do this, we may first have to perform a coordinate transformation, since the Chebychev polynomials exist only on the interval

$[-1, 1]$, whereas in the physical problem of interest the coordinate x may span a completely different interval. In the following we assume that a suitable coordinate transformation has been performed, so that the new coordinate x indeed covers the interval $[-1, 1]$. We then expand the function $f(x)$ as

$$f(x) = \sum_{k=0}^{N} \tilde{f}_k T_k(x). \tag{6.127}$$

As we have said in Section 6.3.1, the beauty of spectral methods lies in the fact that we can express derivatives of functions analytically in terms of derivatives of the basis functions, so in this case

$$\partial_x f(x) = \sum_{k=0}^{N} \tilde{f}_k T_k'(x), \tag{6.128}$$

where the prime denotes a derivative with respect to x. The goal is now to express derivatives of the Chebychev polynomials in terms of Chebychev polynomials themselves. To do that the identities proven in exercise 6.23 will be useful.

Exercise 6.23   Show that derivatives of Chebychev polynomials satisfy the relation

$$T_n'(x) = 2n T_{n-1}(x) + \frac{n}{n-2} T_{n-2}'(x), \qquad n > 2, \tag{6.129}$$

as well as $T_2'(x) = 4T_1(x)$, $T_1'(x) = T_0$ and, evidently, $T_0'(x) = 0$. Also show that

$$\begin{aligned} T_n'(x) &= 2n(T_{n-1}(x) + T_{n-3}(x) + T_{n-5}(x) + \cdots + T_1(x)), & n \text{ even}, \\ T_n'(x) &= 2n(T_{n-1}(x) + T_{n-3}(x) + T_{n-5}(x) + \cdots + T_2(x)) + n T_0(x), & n \text{ odd}. \end{aligned} \tag{6.130}$$

Given that the derivatives $T_n'(x)$ are functions like any other, we can expand them as

$$T_k'(x) = \sum_{l=0}^{N} D_{lk} T_l(x) \tag{6.131}$$

where, according to equation (6.121), the coefficients D_{lk} are the projections of $T_k'(x)$ into $T_l(x)$ given by

$$D_{lk} = (T_l, T_k'). \tag{6.132}$$

Inserting equation (6.130) for $l \geq 0$ and $k \geq 0$ we find

$$D_{lk} = \begin{pmatrix} 0 & 1 & 0 & 3 & 0 & 5 & \cdots \\ 0 & 0 & 4 & 0 & 8 & 0 & \cdots \\ 0 & 0 & 0 & 6 & 0 & 10 & \cdots \\ 0 & 0 & 0 & 0 & 8 & 0 & \cdots \\ 0 & 0 & 0 & 0 & 0 & 10 & \cdots \\ 0 & 0 & 0 & 0 & 0 & 0 & \cdots \\ \vdots & \vdots & \vdots & \vdots & \vdots & \vdots & \ddots \end{pmatrix}. \tag{6.133}$$

Exercise 6.24 Given a function $f(x)$ expanded as in equation (6.127), we can also write its derivative $f'(x)$ as

$$\partial_x f(x) = \sum_{l=0}^{N} \tilde{f}'_l T_l(x). \tag{6.134}$$

Combining equations (6.128) and (6.131) shows that the expansion coefficients $\tilde{f}'_k$ are related to the coefficients $\tilde{f}_k$ by

$$\tilde{f}'_l = \sum_{k=0}^{N} D_{lk} \tilde{f}_k. \tag{6.135}$$

Show that the $\tilde{f}'_k$'s satisfy

$$\tilde{f}'_l = 2(l+1)\tilde{f}_{l+1} + \tilde{f}'_{l+2} \quad l < N-1; \, l > 0. \tag{6.136}$$

This recurrence relation can be used to find the derivative of a function $f(x)$ directly from its spectral coefficients $\tilde{f}_l$.

Inserting equation (6.135) into (6.134) we now have

$$\partial_x f(x) = \sum_{k=0}^{N} \sum_{l=0}^{N} \tilde{f}_k D_{lk} T_l(x), \tag{6.137}$$

demonstrating that we can express the first derivative of $f(x)$ in terms of Chebychev polynomials simply by applying the matrix D_{lk} to the spectral coefficients $\tilde{f}_k$. Luckily, we already have developed all the tools to construct similarly the second derivative of $f(x)$ as it appears in the differential equation (6.126),

$$\partial_x^2 f(x) = \partial_x \left(\sum_{k=0}^{N} \sum_{l=0}^{N} \tilde{f}_k D_{lk} T_l(x) \right) = \sum_{k=0}^{N} \sum_{l=0}^{N} \tilde{f}_k D_{lk} \partial_x T_l(x)$$

$$= \sum_{k=0}^{N} \sum_{l=0}^{N} \sum_{m=0}^{N} \tilde{f}_k D_{lk} D_{ml} T_m(x). \tag{6.138}$$

Defining D_{mk}^2 as the square of the matrix D_{lk},

$$D_{mk}^2 = \sum_{l=0}^{N} D_{ml} D_{lk}, \tag{6.139}$$

equation (6.138) reduces to

$$\partial_x^2 f(x) = \sum_{k=0}^{N} \sum_{m=0}^{N} \tilde{f}_k D_{mk}^2 T_m(x), \tag{6.140}$$

showing that we can construct the *second* derivative of $f(x)$ by applying D_{lk} to the spectral coefficients $\tilde{f}_k$ *twice*.

Inserting equation (6.140) into the differential equation (6.126) then yields

$$\sum_{k=0}^{N} \sum_{m=0}^{N} \tilde{f}_k D_{mk}^2 T_m(x) = s(x).$$ (6.141)

In a pseudo-spectral method we now evaluate this equation at the $N + 1$ collocation points x_i. Defining

$$A_{ki} \equiv \sum_{m=0}^{N} D_{mk}^2 T_m(x_i),$$ (6.142)

we can write the resulting $N + 1$ equations as the simple matrix equation

$$\sum_{k=0}^{N} A_{ki} \tilde{f}_k = s_i,$$ (6.143)

where $s_i = s(x_i)$, hence

$$\tilde{f}_k = \sum_{i=0}^{N} A_{ki}^{-1} s_i.$$ (6.144)

We thus have reduced the problem to matrix inversion.

The attentive reader will recall that for the finite-difference methods in Section 6.2.2 we also reduced the elliptic differential equation into a set of linear equations (e.g., equation 6.37). However, to achieve the same accuracy in spectral and finite-difference methods we often need a much smaller number N of basis functions than the number N of grid points. For many applications spectral methods converge *exponentially* with increasing N.[32] The matrix A_{ki} is therefore much smaller than the typical matrices encountered for finite-difference methods and can be inverted much more easily and faster.

So far we have ignored boundary conditions. They can be accounted for by replacing some of the equations in (6.143) with conditions that enforce the boundary conditions. Imagine, for example, a boundary condition

$$f(1) = 0.$$ (6.145)

Since all Chebychev polynomials satisfy $T_n(1) = 1$, we would then have to require

$$\sum_{k=0}^{N} \tilde{f}_k = 0.$$ (6.146)

This constraint can be enforced by replacing the row in A_{ki} that corresponds to the collocation point $x_i = 1$ with 1's.

For elliptic equations containing nonlinear terms we can construct the solution iteratively as we described towards the end of Section 6.2.2 for finite differencing. For an equation

---

[32] See, e.g., Press *et al.* (2007).

of the form (6.54), for example, we can linearize the equation around the approximate solution f^N after N iteration steps, find the linear equation (6.60) for the next correction δf, and then solve this linear equations for δf with the techniques described above.

6.3.5 Initial value problems

Spectral methods applied to initial value problems typically treat the space and time coordinates differently and expand only the space-dependence into basis functions. This choice is quite similar, in fact, to the idea behind the method of lines which we discussed at the end of Section 6.2.3 in the context of finite difference methods. This approach again converts partial differential equations into a set of ordinary differential equations in time, in this case for the spectral expansion coefficients.

As in Section 6.2.3, consider the model advective equation

$$\partial_t u + v \partial_x u = 0, \tag{6.147}$$

where the function $u = u(t, x)$ depends on both time and space. Using Chebychev polynomials we expand this function as

$$u(t, x) \simeq u^{(N)} = \sum_{k=0}^{N} \tilde{u}_k(t) T_k(x), \tag{6.148}$$

which, when inserted into equation (6.147), yields

$$\partial_t \sum_{k=0}^{N} \tilde{u}_k(t) T_k(x) + v \sum_{k=0}^{N} \tilde{u}_k(t) \partial_x T_k(x) = 0. \tag{6.149}$$

We again have to evaluate this equation $N + 1$ times to derive $N + 1$ equations for the $N + 1$ spectral coefficients. In a Galerkin method we would take the scalar product of the equation with the basis functions $T_l(x)$ themselves, which would yield linear equations for the spectral coefficients $\tilde{u}_k(t)$. In numerical relativity, pseudo-spectral methods are more commonly used, for which we evaluate equation (6.149) at the $N + 1$ collocation points X_i given by equation (6.119). We then find equations for the functions u_i according to

$$\partial_t u_i(t) = -v \sum_{k=0}^{N} \tilde{u}_k(t) T_k'(X_i) = -v \sum_{l=0}^{N} \tilde{u}_l'(t) T_l(X_i), \tag{6.150}$$

where we have expressed the derivatives $T_k'(X_i)$ in terms of the Chebychev polynomials themselves using equations (6.131) and (6.135).

We can evaluate the right-hand side of equation (6.150) as follows. For a set of functions u_i (at a certain time t) use equation (6.122) to find the spectral coefficients $\tilde{u}_k$, then find the $\tilde{u}_k'$ (for example from the recurrence relation 6.135), and finally perform the sum in equation (6.150). Using Chebychev polynomials both the first and last step can be carried out with fast Fourier transforms.

Since we have expressed the spatial derivative of u in terms of derivatives of the basis functions $T_k(x)$, the equations (6.150) can now be treated like ordinary differential equations for the u_i, and a variety of methods can be used for their integrations. Given that we have invested a fair amount of effort in the accurate representation of the spatial derivatives, it would be a waste not to treat the time derivatives with some care as well. It is therefore reasonably common to integrate these equations with an explicit fourth-order Runge–Kutta method. As for the explicit finite difference methods discussed in Section 6.2.3 the time step Δt is limited by a Courant stability criterion. For the uniform-grid methods in Section 6.2.3 we found that typically $\Delta t \sim \Delta x \sim N^{-1}$. Here, however, the collocation points cluster near the domain boundaries – from equation (6.119), the distance Δx between the collocation points X_0 and X_1, for example, is $1 - \cos(\pi/N) \sim N^{-2}$ – which reduces the limit to $\Delta t \sim N^{-2}$. This is less severe than it may sound, since N is typically much smaller for spectral methods than for finite difference methods, and if needed this problem can also be avoided by implementing implicit methods.

6.3.6 Comparison with finite-difference methods

It may be useful to include a brief comparison between the respective advantages and dis-advantages of finite-difference and spectral methods. There are other methods for solving partial differential equations, of course, like finite element, Monte Carlo and variational methods, but since these have not been adopted widely in numerical relativity we omit a discussion of them.

The most attractive feature of spectral methods is the fact that, for many situations, solutions converge exponentially with the number of basis functions. That means that for a fixed allocation of computational resources, spectral methods can often achieve a much higher accuracy than finite difference methods, which yield solutions that converge only as some power of the number of grid points.

One of the disadvantages of spectral methods is that they can work well only when the solution functions are well represented by the basis functions. In particular this means that the solutions should be smooth, meaning that the functions and all their derivatives must be continuous (unless the adopted basis functions are chosen to reflect any known discontinuous behavior). If this is not the case, so-called "Gibbs phenomena" may appear (i.e., spurious oscillations of the spectral solution near discontinuities), and this behavior adversely affects the convergence of the method.

As a consequence, spectral methods have been most successful for situations in which the solutions are indeed smooth. One example in the context of numerical relativity are initial data for binary black holes (see Chapter 12), for which all gravitational fields are expected to be smooth outside of the black holes. For binary neutron star initial data the situation is already more complicated, since the matter variables may have discontinuous derivatives on the stellar surfaces. This problem can be circumvented by introducing "surface-fitting" coordinate patches. The stellar interiors and exteriors are then handled on different computational domains. In each one of them the solutions are smooth, and the different patches are glued together with suitable boundary conditions on the stellar surfaces. This approach works well as long as the surface does not develop cusps, which may result in Gibbs phenomena. This may happen, for example, just before the star is tidally disrupted by a binary companion. We will discuss techniques and results for binary neutron star initial data in Chapter 15, and for black hole-neutron star binaries in Chapter 17.1.

The most attractive features of finite-difference methods are their ease of implementation and their robustness. In general, it is reasonably straightforward to implement at least a low-order finite-difference algorithm, and in general these algorithms are reasonably robust, and are less sensitive to the properties of the solutions than spectral methods. These observations probably explain why, with some notable exceptions, most dynamical simulations of binary mergers have been carried out with finite difference simulations. This is especially true of simulations involving fluid stars, where the presence of discontinuous shocks proves particularly challenging for spectral methods. We will describe simulations of binary black holes in Chapter 13, for binary neutron stars in Chapter 16, and for black hole-neutron star binaries in Chapter 17.2, where we will summarize some of the different methods used to perform them.

6.4 Code validation and calibration

Before closing this chapter we should briefly discuss some general strategies for testing numerical codes. One obvious code test is to use the code to treat a case in which an analytical solution, or at least some very accurate numerical solution, is available for comparison. For example, a code that is designed to evolve multiple spinning black holes could be used to simulate a single, stationary black hole. The results could then be compared to a stationary Kerr spacetime. Even if we were careful to employ only gauge-invariant quantities in performing such comparisons, we inevitably would find that the numerical results do *not* agree exactly with the analytical results. The reason for the discrepancy would be that, with very few exceptions, any numerical calculation will always have some truncation error. Comparing a *single* simulation with an analytical solution is therefore not very meaningful, because it would not be possible to distinguish a deviation caused by a coding error from a truncation error. A more meaningful test, one that can distinguish between coding mistakes and truncation error, is performing a *sequence* of simulations and using the sequence to check the *convergence* of the results to an analytical solution. The point is that the truncation error decreases in a predictable way with increasing numerical resources, while coding mistakes do not. We have already relied on such a convergence test in Chapter 5.2.4 to calibrate a relativistic MHD code.

Focusing on finite difference methods for the remainder of the section, we will denote the numerical solution at a point (t, x), achieved with a grid spacing h, as $u_h(t, x)$. We assume that we can express this solution as a Taylor expansion about the analytical solution

$u(t, x)$, in which case we may write

$$u_h(t, x) = u(x, t) + h E_1 + h^2 E_2 + h^3 E_3 + \mathcal{O}(h^4), \qquad (6.151)$$

where the error terms E_i are independent of the grid spacing h. For concreteness, imagine that we constructed a second-order scheme, in which case E_1 should be zero and the numerical error should scale with h^2,

$$u_h(t, x) - u(x, t) = h^2 E_2 + h^3 E_3 + \mathcal{O}(h^4). \qquad (6.152)$$

Now consider redoing the same calculation with a higher resolution, for example with a new grid spacing $h/2$. The new error should then be

$$u_{h/2}(t, x) - u(x, t) = \frac{1}{4} h^2 E_2 + \frac{1}{8} h^3 E_3 + \mathcal{O}(h^4), \qquad (6.153)$$

or

$$4(u_{h/2}(t, x) - u(x, t)) = h^2 E_2 + \frac{1}{2} h^3 E_3 + \mathcal{O}(h^4). \qquad (6.154)$$

Doubling the resolution again we find

$$16(u_{h/4}(t, x) - u(x, t)) = h^2 E_2 + \frac{1}{4} h^3 E_3 + \mathcal{O}(h^4). \qquad (6.155)$$

As we increase the resolution, the higher order terms keep decreasing, so that the right hand sides converge to $h^2 E_2$ (where h is the original grid spacing). In a convergence test to an analytical solution we test this behavior by plotting the rescaled errors

$$2^{2k}(u_{h/k}(t, x) - u(t, x)) \rightarrow h^2 E_2, \qquad (6.156)$$

which will converge only if the implementation is indeed second-order accurate with $E_1 = 0$. This is a very strong test, since many coding mistakes and typos will result in errors that, even if they are small, do not converge away.

For a convergence test to an analytical solution we need to compare the numerical solution for at least *three* different resolutions. For the two finer resolutions the rescaled errors should then be closer to each other than for the two coarser resolutions. In fact, many second-order schemes are symmetric, in which case the third-order error terms E_3 vanish also, and the rescaled errors converge to $\mathcal{O}(h^4)$. As an example, we show in Figure 6.12 a test for a code designed to construct binary black hole-neutron star quasiequi-librium initial data.[33] In this test the code, which solves partial differential elliptic equations in three spatial dimensions, is shown to converge, in the absence of a black hole, to the Oppenheimer–Volkoff solution for a spherical equilibrium star (see Chapter 1.4). The latter is not an analytical solution, but as a solution to a coupled set of ordinary differential equations in one dimension it can be computed to sufficiently high accuracy so serve as a "semi-analytical" solution in this test.

---

[33] Baumgarte *et al.* (2004); see Chapter 17.1.

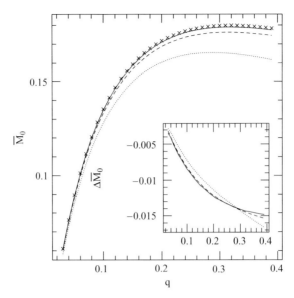

Figure 6.12 The rest mass $\bar{M}_0$ of a spherical $n = 1$ polytrope as a function of a density parameter q for three different grid resolutions h (dotted line), $h/2$ (dashed line) and $h/4$ (solid line), computed with a code that solves elliptic equations in three dimensions for quasiequilibrium binaries. The crosses mark the high-accuracy solution obtained by solving the Oppenheimer–Volkoff ordinary differential equations for an equilibrium star in spherical symmetry. The inset shows the rescaled numerical errors, which clearly converge to a single line, establishing second-order convergence of the code. [From Baumgarte *et al.* (2004).]

Interestingly, neither an analytical nor a semi-analytical solution is needed to test the convergence of a code. In a *self-consistent convergence test* we can eliminte the unknown analytical solution $u(t, x)$ in (6.152) by subtracting (6.153)

$$u_h - u_{h/2} = \frac{3}{4}h^2 E_2 + \frac{7}{8}h^3 E_3 + \mathcal{O}(h^4). \qquad (6.157)$$

We similarly find

$$4(u_{h/2} - u_{h/4}) = \frac{3}{4}h^2 E_2 + \frac{1}{2}\frac{7}{8}h^3 E_3 + \mathcal{O}(h^4). \qquad (6.158)$$

The higher-order error terms again keep decreasing, and the rescaled differences approach

$$2^{2k}(u_{h/k}(t, x) - u_{h/(2k)}(t, x)) \rightarrow \frac{3}{4}h^2 E_2. \qquad (6.159)$$

This means that we can again check the convergence of our code, but this time completely independently of any analytical solution. Clearly, this is a very powerful tool. Often we can use the convergence to an analytical solution to test codes in certain regimes for which an analytical solution can be found. But these tests may or may not test all aspects of the computational algorithm. Self-consistent convergence tests, on the other hand, can be used to test all parts of a code. The one down-side of a self-consistent convergence test, compared to the convergence to an analytical solution, is that we now need at least *four* different grid resolutions, so that we can form three pairs whose convergence we can test.[34]

> **Exercise 6.25** Sometimes it is useful to "invent" an analytic solution where none is known in order to perform a convergence test on a code. Consider the nonlinear, flat-space elliptic equation (6.54) for the function $f(r)$:
>
> $$\nabla^2 f = f^n g, \qquad (6.160)$$
>
> subject to boundary conditions
>
> $$\partial_r f = 0, \quad r = 0, \qquad (6.161)$$
>
> $$\partial_r (rf) = 0, \quad r = \infty, \qquad (6.162)$$
>
> where $r^2 = x^2 + y^2 + z^2$. Take $n = 4$ for definiteness. Here we will write and test a code to solve this equation in one dimension (spherical symmetry), two dimensions (axisymmetry), or three dimensions.
> (a) Choose the arbitrary analytic function
>
> $$f = \frac{r^2}{1 + r^3}, \qquad (6.163)$$
>
> constructed to satisfy the boundary conditions, substitute it into equation (6.160) and solve for g. Given this analytic source function g we now have an exact solution f to equation (6.160) that satisfies the appropriate boundary conditions.
> (b) Use the analytic source function g derived in part (a) to solve equation (6.160) by finite-differencing. You may need to linearize the equation first, as discussed at the end of Section 6.2.2, and iterate the solution.
> (c) Perform a convergence test to check your code by comparing with the analytic solution. Does your code converge at the expected convergence rate?

---

[34] We also point out that any typo or mistake in an analytic equation that propagates into a high-order code will not be discovered by a self-consistent convergence test. Only by comparison with an analytic solution can a convergence test alone reveal such an error.

7 Locating black hole horizons

Black holes are characterized by the horizons surrounding them. Clearly, then, the numerical simulation of black holes requires the ability to locate and analyze black hole horizons in numerically generated spacetimes. In this chapter we first review different concepts of horizons in asymptotically flat spacetimes, and then discuss how these horizons can be probed numerically.

7.1 Concepts

Several different notions of horizons exist in general relativity. The defining property of a black hole is the presence of an *event horizon* (Section 7.2), but, as we will see, *apparent horizons* (Section 7.3) also play an extremely important role in the context of numerical relativity. In addition, the concepts of *isolated* and *dynamical horizons* (Section 7.4) serve as useful diagnostics in numerical spacetimes containing black holes.

A black hole is defined as a region of spacetime from which no null geodesic can escape to infinity. The surface of a black hole, the *event horizon*, acts as a one-way membrane through which light and matter can enter the black hole, but once inside, can never escape. It is the boundary in spacetime separating those events that can emit light rays that can propagate to infinity and those which cannot. More precisely, the event horizon is defined as the boundary of the causal past of future null infinity.[1] It is a $2 + 1$ dimensional hypersurface in spacetime formed by those outward-going, future-directed null geodesics that neither escape to infinity nor fall toward the center of the black hole. The event horizon is a gauge-invariant entity, and contains important geometric information about a black hole spacetime.

The intersection $\mathcal{H}$ of the event horizon with a $t = constant$ spatial hypersurface Σ, i.e., the spatial "snapshot" of the horizon at the instant of time associated with Σ, forms a closed, two-dimensional surface, whose proper surface area we denote as $\mathcal{A}$. The *area theorem*[2] of classical general relativity states that this surface area can never decrease in time,

$$\delta \mathcal{A} \geq 0, \tag{7.1}$$

---

[1] See, e.g., Hawking and Ellis (1973); Wald (1984).
[2] Hawking (1971, 1972, 1973).

as long as all matter satisfies the null energy condition.[3] In the collision and coalescence of two or more black holes, the surface area of the remnant black hole must be greater than the sum of the progenitor black holes.

The fact that the event horizon area cannot decrease motivates the definition of the *irreducible mass*[4]

$$M_{\mathrm{irr}} \equiv \left(\frac{\mathcal{A}}{16\pi} \right)^{1/2}. \tag{7.2}$$

It is possible to extract energy and angular momentum from a rotating Kerr black hole. While such an interaction can reduce the black hole's mass, it cannot reduce its area, according to the area theorem. The definition (7.2) then implies that the irreducible mass of the black hole cannot decrease, which motivates its name.

> **Exercise 7.1** Verify that for a Schwarzschild black hole the irreducible mass M_{irr} is equal to the ADM mass M_{ADM}.

The area theorem can be used to place a strict upper limit on the amount of energy that is emitted in gravitational radiation in black hole collisions.

> **Exercise 7.2** Consider two widely separated, nonrotating black holes of masses M_1 and M_2, initially at rest with respect to some distant observer. Use the area theorem to find an upper limit on the energy emitted in gravitational radiation that arises from the head-on collision of the two black holes. Verify that for equal mass black holes at most 29% of the total initial energy can be emitted in gravitational radiation.

In Chapter 13 we will find that considerably less energy is emitted in collisions of black holes than the upper limit allowed by the area theorem.

Given the irreducible mass M_{irr} and the angular momentum J of an isolated, stationary black hole, we can compute the *Kerr mass* M ($= M_{\mathrm{ADM}}$) from

$$M^2 = M_{\mathrm{irr}}^2 + \frac{1}{4} \frac{J^2}{M_{\mathrm{irr}}^2}. \tag{7.3}$$

Solving for M_{irr} we find

$$M_{\mathrm{irr}}^2 = \frac{M^2}{2} \left(1 + \sqrt{1 - \frac{J^2}{M^4}} \right), \tag{7.4}$$

which implies that we have $M^2/M_{\mathrm{irr}}^2 \leq 2$ for Kerr black holes, with equality in the extreme Kerr limit when $J = M^2$.

While the event horizon has some very interesting geometric properties, its global nature makes it very difficult to locate in a numerical simulation. The reason is that knowledge of

---

[3] The null energy condition requires that $T_{ab}k^a k^b \geq 0$ for all null vectors k^a. For a perfect gas, this condition requires $\rho + P \geq 0$.

[4] Christodoulou (1970); Christodoulou and Ruffini (1971); see also Chapter 1.2.

the entire future spacetime is required to decide whether or not any particular null geodesic will ultimately escape to infinity. In numerical simulations an event horizon can be found only "after the fact", i.e., after the evolution has proceeded long enough to have settled down to a stationary state.

Locating event horizons in "post-processing" may be sufficient for diagnostic purposes, i.e., for analyzing the geometrical and physical consequences of a black hole simulation after it is completed, but it does not allow us to locate the black holes during the course of a numerical simulation. The later can be important, and is sometimes essential, for allowing the simulation to continue in the presence of one or more black holes. The spacetime singularities inside the black holes must be excluded from the numerical grid, since they would otherwise spoil the numerical calculation. Several of the following chapters treat simulations in which black holes are present and there we will discuss several different strategies for avoiding black hole singularities numerically. One approach is based on the realization that, by definition, the interior of a black hole is causally disconnected from, and hence can never influence, the exterior. This fact suggests that we may "excise", i.e., remove from the computational domain, the spacetime region inside the event horizon.[5] Black hole "excision" requires at least approximate knowledge of the location of the horizon at all times during the evolution, so the construction of the event horizon after the fact is not sufficient.

The concept of *apparent horizons* allows us to locate black holes during the evolution. The apparent horizon is defined as the outermost smooth 2-surface, embedded in the spatial slices Σ, whose outgoing future null geodesics have zero expansion everywhere. We will explain this notion in much greater detail in Section 7.3 below. As we will see, the apparent horizon can be located on each slice Σ, when it exists, and is therefore a local (in time) concept. The singularity theorems of general relativity[6] tell us that if an apparent horizon exists on a given time slice, it must be inside a black hole event horizon.[7] This theorem makes it safe to excise the interior of an apparent horizon from a numerical domain. Note, however, that absence of proof is not proof of absence: the absence of an apparent horizon does not necessarily imply that a black hole is absent. One example can be found in the Oppenheimer–Snyder collapse of spherical dust to a black hole as constructed in Chapter 1.4; we will return to this example in Section 7.3.1. It is also possible to construct slicings of the Schwarzschild geometry in which no apparent horizon exists.[8] Also, it is straightforward to show that apparent horizons do not form during spherical collapse in polar slicing.[9]

These examples demonstrate the gauge-dependent nature of the apparent horizon. Nevertheless, the usual expectation when performing a black hole a simulation is that, except

---

[5] Unruh (1984), as quoted in Thornburg (1987); see also Chapters 13.1.1 and 14.2.3.
[6] See Hawking and Ellis (1973); Wald (1984) for an introduction.
[7] This statement is not necessarily true in other theories of gravity. In Brans–Dicke theory, for example, apparent horizons may exist outside of event horizons; see Scheel *et al.* (1995b) for a numerical example.
[8] Wald and Iyer (1991)
[9] See exercises 7.13 and 8.10.

when choosing a special slicing in which an apparent horizon is known to be absent, an apparent horizon will eventually appear on the slice whenever a black hole is present. The fact that the apparent and event horizons always coincide in a stationary spacetime promotes this expectation.

Now consider constructing the worldtube H formed by stacking together apparent horizons on different spatial slices Σ. In general, H can make discrete jumps and does not need to be continuous (see Chapter 1.4 and Figure 1.3). When matter or radiation falls into the black hole the horizon H expands. When the black hole is isolated, however, and absorbs no more matter or radiation, H becomes a null-surface. In this regime H can be described within the *isolated horizon* framework.[10] Using information on H, this formalism provides a coordinate-independent definition of the black hole mass and angular momentum, as we will discuss below.

7.2 Event horizons

Event horizons are spanned by outgoing light rays that neither reach future null infinity nor hit the black hole singularity. As a nontrivial but analytical example, we located the event horizon in the Oppenheimer–Snyder collapse of a dust sphere to a black hole in Chapter 1.4. In principle, knowledge of the entire future evolution of a spacetime is necessary to determine the fate of outgoing light rays and thereby locate event horizons. In practice, however, event horizons can be identified fairly accurately after a finite evolution time in a numerical simulation, provided the spacetime has settled down to a nearly stationary state.

An obvious approach to locating an event horizon in this situation is to evolve null geodesics,[11] whose worldlines are governed by the equation

$$\frac{d^2 x^a}{d\lambda^2} + {}^{(4)}\Gamma^a_{bc} \frac{dx^b}{d\lambda} \frac{dx^c}{d\lambda} = 0, \tag{7.5}$$

where λ is an affine parameter. Splitting this second-order equation into two first-order equations and substituting $3 + 1$ metric quantities yields

$$\begin{aligned} \frac{dp_i}{d\lambda} &= -\alpha \partial_i \alpha (p^0)^2 + \partial_i \beta^k p_k p^0 - \tfrac{1}{2} \partial_i \gamma^{lm} p_l p_m \\ \frac{dx^i}{d\lambda} &= \gamma^{ij} p_j - \beta^i p^0, \end{aligned} \tag{7.6}$$

where we have used $p^i = dx^i/d\lambda$ and $p^0 = (\gamma^{ij} p_i p_j)^{1/2}/\alpha$ (which enforces $g^{ab} p_a p_b = 0$).

Exercise 7.3 Derive equations (7.6).

In a numerical spacetime, the lapse α, the shift β^i and the spatial metric γ_{ij} are known on the computational grid, so that light rays can be ejected in different directions p^i from

---

[10] Ashtekar *et al.* (2000); Ashtekar and Krishnan (2004).
[11] See, e.g., Hughes *et al.* (1994).

every point x^a in spacetime and their geodesic trajectories tracked. The search for an event horizon can be expedited by knowing the location of the apparent horizons. If all light rays sent out from an event x^a end up inside an apparent horizon (which is always located inside an event horizon) the event must reside inside the event horizon as well. If, on the other hand, at least one light ray sent out from the event escapes to large separations, it is not inside an event horizon. By distinguishing events in this way, the ejection and propagation of light rays from various points in spacetime can delineate the location of the event horizon.

In practice it is more expeditious to integrate null geodesics *backwards* in time.[12] Future directed light rays diverge away from the event horizon, either toward the interior of the black hole or toward future null infinity. By contrast, backwards propagating rays converge on the event horizon, which thus acts as an "attractor" for these rays. This method is particularly efficient if one can identify a finite region in spacetime within which the event horizon is expected to reside. It is then sufficient to integrate light rays from events residing in this region at late times, and they will be attracted to the event horizon.

This method becomes quite transparent in spherical symmetry, where we can often find the trajectories of "outgoing" null geodesics analytically. The label "outgoing" is somewhat misleading, since inside the event horizon all worldlines propagate to smaller areal radius. For example, for a Schwarzschild black hole in Kerr–Schild coordinates, we noted in exercise 6.3 that outgoing null geodesics must satisfy

$$t - r = 4M \ln |r/2M - 1| + constant. \tag{7.7}$$

As shown in Figure 7.1, tracing these geodesics backwards in time inevitably brings us to the event horizon at $r = 2M$.

In nontrivial applications, this technique has been used to probe the topology of the event horizon arising during the head-on collision and merger of two black holes, as well as the collapse of a rotating toroidal cluster to a toroidal black hole. The "pair of pants" that the event horizon forms in a spacetime diagram for the collapse of two collisionless clusters to two black holes, followed by their head-on collision, is depicted in Figure 10.5. We postpone a discussion of these simulations until Chapter 10.2 and 10.4.

Instead of integrating individual null geodesics, we can also consider constructing a $2 + 1$ dimensional null hypersurface enclosing a bundle of outgoing null geodesics in spacetime.[13] We can define such a null hypersurface as a level surface of some function $f(t, x^i)$, say $f(t, x^i) = 0$. Given that the normal vector $\partial_a f$ to such a null hypersurface must itself be null,[14] the function f must satisfy

$$g^{ab} \partial_a f \partial_b f = 0. \tag{7.8}$$

---

[12] See Hughes *et al.* (1994); Anninos *et al.* (1995); Libson *et al.* (1996).
[13] See Anninos *et al.* (1995); Libson *et al.* (1996).
[14] See, e.g., Wald (1984), p. 65.

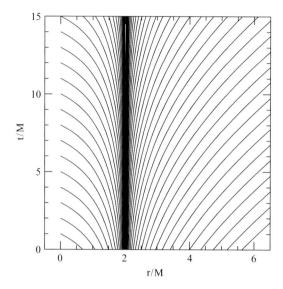

Figure 7.1 Worldlines of outgoing null geodesics in a Schwarzschild spacetime in Kerr–Schild coordinates (see exercise 6.3). Traced backwards in time, these rays are strongly attracted to the event horizon at $r/M = 2$. These worldlines coincide with contours of the function f introduced in equation (7.8).

Solving for $\partial_t f$ we then find an evolution equation for f,

$$\partial_t f = \frac{-g^{ti}\partial_i f + \sqrt{(g^{ti}\partial_i f)^2 - g^{tt} g^{ij} \partial_i f \partial_j f}}{g^{tt}}, \tag{7.9}$$

where we have chosen the positive root so that the surface is generated by outgoing null geodesics. Expressing the spacetime metric in terms of the lapse, shift and spatial metric we find

$$\partial_t f = \beta^i \partial_i f - \left(\alpha^2 \gamma^{ij} \partial_i f \partial_j f\right)^{1/2}. \tag{7.10}$$

Exercise 7.4 Derive equation (7.10) from (7.8).

It is particularly easy to use equation (7.10) to locate an event horizon if, at late time, we can bracket it by two null hypersurfaces, e.g., S_1 and S_2. Suppose each one of these two hypersurfaces is defined by a certain value, say zero, of two functions f_1 and f_2. Knowing the location of the apparent horizon at late times again facilitates the choice of S_1 and S_2. We can use equation (7.10) to evolve f_1 and f_2 backwards in time, and find S_1 and S_2 at earlier times by locating $f_1 = 0$ and $f_2 = 0$. The two hypersurfaces will converge on the event horizon.

In spherical symmetry it is easy to find the general solution for f, which illustrates why this approach works. By construction, f is constant along outgoing light rays. In Kerr–Schild coordinates, outgoing light rays travel along the characteristics (7.7). Any arbitrary

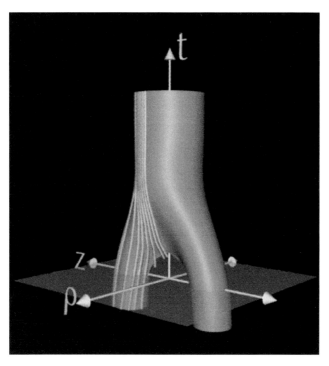

Figure 7.2 The "pair of pants" formed by the event horizon in a spacetime diagram depicting the head-on collision of two "eternal" black holes. [From Matzner *et al.* (1995).]

function

$$f = f(t - r - 4M \ln |r/2M - 1|) \tag{7.11}$$

must therefore be a solution to equation (7.10).

> **Exercise 7.5** Verify that any function $f = f(t - r - 4M \ln |r/2M - 1|)$ is a solution to equation (7.10).

This technique has been used to construct the event horizon for yet another scenario describing the head-on collision of two black holes. We will also discuss this scenario, which involves "eternal" black holes, in Chapter 10.2, but here, in anticipation, we show a spacetime diagram depicting the event horizon in Figure 7.2.

7.3 Apparent horizons

Consider a smooth, closed, 2-dimensional hypersurface S residing in Σ. By construction, S is spatial. Let s^a be its outward pointing unit normal lying in Σ. Evidently s^a then satisfies $s_a s^a = 1$ and $s^a n_a = 0$. Just as the spacetime metric g_{ab} induces the spatial metric γ_{ab} on

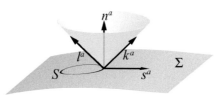

Figure 7.3 A smooth, 2-dimensional, hypersurface S is embedded in Σ. The unit, outward-pointing normal to S lying in Σ is s^a, and the normal to Σ is n^a. The outgoing and ingoing null tangent vectors k^a and l^a are constructed as linear combinations of n^a and s^a.

Σ (see Chapter 2.3), the metric γ_{ab} induces a 2-dimensional metric m_{ab} on S given by

$$m_{ab} = \gamma_{ab} - s_a s_b = g_{ab} + n_a n_b - s_a s_b. \tag{7.12}$$

For each point on S we can now construct a pair of future-pointing null geodesics whose projection on Σ is orthogonal to S. Up to an overall factor, the tangents k^a and l^a to these geodesics on S are

$$k^a \equiv \frac{1}{\sqrt{2}}(n^a + s^a) \quad \text{and} \quad l^a \equiv \frac{1}{\sqrt{2}}(n^a - s^a) \tag{7.13}$$

(see Figure 7.3). By construction we have $k_a k^a = 0$ and $m_{ab}k^a = 0$ as well as $l_a l^a = 0$ and $m_{ab}l^a = 0$, and we have chosen the normalization so that $k^a l_a = -1$. We call k^a the tangent to the "outgoing" and l^a the tangent to the "ingoing" null geodesic. These names are again somewhat misleading, since inside a black hole both tangents point toward the black hole's interior. Combining equations (7.12) and (7.13) we can express m_{ab} in terms of k^a and l^a as

$$m_{ab} = g_{ab} + k_a l_b + l_a k_b. \tag{7.14}$$

The expansion of the outgoing null geodesics orthogonal to S is

$$\boxed{\Theta = m^{ab}\nabla_a k_b.} \tag{7.15}$$

Note that k^a is only defined on S. The projection with m^{ab} ensures that only derivatives tangent to S enter this expression, so that no knowledge of k^a away from S is required in this definition of the expansion.

Exercise 7.6 Show that the definition (7.15) is equivalent to

$$\Theta = \nabla_a k^a \tag{7.16}$$

if we assume that the null tangent vectors k^a are affinely parametrized, i.e., $k^a \nabla_a k^b = 0$, in a neighborhood of S.

We now define an *outer-trapped surface* as a 2-dimensional surface S embedded in Σ on which the expansion Θ of the outgoing null geodesics orthogonal to S is negative everywhere.[15] We further define a *trapped region* as any region of Σ that contains outer-trapped surfaces.[16] Finally, we define the *apparent horizon* as the outer boundary of any connected trapped region. This definition makes the apparent horizon a *marginally outer-trapped surface* on which the expansion of outgoing null geodesics vanishes

$$\boxed{\Theta = 0.} \tag{7.17}$$

Before proceeding with the formal derivation it may be useful to illustrate the above concepts with a simple example in spherical symmetry. Without loss of generality we can write the spacetime line element in spherical polar coordinates in the form

$$ds^2 = -(\alpha^2 - A^2\beta^2)dt^2 + 2A^2\beta\, dr\, dt + A^2 dr^2 + B^2 r^2 (d\theta^2 + \sin^2\theta\, d\phi^2). \tag{7.18}$$

The only nonvanishing component of the shift vector β^i is the radial component, which we call β, and we may further assume that α, β, A and B are functions of t and r only.

Consider a spherical surface S centered on the origin. The spatial normal vector s^i to S must then be radial and take the form

$$s^a = (0, A^{-1}, 0, 0) \tag{7.19}$$

and hence $s_a = (A\beta, A, 0, 0)$. We can now construct the outgoing null normal

$$k_a = \frac{1}{\sqrt{2}}(A\beta - \alpha, A, 0, 0) \quad \text{and} \quad k^a = \frac{1}{\sqrt{2}}(\alpha^{-1}, A^{-1} - \alpha^{-1}\beta, 0, 0), \tag{7.20}$$

as well as the induced metric on S

$$m_{ab} = \text{diag}(0, 0, B^2 r^2, B^2 r^2 \sin^2\theta). \tag{7.21}$$

Computing the expansion Θ we find

$$\Theta = \frac{\sqrt{2}}{rB}\left(\frac{1}{\alpha}\partial_t(Br) + \left(\frac{1}{A} - \frac{\beta}{\alpha}\right)\partial_r(Br)\right). \tag{7.22}$$

Exercise 7.7 Derive equation (7.22).

So far this result is not too enlightening. It turns out, however, that the expression in parentheses is proportional to the rate of change of the areal radius along k^a. The areal radius is $R_{\text{area}} = Br$, and its directional derivative along k^a is therefore

$$k^a \nabla_a R_{\text{area}} = k^t \partial_t(Br) + k^r \partial_r(Br) = \frac{1}{\sqrt{2}}\left(\frac{1}{\alpha}\partial_t(Br) + \left(\frac{1}{A} - \frac{\beta}{\alpha}\right)\partial_r(Br)\right). \tag{7.23}$$

---

[15] E.g. Hawking and Ellis (1973); Wald (1984).

[16] Some authors define a trapped region in terms of *trapped surfaces* as opposed to outer-trapped surfaces. For trapped surfaces the expansion of the ingoing null geodesics orthogonal to S is required to be negative, $\Theta_{(l)} < 0$, in addition to $\Theta < 0$. For the purposes of this section this distinction is unimportant.

Combining equations (7.22) and (7.23) we have

$$\Theta = \frac{2}{R_{\text{area}}} k^a \nabla_a R_{\text{area}} = \frac{1}{4\pi R_{\text{area}}^2} k^a \nabla_a (4\pi R_{\text{area}}^2). \tag{7.24}$$

Equation 7.24 shows that the expansion Θ measures the fractional change of the cross-sectional area of a bundle of outgoing null rays with tangent vectors k^a, or, equivalently, the fractional change in the area of an outward spherical flash of light.[17]

Exercise 7.8   Show that

$$\Theta = \frac{1}{m^{1/2}} \mathcal{L}_{\mathbf{k}} m^{1/2} = \mathcal{L}_{\mathbf{k}} \ln m^{1/2}, \tag{7.25}$$

where m is the determinant of the induced metric m_{ab}. Since $m^{1/2} d^2 x$ is the proper area element of a cross-section of a bundle of null rays k^a, it can then be shown that Θ measures its fractional change of cross-sectional area in general, and not only in spherical symmetry.

With this result the definition of an apparent horizon becomes very simple to apply in spherical symmetry. The area $\mathcal{A}$ of a spherical flash of light rays emitted radially outward will propagate instantly to a larger area if it is emitted outside an apparent horizon, to a smaller area if emitted inside an apparent horizon, and remain constant if emitted on the apparent horizon.

Consider, for example, a Schwarzschild black hole in isotropic coordinates (see, e.g., Table 2.1), for which $A = B = \psi^2 = (1 + M/(2r))^2$, $\beta = 0$, and $\partial_t(Br) = 0$. According to equation (7.22) the expansion Θ then vanishes when $\partial_r(Br) = 0$, which occurs at

$$r = \frac{M}{2}. \tag{7.26}$$

In exercise 3.4 we have seen that this isotropic radius corresponds to an areal radius of $R_{\text{areal}} = 2M$. Not surprisingly, the apparent horizon coincides with the event horizon in this static spacetime.

Exercise 7.9   Verify that for a Schwarzschild black hole in Schwarzschild, Kerr–Schild and Painlevé–Gullstrand coordinates the apparent horizon is located at $R_{\text{areal}} = 2M$.

Returning to our general derivation, we would like to bring the expansion (7.15) into a form that is more suitable for evaluation in a $3 + 1$ numerical simulation and write it in terms of 3-dimensional spatial objects. Substituting equation (7.13) into (7.15) we have

$$\sqrt{2}\, m^{ab} \nabla_a k_b = m^{ab} \nabla_a (n_b + s_b) = m^{ij}(D_i s_j - K_{ij}) \tag{7.27}$$

or, using equation (7.12) in $m^{ij} K_{ij}$, we obtain

$$\sqrt{2}\, m^{ab} \nabla_a k_b = D_i s^i - K + s^i s^j K_{ij}. \tag{7.28}$$

---

[17] See also the discussion in Poisson (2004), Section 2.4.

Here we also have used the relation $s^a s_a = 1$ as well as the definition (2.49) of the extrinsic curvature. The apparent horizon condition is therefore

$$0 = \sqrt{2}\,\Theta = m^{ij}(D_i s_j - K_{ij}) \tag{7.29}$$

or equivalently

$$0 = \sqrt{2}\,\Theta = D_i s^i - K + s^i s^j K_{ij}. \tag{7.30}$$

Exercise 7.10 Show that with the conformal rescalings (3.5), (3.35) and $s^i = \psi^{-2}\bar{s}^i$ the apparent horizon condition (7.30) can be written as

$$\bar{D}_i \bar{s}^i + 4\bar{s}^i \bar{D}_i \ln\psi - \frac{2}{3}\psi^2 K + \psi^{-4}\bar{A}_{ij}\bar{s}^i \bar{s}^j = 0. \tag{7.31}$$

It is often useful in the search for apparent horizons to characterize the horizon as a level surface of a scalar function, e.g.,

$$\tau(x^i) = 0. \tag{7.32}$$

We can then write the unit normal s^i as

$$s^i = \lambda D^i \tau, \quad \text{or} \quad s_i = \lambda D_i \tau = \lambda \partial_i \tau, \tag{7.33}$$

where λ is the normalization factor

$$\lambda \equiv \left(\gamma^{ij} D_i \tau D_j \tau\right)^{-1/2}. \tag{7.34}$$

In the expression

$$m^{ij} D_i(\lambda D_j \tau) = m^{ij}\lambda D_i D_j \tau + m^{ij}(D_i\lambda)(D_j\tau) \tag{7.35}$$

the second term vanishes, since $D_j\tau$ is proportional to s_j, which vanishes when contracted with m^{ij}. Substituting equation (7.33) into (7.29) therefore yields

$$0 = m^{ij}(\lambda D_i D_j \tau - K_{ij}) = m^{ij}(\lambda \partial_i \partial_j \tau - s_k \Gamma^k_{ij} - K_{ij}). \tag{7.36}$$

A particularly useful form of the level function τ is

$$\tau(x^i) = r_C(x^i) - h(\theta, \phi), \tag{7.37}$$

where r_C is the coordinate separation between the point x^i and some fiducial point C^i inside the $\tau = 0$ surface, and where θ and ϕ are spherical polar coordinates centered on C^i. In the following we will assume that C^i is the origin of the coordinate system, so that $C^i = 0$, but the generalization to a point that does not coincide with the origin is straightforward. The function h then measures the coordinate distance from the origin to the $\tau = 0$ surface in the (θ, ϕ) direction.

The first derivative of equation (7.37) is

$$\partial_i \tau = \sigma_i - \partial_i h, \tag{7.38}$$

where $\sigma_i \equiv \partial_i r_C$ is the unit vector in the (θ, ϕ) direction. In spherical polar coordinates r_C is simply r, so that $\sigma_i = (1, 0, 0)$. Since $\partial_i \sigma_j = 0$, the apparent horizon condition (7.36) reduces to

$$0 = \sqrt{2}\Theta = m^{ij}(\lambda \partial_i \partial_j h - s_k \Gamma^k_{ij} - K_{ij}) \qquad \text{(spherical polar coordinates).} \qquad (7.39)$$

> **Exercise 7.11** Show that in Cartesian coordinates the apparent horizon condition (7.36) becomes
>
> $$0 = \sqrt{2}\Theta = m^{ij}\left(\frac{\lambda}{r_C}(\delta_{ij} - \sigma_i \sigma_j) + \lambda \partial_i \partial_j h - s_k \Gamma^k_{ij} - K_{ij}\right) \qquad \text{(Cartesian coordinates).}$$
>
> $$(7.40)$$

We have now expressed the apparent horizon condition as a second-order partial differential equation for the function h that measures the horizon's coordinate distance from the origin. The principal part of the equation, $m^{ij}\partial_i \partial_j h$, involves a Laplacian with respect to the 2-dimensional metric m_{ij} on the surface S. Finding solutions to these elliptic equations, and hence locating apparent horizons, is in general not at all a trivial matter, in particular since the normal vector s^i also contains derivatives of h. For the remainder of this section we will discuss strategies for locating apparent horizons on spatial hypersurfaces exhibiting spherical symmetry, axisymmetry, and without any special symmetry.

7.3.1 Spherical symmetry

In spherical symmetry h is a constant and does not depend on θ or ϕ. The second derivatives in the differential equation (7.39) therefore vanish, and $s^i = \lambda \sigma^i$ can be constructed algebraically from the metric. The expansion Θ then reduces to the algebraic expression

$$\Theta = -\frac{1}{\sqrt{2}} m^{ij}(s_k \Gamma^k_{ij} + K_{ij}) \qquad (7.41)$$

which only depends on radius r. Finding an apparent horizon simply amounts to finding a root of this function. This simplification does not come as a surprise, of course, since we have seen before that apparent horizons can be located quite easily in spherical symmetry.

> **Exercise 7.12** Verify that for a metric of the form (7.18) the expansion (7.41) yields (7.22).

> **Exercise 7.13** (a) Show that in spherical symmetry and isotropic coordinates (i.e., $A = B$ in the metric 7.18) the apparent horizon condition reduces to
>
> $$1 + r\frac{\partial_r A}{A} - \frac{Ar}{2}K_T = 0, \qquad (7.42)$$
>
> where K_T is defined by equation (4.63).
> (b) Use equation (7.42) to argue that apparent horizons do not form in polar slicing, provided $dr_s/dr \neq 0$, where r is the isotropic radius and r_s is the areal radius.

In Chapter 1.4 we presented an instructive example in which an apparent horizon arises during stellar collapse, namely the Oppenheimer–Snyder collapse of a homogeneous dust sphere to a Schwarzschild black hole. The appearance of an apparent horizon is slicing-dependent; in this example geodesic (Gaussian normal) slicing was employed. The adopted slicing and coordinate conditions yield a spacetime that is completely analytic, including the location of the trapped surfaces, the apparent horizon and the event horizon; see Figure 1.3. The figure shows that trapped surfaces can appear suddenly and in a discontinuous manner, but this feature depends on the slicing. In polar slicing, for example, no trapped surfaces arise, as demonstrated in Exercise 7.13.[18] But it is already clear from this figure that the absence of an apparent horizon does not imply the absence of an event horizon. Consider, for example, a spatial slice at $\tau = 10M$. This slice penetrates the event horizon, and hence the interior of the nascent black hole, but nowhere does this slice exhibit any trapped surfaces or an apparent horizon.

7.3.2 Axisymmetry

In axisymmetry, the function h is allowed to depend on θ, $h = h(\theta)$. In equation (7.39) only derivatives with respect to θ give a nonzero contribution, so that the apparent horizon condition now becomes an ordinary differerential equation for $h(\theta)$. This differential equation is still quite complicated (recall that both $s_i = \lambda(\sigma_i - \partial_i h)$ and λ contain derivatives of h), but we will bring it into a tractable form as follows.

Adopting spherical polar coordinates, we may define

$$m_i \equiv \partial_i \tau = \sigma_i - \partial_i h = (1, -\partial_\theta h, 0), \tag{7.43}$$

so that $s_i = \lambda m_i$. At each point $x^i \equiv (h, \theta, \phi)$ we introduce the tangent vector

$$u^i \equiv \partial_\theta x^i = (\partial_\theta h, 1, 0) \tag{7.44}$$

(see Figure 7.4). Since for these vectors the ϕ components vanish, it is also useful to introduce capital letter indices A, B, ... that run only over r and θ.

The vector u^A is the tangent vector to the $r = h$ surface in the θ direction, and we can compute the arc length s of its integral curves from

$$\left(\frac{ds}{d\theta} \right)^2 = \gamma_{AB} u^A u^B. \tag{7.45}$$

We now ask the reader to derive two useful relationships.

Exercise 7.14   Show that

$$m^{AB} = \frac{\lambda^2}{\gamma^{(2)}} u^A u^B, \tag{7.46}$$

---

[18] See Petrich *et al.* (1986) for Oppenheimer–Snyder collapse in polar slicing and Chapter 8.2 for other examples of spherical collapse in polar slicing.

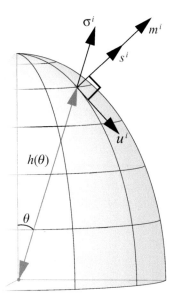

Figure 7.4 A schematic drawing of the surface $r = h(\theta)$, together with the vectors s^a, σ^a, m^a and u^a, defining an apparent horizon in axisymmetry.

where $\gamma^{(2)} = \gamma_{rr}\gamma_{\theta\theta} - \gamma_{r\theta}\gamma_{r\theta}$ is the determinant of the metric induced in the (r, θ) hypersurface.

Exercise 7.15 Show that

$$\frac{ds}{d\theta} = \frac{(\gamma^{(2)})^{1/2}}{\lambda}. \tag{7.47}$$

An immediate consequence of equation (7.46) is that the first term in equation (7.39) reduces to

$$m^{ij}\lambda\partial_i\partial_j h = \frac{\lambda^3}{\gamma^{(2)}}\partial_\theta\partial_\theta h. \tag{7.48}$$

We now make the assumption that $\gamma_{r\phi} = \gamma_{\theta\phi} = 0$, which can always be arranged in axisymmetry. The contractions in equations (7.39) then decouple into (r, θ) and ϕ terms; for example, $m^{ij}K_{ij} = m^{AB}K_{AB} + \gamma^{\phi\phi}K_{\phi\phi}$. With equations (7.46) and (7.47) and the relation $s_i = \lambda m_i$ the apparent horizon condition (7.39) becomes[19]

$$\partial_\theta^2 h = -\Gamma^A_{BC}m_A u^B u^C - \left(\frac{ds}{d\theta}\right)^2 \gamma^{\phi\phi}\Gamma^A_{\phi\phi}m_A$$

$$- (\gamma^{(2)})^{-1/2}\left(\frac{ds}{d\theta}\right)u^A u^B K_{AB} - (\gamma^{(2)})^{-1/2}\left(\frac{ds}{d\theta}\right)^3 \gamma^{\phi\phi}K_{\phi\phi}. \tag{7.49}$$

Derivatives $\partial_\theta h$ are still hidden in u^A, m_A and $ds/d\theta$, but we have eliminated λ, which depends on $\partial_\theta h$ in a rather complicated way.

---

[19] See Eppley (1977), whose notation we adopt, for an alternative derivation originally due to D. Eardley.

As anticipated, this equation is an ordinary differential equation for the location $r = h$ of the apparent horizon as a function of θ. The coefficients in this equation have to be evaluated at the location $h(\theta)$. Exercise 7.16 provides an example for which this equation assumes a much simpler form.

> **Exercise 7.16** Consider an axisymmetric spacetime at a moment of time symmetry ($K_{ij} = 0$) that is described by the conformally-flat 3-metric
>
> $$ds^2 = \psi^4 (dr^2 + r^2(d\theta^2 + \sin^2\theta \, d\phi^2)), \qquad (7.50)$$
>
> where $\psi = \psi(r, \theta)$. Show that equation (7.49) defining the apparent horizon surface $r = h(\theta)$ reduces to
>
> $$\partial_\theta^2 h = -\left(\frac{(\partial_\theta h)^3}{h^2} + \partial_\theta h\right)\left(\frac{4\partial_\theta \psi}{\psi} + \cot\theta\right)$$
>
> $$+ (\partial_\theta h)^2\left(\frac{3}{h} + \frac{4\partial_r \psi}{\psi}\right) + 2h + h^2\frac{4\partial_r \psi}{\psi}. \qquad (7.51)$$

> **Exercise 7.17** Verify that for a Schwarzschild black hole in isotropic coordinates the apparent horizon condition (7.51) recovers equation (7.26), $h = M/2$.

Boundary conditions for equation (7.49) arise from the requirement that the surface S be smooth (e.g., no cusps at poles or equator). This requirement implies that the first derivative $\partial_\theta h$ has to vanish at the poles $\theta = 0$ and $\theta = \pi$, and, in the case of equatorial symmetry, at $\theta = \pi/2$.

> **Exercise 7.18** The term $\cot\theta$ in equation (7.51) is irregular at the poles. Use a Taylor expansion of $h(\theta)$ about $\theta = 0$ to show that, to lowest order in θ, the differential equation (7.51) near the poles can be written as
>
> $$\partial_\theta^2 h = h + 2h^2\frac{\partial_r \psi}{\psi}, \qquad (7.52)$$
>
> demonstrating that the coordinate singularity there can be removed.

Using the above formalism we can now find the apparent horizon for the boosted black hole solution that we discussed in Chapter 3.2.

> **Exercise 7.19** (a) Consider the boosted black hole solution of Chapter 3.2 and find the location of the apparent horizon to leading order in the correction due to P. Specifically, write h as $h = h_0 + h_P + \mathcal{O}(P^2)$ where, from exercise 7.17, $h_0 = \mathcal{M}/2$, and show that to linear order in P equation (7.49) reduces to[20]
>
> $$\partial_\theta^2 h_P - h_P + (\cot\theta)\partial_\theta h_P - \frac{3P}{16}\cos\theta = 0, \qquad (7.53)$$
>
> where $P = (P^i P^j \gamma_{ij})^{1/2}$ is the magnitude of the 3-momentum P^i, and where we have taken P^i to be aligned with the polar axis.

---

[20] See Appendix A.2 of Dennison *et al.* (2006); see also Cook and York, Jr. (1990).

(b) Show that this equation, together with the boundary conditions at the poles, is solved by

$$h_P = -\frac{P}{16}\cos\theta. \qquad (7.54)$$

Hint: Compute $K_{ij} = \psi^{-2}\bar{A}_{ij}$ from equation (3.80); also note that, according to exercise 3.13, corrections to ψ enter only at second order in P, so that they can be neglected here.

To lowest order in P, the boost of the black hole therefore results in a translation of the apparent horizon in the direction opposite to the boost. We can now use this result to compute the irreducible mass (7.2) of this boosted black hole – at least approximately. The irreducible mass (7.2) is defined in terms of the proper area of the event horizon, and we, of course, have only located the apparent horizon. For stationary spacetimes the two horizons coincide, and for small deviations from stationarity, as is the case here, the assumption is that the two are very close to each other. Making this assumption, we will approximate the irreducible mass by its value in terms of the apparent horizon.

Exercise 7.20   (a) Show that, for a conformally flat 3-metric in axisymmetry (equation 7.50), the proper area of the black hole apparent horizon is given by

$$A = \int_0^{2\pi}\int_0^\pi \psi^4 h^2 \left(1 + \left(\frac{\partial_\theta h}{h}\right)^2\right)^{1/2}\sin\theta\,d\theta\,d\phi. \qquad (7.55)$$

Hint: Treat the horizon as a "surface of revolution" $r = h(\theta)$ about the z-axis and use elementary calculus, or use the results of Appendix C.
   (b) Expand both h and ψ to second order in P, i.e., write

$$h = h_0 + h_P + h_{P^2} = \frac{M}{2} + h_P + h_{P^2} \qquad (7.56)$$

and

$$\psi = \psi_0 + \frac{P^2}{M^2}\psi^{(2)} = 1 + \frac{M}{2h} + \frac{P^2}{M^2}\psi^{(2)} \qquad (7.57)$$

to show that

$$\psi^4 h^2 = 16\left(\frac{M^2}{4} + \frac{h_P^2}{2} + \frac{P^2}{2}\psi^{(2)}\right). \qquad (7.58)$$

Notice that the second-order perturbation of the horizon, h_{P^2}, has cancelled out of this expression, so that the results of exercises 3.13 and 7.19 are sufficient to compute the irreducible mass to second order.
   (c) Now insert (7.58) into (7.55) to show that, to second order in P,

$$A = 16\pi M + 16\pi P^2 \int_0^\pi \psi^{(2)}\sin\theta\,d\theta + 16\pi \int_0^\pi h_P^2 \sin\theta\,d\theta$$

$$+ \pi \int_0^\pi 16(\partial_\theta h_P)^2 \sin\theta\,d\theta. \qquad (7.59)$$

Finally insert the results of exercises 3.13 and 7.19, carry out the integrations, and use (7.2) to show that[21]

$$M_{\text{irr}} = \mathcal{M} \left(1 + \frac{P^2}{8\mathcal{M}^2} \right). \tag{7.60}$$

The result of exercise 7.20 now helps to resolve an issue that we encountered in exercise 3.24, where we found the expression

$$M_{\text{ADM}} = \mathcal{M} + \frac{5}{8} \frac{P^2}{\mathcal{M}} + \mathcal{O}(P^4) \tag{7.61}$$

for the ADM mass of this boosted black hole solution (see equation 3.155). At first sight, this result seems surprising because we might have expected a factor of $1/2$ instead of $5/8$ in front of the second term in order to obtain the correct Newtonian expression for the kinetic energy (see also the paragraph following exercise 3.24). The culprit is the mass parameter $\mathcal{M}$, which, in general, is not a physical measure of the black hole mass. Now we can use equation (7.60) to replace $\mathcal{M}$ with M_{irr} in equation (7.61). This yields

$$M_{\text{ADM}} = M_{\text{irr}} + \frac{1}{2} \frac{P^2}{M_{\text{irr}}} + \mathcal{O}(P^4), \tag{7.62}$$

which is very reassuring. This example illustrates that the ADM mass of a boosted, nonspinning black hole differs from its irreducible mass by its kinetic energy, which is all the mass-energy that can be extracted from such a black hole.

Following these examples we now return to a more general treatment of the apparent horizon in axisymmetry. We have reduced the problem of finding apparent horizons in axisymmetry to a standard two point boundary value problem that can be solved with standard numerical techniques, e.g., spectral methods, the shooting method, or relaxation methods.[22]

In a variant of the spectral method we can write the location $h(\theta)$ in a power series in $\cos\theta$,

$$h(\theta) = \sum_{n=0}^{n_{\max}} c_n \cos^n \theta, \tag{7.63}$$

which automatically satisfies the boundary conditions at the poles. Inserting this expansion into equation (7.49) yields an equation for the coefficients c_n that has to hold for all θ, say

$$P(c_0, c_1, \ldots, c_n, \theta) = 0. \tag{7.64}$$

We could solve this equation by evaluating it at a certain set of points, which yields a set of equations for the c_n that can then be solved by matrix inversion (see Chapter 6.3). In an

---

[21] See Appendix A.3 of Dennison *et al.* (2006).
[22] See, e.g., Press *et al.* (2007).

alternative iterative method we can vary each c_n, keeping all other coefficients constant, until the integral $\int P^2 d\theta$ assumes a minimum.[23] An apparent horizon will have been located provided this integral becomes zero to within a desired tolerance.

Alternatively we can adopt a simple "shooting" technique. Starting at the pole $\theta = 0$, we can set $\partial_\theta h = 0$ and take a trial value $h(0) = h_0$ as initial data and integrate equation (7.49) either to the equator (assuming equatorial symmetry) or to the other pole (in the absence of equatorial symmetry). We can then vary h_0 from values much larger than the suspected location of the apparent horizon to values much smaller than this value and search for a sign change of $\partial_\theta h$ at the equator or the other pole. If there is no sign change, there is no apparent horizon. If there is a sign change, we can iterate over h_0 until $\partial_\theta h = 0$, which indicates that we have located the apparent horizon.[24]

Finally, the apparent horizon condition (7.49) can be finite differenced and then solved by relaxation.[25]

An interesting application of apparent horizon finders in axisymmetry is to examine the horizons in Brill–Lindquist initial data for two black holes (see Chapter 3.1). For a moment of time symmetry ($K_{ij} = 0$), a vacuum solution to the constraint equations describing two equal-mass black holes momentarily at rest with respect to each other is given by

$$\gamma_{ij} = \psi^4 \eta_{ij} \tag{7.65}$$

where

$$\psi = 1 + \frac{\mathcal{M}}{2r_1} + \frac{\mathcal{M}}{2r_2} \tag{7.66}$$

(see equation (3.24) and the related discussion). We can choose the two black holes to be aligned along the z-axis and the origin to lie midway between them. In this case we have $r_{1,2} = (x^2 + y^2 + (z \pm z_0)^2)^{1/2}$, where z_0 is the coordinate separation between each singularity and the origin.

For large separations $z_0 \gg \mathcal{M}$ each black hole is surrounded by its own "disjoint" apparent horizon. For small separations there is also a "common" apparent horizon that surrounds both black holes. It is of interest to locate the critical separation z_{crit} at which this common horizon first appears. Numerical integrations give[26] $z_{crit} = 0.767\mathcal{M}$, at which point the common horizon has an area of $\mathcal{A} = 62.5\pi\mathcal{M}^2$.

7.3.3 General case: no symmetry assumptions

In the absence of any symmetry assumptions the operator $m^{ij}\partial_i\partial_j$ that acts on h in equation (7.39) or (7.40) is a 2-dimensional Laplace operator with respect to the metric

---

[23] Eppley (1977).

[24] Nakamura *et al.* (1988); Shapiro and Teukolsky (1992b). Shapiro and Teukolsky (1992b) reintroduce the parameter s (arc length) and integrate ordinary differential equations for $h(s)$ and $\theta(s)$ simultaneously, in lieu of equation (7.49).

[25] Cook and York, Jr. (1990).

[26] Brill and Lindquist (1963); Čadež (1974); Bishop (1982).

m_{ij} induced on the 2-surface S. We can use several different numerical techniques to integrate either one of these partial differential equations, including spectral methods, finite difference methods, and a flow method in which the elliptic operator is converted into a parabolic operator.

We implement a spectral method by expanding the location of the horizon $r = h(\theta, \phi)$ in spherical harmonics,[27]

$$h(\theta, \phi) = \sum_{l=0}^{l_{max}} \sum_{m=-l}^{l} a_{lm} Y_{lm}(\theta, \phi), \tag{7.67}$$

and then finding an iterative algorithm to determine the expansion coefficients a_{lm}. Spherical harmonics are eigenfunctions of the *flat-space* Laplace operator L^2 on a 2-sphere, i.e., $L^2 Y_{lm} = -l(l+1)Y_{lm}$. However, they are not eigenfunctions of the operator $m^{ij}\partial_i\partial_j$ appearing in the apparent horizon equation (7.39) or (7.40). Nevertheless, we can take advantage of the eigenfunction property of the spherical harmonics by recasting the apparent horizon condition into an equation for $L^2 h$, where

$$L^2 h \equiv \partial_\theta \partial_\theta h + \cot\theta \, \partial_\theta h + \sin^{-2}\theta \, \partial_\phi \partial_\phi h. \tag{7.68}$$

Suppose then that we rewrite the apparent horizon condition $\Theta = 0$ as

$$L^2 h = \rho \, \Theta + L^2 h, \tag{7.69}$$

where we substitute equation (7.39) for Θ. Let us choose the scalar function ρ so that the partial derivative $\partial_\theta \partial_\theta h$, which appears in both Θ and $L^2 h$, exactly cancels on the right-hand side of equation (7.69).[28] Substituting equation (7.67) and using the eigenvalue equation for Y_{lm} on the left-hand side of equation 7.69, we can then multiply both sides of this equation by Y_{lm}^* and integrate over S to obtain

$$-l(l+1)a_{lm} = \int_S Y_{lm}^* (\rho \, \Theta + L^2 h) \, d\Omega, \tag{7.70}$$

where $d\Omega = \sin\theta \, d\theta \, d\phi$, and where we have used the orthogonality property of the spherical harmonics. A priori, this new equation is not very helpful, since the right hand side must be evaluated on S, whose location depends on the very coefficients a_{lm} that we are trying to determine. However, equation (7.70) can be used to establish an iteration procedure by which the integral on the right hand side is evaluated using a previous guess for the set a_{lm}^n in order to determine a new, improved set, a_{lm}^{n+1}. Clearly this algorithm works only for $l \geq 1$, while for $l = m = 0$ the left-hand side of equation (7.70) vanishes. In this case we may consider the integral on the right-hand side a function of the coefficient a_{00} through Θ, and vary this coefficient until a root of the right-hand side has been located. This determines the improved value a_{00}^{n+1}.

---

[27] See, e.g., Nakamura *et al.* (1984, 1985); Kemball and Bishop (1991); Gundlach (1998)
[28] See Gundlach (1998) for a generalization.

In an alternative approach we can determine the expansion coefficients a_{lm} with the help of a multidimensional minimization method.[29] Consider the integral of Θ^2 over S,

$$S = \int_S \Theta^2 d\sigma. \tag{7.71}$$

Since Θ depends on h, and h is expressed in terms of the expansion coefficients a_{lm}, S is also a function of these coefficients. We can then use a standard minimization method, for example Powell's method or a Davidson–Fletcher–Powell algorithm,[30] to vary the a_{lm} until a mimimum of S has been found. An apparent horizon has been located if S can be brought sufficiently close to zero within a specified tolerance.

We can also solve the apparent horizon condition (7.39) or (7.40) with finite difference methods.[31] We can again write the condition as in equation (7.69), but now cover S with a finite difference grid (θ_i, ϕ_j), on which $h(\theta, \phi)$ is represented as $h_{i,j}$. The operator (7.68) can then be represented as

$$(L^2 h)_{i,j} = \frac{h_{i+1,j} - 2h_{i,j} + h_{i-1,j}}{(\Delta\theta)^2} + \cot\theta_i \frac{h_{i+1,j} - h_{i-1,j}}{2\Delta\theta} +$$

$$\sin^{-2}\theta_i \frac{h_{i,j+1} - 2h_{i,j} + h_{i,j-1}}{(\Delta\phi)^2}. \tag{7.72}$$

As before, we can solve equation (7.69) using an iterative algorithm. On the right-hand side, the operator L^2 acting on h (equation 7.72) can be evaluated for a previous set of values $h^n_{i,j}$. On the left-hand side, the same operator acts on the new values $h^{n+1}_{i,j}$. Evaluating equation (7.69) at all gridpoints (θ_i, ϕ_j) then yields a coupled set of linear equations for the $h^{n+1}_{i,j}$, which can be solved with standard techniques of matrix inversion. Using the representation (7.69) which employs L^2 simplifies the matrix that needs to be inverted, but alternatively we can finite difference and invert the operator $m^{ij}\partial_i\partial_j$ directly, without employing equation (7.69).[32]

Yet another method that has been used for the locating apparent horizons in three spatial dimensions is the curvature flow method.[33] This method is related to solving an elliptic equation by converting it into a parabolic equation in an artificial "time" coordinate – we have described a similar approach for solving the maximal slicing condition in Chapter 4.2. During evolution in "time", the solution of such a parabolic problem settles down to equilibrium, which furnishes the solution to the original elliptic equation. In this way, we can deform a trial surface S according to

$$\frac{\partial x^i}{\partial\lambda} = -s^i\Theta, \tag{7.73}$$

---

[29] See Libson *et al.* (1996); Baumgarte *et al.* (1996); Anninos *et al.* (1998) who expand h in terms of symmetric traceless tensors instead of spherical harmonics, which is completely equivalent, but more convenient in Cartesian coordinates.
[30] See, e.g., Press *et al.* (2007).
[31] Shibata (1997); also Shibata and Uryū (2000).
[32] See, e.g., Thornburg (1996); Huq *et al.* (2002); Schnetter (2003).
[33] Tod (1991).

where λ is the "time" parameter. For time-symmetric data with $K_{ij} = 0$ we have $\sqrt{2}\Theta = D_i s^i$, so that the expansion becomes proportional to the trace of the extrinsic curvature of S in Σ, $D_i s^i$. The apparent horizon then satisfies $D_i s^i = 0$ and is therefore a minimal surface,[34] for which this method is known to converge. For general data, the flow equation (7.73) is no longer guaranteed to converge, but numerical experience shows that it typically does.[35]

Various variations and combinations of the above methods have been implemented, but it is not clear whether any one of these methods is preferable to the others for all situations.[36] We will encounter several examples of apparent horizon identification in nonaxisymmetric, $3 + 1$ dimensional spacetimes when we discuss simulations of binary black holes in Chapters 12 and 13 and black hole–neutron star binaries in Chapter 17.

7.4 Isolated and dynamical horizons

The formalism of isolated and dynamical horizons combines in many respects the different advantages of event and apparent horizons as black hole diagnostics. Like apparent horizons, isolated and dynamical horizons are quasilocal and do not require global knowledge of the spacetime. Like event horizons, but unlike apparent horizons, they furnish insight into the evolution of a black hole. For example, isolated and dynamical horizons provide a framework for a quasilocal formulation of black hole thermodynamics. This framework furnishes a useful diagnostic of the physical properties (e.g., the mass and spin) of any black hole that may be present in a numerical simulation. It also provides natural boundary conditions for initial data describing black holes in quasiequilibrium.

Here we will restrict our discussion of isolated and dynamical horizons to a brief and qualitative sketch of their definitions and properties. We shall refer the reader to the literature for a more comprehensive discussion and proofs.[37]

Consider a sequence of apparent horizons S on neighboring spatial slices Σ. Since apparent horizons can "jump" discontinuously on neighboring slices,[38] the resulting worldtube H may not be continuous. Let us disregard these jumps, and instead focus on smooth sections of H. The worldtube H can then be either spacelike or null. If matter or radiation is falling into the horizon, the black hole is growing in mass, its horizon is expanding, H is spacelike, and we call it a *dynamical horizon*. If no matter or radiation is falling into the black hole, H becomes null and we call it a *nonexpanding horizon*. The definition of an *isolated horizon* requires some additional mathematical structure, but for our purposes it

---

[34] See discussion in Chapter 4.2.
[35] See also Gundlach (1998); Shoemaker *et al.* (2000) for implementations of related methods.
[36] See Baumgarte and Shapiro (2003c) for more details.
[37] See the review article by Ashtekar and Krishnan (2004) and Dreyer *et al.* (2003) for discussion, proofs and references, and Schnetter *et al.* (2006) for an introduction in the context of numerical relativity.
[38] See Chapter 1.4 and Figure 1.3 for an example.

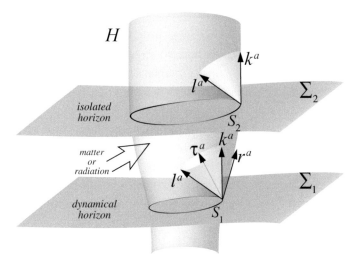

Figure 7.5 Spacetime diagram distinguishing an isolated *vs.* a dynamical horizon (see text for details).

is sufficient to identify a nonexpanding horizon with an isolated horizon.[39] The concepts of dynamical and isolated horizons are illustrated in Figure 7.5.

First consider dynamical horizons. Since H is spacelike, we can apply, at least locally, the same formalism that we developed to describe the spatial slices Σ in Chapter 2. In analogy with the normal vector n^a on Σ, we can define τ^a as a timelike unit vector that is normal to H and satisfies $g_{ab}\tau^a\tau^b = -1$. We can then define the spatial metric $q_{ab} = g_{ab} + \tau_a\tau_b$ induced on H (*cf.* equation 2.27), as well as the extrinsic curvature $K_{ab}^H = -q_a{}^c q_b{}^d \nabla_c \tau_d$ (*cf.* equation 2.49). Evidently, q_{ab} and K_{ab}^H have to satisfy the Hamiltonian constraint (2.132) and the momentum constraint (2.133). We also define a spatial normal on cross sections S of H in analogy to s^a; while s^a is the normal to S in Σ (see Figure 7.3), we now define r^a as the normal to S in H (see Figure 7.5). We also construct outgoing and ingoing null vectors k^a and l^a in terms of τ^a and r^a, in analogy with equation (7.13).

We can now use this machinery to compute the angular momentum contained within S; i.e., the spin of the black hole. To do so we need a Killing vector field ϕ^a of q_{ab}, which is defined on H and is tangent to all cross sections S.[40] Similar to the angular momentum integral (3.187) we can define the angular momentum as

$$J_S = \frac{1}{8\pi} \oint_S K_{ab}^H \phi^a dS^b, \qquad (7.74)$$

where $dS^b = d^2x \sqrt{m} r^b$ and where m is the determinant of the induced metric on S. As it turns out, we may replace K_{ab}^H in the integrand with K_{ab} (the extrinsic curvature of Σ) and

---

[39] See Ashtekar *et al.* (2000); Ashtekar and Krishnan (2003) for precise definitions.
[40] In the absence of an exact Killing vector field, an approximate angular momentum can also be computed from an approximate Killing vector field; see, e.g., Cook and Whiting (2007).

r^a in the surface element with s^a, so that we can compute the angular momentum directly in terms of $3 + 1$ quantities.

Having defined the angular momentum J_S, it can be shown that the horizon mass is given by

$$\boxed{M_S = \frac{1}{2R_S}\sqrt{R_S^4 + 4J_S^2},}$$ (7.75)

where $R_S = (\mathcal{A}_S/4\pi)^{1/2}$ is the areal radius of the horizon. This result is identical to that found for Kerr black holes, but derived independently within the framework of dynamical horizons.

It is also possible to derive thermodynamic laws, and to relate the changes in horizon mass and angular momentum between two intersections S_1 and S_2 of H to the matter and radiation flux across H between S_1 and S_2.[41]

Now turn to isolated horizons. When no more matter or radiation fall into the black hole, the outgoing null vector k^a becomes tangent to H, which is now a null surface. The induced metric on H is then degenerate;[42] in fact, we may view the 2-metric m_{ab} induced on the horizon S (see equation 7.14) as a degenerate 3-metric on H.

Since k^a is tangent to H, equation (7.25) implies that the horizon area,

$$\mathcal{A}_S = \oint_S m^{1/2} d^2x,$$ (7.76)

must remain constant. Several other interesting properties follow, and in particular we can still measure the angular momentum and mass as in equations (7.74) and (7.75).

In anticipation of boundary conditions that we will use for the numerical construction of initial data for black holes in equilibrium or quasiequilibrium (see Chapter 12.3) we shall discuss one additional property of isolated horizons in more detail. By construction, the expansion Θ vanishes everywhere on H. For isolated horizons, k^a is tangent to H, so we must also have

$$\mathcal{L}_k\Theta = 0.$$ (7.77)

Before we proceed it is useful to define the projection of the gradient of k^a as

$$\Theta_{ab} = m_a{}^c m_b{}^d \nabla_c k_d.$$ (7.78)

We now decompose Θ_{ab} into its trace, symmetric-trace-free and antisymmetric parts as

$$\Theta_{ab} = \frac{1}{2}\Theta\, m_{ab} + \sigma_{ab} + \omega_{ab},$$ (7.79)

---

[41] Ashtekar and Krishnan (2003).

[42] We call a metric q_{ab} induced on a surface H degenerate if it has a degenerate direction, meaning that there exists a vector X^a tangent to H so that $q_{ab}X^a = 0$. A surface H is null if the metric induced on the surface is degenerate.

where

$$\Theta = m^{ab}\Theta_{ab} \tag{7.80}$$

is the expansion scalar, with which we are well acquainted by now,

$$\sigma_{ab} = \Theta_{(ab)} - \frac{1}{2}\Theta\, m_{ab} \tag{7.81}$$

is the shear tensor, and

$$\omega_{ab} = \Theta_{[ab]} \tag{7.82}$$

is the rotation tensor.

Raychaudhuri's equation applied to null geodesics with tangent vectors k^a gives[43]

$$\mathcal{L}_k\Theta = -\frac{1}{2}\Theta^2 - \sigma^{ab}\sigma_{ab} + \omega^{ab}\omega_{ab} - R_{ab}k^a k^b. \tag{7.83}$$

As we have discussed, on isolated horizons both Θ and $\mathcal{L}_k\Theta$ vanish. Since the vectors k^a are hypersurface orthogonal the rotation tensor ω_{ab} also must vanish.[44] Invoking Einstein's equation on the horizon in the absence of matter, we see that the last term in equation (7.83) must be zero as well. That leaves $\sigma^{ab}\sigma_{ab} = 0$, which implies the relation

$$\sigma_{ab} = 0 \tag{7.84}$$

on isolated horizons. As we will see in Chapter 12.3.2, this property of isolated horizons will provide a boundary condition for the numerical construction of black holes in (quasi)equilibrium.

---

[43] See equation (2.39) in Section 2.4.3 of Poisson (2004).
[44] See exercise 2.10 and Section 2.4.3 in Poisson (2004).

8 Spherically symmetric spacetimes

Turn now from the most general spacetime in full $3 + 1$ dimensions to the special case of spherical symmetry. Why should we do this? Actually, spherical systems provide very useful computational, physical and astrophysical insight and working with them serves multiple purposes. The field equations reduce to $1 + 1$ dimensions – variables may be written as functions of only two parameters, a time coordinate t and a suitable radial coordinate r – and are much simpler to solve in spherical symmetry. Solving them is a very cost-effective way of probing dynamical spacetimes with strong gravitational fields, including spacetimes with black holes. After all, nonrotating stars and black holes are themselves spherical, so many important aspects of gravitational collapse, including black hole formation and growth, can be studied in spherical symmetry. For example, the numerical study of spherically symmetric collapse to black holes led to the discovery of critical phenomena in black hole formation. The simplification in the equations, together with the reduction in the number of spatial dimensions, means that the system of spherical equations can be solved more quickly, in terms of both human input and computer time, and with much higher accuracy, than the set required for more general spacetimes. As a result, tackling problems in spherical symmetry provides an excellent starting point for learning how to do numerical relativity. It also serves as a convenient laboratory for experimenting with different gauge choices (coordinates) and for generating high precision, test-bed solutions for numerical codes designed to work in higher dimensions.

Having said all of this, we must bear in mind that there are important features of dynamical spacetimes that will be missed when we restrict our attention to spherical symmetry. Rotation cannot be treated in spherical symmetry. Spinning stars, star clusters and black holes, rotational instabilities in stars and star clusters, relativistic effects induced by the dragging of inertial frames – none of these features are present in spherical symmetry. Moreover, gravitational radiation cannot be generated in spherical spacetimes: Birkhoff's theorem forbids it. We thus will have to postpone studying rotation and gravitational wave generation until we relax the restriction to spherical symmetry and advance to axisymmetry. In axisymmetry, spacetime has $2 + 1$ dimensions – two spatial coordinates, e.g., r and a polar angle θ, plus t are necessary to specify the value of any function. Axisymmetry represents the lowest dimensionality at which rotation and gravitational radiation arise in asymptotically flat spacetimes.[1]

The reduction in the number of degrees of freedom characterizing a spherical space-time is reflected in the reduced number of nontrivial components of the 3-metric γ_{ij} that must be determined by solving the $3 + 1$ equations. In general, there are six independent components of γ_{ij}: one component can be accounted for by the conformal factor ψ, which is fixed by the Hamiltonian constraint (e.g., equation 3.37), leaving five components of the conformal 3-metric $\bar{\gamma}_{ij}$. Three of these components are related to the three spatial coordinates and may be removed altogether by a suitable choice of the components of the shift 3-vector β^i. The remaining two components characterize the two dynamical degrees of freedom associated with a gravitational field, namely, the two polarization states of a gravitational wave. Axisymmetric systems with rotation may give rise to both dynamical degrees of freedom. However, if such a system happens to be stationary and nonradiating, like a Kerr black hole or a rotating equilibrium star, neither of these dynamical degrees of freedom are present. Axisymmetric systems without rotation can possess only one gravitational degree of freedom. Moreover, if such a system is static and nonrotating, as in the case of an oblate equilibrium star cluster, none are present. In spherical symmetric spacetimes, neither dynamical degrees of freedom are allowed, ever. The spacetime is nonradiating no matter how violently it may be changing with time.

Choices! Choices!

The high degree of symmetry permits us to write the 3-metric of a spherical spacetime in the general form

$$dl^2 = A\,dr^2 + Br^2(d\theta^2 + \sin^2\theta\,d\phi^2), \tag{8.1}$$

where A and B are functions only of t and r. To specify the full spacetime 4-metric we also require the gauge functions $\alpha(t, r)$ and $\beta^r(t, r)$. As we discussed above, we really require only one nontrivial 3-metric coefficient to fix the 3-geometry.[2] But we have the gauge freedom inherent in our choice of the radial shift function $\beta^r(t, r)$ to cast the spatial 3-metric in alternative forms. There are a couple of different strategies we could adopt in choosing the shift, as we discussed in Chapter 4.4. On the one hand, we could use the shift to

---

[1] Both rotation and gravitational wave generation can arise in $1 + 1$ dimensional spacetimes describing *infinite*, axisymmetric cylinders; see, e.g., Piran (1979) for a discussion and references. Plane gravitational waves of infinite extent can propagate in $1 + 1$ dimensional spacetimes with planar symmetry; see, e.g., Bondi *et al.* (1959) and Ehlers and Kundt (1962) for some exact solutions and Centrella and Wilson (1983, 1984) and Anninos *et al.* (1989) for numerical simulations.

[2] Actually, we know from the Painlevé–Gullstrand metric for a Schwarzschild spacetime (see Table 2.1) that we can have $A = B = 1$, so that *no* nontrivial components are needed to specify the 3-metric in some cases.

reduce the number of nontrivial 3-metric functions to the minimum required number, one. Thus we could set $B = 1$ in equation (8.1), in which case the radial coordinate r is the areal or circumferential radius commonly referred to as the Schwarzschild radial coordinate: $r = r_s = (\mathcal{A}/4\pi)^{1/2} = (\mathcal{C}/2\pi)$, where $\mathcal{A}$ is the proper area and $\mathcal{C}$ is the circumference of a sphere centered at $r_s = 0$. This is the "radial gauge" of Chapter 4.4. Alternatively, we could set $A = B$, whereby $r = \bar{r}$ is an isotropic radial coordinate and we have the "isotropic gauge". The later choice also serves to globally minimize the distortion on the grid, and is a special case of the minimum distortion gauge condition, as discussed in Chapter 4.5. On the other hand, we could employ the shift vector to accomplish a different task, like simplifying the matter field rather than the gravitational field. For example, we might use the shift to maintain comoving coordinates, whereby the matter remains at rest with respect to the spatial coordinates, with its 4-velocity satisfying $u^a \propto (\partial/\partial t)^a$ at all times.[3] Finally, we could simply get rid of the radial shift altogether, setting $\beta^r = 0$. While this may appear to be a simplification, the price we pay is that we must now solve for both A and B, except in special cases where $(\partial/\partial t)^a$ is a Killing vector and the metric functions do not change with time.

Given everything we have discussed so far, we still have left the freedom to choose the lapse function α to specify the time slicing. We could simply set $\alpha = 1$, but we have seen in Chapter 4.1 that such a choice ("geodesic slicing") often leads to fatal coordinate singularities in a numerical simulation. The appearance of a black hole raises a special concern: the computational domain must avoid the physical (curvature) singularity inside the hole at all costs, least the metric functions blow up and cause the code to crash before the evolution is complete. One way to accomplish this is to choose a geometric "singularity avoiding" lapse function, like maximal slicing or polar slicing. Another way is to choose a "horizon penetrating" lapse condition that enables our spatial hypersurfaces to cross the apparent horizon without encountering coordinate singularities in any of the metric functions. That way we can employ black hole excision techniques, removing the central singularity of the black hole and the surrounding neighborhood from the computational domain altogether and replacing it, if needed, with a simple boundary condition at or just inside the apparent horizon, where the metric is well-behaved. Yet another approach employs the "moving puncture" method and gauge conditions, which we will discuss in greater detail in Chapter 13.1.3.

All of the gauge choices discussed above have been utilized numerically at one time or another. Several of them will be explored in the examples which follow, where we will solve Einstein's equations in $1 + 1$ dimensions to construct spherical spacetimes. We will begin with the simplest nontrivial spacetime – a single, isolated Schwarzschild black hole – and work our way through some more complicated examples.

---

[3] See Taub (1978), Section 15, for a shift prescription that maintains comoving coordinates for fluids in *arbitrary* dimensions and Eardley and Smarr (1979) for some numerical examples involving dust in spherical symmetry.

8.1 Black holes

Here we shall construct the spacetime for a nonrotating, vacuum black hole by solving the $1 + 1$ equations. Starting from the same initial conditions, let us see how different lapse and shift conditions lead to different foliations of spacetime. That is, let us see how the resulting $t = constant$ spatial slices differ, both in their "upward" climb in a spacetime diagram, which measures the rate at which proper time as measured by a normal observer n^a advances from one slice to the next at different spatial locations, and in their "lateral" extent, which measures the degree to which the slicing covers the interior of the black hole at any time.

Familiar gauge choices

Before we solve the $1 + 1$ equations formally to construct some examples of spherical black hole spacetimes, let us anticipate the results we should rediscover if we adopt a few familiar, analytic gauge conditions for α and β^r together with suitable initial data. In all of these examples, the lapse and shift form a Killing lapse and shift, as discussed at the end of Chapter 2.7, so that all metric functions remain independent of time.

Begin with a case for which we specify the initial data on the time-symmetry slice in the Kruskal–Szekeres diagram (see Figure 1.1): the spacelike $v = 0$ surface covering $0 \leq u \leq \infty$ in the $u - v$ plane, or, equivalently, the $t_s = 0$ slice from $2M \leq r_s \leq \infty$ in standard Schwarzschild time and radial coordinates, t_s and r_s. Then we know that the familiar gauge choice

$$\alpha(r_s) = (1 - 2M/r_s)^{1/2}, \quad \beta^{r_s}(r_s) = 0, \tag{8.2}$$

gives the usual static solution for the 3-metric in Schwarzschild coordinates,

$$A(r_s) = \frac{1}{1 - 2M/r_s}, \quad B(r_s) = 1. \tag{8.3}$$

Slices of $t_s = constant$ in this foliation are shown in Figure 8.1.

The limiting slice asymptotes to the event horizon at $r_s = 2M$ as $t_s \to \infty$; no slice ever penetrates the horizon. Thus, standard Schwarzschild time slicing is *singularity avoiding* but it is *not horizon penetrating*. A simple radial coordinate transformation can be used to express this solution in terms of an isotropic radial coordinate $\bar{r}$:

$$\alpha(\bar{r}) = \frac{1 - M/2\bar{r}}{1 + M/2\bar{r}}, \quad \beta^{\bar{r}}(\bar{r}) = 0, \tag{8.4}$$

whereby

$$A(\bar{r}) = B(\bar{r}) = \left(1 + \frac{M}{2\bar{r}}\right)^4. \tag{8.5}$$

This is not a different foliation, but rather the same time slicing as Schwarzschild, only expressed in terms of a different radial coordinate.

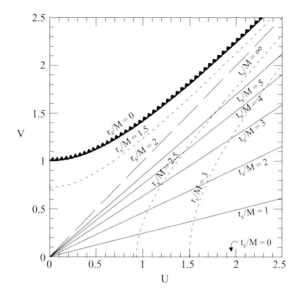

Figure 8.1 Slices of constant Schwarzschild time t_s for a spherical black hole, plotted in a Kruskal–Szekeres diagram. The dashed line denotes the event horizon at $r_s/M = 2$ and the sawtooth curve denotes the central singularity at $r_s/M = 0$.

By contrast with Schwarzschild time slicing, Kerr–Schild (or ingoing Eddington–Finkelstein) slicing *is* horizon penetrating, but it is not singularity avoiding. These properties make Kerr–Schild slicing (or a lapse condition close to it), a good candidate for black hole excision. As in the case of Schwarzschild, the Kerr–Schild solution is analytic and static. However, none of the Kerr–Schild $t_{KS} = constant$ slices are time-symmetric, since the diagonal components of $K_{ij}(r_s)$ never vanish. So we cannot expect the time-symmetric slice in the Kruskal–Szekeres diagram to provide initial data for this solution. Instead, we need data on a $t_{KS} = constant$ slice, any one of which will suffice, since the 3-geometry is still static in this time coordinate. The Kerr–Schild gauge conditions are

$$\alpha(r_s) = \left(\frac{r_s}{r_s + 2M} \right)^{1/2}, \quad \beta^{r_s}(r_s) = \frac{2M}{r_s + 2M}. \tag{8.6}$$

and the 3-metric for this solution is given by

$$A(r_s) = 1 + \frac{2M}{r_s}, \quad B(r_s) = 1. \tag{8.7}$$

To locate the $t_{KS} = constant$ time slices on a Kruskal–Szekeres diagram we need to express u and v as functions of t_{KS} and r_s in regions I and II of the diagram. We find, up to an arbitrary additive constant absorbed in t_{KS},

$$u = \tfrac{1}{2} \left[e^{(t_{KS}+r_s)/4M} + e^{(t_{KS}-r_s)/4M}(r_s/2M - 1) \right],$$
$$v = \tfrac{1}{2} \left[e^{(t_{KS}+r_s)/4M} - e^{(t_{KS}-r_s)/4M}(r_s/2M - 1) \right]. \tag{8.8}$$

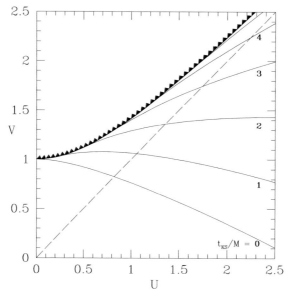

Figure 8.2 Slices of constant Kerr–Schild (ingoing Eddington–Finkelstein) time of a Schwarzschild black hole, plotted on a Kruskal–Szekeres diagram. The dashed line denotes the event horizon at $r_s/M = 2$ and the sawtooth curve denotes the central singularity at $r_s/M = 0$.

Exercise 8.1   Verify equation (8.8). Begin by matching the Schwarzschild and Kerr–Schild line elements to show that

$$t_{KS} = t_s + 2M \ln |r_s/2M - 1| + C, \qquad (8.9)$$

where C is a constant which can be set equal to zero. Then use this equation together with the transformations between Kruskal–Szekeres and Schwarzschild coordinates, in regions I and II of the Kruskal–Szekeres diagram,

$$(r_s/2M - 1)e^{r_s/2M} = u^2 - v^2 \qquad (8.10)$$

and

$$t_s = \begin{cases} 4M \tanh^{-1}(v/u), & \text{in region I,} \quad r_s > 2M, \\ 4M \tanh^{-1}(u/v), & \text{in region II,} \quad r_s < 2M, \end{cases} \qquad (8.11)$$

to derive

$$t_{KS} = 4M \ln(u + v) - r_s, \qquad (8.12)$$

which holds in both regions. Finally, use equations (8.10) and (8.12) to derive equation (8.8).

Using equation (8.8), we plot slices of constant t_{KS} in Figure 8.2. We see from the figure that all of the slices penetrate the horizon, and all of them hit the central singularity. So Kerr–Schild time slicing is horizon penetrating but not singularity avoiding. This time slicing, and slicings with lapse functions close to it, provide a good gauge for performing black hole excision.

Maximal slicing

Maximal slicing has the advantage that it is both horizon penetrating and singularity avoiding. As a result, it has been adopted in many numerical simulations involving black holes. Typically, imposing the maximal slicing condition does not admit an exact solution to the $1 + 1$ equations. Remarkably, for the case of a single, nonrotating black hole in the radial gauge, analytic solutions for a maximal spacetime do exist, even in the presence of a radial shift. We already derived an entire family of time-independent solutions in Chapter 4.2 (see equations 4.23–4.25). Here we will construct another maximal slicing of the same spacetime, one that gives a time-dependent or "dynamical slicing" of a Schwarzschild black hole.[4] This spacetime is also analytic, if by "analytic" we allow 1-dimensional quadratures. The metric for this solution has some generic features that characterize the metric that develops at late times during stellar collapse to a black hole when maximal slicing is employed. Not only will our examination of this solution be useful to illustrate how maximal slicing works, but our reconstruction of the solution will provide a convenient opportunity to review the typical steps required to build a spacetime in the $3 + 1$ formalism, at least in spherical symmetry.

We start with the line element in the form

$$ds^2 = -(\alpha^2 - \beta^2/A)d\bar{t}^2 + 2\beta d\bar{t}dr + Adr^2 + r^2(d\theta^2 + \sin^2\theta d\phi^2), \qquad (8.13)$$

where r is the Schwarzschild radial coordinate (we drop the subscript "s" in this section), $\beta = \beta_r = A\beta^r$, $\bar{t}$ is the maximal time coordinate, and the functions α, β and A depend only on $\bar{t}$ and r. Given this form of the metric we compute all the 3-dimensional Christoffel symbols, which we will need in the evaluation of the standard $3 + 1$ or ADM equations. We find

$$\Gamma^r_{rr} = \partial_r A/(2A), \quad \Gamma^r_{\theta\theta} = -r/A, \quad \Gamma^r_{\phi\phi} = -r\sin^2\theta/A,$$
$$\Gamma^\theta_{r\theta} = \Gamma^\theta_{\theta r} = 1/r, \quad \Gamma^\theta_{\phi\phi} = -\sin\theta\cos\theta, \qquad (8.14)$$
$$\Gamma^\phi_{\theta\phi} = \Gamma^\phi_{\phi\theta} = \cot\theta, \quad \Gamma^\phi_{r\phi} = \Gamma^\phi_{\phi r} = 1/r,$$

with the remaining coefficients equal to zero.

Exercise 8.2 For practice, obtain the Christoffel symbols displayed in equation (8.14) by employing the Lagrangian

$$L = A\dot{r}^2 + r^2\dot{\theta}^2 + r^2\sin^2\theta\dot{\phi}^2 \qquad (8.15)$$

to write down the 3-dimensional Euler–Lagrange equations and then matching them to the 3-dimensional geodesic equation

$$\ddot{x}^i + \Gamma^i_{jk}\dot{x}^j\dot{x}^k = 0, \qquad (8.16)$$

to get the nonvanishing symbols.

---

[4] Estabrook *et al.* (1973); our discussion is patterned after their treatment.

Next we insert the Christoffel symbols in equation (2.142) to calculate the nonvanishing components of the 3-dimensional Riemann tensor, R_{ij}, obtaining

$$R_{rr} = \partial_r A/(rA), \quad R_{\theta\theta} = R_{\phi\phi}/\sin^2\theta = 1 - 1/A + r\partial_r A/(2A^2). \tag{8.17}$$

We now can get $R = R^i{}_i$, required to solve the Hamiltonian constraint equation (2.132):

$$R = 2\partial_r A/(rA^2) + 2(1 - 1/A)/r^2. \tag{8.18}$$

The nonvanishing components of the extrinsic curvature may be calculated from equation (2.134), yielding

$$K_{rr} = -(\partial_{\bar{t}} A + \beta\partial_r A/A - 2\partial_r\beta)/(2\alpha), \quad K_{\theta\theta} = K_{\phi\phi}/\sin^2\theta = r\beta/(\alpha A). \tag{8.19}$$

Maximal slicing requires $K = K^i{}_i = 0$, in which case equation (8.19) implies

$$K_{rr} = -2\beta/(\alpha r), \quad K_{ij}K^{ij} = 6(\beta/\alpha Ar)^2 \tag{8.20}$$

and

$$-\partial_{\bar{t}}\ln A + (\beta/A)\partial_r\ln(\beta^2 r^4/A) = 0. \tag{8.21}$$

The Hamiltonian constraint (2.132) reduces to

$$R = K_{ij}K^{ij}, \tag{8.22}$$

which, inserting equations (8.18) and (8.20), yields

$$3\beta^2/(\alpha^2 A) = A - 1 + r\partial_r A/A. \tag{8.23}$$

When combined with equation (8.20) for K_{rr}, the radial component of the momentum constraint (2.133) may be evaluated to give

$$\partial_r\ln(\beta r^2/A\alpha) = 0. \tag{8.24}$$

Maximal slicing also requires $\partial_{\bar{t}} K = 0$, in which case equation (2.137), combined with equation (8.22), gives $D^2\alpha = \alpha R$. Substituting equation (8.18) in the right-hand side and expanding the derivative on the left-hand side yields an equation for the lapse,

$$\partial_r\partial_r\alpha + 2\partial_r\alpha/r - (\partial_r\ln A)\partial_r\alpha/2 = 2\alpha(A - 1 + r\partial_r\ln A)/r^2. \tag{8.25}$$

Finally, the evolution equation (2.135) for K_{rr} gives

$$\begin{aligned}
\partial_{\bar{t}}\ln(\beta/\alpha) = {} & (3\beta/A + \alpha^2 A/\beta - \alpha^2/\beta)/r \\
& + 3(\partial_r\beta)/A + (\alpha^2/\beta - 4\beta/A)(\partial_r\ln A)/2 \\
& - (\beta/A + \alpha^2/\beta)\partial_r\ln\alpha,
\end{aligned} \tag{8.26}$$

where we have used equation (8.20) to replace K_{rr}, equation (8.17) for R_{rr}, and equation (8.25) for $\partial_r\partial_r\alpha$.

We have now assembled the basic equations and are ready to integrate them to obtain the solution. Note that in spherical symmetry, which contains no dynamical degrees of freedom (i.e., no gravitational waves), it is not necessary to solve an evolution equation either for the 3-metric variable A or the extrinsic curvature variable $K^r{}_r$. These longitudinal quantities are determined on any time slice entirely by the Hamiltonian and momentum constraint equations. For example, we may solve the constraint equation (8.22) for A in lieu of the evolution equation (8.21). For this reason we say that in spherical symmetry the gravitational field evolution is completely *constrained*.

Equation (8.24) immediately gives

$$\beta = \alpha A T / r^2, \tag{8.27}$$

where $T = T(\bar{t})$ is a constant of integration that can be a function of $\bar{t}$ only. Substituting this result into equation (8.23) and integrating yields

$$A = \frac{1}{1 - 2M/r + T^2/r^4}, \tag{8.28}$$

where the new constant of integration M is again a function of $\bar{t}$ only. By going to large r we we will be able to identify M with the total mass-energy of the black hole, once we show that it is independent of $\bar{t}$. Taking a time derivative of equation (8.28) and using equations (8.21) and (8.27) yields

$$\partial_r(\alpha A^{1/2}) = A^{3/2}(r\,\partial_{\bar{t}} M / T - \partial_{\bar{t}} T / r^2), \tag{8.29}$$

which we will integrate for the lapse shortly. Substituting equation (8.27) into equation (8.26), and using the time derivative of equation (8.28) to help evaluate the left-hand side shows, after some algebra, that $\partial_{\bar{t}} M = 0$, hence M is constant, as anticipated. We now integrate equation (8.29) to get the lapse function, using equation (8.28) and the condition that $\alpha \to 1$ as $r \to \infty$:

$$\alpha = (1 - 2M/r + T^2/r^4)^{1/2} \left[1 + \frac{\partial_{\bar{t}} T}{M} \int_0^{M/r} dx\,(1 - 2x + T^2 x^4/M^4)^{-3/2} \right]. \tag{8.30}$$

Using equations (8.23) and (8.27) one can check that equation (8.25), which we have not needed to solve for the lapse, is automatically satisfied by equation (8.30).

So far, the constant of integration $T = T(\bar{t})$ is undetermined. We can now pick out a particular maximal slicing of Schwarzschild by determining this function. Setting $T = 0$, for example, we recover the familiar $t_s = constant$ slices, for which equations (8.28) and (8.30) reduce to equations (8.3) and (8.2) with zero shift.

> **Exercise 8.3** Check that the static Schwarzschild metric (8.3) and (8.2) satisfies equation (8.25).

If we set T to a constant, say $T = C$, we recover the family of time-independent maximal slicings of Schwarzschild, equations (4.23)–(4.25), that we derived in Chapter 4.2 with the help of an alternative "height-function" approach. Evidently, there exist many different maximal slices of Schwarzschild.

How is it possible that the same maximal slicing condition for the lapse, equation (8.25), a second-order partial differential equation, is satisfied for different solutions? The answer lies in the different *inner boundary condition* imposed on the lapse on the black hole throat. Setting $\alpha = 0$ on the Einstein–Rosen bridge at the center of the throat leads to standard *static* time slicing with $T = 0$. By contrast, setting $\partial_r \alpha = 0$ to require "smoothness" across the throat, results in a *dynamical* time slicing solution, the derivation of which we have nearly completed. In obtaining equation (8.30) for the lapse by integrating equation (8.29) rather than by integrating equation (8.25), we simply postponed having to specify inner boundary conditions on the throat.

The dynamical maximal time-slicing solution for the spacetime is given by equations (8.27), (8.28) and (8.30). However, there still remains one unknown in these equations: the function T and its dependence on $\bar{t}$. Specifying this function will now require that we explicitly impose an inner boundary condition on the black hole throat, and, as mentioned above, we shall impose "smoothness" across the throat to get the desired solution. The simplest way to do this is to transform to standard Schwarzschild coordinates, (t, r, θ, ϕ), where $t = t(\bar{t}, r)$ but where the spatial coordinates are the same. Matching the line element (8.13) with the standard Schwarzschild line element gives

$$\partial_{\bar{t}} t = \alpha A^{1/2}, \tag{8.31}$$

$$\partial_r t = A^{1/2} T / \left[r^2 (2M/r - 1) \right]. \tag{8.32}$$

Exercise 8.4   Verify equations (8.31) and (8.32).

Integrating equation (8.32) yields

$$t/M = (T/M^2) \int_{M/r}^{X(T)} dx (1 - 2x + T^2 x^4/M^4)^{-1/2} (2x - 1)^{-1}, \tag{8.33}$$

where $X(T)$ is determined by differentiating equation (8.33) with respect to $\bar{t}$, substituting into equation (8.31), and using equation (8.30) to get

$$\frac{dX}{dT} = T^{-1}(2X - 1)(1 - 2X + T^2 X^4/M^4)^{1/2} \left[\frac{M}{\partial_{\bar{t}} T} + \int_0^X dx (1 - 2x + T^2 x^4/M^4)^{-3/2} \right]. \tag{8.34}$$

Now we can impose the "smoothness" requirement at the center of the Einstein–Rosen bridge along $t = 0$ for $r < 2M$ (i.e., along $u = 0$, $v > 0$, in Kruskal–Szekeres coordinates). Smoothness requires that, when viewed in a Kruskal–Szekeres diagram, each $\bar{t} = $ constant slice intersect the center of the bridge in the normal direction, so that as $t \to 0$, we have that $\partial_r t \to \infty$ as r approaches its minimum value $r_{min} = r_{min}(\bar{t})$ along the slice (see Figure 8.3). Demanding that this condition be satisfied by equations (8.32) and (8.34) requires that r_{min} be equal to $M/X(T)$ for that value of $X(T)$ that is the smaller of the two real roots of the

Table 8.1 Parameters on selected maximal slices.

$\bar{t}/M$	T/M^2	$\partial_{\bar{t}}T/M$	X	r_{min}/M
0	0	1	$\frac{1}{2}$	2
1	0.8104	0.5350	0.5249	1.9050
2	1.1384	0.1791	0.5670	1.7638
3	1.2460	0.0585	0.6019	1.6615
4	1.2814	0.0193	0.6263	1.5965
5	1.2931	0.0065	0.6422	1.5570
6	1.2970	0.0022	0.6521	1.5334
7	1.2984	0.0007	0.6581	1.5195
8	1.2988	0.0002	0.6617	1.5114
∞	$\frac{3\sqrt{3}}{4}$	0	$\frac{2}{3}$	$\frac{3}{2}$

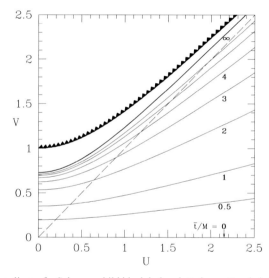

Figure 8.3 Maximal time slices of a Schwarzschild black hole, plotted on a Kruskal–Szekeres diagram. The dashed line denotes the event horizon at $r_s/M = 2$, the sawtooth curve denotes the central singularity at $r_s/M = 0$ and the darker solid line shows the limiting slice at $r_s = 3M/2$.

fourth-order polynomial equation $1 - 2x + T^2x^4/M^4 = 0$. It turns out that equation (8.34) is automatically satisfied when $X(T)$ is a root of this polynomial.

We now can get the desired relation between T and $\bar{t}$. Setting $\bar{t} = t$ at $r = \infty$ in equation (8.33) gives

$$\bar{t}/M = (T/M^2) \int_0^{X(T)} dx(1 - 2x + T^2x^4/M^4)^{-1/2}(2x - 1)^{-1}. \qquad (8.35)$$

For each allowed value of T, where $0 \leq T/M^2 \leq 3\sqrt{3}/4$, the polynomial equation determines $X(T)$ and equation (8.35) gives the value of $\bar{t}$ parametrizing the maximal hypersurface. Values of key parameters on select hypersurfaces are listed in Table 8.1. It

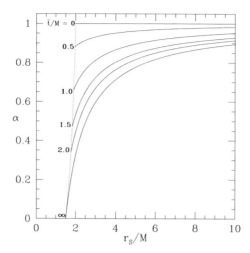

Figure 8.4 The lapse α as a function of areal radius r_s on selected maximal slices.

is relevant to note that the integral in equation (8.35) diverges when the two real roots of the fourth-order polynomial coincide. This occurs at $T/M^2 = 3\sqrt{3}/4$, for which one finds $\bar{t} = \infty$, $X = 2/3$, $r_{\min}/M = 3/2$ and $\alpha = 0$. Thus the foliation of the black hole spacetime terminates at this *limit slice*, beyond which maximal slicing cannot probe. Encountering a limit slice is a characteristic of maximal time slicing and even occurs in nonvacuum scenarios when using maximal slicing to follow the collapse of a star or star cluster to a black hole. Given that all of the slices, including the limit slice, avoid hitting the central singularity, while all events in the domain of outer communications are covered, such a foliation of the black hole spacetime is "good news" from a numerical point of view. Otherwise, a simulation designed to build the spacetime numerically would terminate due to computer overflows upon encountering the central singularity. In this situation the numerical evolution would come to an abrupt end, even if one were interested only in the black hole exterior.[5]

   In Figures 8.4–8.6 we plot the profiles of α, β and A on selected maximal hypersurfaces. The "collapse of the lapse" near the center of the Einstein–Rosen bridge inside $r = 2M$ that we expect with maximal slicing at late times[6] is evident in Figure 8.4. The "bad news" from a numerical standpoint is that metric functions like A blow up at the center of the bridge at $r_{\min}$. The metric coefficients in Kerr–Schild slicing, discussed in the previous section, do not blow up on the slice until reaching the central singularity (*cf.* equation 8.7). However, Kerr–Schild slices do hit the central singularity, which would be disastorous in a numerical simulation. So a better solution to constructing the spacetime numerically would be to eliminate at least part of the black hole interior – after all, the black hole exterior

---

[5] However there do exist cases, e.g., extremely inhomogeneous dust balls, for which maximal slicing fails, hitting the central singularity before covering the domain of outer communications of the resulting black hole; Eardley and Smarr (1979).

[6] Recall our discussion in Chapter 4.2.

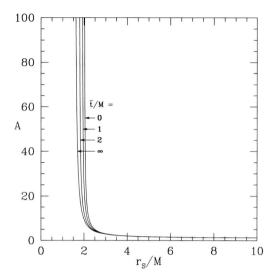

Figure 8.5 The radial metric coefficient $A(=\gamma_{rr})$ as a function of areal radius r_s on selected maximal slices.

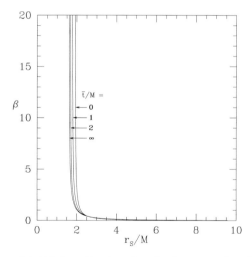

Figure 8.6 The radial shift β as a function of areal radius r_s on selected maximal slices.

cannot be affected by the interior, so there should be no need to compute the interior. This black hole *excision* technique[7] does not rely on any particular slicing condition in the exterior, and can be used as long as the grid functions remain sufficiently well-behaved outside the excision surface. We will discuss some examples in Chapters 13 and 14. An alternative approach might use the *moving-puncture* approach,[8] which adopts a particularly well suited set of coordinates that avoids the spacetime singularities without developing the

[7] See Chapters 13.1.2 and 14.2.3.
[8] See Chapter 13.1.3.

pathologies described above. Examples of moving-puncture simulations will be described in Chapters 13 and 17.

8.2 Collisionless clusters: stability and collapse

Collisionless particles provide a convenient matter source for developing numerical relativity algorithms in general, and for problems involving relativistic matter in particular. By the same token, numerical relativity provides a powerful tool for studying the physical behavior and astrophysical implications of self-gravitating clusters of relativistic, collisionless particles.

From a technical point of view, working with collisionless matter has some advantages over fluid systems for designing and testing numerical relativity schemes. The collisionless matter equations can be represented as ordinary differential equations (geodesic equations; recall Chapter 5.3 and equation 5.222) and thus are straightforward to integrate, while hydrodynamical equations are partial differential equations and require more subtle handling. Furthermore, collisionless matter is not subject to shocks or other discontinuities that often require special treatment, or sophisticated schemes, in numerical simulations to give reliable results.

Numerical relativity simulations have resolved a number of longstanding issues in relativistic stellar dynamics, issues which had been unresolved even for spherically symmetric systems. These results have far wider applicability than stellar dynamics, in that the matter fields need not be identified with stars *per se* but with any gas of self-gravitating, collisionless particles, as emphasized in Chapter 5.3. As an example of such an issue, consider that, in contrast to the situation for spherical *fluid* stars in general relativity, there exists only sufficient, but not necessary, criteria for the dynamical radial stability of a spherical *collisionless* cluster. Specifically, linear perturbation theory, using trial functions in a variational principle, had demonstrated that along a one-parameter sequence of equilibrium clusters parametrized by z_c, the central redshift of the configuration, the onset of instability occurs *near* the point of maximum fractional binding energy ($E_b \equiv (M_0 - M)/M_0$, where M is the total mass-energy and M_0 is the rest-mass), independent of the nature of the equilibrium models.[9] In typical models a turning point in the relativistic binding energy curve occurs at high redshift, $z_c \approx 0.5$. Subsequently, a theorem was then proven rigorizing this result, but it is restricted, stating only that that the equilibrium configurations are stable *at least* up to the first maximum of the fractional binding energy along the sequence.[10] This contrasts with the situation for a spherical fluid equilibria (stars) in general relativity, for which the binding energy maximum, equivalent to the "turning point" along the M *vs.* ρ_c curve (see Chapter 1.3), identifies precisely the onset of radial instability. Fully nonlinear, time-dependent simulations furnish strong numerical evidence that the turning point in the binding energy curve does in fact signal the the onset of dynamical radial

---

[9] Ipser and Thorne (1968); Ipser (1969a,b); Fackerell (1970).
[10] Ipser (1980).

instability along an equilibrium sequence of collisionless clusters. Numerical relativity has thus "discovered" a theorem awaiting a formal proof.[11]

By far the most interesting and important results to emerge from the numerical simulations deal with the nonlinear evolution and final fate of unstable clusters. They corroborate earlier speculation[12] that unstable clusters inevitably undergo catastrophic collapse to black holes. They also demonstrate that when the nearly homogeneous core of a centrally concentrated ("core-halo") cluster undergoes collapse, a mass much larger than the core ultimately forms the central black hole. This occurs because of the so-called "avalanche effect", whereby the black hole formed from the collapse of the initial core grows by capture of lower-angular momentum stars that orbit at increasingly larger apocenter from the center but wander close to the hole at pericenter. At the end of the collapse, the cluster settles into a new stationary state consisting of a massive, nearly Newtonian halo in orbit about a central black hole. This numerical example provides one viable scenario for forming a supermassive black hole.

Relativistic "violent relaxation", or "dynamical phase mixing", has also been explored by simulations. In violent relaxation, collective fluctuations in the time-varying gravitational field lead to particle thermalization.[13] In general relativity, a bound, nonequilibrium collisionless cluster can either achieve virial equilibrium via violent relaxation or else collapse to a black hole.

In the sections that follow we will sketch exactly how some of the numerical calculations of collisionless clusters have been performed. The key computational challenge involves black hole formation and singularity avoidance in cases of catastrophic cluster collapse. We will focus on a few of the ways that this challenge has been overcome successfully.

8.2.1 Particle method

Physical picture

Consider any spherical surface drawn in the interior of a spherical distribution of particles (see Figure 8.7). The surface is densely and uniformly covered with an infinite number of particles, each with an infinitesimal rest mass. Particles move both in the radial and transverse directions, but to preserve spherical symmetry, their transverse motion must be isotropic. Accordingly, even as individual particles may have nonzero angular momentum about the cluster center, the total angular momentum summed over all the particles is strictly zero.

The restriction to spherical symmetry reduces the number of degrees of freedom in phase space that we need to consider. In coordinate space the only nontrivial dynamical variable

---

[11] For numerical simulations in spherical symmetry, see Shapiro and Teukolsky (1985b,a,c, 1986) and, for a review and references, Shapiro and Teukolsky (1992a). The discussion in this section is drawn from the later two references and Rasio *et al.* (1989b). See also Olabarrieta and Choptuik (2002).

[12] Zel'dovich and Podurets (1965); see also Fackerell *et al.* (1969).

[13] This was first discussed in Newtonian gravitation by Lynden-Bell (1967). See also Binney and Tremaine (1987) for discussion.

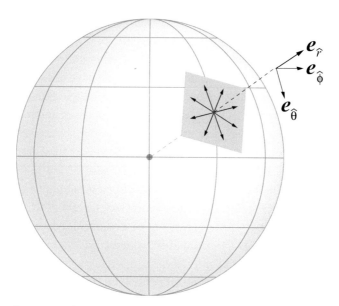

Figure 8.7 Schematic representation of the distribution of particles in a spherical cluster (see text). [After Shapiro and Teukolsky (1992a).]

is the radius r of a particle. In velocity space the only nontrivial dynamical variables are the radial and transverse velocities, $u^{\hat{r}}$ and $u^{\perp}$.

Newtonian limit

Prior to describing the relativistic problem, it is useful to consider how a spherical cluster of collisionless particles can be evolved in Newtonian physics. In this limit, the metric reduces to

$$ds^2 = -(1 + 2\Phi)dt^2 + dr^2 + r^2 d\Omega^2, \tag{8.36}$$

where Φ is the Newtonian potential. The particles move according to the geodesic equation, which is simply Newton's law of motion,

$$\frac{dx^i}{dt} = u^i, \qquad \frac{du^i}{dt} = -\nabla\Phi, \tag{8.37}$$

where x^i and u^i are the position and velocity 3-vectors of each particle. In spherical symmetry these equations simplify to

$$\frac{dr}{dt} = u_r,$$

$$\frac{du_r}{dt} = -\partial_r \Phi + \frac{u_\phi^2}{r^3},$$

$$u_\theta = 0 \quad (\text{orbit confined to the plane } \theta = \frac{\pi}{2}),$$

$$\frac{du_\phi}{dt} = 0 \quad (\text{conservation of angular momentum}). \tag{8.38}$$

These equations are integrated forward in time for every particle for a small time step. The new particle positions yield the rest-mass density ρ at the new time:

$$\rho = \sum_{\text{all particles}} mn, \tag{8.39}$$

where m is the particle rest mass and n is the number density. The rest-mass density serves as the (only) source term for the gravitational field equation, which in this case is just Poisson's equation,

$$\nabla^2 \Phi = 4\pi\rho. \tag{8.40}$$

Solving this equation gives the self-consistent gravitational field at the new time. The new potential is then inserted in the particle equations of motion and the process is repeated for another time step. This approach to evolving a self-gravitating, collisionless N-body system is known as a *mean-field, particle simulation scheme*, and is well studied in the Newtonian domain.

The general relativistic formulation

A mean-field, particle simulation scheme for general relativity can be constructed in a similar fashion to the Newtonian scheme described in the previous section, as shown by Shapiro and Teukolsky (1985b,a,c, 1986, 1992b). The matter is evolved by N-body particle simulation and provides the source for the mean gravitational field, or metric. The metric is determined by solving the standard $3 + 1$ (ADM) equations.

For a spherical spacetime, we shall adopt the metric in isotropic form,

$$ds^2 = -(\alpha^2 - A^2\beta^2)dt^2 + 2A^2\beta drdt + A^2(dr^2 + r^2 d\theta^2 + r^2 sin^2\theta d\phi^2). \tag{8.41}$$

Here we have set $\beta = \beta^r$, the contravariant radial component of the shift vector, the only nonvanishing component.

The stress-energy tensor for the matter is given by equation (5.221) and the equations of motion ($\nabla_a T^{ab} = 0$) are the geodesic equations for each particle, given by equations (5.223) and (5.224), with normalization condition (5.225). As in Newtonian gravitation, these equations simplify considerably in spherical symmetry, yielding

$$\frac{dr}{dt} = \frac{\alpha u_r}{A^2(\alpha u^0)} - \beta,$$

$$\frac{du_r}{dt} = -(\alpha u^0)\partial_r\alpha + u_r\partial_r\beta + \frac{u_r^2}{u^0}\frac{\partial_r A}{A^3} + \frac{u_\phi^2}{u^0}\left(\frac{1}{r^3 A^2} + \frac{\partial_r}{r^2 A^3}\right), \tag{8.42}$$

$$u_\theta = 0,$$

$$\frac{du_\phi}{dt} = 0.$$

The normalization condition $u_a u^a = -1$ gives

$$\alpha u^0 = \left(1 + \frac{u_r^2}{A^2} + \frac{u_\phi^2}{r^2 A^2} \right)^{1/2}. \qquad (8.43)$$

Exercise 8.5   Verify equations (8.42) and (8.43). Show that they reduce to equations (8.38) in the Newtonian limit, where $A \to 1, \alpha \to 1, \partial_r \alpha \to \partial_r \Phi, \beta \to 0$ and $\alpha u^0 \to 1$.

Hence, given the metric at any time t, equations (8.42) and (8.43) are integrated forward in time for the new particle positions and 4-velocities at $t + \Delta t$.

Given the particle positions and velocities, the matter source terms appearing in the field equations can be calculated according to equations (5.226)–(5.229). For spherical systems, they yield

$$\rho = \sum_A m_A n_A W^2, \qquad (8.44)$$

$$S_r = \sum_A m_A n_A W u_r^A, \qquad (8.45)$$

$$S_{rr} = \sum_A m_A n_A u_r^A u_r^A, \qquad (8.46)$$

$$S = \rho - \sum_A m_A n_A, \qquad (8.47)$$

where we regard each particle as being in its own category A, so that the sum over A is a sum over particles (recall $W = \alpha u_A^0$). The number density n_A can then be calculated according to equation (5.230), which in spherical symmetry can be taken to be

$$n_A = \frac{1}{4\pi W A^3 r^2 \Delta r}. \qquad (8.48)$$

Here we treat the particle as smeared out over a zone size Δr, which is the radial grid size.

In writing down the field equations we want to choose a suitable time coordinate, i.e., a lapse, that allows us to integrate forward in time without encountering singularities in the case of black hole formation. Based on our discussion in Section 8.1, where we constructed the spacetime for a vacuum black hole, we suspect that *maximal slicing* might provide a reasonable choice. The maximal slicing condition requires $K = 0 = \partial_t K$ (see Chapter 4.2), where we recall $K \equiv K_i{}^i$ is the trace of the extrinsic curvature. Taking the trace of the evolution equation for $\partial_t K_{ij}$, i.e., equation (2.137), then leads to a linear elliptic equation for α,

$$\partial_r \left(A r^2 \partial_r \alpha \right) = \alpha A^3 r^2 \left[\frac{3}{2} (K^r{}_r)^2 + 8\pi \rho + 4\pi T \right] \quad \text{(maximal slicing)}, \qquad (8.49)$$

where $T = S - \rho$ is the trace of T_{ab}. Alternatively, we might try *polar slicing*, for which the trace of the part of K_{ij} transverse to the radial direction is set to zero: $K_T \equiv K^\theta{}_\theta + K^\phi{}_\phi = 0$. As we have seen in exercise 6.14, and as we will rederive in exercise 8.10, polar slicing avoids regions of spacetime containing trapped surfaces in spherical symmetry, suggesting that it might have better singularity avoidance than maximal slicing when a black hole forms. The lapse equation for polar slicing is a simple quadrature,

$$\alpha = \left[1 - \left(\frac{M}{2r_{\max}} \right)^2 \right] \exp \left[\frac{1}{2} \int_{r_{\max}}^{r} \frac{r(\partial_r A/A)^2 + 8\pi r S_{rr}}{1 + r \partial_r A/A} dr \right] \quad \text{(polar slicing)}, \quad (8.50)$$

where $r_{\max}$ is an arbitrary radius in the vacuum exterior (see exercise 8.8). Both maximal and polar slicings have been used in simulations of collisionless clusters.

It is convenient to use the quantity K_T to express equations that are valid in both maximal and polar gauges:

$$K_T = \begin{cases} 0 & \text{(polar slicing)}, \\ -K^r{}_r & \text{(maximal slicing)}. \end{cases} \quad (8.51)$$

Then the evolution equation (2.134) for the metric coefficient A is

$$\partial_t A = \beta \left(\partial_r A + \frac{A}{r} \right) - \frac{1}{2} \alpha A K_T. \quad (8.52)$$

The Hamiltonian constraint (2.132) becomes a nonlinear elliptic equation for A,

$$\frac{1}{r^2} \partial_r \left(r^2 \partial_r A^{1/2} \right) = -\frac{1}{4} A^{5/2} \left(8\pi\rho + \frac{3}{4} K_T^2 \right). \quad (8.53)$$

We used up our spatial coordinate freedom by choosing the 3-metric to be isotropic. This choice automatically leads to a condition on the shift, β. This shift condition may be derived by using the definition of K_{ij}, equation (2.134), to evaluate $\alpha(K_{rr} - K_{\theta\theta})$, which yields

$$\partial_r \left(\frac{\beta}{r} \right) = \frac{\alpha}{r} (K^r{}_r - K^\theta{}_\theta), \quad (8.54)$$

or

$$\beta = -r \int_r^\infty \alpha (K^r{}_r - \frac{1}{2} K_T) \frac{dr}{r} \quad (8.55)$$

(cf. exercise 4.12).

Exercise 8.6 In spherical symmetry, a "first-order form" of the *minimal distortion* gauge (cf. equation 4.70) may be written

$$(L\beta)_{ij} = 2\alpha(K_{ij} - \frac{1}{3} K \gamma_{ij}), \quad (8.56)$$

where the vector gradient $(L\beta_{ij})$ of the vector β^i is defined in equation (4.67).[14] Show that equation (8.54) is consistent with condition (8.56).

---

[14] For spherical systems, the first-order form of the minimum distortion condition is sufficient, since there are no radiative modes and only longitudinal shear is present. See equations (4.71) or (4.74) for the second-order form of the minimal distortion condition.

The momentum constraint (2.133) determines the radial component of the extrinsic curvature:

$$D_i K^i{}_r - D_r K = 8\pi S_r, \tag{8.57}$$

which yields

$$
K^r{}_r =
\begin{cases}
(8\pi/(A^3 r^3)) \int_0^r A^3 r^3 S_r dr & \text{(maximal slicing)}, \\[2ex]
4\pi r S_r / (1 + r \partial_r A / A) & \text{(polar slicing)}.
\end{cases}
\tag{8.58}
$$

The relevant field equations have now all been assembled. We point out again that in spherical symmetry, which contains no dynamical degrees of freedom, it is not necessary to solve an evolution equation either for the one nontrivial 3-metric coefficient A or the one independent extrinsic curvature variable $K^r{}_r$. These quantities on any time slice are determined entirely by the Hamiltonian and momentum constraint equations. Hence, we may solve the Hamiltonian constraint equation (8.53) for A in lieu of the evolution equation (8.52), and similarly for $K^r{}_r$. The resulting approach constitutes a completely constrained evolution scheme.

Exercise 8.7 Derive field equations (8.52)–(8.58). To do so you will first need to use the isotropic metric (8.41) to calculate the following quantities:
(a) $\Gamma^i{}_{jk}$ for all i, j, k,
Hint: Use equation (3.7) with $\bar{\gamma}_{ij} = (1, r^2, r^2\sin^2\theta)$ to obtain

$$
\begin{aligned}
&\Gamma^r{}_{rr} = \partial_r A / A, &&\Gamma^r{}_{\theta\theta} = -(r + r^2 \partial_r A / A), \\
&\Gamma^\theta{}_{r\theta} = \Gamma^\phi{}_{r\phi} = 1/r + \partial_r A / A, &&\Gamma^\theta{}_{\phi\phi} = -\sin\theta\cos\theta, \\
&\Gamma^r{}_{\phi\phi} = -r\sin^2\theta - r^2\sin^2\theta\,\partial_r A / A, \\
&\Gamma^\phi{}_{\theta\phi} = \cot\theta.
\end{aligned}
$$

(b) R_{ij},
Hint: Use equation (3.10) to find

$$R^r{}_r = -2A^{-3}[\partial_r \partial_r A + \frac{1}{r}\partial_r A - \frac{1}{A}(\partial_r A)^2],$$

$$R^\theta{}_\theta = -A^{-3}[\partial_r \partial_r A + \frac{3}{r}\partial_r A] = R^\phi{}_\phi.$$

(c) R,
(d) K_{ij},
Hint: Use equation (2.135) to show

$$K_{rr} = -[\partial_t A^2 - 2\partial_r \beta_r + 2\beta_r \partial_r A / A]/(2\alpha),$$

$$K_{\theta\theta} = -[\partial_t A^2 - 2\beta_r(1/r + \partial_r A / A)]r^2/(2\alpha)$$

$$= K_{\phi\phi}\sin^{-2}\theta,$$

where $\beta_r = A^2\beta$.
(e) $K = K^i{}_i$.

Exercise 8.8 (a) Derive the maximal slicing condition (8.49).
(b) Derive the polar slicing condition (8.50).
Hint: Define $d/dt \equiv \partial_t - \beta^i \partial_i$. Then show that

$$\frac{dK_j{}^i}{dt} = -\partial_k \beta^i K_j{}^k + \partial_j \beta^k K_k{}^i - D_j D^i \alpha + \alpha [R_j{}^i + K K_j{}^i + 4\pi T \delta_j{}^i - 8\pi S_j{}^i].$$
(8.59)

Set $K_T = K_\theta{}^\theta + K_\phi{}^\phi$, use equation (8.59), and demand that $K_T = 0 = dK_T/dt$.

To solve the field equations uniquely we need boundary conditions. Outer boundary conditions are obtained by matching the metric to the isotropic form of the static Schwarzschild metric as $r \to \infty$. In fact, it is easy to show that in polar slicing, the metric takes the isotropic Schwarzschild form *everywhere* in the vacuum exterior. Set the right hand side of equation (8.53) equal to zero and integrate. Requiring that A tend asymptotically to the isotropic value at large r yields

$$A = \left(1 + \frac{M}{2r}\right)^2 \quad \text{(polar slicing, exterior)}$$
(8.60)

for all r outside the matter. Substituting this result into equation (8.50), with $S_{rr} = 0$, gives

$$\alpha = \frac{1 - M/(2r)}{1 + M/(2r)} \quad \text{(polar slicing, exterior)},$$
(8.61)

which is the familiar isotropic lapse function. Equation (8.58) shows that $K^r{}_r = 0$ everywhere outside the matter, and hence by equation (8.55), $\beta = 0$ as well. Thus in polar slicing one need only integrate Einstein's equations *inside* the matter and match to the standard Schwarzschild metric in isotropic coordinates at the matter surface. By contrast, in maximal slicing, the nonzero β and $K^r{}_r$ require the metric to be integrated to reasonably large values of $r \gg M$ in the vacuum exterior in order to match to the standard Schwarzschild line element. For maximal slicing at large r matching yields $A \to 1 + const/r$ to leading order in $1/r$, and similarly for α. This behavior can be imposed by setting a Robin boundary condition at the outermost grid point, $\partial_r [r(A - 1)] = 0$, and likewise for α. Equation (8.58) gives $K^r{}_r \sim r^{-3}$ as $r \to \infty$, so that the asymptotic solution to equation (8.55) is

$$\beta = \frac{r K^r{}_r}{2}(1 + \mathcal{O}(1/r)) \quad \text{(maximal slicing, large } r).$$
(8.62)

The boundary conditions for the field variables at the center of the cluster are chosen to enforce *regularity*, which is appropriate for a matter profile that is smooth near the origin and exhibits no spikes (e.g. $\partial_r \rho = 0 = S^r$, etc.):

$$0 = \partial_r \alpha = \partial_r A = \beta = K^r{}_r \quad (r \to 0).$$
(8.63)

Solving the Hamiltonian equation (8.53) numerically for $A^{1/2}$ by finite-differencing in r to replace the second-order differential operator requires linearization of the right-hand side. The resulting set of coupled, linear equations, differenced to second order in the grid spacing Δr, typically takes on a tridiagonal form, for which simple recipes exist for inverting the matrix. Linearization requires iteration of the solution until convergence.[15] Using the evolution equation for A to provide a good initial guess on the new time slice, given its value on the previous slice, usually accelerates the convergence. The maximal slicing condition (8.49) yields a similar tridiagonal set of equations, but this equation is linear in α, so no iteration following inversion is necessary. The other field equations may be solved either algebraically or by simple one-dimensional, numerical quadrature.

Since there are no field evolution equations to solve in a completely constrained code, and only ODEs for the matter, the time step Δt between computational time slices is not subject to a hyperbolic Courant condition (e.g., the light-travel time across a radial grid spacing; see Chapter 6.2.3). Rather, it is determined by the time scale for the matter source terms to change, i.e., the dynamical time scale of the system. A reasonable time-step criterion that has worked in practice is

$$\Delta t \approx \min \left[\frac{q}{\alpha} \left(\frac{3\pi}{32\rho} \right)^{1/2} \right],$$ (8.64)

where $q \lesssim 0.1$ is a constant. Here the quantity in parantheses is the free-fall collapse time for a homogeneous dust ball at the local density ρ. The insertion of α in the denominator accounts for the slowing down of *proper* time with respect to *coordinate* time as the lapse function falls to zero. Equation (8.64) still guarantees that the matter source terms change only a small amount between time steps, because they change only when proper time as measured by a normal observer $(d\tau = \alpha dt)$ increases by a significant fraction of the local free-fall time.

Diagnostics

As in any simulation, one can identify a number of nontrivial diagnostics that provide a check on the accuracy of the numerical simulation. For example, the total mass-energy of the configuration $M = M_{\mathrm{ADM}}$, which can be computed as in equation (3.145) or (3.147), must remain constant in time. In fact, in a spherical spacetime, one can use the conservation of mass-energy to prove additional constraints on the metric, as suggested by exercise (8.9).

> **Exercise 8.9**   Everywhere in the vacuum exterior of a spherical spacetime, M may be expressed invariantly as
>
> $$M = \left(\frac{\mathcal{A}}{16\pi} \right)^{1/2} \left(1 - \frac{\nabla_a \mathcal{A} \nabla^a \mathcal{A}}{16\pi \mathcal{A}} \right),$$ (8.65)

---

[15] See Chapter 6.2.2.

where $\mathcal{A}$ is the proper area of a 2-sphere of radius r.[16] Evaluate equation (8.65) to show that in the coordinates adopted in this section we have

$$\mathcal{A} = 4\pi A^2 r^2 \tag{8.66}$$

and

$$M = \frac{1}{2} A r \left[1 + \frac{1}{4}(Ar K_T)^2 - \left(1 + \frac{r}{A}\partial_r A\right)^2 \right]. \tag{8.67}$$

The expression appearing on the right-hand side of equation (8.67) must therefore be constant in space and time throughout the vacuum exterior and must equal the initial mass-energy of the configuration. This provides a useful self-consistency check on the integrations. In the case of polar slicing, where A is given by equation (8.60), the identity is satisfied trivially. The check in this case thus reduces to seeing whether the expression for M formed from the gradient of A is continuous at the matter surface.

A further check is available for stable, equilibrium clusters. Since the metric is static in such a case, each particle has a conserved energy, $E = -p_0$. It is thus useful to check for the conservation of the total particle "energy",

$$-E_0 \equiv \sum_A m_A u_0{}^A. \tag{8.68}$$

In cases where the cluster collapses to a black hole, it is useful to follow the growth of the black hole *event* horizon. This can be done easily in spherical symmetry simply by calculating the trajectories of outgoing radial null rays emitted from various spacetime points. The geodesic equation for the jth ray is

$$\frac{dr_j}{dt} = \frac{\alpha(t, r_j)}{A(t, r_j)} - \beta(t, r_j). \tag{8.69}$$

The location of the event horizon is then found by finding pairs of null rays emitted from the same radius at slightly different times, one of which escapes to infinity and one of which is pulled back into the hole.[17]

Inside the black hole there can be *trapped surfaces*, i.e., regions where the cross-sectional area of an outgoing bundle of null rays immediately converges.[18] In spherical symmetry this condition can be expressed simply by the equation

$$\frac{d\mathcal{A}}{dt} \leq 0, \tag{8.70}$$

where d/dt is the total derivative along the null ray and $\mathcal{A}$ is given by equation (8.66) for a radially propagating bundle of null rays spanning a 2-sphere. Using equations (8.52) and

[16] See Lightman *et al.* (1975), problem 16.10.
[17] See Chapter 7.2.
[18] Recall Chapter 7.3.

(8.69), we find that condition (8.70) becomes

$$1 + r(\partial_r A)/A - ArK_T/2 \le 0. \tag{8.71}$$

Recall that the *apparent* horizon is the outer boundary of the region of trapped surfaces and occurs where equality holds in equations (8.70) and (8.71). In polar slicing, where $K_T = 0$, trapped surfaces do not form, as shown in exercise 8.10.[19]

> **Exercise 8.10** Show that the existence of trapped surfaces in polar slicing would be equivalent to the condition
>
> $$\frac{dr_s}{dr} \le 0, \tag{8.72}$$
>
> where $r_s = Ar$ is the Schwarzschild areal radial coordinate. Thus argue that in a nonpathological spacetime where r_s is a monotonic increasing function of r, no trapped surfaces are encountered.

For collapse to a black hole, trapped surfaces are generally found in maximal slicing, but the polar slices avoid these regions. Thus polar slicing has a somewhat stronger "singularity avoidance" property than maximal slicing.

α-Freezing

Integrating the above system of equations yields very accurate numerical spacetimes for the most part, as we shall illustrate below. However, some simulations of collapsing clusters become inaccurate before the exterior spacetime surrounding the growing, central black hole reaches a final stationary state. This problem can be particularly severe for clusters with appreciable central mass concentration – so-called "extreme core-halo configurations". Such clusters are characterized by enormous dynamic range, with orbital timescales in the central core much shorter than those in the outer halo. A "seed" black hole forms at the center well before the bulk of the matter in the outer regions has had time to evolve significantly. Determining what fraction of the total cluster mass ultimately forms a black hole and what fraction remains outside in orbit about the hole can prove challenging in such cases. Yet the outcome may shed light on plausible mechanisms for forming supermassive massive black holes in the cores of collisionless clusters arising in nature, like dense star clusters or dark matter halos.

The traditional way of attacking this issue is to find coordinate conditions (i.e., lapse and shift functions) that make the problem trackable. Figure 8.8 illustrates the main effect responsible for the inability of the choices discussed in this section to track the late-time evolution of extreme core-halo configurations. Isotropic coordinates, while preventing "spikes" from forming in the radial metric component near horizons, lead to considerable *grid stretching* all along the black hole throat. Grid stretching arises because the

---

[19] See also exercise 7.13, where the result is derived by a more formal route.

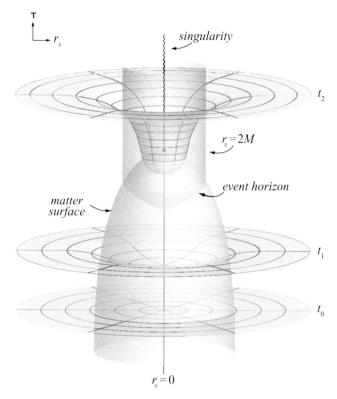

Figure 8.8 Schematic spacetime diagram of the collapse of a collisionless cluster to a black hole. The vertical axis measures proper time τ measured by a normal observer and the horizontal axis measures areal (Schwarzschild) radius r_s. A singularity forms at the center in a finite proper time. The time slicing chosen here avoids the singularity by "slowing down" the advance of proper time near the center, hence the warping there of the spatial hypersurfaces with increasing coordinate time t. Spherical polar coordinate lines appearing on each slice mark the spatial grid, with the radial grid based on an isotropic coordinate r. When the spatial metric on the time slice t has the "radial" form $A_s^2 dr_s^2 + r_s^2 d\Omega^2$, a spike develops in A_s near the event horizon, as in the usual static Schwarzschild geometry. When the isotropic metric $A^2 dr^2 + A^2 r^2 d\Omega^2$ is used, no spike appears. However, considerable radial grid must be expended along the throat near the horizon to determine the metric accurately ("grid stretching"). [After Shapiro and Teukolsky (1986).]

isotropic radial coordinate r must span many decades along the black hole throat. The metric field also varies rapidly on the throat. Consequently, it is necessary to cover the throat with an increasing number of grid points to determined the metric accurately in a numerical calculation. But given finite computational resources, there are only a finite number of radial grid points available to cover the throat. The result is that the growing numerical inaccuracies attributed to grid stretching ultimately force the integrations to terminate.

In addition, maximal time slicing, although successful in holding back the advance of proper time at the center and thus postponing the appearance of singularities, causes the lapse function α to decay and the spatial conformal factor A to increase exponentially at late times near the center. This can eventually cause underflows and overflows in the computed metric components whenever a black hole forms, forcing the integrations to terminate before the exterior spacetime reaches a final stationary state.

Several techniques have been developed that allow us to evolve black holes for very long times, without the occurence of the above problems. One such technique is black hole excision;[20] another is the "moving-puncture" approach.[21] However, in the case of collisionless particles undergoing spherical collapse to a black hole, there is a very simple and elegant choice of coordinates and matter variables that provide accurate solutions for arbitrary late times. The idea behind this choice is easily understood. At late times during the collapse the lapse goes to zero exponentially with t near the cluster center (the "collapse of the lapse"; see Chapter 4.2). Hence the proper time interval $d\tau = \alpha dt$ measured by a normal observer near the cluster center goes to zero for a finite interval of coordinate time dt. Thus we expect that any physical quantity measured by such an observer will freeze, i.e., become constant with t, at late times, because there is no advance of proper time in that observer's reference frame.

As an example, consider the radial component of the velocity of a particle, $v^{\hat{r}}$, where the caret denotes an orthonormal component measured by the normal observer n^a. We have

$$v^{\hat{r}} = u^{\hat{r}}/u^{\hat{0}} = -u_a e^a_{(\hat{r})}/u_a e^a_{(\hat{0})}, \qquad (8.73)$$

where $e^a_{(\hat{0})} = n^a$ is the orthonormal time basis vector and $e^a_{(\hat{r})} = e^a_r/A$ is the orthonormal radial basis vector. Evaluating equation (8.73) yields

$$v^{\hat{r}} = (u_r/A)/\alpha u^0 \qquad (8.74)$$

and similarly

$$v^{\hat{\phi}} = (u_\phi/Ar)/\alpha u^0. \qquad (8.75)$$

Substituting equations (8.74) and (8.75) into (8.43) gives

$$\alpha u^0 = \left[1 - (v^{\hat{r}})^2 - (v^{\hat{\phi}})^2\right]^{-1/2}. \qquad (8.76)$$

Since $v^{\hat{r}}$ and $v^{\hat{\phi}}$ must freeze at late times, so must αu^0, and since u_ϕ is a constant of the motion, we learn from equations (8.74) and (8.75) that the areal radius of a particle $r_s = Ar$ and the ratio u_r/A also freeze.

In this fashion it is straightforward to determine which quantities freeze and which do not as $\alpha \to 0$. We find that, as functions of r_s, the source functions ρ and S freeze, while S_r and S_{rr} do not. The quantities β, A and r (the isotropic radius of a particle) do not freeze. Equation (8.42) thus shows that even when $\alpha \to 0$, the shift continues to drive changes in r. The result is grid stretching.

---

[20] See Chapters 13.1.2 and 14.2.3 *ff.*
[21] See Chapter 13.1.3.

To overcome grid stretching one needs to take advantage of α-freezing and recast the equations in terms of the freezing variables. In either polar or maximal time slicing, the equations of motion of a particle become

$$\frac{dr_s}{dt} = -\frac{1}{2}\alpha r_s K_T + \alpha \frac{(u_r/A)}{\alpha u^0} \left(1 - \frac{r_s \partial_{r_s} A}{A}\right)^{-1}, \tag{8.77}$$

$$\frac{d(u_r/A)}{dt} = \left(-(\alpha u^0)\partial_{r_s}\alpha + \alpha \frac{u_\phi^2}{r_s^3(\alpha u^0)}\right)\left(1 - \frac{r_s \partial_{r_s} A}{A}\right)^{-1} + \alpha K^r{}_r(u_r/A). \tag{8.78}$$

Since every term on the right-hand sides of the above equations contains an explicit factor of α or $\partial_{r_s}\alpha$, we see that

$$\frac{dr_s}{dt} \to 0, \quad \frac{d(u_r/A)}{dA} \to 0 \quad \text{as} \quad \alpha \to 0. \tag{8.79}$$

This provides a formal proof of the freezing of the particle motion at late times near the center of a collapsing configuration in either maximal or polar slicing.[22] In the next section we will present results of simulations, some of which utilize freezing variables.

Numerical calculations

As a test of the overall mean-field, particle simulation scheme, consider the results obtained for Oppenheimer–Snyder collapse, i.e., the collapse from rest of a spherical, homogeneous dust ball to a Schwarzschild black hole, as discussed in Chapter 1.3. The interior is given by the familiar Friedmann solution for a closed universe, which is known in closed analytic form in Gaussian normal (geodesic) time and comoving spatial coordinates. The interior must be matched smoothly onto the vacuum exterior, which is just the static, vacuum Schwarzschild metric. For comparison with a numerical simulation, it is necessary to express the solution in the same coordinate system used in the simulation. The transformation of the Oppenheimer–Snyder solution to both maximal and polar time slicing and isotropic spatial coordinates has been carried out[23] in order to compare numerical with analytic solutions in these gauges.

In Figure 8.9 the lapse profile is plotted on selected maximal time slices during the collapse from an initial areal radius $R/M = 10$. The agreement between the analytic and numerical solutions is good, even after the black hole forms at $t/M \approx 40$. Freezing-variables were not required for this example. A spacetime diagram is plotted in Figure 8.10, showing that numerical code tracks the matter worldlines fairly reliably. It is also reassuring that the event horizon appears first at the origin and grows monotonically outwards, remaining stationary at $r_s/M = 2$ once the last particle crosses inside. Not surprisingly, the stellar surface approaches a limit surface at $r_s/M = 3/2$; this is the same limit surface

---

[22] For the detailed form of the field equations using freezing variables, see Shapiro and Teukolsky (1986).
[23] Petrich *et al.* (1985, 1986).

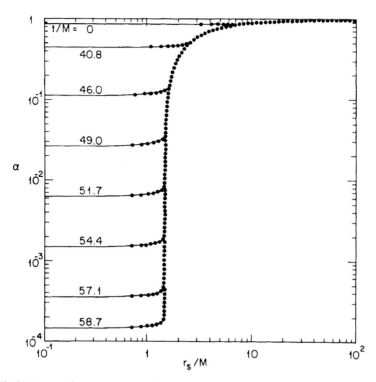

Figure 8.9 The lapse α as a function of areal radius r_s on selected maximal time slices for Oppenheimer–Snyder collapse from $R = 10M$. The solid lines are the results of exact integrations. The dots are the results obtained with the numerical code. [From Shapiro and Teukolsky (1985b).]

that we found when we evolved a vacuum Schwarzschild spacetime in maximal time slicing in Section 8.1. Maximal slicing is successful in holding back the collapse and preventing the formation of central singularity, but not before the matter collapses deep inside the event horizon. By contrast, in polar slicing, the matter surface asymptotes to $r_s/M = 2$ and no trapped surface forms, as depicted in Figure 8.11.

A more interesting case is the evolution of a collisionless ensemble of identical particles initially in equilibrium and described by a truncated, isothermal Maxwell–Boltzmann distribution function,

$$f(E) = \begin{cases} K \exp(-E/T), & E \leq E_{\max}, \\ 0, & E > E_{\max}. \end{cases} \tag{8.80}$$

Here K is a normalization constant, T is a constant "temperature", $E = -p_0$ is the energy of a particle, and $E_{\max}$ is the maximum allowed energy, corresponding to a particle momentarily at rest at the surface of the configuration, $r_s = R$. The energy cutoff guarantees that all particles are restricted to a finite region of space.

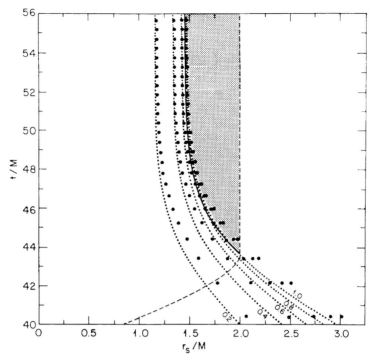

Figure 8.10 Spacetime diagram at late times for Oppenheimer–Snyder collapse in *maximal* slicing from $R = 10M$. The dotted lines are worldlines of Lagrangian matter elements from exact integrations. Each worldline is labeled by the fixed interior rest-mass fraction. The dots are points for the corresponding elements obtained with the numerical code. The dashed line is the event horizon. The shaded area is the region of trapped surfaces. Its outer boundary, the apparent horizon, coincides with the event horizon. Its inner boundary is just inside the surface of the matter. [From Shapiro and Teukolsky (1985b).]

Relativistic star clusters in equilibrium with truncated Maxwell–Boltzmann distributions have been studied extensively in the literature.[24] Prior to the development of numerical relativity such studies were restricted to the construction of static equilibria and depended on linear perturbation theory employing trial functions to analyze stability. The precise point of onset of dynamical instability along parametrized sequences of equilibria could not be determined by this approach. However, by using these models as initial data in a relativistic mean-field, particle simulation scheme, the point of onset of dynamical instability can be rigorously identified and, more significantly, the full nonlinear evolution of unstable systems can be tracked and their final fate ascertained.

As an example, consider the evolution of a Maxwell–Boltzmann cluster with areal radius $R/M = 9.2$ and central gravitational redshift $z_c = 0.52$ along a one-parameter equilibrium sequence defined by $E_{\max} = m - T/2$, where m is the mass of each particle, assumed identical. This cluster is only moderately centrally condensed, with the ratio of

---

[24] See, e.g., Zel'dovich and Podurets (1965); Ipser (1969b); Misner *et al.* (1973); Katz *et al.* (1975) for models.

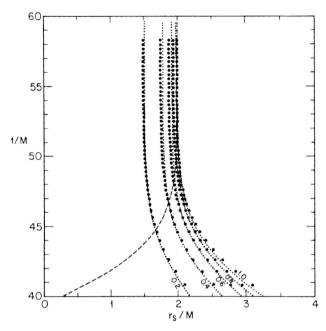

Figure 8.11 Spacetime diagram at late times for Oppenheimer–Snyder collapse in *polar* slicing from $R = 10M$. The labelling is the same as in Figure 8.10. No trapped surface forms in this slicing. [From Shapiro and Teukolsky (1986).]

mean-to-central mass-energy density given by $\langle \rho \rangle / \rho_c = 0.072$. It lies just beyond the point of onset of instability, which is near $z_c = 0.42$, as determined by numerical simulations. The collapse of this cluster has been followed using both maximal and polar time slicing and isotropic radial coordinates.[25] Snapshots of the configuration are shown at selected maximal times during the collapse in Figure 8.12.

By solving equation (8.69) and identifying pairs of outgoing null rays emitted at neighboring points in spacetime, one of which escapes to infinity and the other of which is pulled back by the black hole, the location of the black hole event horizon was determined. By solving equation (8.71) in maximal slicing, the region of trapped surfaces and the apparent horizon were identified. All the matter collapses inside the black hole and the surface approaches a limit surface at $r_s/M = 1.5$ in this time slicing, As expected, both horizons merge and remain locked at $r_s/M = 2$ once the last particle is captured. In polar slicing the collapsing surface asymptotes to radius $r_s/M = 2$ at late times and no trapped surface forms. A spacetime diagram showing the late-time behavior is shown in Figure 8.13 for maximal slicing and in Figure 8.14 for polar slicing. Once again, freezing variables were not required for this integration.

A more computationally challenging example is provided by the collapse of an unstable equilibrium cluster with an extreme core-halo profile. Consider a relativistic, collisionless,

---

[25] Shapiro and Teukolsky (1985a, 1986).

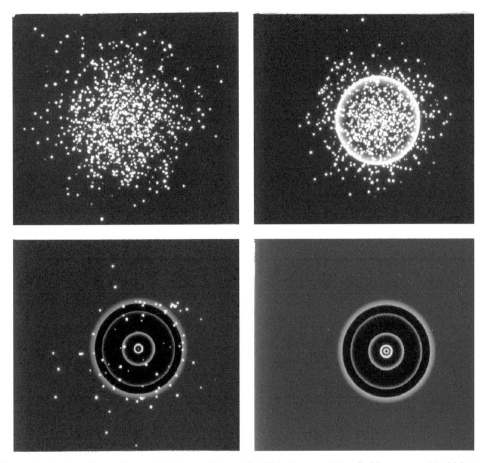

Figure 8.12 The collapse of a marginally unstable gas of collisionless particles of arbitrary mass M which at $t = 0$ obeys a truncated, isothermal Maxwell–Boltzmann distribution function with an areal radius $R/M = 9$. Spherical flashes of light are used to probe the spacetime geometry; at late times the light rays are trapped by the gravitational field. Their trajectories help locate the black hole event horizon, which in this example eventually reaches $r_s/M = 2$ and encompasses all the matter. [From Shapiro and Teukolsky (1988).]

spherical polytrope of index $n = 4$ with a central redshift $z_c = 0.50$. The phase space distribution function for such a configuration has the form[26]

$$f(E) = \begin{cases} K\,(E/E_{\max})^{-5}[1 - (E/E_{\max})^2]^{5/2}, & E \leq E_{\max}, \\ 0, & E > E_{\max}, \end{cases} \qquad (8.81)$$

where the variables have the same definitions as in equation (8.80). In this case, the relativistic core contains only 0.5% of the total rest-mass. The remainder of the mass

---

[26] Fackerell (1970).

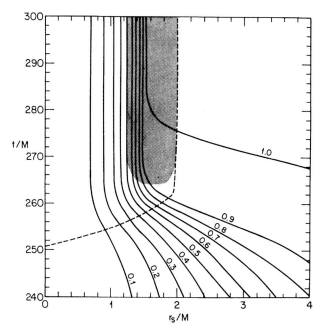

Figure 8.13 Spacetime diagram in *maximal* slicing for the Maxwell–Boltzmann cluster depicted in Figure 8.12. The solid lines are worldlines of fictitious Lagrangian matter tracers labeled by the fixed interior rest-mass fraction. The dashed line is the event horizon. The shaded area is the region of trapped surfaces. The event and apparent horizons are both numerically asymptote to $r_s = 2M$ to high accuracy. [From Shapiro and Teukolsky (1986).]

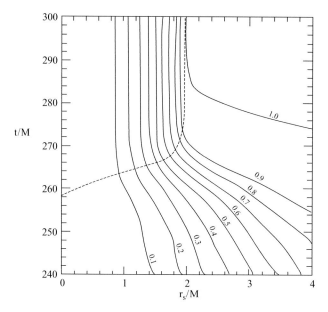

Figure 8.14 Spacetime diagram in *polar* slicing for the Maxwell–Boltzmann cluster depicted in Figures 8.12 and 8.13. Labeling is the same as in Figure 8.13. No trapped surface forms in polar slicing. The event horizon and matter surface are both numerically asymptote to $r_s = 2M$ to high accuracy. [From Shapiro and Teukolsky (1986).]

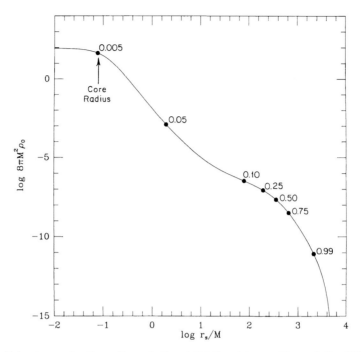

Figure 8.15 Initial rest-mass density profile ρ_0 for the relativistic polytrope with $n = 4$, $\Gamma = 5/4$, and central redshift $z_c = 0.50$. Dots are located at points at which the interior rest-mass fraction has the values shown. This extreme core-halo configuration has a highly relativistic core and an extensive Newtonian halo. A the end of the simulation, the black hole which forms at the center contains a fraction 0.05 of the total rest-mass, approximately 10 times the initial core mass. [From Shapiro and Teukolsky (1986).]

resides in a nearly Newtonian halo extending out to $R/M \approx 5000$. The ratio of the mean-to-central density in the cluster is tiny: $\langle \rho \rangle / \rho_c = 4.0 \times 10^{-13}$; see Figure 8.15.

The cluster is dynamically unstable. During its collapse, approximately 5% of the cluster mass, considerably more mass than the core, forms a central black hole. By the end of collapse, the cluster settles into a new equilibrium state consisting of a massive Newtonian halo of particles in orbit about a central black hole. Such a centrally condensed system could provide a plausible model of a galaxy containing a supermassive black hole.[27] The catastrophic collapse of an unstable, collisionless gas like the one evolved here might even provide a viable formation mechanism for such a supermassive black hole.[28] Supermassive black holes are believed to be the engines that power quasars and active galactic nuclei and their formation and growth remains one of the major puzzles of cosmological structure formation.

---

[27] Most galaxies containing a bulge, including the Milky Way, are observed to contain a central supermassive black hole; see, e.g., Ho (2004) for reviews and references.

[28] Zel'dovich and Podurets (1965); Rees (1984); Shapiro and Teukolsky (1985c); Balberg and Shapiro (2002).

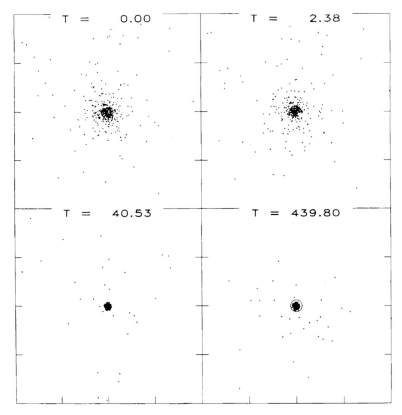

Figure 8.16 Snapshots of the central regions $r_s/M \leq 2$ during the collapse of the extreme core-halo configuration shown in Figure 8.15. The markers on frames are placed at intervals of $\Delta r_s = 1M$. The times are labeled by $T \equiv t/M$. The circle in the last frame shows the event horizon at $r_s/M = 0.1$. The collapse of the innermost regions to a black hole is evident. By $t/M \approx 40$, the cluster reaches a quasistationary state consisting of central black hole surrounded by a massive halo of orbiting particles. [From Shapiro and Teukolsky (1986).]

Freezing-variables, designed to handle configurations with large dynamic range, are required to follow the evolution of this cluster reliably.[29] Maximal slicing is not adequate to hold back the collapse, as the value of the central radial metric coefficient increases above $A_c \approx 10^{49}$ and the central lapse falls below $\alpha_c \approx 10^{-70}$, leading to inaccuracies due to excessive grid stretching. Polar slicing with freezing variables proves necessary, in part because A_c grows much more slowly in this gauge.

Highlights of the collapse are summarized in Figures 8.16–8.18. Snapshots of the imploding central parts of the cluster inside $r_s/M = 2$ are shown at selected times in Figure 8.16. There is little evolution in the outer Newtonian halo where most of the mass resides. However, the growing black hole is effective in "sweeping clean" the innermost

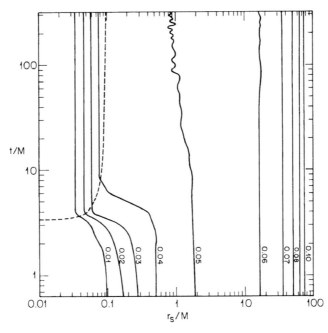

Figure 8.17 Spacetime diagram in polar slicing for the relativistic polytrope shown in Figure 8.16. Labeling is the same as in Figure 8.13. The event horizon asymptotes to $r_s/M = 0.1$, at which point it encompasses 5% of the total cluster rest-mass. [From Shapiro and Teukolsky (1986).]

region in and around the core. The cluster settles into a stationary equilibrium state after $t/M = 40$. A spacetime diagram for the collapse is shown in Figure 8.17.

In Figure 8.18 the trajectories of four typical particles orbiting near the cluster center are plotted. The simulation is performed with $N_{tot} = 7198$ particles; at $t/M = 0$, a fraction $N/7198$ of the total cluster rest-mass resides inside particle N. Particle 333, initially in an elliptical-like orbit near $r_s/M = 1$, moves along an inward spiral and is captured by the black hole after two orbital periods. It illustrates the "avalanche-instability" whereby the mass interior to this particle at pericenter grows sufficiently in two periods so that the particle, initially outside the collapsing core, eventually finds itself on a capture orbit. Particle 338 moves in a nearly elliptical orbit that extends out to $r_s/M = 1.5$ and exhibits large perhelion precession about the central hole. Pericenter for this particle is one of the closest of all the ambient particles that do not get captured. It is located at $r_s/M = 0.25$ or $r_s/M_H \approx 5$. This result is consistent with the fact that particles which orbit a stationary Schwarzschild black hole inside $r_s/M = 4$ are inevitably captured.[30] It is also satisfying that the pericenter position of this marginally stable orbit remains stationary with time, further confirming that, by the time the integrations terminate at $t/M = 439$, the cluster achieves a new dynamic equilibrium about a stationary central black hole. Particle 340 moves nearly unperturbed in a circular orbit at $r_s/M = 1$.

---

[30] See, e.g., Misner *et al.* (1973).

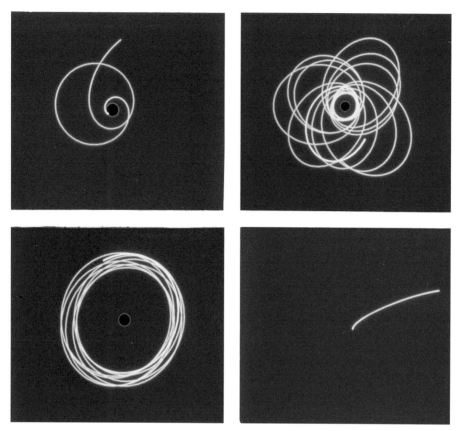

Figure 8.18 Orbital trajectories of four typical particles near the cluster center. In the *top left* panel the particle is initially in an elliptical-like orbit, but spirals into the black hole after about two orbits. In the *top right* panel the particle moves in a nearly elliptical orbit, exhibiting large perihelion precession about the central hole. In the *bottom left* panel the particle moves essentially unperturbed in a nearly circular orbit. In the *bottom right* panel the particle falls nearly radially into the black hole. [From Shapiro and Teukolsky (1988).]

Exercise 8.11 Interpret the behavior of particle 340 in light of Birkhoff's theorem.

Particle 410 falls nearly radially from $\approx 7M$ into the black hole, illustrating how the mass of the black hole grows beyond that of the collapsing core.

Finally, consider Figure 8.19 showing the fractional binding enery E_b/M_0 *vs.* central redshift along an equilibrium sequence of $n = 4$ polytropes. The onset of instability as determined by linear perturbation theory using trial functions ($\omega^2 < 0$) occurs well beyond the turning point. However, numerical integrations with the mean-field, particle simulation scheme show that all configurations beyond the first turning point are dynamically unstable. From this result and simulations involving other examples of parametrized sequences of

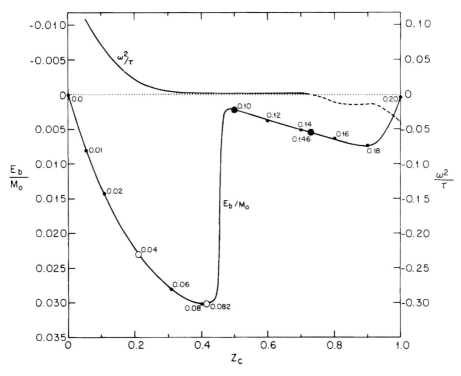

Figure 8.19 The fractional binding energy E_b/M_0 and oscillation frequency ω^2 in units of $\tau = \rho_c^* - P_c$ are plotted as functions of central redshift z_c for the $n = 4$ polytropic sequence. Clusters along the sequence are labeled by their value of the relativity parameter α as defined in Fackerell (1970). The dynamical fates of four models along the sequence, as determined by numerical simulation, are as follows: open large circles indicate stable equilibrium; filled large circles indicate collapse to a black hole. [From Shapiro and Teukolsky (1985a).]

collisionless clusters, one concludes that the turning point on the binding energy curve does signify the onset of dynamically instability, as in fluid systems.

8.2.2 Phase space method

While the mean-field, particle simulation scheme discussed in the previous section determines particle positions and velocities, it does not provide the phase space distribution function f. As discussed in Chapter 5.3 a method to solve the relativistic Vlasov equation in spherical symmetry directly for f by exploiting Liouville's theorem has been developed.[31] The key idea for determining f is to implement equation (5.231). Knowing f determines the matter source functions. The gravitational field equations are the same as described in the previous section. The virtue of the phase space method is that it accurately tracks the

---

[31] Rasio *et al.* (1989b).

increasingly complicated fine-grained structure of the distribution function due to phase mixing. For determining the global behavior of a collisionless system, the mean-field particle-simulation scheme is quite adequate. However, to obtain a detailed description of the phase space distribution, especially when there is counter-streaming and phase mixing, the direct phase space method is necessary.

Adopt the form for the line element given in equation (8.41). Again assume for simplicity that the gas consists of a single species of mass m. As coordinates in phase space, choose the radial velocity u_r, the specific angular momentum j given by

$$j \equiv \sqrt{u_\theta^2 + \frac{u_\phi^2}{\sin^2 \theta}}, \tag{8.82}$$

and the angle

$$\psi \equiv \tan^{-1}\left(\frac{u^{\hat{\theta}}}{u^{\hat{\phi}}}\right) = \tan^{-1}\left(\frac{u_\theta \sin \theta}{u_\phi}\right), \tag{8.83}$$

measuring the orientation of the transverse velocity. Here the carets denote orthonormal components. In spherical symmetry, f cannot depend on ψ, and j is a conserved quantity, so that the Boltzmann equation (5.217) reduces to

$$\frac{\partial f}{\partial t} + \left(\frac{dr}{dt}\right)\frac{\partial f}{\partial r} + \left(\frac{du_r}{dt}\right)\frac{\partial f}{\partial u_r} = 0 \tag{8.84}$$

since $\partial f/\partial \psi = 0$ and $dj/dt = 0$. Here we have taken $f = f(t, x^j, u_j) \equiv dN/(d^3 V_x d^3 V_u)$, where $d^3 V_u = d^3 V_p/m^3$ and where $d^3 V_p$ is given by equation (5.213).

> **Exercise 8.12** Show that for a spherical metric given by equation (8.41) $(-g)^{1/2} = A^3 \alpha r^2 \sin \theta$ and
>
> $$d^3 V_u = \frac{du_r du_\theta du_\phi}{A^3 r^2 \alpha u^0 \sin \theta} = \frac{\pi}{A^3 r^2 \alpha u^0} dj^2 du_r, \tag{8.85}$$
>
> where the integral over ψ has been carried out in the last expression.

In spherical symmetry, we can thus write $f = f(t, r, u_r, j)$. The coefficients dr/dt and du_r/dt in equation (8.84) are given by geodesic equations (8.42) and (8.43). The field equations are unchanged from those described in the previous section. The matter source terms are derived from f as in equations (5.207)–(5.210), using equation (5.213). For example, the quantity $\rho(t, r)$ can be computed from

$$\rho = m \int (\alpha u^0)^2 f d^3 V_u = \frac{\pi m}{A^3 r^2} \int_0^\infty dj^2 \int_{-\infty}^\infty du_r f \left(1 + \frac{u_r^2}{A^2} + \frac{j^2}{A^2 r^2}\right)^{1/2}. \tag{8.86}$$

Note that one must be careful when evaluating the source terms at $r = 0$, where $f \neq 0$ only for $j = 0$. The coordinate singularity at $r = 0$ in these quadratures is eliminated by recasting the $d^3 V_u$ in terms of velocity components in an orthonormal frame: $d^3 V_u = (du_{\hat{x}} du_{\hat{y}} du_{\hat{z}})/u^{\hat{0}}$. The point $r = 0$ is the center of symmetry, in which case the distribution function is isotropic and can only depend on the magnitude $u \equiv (u_{\hat{x}}^2 + u_{\hat{y}}^2 + u_{\hat{z}}^2)^{1/2}$ of the velocity. In this case we can write $d^3 V_u = (4\pi u^2 du)/(1 + u^2)^{1/2}$, which gives finally

$$\rho_c = 4\pi m \int_0^\infty du f_c(u) u^2 (1 + u^2)^{1/2}. \tag{8.87}$$

Here the subscript c indicates a value at $r = 0$ and $f_c(u) \equiv f_c(t, r = 0, j = 0, u_r = A_c u)$.

Recall from our discussion in Chapter 5.3 that equation (8.84) is solved by applying Liouville's theorem in the form of equation (5.231). Specifically, f at any time t is evaluated numerically by integrating a trajectory backward in time, from a point (r, u_r, j) in phase space to $t = 0$, where f is specified via the initial data. This trajectory is constructed by integrating ordinary differential equations (8.42) using equation (8.43). The right-hand sides of these equations involve the values of the fields and their derivatives, which therefore must be stored on a radial grid. The time step is chosen according to equation (8.64). A good check on the method is provided by calculating the total mass-energy of the system via equation (8.67), which should be conserved in time.

As an example, consider the evolution in maximal slicing of the same truncated, isothermal Maxwell–Boltzmann distribution function treated in the previous section by the particle simulation method.[32] The point of onset of instability along the equilibrium sequence is found to be at the same point as before, i.e., the turning point in the binding energy at $z_c = 0.42$. The spacetime diagram for the unstable collapse of a collisionless cluster at $z_c = 0.52$ is in excellent agreement with the diagram plotted in Figure 8.13. The unique character of the phase-space method is revealed in Figure 8.20.

Contours of f are plotted on two different slices of fixed j. The phase space coordinates in the diagram are the *freezing* variables r_s and $v^{\hat{r}} = u_r/(A\alpha u^0)$ measured by a normal observer. The expectation is that f expressed in these variables should exhibit a steady configuration at late times when $\alpha \to 0$. This is what is found: once all the matter with a given j collapses inside the horizon, the distribution function evolves very slowly toward a final static structure.

The phase space method has also been used to demonstrate that it is possible to construct stable relativistic star clusters with *arbitrarily* large central redshifts.[33] Prior to this, it had generally been believed that *all* clusters with $z_c \gtrsim 0.5$ would be dynamically unstable.

8.3 Fluid stars: collapse

The most important problem to date involving numerical simulations of fluid stars in spherical symmetry has been the supernova problem. Here the collapse of the degenerate stellar core of a massive star at the endpoint of stellar evolution is believed to lead to the formation of a neutron star, accompanied by an explosion of the more massive stellar envelope. For the most massive stars, core collapse can lead to the formation of a black

[32] Rasio *et al.* (1989b).
[33] Rasio *et al.* (1989a).

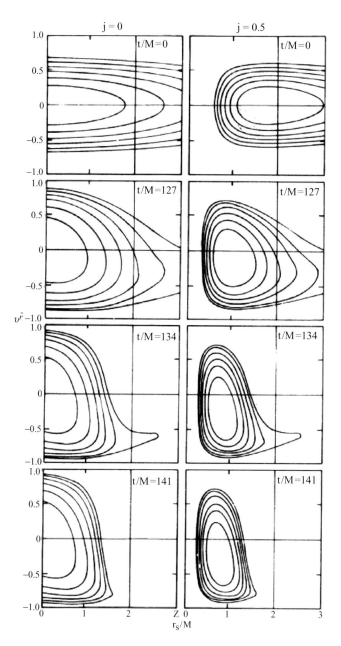

Figure 8.20 Time evolution of the distribution function f for the Maxwell–Boltzmann cluster. The areal radius r_s (in units of M) and the radial velocity $v^{\hat{r}}$ are used as phase space coordinates. Each plane $(r_s, v^{\hat{r}})$ is a 2-dimensional slice taken from the 3-dimensional phase space by setting the particle angular momentum j equal to a constant. The series on the left corresponds to $j = 0$, while that on the right corresponds to $j = 0.5 M^2$, representing the angular momentum of a "typical particle" in this cluster $(0 \leq j \lesssim M^2$ for $f \neq 0)$. Lines of constant f are shown, equally spaced between 0 and its maximum value on the slice. Note that for $j \neq 0$, the matter never reaches $r_s = 0$. In the final plots, the entire mass of the cluster has collapsed inside an event horizon at $r_s = 2M$. [From Rasio *et al.* (1989b).]

hole instead of a neutron star, with or without an explosion. The demarcation line between collapse to a neutron star and collapse to a black hole is still uncertain, and all the detailed microphysical processes (e.g., the "hot" nuclear equation of state; neutrino production mechanisms and transport, etc.) and all the important macrophysical effects (e.g., general relativity, rotation, magnetic fields, convection, etc.) that make up a realistic simulation are yet to be fully incorporated in a totally rigorous fashion.[34] One thing is certain: during stellar core collapse, the gravitational field becomes strong and fluid velocities approach the speed of light, hence a reliable calculation requires a fully relativistic treatment. The first relativistic treatment was the pioneering work of May and White (1966, 1967). Their work represented an important milestone in computational astrophysics and helped launch numerical relativity. Their code was based on the formulation of Misner and Sharp (1964) for spherically symmetric gravitational collapse.[35] This formulation has the desirable feature that the equations take the form of Newtonian *Lagrangian* hydrodynamics plus relativistic corrections. Hence all the machinery and expertise for handling Lagrangian hydrodynamics in Newtonian theory could be taken over to the relativistic case. However, one fundamental problem with the Misner–Sharp formalism and the coordinate system on which it is based is that collapse to a black hole cannot be followed once the black hole forms, because the equations become singular. This means that we are unable to follow the fate of the outer layers of the star when the inner core has formed a black hole.

Schemes that avoid a singularity during spherical fluid collapse to a black hole have been constructed by many groups over the years.[36] The essential feature of these codes is a different choice of time coordinate from that of Misner and Sharp, which allows the evolution to be followed to late times, without encountering singularities. Indeed, we have already shown in Section 8.1 how different time slicings can be chosen to avoid singularities when evolving a vacuum black hole spacetime, and we then demonstrated in Section 8.2 that the situation is very similar in the presence of collisionless matter. For the most part, analogous schemes for fluid matter work on a fixed *Eulerian* spatial mesh and adapt the Eulerian equations of relativistic hydrodynamics discussed in Chapter 5.2 to spherical symmetry. In all of these schemes the equations depart much more from Newtonian hydrodynamics than does the Misner–Sharp formulation. More significantly, it is usually the case that greater computational effort is required in an Eulerian formulation to attain the accuracy of a Lagrangian formulation. The reason is that the spatial grid in a Lagrangian scheme follows the fluid elements, so the entire fluid is automatically and completely covered by the same fixed number of grid points that covered the fluid at the initial time. Besides the extra effort required to make the hydrodynamics competitive with

---

[34] For a general discussion and references, see Shapiro and Teukolsky (1983); Arnett (1996); Janka *et al.* (2007); Burrows *et al.* (2007).

[35] The discussion in this section closely follows Baumgarte *et al.* (1995).

[36] See, e.g., Wilson (1979), Shapiro and Teukolsky (1980); Schinder *et al.* (1988); Mezzacappa and Matzner (1989) for early work.

a Lagrangian scheme, an Eulerian scheme suffers an additional penalty in treating the field equations. Specifically, a relativistic Eulerian code typically requires an exterior grid extending to large distances outside the star in order to impose boundary conditions on the asymptotically flat gravitational field. By contrast, a spherical relativistic Lagrangian code only needs a grid that covers the matter.[37] On the other hand, Lagrangian codes are more difficult to extend to more than one spatial dimension, so they are mainly useful for spherical systems.[38]

The problem of singularities is irrelevant if collapse always leads to the formation of a neutron star, in which case the Misner–Sharp formulation is completely adequate. However, we expect that the collapse of a very massive or supermassive star produces a black hole. For such cases it is desirable to have a scheme that can handle black hole formation but with the advantages of the Misner–Sharp formulation. Such a code was developed by Baumgarte *et al.* (1995) based on the formulation of Hernandez and Misner (1966), which uses *retarded* time as a coordinate instead of the usual Schwarzschild time that appears in the Misner–Sharp equations. This feature prevents the computational grid following the fluid from penetrating inside any black hole that may form and encountering a singularity.

In the next section we shall review the widely used Misner–Sharp formulation for spherical hydrodynamics and show how its main deficiencies can be easily removed by adopting the closely related Hernandez–Misner formulation. Both formulations yield Lagrangian simulation schemes. We postpone a discussion of Eulerian treatments to Chapters 14, 16 and 17, where we describe some important applications of relativistic hydrodynamics in *nonspherical* spacetimes.[39]

8.3.1 Misner–Sharp formalism

The Lagrangian equations of relativistic hydrodynamics in spherical symmetry were first derived by Misner and Sharp (1964) and have been used in many numerical calculations, including those of May and White (1966, 1967).[40] A straightforward re-derivation of the equations is contained in Misner *et al.* (1973)[41] so we simply summarize the results below.

The line element is written in a diagonal form

$$ds^2 = -e^{2\phi(t,A)}dt^2 + e^{\lambda(t,A)}dA^2 + R^2(t,A)d\Omega^2, \tag{8.88}$$

where R is the circumferential radius. In the parlance of $3 + 1$, the lapse is e^ϕ and the shift is zero. We can think of each spherical shell of matter as labeled by a parameter A and

---

[37] Recall that no gravitational waves are generated in spherical symmetry, so there is no need to track their propagation outside the star.

[38] Exceptions are Lagrangian fluid codes based on the SPH method, which are straightforward to construct for multidimensional spacetimes; see Chapters 5.2, 16 and 17. See also Taub (1978) for shift prescriptions that maintain comoving (i.e., Lagrangian) coordinates in arbitrary dimensions.

[39] For a robust Eulerian relativistic hydrodynamics scheme specifically adapted to spherical symmetry, see Wilson (1979).

[40] See also van Riper (1979).

[41] Misner *et al.* (1973), exercise 32.7.

$R(t, A)$ as the worldline of the shell with label A. The comoving radial coordinate A can be chosen to be the rest-mass (e.g., baryon number) enclosed within R.

The matter is assumed to be a perfect fluid, for which the stress-energy tensor is given by equation (5.4) with specific enthalpy given by equation (5.5). In a comoving (Lagrangian) coordinate frame, the fluid 4-velocity takes the form

$$u^a = (e^{-\phi}, 0, 0, 0). \tag{8.89}$$

It is useful to define the quantities

$$m = 4\pi \int_0^A \rho_0(1 + \epsilon)R^2(\partial_A R)dA, \tag{8.90}$$

$$U = e^{-\phi}\partial_t R, \tag{8.91}$$

$$\Gamma = e^{-\lambda/2}\partial_A R. \tag{8.92}$$

Here m can be interpreted as the gravitational mass inside A, U is the coordinate velocity (rate of change of R along a fluid worldline with respect to the proper time of that fluid element), and Γ is simply a more convenient form for the radial metric function.[42] Note that since U is only a *coordinate* velocity, its magnitude may exceed unity.

In terms of these variables the equations of motion and the Einstein field equations can be written as

$$\partial_t U = -e^{\phi}\left(\frac{4\pi\Gamma R^2}{h}\partial_A P + \frac{m + 4\pi R^3 P}{R^2}\right), \tag{8.93}$$

$$\partial_t m = -e^{\phi}4\pi R^2 PU, \tag{8.94}$$

$$\partial_A \phi = -\frac{1}{\rho_0 h}\partial_A P, \tag{8.95}$$

$$\Gamma = (1 + U^2 - 2m/R)^{1/2}, \tag{8.96}$$

$$\rho_0 = \frac{\Gamma}{4\pi R^2 \partial_A R}. \tag{8.97}$$

Differentiating equation (8.90) and using equation (8.97) yields

$$\partial_A m = (1 + \epsilon)\Gamma. \tag{8.98}$$

Also, equations (8.92) and (8.97) can be combined to give

$$e^{-\lambda/2} = 4\pi\rho_0 R^2. \tag{8.99}$$

This system of equations still has to be supplemented with the first law of thermodynamics

$$\partial_t \epsilon = -P\partial_t\left(\frac{1}{\rho_0}\right), \tag{8.100}$$

---

[42] Following convention, we use the symbol Γ in this section to denote the right-hand side of equation (8.92), and γ to denote the adiabatic index of a perfect gas.

as well as an equation of state of the form $P = P(\rho_0, \epsilon)$. For a γ-law EOS we have, as usual,

$$P = (\gamma - 1)\rho_0 \epsilon. \tag{8.101}$$

The appropriate boundary conditions are

$$
\begin{aligned}
&R = 0, \quad U = 0, \quad \Gamma = 1, \quad m = 0 \quad \text{at the origin, } A = 0, \\
&P = 0, \quad e^\phi = 1 \quad\quad\quad\quad\quad\quad\quad \text{at the surface, } A = A_{\text{total}}.
\end{aligned}
\tag{8.102}
$$

The choice of the boundary condition for ϕ is somewhat arbitrary; the choice made here ensures that the coordinate time t agrees with the proper time on the surface of the star. The above equations define an initial boundary-value problem for initial data $U(R)$, $\rho_0(R)$ and $\epsilon(R)$, which have to be chosen such that $1 + U^2 - 2m/R > 0$.

> **Exercise 8.13** Derive the Newtonian limit of the above equations. Specifically, take the limits $U \ll 1$, $\epsilon \ll 1$, $P/\rho_0 \ll 1$, and $m/R \ll 1$ to show
>
> $$\Gamma = 1 = \partial_A m. \tag{8.103}$$
>
> Hence argue that equation (8.103), together with the boundary conditions, implies that the rest-mass A and the gravitational mass m are the same in the Newtonian limit. Also argue that equation (8.94) implies that m is now a constant of the motion. Show that
>
> $$\partial_m \phi = -\frac{1}{\rho_0} \partial_m P, \tag{8.104}$$
>
> from which we conclude that in the Newtonian limit the metric function ϕ approaches the Newtonian potential. Argue that $\phi \ll 1$, or $e^\phi \approx 1$, hence all the evolution equations can be written independently of ϕ. Consequently, show that equation (8.91) becomes
>
> $$\partial_t R = U, \tag{8.105}$$
>
> while equation (8.93) yields the Lagrangian equation of motion for Newtonian spherical hydrodynamics,
>
> $$\partial_t U = -\left(4\pi R^2 \partial_m P + \frac{m}{R^2}\right), \tag{8.106}$$
>
> and equation (8.97) becomes
>
> $$\rho_0 = \frac{1}{4\pi R^2 (\partial_m R)}. \tag{8.107}$$

As is apparent from exercise (8.13), an obvious benefit of this coordinate system is that the relativistic equations are very close to the corresponding Newtonian ones. The meaning and interpretation of the variables can be carried over directly from the Newtonian theory. Most important, an existing Newtonian code can easily be upgraded to a fully relativistic one simply by adding a few terms and equations.

On the other hand, this coordinate system has a severe drawback. If a configuration collapses to a black hole, a coordinate singularity arises which prevents any further evolution.

Exercise 8.14 Explore what happens to the lapse function $\alpha = e^\phi$ in comoving Misner–Sharp coordinates.

(a) Consider first the case of dust collapse, where $P = 0$. Show that $\alpha = 1$ (geodesic slicing), and recall the discussion of Chapter 4.1 regarding the ultimate appearance of coordinate singularities when using this slicing.

(b) Now treat the situation with pressure gradients. Use the Euler equation for the fluid acceleration a^a in the form

$$\rho_0 h a_a = -[\nabla_a P + (u^b \nabla_b P) u_a] \tag{8.108}$$

to show that in comoving coordinates

$$D_i \ln \alpha = -\frac{1}{\rho_0 h} D_i P. \tag{8.109}$$

Combine equation (8.109) with the second law of thermodynamics for adiabatic flow to obtain

$$D_i \ln \alpha = -\frac{1}{h} D_i h, \quad \text{or} \quad \alpha \propto \frac{1}{h}. \tag{8.110}$$

Comment on the likely consequences of equation (8.110) for calculations of collapse to black holes.

The singularity problem motivated Hernandez and Misner to introduce a null coordinate u and transform the above equations to what they called "observer time coordinates". The virtue is that these coordinates always stay outside event horizons. Not only are there no coordinate singularities, but also these coordinates never encounter the physical curvature singularity at the center of the black hole. We discuss this formalism in the next section.

8.3.2 The Hernandez–Misner equations

Here we summarize the transformations of the equations to observer time coordinates.[43] The idea is to find a coordinate system in which the time coordinate t is replaced by a null coordinate u, which is constant along outgoing light rays. In addition, u can be scaled so that it measures the time of an observer at infinity ("observer time coordinates"). For a configuration that collapses to a black hole, no light ray from inside the event horizon will, by definition, ever reach spatial infinity. Therefore no event inside an event horizon corresponds to finite "observer time" u; in fact, the event horizon itself is the surface $u \to \infty$.

We introduce the outgoing null coordinate u (outgoing Eddington–Finkelstein coordinate) by

$$e^\psi du = e^\phi dt - e^{\lambda/2} dA, \tag{8.111}$$

---

[43] Details may be found in Hernandez and Misner (1966) or Baumgarte *et al.* (1995).

whereby the line element (8.88) takes the form

$$ds^2 = -e^{2\psi}du^2 - 2e^{\psi}e^{\lambda/2}du\,dA + R^2 d\Omega^2. \tag{8.112}$$

We note that the lapse function is now given by e^{ψ}. The equations of the last section can be transformed to this new coordinate system by transforming the differential operators. Equation (8.111) gives

$$e^{-\psi}\partial_u\Big|_A = e^{-\phi}\partial_t\Big|_A, \tag{8.113}$$

while the chain rule for partial differentiation gives the new spatial derivative in terms of the old ones according to

$$e^{-\lambda/2}\partial_A\Big|_u = e^{-\lambda/2}\partial_A\Big|_t + e^{-\phi}\partial_t\Big|_A. \tag{8.114}$$

The definition of coordinate velocity U in equation (8.91) becomes

$$U = e^{-\psi}\partial_u R. \tag{8.115}$$

Treating the spatial derivative on the right hand side of equation (8.93) with care yields[44]

$$\partial_u U = -\frac{e^{\psi}}{1-v_s^2}\left(\frac{4\pi\Gamma R^2}{h}\partial_A P + \frac{m+4\pi R^3 P}{R^2}\right)$$
$$-\frac{e^{\psi}v_s^2}{1-v_s^2}\left(4\pi\rho_0 R^2\partial_A U + \frac{2U\Gamma}{R}\right), \tag{8.116}$$

where v_s is the speed of sound given by

$$v_s^2 = \partial_{\rho^*}P\big|_s = \frac{1}{\rho_0^2 h}\left[P\,\partial_\epsilon P\big|_{\rho_0} + \rho_0^2\,\partial_{\rho_0}P\big|_\epsilon\right], \tag{8.117}$$

and where $\rho^* = \rho_0(1+\epsilon)$ is the total comoving mass-energy density. For a γ-law EOS the speed of sound can be written

$$v_s^2 = \frac{\gamma-1}{\rho_0 h}(P + \rho_0\epsilon) = (\gamma-1)\frac{h-1}{h}. \tag{8.118}$$

Exercise 8.15 Derive equation (8.118) for the sound speed from equation (8.117).

The remaining Einstein equations (8.94), (8.92), and (8.97) now become

$$\partial_u m = -e^{\psi}4\pi R^2 PU \tag{8.119}$$

$$\Gamma = (1 + U^2 - 2m/R)^{1/2} \tag{8.120}$$

$$\rho_0 = \frac{\Gamma+U}{4\pi R^2\partial_A R}. \tag{8.121}$$

---

[44] See Appendix B of Baumgarte *et al.* (1995).

An integration factor e^ϕ was inserted in equation (8.111) to make du a perfect differential. This requirement yields a differential equation to replace equation (8.95):

$$\partial_A \psi - \frac{1}{\Gamma}\partial_A U + \frac{m}{4\pi\rho_0 R^4 \Gamma} + \frac{P}{\rho_0 \Gamma R}. \tag{8.122}$$

Using equation (8.114) to replace the spatial derivative in equation (8.98), as well as equation (8.94), yields

$$\partial_A m = (1+\epsilon)\Gamma - \frac{PU}{\rho_0}. \tag{8.123}$$

Once again, the equations have to be supplemented by the first law of thermodynamics for adiabatic flow,

$$\partial_u \epsilon = -P \partial_u \left(\frac{1}{\rho_0}\right), \tag{8.124}$$

together with an EOS, $P = P(\rho_0, \epsilon)$.

The boundary conditions are mostly the same as in the Misner–Sharp scheme described in the last section:

$$
\begin{array}{llll}
R = 0, \; U = 0, & \Gamma = 1, & m = 0 & \text{at the origin, } A = 0, \\
P = 0, \; e^\psi = \Gamma + U & & & \text{at the surface, } A = A_{\text{total}}.
\end{array} \tag{8.125}
$$

> **Exercise 8.16** Show that the boundary condition for ψ matches the interior u to the exterior, where observer time coordinates reduce to outgoing Eddington–Finkelstein coordinates. Begin by noting that for an observer comoving with the fluid at the surface of the configuration,
>
> $$ds^2 = -e^{2\phi}dt^2 = -dt^2. \tag{8.126}$$
>
> Next, express the exterior spacetime at the surface in Schwarzschild coordinates,
>
> $$ds^2 = -(1 - 2M/R)dt_s^2 + (1 - 2M/R)^{-1}dR^2 + R^2 d\Omega^2, \tag{8.127}$$
>
> to find dt_s/dt, then express du in terms of the same coordinates,
>
> $$du = dt_s - (1 - 2M/R)^{-1}dR, \tag{8.128}$$
>
> to find du/dt. Combine with equation (8.111) to get the desired result.

Note that the boundary condition for ψ implies that u measures the proper time of a stationary observer at infinity (*cf.* equations 8.127 and 8.128).

The Newtonian limit of the above equations results in the same equations found in the Newtonian limit of the Misner–Sharp equations and derived in exercise (8.13). The result is not surprising, since the Newtonian limit corresponds to setting $c = \infty$, in which case the light cones along which $u = constant$ open up and coincide with $t = constant$ surfaces. In the Newtonian limit, ψ disappears from the equations, as did ϕ in the previous case.

Box 8.1   The Misner–Sharp and Hernandez–Misner equations

For an easy comparison we list the equations of Misner and Sharp (1964) in the left column and those of Hernandez and Misner (1966) in the right column:

$$\partial_t U = -e^\phi \left[\frac{4\pi \Gamma R^2}{h} \partial_A (P + P_{\text{vis}}) \right.$$
$$\left. + \frac{m + 4\pi R^3 (P + P_{\text{vis}})}{R^2} \right]$$

$$\partial_u U = -\frac{e^\psi}{1 - v_s^2} \left[\frac{4\pi \Gamma R^2}{h} \partial_A (P + P_{\text{vis}}) \right.$$
$$\left. + \frac{m + 4\pi R^3 (P + P_{\text{vis}})}{R^2} \right]$$
$$- \frac{e^\psi v_s^2}{1 - v_s^2} \left(4\pi \rho_0 R^2 \partial_A U + 2U\Gamma/R \right)$$
$$+ \frac{1}{1 - v_s^2} \frac{\Gamma}{\rho_0 h} \partial_u P_{\text{vis}}$$

$$\partial_t R = e^\phi U$$

$$\partial_u R = e^\psi U$$

$$\partial_t m = -e^\phi 4\pi R^2 (P + P_{\text{vis}}) U$$

$$\partial_u m = -e^\psi 4\pi R^2 (P + P_{\text{vis}}) U$$

$$\Gamma = (1 + U^2 - 2m/R)^{1/2}$$

$$\Gamma = (1 + U^2 - 2m/R)^{1/2}$$

$$\rho_0 = \frac{\Gamma}{4\pi R^2 \partial_A R}$$

$$\rho_0 = \frac{\Gamma + U}{4\pi R^2 \partial_A R}$$

$$\partial_t \epsilon = -(P + P_{\text{vis}}) \partial_t (1/\rho_0)$$

$$\partial_u \epsilon = -(P + P_{\text{vis}}) \partial_u (1/\rho_0)$$

$$h = 1 + \epsilon + (P + P_{\text{vis}})/\rho_0$$

$$h = 1 + \epsilon + (P + P_{\text{vis}})/\rho_0$$

$$\partial_A m = (1 + \epsilon)\Gamma$$

$$\partial_A m = (1 + \epsilon)\Gamma - (P + P_{\text{vis}})U/\rho_0$$

$$\partial_A \phi = -\frac{1}{\rho_0 h} \partial_A (P + P_{\text{vis}})$$

$$\partial_A \psi = \frac{1}{\Gamma} \partial_A U + \frac{m}{4\pi \rho_0 R^4 \Gamma} + \frac{P + P_{\text{vis}}}{\rho_0 \Gamma R}$$

$$v_s^2 = \frac{1}{\rho_0^2 h} \left((P + P_{\text{vis}}) \left. \frac{\partial P}{\partial \epsilon} \right|_{\rho_0} + \rho_0^2 \left. \frac{\partial P}{\partial \rho_0} \right|_\epsilon \right)$$

We list the equations in the order in which they are typically evaluated in a numerical scheme. For completeness we explicitly include an artificial viscosity term P_{vis} (see equation 5.24).

A summary of the two sets of equations appears in Box 8.1. All relevant equations are listed next to each other and in the order in which they should be evaluated in a numerical scheme.[45]

We remark that the $u = constant$ surfaces are not characteristics of the Hernandez–Misner equations. This is because there is no gravitational radiation in spherical symmetry. Therefore we are still dealing with a Cauchy problem for the fluid evolution, rather than a characteristic initial value problem, and both the Misner–Sharp and Hernandez–Misner equations can be treated in the same way.

---

[45] For a finite difference version of the Misner–Sharp equations, see van Riper (1979) and for the Hernandez–Misner equations, see Baumgarte *et al.* (1995).

It is clear that very few modifications are needed to transform the Misner–Sharp equations to observer time coordinates. Hence a code using Misner–Sharp equations can be rewritten in observer time coordinates simply by adding a few terms and using ψ instead of ϕ. The advantage is that the revised code can handle black holes without breaking down.

There are two very attractive properties of observer time coordinates. One is that the coordinate u immediately corresponds to the time at which a distant observer would see a certain event, as for example a gamma-ray burst (GRB) in a supernovae explosion. The other one is that the global structure of spacetime is conveniently "hard-wired" into the integration scheme. This means that there is no need to search for apparent horizons (actually, apparent horizons never appear in observer time coordinates because they are always inside event horizons) or to track null rays in order to locate event horizons. In this case the event horizon can be found simply by looking for events at the which the lapse function e^{ψ} becomes exceedingly small. The lapse function plummets for every fluid element approaching the event horizon and essentially causes its further evolution to cease. For a typical application involving collapse to a black hole, one can terminate the evolution if and when e^{ψ} drops below, say 10^{-3} at the outermost shell. By then, the lapse in the center can be considerably smaller and can reach machine underflow.

The only subtlety that arises in using observer time coordinates has to do with the implementation of initial data. It is usually convenient to specify initial data on a spatial $t = constant$ hypersurface instead of a null hypersurface. Consequently, for typical applications, the implementation of initial data occurs in two stages. First, initial data are given on a $t = constant$ surface. These are then evolved using a Misner–Sharp scheme. During this evolution, a null geodesic is sent out from the center of the configuration, and the data on its path are stored. When the null ray arrives at the surface, this stage of the evolution can be stopped and the data on the ray's path can now be used as initial data on a $u = constant = 0$ surface, at which point the Hernandez–Misner scheme takes over for the rest of the evolution.

As an example of the Hernandez–Misner scheme at work, we show in Figure 8.21 "testbed" results for the collapse of a homogeneous, dust ball initially at rest, i.e., Oppenheimer–Snyder (OS) collapse. As we noted earlier, OS collapse provides one of the few highly nonlinear, dynamical examples in general relativity for which the solution is known analytically (see Chapter 1.4). However, as we noted in previous cases, before this analytic solution can be compared with the numerical results, the solution must be transformed to the coordinate system adopted in the numerical approach. This transformation has been carried out[46] for null coordinates and the comparison with a numerical simulation in these coordinates is shown in the figure.

A more interesting result is shown in Figure 8.22 for the collapse of a $1.4 M_{\odot}$, $n = 3$ polytrope with adiabatic index $\gamma = 4/3$ and initial central density $\rho_{0c} = 10^{12}\, \mathrm{g\, cm^{-3}}$. Plotted in the figure is a spacetime diagram for the late time evolution of a configuration in

---

[46] Baumgarte *et al.* (1995).

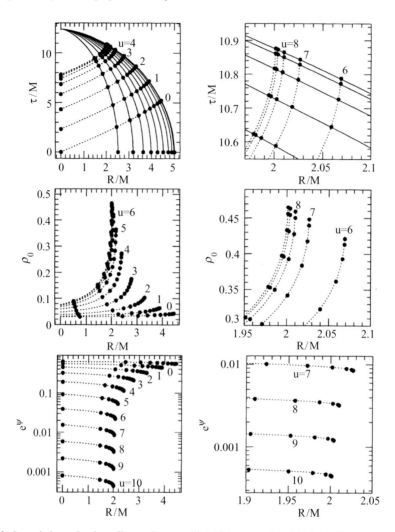

Figure 8.21 Oppenheimer–Snyder collapse of a star with initial compaction $R/M = 5$. Shown are a spacetime diagram with worldlines of selected mass shells (top row), rest-density profiles (middle row) and lapse function profiles (bottom row) at different observer times u. The analytic solution is represented by solid lines (worldlines of mass shells) and dotted lines (lines of constant u), while numerical results are represented by dots. [From Baumgarte *et al.* (1995).]

which the initial pressure was slightly reduced everywhere by a factor $d = 0.9946$ below the equilibrium value at $t = 0$.[47] In general relativity, all spherical equilibrium stars with $\gamma = 4/3$ are unstable to radial collapse,[48] so it is not surprising that the star undergoes an implosion. This case serves as a crude model for core collapse in a nonrotating, massive star at the endpoint of stellar evolution (at least prior to reaching nuclear densities). By simply

---

[47] This case was first studied by van Riper (1979).
[48] See, e.g., Shapiro and Teukolsky (1983), Chapter 6.9.

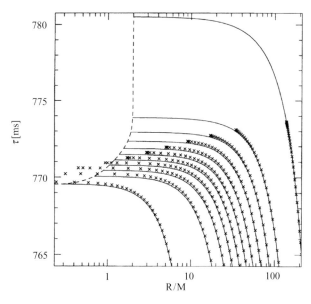

Figure 8.22 Comparison of simulations of the collapse of a slightly pressure-depleted $1.4M_\odot$ core with $\rho_0 = 10^{12}$ g cm^{-3} and $\gamma = 4/3$. Plotted is a spacetime diagram of the late-time evolution showing the worldlines of Lagrangian mass shells obtained with a Hernandez–Misner code (solid lines) and a Misner–Sharp code (crosses). The dashed line marks the event horizon. [From Baumgarte *et al.* (1995).]

rescaling the mass, the calculation also provides a good model for the collapse of a radially unstable, nonrotating, supermassive star.[49] Results are shown both for a Misner–Sharp simulation and a Hernandez–Misner simulation. The agreement in the time of collapse is within 0.05%. We note, however, that the Misner–Sharp simulation penetrates the event horizon and breaks down shortly after the formation of an apparent horizon, while the Hernandez–Misner simulation covers the entire spacetime outside the event horizon.[50]

8.4 Scalar field collapse: critical phenomena

Initial data sets for relativistic collapse typically divide themselves into two groups: those that lead to black hole formation and those that do not. One of the most important triumphs of numerical relativity has been the discovery by Choptuik that at the threshold of black hole formation, gravitational collapse solutions in general relativiy exhibit "critical behavior".[51] Specifically, collapse solutions for initial data at black hole threshold exhibit universality,

---

[49] In core collapse of a massive star, the EOS is dominated by relativistic, degenerate electrons above nuclear densities, while in a supermassive star, the EOS is dominated by thermal radiation pressure. In both cases, $\gamma \approx 4/3$.

[50] For more realistic examples of stellar core collapse employing the Hernandez–Misner formulation, including neutrino emission, see Baumgarte *et al.* (1996).

[51] Choptuik (1993); as reported here, the discovery came about as the direct result of a question posed to Choptuik in 1987 by Christodoulou: "will black hole formation turn on at *finite* or *infinitesimal* mass for a generic interpolating family at threshold?"

scaling and self-similarity in analogy with critical behavior in statistical mechanics. The existence of such "critical phenomena" are interesting for revealing the surprising structure and simplicity that underlies collapse solutions in general relativity and for providing insight into cosmic censorship and the dynamical character of Einstein's field equations. Following their discovery, critical phenomena have been pursued by numerous numerical simulations and also by perturbation treatments which utilize the formalism and techniques of dynamical systems theory and the renormalization group.[52]

Remarkably, critical phenomena can be explored in some detail by working with one of the simplest examples of gravitational collapse: the implosion of a massless scalar field in spherical symmetry. In fact, it was this model that was probed by Chopuik in his original analysis. The existence of critical behavior in such a system arises from the generic competition between two dynamical effects: the kinetic energy of the field, which tends to disperse it to infinity, and its gravitational potential energy, which, if sufficiently strong, can trap some of the mass-energy in a black hole. Choptuik realized that for any family of initial data, the dynamical competition could be controlled by tuning a parameter in the initial conditions (e.g., the amplitude of the initial field). If the parameter p is less than some critical (threshold) value p^*, the scalar field disperses completely, while if $p > p^*$, a black hole forms.

To examine quantitatively what happens for p in the neighborhood of p^*, we first need to evolve the system numerically. Toward this end, we assemble the full set of equations for evolving a massless scalar field in spherical symmetry in the next section. This is a good exercise in working with the equations of numerical relativity and then applying them to address a fundamental physics issue.

Basic equations

Suppose we start with the spherical metric in the form given by equation (8.13). Here r is the areal coordinate, whereby the surface area of a 2-sphere of constant t and r is $4\pi r^2$. Adopt polar time slicing, for which

$$K = K^r{}_r, \qquad K^\theta{}_\theta = K^\phi{}_\phi = 0. \qquad (8.129)$$

According to equation (8.19), the requirement of polar slicing implies $\beta = 0$, i.e., the shift must be zero everywhere. Setting $A = a^2$ we can thus cast the metric in the simple diagonal form

$$ds^2 = -\alpha^2(t, r)dt^2 + a^2(t, r)dr^2 + r^2(d\theta^2 + \sin^2\theta \, d\phi^2), \qquad (8.130)$$

which serves as the starting point of Choptuik's analysis. We need to find two field equations to determine the two functions α and a in the metric. We can obtain an equation for a from

---

[52] For excellent reviews and references, see Gundlach (2000, 2003) and Choptuik (1998), from which much of the discussion in this section is drawn

the Hamiltonian contraint, equation (2.132), which simplifies to $R = 16\pi\rho$ in the adopted slicing, since $K^2 = K_{ij}K^{ij}$. Using equation (8.18) to evaluate R, we arrive at

$$\frac{1}{a}\frac{da}{dr} + \frac{a^2 - 1}{2r} = 4\pi\rho r a^2. \tag{8.131}$$

The slicing condition $\partial_t K_{\theta\theta} = 0$ yields the equation for α. Using equation (2.135), we may write

$$0 = \partial_t K_{\theta\theta} = \alpha R_{\theta\theta} - D_\theta D_\theta \alpha - 8\pi\alpha \left[S_{\theta\theta} - \frac{1}{2}r^2(S - \rho) \right]. \tag{8.132}$$

Using equation (8.17) to evaluate $R_{\theta\theta}$, and using $D_\theta D_\theta \alpha = r\partial_r\alpha/a^2$, yields the desired equation for α,

$$\frac{1}{\alpha}\frac{d\alpha}{dr} - \frac{1}{a}\frac{da}{dr} - \frac{a^2 - 1}{r} = -8\pi\frac{a^2}{r}\left[S_{\theta\theta} - \frac{1}{2}(S - \rho) \right]. \tag{8.133}$$

The matter source is given by the stress-energy tensor for a massless, noninteracting, real scalar field,

$$T_{ab} = \nabla_a\varphi\nabla_b\varphi - \frac{1}{2}g_{ab}g^{cd}\nabla_c\nabla_d\varphi \tag{8.134}$$

(*cf.* equation 5.232). Defining the auxiliary scalar field variables Φ and Π according to

$$\Phi(t, r) \equiv \partial_r\varphi(t, r), \tag{8.135}$$

$$\Pi(t, r) \equiv \frac{a}{\alpha}\partial_t\varphi(t, r), \tag{8.136}$$

and referring to equation (2.138), we can express the matter source terms as follows:

$$\rho = (\Phi^2 + \Pi^2)/(2a^2), \tag{8.137}$$

$$j_r = -(\Phi\Pi)/a, \tag{8.138}$$

$$S^r_r = \rho, \tag{8.139}$$

$$S^\theta_\theta = S^\phi_\phi = (\Pi^2 - \Phi^2)/(2a^2), \tag{8.140}$$

$$S = (3\Pi^2 - \Phi^2)/(2a^2). \tag{8.141}$$

The field equations (8.131) and (8.133) thus reduce to

$$\frac{1}{a}\frac{da}{dr} + \frac{a^2 - 1}{2r} - 2\pi r(\Pi^2 + \Phi^2) = 0, \tag{8.142}$$

$$\frac{1}{\alpha}\frac{d\alpha}{dr} - \frac{1}{a}\frac{da}{dr} - \frac{a^2 - 1}{r} = 0. \tag{8.143}$$

Exercise 8.17   Derive an evolution equation for a. Use equation (8.19) for K_{rr}, together with the momentum constraint (2.133), to show

$$\partial_t a = 4\pi r \alpha \Phi \Pi. \tag{8.144}$$

It is thus not necessary to solve any evolution equations for the gravitational field, like equation (8.144), but instead one can integrate the first order ODE equations (8.142) and (8.143), (really radial quadratures), at each new time slice. Such a "constrained" scheme typically produces a more stable and accurate integration; equation (8.144) will be satisfied automatically. The matter field, however, must be evolved. The equation of motion is the massless Klein–Gordon equation (5.235). In terms of the auxiliary variables, this wave equation may be cast as two first-order equations,

$$\partial_t \Phi = \partial_r \left(\frac{\alpha}{a} \Pi \right), \tag{8.145}$$

$$\partial_t \Pi = \frac{1}{r^2} \partial_r \left(r^2 \frac{\alpha}{a} \Phi \right). \tag{8.146}$$

Exercise 8.18   Derive the two first-order equations above.

A useful geometric diagnostic is the mass function $m(t, r)$ which in spherical symmetry can defined invariantly by

$$1 - \frac{2m(t, r)}{r} \equiv \nabla_a r \nabla^a r = a^{-2}, \tag{8.147}$$

(*cf.* exercise 8.9). In the limit $r \to \infty$, m approaches the ADM mass ($=$ total mass-energy) of the spacetime. As we have discussed, polar slices cannot cross apparent horizons. However, black hole formation is definitely signaled by $2m/r \to 1$ at some areal radius $r = R_{BH}$; at this radius, the mass of the final black hole can be calculated from $M_{BH} = R_{BH}/2$.

Boundary conditions on the field and matter variables are now straightforward to specify. Equations (8.142) and (8.143) can be integrated outward from $r = 0$, where we can set $a = 1$ and $\alpha = 1$. The boundary value for a guarantees regularity at the origin according to equation (8.147). The boundary value for α makes the coordinate time t the proper time measured by a normal observer at the origin, which is a convenient parametrization of the $t = constant$ hypersurfaces. For the scalar field, regularity at the origin requires $\partial_r \varphi = 0$. At large radii, where the spacetime is asymptotically flat, we can impose outward spherical wave boundary conditions, e.g., $r\varphi(t, r) = f(t - r)$ for some function f. An equivalent way to write this condition is $\partial_r(r\varphi) + \partial_t(r\varphi) = 0$ as $r \to \infty$, which is simple to impose.

Exercise 8.19   Translate the inner and outer boundary conditions on $\varphi(t, r)$ to boundary conditions on $\Phi(t, r)$ and $\Pi(t, r)$.

We have now assembled all of the relevant equations required in Choptuik's original analysis.

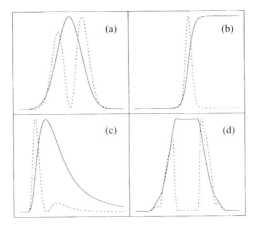

Figure 8.23 Typical initial profiles of the scalar field φ (solid lines) and the radial mass-energy density dm/dr (dotted lines) for four families of initial data considered by Choptuik. [From Choptuik (1993).]

Exercise 8.20 Show that the rescaling

$$t \to kt, \quad r \to kr, \quad \varphi \to \varphi, \quad \Pi \to k^{-1}\Pi, \quad \Phi \to k^{-1}\Phi, \quad (8.148)$$

transforms one solution into another for any positive k.

Numerical results

Suppose we now consider one-parameter families of massless scalar field initial data and evolve them for many different values of the parameter p. For example, one of the many families considered by Choptuik consisted of ingoing Gaussian wave packets $\varphi(0, r) = \varphi_0 r^3 \exp(-[(r - r_0)/\delta]^q)$, with the parameter p taken variously to be the amplitude φ_0, the centroid r_0, the width δ and the power-law q. A typical profile is sketched in Figure 8.23.

Exercise 8.21 Consider the evolution of a massless spherical scalar field in flat space for a specified initial packet $\varphi(0, r)$. Derive the initial auxiliary scalar fields $\Phi(0, r)$ and $\Pi(0, r)$ in terms of $\varphi(0, r)$ by requiring that the initial data yield a purely *ingoing* spherical wave.
Hint: Argue that the solution must be of the form $\varphi(t, r) = g(t + r)/r$ for some function $g(x)$.

The parameter p is a measure of the strength of the gravitational interaction. Strong interactions (high p, say) lead to black hole formation; weak interactions (low p) do not and lead to dispersal instead.[53] The critical transition value p^* is found empirically.

The first significant result to emerge from Choptuik's study is that for generic families of initial scalar field data, the black hole masses are well-fit by the scaling

---

[53] Christodoulou (1986, 1991, 1993) established that weak spherical scalar waves disperse and strong waves form black holes and that these are the only two final states.

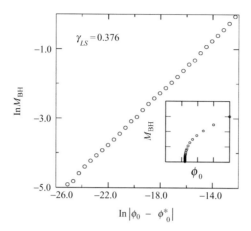

Figure 8.24 Typical evidence for mass scaling in the collapse of a spherical massless scalar field. The initial data consists of a one-parameter family of ingoing Gaussian pulses of scalar field in which the amplitude φ_0 is varied. The mass-scaling exponent, $\gamma_{LS} \approx 0.376$, is determined from a least-squares fit. The inset shows that the transition is Type II. For Gaussian initial data, the scaling persists well beyond the critical limit, $p \to p^*$. [From Choptuik (1998).]

relation

$$M_{\mathrm{BH}} = C|p - p^*|^{\gamma}, \qquad (8.149)$$

where C is a constant that depends on the family but where the exponent γ is universal, $\gamma \simeq 0.37$, independent of family. The evidence is displayed in Figure 8.24.

Black hole formation turns on at infinitesimal mass in the case of a massless scalar field, and this situation is designated a Type II critical phenomenon. By contrast, in a Type I transition, black holes first appear at finite mass.[54] The designations are in analogy with first and second order phase transitions in statistical mechanics. The two possibilities are distinguished schematically in Figure 8.25.

The second surprising result was the appearance of *universality*, the phenomenon whereby for a finite length of time in a finite region of space, the spacetime generated by all families of near-critical initial data approach the same solution. This universal critical solution to the equations of motion is approached by all initial data that are close to black hole threshold, on both sides of the critical solution, for any one-parameter family. The solution is determined up to an overall scale factor depending on the family. The critical solution is obviously unstable, as the slightest perturbation will result either in black hole formation or complete dispersal. When the evolution chooses one of the two routes, the universal phase ends.

The third finding revealed by Chotpuik's simulations is *scale-echoing*. This phenomenon is a form of discrete self-similarity, whereby as one tunes closer and closer to the critical

[54] The collapse of a massive scalar field is a Type I phenomenon, as is the collapse of collisionless matter.

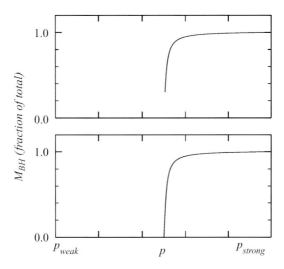

Figure 8.25 Schematic illustration of possible black-hole threshold behavior. The top panel represents a Type I ("first-order") transition, where the "order parameter" M_{BH} is nonzero at threshold. The bottom panel shows a Type II ("second order") transition, where M_{BH} is infinitesimal at the critical point. [From Choptuik (1998).]

solution, the dynamical character of the solution repeats itself at ever-decreasing time and length scales related by a factor of e^Δ, where $\Delta \simeq 3.44$, so that $e^\Delta \simeq 30$. Mathematically, the critical solution $\varphi^*(t, r)$ and associated spacetime satisfy

$$t' = e^{-n\Delta}t \tag{8.150}$$

$$r' = e^{-n\Delta}r \tag{8.151}$$

$$ds'^2 = e^{-2n\Delta}ds^2 \tag{8.152}$$

$$\varphi^*(t', r') = \varphi^*(t, r) \tag{8.153}$$

$$(n = 1, 2, 3 \ldots). \tag{8.154}$$

The dimensionless critical exponent γ and echo period Δ are two interesting constants whose origins are still somewhat of a mystery.

The behavior discovered by Choptuik has been found in other forms of matter coupled to gravity. It has even been demonstrated in the collapse of axisymmetric gravitational waves in pure vacuum spacetimes.[55] The existence of the critical exponent and echo period appears to be universal, but their values depend on the type of matter.

A significant consequence of Type II critical solutions is that they produce *naked singularities*: a critical solution results in a point of infinite curvature which is visible to observers at future null infinity. For example, the spacetime scalar curvature $^{(4)}R$ grows without bound near $r = 0$ in a precisely critical massless scalar field evolution.

[55] Abrahams and Evans (1993); $\gamma \approx 0.37$, $\Delta \approx 0.6$

Exercise 8.22   Show that $^{(4)}R$ is easily evaluated for a massless scalar field as

$$^{(4)}R = 8\pi \nabla^a \varphi \nabla_a \varphi = 8\pi (\Phi^2 - \Pi^2)/a^2. \qquad (8.155)$$

Thus Type II critical solutions provide counterexamples to the Cosmic Censorship Conjecture,[56] which asserts that the collapse of well-behaved initial data does not result in a naked singularity. While it is always possible to restate the conjecture so that it applies only to the collapse of *generic* initial data, and thereby rule out these critical solutions on technical grounds, these counterexamples do highlight why is has been so difficult to fashion a rigorous proof.[57]

---

[56] Penrose (1969).
[57] See Berger (2002) for discussion and references; see also Chapter 10.1.

9 Gravitational waves

Spherically symmetric spacetimes, which we discussed in Chapter 8, do not admit gravitational radiation. Once we relax this symmetry restriction, as we shall do in the following chapters, we will encounter spacetimes that do contain gravitational radiation. In fact, simulating promising sources of gravitational radiation and predicting their gravitational wave signals are among the most important goals of numerical relativity. These goals are especially urgent in light of the new generation of gravitational wave laser interferometers which are now operational. A book on numerical relativity therefore would not be complete without a discussion of gravitational waves.

In this chapter we review several topics related to gravitational waves. We start in Section 9.1 with a discussion of linearized waves propagating in nearly Minkowski spacetimes and the role that these waves play even in the case of nonlinear sources of gravitational radiation. In Section 9.2 we survey plausible sources of gravitational waves, highlighting those that seem most promising from the perspective of gravitational wave detection. We briefly describe some of the existing and planned gravitational wave detectors in Section 9.3. Finally, in Section 9.4 we make contact with numerical relativity, and review different strategies that have been employed to extract gravitational radiation data from numerical relativity simulations.

9.1 Linearized waves

Most of this book deals with strong-field solutions of Einstein's equations, including black holes, neutron stars, and binaries containing these objects. As long as these solutions are dynamical and nonspherical, they will emit gravitational radiation. In the near-field region of such sources, the gravitational fields consist of a combination of longitudinal and transverse (i.e., radiative) components that cannot be disentangled unambigiously. As the transverse fields propagate away from their sources, however, they will reach an asymptotic region in which they can be modeled as a linear perturbation of a nearly Minkowski spacetime. These linearized gravitational waves carry information about the nature of the nonlinear sources that generated them. It is these linearized waves that are measured by gravitational wave detectors. The goal in simulating astrophysically promising sources of gravitational radiation is therefore to predict the emitted gravitational waveforms that reach this asymptotic regime. We will discuss this numerical "extraction" of gravitational waveforms in Section 9.4. Before doing so, however, we begin with a review of some of the basic properties of linearized gravitational waves.

9.1.1 Perturbation theory and the weak-field, slow-velocity regime

A detailed discussion and derivation of the generation and propagation of gravitational waves in the weak-field, slow-velocity regime of general relativity can be found in any of the standard textbooks on general relativity. We have already provided a brief summary in Chapter 1.1; here we will review the main results and use them in several applications.

Consider a small perturbation h_{ab} of a known "background" solution to Einstein's equations. In principle the background could be any solution, but here we are interested in waves propagating in a nearly Minkowski spacetime, for which the metric becomes[1]

$$g_{ab} = \eta_{ab} + h_{ab}, \quad |h_{ab}| \ll 1. \tag{9.1}$$

It is convenient to introduce the "trace-reversed" perturbation

$$\bar{h}_{ab} \equiv h_{ab} - \frac{1}{2}\eta_{ab}h_c^{\ c}. \tag{9.2}$$

We can now exploit our coordinate freedom to impose the "Lorentz gauge" condition,

$$\nabla_a \bar{h}^{ab} = 0, \tag{9.3}$$

in which case Einstein's equations in *vacuum* reduce to the wave equation

$$\boxed{\Box \bar{h}_{ab} \equiv \nabla^c \nabla_c \bar{h}_{ab} = 0 \quad \text{(vacuum)}.} \tag{9.4}$$

As it turns out, the Lorentz-gauge condition (9.3) does not determine $\bar{h}_{ab}$ uniquely, since we can introduce further infinitesimal gauge transformations that leave this condition unchanged. We can therefore use this remaining gauge freedom to impose further conditions on the perturbations $\bar{h}^{ab}$. Particularly useful is the transverse-traceless or "TT" gauge, in which

$$\bar{h}_{a0}^{TT} = 0, \quad \bar{h}^{TT a}_{\ \ a} = 0. \tag{9.5}$$

The first condition implies that the only nonzero components of $\bar{h}_{ab}^{TT}$ are purely spatial. The second condition implies that $h_c^{\ c} = 0$, so that, according to equation (9.2), the trace-reversed metric perturbations $\bar{h}_{ab}$ are identical to the original perturbations h_{ab}, and we are entitled to drop the bars whenever we write down results in the TT gauge. In the TT gauge the time-space components of the 4-dimensional Riemann tensor are simply expressed in terms of the metric perturbations:

$$^{(4)}R_{i0j0} = -\frac{1}{2}\ddot{h}_{ij}^{TT}. \tag{9.6}$$

---

[1] Departing from the notation in most of the rest of the book we shall adopt Cartesian coordinates here, whereby $\eta_{ab} = \text{diag}(-1, 1, 1, 1)$.

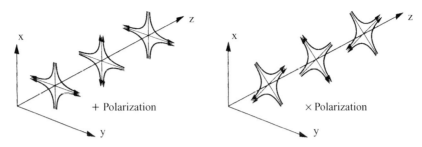

Figure 9.1 Lines of force associated with the two polarization states h_+ and $h_\times$ of a linear plane gravitational wave traveling in vacuum in the z-direction. [From Abramovici *et al.* (1992).]

Here the double dot denotes the second time derivative. Equation (9.6) exhibits the special role that the metric perturbations in the TT gauge play in providing a direct measure of the Riemann tensor, which determines the gravitational tidal field. Together, the conditions (9.3) and (9.5) provide eight constraints on the originally ten independent components of h_{ab}.[2] The remaining two degrees of freedom correspond to the two possible polarization states of gravitational radiation. It is often convenient to express these two polarization states in terms of the two polarization tensors e_{ab}^+ and $e_{ab}^\times$. For a linear plane wave propagating in vacuum in the z-direction, for example, we have

$$e_{xx}^+ = -e_{yy}^+ = 1, \quad e_{xy}^\times = e_{yx}^\times = 1, \quad \text{all other components zero.} \quad (9.7)$$

A general gravitational wave is then specified by two dimensionless amplitudes h_+ and $h_\times$ as

$$h_{jk}^{TT} = h_+ e_{ij}^+ + h_\times e_{ij}^\times. \quad (9.8)$$

The effect of a passing gravitational wave on two nearby, freely-falling test particles at a spatial separation ξ^i is to change their separation according to

$$\ddot{\xi}_i = \frac{1}{2} \ddot{h}_{ij}^{TT} \xi^j. \quad (9.9)$$

Equation (9.9) is a direct consequence of the equation of geodesic deviation applied to the two particles, combined with equation (9.6). The relative strain $\delta\xi/\xi$ between these two particles is therefore proportional to the gravitational wave amplitude, which explains why h_{ij}^{TT} is sometimes called the gravitational wave strain. In Figure 9.1 we show the lines of force associated with the two polarization states h_+ and $h_\times$. For the h_+ polarization, for example, particles separated along the x-direction (with coordinate axes chosen as in the

---

[2] Superficially, it appears as if the conditions (9.3) and (9.5) provided nine conditions, not eight. However, one of the four equations $\bar{h}^{TT}{}_{a0} = 0$ in (9.5) is redundant with one of the conditions in (9.3), reducing the number of conditions to eight. See, for example, Section 35.4 in Misner *et al.* (1973) for a more detailed discussion.

figure), will be pulled apart while particles along the y-direction are pushed together, and vice versa a half-cycle later.

Consider now the generation of gravitational radiation from a *weak-field, slow-velocity* source.[3] For such a source, Einstein's equations reduce to[4]

$$\Box \bar{h}_{ab} = \nabla^c \nabla_c \bar{h}_{ab} = -16\pi\, T_{ab} \quad \text{(weak-field, slow-velocity)}. \tag{9.10}$$

Imposing an outgoing-wave boundary condition, we can solve equation (9.10) with the help of a Green function to obtain the integral equation,

$$\bar{h}_{ab}(t, x^i) = 4 \int d^3 x' \frac{T_{ab}(t - |x^i - x'^i|, x'^i)}{|x^i - x'^i|}. \tag{9.11}$$

Here $t - |x^i - x'^i|$ is the retarded time; it implies that $\bar{h}_{ab}$ at the event (t, x^i) depends on the stress-energy tensor T_{ab} on the event's past light cone.

In the wave zone at large distances r from the source we may expand the retarded integral in equation (9.11) in powers of x'^i/r to find

$$\bar{h}_{ij}(t, x^k) = \frac{4}{r} \int d^3 x'\, T_{ij}(t - r, x'^k). \tag{9.12}$$

We now use the tensor virial theorem[5]

$$\int d^3 x'\, T^{ij} = \frac{1}{2} \frac{d^2}{dt^2} \int d^3 x'\, T^{tt} x'^i x'^j, \tag{9.13}$$

together with the expression $T^{tt} \approx \rho_0$, valid for Newtonian sources, and the definition of the second moment of the mass distribution,

$$I^{ij}(t) \equiv \int d^3 x'\, \rho_0(t, x'^k) x'^i x'^j, \tag{9.14}$$

to rewrite (9.12) as

$$\bar{h}_{ij}(t, x^k) = \frac{2}{r} \ddot{I}_{ij}(t - r). \tag{9.15}$$

It is then convenient to define the *reduced quadrupole moment* as the traceless part of the second moment of the mass distribution,

$$\mathcal{I}_{ij} \equiv I_{ij} - \frac{1}{3}\eta_{ij} I, \tag{9.16}$$

where $I = I^a_a$. The reduced quadrupole moment has a more transparent physical meaning than the second moment of the mass distribution, since it appears in the near-zone expansion

---

[3] Weak-field means $\Phi \ll 1$, where Φ is the Newtonian potential of the source; slow-velocity means $v \ll 1$, where v is the characteristic speed of the source and its internal components.

[4] In what follows, adopting the weak-field, slow-velocity limit allows us to neglect the stress-energy pseudotensor t_{ab} that must be added to T_{ab} on the right hand side of equation (9.10) in the general case; see Misner *et al.* (1973), Section 36.10.

[5] This can be proven by using $\partial_t^2 T^{tt} = -\partial_t \partial_k T^{tk} = \partial_k \partial_l T^{lk}$ (a consequence of successive application of the conservation of energy-momentum in the weak-field limit, $\partial_a T^{ab} = 0$) and repeated integration by parts.

of the Newtonian gravitational potential,

$$\Phi = -\left(\frac{M}{r} + \frac{d_i x^i}{r^3} + \frac{3}{2} \frac{\mathcal{I}_{ij} x^i x^j}{r^5} + \cdots \right). \tag{9.17}$$

Here M is the total mass of the system, and d_i is its dipole moment.

Finally, to bring equation (9.15) into the TT gauge, we need to project out its transverse-traceless part. Using the projection operator

$$P_i{}^j \equiv \eta_i{}^j - n_i n^j, \tag{9.18}$$

where $n^i = x^i / r$ is a unit vector pointing in the wave's local direction of propagation, this can be accomplished by defining

$$\mathcal{I}_{ij}^{TT} \equiv \left(P_i{}^k P_j{}^l - \frac{1}{2} P_{ij} P^{kl} \right) \mathcal{I}_{kl}. \tag{9.19}$$

We then have

$$h_{ij}^{TT}(t, x^k) = \frac{2}{r} \ddot{\mathcal{I}}_{ij}^{TT}(t - r) \quad \text{(weak-field, slow-velocity).} \tag{9.20}$$

since the TT parts of I_{ij} and $\mathcal{I}_{ij}$ are identical. We emphasize that equation (9.20), referred to as the "quadrupole approximation", holds only for a weak-field, slow-velocity source. But a majority of the most promising sources of detectable gravitational waves are characterized by strong fields and high velocities, for which numerical relativity is required to determine the waveforms. Nevertheless, equation (9.20) often provides a useful first approximation even in these cases. It also serves as a good check on numerical calculations in the appropriate limit.

Exercise 9.1 A distant observer at (r, θ, ϕ) in the wave zone measures a gravitational wave from a radiating source at the origin. Decompose the waveform into its two polarization modes as in equation (9.8), whereby the 3-metric in the wave zone may be written

$$dl^2 = dr^2 + (1 + h_+) r^2 d\theta^2 + (1 - h_+) \sin^2\theta d\phi^2 + 2h_\times \sin\theta d\theta d\phi. \tag{9.21}$$

(a) Show that the two polarization modes are related to $\mathcal{I}_{ij}$ according to

$$h_+ = \frac{1}{r} (\ddot{\mathcal{I}}_{\hat{\theta}\hat{\theta}} - \ddot{\mathcal{I}}_{\hat{\phi}\hat{\phi}}), \tag{9.22}$$

$$h_\times = \frac{1}{r} \ddot{\mathcal{I}}_{\hat{\phi}\hat{\phi}}, \tag{9.23}$$

where $\mathcal{I}_{\hat{i}\hat{j}}$ are the components of the reduced quadrupole moment in an orthonormal basis $(\mathbf{e}_{\hat{r}}, \mathbf{e}_{\hat{\theta}}, \mathbf{e}_{\hat{\phi}})$.

(b) Show that if we relate spherical polar coordinates to Cartesian coordinates (x, y, z) in the usual way, the two polarization modes can be expressed in terms of the components of $\mathcal{I}_{ij}$ in a Cartesian basis according to

$$h_+ = \frac{1}{r}\left[\frac{\ddot{\mathcal{I}}_{xx} - \ddot{\mathcal{I}}_{yy}}{2}(1 + \cos^2\theta)\cos(2\phi) + \ddot{\mathcal{I}}_{xy}(1 + \cos^2\theta)\sin(2\phi)\right.$$
$$\left. + \left(\ddot{\mathcal{I}}_{zz} - \frac{\ddot{\mathcal{I}}_{xx} + \ddot{\mathcal{I}}_{yy}}{2}\right)\sin^2\theta\right],\tag{9.24}$$

$$h_\times = \frac{1}{r}\left[-\frac{\ddot{\mathcal{I}}_{xx} - \ddot{\mathcal{I}}_{yy}}{2}\cos\theta\,\sin(2\phi) + \ddot{\mathcal{I}}_{xy}\cos\theta\,\cos(2\phi)\right].\tag{9.25}$$

Exercise 9.2 Consider a Newtonian binary consisting of two point masses m_1 and m_2 in circular orbit about each other. Choose coordinates so that the binary orbits in the xy plane and is aligned with the x-axis at $t = 0$.
(a) Compute the reduced quadrupole moment in the center-of-mass frame to show that it can be written as

$$\mathcal{I}_{ij} = \frac{1}{2}\mu R^2 \begin{pmatrix} \cos 2\Omega t + 1/3 & \sin 2\Omega t & 0 \\ \sin 2\Omega t & -\cos 2\Omega t + 1/3 & 0 \\ 0 & 0 & -2/3 \end{pmatrix},\tag{9.26}$$

where Ω is the orbital angular velocity, $\mu = m_1 m_2/(m_1 + m_2)$ the reduced mass, and R the binary separation.
(b) Show that a distant observer situated at (r, θ, ϕ), where, without loss of generality, $\phi = 0$, measures the waveforms to be

$$h_+ = -\frac{2}{r}\Omega^2\mu R^2(1 + \cos^2\theta)\cos 2\Omega(t - r),$$

$$h_\times = -\frac{4}{r}\Omega^2\mu R^2\cos\theta\sin 2\Omega(t - r).$$

Thus the magnitude h of the waveform amplitude is of order

$$h \simeq \frac{4\mu\Omega^2 R^2}{r} = \frac{4\mu M^{2/3}\Omega^{2/3}}{r} = \frac{4}{r}\frac{\mu M}{R},\tag{9.27}$$

where the total mass is given by $M = m_1 + m_2$ and where we have used Kepler's law $\Omega^2 = M/R^3$. Note that the frequency of the wave is twice the frequency of the orbit in the quadrupole approximation.

Gravitational radiation carries energy, momentum, and angular momentum. To see this formally[6] we need to expand Einstein's equations to second order. In vacuum, the resulting second-order equations contain two different kinds of terms – terms that are linear in the second-order perturbations of the metric, and terms that are quadratic in the first-order perturbations h_{ij}. Upon suitable averaging, the latter constitute a source term for the background curvature. This source term thus defines an "effective" stress-energy tensor

---

[6] The proceedure is know as the "shortwave approximation"; see Misner *et al.* (1973), Sections 35.13–35.15.

for the gravitational waves, T_{ab}^{GW}. In a (nearly) Minkowski frame of linearized theory, this effective stress-energy tensor reduces to

$$T_{ab}^{\text{GW}} = \frac{1}{32\pi} \langle \partial_a h_{ij}^{TT} \partial_b h^{TTij} \rangle, \tag{9.28}$$

where $\langle \rangle$ denotes an average over several wavelengths. We can now use this stress-energy tensor to compute the energy, momentum and angular momentum carried off by the gravitational waves. The outgoing energy flux in the radial direction, for example, is given by the T_{GW}^{tr} component of equation (9.28). To find the gravitational wave luminosity, which is equal and opposite the energy change of the source, we integrate the energy flux passing through a large sphere surrounding the source,

$$L_{\text{GW}} \equiv -\frac{dE}{dt} = -\lim_{r\to\infty} \int T_{tr}^{\text{GW}} r^2 d\Omega. \tag{9.29}$$

We can express the radiated energy in terms of the wave amplitudes h_+ and $h_\times$ of the outgoing gravitational radiation according to

$$L_{\text{GW}} = -\frac{dE}{dt} = \lim_{r\to\infty} \frac{r^2}{16\pi} \int \langle \dot{h}_+^2 + \dot{h}_\times^2 \rangle d\Omega, \tag{9.30}$$

where we have used the fact that $\partial_r h_{ij}^{TT} = -\partial_t h_{ij}^{TT}$ for radially outgoing radiation. Similarly, we can find the loss of angular momentum due to radiation,[7]

$$\frac{dJ_i}{dt} = \lim_{r\to\infty} \frac{r^2}{16\pi} \int \text{Re}\langle \dot{H}^* \hat{J}_i H \rangle d\Omega, \tag{9.31}$$

where $H \equiv h_+ - ih_\times$ and where the operator $\hat{J}_i$ is defined as

$$\hat{J}_x = -\sin\phi\partial_\theta - \cos\phi(\cot\theta\partial_\phi + 2i\csc\theta) \tag{9.32}$$
$$\hat{J}_y = \cos\phi\partial_\theta - \sin\phi(\cot\theta\partial_\phi + 2i\csc\theta) \tag{9.33}$$
$$\hat{J}_z = \partial_\phi. \tag{9.34}$$

In particular, this relation gives

$$\frac{dJ_z}{dt} = \lim_{r\to\infty} \frac{r^2}{16\pi} \int \langle \partial_t h_+ \partial_\phi h_+ + \partial_t h_\times \partial_\phi h_\times \rangle d\Omega. \tag{9.35}$$

Finally, the loss of linear momentum due to radiation is

$$\frac{dP^i}{dt} = \lim_{r\to\infty} -\frac{r^2}{16\pi} \int \frac{x^i}{r} \langle \dot{h}_+^2 + \dot{h}_\times^2 \rangle d\Omega. \tag{9.36}$$

[7] See, e.g., Ruiz *et al.* (2008a).

In all the above expressions the quantities E, J and P^i refer to the source, and their changes are equal and opposite to the corresponding quantities carried off by the waves. We note that *equations (9.30)–(9.36) apply even in the case of strong-field sources*, since the waves they generate constitute small perturbations to the nearly flat background spacetime once they reach large distances from the source, where these fluxes are evaluated.

In the case of a weak-field, slow-velocity source, we can express the radiated energy in terms of the source's reduced quadrupole moment. Inserting equation (9.20) into (9.29) and integrating yields[8]

$$L_{\text{GW}} = -\frac{dE}{dt} = \frac{1}{5}\langle \dddot{\mathcal{I}}_{jk} \dddot{\mathcal{I}}^{jk}\rangle. \tag{9.37}$$

This energy loss can also be modeled with a corresponding radiation-reaction force F_i^{react} that can be written as the gradient of a radiation-reaction potential Φ^{react},

$$F_i^{\text{react}} = -m\,D_i\Phi^{\text{react}}, \quad \Phi^{\text{react}} = \frac{1}{5}\mathcal{I}_{jk}^{(5)}x^j x^k. \tag{9.38}$$

Here the superscript (5) denotes a fifth time derivative. To show that equations (9.37) and (9.38) are consistent with each other we compute the energy loss of a swarm of particles, labeled by an index A, as[9]

$$\frac{dE}{dt} = \sum_A v_A^i F_i^{\text{react}} = -\sum_A m_A v_A^i D_i\Phi^{\text{react}} = -\frac{2}{5}\mathcal{I}_{ij}^{(5)}\sum_A m_A v_A^i x_A^j$$

$$= -\frac{1}{5}\mathcal{I}_{ij}^{(5)}\frac{d}{dt}\sum_A m_A x_A^i x_A^j = -\frac{1}{5}\mathcal{I}_{ij}^{(5)}\mathcal{I}^{ij(1)}. \tag{9.39}$$

In the last step we have been able to convert an I^{ij} into an $\mathcal{I}^{ij}$ since $\mathcal{I}_{ij}\eta^{ij} = 0$. Taking an average over several wave cycles we can integrate by parts twice to convert $\mathcal{I}_{ij}^{(5)}\mathcal{I}^{ij(1)}$ into $\mathcal{I}_{ij}^{(3)}\mathcal{I}^{ij(3)}$, which yields equation (9.37). We can perform a very similar calculation to compute the loss of angular momentum, obtaining

$$\frac{dJ_i}{dt} = -\frac{2}{5}\epsilon_{ijk}\langle \ddot{\mathcal{I}}^{jm}\, \dddot{\mathcal{I}}^k{}_m\rangle. \tag{9.40}$$

The so-called "quadrupole formulae" (9.37) and (9.40) can be used to calculate the loss of energy and angular momentum from any weak-field, slow-velocity source. As an example, we return in exercise 9.3 to the point-mass binary system of exercise 9.2.

Exercise 9.3   Revisit the point-mass binary of exercise 9.2 and evaluate (9.37) and (9.40) to find

$$\frac{dE}{dt} = -\frac{32}{5}\mu^2 R^4 \Omega^6 \tag{9.41}$$

---

[8] Before the integration can be carried out, the transverse-traceless components of $\mathcal{I}_{ij}^{TT}$ in equation (9.20) have to be expanded in terms of $\mathcal{I}_{ij}$, whereby the normal vector $n^i = x^i/r$ in the projection operator (9.18) introduces a (quadrupolar) angular dependence. See, e.g., Carroll (2004), page 313, for details.

[9] See, e.g., Burke (1971) as well as the discussion in Shapiro and Teukolsky (1983).

and

$$\frac{dJ_z}{dt} = -\frac{32}{5}\mu^2 R^4 \Omega^5. \tag{9.42}$$

The loss of energy and angular momentum due to gravitational radiation causes the binary orbit to shrink, i.e., gravitational radiation drives binary inspiral. The inspiral and merger of binaries containing compact stars is a very promising source of gravitational waves for detection by laser interferometers (see Section 9.2). Consequently, compact binary inspiral and merger is a central theme in this book, and we shall return to this topic on several occasions.[10]

Exercise 9.3 provides an example of the remarkable relationship

$$dE = \Omega \, dJ, \tag{9.43}$$

which holds under quite general conditions involving relativistic systems in quasistationary equilibrium undergoing secular changes. We will encounter this relation several times later in this book. For example, in exercise 12.2, we will see that equation (9.43) guarantees that a circular binary undergoing slow inspiral due to gravitational wave emission will remain in a (nearly) circular orbit.

9.1.2 Vacuum solutions

Return now to Einstein's linearized equations (9.4) in vacuum,

$$\Box \bar{h}_{ab} = 0. \tag{9.44}$$

Just as in electrodynamics, this type of equation admits simple plane wave solutions of the form

$$\bar{h}_{ab} = \text{Re}\left(A_{ab} e^{ik_c x^c}\right). \tag{9.45}$$

Here $k_a = (\omega, k^i)$ is a 4-dimensional wave vector, $x^a = (t, x^i)$ denote the inertial coordinates of a point in spacetime and A_{ab} is a constant tensor representing the wave amplitude. Einstein's equations (9.44) then demand that k^a be a null vector,

$$k_a k^a = 0, \tag{9.46}$$

whereby $\omega = |k^i|$. This dispersion relation implies that gravitational waves propagate at the speed of light. The Lorentz condition (9.3) requires

$$k^a A_{ab} = 0, \tag{9.47}$$

implying that gravitational waves are transverse. The above results are quite general, since we can always decompose an arbitrary, linear gravitational wave propagating in a nearly flat, vacuum spacetime into a superposition of the plane wave solutions (9.45).

---

[10] See Chapter 12.1 for a more detailed overview of binary inspiral.

From a numerical point of view the plane wave solutions found above are not the most useful. Most numerical simulations treat spacetimes with finite, bounded sources, for which the waves propagate radially outward at large distance. Moreover such spacetimes approach asymptotic flatness at least as fast as r^{-1}. Clearly spacetimes containing with plane waves (9.45) do not share these properties. More useful for simulation purposes are multipole expansions of linear, vacuum solutions to equation (9.45) expressed in terms of tensor spherical harmonics. Tensor spherical objects are defined in Appendix D.

When working in spherical coordinates, it is more convenient to express the two polarization states of gravitational radiation in terms of *even-* and *odd-parity modes* rather than the $+$ and $\times$ modes of equation (9.7) An even-parity mode (l, m) has parity $(-1)^l$ under space inversion $(\theta, \phi) \to (\pi - \theta, \phi + \pi)$, while an odd-parity mode has parity $(-1)^{l+1}$. Evidently, depending on the index l, both odd- and even-parity modes can have odd or even parity. To choose a less confusing name, some authors therefore refer to the even-parity modes as *polar* and the the odd-parity modes as *axial*.

Of particular interest is the quadrupole $l = 2$ wave, which often is the dominant mode for many sources of gravitational radiation. For even-parity or polar quadrupole modes, the metric takes the form[11]

$$ds^2 = -dt^2 + (1 + Af_{rr})\, dr^2 + (2Bf_{r\theta})r\, dr\, d\theta + (2Bf_{r\phi})r \sin\theta dr\, d\phi$$
$$+ (1 + Cf_{\theta\theta}^{(1)} + Af_{\theta\theta}^{(2)})r^2 d\theta^2 + [2(A - 2C)f_{\theta\phi}]r^2 \sin\theta d\theta d\phi$$
$$+ (1 + Cf_{\phi\phi}^{(1)} + Af_{\phi\phi}^{(2)})r^2 \sin^2\theta d\phi^2. \tag{9.48}$$

Here the coefficients A, B and C can be constructed from an arbitrary function $F(x)$, where we have $x = t - r$ for an outgoing solution or $x = t + r$ for an ingoing solution. In the general case, we may take

$$F = F_1(t - r) + F_2(t + r) \tag{9.49}$$

and define

$$F^{(n)} \equiv \left[\frac{d^n F_1(x)}{dx^n}\right]_{x=t-r} + (-1)^n \left[\frac{d^n F_2(x)}{dx^n}\right]_{x=t+r}. \tag{9.50}$$

In terms of $F(x)$ and its derivatives we then have

$$A = 3\left[\frac{F^{(2)}}{r^3} + \frac{3F^{(1)}}{r^4} + \frac{3F}{r^5}\right], \tag{9.51}$$

$$B = -\left[\frac{F^{(3)}}{r^2} + \frac{3F^{(2)}}{r^3} + \frac{6F^{(1)}}{r^4} + \frac{6F}{r^5}\right], \tag{9.52}$$

$$C = \frac{1}{4}\left[\frac{F^{(4)}}{r} + \frac{2F^{(3)}}{r^2} + \frac{9F^{(2)}}{r^3} + \frac{21F^{(1)}}{r^4} + \frac{21F}{r^5}\right]. \tag{9.53}$$

---

[11] Teukolsky (1982); linear, vacuum quadrupole waves in this representation are sometimes referred to as "Teukolsky waves". For generalization to all multipoles, see Rinne (2008b).

The angular functions f_{ij} in the metric (9.48) depend on the axial parameter m. We list these functions in the order $m = \pm 2, \pm 1, 0$, with the functions corresponding to the upper sign displayed on top of those corresponding to the lower sign:

$$f_{rr} = \sin^2\theta \begin{pmatrix} \cos 2\phi \\ \sin 2\phi \end{pmatrix}, \quad 2\sin\theta\cos\theta \begin{pmatrix} \cos\phi \\ \sin\phi \end{pmatrix}, \quad 2 - 3\sin^2\theta,$$

$$f_{r\theta} = \sin\theta\cos\theta \begin{pmatrix} \cos 2\phi \\ \sin 2\phi \end{pmatrix}, \quad (\cos^2\theta - \sin^2\theta)\begin{pmatrix} \cos\phi \\ \sin\phi \end{pmatrix}, \quad -3\sin\theta\cos\theta,$$

$$f_{r\phi} = \sin\theta \begin{pmatrix} -\sin 2\phi \\ \cos 2\phi \end{pmatrix}, \quad \cos\theta \begin{pmatrix} -\sin\phi \\ \cos\phi \end{pmatrix}, \quad 0,$$

$$f_{\theta\theta}^{(1)} = (1 + \cos^2\theta)\begin{pmatrix} \cos 2\phi \\ \sin 2\phi \end{pmatrix}, \quad 2\sin\theta\cos\theta \begin{pmatrix} -\cos\phi \\ -\sin\phi \end{pmatrix}, \quad 3\sin^2\theta, \tag{9.54}$$

$$f_{\theta\theta}^{(2)} = \begin{pmatrix} -\cos 2\phi \\ -\sin 2\phi \end{pmatrix}, \quad 0, \quad -1,$$

$$f_{\theta\phi} = \cos\theta \begin{pmatrix} \sin 2\phi \\ -\cos 2\phi \end{pmatrix}, \quad \sin\theta \begin{pmatrix} -\sin\phi \\ \cos\phi \end{pmatrix}, \quad 0,$$

$$f_{\phi\phi}^{(1)} = -f_{\theta\theta}^{(1)}$$

$$f_{\phi\phi}^{(2)} = \cos^2\theta \begin{pmatrix} \cos 2\phi \\ \sin 2\phi \end{pmatrix}, \quad 2\sin\theta\cos\theta \begin{pmatrix} -\cos\phi \\ -\sin\phi \end{pmatrix}, \quad 3\sin^2\theta - 1.$$

Exercise 9.4 Consider a superposition of ingoing and outgoing waves with

$$F_1(x) = -F_2(x). \tag{9.55}$$

(a) Show that if $F_1(x)$ is an odd function in x, then $t = 0$ corresponds to a moment of time symmetry (i.e., show that $K_{ij} = 0$ at $t = 0$).
(b) Now consider the specific choice

$$F_1(x) = -F_2(x) = \mathcal{A}x e^{-(x/\lambda)^2}, \tag{9.56}$$

where $\mathcal{A}$ is an amplitude and λ measures the spatial extent of the wave packet. Perform an expansion around $r = 0$ to show that the waveform is regular (i.e., nonsingular) at the origin.
Hint: It may be helpful to use an algebraic package like Mathematica, or else to restrict the analysis to $t = 0$.

To translate to the h_+ and $h_\times$ polarizations, we can introduce a local orthonormal coordinate system and identify

$$h_+ = h_{\hat\theta\hat\theta}^{TT} = C f_{\theta\theta}^{(1)} \tag{9.57}$$

and

$$h_\times = h_{\hat\theta\hat\phi}^{TT} = -2C f_{\theta\phi}. \tag{9.58}$$

Here we have kept only those terms that fall off with $1/r$, dropping higher-order terms.

Exercise 9.5 Use equation (9.30), without averaging over wave cycles, to show that the total energy emitted by an even-parity $l = 2$, $m = 0$ wave is

$$\frac{dE}{dt} = -\frac{6}{5}r^2\dot{C}^2 = -\frac{3}{40}\left(F^{(5)}\right)^2, \tag{9.59}$$

provided the $F^{(4)}/r$ term constitutes the dominant contribution to C at large r.

Similarly, the odd-parity or axial metric takes the form

$$ds^2 = -dt^2 + dr^2 + (2Kd_{r\theta})r\,dr\,d\theta + (2Kd_{r\phi})r\sin\theta\,dr\,d\phi + (1 + Ld_{\theta\theta})r^2d\theta^2$$
$$+ (2Ld_{\theta\phi})r^2\sin\theta\,d\theta\,d\phi + (1 + Ld_{\phi\phi})r^2\sin^2\theta\,d\phi^2. \tag{9.60}$$

Here we construct the coefficients K and L from a function

$$G = G_1(t - r) + G_2(t + r) \tag{9.61}$$

and its derivatives

$$G^{(n)} \equiv \left[\frac{d^n G_1(x)}{dx^n}\right]_{x=t-r} + (-1)^n\left[\frac{d^n G_2(x)}{dx^n}\right]_{x=t+r} \tag{9.62}$$

according to

$$K = \frac{G^{(2)}}{r^2} + \frac{3G^{(1)}}{r^3} + \frac{3G}{r^4}, \tag{9.63}$$

$$L = \frac{G^{(3)}}{r} + \frac{2G^{(2)}}{r^2} + \frac{3G^{(1)}}{r^3} + \frac{3G}{r^4}. \tag{9.64}$$

The angular functions d_{ij} are again listed in the order $m = \pm 2, \pm 1, 0$, yielding

$$d_{r\theta} = 4\sin\theta\begin{pmatrix}\cos 2\phi\\\sin 2\phi\end{pmatrix}, \quad -2\cos\theta\begin{pmatrix}\cos\phi\\\sin\phi\end{pmatrix}, \quad 0,$$

$$d_{r\phi} = -4\sin\theta\cos\theta\begin{pmatrix}\sin 2\phi\\-\cos 2\phi\end{pmatrix}, \quad -2(\cos^2\theta - \sin^2\theta)\begin{pmatrix}\sin\phi\\-\cos\phi\end{pmatrix}, \quad -4\cos\theta\sin\theta,$$

$$d_{\theta\theta} = -2\cos\theta\begin{pmatrix}\cos 2\phi\\\sin 2\phi\end{pmatrix}, \quad -\sin\theta\begin{pmatrix}\cos\phi\\\sin\phi\end{pmatrix}, \quad 0,$$

$$d_{\theta\phi} = (2 - \sin^2\theta)\begin{pmatrix}\sin 2\phi\\-\cos 2\phi\end{pmatrix}, \quad \cos\theta\sin\theta\begin{pmatrix}\sin\phi\\-\cos\phi\end{pmatrix}, \quad -\sin^2\theta,$$

$$d_{\phi\phi} = 2\cos\theta\begin{pmatrix}\cos 2\phi\\\sin 2\phi\end{pmatrix}, \quad \sin\theta\begin{pmatrix}\cos\phi\\\sin\phi\end{pmatrix}, \quad 0. \tag{9.65}$$

As we have seen in exercise 9.4, we can construct superpositions of ingoing and outgoing waves in vacuum that at $t = 0$ are time symmetric with $K_{ij} = 0$. If the amplitude $\mathcal{A}$ of the wave is small ($\mathcal{A} \ll 1$), the metric for the complete spacetime is given by the linearized solutions (9.48) or (9.60) to equation (9.44). For a nonlinear wave, the $3 + 1$

equations must be solved numerically to determine the spacetime. However, a time-symmetric waveform at $t = 0$ automatically satisfies the momentum constraint (2.133). To completely specify the initial data for such a nonlinear wave, we only need to solve the Hamiltonian constraint (2.132).[12] Linearized waves provide useful analytic solutions for testing initial value routines and $3 + 1$ vacuum evolution codes designed to handle nonlinear problems.

9.2 Sources

The gravitational waves that we hope to observe on Earth have exceedingly small amplitudes.[13] For a binary, for example, we can evaluate equation (9.27) to estimate the typical gravitational wave strain h. For equal-mass binaries we have $\mu = M/4$, so

$$h \simeq \frac{4}{r} \frac{\mu M}{R} \simeq 5 \times 10^{-20} \left(\frac{1 \text{ Mpc}}{r} \right) \left(\frac{M}{M_\odot} \right) \left(\frac{M}{R} \right). \tag{9.66}$$

For a binary of stellar-mass black holes with $M = 10 M_\odot$, for example, located in the Virgo cluster at a distance of about 20 Mpc, at a binary separation of about[14] $R = 6M$, we see that the strain h is smaller than about 10^{-20}. Similar estimates hold for other stellar sources of gravitational radiation. Clearly it is a formidable challenge to observe this radiation, and we will return to this issue in Section 9.3. In the meantime we recognize that for an astrophysical object of a given mass M to emit strong gravitational radiation, its compaction M/R needs to be large. This requirement singles out compact objects as the most promising stellar sources of gravitational radiation.

Of course, there may also be other astrophysical sources of gravitational radiation that are not related to stars. Processes in the early Universe, for example, may generate gravitational radiation that may still be observable today, not unlike the cosmic microwave background. Potential origins of such a radiation include primordial fluctuations in the Universe's geometry, amplified during inflation, and phase transitions. String theory also predicts possible mechanisms for the generation of gravitational radiation, including vibrating cosmic strings and the condensation of branes. Finally, one might suspect that there are sources of detectable gravitational radiation of which we are unaware at the time of the writing of this book. Every time another form of radiation has opened up a new window for us to view the Universe, we have been surprised by some unexpected observations. These observations, in turn, have led to a new and deeper understanding of the Universe. It is entirely possible that nature will reward us with a similar surprise when we are able to observe and measure gravitational radiation at sufficiently high sensitivities.

---

[12] Time-symmetric, nonlinear wave solutions constructed in this way in axisymmetry are often refered to as *Brill waves* (see exercise 3.5).

[13] Our discussion in this section is drawn from Flanagan and Hughes (2005), who provide a more detailed overview as well as numerous references.

[14] As we will see in Chapters 12 and 13, this binary separation is close to the "innermost stable circular orbit", at which the binary inspiral becomes increasingly rapid and decreasingly circular.

Different sources of gravitational radiation emit waves at vastly different frequencies. It is useful to divide this large spectrum of gravitational radiation into different frequency bands. The *high frequency band* includes frequencies in the range $1 \text{ Hz} \lesssim f \lesssim 10^4$ Hz, the *low frequency band* covers the range $10^{-5} \text{ Hz} \lesssim f \lesssim 1$ Hz, the *very low frequency band* spans $10^{-9} \text{ Hz} \lesssim f \lesssim 10^{-7}$ Hz, and the *ultra low frequency band* covers $10^{-18} \text{ Hz} \lesssim f \lesssim 10^{-13}$ Hz.

The partition into these different bands is motivated in part by the means by which we detect them, as we will discuss below. It is also motivated by the different classes of potential sources. To distinguish the later, we shall estimate the characteristic gravitational wave frequency from a stellar object of mass M, radius R and compaction M/R by[15]

$$f \simeq \frac{1}{M}\left(\frac{M}{R}\right)^{3/2} \simeq 2 \times 10^5 \text{ Hz} \left(\frac{M_\odot}{M}\right)\left(\frac{M}{R}\right)^{3/2}. \tag{9.67}$$

The highest frequency sources are compact objects with large compactions (black holes or neutron stars) and small masses; stellar-mass compact objects fall into this category. These sources fall into the high frequency band. Objects with either larger masses (supermassive black holes) or smaller compactions (white dwarfs, or binaries with large binary separation) radiate in the low frequency band. Other stellar sources may radiate with even lower frequencies; however, the strain of their radiation is so weak that this radiation is not interesting from an observational perspective. That leaves nonstellar sources as the most interesting sources of gravitational radiation in the very low and ultra low frequency bands.

9.2.1 The high frequency band

The high frequency band includes frequencies in the approximate range $1 \text{ Hz} \lesssim f \lesssim 10^4$ Hz. This frequency band is observable with the new generation of ground-based gravitational wave interferometers, which we will discuss in Section 9.3. The upper limit of this band corresponds to the highest frequency of gravitational radiation that we may expect from stellar sources. As discussed above, we can estimate this limit from equation (9.67) for stellar-mass compact objects with large compactions, i.e., neutron stars or stellar-mass black holes. The lower limit of this band is set by our ability to dectect gravitational radiation with ground-based interferometers. For smaller frequencies, it becomes increasingly difficulty to decouple the signal from the noise arising from low-frequency vibrations, both mechanical and gravitational, on the ground. Curiously, the high frequency band coincides more or less with the audible range of human hearing; if converted to sound, we would be able to hear the gravitational radiation in this band.

Promising sources in this frequency band include compact, stellar-mass binaries, stellar core collapse, rotating neutron stars, and stochastic backgrounds.[16]

---

[15] For binaries this relation follows directly from Kepler's third law. For other objects we can estimate their characteristic frequency from their dynamical time scale: $f \simeq 1/\tau_{\text{dyn}} \simeq \sqrt{\rho}$, which again results in equation (9.67).

[16] We refer to Cutler and Thorne (2002), both for a more detailed discussion of these and other sources and a perspective on the possibility of their detection.

Compact binaries

Compact binaries – binary neutron stars, binary black holes, and binary black hole-neutron stars – are probably the most promising sources of gravitational radiation for ground-based gravitational wave detectors. Very simple estimates of the emitted gravitational wave signals can be obtained from exercises 9.2, 9.3 and 12.3. More accurate predictions applicable for close, mildly relativistic orbits require post-Newtonian calculations; see Appendix E for a brief summary of these calculations. The highly relativistic late inspiral, plunge and merger phases of the orbit require full numerical simulations for the evolution and emitted gravitational radiation. Such numerical simulations will be described in detail in Chapters 12–13 and 15–17.

Compact binaries are known to exist, at least in the case of binary neutron stars. Since the discovery by Hulse and Taylor (1975) of PSR 1913+16, a binary neutron star system containing a radio pulsar, a number of similar systems have been identified.[17] While several of these binaries will coalesce within a Hubble time[18] none have small enough binary separation to be observable with the current generation of gravitational wave detectors. However, for some binaries the gravitational wave back-reaction is strong enough for us to measure its effect on the binary orbit. By monitoring the orbit of PSR 1913+16 by radio pulsar techniques, Hulse and Taylor were able to confirm for the first time the validity of equation (9.37) and, hence, the rate at which gravitational radiation carries off energy as predicted by general relativity in the weak-field, slow-velocity limit.

We can use a statistical analysis based on the known sample of observed binary neutron stars to estimate the rate of binary neutron star coalescence per Milky Way-type galaxy. Since this sample is rather small, the estimates are not very rigorous, but presumably they improve with each new discovery of a binary neutron star system.[19] To date, no binary black hole or black hole–neutron star system has been discovered, so we cannot perform such an analysis to estimate their merger rates.

An alternative way of estimating compact binary coalescence rates is to model the evolution of stellar populations. These "population synthesis" calculations rely on theoretical models for the late stages of stellar evolution and the formation of compact binaries and have significant uncertainties. Observational constraints can sharpen these population synthesis estimates, yielding[20] merger rates of binary neutron stars of approximately 10^{-5}–10^{-4} per year per Milky Way-type galaxy, 10^{-6}–10^{-5} for binary black hole-neutron stars, and 10^{-7}–10^{-6} for binary black holes. Thus, to have an appreciable detection rate, gravitational wave detectors must be able to observe mergers out to sufficiently large distances so that the volume they survey contains many galaxies. We shall discuss the sensitivities of the current generation of detectors in Section 9.3.

---

[17] See, e.g., Stairs (2004) for a review.
[18] At least seven such neutron star binaries have been discovered via radio pulsar observations at this time. They are J0737−3039 (a double pulsar), J1518+4904, B1534+12, J1811−1736, J1829+2456, B1913+16 (the Hulse–Taylor binary), and B2127+11C; see Stairs (2004).
[19] See Burgay *et al.* (2003).
[20] Kalogera *et al.* (2007).

One payoff of measuring gravitational waves from compact binaries is that we can use the wave data to perform "stellar spectroscopy". Specifically, once the data are analyzed using theoretical templates and "matched-filtering" techniques (see Section 9.3), they can provide direct measurements of the masses and spins of the inspiraling companions. These gravitational wave measurements are the only way we can obtain these parameters for binary black holes, since they emit no other form of radiation. Unlike other means used to estimate the masses and spins of black holes in gaseous environments, gravitational wave measurements from binary inspiral do not have the complications and uncertainties associated with the modeling of nonvacuum physics near black holes, such as turbulent MHD accretion or electromagnetic radiation transport. The accuracy to which binary parameters can be determined from a gravitational wave detection depends on several factors, including the masses of the binary companions themselves.[21] For a "typical" observation of a binary black hole system consisting of two 10 $M_\odot$ black holes with ground-based gravitational wave detectors (see Section 9.3.1), an analysis of the the lowest-order quadrupole radiation emitted during the inspiral phase[22] is sufficient to determine the so-called *chirp* mass $(m_1 m_2)^{3/5}(m_1 + m_2)^{-1/5}$ to within a fraction of a percent.[23] Determining the individual masses m_1 and m_2 requires additional information, for example the reduced mass, which appears at the next post-Newtonian order. Unfortunately, the effects of black hole spin also appear in these post-Newtonian terms. Measuring these quantities individually therefore requires some way of breaking the degeneracy. At least in principle, one approach may involve observations of the binary merger. As we will see in Chapter 13, the black hole spins do have an important effect on the dynamics of the binary black hole mergers; matching observed gravitational wave signals from the merger phase to numerical simulations may therefore hold the key to determining the individual black hole parameters.

The expectation that compact binaries are significant sources of detectable gravitational radiation, together with their unique role as fundamental probes of general relativity, make compact binaries of crucial importance for gravitational wave physics and astronomy. This is why we devote so much attention to these objects in this book.

Stellar core collapse

Stellar core collapse, which occurs in a Type II supernova, for example, may also emit a strong gravitational wave signal. Core collapse will lead to the formation of a neutron star or, for the most massive progenitors, a black hole. Such an event involves a large amount of mass moving very rapidly in a very small volume. Just how much gravitational

---

[21] The mass of the binary affects the frequency range of the emitted gravitational wave pattern (equation 9.67), and hence it determines whether or not the signal is received in a part of spectrum at which gravitational wave detectors are sensitive; compare the gravitational wave "noise-curves" in Section 9.3.

[22] See Chapter 12.1, including Figure 12.1 and exercise 12.3.

[23] See, e.g., Cutler and Flanagan (1994); Cutler and Thorne (2002).

radiation is emitted nevertheless depends on how much the collapse deviates from spherical symmetry. If the progenitor star has very little rotation, then the collapse will proceed almost spherically, and only a very small amount of gravitational radiation will be emitted. If the progenitor star rotates more rapidly, however, then the collapsing core will be deformed and this will induce a time-varying quadrupole moment and, hence, a potentially significant burst of gravitational radiation.

A rotating star in equilibrium is oblate – even the Earth has a slightly larger radius at the equator than at the poles. The degree of oblateness increases as the ratio of the rotational kinetic energy $T = I\Omega^2/2 = J^2/(2I)$ to the gravitational binding energy $|W|$ increases. Here I is the moment of inertia, $J = I\Omega$ the angular momentum, and W is the gravitational potential energy of the star.[24] During the collapse the mean radius of the core, R, decreases. Since $I \propto R^2$ and $|W| \propto 1/R$, we find

$$\frac{T}{|W|} \propto \frac{1}{R}. \tag{9.68}$$

The above relationship implies that, assuming conservation of angular momentum, the ratio $T/|W|$ increases during the collapse. Even a modestly oblate progenitor star will therefore become increasingly oblate following collapse. Such a collapse leads to a gravitational wave burst signal. It would be particularly exciting to observe such a signal in conjunction with an electromagnetic or a neutrino counterpart.

It is also possible that the collapsed core does not remain axisymmetric. If the ratio $T/|W|$ becomes sufficiently large, nonaxisymmetric instabilities may be triggered. Best understood is the so-called bar-mode instability (see Chapters 14.2.2 and 16.3), which deforms the star into a triaxial ellipsoid, i.e., the shape of an American football. Typically, the critical value of $T/|W|$ for this to happen is about 0.14, which is quite large. However, other instabilities – including a single-arm $m = 1$ spiral mode or the r-mode – may develop at smaller values of $T/|W|$. If a bar-mode instability develops in a neutron star remnant following core collapse, the remnant would emit a periodic gravitational wave signal in addition to the initial burst. Depending on its strength, the periodic signal may be much easier to identify via gravitational wave observations, since the periodicity can be exploited to build up a higher signal-to-noise ratio in the output of a gravitational wave detector.

Finally, we mention the possibility of a neutron star remnant ultimately undergoing collapse to a black hole (see Chapter 14.2.1). A number of different scenarios suggest how such a "delayed collapse" could be triggered – by phase transitions as a newly-formed neutron star cools, by accretion fall-back of the stellar mantle, or by changes in the angular velocity profile of the remnant induced by magnetic braking or viscosity. If the remnant is rotating, delayed collapse will result in a delayed burst of gravitational waves, and this

---

[24] For an incompressible, rotating Newtonian star $T/|W| \sim e^2$ for small $e^2 \ll 1$, where $e^2 \equiv 1 - R_{pol}^2/R_{eq}^2$, R_{pol} is the polar radius and R_{eq} is the equatorial radius. See Shapiro and Teukolsky (1983), equation (7.3.24), for the exact expression, as well as Chapters 7.3 and 16.8 in that text for further discussion.

burst will be followed by emission from the quasinormal ringing of the black hole as it settles into stationary equilibrium.

Rotating neutron stars

Rotating neutron stars are also promising sources of gravitational radiation in the high frequency band. We have observed many rotating neutron stars as radio and X-ray pulsars, so we know not only that they exist, but also where they are located. Also, such a star emits a periodic gravitational wave signal at a nearly constant, known frequency, which serves to increase the signal-to-noise ratio as mentioned above.

A perfectly axisymmetric, rotating neutron star in stationary equilibrium emits no gravitational radiation. For the star to radiate gravitational waves, its axisymmetry has to be broken. A number of different mechanisms have been suggested that could accomplish this: an ultra-strong internal magnetic field that is not aligned with the axis of rotation, accretion pile-up from a companion star, or a small "mountain" that emerges on the crust from a starquake.[25] It is difficult to predict how large a deformation to expect from these mechanisms, making it somewhat uncertain how strong we may expect the gravitational wave signal to be. Current gravitational wave detectors already provide useful upper limits on the gravitational wave emission from a few rapidly rotating pulsars, and hence, on their nonaxisymmetric deformation.[26]

Stochastic backgrounds

Random, independent and uncorrelated sources that we cannot resolve observationally give rise to so-called "stochastic background" radiation. Stochastic backgrounds include the gravitational radiation that may have been produced in the early Universe, for example by phase transitions, from quantum fluctuations arising during inflation or from cosmic strings.[27] Stochastic background radiation could exist at all frequencies, and thus spans all the frequency bands considered here. Physically useful upper limits on stochastic background radiation in the high frequency regime already can be established with current gravitational wave detectors.[28]

9.2.2 The low frequency band

The low frequency band between 10^{-5} Hz $\lesssim f \lesssim 1$ Hz cannot be observed with ground-based detectors, since their coupling – both mechanically and gravitationally – to a host of

---

[25] See, e.g., Shapiro and Teukolsky (1983), Chapter 10.5 and 10.11 for discussion and references.
[26] See The LIGO Scientific Collaboration: B. Abbott *et al.* (2007).
[27] See, e.g., Peacock (1999) and Flanagan and Hughes (2005) for further discussion and references.
[28] See The LIGO Scientific Collaboration: B. Abbott (2007).

Earth-related sources of noise is too strong. This leaves space-based observatories as the best means of detecting such radiation, and we will describe one such planned instrument, LISA, in Section 9.3.

As we can infer from equation (9.67), promising stellar sources in the low frequency regime include both high compaction, supermassive objects, such as a supermassive black hole binaries, and low compaction, stellar-mass objects, such as white dwarf binaries. As discussed above, there also may be stochastic background radiation present in the low frequency band, but here we will focus only on stellar sources.

Supermassive black holes

Supermassive black holes ($M \sim 10^6$–$10^9 M_\odot$) are believed to be the engines that power quasars and active galactic nuclei. Most, if not all, nearby bulge galaxies host a supermassive black hole at their centers. Black holes can grow to supermassive scale in the early Universe by a combination of accretion and binary mergers.[29] Because supermassive black hole masses are well-correlated with the velocity dispersions of stars in the central bulges of their host galaxies, as well as with the total bulge masses, it appears that supermassive black hole formation and growth are intimately connected with galaxy formation.[30]

Binaries containing supermassive black holes are promising sources of quasiperiodic, low frequency gravitational radiation. They can be divided into two broad categories. One category consists of two supermassive black holes of comparable mass in binary orbit. The other category consists of a supermassive black hole with a stellar-mass companion, sometimes referred to as an "extreme mass ratio inspiral", or EMRI, binary.

Binaries containing two supermassive black holes of comparable mass can form when two galaxies collide and merge. Galaxy mergers are believed to provide an important mechanism for the hierachical build-up of large-scale structure in the early Universe. Such mergers continue to be observed in the present epoch. Thus, the formation of massive and supermassive binary black holes may not be uncommon.[31]

From a computational point of view, the inspiral and coalescence of two supermassive black holes is identical to that of two stellar-mass black holes. We will discuss general relativistic simulations of binary black hole coalescence in detail in Chapter 13. LISA, or a detector like it, may be able to observe supermassive black hole mergers up to fairly high redshift (possibly $z \sim 5$–10), whereby we can use gravitational waves to probe structure formation in the early Universe and provide insight into how supermassive black holes form in the first place.

---

[29] One quasar, QSO SDSS 1148+5251, observed at redshift $z = 6.43$ (Fan, X. *et al.* (2003)), is believed to host a supermassive black hole that has grown to a mass of about $10^9 M_\odot$ within 0.9 Gyr after the Big Bang; see Shapiro (2005) and references therein.

[30] See articles in Ho (2004) for an overview and references.

[31] See Sesana *et al.* (2007) for the tentative implications for LISA.

EMRI binaries form when a supermassive black hole captures a smaller, stellar-mass object. These binaries are promising sources of gravitational radiation if the binary is compact, i.e., if the stellar-mass object is a black hole, neutron star or white dwarf. For these compact EMRI binaries, tidal effects on the low-mass companion are small and the stars are completely intact when they are swallowed by the supermassive black hole. The challenge in modeling these binaries lies in correctly accounting for the gravitational radiation reaction force, which causes a gradual deviation of the object's orbit from a pure geodesic. Observing such a binary may provide our best opportunity to map out the strong-field geometry of a Kerr black hole. Tentative estimates suggest that LISA may be able to observe at least several of these systems per year.

We also mention the possibility that LISA may detect intermediate-mass black holes with masses somewhere between stellar-mass and supermassive black holes (i.e., $M \sim 10^2$–$10^4 M_\odot$). There have been some tentative observational suggestions that such black holes may exist.[32] Since binaries involving intermediate-mass black holes would also radiate in the low frequency band, gravitational wave observations of, say, a binary containing one or more intermediate-mass black hole could provide unambiguous evidence of their existence.

White dwarf binaries

White dwarf binaries constitute a well established and understood group of weak-field sources in the low frequency band. These systems emit gravitational radiation that is strong enough to be detected by proposed detectors like LISA, but weak enough that the radiation reaction force on the binary orbit is very small, making the signal almost perfectly periodic over typical observation timescales. In fact, these sources are so well understood that they can be used to calibrate LISA once the instrument is operational.

From a "new discovery" point of view, these sources are not so exciting as some of the strong-field sources we have described. Since we already understand these binaries fairly well, there is not much we can learn about them from gravitational radiation. We may nevertheless discover many more of these binaries, and, in addition, there are some aspects of known binaries that we may be able to probe better with the help of gravitational waves. For example, the orbit's inclination angle, which is difficult to determine from other observations, can be deduced from the relative amplitudes of the polarization waveforms h_+ and $h_\times$.

It is expected that LISA will be able to detect so many periodic binary sources that they will form a stochastic background noise. This background is likely to exceed the instrument's intrinsic noise at low frequencies $f \lesssim 10^{-3}$ Hz.

9.2.3 The very low and ultra low frequency bands

The timescales associated with the very low frequency band, 10^{-9} Hz $\lesssim f \lesssim 10^{-7}$ Hz, correspond to a few months to a few decades. These limits are set by the observational

---

[32] See van der Marel (2004) and references therein.

technique – pulsar timing – that has been used already to establish upper limits on the gravitational radiation in this band. Radio pulsars are extremely precise clocks, but the pulses' arrival times on Earth may be modulated by the presence of gravitational waves. Making these measurements sufficiently accurate requires integrating the pulsar data for at least a few months, which sets the higher limit of the frequency band; on the other hand, we have observed pulsars only for a few decades, which sets the lower limit. Pulsar timing analysis has been used to establish upper limits both on stochastic background radiation and on supermassive black hole binaries that have such large masses or binary separations that they radiate in the very low frequency band.

The ultra low frequency band spans frequencies of approximately 10^{-18} Hz $\lesssim f \lesssim 10^{-13}$ Hz, or wavelengths that are comparable to the Universe's Hubble length. Waves in this frequency band may be generated by quantum fluctuations in the early Universe and are amplified during inflation. Measuring the amplitude of this gravitational radiation would therefore probe the inflation epoch and help discriminate between competing models. Gravitational waves in this frequency range leave an imprint in the cosmic microwave background, and may therefore be detected indirectly by analyzing the cosmic microwave background and its polarization.[33]

9.3 Detectors and templates

In many ways, gravitational wave detection is more like hearing than seeing.[34] In most other astronomical observations we detect photons, which behave very differently from their gravitational analogs. Photons typically have wavelengths that are much shorter than the emitting object, so that we can create images. Gravitational waves, on the other hand, have wavelengths that are larger than or at least comparable to the size of the emitting object. That means that we cannot use gravitational waves to create an image of the emitting object. In analogy to hearing we cannot even locate a gravitational wave source in the sky with just one detector. This makes it so important to operate a number of different gravitational wave detectors, spread far apart over the Earth or in space.

Photons are also emitted incoherently from very small regions within the emitting object, usually from atoms or electrons, and we therefore observe the radiation's intensity – which measures the time-average of the square of the wave amplitude – rather than the individual waveform itself. By contrast, gravitational waves are created coherently by the bulk motion of the emitting object, and we observe the gravitational waveform directly. This difference has two very important consequences.

The first consequence is related to the fact that the wave amplitude falls off with one over the distance from the emitting object, while the intensity falls of with one over the square of the distance. That means that an increase of a factor of 2, say, in the sensitivity of

---

[33] See, e.g., Smith *et al.* (2006).
[34] Flanagan and Hughes (2005) explore this analogy in detail.

our gravitational wave detector doubles the distance out to which we can observe certain objects. The total observable volume of the Universe then increases by a factor of eight – meaning that doubling the sensitivity increases the expected event rate by almost a factor of 10! For instruments that measure the intensity of electromagnetic radiation, the corresponding increase in the event rate is smaller by the square root of eight.

The second consequence is that gravitational wave astronomy benefits from theoretically predicted waveforms. Gravitational wave detectors measure gravitational amplitudes directly, as a function of time, and this measurement can be compared with theoretical models. Clearly such a comparison will be necessary for the physical interpretation of any observed signal. In addition, "matched-filtering" techniques – in which the noisy output of the detector is compared with a catalog of theoretical gravitational waveform templates – dramatically increase the likelihood of identifying a particular signal. Using this technique, the distance out to which an object can be observed increases approximately with the square root of the number of wave cycles – which again may increase the effective event rate significantly.

The first gravitational wave detector was a bar detector constructed by Joseph Weber, one of the pioneers of gravitational wave astronomy. Several bar detectors are still operational. These detectors have a high sensitivity only in a very narrow frequency range. While these detectors are probably not as promising as the new generation of gravitational wave interferometers, they may provide some useful and important complementary information.

Gravitational wave interferometers currently come in two different types: ground-based and space-based. As we discussed in Section 9.2, the ground-based detectors are designed to observe gravitational wave sources in the high frequency band (Section 9.2.1), while space-based detectors will be able to detect sources in the low frequency band (Section 9.2.2).

9.3.1 Ground-based gravitational wave interferometers

Ground-based gravitational wave detectors are giant Michelson–Morley interferometers. As illustrated in Figure 9.2, the basic idea is to send a laser beam on a beam splitter that splits the beam into two. The two beams are then sent down two orthogonal vacuum tubes, or "arms", reflected by mirrors at their end, and reassembled upon returning to the beam splitter, possibly after several return trips through the arms. A passing gravitational wave will distort the relative length of the two arms according to equation (9.9),[35] and will therefore modify the interference pattern of the two returning light beams.

The challenge, of course, lies in the fact that gravitational radiation is so weak. As we have seen from equation (9.66), even a reasonably optimistic estimate predicts a gravitational wave strain in the order of $h \sim 10^{-20}$. From (9.9) we then have $\delta\xi \sim 10^{-20}\xi$; inserting a few kilometers for ξ (which is the arm length for the largest ground-based detectors),

---

[35] Recall the effect of the two gravitational wave polarizations illustrated in Figure 9.1.

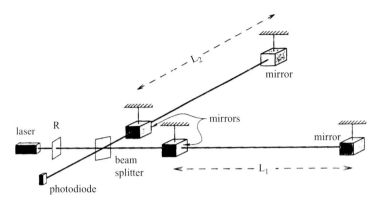

Figure 9.2 Schematic diagram of a Michelson–Morley interferometer as used in laser interferometer gravitational wave detectors. [From Abramovici *et al.* (1992).]

we still end up with a dauntingly small distortion, similar in size to a small fraction of the nucleus of the hydrogen atom. To measure such a small effect is an extraordinary proposition indeed! Undeterred by these challenges a generation of gravitational wave pioneers have not only designed the necessary instruments, but also demonstrated that these instruments are capable of detecting such exceedingly small distortions.

Several ground-based detectors are currently either operational or in the planning phase. The Laser Interferometer Gravitational-wave Observatory – LIGO – in the US operates three interferometers: one interferometer with an arm length of four kilometers in Livingston, Lousiana, and a pair of interferometers with arm lengths of two and four kilometers in Hanford, Washington. Figure 9.3 shows an areal view of the Hanford facility. VIRGO is a joint French–Italian instrument, located in Pisa, Italy, with an arm length of 3 km. GEO600 is a detector with an arm length of 600 m, operated by a English–German collaboration near Hannover, Germany. While this instrument has a slightly smaller arm length, it uses a very advanced interferometer technology that makes its sensitivity comparable to those with longer arm lengths. It is also being used as a test-bed for advanced technologies to be used in the next generation of larger detectors. TAMA300, a 300-m interferometer located close to Tokyo, Japan, was the first operational, large-scale gravitational wave interferometer, and a successor with a longer arm length is being planned. The only gravitational wave interferometer in the southern hemisphere, ACIGA, is being constructed near Perth in Australia. The current instrument is an 80-meter interferometer, but the hope is that it can be extended to a multi-kilometer instrument some time in the future.

Figure 9.4 shows noise curves $\tilde{h}(f)$ for the LIGO instruments, together with the signal strengths $\tilde{h}_s(f)$ for several different gravitational wave sources that we discussed in Section 9.2. Ignoring for simplicity the exact definition of these quantities[36] we merely note that a point on the signal curve that lies above a noise curve in the graph can be

---

[36] See Cutler and Thorne (2002).

Figure 9.3 The LIGO observatory in Hanford, Washington. [From LIGO website, `http://www.ligo-wa.caltech.edu.`]

detected with the corresponding instrument with a false-alarm probability of less than one percent.

The upper curve in Figure 9.4 is the design noise curve for LIGO-I, which already has been achieved by the LIGO collaboration. The graph then indicates that with this instrument we would be able to detect a black hole binary inspiral of two 10 $M_\odot$ black holes at a distance of 100 Mpc. Whether or not we will indeed observe such an inspiral with this generation of gravitational wave interferometers depends primarily on nature's generosity, i.e., on whether or not such an inspiral and merger happens to occur within this distance while the detectors are operating.

Figure 9.4 also includes two noise curves for the next-generation of detector, LIGO-II or Advanced LIGO. One of these curves, labeled "WB LIGO-II", is basically the improved "wide-band" version of LIGO-I with reduced noise sources. However, LIGO-II will also have the capability of fine-tuning the noise-curve to a narrow band of promising sources. The figure shows an example, labeled "NB LIGO-II", that is fine-tuned to typical frequencies in low-mass X-ray binaries. Clearly, LIGO-II should be able to detect many more potential gravitational wave sources than its predecessor LIGO-I.

9.3.2 Space-based detectors

As we have seen in Section 9.2, gravitational waves in the low frequency band cannot be observed with ground-based detectors. Observing gravitational radiation in this frequency

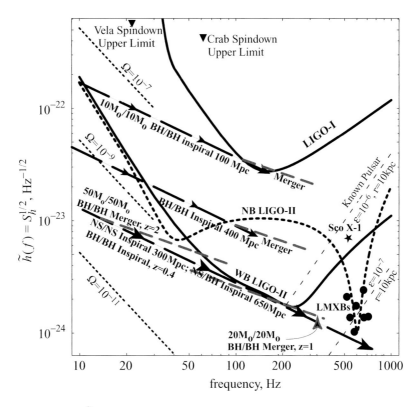

Figure 9.4 The noise curve $\tilde{h}(f)$ for several planned LIGO Interferometers. Also included is the signal strength $\tilde{h}_s(f)$ for a number of promising sources in the high frequency band (see Section 9.2.1). Whenever a point on the signal curve lies above the interferometer's noise curve, the signal is detectable with a false alarm probability of less than one percent. [From Cutler and Thorne (2002).]

band therefore requires space-based instruments. One such instrument, the Laser Interferometer Space Antenna – LISA – is currently being planned as a joint venture between the US space agency NASA and its European counterpart ESA.

As currently envisioned, LISA will be made up of three spacecraft located at the corners of a triangle with sides of length of about 5×10^6 km. Each spacecraft houses two lasers that send laser beams along the sides of the triangle. In essence, each side of the triangle functions as the arm of a large interferometer. As illustrated in Figure 9.5, the instrument as a whole will be placed in a solar orbit, trailing the Earth in its orbit about the Sun by about $20°$. To maintain an approximately equal distance between the spacecraft over the course of a year, each spacecraft is placed into a slightly eccentric orbit, with a phase difference of $120°$ between each one of them. The resulting motion of an individual spacecraft is illustrated in the bottom panel of Figure 9.5. The whole configuration therefore exhibits a "rolling" motion, which turns out to be very useful for locating sources in the sky.

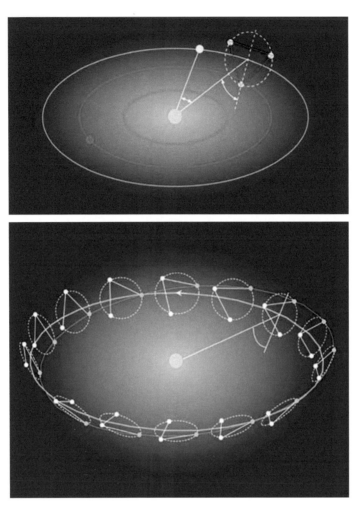

Figure 9.5 LISA's orbit about the Sun. The top image shows how the triangle of three LISA spacecrafts trails Earth's orbit about the Sun by $20°$. The bottom image follows one of the three satellites' orbit about the Sun, demonstrating how the triangle's orientation changes during the course of a year. [From LISA web site, `http://lisa.nasa.gov.`]

In Figure 9.6 we show the projected noise curve for LISA, together with some of the potential low frequency band sources that we discussed in Section 9.2.2. At low frequencies, the noise arising from unresolved white dwarf binaries is expected to exceed LISA's intrinsic instrument noise. Evidently LISA has great potential to detect and discover extremely interesting objects.

Another space-based detector is the gravitational wave antenna DECIGO (Deci-hertz Interferometer Gravitational Wave Observatory), proposed by scientists in Japan. This instrument will bridge the gap between LISA and terrestrial detectors like LIGO, as it is most sensitive to gravitational waves in the frequency range $f \sim 0.1$–10 Hz.

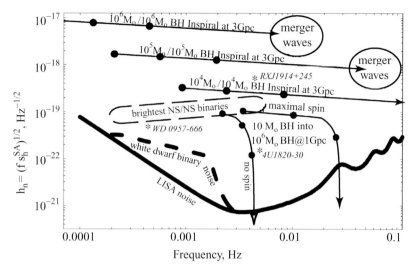

Figure 9.6 Projected noise curve $h_n(f)$ for LISA, together with a number of promising souces of gravitational radiation in the low frequency band (see Section 9.2.2). The thick line is LISA's instrument noise; the thick-dashed curve is the noise curve incorporating the expected stochastic background of gravitational radiation from unresolved white-dwarf binaries. Wave strengths above these curves have sufficiently high signal-to-noise ratio that they should be detectable using optimal signal processing. [From Cutler and Thorne (2002).]

9.4 Extracting gravitational waveforms

We now turn to the topic that is most important from a numerical relativity perspective: the extraction of gravitational waveforms in a numerical simulation. One of the major motivations for performing numerical relativity simulations is the accurate calculation of gravitational waveforms from promising sources in order that these theoretically computed signals can be compared with observational data from gravitational wave detectors. As we discussed in Section 9.3, such a comparison will be needed not only for the physical interpretation of any observed data, but also to increase significantly the liklihood of a detection in the first place. The latter can be accomplished by employing a "matched-filtering" technique that compares observed signals with templates of theoretical waveforms.

Far from the sources, gravitational radiation is weak and can be described very conveniently in terms of the formalism introduced in Section 9.1. In particular, the wave information can be expressed in terms of two polarization amplitudes in the TT gauge. Numerical relativity simulations, however, focus on the strong-field regime of the sources, and compute a $3 + 1$ spacetime metric decomposed in terms of a lapse function, shift vector and spatial 3-metric. The functional form of this spacetime metric is strongly dependent on the chosen coordinate system. Under some special gauge conditions, we can directly read off the wave polarization amplitudes h_+ and $h_\times$ from the evolved metric.[37] In general,

---

[37] See Shibata (1999a,b) for an example.

however, it is not trivial to extract, in a gauge-invariant way, the linearized wave quantities that we introduced in Section 9.1.

In this section we focus on two different strategies for this gravitational wave extraction.[38] The first approach, described in Section 9.4.1, employs the gauge-invariant Moncrief formalism, and involves decomposing the metric into spherical harmonics and then combining the latter in a gauge-invariant way. The second method, summarized in Section 9.4.2, is based on the Newman–Penrose formalism and uses Weyl scalars, which are computed from a projection of the Weyl tensor onto a null tetrad.

9.4.1 The gauge-invariant Moncrief formalism

The Moncrief formalism is based on a perturbative decomposition of the metric.[39] In particular, we assume that at a large distance from any source we can decompose the spacetime metric g_{cd} into a background metric g_{cd}^B, which we take to be the Schwarzschild metric, and a perturbation h_{cd},

$$g_{cd} = g_{cd}^B + h_{cd}. \tag{9.69}$$

The basic idea is then the following: As we did for the linearized wave solutions (9.48) and (9.60), we split the perturbative metric h_{cd} into even- and odd-parity parts h_{cd}^e and h_{cd}^o, and then decompose both in terms of spherical harmonics,

$$h_{cd} = \sum_{l=2}^{\infty} \sum_{m=-l}^{l} (h_{cd}^{e\,lm} + h_{cd}^{o\,lm}). \tag{9.70}$$

For each mode we can form a gauge-invariant *Moncrief function* that satisfies a certain wave equation. From these Moncrief functions we can then determine the gravitational wave forms h_+ and $h_\times$.

The decomposition into spherical harmonics involves scalar, vector and tensor spherical harmonics, which we review in Appendix D. As a brief reminder, any scalar on the unit sphere can be decomposed into scalar spherical harmonics Y_{lm}. To represent a vector on the unit sphere, we need two basis vectors E_a^{lm} and S_a^{lm}, which we can express in terms of derivatives of the scalar spherical harmonics on the unit sphere. To represent a symmetric, traceless, rank-2 tensor on the unit sphere we require two basis tensors, Z_{ab}^{lm} and S_{ab}^{lm}, which we can express in terms of second derivatives of the scalar spherical harmonics. As we might expect, the decomposition of the even-parity perturbations will involve those spherical harmonics with even parity (namely the scalar spherical harmonics Y_{lm}, the vector spherical harmonics E_a^{lm}, and the tensor spherical harmonics Z_{ab}^{lm}), while the decomposition

---

[38] We are greatly indepted to Y. T. Liu, who provided useful notes for this section.
[39] In this section we will present most results without proof; we refer to the original paper by Moncrief (1974), as well as the review by Nagar and Rezzolla (2005), for more details.

of the odd-parity perturbations only involves odd-parity spherical harmonics (the vector spherical harmonics S_a^{lm}, and the tensor spherical harmonics S_{ab}^{lm}).

We now alert the reader to a departure from our general notation convention. Instead of separating spatial indices from spacetime indices, it is more convenient, in the context of the spherical coordinates used in this section, to separate the angular coordinates θ and ϕ from the remaining spacetime coordinates r and t. Following this convention, we will label the former (θ and ϕ) with lower-case letters $a, b, \ldots$, and the latter (t and r) with upper-case letters $A, B, \ldots$ In fact, we have already adopted this convention for the vector and tensor spherical harmonics in the previous paragraph, which only have θ and ϕ components. Similarly, the 2-dimensional metric σ_{ab} on the unit sphere S^2, given by

$$\sigma_{ab}dx^a dx^b = d\theta^2 + \sin^2\theta d\phi^2, \qquad (9.71)$$

has only angular components.

Before discussing the specific decompositions of the even- and odd-parity perturbations we can already anticipate their general character. The completely angular parts of the perturbative metric, h_{ab}^{lm}, will involve tensor spherical harmonics. For the odd-parity perturbations this can only be the S_{ab}^{lm}, but for the even-parity perturbations this could be a combination of the tensor spherical harmonics Z_{ab}^{lm} and a scalar spherical harmonic multiplying the 2-dimensional metric, $Y_{lm}\sigma_{ab}$. Similarly, we will express the mixed angular-nonangular parts of the metric, h_{Ab}^{lm}, as vector spherical harmonics. For the even-parity perturbations this will involve the E_a^{lm}; for the odd-parity parturbations the S_a^{lm}. Finally, the nonangular parts of the metric, h_{AB}^{lm}, decompose like scalars. Since we can express scalars only in terms of the even-parity scalar spherical harmonics Y_{lm}, these terms can be nonzero only for the even-parity perturbations.

Even-parity (polar) modes

As we discussed above, we can express the nonangular parts of the perturbative metric, $h_{AB}^{e\,lm}$, in terms of scalar spherical harmonics. We denote these components as[40]

$$h_{tt}^{e\,lm} = \left(1 - \frac{2M}{r}\right) H_{0lm}(t,r)Y_{lm}(\theta,\phi), \qquad (9.72)$$

$$h_{tr}^{e\,lm} = H_{1lm}(t,r)Y_{lm}(\theta,\phi), \qquad (9.73)$$

$$h_{rr}^{e\,lm} = \frac{H_{2lm}(t,r)}{1 - 2M/r}Y_{lm}(\theta,\phi), \qquad (9.74)$$

where H_{0lm}, H_{1lm} and H_{2lm} are expansion coefficients. The mixed angular-nonangular parts of the metric, $h_{Ab}^{e\,lm}$, are decomposed into electric-type vector spherical

---

[40] Here we follow the notation of Shibata (1999a).

harmonics E_a^{lm},

$$h_{Ab}^{e\,lm} = h_{Alm}(t,r)E_b^{lm}(\theta,\phi), \tag{9.75}$$

where we will use labels 0 for $A = t$ and 1 for $A = r$ in the expansion coefficients h_{Alm}. Finally, we decompose the angular part of the metric $h_{ab}^{e\,lm}$ according to

$$h_{ab}^{e\,lm} = r^2 \left(K_{lm}(t,r)\sigma_{ab}(\theta)Y_{lm}(\theta,\phi) + 2G_{lm}(t,r)Z_{ab}^{lm}(\theta,\phi) \right), \tag{9.76}$$

where the expansion coefficient K_{lm} captures the trace and G_{lm} the traceless part of $h_{ab}^{e\,lm}$. In matrix form, the perturbative metric then appears as

$$h_{\mu\nu}^{e\,lm} = \begin{pmatrix} \Lambda H_{0lm}Y_{lm} & H_{1lm}Y_{lm} & h_{0lm}E_\theta^{lm} & h_{0lm}E_\phi^{lm} \\ \text{sym} & H_{2lm}Y_{lm}/\Lambda & h_{1lm}E_\theta^{lm} & h_{1lm}E_\phi^{lm} \\ \text{sym} & \text{sym} & r^2(K_{lm}Y_{lm} + G_{lm}W_{lm}) & r^2 G_{lm}X_{lm} \\ \text{sym} & \text{sym} & \text{sym} & r^2 \sin^2\theta(K_{lm}Y_{lm} - G_{lm}W_{lm}) \end{pmatrix}, \tag{9.77}$$

where we have abbreviated $\Lambda = 1 - 2M/r$, and used equation (D.25) to express Z_{ab}^{lm} in terms of the functions X_{lm} and W_{lm} (see equations D.10 and D.11). We have also denoted components that can be inferred from symmetry with a *sym*.

A numerical simulation will result in a spacetime metric. In most cases, this space-time metric is decomposed into the lapse α, the shift β^i, and the spatial metric γ_{ij}. We now would like to express the asymptotic metric in the form of equation (9.77). To do so, we compute the expansion coefficients appearing in equation (9.77) using the orthogonality relations for the spherical harmonics (see Appendix D), evaluating the surface integrals at a large radius. In particular, we find the following expressions for the spatial components:

$$H_{2lm} = \int \left(1 - \frac{2M}{r} \right) \gamma_{rr} Y_{lm}^* d\Omega \tag{9.78}$$

$$h_{1lm} = \frac{1}{l(l+1)} \int \sigma^{ab}(E_b^{lm})^* \gamma_{ra} d\Omega$$
$$= \frac{1}{l(l+1)} \int \left((\partial_\theta Y_{lm})^* \gamma_{r\theta} + (\partial_\phi Y_{lm})^* \frac{\gamma_{r\phi}}{\sin^2\theta} \right) d\Omega \tag{9.79}$$

$$K_{lm} = \frac{1}{2r^2} \int \gamma_{cd}\sigma^{cd} Y_{lm}^* d\Omega = \frac{1}{2r^2} \int \gamma_+ Y_{lm}^* d\Omega \tag{9.80}$$

$$G_{lm} = \frac{1}{(l-1)l(l+1)(l+2)r^2} \int \gamma_{cd}(Z_{lm}^{cd})^* d\Omega$$
$$= \frac{1}{2(l-1)l(l+1)(l+2)r^2} \int \left(\gamma_- W_{lm}^* + \frac{2\gamma_{\theta\phi}}{\sin^2\theta} X_{lm}^* \right) d\Omega, \tag{9.81}$$

where we have used the abbreviation

$$\gamma_\pm \equiv \gamma_{\theta\theta} \pm \frac{\gamma_{\phi\phi}}{\sin^2\theta}. \tag{9.82}$$

Exercise 9.6 Show that for an even-parity, $l = 2, m = 0$ linearized wave represented by equation (9.48), the coefficients (9.78)–(9.81) are given by

$$H_{2\,20} = 4A\sqrt{\frac{\pi}{5}} \tag{9.83}$$

$$h_{1\,20} = 2rB\sqrt{\frac{\pi}{5}} \tag{9.84}$$

$$K_{20} = -2A\sqrt{\frac{\pi}{5}} \tag{9.85}$$

$$G_{20} = (2C - A)\sqrt{\frac{\pi}{5}}. \tag{9.86}$$

Hint: Instead of inserting the metric coefficients defined by (9.48) into the quadratures (9.78)–(9.81) and carrying out the integrations, it may be easier to express the angular dependence in the metric coefficients in terms of spherical harmonics, and then use their orthogonality relations to find the integrals.

We can now combine the coefficients (9.78)–(9.81) to form the gauge-invariant Moncrief function

$$R_{lm} = \frac{r[l(l+1)k_{1lm} + 4(1 - 2M/r)^2 k_{2lm}]}{l(l+1)[(l-1)(l+2) + 6M/r]}, \tag{9.87}$$

where the functions

$$k_{1lm} = K_{lm} + l(l+1)G_{lm} + 2\left(1 - \frac{2M}{r}\right)\left(r\partial_r G_{lm} - \frac{h_{1lm}}{r}\right), \tag{9.88}$$

$$k_{2lm} = \frac{H_{2lm}}{2(1 - 2M/r)} - \frac{1}{2\sqrt{1 - 2M/r}}\partial_r\left(\frac{r[K_{lm} + l(l+1)G_{lm}]}{\sqrt{1 - 2M/r}}\right) \tag{9.89}$$

are gauge-invariant themselves. In a Schwarzschild spacetime, the function R_{lm} satisfies the famous *Zerilli equation*[41]

$$\partial_t^2 R_{lm} - \partial_{r_*}^2 R_{lm} + V_l^{(e)} R_{lm} = 0. \tag{9.90}$$

Here $V_l^{(e)}$ is the Zerilli potential

$$V_l^{(e)}(r) = \left(1 - \frac{2M}{r}\right) \tag{9.91}$$

$$\times \frac{l(l+1)(l-1)^2(l+2)^2 r^3 + 6(l-1)^2(l+2)^2 Mr^2 + 36(l-1)(l+2)M^2 r + 72M^3}{r^3\left((l-1)(l+2)r + 6M\right)^2},$$

---

[41] Zerilli (1970) and references therein.

and r_* denotes the *tortoise coordinate*

$$r_* = r + 2M \ln \left(\frac{r}{2M} - 1 \right). \tag{9.92}$$

Exercise 9.7   Revisiting the even-parity, $l = 2$, $m = 0$ linearized wave of exercise 9.6, show that the Moncrief function R_{20} is given by

$$R_{20} = \frac{r}{6} \sqrt{\frac{\pi}{5}} \, (r \partial_r A - 6A - 6B + 12C). \tag{9.93}$$

From the Moncrief functions R_{lm} we can determine the asymptotic gravitational wave amplitudes h_+ and $h_\times$. We postpone displaying these expressions (see equation 9.106) until after we have introduced the odd-parity (or axial) modes.

Odd-parity (axial) modes

Recall that the only odd-parity spherical harmonics are the magnetic-type vector spherical harmonics S_a^{lm} and the tensor harmonics S_{ab}^{lm}. Accordingly, the nonangular parts of the metric cannot have an odd-parity perturbation, and we must have

$$h_{AB}^{o\,lm} = 0. \tag{9.94}$$

The mixed angular-nonangular parts can be expanded in terms of the vector spherical harmonics S_a^{lm},

$$h_{tc}^{o\,lm} = V_{lm}(t,r) S_c^{lm}(\theta, \phi) \tag{9.95}$$

$$h_{rc}^{o\,lm} = C_{lm}(t,r) S_c^{lm}(\theta, \phi), \tag{9.96}$$

and the completely angular part in terms of the tensor spherical harmonics S_{ab}^{lm},

$$h_{cd}^{o\,lm} = -2r^2 D_{lm}(t,r) S_{cd}^{lm}(\theta, \phi). \tag{9.97}$$

In matrix form, the perturbative matrix then appears as

$$h_{\mu\nu}^{o\,lm} = \begin{pmatrix} 0 & 0 & -V_{lm} S_\phi^{lm} / \sin\theta & V_{lm} \sin\theta \, S_\theta^{lm} Y_{lm} \\ 0 & 0 & -C_{lm} S_\phi^{lm} / \sin\theta & C_{lm} \sin\theta \, S_\theta^{lm} Y_{lm} \\ \text{sym} & \text{sym} & r^2 D_{lm} X_{lm} / \sin\theta & -r^2 D_{lm} W_{lm} \sin\theta \\ \text{sym} & \text{sym} & \text{sym} & -r^2 D_{lm} X_{lm} \sin\theta \end{pmatrix}, \tag{9.98}$$

where we have used equation (D.25) to express S_{ab}^{lm} in terms of the functions X_{lm} and W_{lm} (see equations D.10 and D.11).

As for the even-parity modes, we can find the expansion coefficients V_{lm}, C_{lm} and D_{lm} for a given spacetime metric from a surface integration at large radius. Using the orthogonality relations in Appendix D we find

$$
\begin{aligned}
C_{lm} &= \frac{1}{l(l+1)} \int \sigma^{ab} (S_b^{lm})^* \gamma_{ra} d\Omega \\
&= -\frac{1}{l(l+1)} \int \frac{1}{\sin\theta} \left((\partial_\phi Y_{lm})^* \gamma_{r\theta} - (\partial_\theta Y_{lm})^* \gamma_{r\phi} \right) d\Omega
\end{aligned}
\tag{9.99}
$$

and

$$
\begin{aligned}
D_{lm} &= -\frac{1}{(l-1)l(l+1)(l+2)r^2} \int \gamma_{cd} (S_{lm}^{cd})^* d\Omega \\
&= \frac{1}{(l-1)l(l+1)(l+2)r^2} \int \frac{1}{\sin\theta} \left(\gamma_- X_{lm}^* - \gamma_{\theta\phi} W_{lm}^* \right) d\Omega.
\end{aligned}
\tag{9.100}
$$

We now combine the coefficients C_{lm} and D_{lm} to form the odd-parity, gauge-invariant Moncrief function

$$
Q_{lm} = \frac{1}{r} \left(1 - \frac{2M}{r} \right) (C_{lm} + r^2 \partial_r D_{lm}).
\tag{9.101}
$$

In a Schwarzschild spacetime, Q_{lm} satisfies the vacuum *Regge–Wheeler equation*[42]

$$
\partial_t^2 Q_{lm} - \partial_{r_*}^2 Q_{lm} + V_l^{(o)} Q_{lm} = 0,
\tag{9.102}
$$

where the Regge–Wheeler potential is

$$
V_l^{(o)}(r) = \left(1 - \frac{2M}{r} \right) \left(\frac{l(l+1)}{r^2} - \frac{6M}{r^3} \right).
\tag{9.103}
$$

The Moncrief functions Q_{lm} are the odd-parity equivalent of the functions R_{lm} for even-parity modes, and we can now discuss how to extract gravitational radiation from these two functions.

Gravitational wave extraction

We start by splitting the two gravitational wave polarizations h_+ and $h_\times$ into even- and odd-parity parts,

$$
h_+(t, r, \theta, \phi) = h_+^{(o)}(t, r, \theta, \phi) + h_+^{(e)}(t, r, \theta, \phi)
\tag{9.104}
$$

$$
h_\times(t, r, \theta, \phi) = h_\times^{(o)}(t, r, \theta, \phi) + h_\times^{(e)}(t, r, \theta, \phi).
\tag{9.105}
$$

It then turns out that we can express both parts in terms of the gauge-invariant Moncrief functions at spatial infinity $r \to \infty$. For the even-parity modes we have

$$
h_+^{(e)} - i h_\times^{(e)} = \frac{1}{r} \sum_{l=2}^\infty \sum_{m=-l}^l \sqrt{\frac{(l+2)!}{(l-2)!}} R_{lm}(t, r) \, _{-2}Y_{lm}(\theta, \phi),
\tag{9.106}
$$

---

[42] Regge and Wheeler (1957).

where $_{-2}Y_{lm}$ is the $s = -2$ spin-weighted spherical harmonic (see equation D.9), and for the odd-parity modes we find

$$h_+^{(o)} - i h_\times^{(o)} = -\frac{i}{r} \sum_{l=2}^{\infty} \sum_{m=-l}^{l} \sqrt{\frac{(l+2)!}{(l-2)!}} \, q_{lm}(t,r) \, _{-2}Y_{lm}(\theta,\phi). \tag{9.107}$$

Here the functions q_{lm} can be computed from the Moncrief function Q_{lm} by integrating over time,

$$q_{lm}(t,r) = \int_{-\infty}^{t} Q_{lm}(t',r)dt'. \tag{9.108}$$

Using the property (D.6) of spherical harmonics, we can compute the individual even- and odd-parity polarizations to be

$$h_+^{(e)} = \frac{1}{r} \sum_{l=2}^{\infty} \left(R_{l0} W_{l0} + 2 \sum_{m=1}^{l} \mathrm{Re}(R_{lm} W_{lm}) \right) \tag{9.109}$$

$$h_\times^{(e)} = \frac{2}{r} \sum_{l=2}^{\infty} \sum_{m=1}^{l} \mathrm{Re}\left(R_{lm} \frac{X_{lm}}{\sin\theta} \right) \tag{9.110}$$

$$h_+^{(o)} = -\frac{2}{r} \sum_{l=2}^{\infty} \sum_{m=1}^{l} \mathrm{Re}\left(q_{lm} \frac{X_{lm}}{\sin\theta} \right) \tag{9.111}$$

$$h_\times^{(o)} = \frac{1}{r} \sum_{l=2}^{\infty} \left(q_{l0} W_{l0} + 2 \sum_{m=1}^{l} \mathrm{Re}(q_{lm} W_{lm}) \right). \tag{9.112}$$

We are now ready to use this decomposition to compute the change in energy, angular momentum and linear momentum of a source due to gravitational radiation. To compute the energy change, we insert equations (9.109)–(9.112) into equation (9.30) and use the orthogonality relation for spin-weighted spherical harmonics (D.13) to find

$$\frac{dE}{dt} = -\frac{1}{16\pi} \sum_{l=2}^{\infty} \sum_{m=-l}^{l} \frac{(l+2)!}{(l-2)!} (|\dot{Q}_{lm}|^2 + |\dot{R}_{lm}|^2). \tag{9.113}$$

Similarly, we can find the change in angular momentum from equation (9.35),

$$\frac{dJ_z}{dt} = -\frac{i}{16\pi} \sum_{l=2}^{\infty} \sum_{m=-l}^{l} m \frac{(l+2)!}{(l-2)!} (q_{lm}^* Q_{lm} + R_{lm}^* \dot{R}_{lm})$$

$$= \frac{1}{8\pi} \sum_{l=2}^{\infty} \sum_{m=1}^{l} m \frac{(l+2)!}{(l-2)!} \mathrm{Im}(q_{lm}^* Q_{lm} + R_{lm}^* \dot{R}_{lm}). \tag{9.114}$$

We could also compute the change in linear momentum from equation (9.36). The result in terms of Moncrief variables is rather complicated, however, and we refer the reader elsewhere for the relevant expression.[43]

---

[43] See Koppitz *et al.* (2007), equation (3), for the relation.

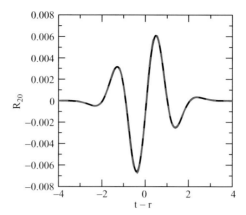

Figure 9.7 The Moncrief function R_{20} at $r = 6.0$ for a $l = 2$, $m = 0$ linear wave. The solid (black) line denotes numerical results, and the dashed (red) line is the analytical value given by equation (9.93).

Exercise 9.8 Returning to the even-parity, $l = 2$, $m = 0$ linearized wave of exercises 9.6 and 9.7, find the wave polarizations h_+ and $h_\times$ from equations (9.109) and (9.110) and compare with the results you would obtain from equations (9.57) and (9.58). Then compute the radiated energy from equation (9.113) and compare with the result of exercise 9.5.

It may be useful to illustrate this procedure with a concrete, numerical example. Consider an even-parity, $l = 2$, $m = 0$ linearized wave, as in exercises 9.4 through 9.8. Choose the function $F(x)$ to have the form of equation (9.56), with amplitude $\mathcal{A} = 0.001$ and spatial extend $\lambda = 1$ (in arbitrary units). Using this wave to specify the initial data ($\gamma_{ij} = g_{ij}$, $K_{ij} = 0$) in a numerical simulation we now evolve it by adopting the BSSN 3 + 1 formalism described in Chapter 11.5. We employ a finite difference implementation that is fourth-order accurate in space and second-order in time (*cf.* Chapter 6.2). We also adopt geodesic slicing with $\alpha = 1$ and $\beta^i = 0$ (Section 4.1), assume equatorial symmetry, and use a spatial grid of $120 \times 120 \times 60$ grid points in the x, y and z directions, respectively.[44] Extracting the Moncrief coefficients at a radius of $r = 6.0$,[45] we can compute the Moncrief function R_{20} as a function of time. In Figure 9.7, we show this function together with its analytical value, which is given by equation (9.93). We could then compute h_+ from equation (9.109), but we shall postpone this step until we can compare this procedure with the Newman–Penrose formalism in Figure 9.8 at the end of the following section.

[44] This is the same gauge as used to express the linearized wave spacetime solutions in equations (9.48) and (9.60), so we can compare the numerical and analytic values of g_{ab} directly. But were we to evolve with a different lapse and shift we would compare gauge-invariant quantities, like the gauge-invariant Moncrief variables, or the polarization amplitudes h_+ and $h_\times$.

[45] This simulation was carried out using a "fisheye" coordinate system that provides radial adaptivity in a simple way (see, e.g., Campanelli *et al.* 2006). The grid-spacing in this coordinate system is $\Delta\bar{x} = \Delta\bar{y} = \Delta\bar{z} = 0.075$, and gravitational waves are extracted at $\bar{r} = 4$, which corresponds to a physical radius of $r = 5.9987$.

Exercise 9.9   In many problems the dominant modes of gravitational wave emission are the even $l = 2$ mass quadrupole modes. Show that in this case the waveform reduces to

$$h_+ = \frac{1}{r}\left[\sqrt{\frac{5}{64\pi}}(R_{22+}\cos 2\phi + R_{22-}\sin 2\phi)(1 + \cos^2\theta) + \sqrt{\frac{15}{64\pi}}R_{20*}\sin^2\theta\right.$$

$$\left. + \sqrt{\frac{5}{16\pi}}(R_{21+}\cos\phi - R_{21-}\sin\phi)\sin\theta\cos\theta\right], \tag{9.115}$$

$$h_\times = \frac{2}{r}\left[\sqrt{\frac{5}{64\pi}}(-R_{22+}\sin 2\phi + R_{22-}\cos 2\phi)\cos\theta\right.$$

$$\left. - \sqrt{\frac{5}{16\pi}}(R_{21+}\sin\phi\hat{A} + R_{21-}\cos\phi)\sin\theta\right], \tag{9.116}$$

where

$$R_{22+} = 2\sqrt{24}\,\text{Re}(R_{22}), \quad R_{22-} = -2\sqrt{24}\,\text{Im}(R_{22}), \tag{9.117}$$

$$R_{20*} = 4\sqrt{3}R_{20}, \tag{9.118}$$

$$R_{21+} = 2\sqrt{24}\,\text{Re}(R_{21}), \quad R_{21-} = 2\sqrt{24}\,\text{Im}(R_{21}). \tag{9.119}$$

Thus, equations (9.115) and (9.116) are the relativistic generalizations of the weak-field, slow-velocity quadrupole expressions (9.24) and (9.25).

9.4.2 The Newman–Penrose formalism

The 4-dimensional Riemann tensor $^{(4)}R_{abcd}$ contains 20 independent components; 10 of these are absorbed in its trace, the Ricci tensor, and the other 10 in its traceless part, the Weyl tensor $^{(4)}C_{abcd}$. In the Newman–Penrose formalism,[46] the 10 independent components of the Weyl tensor are expressed in terms of five complex scalars, $\psi_0, \ldots, \psi_4$, which are sometimes called *Newman–Penrose scalars* and sometimes *Weyl scalars*. These scalars are formed by contracting the Weyl tensor with a complex null tetrad. Unfortunately there is not one unique such null tetrad, and different choices for this tetrad affect the Weyl scalars and their physical interpretation. One of the conceptional issues in the Newman–Penrose formalism is therefore the identification of a suitable null tetrad.[47] It is possible to identify a class of tetrads, called *transverse frames*, in which the "odd" scalars ψ_1 and ψ_3 vanish. In a subset of these frames, the so-called *quasi-Kinnersley frames*, we can then interpret the scalars ψ_0 and ψ_4 as measures of the ingoing and outgoing gravitational radiation, while ψ_2 represents longitudinal, or "Coulombic" parts of the gravitational fields, related to the mass and angular momentum of the spacetime. In this section we are particularly interested in the Weyl scalar ψ_4.

---

[46] Newman and Penrose (1962, 1963).
[47] See, for example, Lehner and Moreschi (2007) for a discussion of some of the related issues.

To construct a null tetrad suitable for our purposes here, we start by choosing two real vectors l^a and k^a that are radially outgoing and ingoing null vectors, respectively. We also construct a complex vector m^a from the two spatial vectors that are orthogonal to l^a and k^a in such a way that the only nonvanishing inner products between these 4-vectors are

$$-l^a k_a = 1 = m^a \bar{m}_a, \tag{9.120}$$

where $\bar{m}^a$ is the complex conjugate of m^a. We then define the Weyl scalar ψ_4 as

$$\boxed{\psi_4 = -^{(4)}C_{abcd} k^a \bar{m}^b k^c \bar{m}^d,} \tag{9.121}$$

where $^{(4)}C_{abcd}$ is the Weyl tensor (1.26).

Exercise 9.10 Show that we may replace the Weyl tensor $^{(4)}C_{abcd}$ with the Riemann tensor $^{(4)}R_{abcd}$ in the definition (9.121) of the Weyl scalar ψ_4.[48]

Some authors adopt a different sign convention in equation (9.121). Further differences exist in the procedure and conventions for constructing specific tetrads. For simplicity we will use a tetrad formed from an orthonormal spherical polar basis according to

$$
\begin{aligned}
l^a &= \frac{1}{\sqrt{2}} \left(e_{\hat{t}}^a + e_{\hat{r}}^a \right) \\
k^a &= \frac{1}{\sqrt{2}} \left(e_{\hat{t}}^a - e_{\hat{r}}^a \right) \\
m^a &= \frac{1}{\sqrt{2}} \left(e_{\hat{\theta}}^a + i e_{\hat{\phi}}^a \right) \\
\bar{m}^a &= \frac{1}{\sqrt{2}} \left(e_{\hat{\theta}}^a - i e_{\hat{\phi}}^a \right).
\end{aligned}
\tag{9.122}
$$

Exercise 9.11 Verify that the tetrad (9.122) satisfies the orthogonality relations (9.120).

We can now verify that ψ_4 indeed provides a measure of outgoing radiation. According to exercise 9.10 we may contract the Riemann tensor $^{(4)}R_{abcd}$ with k^a and $\bar{m}^a$ to compute ψ_4 in equation (9.121), which yields

$$\psi_4 = -\frac{1}{4} \Big(^{(4)}R_{\hat{t}\hat{\theta}\hat{t}\hat{\theta}} - 2i\,^{(4)}R_{\hat{t}\hat{\theta}\hat{t}\hat{\phi}} - 2\,^{(4)}R_{\hat{t}\hat{\theta}\hat{r}\hat{\theta}} + 2i\,^{(4)}R_{\hat{t}\hat{\phi}\hat{r}\hat{\theta}} - {}^{(4)}R_{\hat{t}\hat{\phi}\hat{t}\hat{\phi}} +$$

$$^{(4)}R_{\hat{r}\hat{\theta}\hat{r}\hat{\theta}} + 2i\,^{(4)}R_{\hat{t}\hat{\theta}\hat{r}\hat{\phi}} + 2\,^{(4)}R_{\hat{t}\hat{\phi}\hat{r}\hat{\phi}} - 2i\,^{(4)}R_{\hat{r}\hat{\phi}\hat{r}\hat{\theta}} - {}^{(4)}R_{\hat{r}\hat{\phi}\hat{r}\hat{\phi}} \Big). \tag{9.123}$$

To linear order in small deviations from flat spacetime the Riemann tensor reduces to

$$^{(4)}R_{abcd} = \frac{1}{2} \left(\partial_a \partial_d h_{bc} + \partial_b \partial_c h_{ad} - \partial_b \partial_d h_{ac} - \partial_a \partial_c h_{bd} \right). \tag{9.124}$$

In the TT gauge, the only nonvanishing components of a radially propagating wave are the transverse, angular components $h_{\hat{\theta}\hat{\theta}}^{TT} = -h_{\hat{\phi}\hat{\phi}}^{TT}$ and $h_{\hat{\theta}\hat{\phi}}^{TT} = h_{\hat{\phi}\hat{\theta}}^{TT}$. As in equations (9.57)

---

[48] We may always replace the Weyl tensor with the Riemann tensor in vacuum, where they are identical, but in the definition of ψ_4 we may make this replacement everywhere.

and (9.58), we may identify the former with h_+ and the latter with $h_\times$. Since all terms in equation (9.123) have exactly two angular indices, the only nonzero terms in equation (9.124) are those for which these two angular indices appear in the metric perturbation h_{ab}, and the t and r indices appear in the partial derivatives. For example, the first term in equation (9.123) reduces to

$$^{(4)}R_{\hat{t}\hat{\theta}\hat{t}\hat{\theta}} = -\frac{1}{2}\partial_t^2 h_{\hat{\theta}\hat{\theta}}^{TT} = -\frac{1}{2}\ddot{h}_+. \tag{9.125}$$

For an outgoing wave at large r we know that $h_{ij}^{TT}(t,r,\theta,\phi) = h_{ij}^{TT}(t-r,\theta,\phi)$, hence $\partial_r h_{ij}^{TT} = -\partial_t h_{ij}^{TT}$, so that we can express radial derivatives in terms of time derivatives. Collecting all terms, we then find[49]

$$\boxed{\psi_4 = \ddot{h}_+ - i\ddot{h}_\times.} \tag{9.126}$$

Exercise 9.12 Retrace the steps outlined above to show that for a purely *ingoing* wave the Weyl scalar ψ_4 vanishes.

As claimed above, the Weyl scalar ψ_4 thus provides a measure of outgoing gravitational radiation. We can therefore use ψ_4 to compute gravitational wave signals, as well as the radiated energy and angular and linear momentum. Before proceeding to do that, however, we point out a computational issue.

Computing ψ_4 involves the 4-dimensional Riemann tensor $^{(4)}R_{abcd}$. Many numerical simulations, however, employ a 3+1 formalism that is based on working with 3-dimensional spatial quantities. Before we can compute ψ_4, then, we first have to construct $^{(4)}R_{abcd}$ from these spatial quantities. In principle, we may do this by reversing the steps in Chapter 2.5 and using the Gauss, Codazzi and Ricci equations. More specifically, we can find $^{(4)}R_{abcd}$ from equation (2.61), which expresses $^{(4)}R_{abcd}$ in terms of its spatial and normal projections. We can then substitute equations (2.68), (2.73) and (2.103), which relate these projections to purely spatial quantities. Under some circumstances, this procedure simplifies significantly.

If we may assume that in some asymptotically flat regime the deviations of the lapse from unity and the shift from zero are at most as large as the Riemann tensor itself, then we may approximate the normal vector in equations (2.73) and (2.103) as $n^a \simeq \delta^a_0$. Restricting equations (2.68), (2.73) and (2.103) to spatial indices, and assuming a vacuum, then yields

$$\begin{aligned}
^{(4)}R_{ijkl} &= R_{ijkl} + 2K_{i[k}K_{l]j} \\
^{(4)}R_{0jkl} &= 2\partial_{[k}K_{l]j} + 2K_{m[k}\Gamma^m_{l]j} \\
^{(4)}R_{0j0l} &= R_{jl} - K_{jm}K^m{}_l + KK_{jl}.
\end{aligned} \tag{9.127}$$

The above expressions for $^{(4)}R_{abcd}$ may then be used in equation (9.121) to evaluate ψ_4 at large r in the wave zone on each time slice in a $3+1$ numerical simulation.

---

[49] This expression again depends on conventions. It sometimes appears with the opposite sign in the literature, related to the sign convention in equation (9.121), or with a factor of $1/2$, if a different normalization is used in the tetrad definition (9.122).

Exercise 9.13   Some authors define the ingoing null tetrad vector k^a in equation (9.122) in terms of the normal vector n^a plus a purely spatial, radially outward pointing vector v^a, e.g.,

$$k^a = \frac{1}{\sqrt{2}}(n^a - v^a). \tag{9.128}$$

Show that with this definition, ψ_4 can be expressed as

$$\psi_4 = -\left(\gamma_a{}^p \gamma_b{}^q \gamma_c{}^r \gamma_d{}^{s\,(4)} R_{pqrs} v^a v^c - 2\gamma_a{}^p \gamma_b{}^q \gamma_d{}^{s\,(4)} R_{pqrs} n^s v^a\right.$$
$$\left. + \gamma_b{}^q \gamma_d{}^{s\,(4)} R_{pqrs} n^p n^r\right) \bar{m}^b \bar{m}^d, \tag{9.129}$$

into which we can then substitute equations (2.68), (2.73) and (2.143) for the projections of $^{(4)}R_{abcd}$.

We now return to computing the gravitational wave emission from ψ_4. We can find the gravitational waveforms h_+ and $h_\times$ by integrating the real and imaginary parts of equation (9.126) twice. Knowing h_+ and $h_\times$, we can then compute the radiated energy and momenta by substituting into equations (9.30), (9.35) and (9.36). In particular, we find

$$L_{\mathrm{GW}} = -\frac{dE}{dt} = \lim_{r \to \infty} \frac{r^2}{16\pi} \int d\Omega \left| \int_{-\infty}^{t} dt'\, \psi_4 \right|^2 \tag{9.130}$$

for the gravitational wave luminosity and corresponding loss of energy from the source,

$$\frac{dJ_z}{dt} = \lim_{r \to \infty} \frac{r^2}{16\pi} \int d\Omega \,\mathrm{Re}\left(\left(\int_{-\infty}^{t} dt'\, \psi_4\right)\left(\partial_\phi \int_{-\infty}^{t} dt' \int_{-\infty}^{t'} dt''\, \psi_4^*\right)\right) \tag{9.131}$$

for the angular momentum loss, and

$$\frac{dP^i}{dt} = \lim_{r \to \infty} -\frac{r^2}{16\pi} \int d\Omega \frac{x^i}{r} \left| \int_{-\infty}^{t} dt'\, \psi_4 \right|^2 \tag{9.132}$$

for the linear momentum loss.

For many applications it is useful to decompose ψ_4 into $s = -2$ spin-weighted spherical harmonics

$$\psi_4(t, r, \theta, \phi) = \sum_{l=2}^{\infty} \sum_{m=-l}^{l} \psi_4^{lm}(t, r)\, {}_{-2}Y_{lm}(\theta, \phi) \tag{9.133}$$

(see Appendix D). Using the orthogonality relation (D.13) we can find the expansion coefficients ψ_4^{lm} from

$$\psi_4^{lm} = \int d\Omega\, {}_{-2}Y_{lm}^* \psi_4. \tag{9.134}$$

Exercise 9.14   Compare equation (9.126) with equations (9.106) and (9.107) to show that ψ_4^{lm} is related to the Moncrief variables R_{lm} and Q_{lm} by

$$\psi_4^{lm} = \frac{1}{r}\sqrt{\frac{(l+2)!}{(l-2)!}}\,(\ddot{R}_{lm} - i\dot{Q}_{lm}). \tag{9.135}$$

In terms of the coefficients ψ_4^{lm}, the gravitational wave luminosity (9.130) is

$$L_{\text{GW}} = -\frac{dE}{dt} = \lim_{r \to \infty} \frac{r^2}{16\pi} \sum_{l=2}^{\infty} \sum_{m=-l}^{l} \left| \int_{-\infty}^{t} dt' \, \psi_4^{lm} \right|^2, \tag{9.136}$$

while the loss of angular momentum due to gravitational waves, equation (9.131), becomes

$$\frac{dJ_z}{dt} = \lim_{r \to \infty} \frac{r^2}{16\pi} \sum_{l=2}^{\infty} \sum_{m=-l}^{l} m \, \text{Im} \left(\left(\int_{-\infty}^{t} dt' \psi_4^{lm} \right) \left(\int_{-\infty}^{t} dt' \int_{-\infty}^{t'} dt'' \, \psi_4^{lm*} \right) \right). \tag{9.137}$$

For some purposes it is useful to introduce real functions A_{lm} and B_{lm} whose second time derivative equals the real and imaginary parts of ψ_4^{lm*},

$$\psi_4^{lm} = \ddot{A}_{lm} - i\ddot{B}_{lm}. \tag{9.138}$$

In terms of these, the loss of energy and angular momentum due to waves become

$$L_{\text{GW}} = -\frac{dE}{dt} = \lim_{r \to \infty} \frac{r^2}{16\pi} \sum_{l=2}^{\infty} \sum_{m=-l}^{l} (\dot{A}_{lm}^2 + \dot{B}_{lm}^2) \tag{9.139}$$

and

$$\frac{dJ_z}{dt} = \lim_{r \to \infty} \frac{r^2}{16\pi} \sum_{l=2}^{\infty} \sum_{m=-l}^{l} m(\dot{A}_{lm} B_{lm} - \dot{B}_{lm} A_{lm}). \tag{9.140}$$

Similarly we can express the loss of linear momentum (9.132) in terms of these coefficients, but as in the Moncrief formalism that leads to a rather lengthy expression that we shall omit.[50]

Before closing this chapter we return to the numerical example at the end of Section 9.4.1, namely the $l = 2, m = 0$ even-parity linearized wave. We evolve the same initial data with the same numerical code as described there, but instead of extracting the Moncrief variable R_{20} (see Figure 9.7) we now compute the Weyl scalar ψ_4 at a point $(r, \theta, \phi) = (6.0, 0.79, 0.52)$ (see footnote 45 above). We compare the numerical values with analytical expressions in the left panels of Figure 9.8. In the right panels we also graph the analytic gravitational waveforms for both polarizations, as well as the corresponding numerical waveforms using both the Moncrief formalism (computed from equation 9.109) and the Newman–Penrose formalism (computed from equation 9.126).

Figure 9.8 demonstrates that, for this particular example, the Moncrief formalism leads to a better agreement with the analytical results than the Newman–Penrose formalism (even though both converge to the analytical results as the numerical resolution is increased[51]). This difference can be attributed to two factors. One difference between the two approaches

---

[50] See, e.g., Ruiz *et al.* (2008a).
[51] We also point out that while $h_\times$ computed in Figure 9.8 from the Newman–Penrose formalism is not as close to the analytic value of zero as the value computed by the Moncrief formalism, it is still more than an order of magnitude below the amplitude of the dominant h_+ polarization amplitude for the adopted resolution.

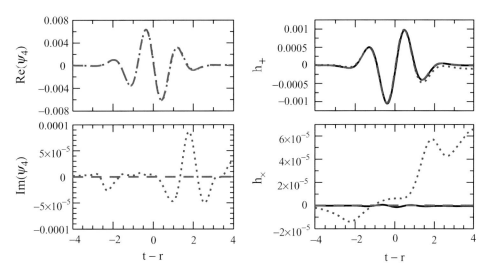

Figure 9.8 The left panels show the Weyl scalar ψ_4 for an even-parity $l = 2$, $m = 0$ linearized wave at a point $(r, \theta, \phi) = (6.0, 0.79, 0.52)$. The dotted (blue) line shows numerical results, and the dashed (red) line analytical values. The right panels show the gravitational waveforms h_+ and $h_\times$ for the same wave. The dashed (red) lines show the analytical waveforms (which are computed in exercise 9.8), the solid (black) lines are the numerical waveforms based on the Moncrief formalism (computed from R_{20}, shown in Figure 9.7, using equation 9.109), and the dotted (blue) line shows the numerical waveforms constructed from the Weyl scalar ψ_4 in equation (9.126).

is that the Moncrief formalism requires taking only a first derivative of the metric (in equations 9.88 and 9.89 for the even-parity modes, or equation 9.101 for the odd-parity modes) to find the Moncrief functions, and only one integration (in equation 9.108) to reconstruct the gravitational waveforms. In the Newman–Penrose formalism, by contrast, we need second derivatives of the metric to compute ψ_4 (which are hidden in the Weyl tensor $^{(4)}C_{abcd}$ in equation 9.121), and then two integrations (in equation 9.126) to compute h_+ and $h_\times$. Not surprisingly, the additional derivative and integration introduce additional numerical error.

Another difference in our particular calculation is that we computed the Moncrief quantities from their modes, and the Newman–Penrose quantities locally, without doing a modal decomposition. Computing the modes involves surface integrals, which effectively filters out some of the numerical noise present in the higher-order modes. Many applications that employ the Newman–Penrose formalism therefore also compute the modes (9.134) and then reconstruct ψ_4 locally from the decomposition (9.133). This procedure yields more accurate results, at least for the dominant, lowest-order modes.

10 | Collapse of collisionless clusters in axisymmetry

As we learned in Chapter 8, where we studied spherical systems, collisionless clusters provide a simple relativistic source for exploring the nature of Einstein's equations and experimenting with numerical techniques to solve them. Once we relax the restriction to spherical symmetry, the spacetimes can exhibit two new dynamical features: rotation and gravitational waves. Not much is known about nonspherical collisionless systems in general relativity, even for stationary equilibria. Some interesting results have emerged by exploiting numerical relativity to investigate the equilibrium structure and collapse of nonspherical rotating and nonrotating clusters in axisymmetry.[1] To highlight the power of the technique, we shall summarize a few of the simulations and their key findings in this chapter.

The examples discussed below are chosen to demonstrate how numerical relativity, quite apart from providing accurate quantitative solutions to dynamical scenarios involving strong gravitational fields, can provide *qualitative* insight into Einstein's equations in those cases where uncertainty still prevails. It can even be helpful as a guide to proving (or disproving) theorems about strong-field spacetimes in those instances where analytic means alone have not been adequate.

10.1 Collapse of prolate spheroids to spindle singularities

It is well-known that classical general relativity admits solutions with singularities, and that such solutions can be produced by the gravitational collapse of nonsingular, asymptotically flat initial data. The *Cosmic Censorship Conjecture* of Penrose[2] states that such singularities will always be clothed by event horizons and hence can never be visible from the outside (no naked singularities). If cosmic censorship holds, then there is no problem with predicting the future evolution outside the event horizon. If it does not hold, then the formation of a naked singularity during collapse would pose a dilemma for general relativity theory. In this situation, one cannot say anything precise about the future evolution of any region of space

---

[1] For calculations of nonspherical equilibrium clusters see, e.g., Shapiro and Teukolsky (1993a,b). For the collapse of nonrotating clusters, see, e.g., Shapiro and Teukolsky (1991a,b), and for a review and additional references, Shapiro and Teukolsky (1992a); see also Abrahams *et al.* (1994, 1995). For collapse of rotating clusters, see, e.g., Shapiro and Teukolsky (1992a); Abrahams *et al.* (1994); Hughes *et al.* (1994); Shapiro *et al.* (1995).

[2] Penrose (1969).

containing the singularity since new information could emerge from it in a completely arbitrary way.

No definitive theorems guarantee that an event horizon always emerges to clothe a singularity. Proving the validity of cosmic censorship remains one of the most outstanding problems in the theory of general relativity.[3] Counter-examples have all been restricted to spherical symmetry and typically involve shell crossing, shell focusing, or self-similarity. We also discussed in Chapter 8.4 how scalar field collapse in spherical symmetry generates a critical solution that produces a naked singularity. A key issue is whether such singularities are accidents of spherical symmetry.

In the absence of general theorems, and prior to simulations of nonspherical collapse, Thorne[4] proposed the *hoop conjecture*: Black holes with horizons form when and only when a mass M ($= M_{\rm ADM}$) gets compacted into a region whose proper circumference in *every* direction is $\mathcal{C} \lesssim 4\pi M$. If the hoop conjecture is correct, aspherical collapse with one or two dimensions appreciably larger than the others might lead to naked singularities.

That such a scenario is at all possible is suggested by the Lin–Mestel–Shu instability, which concerns the collapse of a nonrotating, homogeneous spheroid of collisionless matter in Newtonian gravity.[5] Such a configuration remains homogeneous and spheroidal throughout collapse. However, if the spheroid is slightly oblate at the onset, the configuration ultimately collapses to a flat pancake, while if the spheroid is slightly prolate, it collapses to an infinitesimally thin spindle. While in both cases the density becomes infinite, the formation of a spindle during prolate collapse is particularly severe. The gravitational potential, gravitational force, tidal force, kinetic and potential energies all blow up to infinity in this case.

Might the Lin–Mestel–Shu instability in Newtonian gravitation have relevance to general relativity? In general relativity it is known that *infinite* cylinders *do* collapse to line singularities which, in accord with the hoop conjecture, are not hidden by event horizons.[6] Of course, these configurations are not asymptotically flat, hence they do not qualify as counterexamples to cosmic censorhip. But what about configurations of *finite* size?

Shapiro and Teukolsky (1991a,b) explored this question by employing a mean-field, particle simulation scheme to solve Einstein's equations for the evolution of nonrotating, collisionless matter in axisymmetric spacetimes.[7] Their $2 + 1$ axisymmetric scheme is an extension of their $1 + 1$ spherical code described in Chapter 8.2.[8] The code was tailored to handle cases in which collisionless matter could collapse to a singularity, as in oblate collapse to a flat pancake or prolate collapse to a thin spindle.

---

[3] See Berger (2002) for a discussion and references.
[4] Thorne (1972); see also Misner *et al.* (1973), p. 867.
[5] Lin *et al.* (1965).
[6] Thorne (1972).
[7] See Shapiro and Teukolsky (1991a) for a popular discussion.
[8] For the Newtonian version of this scheme in axisymmetry, with applications, see Shapiro and Teukolsky (1987).

The metric is written in the form

$$ds^2 = -\alpha^2 dt^2 + A^2 (dr + \beta^r dt)^2 + A^2 r^2 (d\theta + \beta^\theta dt)^2 + B^2 r^2 \sin^2 \theta d\phi^2. \qquad (10.1)$$

The full set of gravitational field and matter equations is listed in Appendix F. The code is fully constrained. Maximal slicing and quasi-isotropic spatial coordinates were adopted as the gauge choices. A large battery of test-bed calculations were performed to ensure the reliability of the code. These tests included the propagation of linearized analytic gravitational waves with and without matter sources and nonlinear Brill waves in vacuum spacetimes;[9] maintaining equilibria and identifying the point of onset of radial instability for spherical equilibrium clusters; reproducing Oppenheimer–Snyder collapse of homogeneous dust spheres and the collapse of homogeneous Newtonian spheroids in the weak-field limit.[10] A number of geometric probes were constructed to diagnose the evolving spacetime. For example, the total mass and outgoing radiation energy flux were calculated to monitor mass-energy conservation. To confirm the formation of a black hole, the spacetime was probed for the appearance of an apparent horizon[11] and its area and shape were computed when it is present. To assess the growth of a singularity, the Riemann invariant $I \equiv {}^{(4)}R_{abcd}\, {}^{(4)}R^{abcd}$, which measures the strength of the gravitational *tidal* field, was computed at every spatial grid point. To test the hoop conjecture, the minimum equatorial and polar circumferences outside the matter were determined.

Typical simulations were performed in equatorial symmetry with a spatial grid of 100 radial and 32 angular zones, and with 6000 test particles. A key adaptive grid feature that enabled the simulations to snuggle close to singularities was that the angular grid was allowed to fan and the radial grid was allowed to contract to follow the matter. In addition, the particles were permitted to move on larger time steps than the field variables, which were restricted by the Courant condition. In particular, the particles were only advanced on a time step comparable to the local dynamical time scale, as in equation (8.64). They were thus held frozen until the field variables caught up after being advanced every Courant time step (see, e.g., equation 6.72, with $v = c = 1$).

The code was utilized to track the collapse of nonrotating relativistic prolate and oblate spheroids of various initial sizes and eccentricities. The matter particles comprising the spheroids were taken to be instantaneously at rest at $t = 0$. The technique adopted for constructing exact, time-symmetric prolate spheroids[12] is given in exercise 10.1. In the Newtonian limit, these initial configurations reduce to homogeneous spheroids, but relativistic configurations are inhomogeneous with density increasing outwards. Given a density profile, particles are distributed to sample the initial phase-space distribution function. When the spheroids are large (size $\gg M$ in all directions) the code correctly tracks the Newtonian solutions.

---

[9] Brill waves are gravitational waves resulting from time-symmetric initial data in axisymmetric, vacuum spacetimes; see exercise 3.5. See also Eppley (1977) for numerical evolutions of Brill waves.

[10] Lin *et al.* (1965); Shapiro and Teukolsky (1987).

[11] See Chapter 7.3.

[12] Nakamura *et al.* (1988).

Exercise 10.1 For a homogeneous spheroid in Newtonian gravitation, Poisson's equation $\nabla^2 \Phi_N = 4\pi \rho_N$ relates the potential Φ_N to the rest-mass density ρ_N where

$$\rho_N = \begin{cases} \dfrac{M_N}{4\pi a^2 c/3}, & R^2/a^2 + z^2/c^2 \leq 1, \\ 0, & \text{elsewhere.} \end{cases} \qquad (10.2)$$

Here M_N is the total Newtonian rest mass, a is the equatorial radius, c is the polar radius and R and z are cylindrical coordinates. The solution for the potential is well-known, but is not needed for this exercise. The resulting Newtonian gravitational binding energy for a prolate spheroid of eccentricity $e = (1 - a^2/c^2)^{1/2}$ is given by

$$W_N = -\frac{1}{2} \int \rho_N \Phi_N d^3 x = \frac{3}{10} \frac{M_N^2}{ce} \ln \frac{1+e}{1-e}. \qquad (10.3)$$

(a) Consider a homogeneous spheroid of collisionless particles momentarily at rest in general relativity. To try and minimize the initial radiation content of the space-time, choose a conformally flat spatial metric $\gamma_{ij} = \psi^4 \eta_{ij}$ (i.e., $A = B = \psi^2$ in equation 10.1) and argue that the only nontrivial equation that the initial data must satisfy is the Hamiltonian constraint $^{(3)}R = 16\pi\rho$, where $\rho = T^{ab} n_a n_b$ is the mass density (= rest-mass density in this case), T^{ab} is the stress-energy tensor for collisionless matter and n^a is the normal vector to the initial $t = 0$ hypersurface. Show the constraint condition reduces to

$$\nabla^2 \psi = -2\pi \psi^5 \rho, \qquad (10.4)$$

with the boundary conditions

$$\nabla\psi = 0, \quad \text{at } r = 0, \quad \psi \to 1 + \frac{M}{2r}, \quad \text{as } r \to \infty, \qquad (10.5)$$

where $M = M_{\text{ADM}}$ is the total mass-energy of the configuration.

(b) Choose the density profile ρ according to $2\pi\psi^5\rho \equiv 4\pi\rho_N$, for which the solution to equation (10.4) is immediately given by $\psi = 1 - \Phi_N$. The density ρ is therefore inhomogeneous, increases outward from the center, and is constant on self-similar coordinate spheroidal surfaces. From this conclude that the total rest-mass energy of the spheroid is

$$M_0 = 2M_N + 4W_N, \qquad (10.6)$$

and the total mass-energy is

$$M = 2M_N = \frac{2M_0}{1 + (1 + \alpha M_0)^{1/2}}, \qquad (10.7)$$

where, for a prolate spheroid,

$$\alpha = \frac{6}{5ce} \ln \frac{1+e}{1-e}. \qquad (10.8)$$

(c) Evaluate the Riemann invariant $I = \,^{(4)}R^{abcd}\,^{(4)}R_{abcd} = R^{ijkl} R_{ijkl}$ in Cartesian coordinates to find

$$I = 96\psi^{-12} \left(\partial_i \psi \partial^i \psi \right)^2 - 96\psi^{-11} \partial_j \partial_i \psi \partial^i \psi \partial^j \psi + \psi^{-10} \partial_j \partial_i \psi \partial^j \partial^i \psi. \qquad (10.9)$$

Evaluating equation (10.9) reveals[13] that as $e \to 1$, prolate configurations form spindle singularities located just outside the matter on the axis. When the spheroids are sufficiently compact ($\lesssim M$ in all of its spatial dimensions) solving equation (7.51) shows that there is an apparent horizon; otherwise there is none. A sequence of these momentary static prolate spheroids of fixed rest mass, but increasing eccentricity, foreshadows the evolutionary collapse sequence of Shapiro and Teukolsky (1991b) that we shall now describe.

The left panels of Figure 10.1 show the fate of a typical, highly compact prolate configuration; such a configuration always collapses to a black hole. To appreciate the scale, recall that in isotropic coordinates a Schwarzschild black hole on the initial time slice would have a radius $r = 0.5M$, corresponding to a Schwarzschild radius $r_s = 2M$. The right panels in Figure 10.1 depict the outcome of prolate collapse with the same initial eccentricity but from a larger semi-major axis. Here the configuration collapses to a spindle singularity at the pole without the appearance of an apparent horizon. A search for either a single global horizon centered on the origin, or a small disjoint horizon around the singularity in each hemisphere, comes up empty. The spindle consists of a concentration of matter near the axis in the vicinity of $r \approx 5M$. Figure 10.2 shows the growth of the Riemann invariant I at $r = 6.1M$ on the axis, just outside the matter.[14] Prior to the formation of the singularity, the typical size of I at any exterior radius r on the axis is $\sim M^2/r^6 \ll 1$.[15] With the formation of the spindle singularity, the value of I rises without bound in the region near the pole. The maximum value of I determined numerically is limited only by the resolution of the angular grid: the better the spindle is resolved, the larger the measured value of I before the singularity causes the code (and possibly the spacetime!) to break down. Unlike shell-crossing singularities, where I blows up in the matter interior whenever the matter density is momentarily infinite, the spindle singularity also extends *outside* the matter beyond the pole at $r = 5.8M$ (Figure 10.3). In fact, the peak value of I occurs in the vacuum at $r \approx 6.1M$. Here the *exterior* tidal gravitational field is blowing up, which is not the case for shell crossing.

Probing the spacetime in the vicinity of the singularity suggests that it is is not a point, but rather an extended region which, while including the matter spindle, grows most rapidly in the vacuum exterior above the pole. The local geometry near the spindle exhibits behavior similar to the late-time geometry near the axis along which a naked singularity forms following the collapse of an infinite cylinder. The spatial metric components grow slowly with time, rising to a maximum of $A \approx B \approx 1.7$. The maximum occurs near the origin and is only moderately larger than one, the value at large distance from the spheroid. However, the tidal-field invariant I, which depends on second derivatives of the metric, diverges much more rapidly. This behavior mimics the logarithmic divergence of the metric found along an analytic, prolate sequence of momentary static configurations of increasing

---

[13] Nakamura *et al.* (1988).

[14] The calculation of I during a $3 + 1$ simulation is simplified by decomposing it into spatial field and matter variables on each time slice; see York (1989), equation (109).

[15] Recall that in Schwarzschild geometry, $I = 48M^2/r_s^6$, where r_s is the Schwarzschild areal radius.

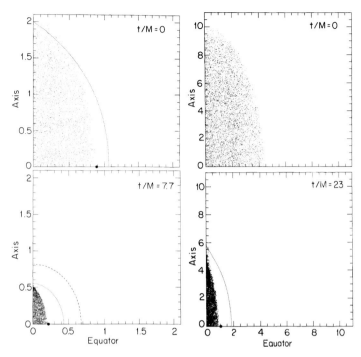

Figure 10.1 *Left panels.* Snapshots of the particle positions at initial and late times for prolate collapse. The positions (in units of M) are projected onto a meridional plane. Initially the semi-major axis of the spheroid is $2M$ and the eccentricity is 0.9. The collapse proceeds nonhomologously and terminates with the formation of a spindle singularity on the axis. However, an apparent horizon (marked by the dashed line) forms to cover the singularity. At $t/M = 7.7$ its area is $\mathcal{A}/16\pi M^2 = 0.98$, close to the expected value of unity (gravitational radiation losses are negligible). Its polar and equatorial circumferences at that time are $\mathcal{C}_{\text{pole}}^{\text{AH}}/4\pi M = 1.03$ and $\mathcal{C}_{\text{eq}}^{\text{AH}}/4\pi M = 0.91$, at later times these circumferences become equal and approach the expected value of unity. The minimum exterior polar circumference is shown by a dotted line and does not coincide with the matter surface. Likewise, the minimum equatorial circumference, which is a circle, is indicated by a solid dot. Here $\mathcal{C}_{\text{eq}}^{\text{min}}/4\pi M = 0.59$ and $\mathcal{C}_{\text{pole}}^{\text{min}}/4\pi M = 0.99$. The formation of a black hole here is thus consistent with the hoop conjecture. *Right panels.* Snapshots of the particle positions at the initial and final times for prolate collapse with the same initial eccentricity as in the left panel but with initial semi-major axis equal to $10M$. The collapse proceeds as in the left panel and terminates with the formation of a spindle singularity on the axis at $t/M = 23$. The minimum polar circumference is $\mathcal{C}_{\text{pole}}^{\text{min}}/4\pi M = 2.8$. There is no apparent horizon, in agreement with the hoop conjecture. This may be a candidate for a naked singularity. [From Shapiro and Teukolsky (1991b).]

eccentricity.[16] Of course, any such description of the singular region and the metric depends on the time slicing and may be different for other choices of time coordinate. In principle the spindle singularity might first occur at the center rather than the pole with a different time slicing.

---

[16] See exercise 10.1.

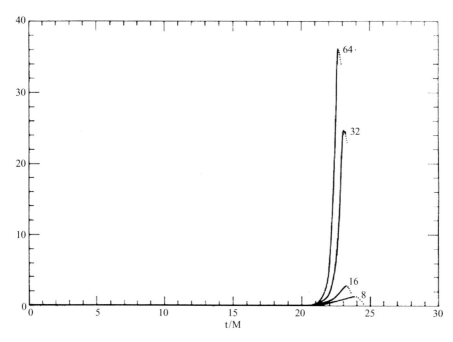

Figure 10.2 Growth of the Riemann invariant I (in units of M^{-4}) *vs.* time for the collapse shown in the right panels of Figure 10.1. The simulation was repeated with various angular grid resolutions. Each curve is labeled by the number of angular zones used to cover one hemisphere. Dots indicate where the singularity has caused the code to become inaccurate. [From Shapiro and Teukolsky (1991b).]

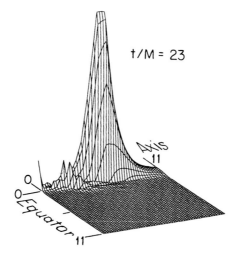

Figure 10.3 Profile of I in a meridional plane for the collapse shown in the right panels of Figure 10.1. For the case shown here employing 32 angular zones to cover one hemisphere, the peak value of I is $24/M^4$ and occurs on the axis just outside the matter. [From Shapiro and Teukolsky (1991b).]

The key question is whether or not this spindle singularity is naked or not. The absence of an apparent horizon does not necessarily imply the absence of a global event horizon, although the converse is true. This point has been emphasized by showing[17] that even Schwarzschild spacetime can be sliced with nonspherical slices that approach arbitrarily close to the singularity without any trapped surfaces. Because such singularities cause numerical integrations to terminate, one cannot map out a spacetime arbitrarily far into the future, which would be necessary to completely rule out the formation of an event horizon.

It may be telling that in the case of collapse from an initially compact state, outward null geodesics are always found to turn around near the singularity, as expected when the singularity resides inside an event horizon. By contrast, for collapse from large radius, outward null geodesics are still propagating freely away from the vicinity of the singularity up to the time the integrations terminate. It remains an open question for future research whether any time slicing can be found which will be more effective in snuggling up to the singularity without actually hitting it. Such a slicing might enable one to confirm whether all outward null geodesics manage to propagate to large distances, thereby determining whether or not the singularities arising from nonrotating, prolate collapse are truly naked.

Vacuum simulations of highly prolate, Brill-wave initial data may shed additional light on the issue of prolate collapse in general relativity. Prolate initial wave packets can be constructed without apparent horizons but with arbitrarily large curvatures.[18] What is the fate of such configurations – black holes or naked singularities? The original attempt[19] to evolve these data used a partially constrained, axisymmetric scheme that could not proceed sufficiently long to settle the question, due to a lack of resolution on the compactified spatial grid. Subsequently, however, a fully constrained scheme, which used a multigrid method for the elliptic equations and fixed mesh refinement, was able to evolve the same initial data for much longer.[20] The same gauge conditions employed by Shapiro and Teukolsky (1991a,b) were used here, but the code was written in cylindrical, rather than spherical polar, coordinates. The longer evolution ultimately revealed an apparent horizon, confirming the formation of a black hole rather than a naked singularity.[21]

10.2 Head-on collision of two black holes

As we know from high energy physics, a good way to to probe the underlying field theory governing the interactions of a relativistic system is to perform a collision experiment. This is the motivation behind the simulations we shall summarize in this section, where we will consider the the head-on collision of two identical clusters of "collisionless" particles

---

[17] Wald and Iyer (1991).

[18] Abrahams *et al.* (1992).

[19] Garfinkle and Duncan (2001).

[20] Rinne (2008a).

[21] But as pointed out by Garfinkle and Duncan (2001), highly prolate Brill waves tend to become less prolate during collapse rather than more, in contrast to prolate collisionless matter spheroids. So the final fates of the two systems may be qualitatively different.

in general relativity.[22] Recall that these particles are called "collisionless" because they interact exclusively by gravity and hence obey the relativistic Vlasov equation. Such particles could represent stars in a cluster, massive neutrinos, axions, or any other form of collisionless matter. Simulations of head-on collisions have been performed for several nonrotating, initial configurations including: (1) a binary consisting of identical spheres of particles, all momentarily at rest; (2) a binary of identical spheres of particles, where each sphere is boosted towards each other; and (3) a binary of identical spheres of particles, where the particles in each sphere are in randomly oriented, circular orbits about their respective centers. In the first two cases, the particles in the two spheres implode towards their respective centers *before* colliding and form two, well-separated, black holes. Such scenarios have been very useful for studying the head-on collision of binary black holes. In the third case, each cluster is constructed to be in near-equilibrium initially, so it does not implode prior to merger. Their collision from rest leads either to coalescence and virialization to a new equilibrium configuration, or to collapse to a black hole. This third collision scenario is the collisionless analog of colliding neutron stars in relativistic hydrodynamics.

> **Exercise 10.2**   Consider the following quantity as a diagnostic of virial equilibrium in a particle simulation,
>
> $$E_0 = -\sum_j (1 + u_0^j), \qquad (10.10)$$
>
> where u_j^a is the 4-velocity of the jth particle and the sum is over all the particles.
> (a) Why does E_0 become constant in time in virial equilibrium?
> (b) What is the meaning of E_0 in the Newtonian limit?
> (c) What other quantities can be evaluated and compared to determine that virial equilibrium has been achieved?
> **Hint:** Consider various mass measures of the configuration.

Shapiro and Teukolsky (1992b) used the same mean-field particle simulation scheme discussed in Section 10.1 and summarized in Appendix F to tackle these scenarios. This scheme adopts the same form of the metric and gauge conditions as before and handles general nonrotating, axisymmetric spacetimes containing collisionless matter. The simulations took the system center of mass in each case to reside in the equatorial plane and then exploited symmetry across the equator. The simulations then employed approximately 250 radial zones and 32 angular zones to cover the spatial grid in the upper hemisphere. The initial data for the two spheres was constructed from prescriptions similar to the one described in exercise 10.1 for single spheroids. The approach described in that exercise was generalized to allow for binary systems, and modified to account for an initial velocity boost in scenario two and for circular particle velocities in scenario three.

---

[22] Shapiro and Teukolsky (1992b).

Topology of the event horizon

The most interesting case is scenario two, the head-on collison of well-separated, boosted spheres in which the particle velocity dispersion in each sphere is initially zero. In this case each cluster collapses on itself to form a black hole prior to merger. Hence this scenario simulates the head-on collision of two nonrotating black holes. The spheres each have an initial coordinate radius $a/M = 0.8$ and their centers are located at $z_0 = \pm 1.4 M$ along the collision axis. Searches for apparent horizons are performed by solving equation (7.51) on each time slice. There are no apparent horizons initially. The spheres are given an inward boost velocity of $v = 0.15$ as measured by a normal observer.

Figure 10.4 shows spatial snapshots of the collision at several different instants of time. Common as well as disjoint horizons appear, as shown in the figures. The event horizon is found[23] by integrating null geodesics backwards in time in the curved background, as discussed in Chapter 7.2. These integrations can only be performed after the evolution has terminated and the global spacetime has been fully determined. Disjoint event horizons appear within the spheres of matter by $t/M \approx 0.13$, much earlier than the formation of the common apparent horizon at $t/M \approx 6.5$ or the disjoint apparent horizons at $t/M \approx 7.1$. Note how the tidal field of each companion causes a distortion in the shape of the nascent black holes and their disjoint event horizons, giving rise to the "hour glass" appearance of the merging holes when they first come into contact. The disjoint event horizons grow towards each other until they coalesce at $t/M \approx 2.18$. By the time the integrations terminate at $t/M \approx 11.7$, all of the matter is encompassed by a single black hole and the exterior spacetime is close to Schwarzschild. The common event and apparent horizons coincide and have an area $\mathcal{A}/16\pi M^2 \approx 1.1$, a polar circumference $\mathcal{C}_{\mathrm{pole}}/4\pi M \approx 1.1$ and an equatorial circumference $\mathcal{C}_{\mathrm{eq}}/4\pi M \approx 1.0$. The deviations of these ratios from unity, their Schwarzschild values, is a measure of the numerical error in the evolution. The total energy in gravitational waves radiated in the collision is $\Delta E/M \approx 3 \times 10^{-4}$.

Figure 10.5 is a spactime diagram of the collision and merger. The time axis is along the vertical direction, while spacelike hypersurfaces (with one of the spatial directions suppressed) are horizontal planes at any instant of time. The collision axis goes from left to right and the black hole horizon is shown as the dark shaded surface. Some of the light rays which generate the horizon ("null generators of the horizon") are shown. Their trajectories were traced by numerically propagating light rays in the background spacetime. The inset shows a closeup view of the formation and merger of the two horizons and how the rays enter the horizons at those early events.

Figure 10.5 is the famous "pair of pants" picture of the event horizon for coalescing black holes that was sketched in general relativity textbooks in the 1970s.[24] The figure shown here, produced over 20 years later, was the first real calculation of such a diagram. Many things were known about the the topology of the merging horizons and the null

---

[23] Hughes *et al.* (1994).
[24] Hawking and Ellis (1973); Misner *et al.* (1973).

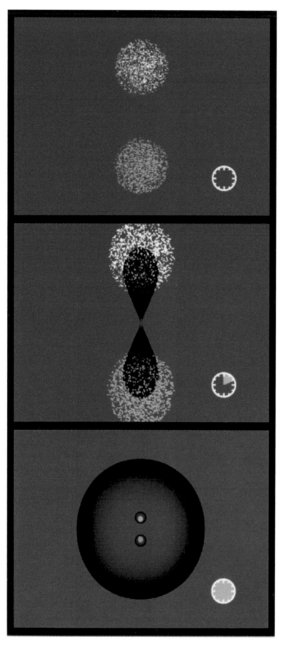

Figure 10.4 Spatial snapshots of the head-on collision of boosted clusters. *Top panel*: Initial data. Each sphere has an initial radius $a/M = 0.8$, an inward boost velocity $v = 0.15$, and their centers are at $z_0/M = \pm 1.4$. Sample particles from the collisionless matter distribution are represented by dots. A clock displaying the elapsed fraction of the total evolution time appears in the corner. *Middle panel*: Horizon coalescence. At time $t/M = 2.18$ the two event horizons coalesce (shaded hour glass). The cluster's centers are at $z/M = \pm 1.07$. The scale of the image has been enlarged by about a factor of 2 over the top panel. *Bottom panel*: Final state. By time $t/M \approx 11.7$ shown here, the event and common apparent horizons have settled down to a quasistationary state and coincide. [From Matzner *et al.* (1995).]

Figure 10.5 Spacetime diagram for the collision depicted in Figure 10.4. The diagram shows some of the null generators of the horizon. The time axis is vertical, the collision axis goes from left to right; one of the spatial dimensions is suppressed. The inset on the left zooms in on the caustic and crossover structure at the birth of the event horizon. [From Matzner *et al.* (1995).]

generators prior to these numerical simulations, but a few important details were not. It was well known that the black hole is spanned by null generators, which can intersect or cross each other only at those points at which they enter the horizon. Once on the horizon, a null generator can never propagate off, nor can it ever cross another null generator. These properties were all understood and nicely corroborated by the simulation. But in addition, the simulation revealed for the first time a line of crossover points for the null generators that extends from the "crotch" on the "pair of pants" down along each inside trouser seam, around each bottom, and continuing a small distance up each outside seam (see Figure 10.5). The points on the outside seam at which the line of crossover terminates are caustics, where the intensity of the intersecting light rays becomes infinite. For a single, isolated cluster undergoing spherical collapse to a black hole, the caustic would arise at the base of the spacetime diagram at the point at which the event horizon first forms, and there would be no crossovers. Here, the gravitational tidal field of the colliding black holes shifts the location of the caustic and produces the line of crossovers. An analysis of the simulation[25] shows that the line of crossovers is spacelike, which means that they cannot be traced by light rays or particles moving slower than c. As this line approaches the caustics on the sides, it becomes asymptotically null.

The topology of the event horizon for merging black holes is somewhat different in the case of vacuum black holes, or "eternal" black holes not formed from matter collapse. Figure 7.2 shows the results of a head-on collision of two nonrotating, vacuum black holes

[25] Matzner *et al.* (1995).

that are initially at rest at finite separation at a moment of time symmetry $(K_{ij} = 0)$.[26] The initial data are the analytic Misner initial data (3.26) discussed in Chapter 3.1.2. The initial separation between horizons L (see equation 3.28) is determined by the parameter μ appearing in the three-metric; for the case shown, $\mu = 2.2$, we have $L/M = 8.92$. The simulation proceeds using maximal slicing for a time $\Delta t \approx 150M$, at which point the merger is complete (note the spacetime diagram in Figure 7.2 goes only up to $t/M = 25$). At late times, the horizon oscillates with decaying amplitude, emitting gravitational wave quasinormal "ringdown" radiation of a vibrating Schwarzschild black hole. The figure shows the event horizon structure from $t/M = 5$ to $t/M = 20$. The event horizon is found by integrating equation (7.10). Figure 7.2 differs from Figure 10.5 in two ways. First, as discussed above, the black holes in Figure 10.5 are "born" from the collapse of matter, while the black holes in Figure 7.2 are eternal. Second, Figure 10.5 plots the coordinate circumference of the event horizon, while Figure 7.2 plots its proper polar circumference. In the latter case, the monotonic increase of the area at early times and its constancy at late times are evident, in compliance with black hole area theorems.

In contrast to black holes filled with matter, the legs in Figure 7.2 do not exhibit cusps, but continue into the past from the times shown in the diagram. As in the case with matter, there does exist a crossover line along the inside seam of the "pair of pants". Note that while the spacetime is symmetric about the initial time, the horizon is not, as it expands monotonically into the future. This is evident in the expansion of the null generators seen in the figure at the initial time.

> **Exercise 10.3**   Can one conclude from the time-symmetry of the spacetime that its past evolution describes a time-reversed collision: ingoing radiation propagates from infinity onto a single, Schwarzschild black hole, bifurcating it into two holes that move apart until coming to a momentary pause, followed by outgoing radiation as the two holes merge?

10.3 Disk collapse

Many astrophysical systems are best represented by infinitesimally thin disks of collisionless matter. Not surprisingly, numerous studies of the dynamical behavior of collisionless disk systems have been performed over the years in Newtonian gravitation.[27] As it happens, thin disks provide particularly simple, but amazingly useful, sources for simulations in numerical relativity. The collapse of an axisymmetric disk of collisionless matter provides the simplest example of matter collapse exhibiting the two most significant and challenging aspects of relativistic gravitation: black hole formation and gravitational radiation generation. Since the matter source resides entirely in the equatorial plane, the matter evolution equations in axisymmetry are 1-dimensional (i.e., they depend on r alone). The

---

[26] Anninos *et al.* (1995) and references therein; Matzner *et al.* (1995).
[27] See, e.g., Hockney and Eastwood (1981); Fridman and Polyachenko (1984); Binney and Tremaine (1987).

source influences the gravitational field, which is 2-dimensional (i.e., a function of r and θ), via "jump conditions" across the equator. When the disk matter is initially at rest, the ensuing collapse provides an interesting analogy to Oppenheimer–Snyder collapse to a black hole in spherical symmetry, but with the added important feature of gravitational radiation production. Since the gravitational field is dynamical in disk collapse, the full machinery of numerical relativity is required to follow the evolution, while spherical Oppenheimer–Snyder collapse is analytic.

The same equations discussed earlier in this chapter to evolve the axisymmetric collapse of isolated clusters and the head-on collision of binary clusters are readily adapted to handle infinitesimally thin disks of collisionless particles.[28] These equations, together with the relevant jump conditions, are summarized in Appendix F. Dynamical simulations that solve these equations have explored different relativistic systems and dynamical effects. The growth of ring instabilities in "cold" equilibrium disks, and their suppression in disks with sufficient velocity dispersion (i.e., "hot" disks), has been studied. Gravitational radiation from oscillating disks, as well as damping of the oscillations by radiation reaction, has been determined. The calculation of gravitational waveforms from disk collapse to black holes has been particularly useful. It has provided another arena for advancing computational machinery – in this case perturbation methods – that can be used in conjunction with numerical simulation data to determine late-time gravitational waveforms when the final state of a dynamical system approaches a Schwarzschild black hole.[29] We will discuss this application below.

We focus on the collapse of a disk of nonrotating, cold matter to a black hole. In this case all the disk matter is instantaneoulsy at rest at $t = 0$ and the spacetime evolution begins at a moment of time symmetry. The construction of appropriate initial data can be accomplished in much the same way that initial data for spacetimes containing relativistic prolate spheroids can be built from their Newtonian counterparts, as in exercise 10.1. For thin disks, however, we need to construct oblate spheroids, taking the limit as their eccentricity goes to unity to get flat configurations. Such a construction is summarized in exercise 10.4.

> **Exercise 10.4** Consider as in exercise 10.1 a homogeneous spheroid in Newtonian gravitation. The potential obeys Poisson's equation, $\nabla^2 \Phi_N = 4\pi\rho_N$, where the density ρ_N is again given by equation (10.2). Now treat an oblate spheroid, for which the eccentricity is defined to be $e = (1 - c^2/a^2)$, and the Newtonian gravitational binding energy is given by
>
> $$ W_N = -\frac{1}{2} \int \rho_N \Phi_N d^3x = \frac{3}{5} \frac{M_N^2}{a} \frac{\arcsin(e)}{e}. \tag{10.11} $$
>
> (a) Construct a relativistic spheroid by setting $2\pi\psi^5\rho \equiv 4\pi\rho_N$ as in exercise 10.1, but now for the mass density ρ inside an oblate spheroid of collisionless particles

[28] Abrahams *et al.* (1994).
[29] Abrahams *et al.* (1995).

momentarily at rest. As in exercise 10.1, the conformal factor satisfies $\psi = 1 - \Phi_N$, where Φ_N is the potential for the corresponding Newtonian spheroid. Show that equations (10.6) and (10.7) apply once again, but for the parameter α given by

$$\alpha = \frac{12}{5ae}\arcsin(e). \tag{10.12}$$

(b) Consider the infinitesimally thin disk of radius a obtained by letting $e \to 1$ in the oblate spheroid considered above. Show that for the homogeneous Newtonian spheroid, the surface density satisfies

$$\sigma_N = \frac{3M_N}{2\pi a^2}\left(1 - \frac{r^2}{a^2}\right)^{1/2}, \tag{10.13}$$

while for the corresponding relativistic disk, the total mass density σ and rest-mass surface density σ_0 are given by

$$\sigma = \sigma_0 = \int_-^+ \rho r \, \sin\theta \, d\theta = \frac{2}{\psi^5}\sigma_N. \tag{10.14}$$

(c) Show that the total mass and rest-mass of the relativistic disk are related by

$$M = \frac{2M_0}{1 + (1 + 6\pi M_0/5a)^{1/2}}. \tag{10.15}$$

The collapse of a thin disk constructed in this fashion is homologous (i.e., self-similar), and the solution is analytic, in Newtonian gravitation.[30] In general relativity, by contrast, one has to integrate the full set of $2 + 1$ equations to determine the spacetime. The simulation summarized here was performed using a spatial grid of 300 radial and 16 angular zones in the upper hemisphere, with the matter source sampled by 24 000 particles. Plots of the particle positions at selected times during the collapse of a disk with initial radius $R_0/M = 1.5$ are shown in Figure 10.6 (recall that in isotropic coordinates, a Schwarzschild black hole has a radius $R_0/M = 0.5$). The location of the apparent horizon, which appears at about $t/M = 4.0$, is also shown. The moving radial mesh algorithm, in which the innermost zones follow the infall of the matter, enables the integrations to continue reliably for a time $\Delta t/M \approx 15$ after the appearance of the horizon. Beyond that, the effect of "grid stretching" along the black hole throat induces numerical inaccuracies. Well before this point, however, all of the matter is well inside the horizon.

Close limit approximation

Even though disk collapse is one of the simplest radiating systems with a matter source, it is not possible to track the evolution long enough with this code to read off the full waveform directly from the simulation. The problem is that a significant amount of the radiation is still in the near-zone, strong-field region at the time the simulation breaks down. This difficulty occurs even though all of the matter has long since disappeared into the black hole. This

---

[30] See Abrahams *et al.* (1994), Section III.A, with $h = 0 = \xi$.

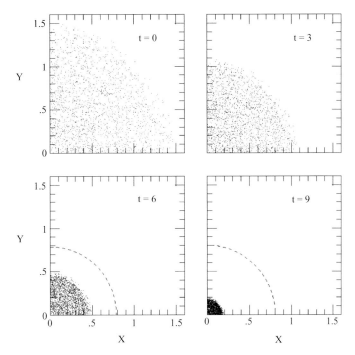

Figure 10.6 Snapshots of the particle positions for the collapse of a cold relativistic disk. Initially the radius is $R_0/M = 1.5$ and the particles are all at rest. The apparent horizon (dashed line) first appears at $t \simeq 4.0M$. The coordinate labels (t, x, y) are in units of M. [From Abrahams *et al.* (1994).]

problem can be overcome in one of several ways. Black hole excision or moving puncture techniques are two approaches; we will discuss these methods in detail in Chapters 13.1 and 14.2.3. Here instead we discuss a different approach that is particularly useful when the final state of a dynamical system is a Schwarzschild black hole, as is the case here. The method is based on black hole perturbation theory and approximates the spacetime at late time as a single perturbed black hole. The perturbation of the black hole is computed on a spatial slice from the numerical metric and extrinsic curvature using gauge-invariant perturbation theory.[31] Waveforms are determined by evolving this perturbation to infinity using the Zerilli equation (9.90).

The method employed here for the collapsing disk is an extension of the "close limit" approach introduced by Price and Pullin (1994) to treat radiation from merging black holes. Price and Pullin considered Misner initial data representing two black holes at a moment of time-symmetry.[32] They realized that when the two holes were sufficiently close, the system could be treated as a single, perturbed black hole. By applying gauge invariant perturbation theory, they calculated this perturbation and evolved it using the Zerilli equation. This allowed them to compute asymptotic waveforms and emitted energies. Remarkably, for

---

[31] Moncrief (1974); see discussion in Chapter 9.4.1.
[32] See Chapter 3.1, equation (3.26) for Misner initial data.

fairly small separations, the energies and waveforms agreed well with the results of full $2 + 1$ numerical simulations, like the ones discussed in the previous section.[33]

The remarkable success of the "close limit" approach can be explained, at least in part, by the presence of the black hole horizon and its use as an inner boundary at which all matter and radiation are purely ingoing. Once the horizon forms and the spacetime settles down to a perturbed Schwarzschild black hole, data obtained by further evolution with a $2 + 1$ numerical routine are no longer required to obtain the final gravitational waveform. The $2 + 1$ simulation can be terminated at a relatively early epoch; instead of continuing the simulation in order to propagate the exterior radiation pulse out to the weak-field extraction regime, a simpler set of black hole perturbation equations can be solved for the waveform, using the fields on the last numerical time slice for initial data. However, this "close limit" approach does require the formation of a black hole during the numerical simulation: its horizon is essential for preventing all further evolution inside the black hole from influencing the spacetime outside. The same approach does not work for, say, an oscillating neutron star, where there is no horizon.

In the perturbation approach we treat the spacetime metric as a static Schwarzschild black hole plus a perturbation (see Chapter 9.4.1). Since for our applications all perturbations have even (i.e., polar) parity, we can restrict the analysis to the even-parity modes and identify, from the perturbations, the gauge-invariant Moncrief functions R_{lm}, as well as their time derivatives, on a spatial slice (see equation 9.87). In axisymmetry, the only nonvanishing modes are those with $m = 0$, R_{l0}. These functions now provide initial data for the Zerilli equation (9.90)

$$\partial_t^2 R_{l0} - \partial_{r_*}^2 R_{l0} + V_l^{(e)} R_{l0} = 0, \tag{10.16}$$

where $r^* = r_s + 2M \ln(r_s/M - 1)$ is the tortoise coordinate, $r_s = r(1 + M/2r)^2$ is the Schwarzschild radius corresponding to (isotropic) radius r, and the Zerilli potential $V_l^{(e)}$ is given by equation (9.91). Given that this equation depends on only one spatial coordinate, it can be integrated on a fine mesh from a small radius very close to the event horizon, say $r_*/M = -500$, to a very large radius, say $r_*/M = 1000$, until the perturbation has propagated out to a large distance, well beyond the peak of the potential just outside the horizon. At large radius, the even-parity gravitational wave amplitude h_+ can be computed from the R_{l0} as in equation (9.109). This perturbation waveform can be compared with the waveform calculated by standard spacelike extraction of the $2 + 1$ radiation data at large distance.[34] In contrast to the perturbation result, the amplitude found from standard extraction is cut short once the $2 + 1$ integrations break down due to, e.g., grid stretching.

We can now mention another advantage of the perturbation method over wave extraction at a finite radius. Integrating the Zerilli equation (10.16) to large radii automatically takes into account the effects of gravitational wave backscatter off the black hole curvature.

---

[33] See also Abrahams and Cook (1994), who extended this technique to initial data representing boosted black holes with a common apparent horizon.

[34] Chapter 9.2.2. The extraction algorithm of Abrahams and Evans (1990) was actually employed for the disk scenario described here.

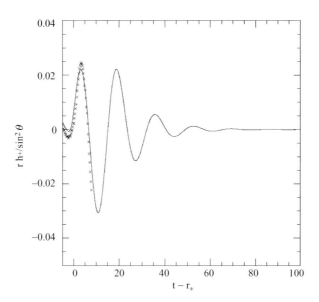

Figure 10.7 The gravitational waveform computed using the perturbation method for disk collapse is plotted as a function of retarded time in units of M. For comparison, the gravitational waveform computed at a radius $r/M = 8$ using a standard spacelike radiation extraction technique is also shown (crosses). [From Abrahams *et al.* (1995).]

These effects can only be incorporated approximately with standard extraction methods, if an integration over a timelike cylinder is used to separate off near-zone effects.

The perturbation and standard extraction waveforms are compared for disk collapse in Figure 10.7.[35] The perturbation results used simulation data at $t/M \approx 4.7$. Since the simulation becomes inaccurate within $t/M \approx 20M$ there is little time for the wave to reach the numerical extraction radius at $r/M = 8$. The standard method can only measure the wave from the initial infall phase and the first black hole oscillation before the simulation breaks down. The two waveforms are in good agreement during this epoch. But the perturbation waveform continues to track the black hole ringdown radiation, exhibiting the dominant quasinormal mode oscillations (of wavelength $\lambda/M \approx 16.8$ for the lowest $l = 2$ mode) that characterize a perturbed Schwarzschild black hole. Simulations that employ either black hole excision or the moving puncture method can track the fully nonlinear evolution of black holes, from formation to ringdown, to arbitrarily late times.

10.4 Collapse of rotating toroidal clusters

Most objects found in nature have spin, including relativistic configurations like neutron stars and black holes. So it is imperative that we learn how to evolve numerical spacetimes with net *rotation*. For an axisymmetric spacetime with rotation, the form of the metric

---

[35] Abrahams *et al.* (1994).

given by equation (10.1) for the nonrotating case must be generalized. The spatial line element in the quasi-isotropic gauge in spherical polar coordinates now becomes

$$dl^2 = A^2(dr^2 + r^2 d\theta^2) + B^2 r^2 (\xi d\theta + \sin\theta d\phi)^2, \qquad (10.17)$$

and all three components of the shift vector, $(\beta^r, \beta^\theta, \beta^\phi)$, may be present. Each of the metric components is a function of t, r and θ, but not ϕ. Once again the variable η represents the even-parity radiative polarization state, as $\eta \to h_+$ at large distances from the source. In the case of rotation, the odd-parity polarization state can also be present. At large distance the variable ξ is a measure of the odd-parity wave amplitude, as $\partial \xi / \partial r \to -\partial h_\times / \partial t$ when there is purely outgoing radiation.

The adopted form of the $2 + 1$ field equations, and the mean-field, particle simulation scheme used to integrate them, are generalizations of the ones used for the nonrotating systems described in the previous sections of this chapter and summarized in Appendix F.[36] The scheme is fully constrained: the Hamiltonian constraint is used to obtain a 3-metric component, and the three momentum constraints are used to compute components of the extrinsic curvature. The two remaining 3-metric components and two extrinsic curvature components are evolved. The shift vector components β^i are used to maintain the quasi-isotropic spatial gauge condition. The lapse function α is determined by the maximal time slicing condition $K = \partial_t K = 0$. The Hamiltonian constraint and lapse equations are elliptic equations, while the shift equations comprise a mixed parabolic-elliptic system of equations.

The above scheme has been used in a number of applications, including a study of the stability of rotating polytropic and toroidal clusters. The formation of Kerr black holes following the collapse of unstable clusters has been demonstrated. The collapse of highly rotating configurations is particularly interesting, since it is impossible to form a Kerr black hole with angular momentum $J/M^2 \geq 1$. In axisymmetry, where angular momentum cannot be radiated, such collapses must result either in new, stationary, nonsingular equilibrium configurations, or Kerr black holes surrounded by rapidly rotating disks, or naked singularities. In a parameter study of collapsing toroidal clusters of varying J/M^2, clusters with $J/M^2 \leq 1$ all collapse to black holes, while those with $J/M^2 \geq 1$ all collapse to new equilibrium configurations. While not unexpected theoretically, it is reassuring to obtain this outcome numerically. We will return to this same issue in Chapter 14.2.3 when we treat collapse of fluid stars.

Toroidal black holes and topological censorship

One of the most interesting outcomes of the simulations of the collapse of rotating toroidal clusters of collisionless particles to Kerr black holes is the emergence of the black hole

---

[36] See Abrahams *et al.* (1994) for the full set of field and collisionless matter equations for *rotating* spacetimes in axisymmetry and a summary of the numerical method.

event horizon as a toroid.[37] The initial matter distribution is based on a solution for a relativistic toroidal cluster in dynamical equilibrium. A discussion of the construction of initial data for rotating equilibrium systems, given a phase space distribution function, is postponed to Chapter 14.1.3. The adopted distribution function for a toroidal cluster is chosen to be of the form

$$f(E, J_z) = g(E)h(J_z), \tag{10.18}$$

where $E = -u_0$ is the energy and $J_z = u_\phi$ is the angular momentum of a particle, both per unit mass. Toroidal clusters with net spin then can be generated most simply by taking

$$g(E) = K\delta(E - E_{max}), \tag{10.19}$$

$$h(J_z) = \delta(J_z - J_0). \tag{10.20}$$

Here the total angular momentum $J = J_{ADM}$ is related to J_0 by $J = (M_0/m)J_0$, where M_0 is the total rest-mass of the configuration and m is the rest-mass of one particle. The quantity K is an arbitrary normalization constant and E_{max} is chosen to be the maximum energy of a particle in a spherical cluster of areal radius R_s, $E_{max} = (1 - 2M/R_s)^{1/2}$.

Here we will discuss the results for a toroidal cluster with an outer circumferential radius of $R_s/M = 4.5$. Collapse was induced by multiplying the angular velocity of each particle by a factor of 0.5, resulting in a *nonequilibrium* cluster with a total angular momentum of $J/M^2 = 0.65$. Re-solving the constraint equations after reducing the angular velocities to get valid initial data, the collapse was followed numerically. The computational mesh consisted of 200 radial zones and 16 angular zones (for one hemisphere) and 3000 particles. The outer boundary of the mesh was placed at $50M$, well outside the matter source.

The toroid initially collapses along the rotation axis to a thin hoop. Then, while undergoing oscillations along the rotation axis, it collapses radially inwards. The final configuration is a stationary Kerr black hole containing all of the matter. The matter distribution and horizons are plotted at selected times in Figures 10.8 and 10.9. The event horizon is determined by tracing null rays backwards in time from the end of the simulation at $t/M = 23.2$, at which point the apparent horizon coincides with the event horizon.[38] The topology of the event horizon is rather remarkable: it initially develops as a toroid, beginning at $t/M = 13.2$. The event horizon is seen to form initially entirely in the vacuum between the origin and the inner edge of the collapsing toroidal cluster. It then expands to fill up the the "doughnut hole", becoming topologically spherical at $t/M \approx 13.5$. At this instant the outer edge of the event horizon has reached the inner edge of the matter toroid. The spactime diagram for the collapse looks quite similar to the diagram plotted in Figure 10.5. The line of crossover points at which light rays enter the horizon, and the cusp formed by rays at the point at which the line of crossovers terminates, are all present as in Figure 10.5.

---

[37] Abrahams *et al.* (1994); Hughes *et al.* (1994).
[38] Hughes *et al.* (1994); see the discussion in Chapter 7.2.

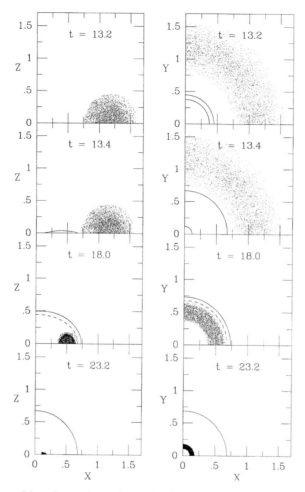

Figure 10.8 Snapshots of the collapse of a rotating toroid of collisionless particles at selected times. The left panels display meridional slices, the right panels show the equatorial plane. The solid line denotes the event horizon, the dashed line the apparent horizon. The earliest times shown occur soon after the appearance of the toroidal event horizon. By the final time shown here, the event and apparent horizons coincide. The coordinate labels (t, x, y, z) are in units of M. [From Hughes *et al.* (1994).]

Of course, some aspects of the collapse noted above are gauge-dependent; the collapse could look quite different in a different time-slicing, for example. Nevertheless, when the results of the numerical simulation were first reported, it appeared that they might be in conflict with black hole theorems regarding "topological censorship". These theorems permit a nonstationary black hole to have the topology of a 2-sphere or a torus,[39] but a torus can persist only for a short time. According to topological censorship,[40] the hole in a

---

[39] Gannon (1976).
[40] Friedman *et al.* (1993); Jacobson and Venkataramani (1995).

Figure 10.9 Three-dimensional views of the collapse of the rotating toroid shown in Figure 10.8. The images are viewed from 45° above the equatorial plane. The first image shows the initial configuration. The remaining three images are at the same times as the last three images in Figure 10.8. The outermost shaded region is the event horizon, the shaded region inside it is the apparent horizon. The scale of the last three images is larger by a factor of 2 over the first image. The clock on the lower right denotes the fraction of time that has elapsed since the onset of the simulation. [From Hughes *et al.* (1994).]

toroidal horizon must close up quickly, before a light ray can pass through. Moreover, the black hole must be a 2-sphere if no null generators enter the horizon at later times.[41]

Once the numerical simulations were studied in detail and the geometry of the transient toroidal event horizon was analyzed by means of a spacetime diagram like Figure 10.5, the results were shown to be completely consistent with the theorems concerning the topology of black hole horizons.[42] At late times, when equilibrium has been reached, the topology is spherical, in agreement with the theorem of Hawking.[43] At early times, the topology is temporarily toroidal. However, the line of crossovers traced out by a point on the inner rim of the torus is *spacelike*. That implies that the "hole" in the torus indeed closes up faster than the speed of light, in compliance with topological censorship. Finally, at intermedidate times, when the horizon is spanned by its full complement of null generators, the rotating black hole has a spherical topology, in accord with the theorems of Browdy and Galloway.

[41] Browdy and Galloway (1995).
[42] Shapiro *et al.* (1995).
[43] Hawking (1972).

11 Recasting the evolution equations

At this point we might suspect that we are all set to carry out dynamical simulations involving relativistic gravitational fields in three spatial dimensions, i.e., dynamical simulations in full $3 + 1$ dimensional spacetimes. After all, we have derived the $3 + 1$ evolution equations for the gravitational fields in Chapter 2, developed techniques for the construction of initial field data in Chapter 3, and discussed strategies for imposing suitable coordinate conditions in Chapter 4. Should the spacetime in which we are interested contain the most common matter sources, we can consult Chapter 5 for the matter source terms and the matter equations of motion. Beyond assembling all of the relevant equations, we have also sketched algorithms for solving them numerically in Chapter 6. If black holes are present, we have derived methods to locate and measure their horizons in Chapter 7; if gravitational waves are generated we have discussed how to extract radiation waveforms numerically in Chapter 9. Finally, we have evolved relativistic systems in spherical symmetry ($1 + 1$ dimensions) in Chapter 8 and in axisymmetry ($2 + 1$ dimensions) in Chapter 10. What else is left to do before going on to perform simulations in full $3 + 1$ dimensions?

Suppose then we were to plunge ahead with the tools currently at our disposal to build a dynamical spacetime in $3 + 1$ dimensions. We could gain some confidence and computational experience by choosing to explore as our first case a simple dynamical system with a known analytic solution. A linear gravitational wave propagating in vacuum would do very nicely. We could grab the initial data for such a wave from Chapter 9.1.2, choose a reasonable set of coordinate conditions from the sample menu provided in Chapter 4 and employ the standard $3 + 1$ (ADM) evolution equations derived in Chapter 2 to evolve the wave.

What would we find? Unfortunately, despite all of our methodical preparation and best numerical implementation, most likely our code would crash after a rather short time! The problem is illustrated in Figure 11.1, which shows the failure of our standard evolution scheme to evolve such a linear wave in a stable fashion. What might we conclude? Simply the following: since even this simple, linear, vacuum spacetime encounters numerical difficulties when evolved with the standard set of $3 + 1$ equations, we can expect that most other dynamical spacetime in $3 + 1$ dimensions will encounter such difficulties.

As it turns out, the standard evolution equations as derived in Chapter 2 are not yet in a form that is suitable for stable numerical integration. Most unconstrained simulations that implement these equations, at least in three spatial dimensions, turn out to be unstable.[1] The failure of these equations can be understood in terms of their mathematical properties, as we will briefly discuss in Section 11.1. Following this summary, we will discuss reformulations of the evolution equations that avoid these problems and have proven more robust and successful numerically. But first, in order to model the shortcomings of the standard evolution equations of Chapter 2 in a simple way and to obtain some guidance on how to fix them, we shall return to Maxwell's equations in Section 11.2. Fortified by the insight provided by this example, we will then present several different reformulations of the $3 + 1$ evolution equations that have proven very successful in many numerical simulations.

11.1 Notions of hyperbolicity

Recall from Chapter 6 that we can cast a hyperbolic partial differential equation – for example the simple wave equation (6.3) – in the first-order form

$$\partial_t \mathbf{u} + \mathbf{A} \cdot \partial_x \mathbf{u} = \mathbf{S}, \tag{11.1}$$

where $\mathbf{u}$ is a solution vector, $\mathbf{S} = \mathbf{S}(\mathbf{u})$ is a source vector, and where we have called the matrix $\mathbf{A}$ the velocity matrix (see equation 6.7). Here we have assumed that the solution depends on t and x only; in more than one spatial dimension we may generalize equation (11.1) to read

$$\partial_t \mathbf{u} + \mathbf{A}^i \cdot \partial_i \mathbf{u} = \mathbf{S}. \tag{11.2}$$

If the solution vector $\mathbf{u}$ has n components, then each matrix $\mathbf{A}^i$ is an $n \times n$ matrix. For the purposes of this section we may ignore the source term and set $\mathbf{S} = 0$.

We call a problem *well-posed* if we can define some norm $\| \dots \|$ so that the norm of the solution vector satisfies[2]

$$\| \mathbf{u}(t, x^i) \| \leq k e^{\alpha t} \| \mathbf{u}(0, x^i) \| \tag{11.3}$$

for all times $t \geq 0$. Here k and α are two constants that are independent of the initial data $\mathbf{u}(0, x^i)$. Stated differently, solutions of a well-posed problem cannot increase more rapidly than exponentially. Clearly, this is a very desirable property, but it is not guaranteed for all hyperbolic systems.

As it turns out, there exist different types of hyperbolicity, and only for some of these types are the equations well-posed. To analyze these hyperbolicity properties, we consider an arbitrary unit vector n^i and construct the matrix $\mathbf{P} = \mathbf{A}^i n_i$, which is sometimes referred

---

[1] The high degree of spatial symmetry in spherical symmetry and even axisymmetry allows for a number of different strategies to obtain stable evolutions, including fully or partially constrained evolution, the use of special coordinate systems, or the introduction of new variables.

[2] Much of this section follows the arguments of Alcubierre (2008a); see also Kreiss and Lorenz (1989) and Alcubierre (2008b).

to as the *principal symbol* or the *characteristic matrix* of the system. Based on the properties of **P** we then distinguish different notions of hyperbolicity. In particular, we call the system[3]

- *symmetric hyperbolic* if **P** can by symmetrized in a way that is independent of n^i,
- *strongly hyperbolic* if, for all unit vectors n^i, **P** has real eigenvalues and a complete set of eigenvectors,
- and *weakly hyperbolic* if **P** has real eigenvalues but not a complete set of eigenvectors.

Simple wave equations, for example, are symmetric hyperbolic. Symmetric hyperbolic systems are automatically strongly hyperbolic. The key result for our purposes states that strongly hyperbolic equations are well-posed, while systems that are only weakly hyperbolic are not.[4]

We are not quite ready yet to analyze the hyperbolicity of the ADM evolution equations (2.134) and (2.135), since our analysis above applies to first-order equations, while the ADM equations are second order in space. One approach to deal with this is to introduce a so-called first-order reduction of the equations, achieved by writing all second derivatives in terms of first derivatives of a new set of auxiliary functions that contain first derivatives of the original variables.[5] An analysis of a first-order reduction of the ADM evolution equations shows that these equations are in fact only weakly hyperbolic.[6] As a consequence, the evolution problem is not well-posed, and we have no reason to expect the solutions – or numerical implementations – to be well-behaved.

Clearly, then, we should try to recast these equations in a form that is strongly hyperbolic. How can we do that without modifying Einstein's equations? The answer lies in the constraint equations. We first observe that the constraints vanish, at least analytically.[7] We are therefore entitled to add multiples of the constraint equations to the evolution equations. Furthermore, the constraints contain up to second derivatives of the gravitational fields, as do the evolution equations. Adding the constraints to the evolution equations therefore affects the appearance of the highest order derivatives, which in turn affects the principal symbol **P** and hence the hyperbolicity. We will see that many reformulations of the evolution equations also involve the introduction of new variables that absorb some of the first derivatives of the gravitational fields.

Before proceeding we add a word of caution. The notion of well-posedness rules out modes that grow faster than exponentially with time.[8] From a numerical perspective,

---

[3] Note that different authors use slightly different conventions.

[4] See, e.g., Kreiss and Lorenz (1989). Note that this analysis focuses on the principal part of the equations only; the source term may also lead to faster-than-exponential growth.

[5] For an alternative approach see, e.g., Gundlach and Martín-García (2006a), who introduce a notion of hyperbolicity that applies to systems that are second order in space.

[6] See, e.g., Kidder *et al.* (2001).

[7] Here we mean that if we move the source terms appearing in equations (3.1) and (3.2) to the left-hand sides, the combination of terms on the left-hand sides must then add up to zero at all times; see, e.g., equations (11.48) and (11.49).

[8] At least in the absence of source terms; see footnote 4 above.

however, exponentially growing modes are still very bad, and can easily terminate a simulation after only a short time. In fact, many of the symmetric or strongly hyperbolic systems that have been introduced over the years lead to such exponentially growing modes in numerical implementations. Unless these modes can be controlled,[9] these systems are not very useful for numerical simulations. From a numerical perspective, then, well-posedness is a necessary but not a sufficient condition.

In the following we will discuss several different approaches that lead to different reformulations of the evolution equations that are both strongly hyperbolic and that have been used successfully in numerical simulations. In the remainder of this chapter we will pursue a more heuristic approach than the one sketched in this section and will marshal mathematical intuition and exploit analogies to motivate the different approaches. In fact, some of the systems discussed below were developed in exactly this way and were discovered to have desirable numerical properties empirically before they were analyzed mathematically and revealed to be strongly hyperbolic. To illustrate some of these heuristic arguments for a simple and familiar example we will begin by discussing Maxwell's equations in Section 11.2.

11.2 Recasting Maxwell's equations

In Chapter 2.2 we have seen that we can bring Maxwell's evolution equations in a Minkowski spacetime into the form

$$\partial_t A_i = -E_i - D_i \Phi \tag{11.4}$$

$$\partial_t E_i = -D^j D_j A_i + D_i D^j A_j - 4\pi j_i \tag{11.5}$$

(see equations 2.11 and 2.12). Recall that A_i is the three-vector potential, the magnetic field satisfies $B_i = \epsilon_{ijk} D^j A^k$ and therefore is automatically divergence-free, ϕ is a gauge potential, and the electric field E_i has to satisfy the constraint equation (2.5),

$$D_i E^i = 4\pi \rho_e. \tag{11.6}$$

In the above equations ρ_e is the electric charge density and j_i the current density.

In Chapter 2.7 we have discussed some of the similarities of Maxwell's equations in the above form with the ADM evolution equations (2.134) and (2.135), namely

$$\partial_t \gamma_{ij} = -2\alpha K_{ij} + D_i \beta_j + D_j \beta_i \tag{11.7}$$

and

$$\partial_t K_{ij} = \alpha(R_{ij} - 2K_{ik}K^k_{\ j} + K K_{ij}) - D_i D_j \alpha - 8\pi\alpha(S_{ij} - \tfrac{1}{2}\gamma_{ij}(S - \rho)) \\ + \beta^k \partial_k K_{ij} + K_{ik}\partial_j \beta^k + K_{kj}\partial_i \beta^k. \tag{11.8}$$

---

[9] See, e.g., Scheel *et al.* (1998); Kidder *et al.* (2001) for examples.

If we identify the vector potential A_i with the spatial metric γ_{ij} and the electric field E_i with the extrinsic curvature K_{ij}, we see that the right-hand sides of both equations (11.4) and (11.7) contain a field variable and a spatial derivative of a gauge variable, while the right-hand sides of both equations (11.5) and (11.8) involve matter sources as well as second spatial derivatives of the second field variable. In equation (11.8) these second derivatives are hidden in the Ricci tensor R_{ij}, which we can write, for example, as

$$R_{ij} = \frac{1}{2}\gamma^{kl}\left(\partial_l\partial_i\gamma_{kj} + \partial_j\partial_k\gamma_{il} - \partial_j\partial_i\gamma_{kl} - \partial_l\partial_k\gamma_{ij}\right) + \gamma^{kl}\left(\Gamma^m_{il}\Gamma_{mkj} - \Gamma^m_{ij}\Gamma_{mkl}\right). \quad (11.9)$$

We can now exploit these similarities by focusing on the simpler Maxwell system of equations to identify some of the computational shortcomings of these forms of the evolution equations.

We first note that equations (11.4) and (11.5) *almost* can be combined to yield a wave equation, which would make the system symmetric hyperbolic. To see this, take a time derivative of equation (11.4) and insert equation (11.5) to form a single equation for the vector potential A_i,

$$-\partial_t^2 A_i + D^j D_j A_i - D_i D^j A_j = D_i \partial_t \Phi - 4\pi j_i. \quad (11.10)$$

On the left-hand side, the second time derivative combines with the Laplace operator $D^j D_j A_i$ to form a wave operator (d'Alembertian). Equations (11.4) and (11.5) would then constitute a wave equation for the components A_i if it weren't for the mixed derivative term $D_i D^j A_j$.

In general relativity the situation is very similar. The Ricci tensor R_{ij} on the right hand side of equation (11.8) contains three mixed derivative terms in addition to the term with a Laplace-like operator acting on γ_{ij}, i.e., $\gamma^{kl}\partial_l\partial_k\gamma_{ij}$. Without these mixed derivative terms the standard ADM equations could be written as a set of wave equations for the components of the spatial metric, which would make them symmetric hyperbolic. As we discussed in Section 11.1 we would like to evolve a system that is well-posed, but the presence of these additional terms may spoil this property.

These considerations suggest that it would be desirable to eliminate the mixed derivative terms. In electrodynamics, three different approaches can be taken to eliminate the $D_i D^j A_j$ term: one can make a special *gauge choice*; one can bring Maxwell's equations into a first-order *symmetric hyperbolic* form; or, one can introduce an *auxiliary variable*. In the remainder of this section we will discuss briefly each one of these three strategies.

11.2.1 Generalized Coulomb gauge

The most straightforward approach to eliminating the undesirable terms is to choose a *gauge* so that the term $D_i D^j A_j$ disappears. Define the quantity

$$\Gamma \equiv D^i A_i. \quad (11.11)$$

To eliminate the offending term, we may then set

$$\Gamma = 0. \tag{11.12}$$

This choice is the familiar *Coulomb gauge* condition.

> **Exercise 11.1**   Show that in the Coulomb gauge the gauge potential Φ satisfies
> Poisson's elliptic equation
>
> $$D^i D_i \Phi = -4\pi \rho_e. \tag{11.13}$$

In general relativity, an analogous approach can be taken by choosing harmonic coordinates, which bring the equations into the form of a wave equation.[10] (The reader may wish to review Chapter 4.3, and, especially, exercise 4.10.) We will return to this strategy in Section 11.3 below.

This approach has disadvantages, too. As exercise 11.1 demonstrates, the choice (11.12) forces us to satisfy a constraint that may be inconvenient to solve. In general relativity, the analogous harmonic gauge condition, $^{(4)}\Gamma^a = 0$, similarly imposes a coordinate choice – the lapse and shift now have to satisfy equations (4.44) and (4.45). These coordinates may not be well-suited to the problem at hand, and may lead to coordinate singularities. One way to regain coordinate freedom is not to set $\Gamma = 0$ as in equation (11.12), but instead to set Γ equal to some yet-to-be-chosen *gauge source function* $H(t, x^i)$,

$$\Gamma = H(t, x^i). \tag{11.14}$$

Retracing the steps in exercise 11.1 we could find out how a particular choice of H affects the gauge potential Φ. Similarly, in general relativity we can relax the harmonic gauge condition with the help of a 4-dimensional gauge source function, $^{(4)}\Gamma^a = H^a(t, x^i)$, as we will see in Section 11.3. This approach is commonly referred to as "generalized harmonic coordinates", and, in analogy, we could refer to the condition (11.14) imposed on Γ as the "generalized Coulomb gauge".

11.2.2 First-order hyperbolic formulations

An alternative, gauge-covariant approach to bringing Maxwell's equations into a *symmetric hyperbolic* form is to take a time derivative of equation (11.5) instead of equation (11.4), which then yields

$$\partial_t^2 E_i = D_i D^j(-E_j - D_j \Phi) - D_j D^j(-E_i - D_i \Phi) - 4\pi \partial_t j_i. \tag{11.15}$$

Using the constraint (11.6) we can eliminate the first term and find a wave equation for E_i,

$$-\partial_t^2 E_i + D_j D^j E_i = 4\pi(\partial_t j_i + D_i \rho_e), \tag{11.16}$$

---

[10] This property was first realized by De Donder (1921) and Lanczos (1922). Many of the early hyperbolic formulations of Einstein's equations were based on this gauge choice, e.g., Choquet-Bruhat (1952, 1962); Fischer and Marsden (1972).

which is symmetric hyperbolic. Interestingly, the gauge fields A_i and Φ have disappeared entirely from this equation.[11]

If desired we can write the wave equation (11.16) as a system of coupled first-order equations by introducing a new set of first-order variables. This is exactly how we found the system (6.6) from the wave equation (6.3) in Section 6.1. We will discuss similar approaches in general relativity in Section 11.4.

11.2.3 Auxiliary variables

A third approach to "fixing" Maxwell's equations is in some ways similar to the first one. We again use the auxiliary variable Γ defined in equation (11.11). However, instead of setting this variable to a predetermined function – given by the gauge source function – we now treat this variable as a new, independent field that we evolve. We can derive an evolution equation for Γ from equation (11.4),

$$\partial_t \Gamma = \partial_t D^i A_i = D^i \partial_t A_i = -D^i E_i - D_i D^i \Phi = -D_i D^i \Phi - 4\pi \rho_e. \quad (11.17)$$

Note that we have used the constraint equation (11.6) in the last equality, similar to the way we used the same constraint to arrive at the wave equation (11.16). Exercise 11.4 below demonstrates how this step affects the properties of constraint violations.

In terms of Γ, the evolution equation (11.5) for E_i becomes

$$\partial_t E_i = -D_j D^j A_i + D_i \Gamma - 4\pi j_i. \quad (11.18)$$

Equations (11.4), (11.17) and (11.18) now constitute the evolution equations in this new formulation, and equations (11.6) and (11.11) are the constraint equations. In this formulation the mixed derivative term $D_i D^j A_j$ has been eliminated without using up any gauge freedom, which is still imposed via the choice of Φ. In Section 11.5 we will introduce an analogous reformulation of the ADM equations.

11.3 Generalized harmonic coordinates

One way to avoid the problems associated with the standard ADM evolution equations (11.7) and (11.8) is to abandon these equations completely. Instead of using a $3 + 1$ decomposition of Einstein's equations as a starting point, we could start with the original 4-dimensional version

$$G_{ab} = 8\pi T_{ab} \quad (11.19)$$

---

[11] To quote from Abrahams and York, Jr., (1997), "the dynamics of electromagnetism have been cleanly separated from the gauge-dependent evolution of the vector and scalar potentials."

in the form $^{(4)}R_{ab} = 8\pi(T_{ab} - (1/2)g_{ab}T)$. Substituting the Ricci tensor as expressed by equation (4.46) yields

$$\frac{1}{2}g^{cd}\partial_d\partial_c g_{ab} - g_{c(a}\partial_{b)}\,^{(4)}\Gamma^c - \,^{(4)}\Gamma^c\,^{(4)}\Gamma_{(ab)c} - 2g^{ed}\,^{(4)}\Gamma^c_{e(a}\,^{(4)}\Gamma_{b)cd} - g^{cd}\,^{(4)}\Gamma^e_{ad}\,^{(4)}\Gamma_{ecb}$$

$$= -8\pi\left(T_{ab} - \frac{1}{2}g_{ab}T\right), \tag{11.20}$$

where we have reintroduced the definition (4.40),

$$^{(4)}\Gamma^a \equiv g^{bc(4)}\Gamma^a_{bc} = -\frac{1}{|g|^{1/2}}\frac{\partial}{\partial x^b}\left(|g|^{1/2}g^{ab}\right) = g^{bc}\nabla_b\nabla_c x^a \tag{11.21}$$

(see exercises 4.8 and 4.10 and the discussion that follows these exercises). The $^{(4)}\Gamma^a$ are contractions of Christoffel symbols, and therefore do not transform like vectors under coordinate transformations. We can now introduce a gauge by setting these quantities equal to some given *gauge source functions* H^a,

$$^{(4)}\Gamma^a = H^a(t, x^i). \tag{11.22}$$

This approach follows very closely our electromagnetic example of Section 11.2.1, and equation (11.22) is the direct analog of (11.14). Just like the choice (11.14) led to the elimination of the "mixed derivatives" terms in Maxwell's equations (11.10), inserting equation (11.22) into Einstein's equations (11.20) yields a nonlinear wave equation for the spacetime metric g_{ab}.[12] After some manipulations this equation can be brought into the form[13]

$$\boxed{\begin{array}{l} g^{cd}\partial_d\partial_c g_{ab} + 2\partial_{(a}g^{cd}\partial_c g_{b)d} + 2H_{(a,b)} - 2H_d\,^{(4)}\Gamma^d_{ab} \\[2mm] \quad + 2\,^{(4)}\Gamma^c_{bd}\,^{(4)}\Gamma^d_{ac} = -8\pi\left(2T_{ab} - g_{ab}T\right). \end{array}} \tag{11.23}$$

Here we have lowered the indices of H^a with the spacetime metric g_{ab}, $H_a \equiv g_{ab}H^b$. For the special choice $H^a = 0$ we recover the harmonic coordinates of Chapter 4.3. More generally we refer to this approach as "generalized harmonic coordinates".[14] This formalism was adopted by Pretorius (2005a,b) in his simulations of binary black hole coalescence and merger, which we will discuss in more detail in Chapter 13.

Identifying equations (11.21) and (11.22) imposes a new, 4-dimensional constraint

$$\mathcal{C}^a \equiv H^a - g^{bc\,(4)}\Gamma^a_{bc} = 0. \tag{11.24}$$

Equation (11.23) can be integrated directly for the spacetime metric g_{ab}. To stabilize the system, it is sometimes necessary to add linear combinations of the constraints (11.24) to the evolution equations (11.23), i.e., it is necessary to add terms proportional to the

---

[12] See Friedrich (1985); Garfinkle (2002). See also the related "Z4" formalism suggested by Bona *et al.* (2003).

[13] See Pretorius (2005b).

[14] This name is somewhat misleading, however, since it does not single out any particular family of coordinate systems. Instead, *any* arbitrary coordinate system can be generated by (11.22) with a suitable choice of H^a.

quantities $\mathcal{C}^a$, which may not be identically zero due to numerical errors.[15] We will discuss why adding constraints can stabilize evolution equations in Section 11.5.

The generalized harmonic approach to determing the metric in a dynamical spacetime differs in several ways from the 3 | 1 formalism, so that it is worthwhile to discuss some aspects of this formalism in some more detail. Most importantly, equation (11.23) is a second-order equation in time for the spacetime metric, while in the usual $3 + 1$ decomposition we integrate coupled equations for spatial metric and the extrinsic curvature that are first order in time. This difference effects both the initial data and the numerical implementation.

Within the $3 + 1$ decomposition, a set of initial data consists of values of the spatial metric γ_{ij} and the extrinsic curvature K_{ij} that satisfy the constraint equations, e.g., (2.132) and (2.133), at one instant of time. Given a choice for the lapse α and the shift β^i, γ_{ij} and K_{ij} can then be integrated forward in time with the evolution equations, e.g., (2.134) and (2.135). Equation (11.23), on the other hand, requires the spacetime metric g_{ab} and its first time derivative at some instant of time t as initial data.

One way of constructing such initial data is the following. We could first solve the contraint equations (2.132) and (2.133), in, say, the conformal thin-sandwich formalism described in Chapter 3.3. The freely specifiable variables then are the conformally related metric $\bar{\gamma}_{ij}$ and its time derivative, together with the trace of the extrinsic curvature K and its time derivative. Solving the equations yields the conformal factor ψ, the lapse α and the shift β^i. This is all the information required to construct the spacetime metric g_{ab} (e.g., equation 2.131). To find the time derivative of g_{ab} we can first evaluate the evolution equation (2.134), which yields $\partial_t \gamma_{ij}$ at $t = 0$. We can find the time derivatives of α and β^i from the condition $^{(4)}\Gamma^a = H^a$, again evaluated at $t = 0$. For the special case $H^a = 0$ this condition yields equations (4.44) and (4.45), which we can solve for the desired time derivatives (see exercise 4.9). If $H^a \neq 0$ these equations have additional source terms, but the derivation is very similar. From the time derivatives of the spatial metric, the lapse and the shift we can finally construct the time derivative of the spacetime metric, which completes the initial data for equation (11.23).

The appearance of second-order time derivatives in equation (11.23) also poses some numerical challenges. In finite difference applications these second derivatives can be handled by choosing a three-level scheme and using a finite difference representation similar to the one described by equation (6.22). The situation is complicated by the presence of the mixed space-time derivatives, which couple the function values at the new time level. To avoid an implicit scheme, the resulting system of equations can be solved iteratively.[16] Alternatively, equation (11.23) can be recast into a system of coupled first-order equations.[17]

---

[15] To stabilize his simulations of binary black holes, Pretorius (2005a) added a combination of these constraints to the evolution equations (11.23), as suggested by Gundlach *et al.* (2005).

[16] Pretorius (2005b).

[17] Lindblom *et al.* (2006).

Another important difference between this approach and $3 + 1$ decompositions is the way in which coordinates are imposed. In the $3 + 1$ decomposition, we choose a coordinate system with the help of the lapse α and the shift β^i. These quantities are directly related to the geometry of the spatial slices Σ, and in Chapter 4 we have seen how geometric considerations can guide the search for coordinate systems with particularly desirable properties. In this "generalized harmonic coordinate" approach, on the other hand, the coordinates are imposed by the gauge source functions H^a. These quantities do not have a direct geometrical meaning, and it is therefore much less clear how to construct a coordinate system with desirable properties. Pretorius (2005a) chose H_t to satisfy a wave equation that includes the lapse α (computed from g^{tt}) as a source, and $H_i = 0$. A priori there is not much reason to expect this choice to lead to a well-behaved coordinate system, but in binary black hole merger simulations it evidently does.

While the wave-like structure of equation (11.23) is very appealing, the uncertainty over how to impose suitable coordinate conditions has served to discourage many researchers from using this formalism. Instead, a number of other formulations of Einstein's equations have been constructed with similarly desirable mathematical properties as equation (11.23) but which, by contrast, impose the gauge conditions via a lapse and shift.

11.4 First-order symmetric hyperbolic formulations

As we discussed in Section 11.1, it could prove quite desirable to cast the evolution equations into a first-order hyperbolic form.[18] The first such formulations were based on harmonic coordinates[19] and used as a starting point the formalism of Section 11.3 with $H^a = 0$. The first formulations that departed from the assumption of harmonic coordinates were based on a spin-frame formalism.[20] Other formulations introduced partial derivatives of the metric and other quantities as new independent variables.[21] In analogy to the electromagnetic example in Section 11.2.2, it is also possible to take a time derivative of equation (11.8) to derive what is sometimes referred to as the "Einstein–Ricci" system of equations.[22] Another system of equations, sometimes called the "Einstein–Bianchi" system, can be derived from the Bianchi identities.[23] The "Einstein–Christoffel" system is constructed by introducing additional "connection" variables.[24] We also mention a so-called "λ"-system that embeds Einstein's equations into a larger symmetric hyperbolic system with the constraint surface of Einstein's equations acting as an attractor of the evolution.[25]

---

[18] See Reula (1998) for a more extensive survey than provided in this section.
[19] Choquet-Bruhat (1952, 1962); Fischer and Marsden (1972).
[20] Friedrich (1981, 1985)
[21] Bona and Massó (1992); Frittelli and Reula (1994).
[22] Choquet-Bruhat and Ruggeri (1983); Abrahams *et al.* (1995); Abrahams and York, Jr., (1997).
[23] Friedrich (1996); Anderson *et al.* (1997).
[24] Anderson and York, Jr. (1999).
[25] Brodbeck *et al.* (1999).

It is beyond the scope of this volume to review all of these formulations. Some have features that are not very desirable numerically, in that they restrict the gauge freedom, introduce extra derivatives of the matter variables, or introduce a large number of auxiliary variables. Only few of these formulations have been implemented numerically, and many of these implementations adopted simplifying symmetry conditions (e.g., spherical symmetry). Some of these implementations display advantages over the standard $3 + 1$ formalism,[26] but others reveal additional problems. For example, a particular equation in the "Einstein–Ricci" system turns out to produce an exponentially growing mode, which can be removed in spherical symmetry, but not in more general $3 + 1$ simulations.[27] Only a few of these systems have been implemented in $3 + 1$ dimensions, including the one of Bona and Massó (1992), and versions of the "Einstein–Christoffel" system of Anderson and York, Jr. (1999).[28] Since this "Einstein–Christoffel" formulation is particularly elegant, we provide a brief summary of this system as an example of a first-order hyperbolic formulation.

Starting with the standard $3 + 1$ or ADM formalism of Chapter 2 we define the new variables[29]

$$f_{kij} \equiv \Gamma_{(ij)k} + \gamma_{ki}\gamma^{lm}\Gamma_{[lj]m} + \gamma_{kj}\gamma^{lm}\Gamma_{[li]m}. \tag{11.25}$$

These functions are now promoted to independent functions. It can then be shown that the evolution equations (11.7) and (11.8) can be rewritten as the coupled system

$$d_t\gamma_{ij} = -2\alpha K_{ij},$$
$$d_t K_{ij} + \alpha\gamma^{kl}\partial_l f_{kij} = \alpha M_{ij}, \tag{11.26}$$
$$d_t f_{kij} + \alpha\partial_k K_{ij} = \alpha N_{kij}.$$

Here we have introduced the abbreviation

$$d_t \equiv \partial_t - \mathcal{L}_\beta, \tag{11.27}$$

and the source terms M_{ij} and N_{ijk} given by

$$
\begin{aligned}
M_{ij} =\ & \gamma^{kl}(K_{kl}K_{ij} - 2K_{ki}K_{lj}) + \gamma^{kl}\gamma^{mn}(4f_{kmi}f_{[ln]j} \\
& + 4f_{km[n}f_{l]ij} - f_{ikm}f_{jln} + 8f_{(ij)k}f_{[ln]m} + 4f_{km(i}f_{j)ln} \\
& - 8f_{kli}f_{mnj} + 20f_{kl(i}f_{j)mn} - 13f_{ikl}f_{jmn}) \\
& - \partial_i\partial_j\ln\bar\alpha - (\partial_i\ln\bar\alpha)(\partial_j\ln\bar\alpha) + 2\gamma_{ij}\gamma^{kl}\gamma^{mn}(f_{kmn}\partial_l\ln\bar\alpha \\
& - f_{kml}\partial_n\ln\bar\alpha) + \gamma^{kl}\Big((2f_{(ij)k} - f_{kij})\partial_l\ln\bar\alpha \\
& + 4f_{kl(i}\partial_{j)}\ln\bar\alpha - 3(f_{ikl}\partial_j\ln\bar\alpha + f_{jkl}\partial_i\ln\bar\alpha)\Big) \\
& - 8\pi S_{ij} + 4\pi\gamma_{ij}T,
\end{aligned} \tag{11.28}
$$

[26] e.g., Bona and Massó (1992).

[27] Scheel et al. (1997, 1998).

[28] The latter have been implemented numerically by Kidder et al. (2001, see also Kidder et al. (2000)) using spectral methods.

[29] Here we adopt the notation of Kidder et al. (2000).

and

$$
\begin{aligned}
N_{kij} = \;& \gamma^{mn}\left(4K_{k(i}f_{j)mn} - 4f_{mn(i}K_{j)k} + K_{ij}(2f_{mnk} - 3f_{kmn})\right) \\
& + 2\gamma^{mn}\gamma^{pq}\Big(K_{mp}(\gamma_{k(i}f_{j)qn} - 2f_{qn(i}\gamma_{j)k}) \\
& + \gamma_{k(i}K_{j)m}(8f_{npq} - 6f_{pqn}) + K_{mn}(4f_{pq(i}\gamma_{j)k} - 5\gamma_{k(i}f_{j)pq})\Big) \\
& - K_{ij}\partial_k \ln\bar\alpha + 2\gamma^{mn}(K_{m(i}\gamma_{j)k}\partial_n \ln\bar\alpha - K_{mn}\gamma_{k(i}\partial_{j)}\ln\bar\alpha) \\
& + 16\pi\gamma_{k(i}j_{j)}.
\end{aligned}
\tag{11.29}
$$

We have also used the "densitized" lapse function

$$
\bar\alpha = \gamma^{-1/2}\alpha
\tag{11.30}
$$

(see equation 3.100) and, in addition to the matter source terms defined in Chapter 2, the 4-dimensional trace of the stress energy tensor,

$$
T = g^{ab}T_{ab}.
\tag{11.31}
$$

The first-order, symmetric hyperbolic ("FOSH") system (11.26) is equivalent to the original set of evolution equations (11.7) and (11.8). Since the f_{kij} are evolved as independent functions, the defining relations (11.25) can be considered as a new set of constraint equations in addition to the usual ones given by equations (2.132) and (2.133). Note also that the source terms M_{ij} and N_{ijk} on the right-hand sides do not contain any derivatives of the fundamental variables (other than of the arbitrary lapse function $\bar\alpha$). Equations (11.26) can be combined to yield a wave equation for the components of the spatial metric γ_{ij} in which the right-hand sides appear as source terms.

Evolutions of a single black hole using a spectral implementation of this system are still unstable, but the lifetime of these simulation can be extended to late times by a generalization of the equations.[30] This generalization involves a redefinition of the independent variables and the addition of new constraints. The above equations can be embedded into a 12-parameter family of strongly hyperbolic formulations. The stability properties of the system depend sensitively on the choice of the free parameters, which can be understood analytically in terms of energy arguments.[31] Given this need for fine-tuning, and given the successes of both the generalized coordinate formalism of Section 11.3 and the BSSN formulation discussed below, this "Einstein–Christoffel" system has lost some of its appeal.

11.5 The BSSN formulation

In our illustration of Section 11.2 employing electromagnetism we have seen that we can also eliminate the "mixed second derivatives" in Maxwell's equations (11.10) by promoting the contraction Γ defined in equation (11.11) to be a new, independent function. In

---

[30] Kidder *et al.* (2001).
[31] See, e.g., Lindblom and Scheel (2002). Also see exercise 11.4 below and the related discussion, which illustrate how the addition of constraints affects the properties of evolution systems.

Sections 4.3 and 11.3 we have seen similarly that we can absorb the mixed second derivatives in the Ricci tensor with the help of the connection functions (11.21). The BSSN
formalism[32] adopts a similar strategy to simplify the three-dimensional, spatial Ricci tensor.[33] In addition, the conformal factor and the trace of the extrinsic curvature are evolved
separately in the BSSN formalism, which follows the philosophy of separating transverse
from longitudinal, or, equivalently, radiative from nonradiative, degrees of freedom. The
later idea was also the basis of the conformal decompositions discussed in Chapter 3 for
the construction of initial data.

To derive this formulation, we begin by writing the conformal factor ψ as $\psi = e^\phi$ so
that we have

$$\bar{\gamma}_{ij} = e^{-4\phi} \gamma_{ij}. \tag{11.32}$$

We then require that the determinant of the conformally related metric $\bar{\gamma}_{ij}$ be equal to that
of the flat metric η_{ij} in whatever coordinate system we are using, i.e.,

$$\phi = \frac{1}{12} \ln\left(\frac{\gamma}{\eta}\right). \tag{11.33}$$

In the following we will adopt a Cartesian coordinate system, so that $\bar{\gamma} = \eta = 1$.

As in equation (3.31) we split off from the extrinsic curvature its trace and conformally
rescale the remaining traceless piece A_{ij}. However, we choose a conformal rescaling that
is different from equation (3.35) and, instead, rescale A_{ij} the same way as we did for the
metric itself,

$$\tilde{A}_{ij} = e^{-4\phi} A_{ij}. \tag{11.34}$$

We will use tildes as opposed to the bars used in Chapter 3 to distinguish between these
different rescalings. Indices of $\tilde{A}_{ij}$ will be raised and lowered with the conformal metric
$\bar{\gamma}_{ij}$, so that $\tilde{A}^{ij} = e^{4\phi} A^{ij}$.

Evolution equations for ϕ and K can now be found by taking the trace of the evolution
equations (2.136) and (2.137), which yields

$$\partial_t \phi = -\frac{1}{6}\alpha K + \beta^i \partial_i \phi + \frac{1}{6}\partial_i \beta^i \tag{11.35}$$

and

$$\partial_t K = -\gamma^{ij} D_j D_i \alpha + \alpha(\tilde{A}_{ij} \tilde{A}^{ij} + \frac{1}{3}K^2) + 4\pi\alpha(\rho + S) + \beta^i \partial_i K. \tag{11.36}$$

Subtracting these equations from the evolution equations (11.7) and (11.8) leaves the
traceless parts of the evolution equations for $\bar{\gamma}_{ij}$ and $\tilde{A}_{ij}$ according to

$$\partial_t \bar{\gamma}_{ij} = -2\alpha \tilde{A}_{ij} + \beta^k \partial_k \bar{\gamma}_{ij} + \bar{\gamma}_{ik}\partial_j \beta^k + \bar{\gamma}_{kj}\partial_i \beta^k - \frac{2}{3}\bar{\gamma}_{ij}\partial_k \beta^k, \tag{11.37}$$

---

[32] Shibata and Nakamura (1995); Baumgarte and Shapiro (1998).
[33] See also Nakamura *et al.* (1987).

and

$$\partial_t \tilde{A}_{ij} = e^{-4\phi} \left(-(D_i D_j \alpha)^{TF} + \alpha(R_{ij}^{TF} - 8\pi S_{ij}^{TF}) \right) + \alpha(K \tilde{A}_{ij} - 2\tilde{A}_{il}\tilde{A}^l_{\ j})$$
$$+ \beta^k \partial_k \tilde{A}_{ij} + \tilde{A}_{ik}\partial_j \beta^k + \tilde{A}_{kj}\partial_i \beta^k - \frac{2}{3}\tilde{A}_{ij}\partial_k \beta^k. \tag{11.38}$$

In the last equation, the superscript TF denotes the trace-free part of a tensor, e.g., $R_{ij}^{TF} = R_{ij} - \gamma_{ij}R/3$. Note that in equations (11.35) through (11.38) the shift terms arise from Lie derivatives $\mathcal{L}_\beta$ of the respective evolution variable appearing on the left-hand side. The divergence of the shift, $\partial_i \beta^i$, appears in the Lie derivative because the choice $\bar{\gamma} = 1$ makes ϕ a tensor density of weight $1/6$, and $\bar{\gamma}_{ij}$ and $\tilde{A}_{ij}$ tensor densities of weight $-2/3$ (see Section A.3).

Exercise 11.2 Derive equations (11.35) through (11.38).

According to equation (3.10) we can split the Ricci tensor into two terms

$$R_{ij} = \bar{R}_{ij} + R_{ij}^\phi, \tag{11.39}$$

where only R_{ij}^ϕ depends on the conformal function ϕ. We can identify the form of R_{ij}^ϕ by inserting $\phi = \ln \psi$ into equation (3.10). We could compute the conformally related Ricci tensor $\bar{R}_{ij}$ by inserting $\bar{\gamma}_{ij}$ into equation (2.143), but that would again introduce the mixed second derivatives that we are trying to avoid. Analogously to the way we introduced a new variable Γ in equation (11.11) to eliminate the mixed derivatives in Maxwell's evolution equations, we can now define "conformal connection functions"

$$\bar{\Gamma}^i \equiv \bar{\gamma}^{jk}\bar{\Gamma}^i_{jk} = -\partial_j \bar{\gamma}^{ij} \tag{11.40}$$

to accomplish the same task in the above evolution equations for the gravitational field. Here the $\bar{\Gamma}^i_{jk}$ are the connection coefficients associated with $\bar{\gamma}_{ij}$, and the last equality holds in Cartesian coordinates when $\bar{\gamma} = 1$. In terms of these conformal connection functions we can now write the Ricci tensor as

$$\bar{R}_{ij} = -\frac{1}{2}\bar{\gamma}^{lm}\partial_m \partial_l \bar{\gamma}_{ij} + \bar{\gamma}_{k(i}\partial_{j)}\bar{\Gamma}^k + \bar{\Gamma}^k\bar{\Gamma}_{(ij)k} + \bar{\gamma}^{lm}\left(2\bar{\Gamma}^k_{l(i}\bar{\Gamma}_{j)km} + \bar{\Gamma}^k_{im}\bar{\Gamma}_{klj}\right). \tag{11.41}$$

The only explicit second-derivative operator acting on $\bar{\gamma}_{ij}$ in this expression involves a Laplacian, $\bar{\gamma}^{lm}\partial_m \partial_l$ – all other second derivatives are absorbed in first derivatives of $\bar{\Gamma}^i$. This derivative structure is similar to the one we encountered in the 4-dimensional Ricci tensor in Sections 4.3 and 11.3, where the only explicit second derivatives on the metric appearing in that quantity managed to form a convenient wave operator (d'Alembertian).

In the generalized harmonic formalism of Section 11.3 we set the 4-dimensional analogs $^{(4)}\Gamma^a$ of the conformal connection functions $\bar{\Gamma}^i$ equal to some specified gauge source function. Here we do not follow this approach but instead promote the $\bar{\Gamma}^i$ to new independent functions, in complete analogy to our treatment of Maxwell's equations in Section 11.2.3. Adopting this approach requires us to derive separate evolution equations for the $\bar{\Gamma}^i$. By analogy with the derivation of equation (11.17) we interchange a partial time and space derivative in the definition (11.40) to obtain

$$\partial_t \bar{\Gamma}^i = -\partial_j \left(2\alpha \tilde{A}^{ij} - 2\bar{\gamma}^{m(j} \partial_m \beta^{i)} + \frac{2}{3} \bar{\gamma}^{ij} \partial_l \beta^l + \beta^l \partial_l \bar{\gamma}^{ij} \right). \tag{11.42}$$

We can now eliminate the divergence of the extrinsic curvature with the help of the momentum constraint (2.133), which then yields the desired evolution equation,

$$\partial_t \bar{\Gamma}^i = -2\tilde{A}^{ij} \partial_j \alpha + 2\alpha \left(\Gamma^i_{jk} \tilde{A}^{kj} - \frac{2}{3} \bar{\gamma}^{ij} \partial_j K - 8\pi \bar{\gamma}^{ij} S_j + 6\tilde{A}^{ij} \partial_j \phi \right)$$
$$+ \beta^j \partial_j \bar{\Gamma}^i - \bar{\Gamma}^j \partial_j \beta^i + \frac{2}{3} \bar{\Gamma}^i \partial_j \beta^j + \frac{1}{3} \bar{\gamma}^{li} \partial_l \partial_j \beta^j + \bar{\gamma}^{lj} \partial_j \partial_l \beta^i. \tag{11.43}$$

Equations (11.35) through (11.38), together with (11.43), form a new system of evolution equations that is equivalent to equations (11.7) and (11.8). Since the $\bar{\Gamma}^i$ are evolved as independent functions, the defining relation (11.40) serves as a new constraint equation, in addition to equations (2.132) and (2.133). We summarize the resulting BSSN formalism in Box 10.1.

Exercise 11.3 Show that the shift terms in equation (11.43) arise from the Lie derivative of $\bar{\Gamma}^i$ along β^i,

$$\mathcal{L}_\beta \bar{\Gamma}^i = \beta^j \partial_j \bar{\Gamma}^i - \bar{\Gamma}^j \partial_j \beta^i + \frac{2}{3} \bar{\Gamma}^i \partial_j \beta^j + \frac{1}{3} \bar{\gamma}^{li} \partial_l \partial_j \beta^j + \bar{\gamma}^{lj} \partial_j \partial_l \beta^i. \tag{11.44}$$

Hint: First show that the conformal connection coefficients transform according to

$$\bar{\Gamma}^{a'} = J^{-W} \frac{\partial x^{a'}}{\partial x^b} \bar{\Gamma}^b + J^{-W} \bar{\gamma}^{ij} \frac{\partial x^{b'}}{\partial x^i} \frac{\partial x^{c'}}{\partial x^j} \frac{\partial x^{a'}}{\partial x^l} \frac{\partial^2 x^l}{\partial x^{b'} \partial x^{c'}} - \frac{1}{2} J^{-W} \bar{\gamma}^{bc} \frac{\partial x^{a'}}{\partial x^b} \partial_c (\ln J^W),$$
$$\tag{11.45}$$

where J is the Jacobian of the transformation and W is the weight of $\bar{\gamma}_{ij}$. The first term is the usual transformation term for a tensor (except for the Jacobian factor), the second term arises because the connection coefficients do not transform like tensors, and the third term appears because $\bar{\gamma}_{ij}$ is a tensor density. Then use definition (A.7) to find equation (11.44) for $W = -2/3$ (*cf.* equations A.36 and A.37).

While the different formulations are equivalent analytically, the difference in performance of numerical implementations is striking. In Figure 11.1 we return to the example of linear gravitational wave propagation cited at the beginning of this chapter. In this example,[34] a small amplitude, time-symmetric, even-parity $l = 2$, $m = 0$ gravitational wave of the kind discussed in Chapter 9.1.2 is evolved with harmonic slicing (see Chapter 4.3), zero shift, and a simple outgoing wave boundary condition. Both the standard $3 + 1$ and BSSN systems give very similar results early on, but the standard system crashes very soon, while the BSSN system remains stable. Similar improvements have been found for many other applications that employ a BSSN scheme, including the propagation of nonlinear

---

[34] Baumgarte and Shapiro (1998); see also a similar example in Shibata and Nakamura (1995).

Box 11.1   The BSSN equations

In the BSSN formulation of the $3+1$ equations the spatial metric γ_{ij} is decomposed into a conformally related metric $\bar{\gamma}_{ij}$ with determinant $\bar{\gamma} = 1$ (assuming Cartesian coordinates) and a conformal factor e^{ϕ},

$$\gamma_{ij} = e^{4\phi} \bar{\gamma}_{ij}. \tag{11.46}$$

We also decompose the extrinsic curvature into its trace and traceless parts and conformally transform the traceless part as we do the metric,

$$K_{ij} = e^{4\phi} \tilde{A}_{ij} + \frac{1}{3}\gamma_{ij} K. \tag{11.47}$$

In terms of these variables the Hamiltonian constraint (2.132) becomes

$$0 = \mathcal{H} = \bar{\gamma}^{ij} \bar{D}_i \bar{D}_j e^{\phi} - \frac{e^{\phi}}{8} \bar{R} + \frac{e^{5\phi}}{8} \tilde{A}_{ij} \tilde{A}^{ij} - \frac{e^{5\phi}}{12} K^2 + 2\pi e^{5\phi} \rho, \tag{11.48}$$

while the momentum constraint (2.133) becomes

$$0 = \mathcal{M}^i = \bar{D}_j (e^{6\phi} \tilde{A}^{ji}) - \frac{2}{3} e^{6\phi} \bar{D}^i K - 8\pi e^{6\phi} S^i. \tag{11.49}$$

The evolution equation (2.136) for γ_{ij} splits into two equations,

$$\partial_t \phi = -\frac{1}{6}\alpha K + \beta^i \partial_i \phi + \frac{1}{6}\partial_i \beta^i, \tag{11.50}$$

$$\partial_t \bar{\gamma}_{ij} = -2\alpha \tilde{A}_{ij} + \beta^k \partial_k \bar{\gamma}_{ij} + \bar{\gamma}_{ik}\partial_j \beta^k + \bar{\gamma}_{kj}\partial_i \beta^k - \frac{2}{3}\bar{\gamma}_{ij}\partial_k \beta^k, \tag{11.51}$$

while the evolution equation (2.135) for K_{ij} splits into the two equations

$$\partial_t K = -\gamma^{ij} D_j D_i \alpha + \alpha \left(\tilde{A}_{ij} \tilde{A}^{ij} + \frac{1}{3}K^2 \right) + 4\pi\alpha(\rho + S) + \beta^i \partial_i K, \tag{11.52}$$

$$\partial_t \tilde{A}_{ij} = e^{-4\phi} \left(-(D_i D_j \alpha)^{TF} + \alpha(R_{ij}^{TF} - 8\pi S_{ij}^{TF}) \right) + \alpha(K \tilde{A}_{ij} - 2\tilde{A}_{il}\tilde{A}^l{}_j)$$
$$+ \beta^k \partial_k \tilde{A}_{ij} + \tilde{A}_{ik}\partial_j \beta^k + \tilde{A}_{kj}\partial_i \beta^k - \frac{2}{3}\tilde{A}_{ij}\partial_k \beta^k. \tag{11.53}$$

In the last equation the superscript TF denotes the trace-free part of a tensor, e.g., $R_{ij}^{TF} = R_{ij} - \gamma_{ij} R/3$. We also split the Ricci tensor into $R_{ij} = \bar{R}_{ij} + R_{ij}^{\phi}$, where R_{ij}^{ϕ} can be found by inserting $\phi = \ln \psi$ into equation (3.10). We express $\bar{R}_{ij}$ in terms of the conformal connection functions $\bar{\Gamma}^i \equiv \bar{\gamma}^{jk}\bar{\Gamma}^i_{jk} = -\partial_j \bar{\gamma}^{ij}$, which yields

$$\bar{R}_{ij} = -\frac{1}{2}\bar{\gamma}^{lm}\partial_m \partial_l \bar{\gamma}_{ij} + \bar{\gamma}_{k(i}\partial_{j)}\bar{\Gamma}^k + \bar{\Gamma}^k \bar{\Gamma}_{(ij)k} + \bar{\gamma}^{lm} \left(2\bar{\Gamma}^k_{l(i}\bar{\Gamma}_{j)km} + \bar{\Gamma}^k_{im}\bar{\Gamma}_{klj} \right). \tag{11.54}$$

The $\bar{\Gamma}^i$ are now treated as independent functions that satisfy their own evolution equations,

$$\partial_t \bar{\Gamma}^i = -2\tilde{A}^{ij}\partial_j \alpha + 2\alpha \left(\bar{\Gamma}^i_{jk}\tilde{A}^{kj} - \frac{2}{3}\bar{\gamma}^{ij}\partial_j K - 8\pi \bar{\gamma}^{ij} S_j + 6\tilde{A}^{ij}\partial_j \phi \right)$$
$$+ \beta^j \partial_j \bar{\Gamma}^i - \bar{\Gamma}^j \partial_j \beta^i + \frac{2}{3}\bar{\Gamma}^i \partial_j \beta^j + \frac{1}{3}\bar{\gamma}^{li}\partial_l \partial_j \beta^j + \bar{\gamma}^{lj}\partial_j \partial_l \beta^i. \tag{11.55}$$

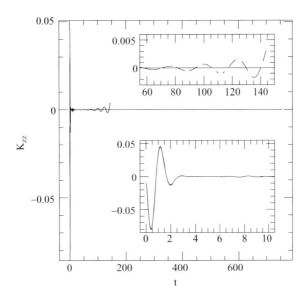

Figure 11.1 Comparison of the evolution of a linearized gravitational wave using the standard $3 + 1$ or ADM equations (dashed lines) and the BSSN equations (solid lines). Shown is the extrinsic curvature component K_{zz} as a function of time. A blow-up of the plot at early times is shown in the bottom insert, where both systems agree very well. The evolution at later times is shown in the top insert, just before the standard $3 + 1$ system crashes. For the BSSN system, which remains stable, the quantity K_{zz} is constructed from $\bar{A}_{zz}$, ϕ and K. [From Baumgarte and Shapiro (1998).]

gravitational waves and the evolution of spacetimes containing black holes and neutron stars. The BSSN system, or a variation closely related to it, is currently the form of the Einstein's equations most commonly used in numerical relativity. With the exception of some that use the generalized harmonic coordinate approach of Section 11.3, most successful simulations of binary black holes, binary neutron stars and binary black hole-neutron stars, as well as stellar collapse, use a version of the BSSN system.

Constraint propagation and damping

The improved numerical stability properties of the BSSN system can be understood in terms of the hyperbolicity criteria discussed in Section 11.1; the BSSN system can be shown to be strongly hyperbolic, and hence well-posed, while the original $3 + 1$ (ADM) formulation is only weakly hyperbolic, and hence ill-posed.[35] We can also gain some heuristic insight into its improved behavior by considering the propagation of the constraints.[36] Linearizing the

---

[35] See, e.g., Sarbach *et al.* (2002); Yoneda and Shinkai (2002); Gundlach and Martin-Garcia (2006a,b), and references therein.
[36] Frittelli (1997).

standard $3+1$ and BSSN equations on a flat Minkowski background reveals[37] that modes that violate the momentum constraint do not propagate in the standard equations. Such zero-speed modes lead to instabilities when nonlinear source terms are included. However, when the momentum constraint, which should be zero if the Einstein equations are satisfied exactly, is added to the $\bar{\Gamma}^i$ evolution equation (11.43) of the BSSN system, the momentum constraint violating modes now propagate with nonzero speed. Consequently, rather than growing locally, as in the standard system, constraint violations can now propagate off the numerical grid,[38] thereby stabilizing the simulation.

To illustrate this effect in a simple setting, let us return to Maxwell's equations and measure the violation of the constraint (11.6) by introducing the variable $\mathcal{C}$ defined as

$$\mathcal{C} \equiv D^i E_i - 4\pi \rho_e. \tag{11.56}$$

By differentiating the continuity equation $D_i j^i + \partial_t \rho_e = 0$ and using the evolution equation (11.5) it is easy to show that the time derivative of this constraint variable vanishes identically,

$$\partial_t \mathcal{C} = \partial_t (D^i E_i - 4\pi \rho_e) = D^i \partial_t E_i - 4\pi \partial_t \rho_e$$
$$= -D^i D^j D_j A_i + D^i D_i D^j A_j - 4\pi (D^i j_i + \partial_t \rho_e) = 0. \tag{11.57}$$

This result indicates that any violation of the constraint (i.e., $\mathcal{C} \neq 0$) will persist and not propagate away.

The situation is different in the BSSN-like reformulation of Maxwell's equations of Section 11.2.3. In fact, we can use this example to show the effect of adding the constraint (11.6) in the evolution equation (11.17). Instead of using the constraint to completely replace the divergence $D^i E_i$ as we did in deriving the Γ-evolution equation (11.17), consider adding the constraint violation parameter $\mathcal{C}$ times some constant a^2 (which we are always allowed to do, since constraint violations are supposed to vanish),

$$\partial_t \Gamma = \partial_t D^i A_i = D^i \partial_t A_i = -D^i E_i - D_i D^i \Phi + a^2 \mathcal{C}$$
$$= (a^2 - 1) D^i E_i - D_i D^i \Phi - 4\pi a^2 \rho_e. \tag{11.58}$$

With $a^2 = 1$ we recover equation (11.17). We can now reconsider the propagation of the constraint violations, i.e., the time evolution of the parameter $\mathcal{C}$. In deriving equation (11.57) we used the evolution equation (11.5), but this evolution equation is no longer part of the new BSSN-like evolution system consisting of equations (11.4), (11.17) and (11.18). For this system, we can show that the constraint violation parameter now satisfies a wave equation.[39]

[37] Alcubierre *et al.* (2000).
[38] Provided that appropriate boundary conditions are imposed.
[39] See Knapp *et al.* (2002), who demonstrate this effect in a numerical example.

Exercise 11.4 Show that in the system (11.4), (11.17) and (11.18) constraint violations measured by the parameter $\mathcal{C}$ defined in equation (11.56) satisfy the wave equation

$$(-\partial_t^2 + a^2 D_i D^i)\mathcal{C} = 0. \tag{11.59}$$

According to equation (11.59), constraint violations measured by $\mathcal{C}$ now propagate like waves with a characteristic speed a. For $a = 0$, the constraints again do not propagate, but with $a = 1$ they propagate with the speed of light. We thus see that adding the constraint in equation (11.17) was crucial in achieving this property, in direct analogy to the results found for the linearized Einstein equations.

If the condition $\mathcal{C} = \partial_t \mathcal{C} = 0$ holds initially, then the two systems are equivalent analytically, since both will guarantee that $\mathcal{C} = 0$ in the domain of dependence of the initial spatial hypersurface. However, the two systems behave very differently numerically, since any numerical (e.g., roundoff) error will lead to a constraint violation $|\mathcal{C}| > 0$, which will then evolve differently in the two systems.

The addition of the constraints to the evolution equations, which comes under the general category of *constraint damping* or *constraint sweeping*, is by no means unique to the BSSN system. We have already discussed in Section 11.1 that adding constraints to the evolution equations affects the principal part of the system, and hence its well-posedness. This technique has been applied in the context of MHD, as we mentioned in Chapter 5.2.4. It was pointed out early on[40] that constraint violations can be controlled by adding additional terms to the standard $3 + 1$ evolution equations. The importance of the propagation of constraints in unconstrained evolution calculations was demonstrated and shown to be linked to the addition of the Hamiltonian constraint to the evolution equations.[41] Several investigators have also experimented with adding the momentum constraints to the standard equations and have found stabilizing effects.[42] We have already mentioned that in the presence of black holes, the generalized harmonic approach summarized in Section 11.3 requires the addition of constraints. It also turns out that in some situations the stability properties of the basic BSSN system itself can be enhanced further by explicitly adding Hamiltonian and/or momentum constraints.[43] We will describe some of these enhancements when we discuss a few specific applications that employ the BSSN formalism in later chapters.[44]

---

[40] Detweiler (1987).
[41] Frittelli (1997); see also Kelly *et al.* (2001).
[42] See, e.g., Yoneda and Shinkai (2001).
[43] See Yoneda and Shinkai (2002); Yo *et al.* (2002); Duez *et al.* (2004); Marronetti (2005, 2006).
[44] See, e.g., Chapter 14.2.3.

12 Binary black hole initial data

In this chapter we wish to construct initial data for *quasi*equilibrium binary black holes. That is, we seek solutions corresponding to two black holes in stable, nearly circular orbit about each other. In contrast to Newtonian theory, a stellar binary in general relativity can never be in *strict* equilibrium, with the companions moving in exactly circular orbits at constant separation for all time. Instead, gravitational radiation emission inevitably leads to loss of orbital energy and angular momentum, causing the orbit to decay. The resulting trajectory then traces out an inspiral rather than a perfect circle. For sufficiently large separations, the binary motion is nearly Newtonian, hence the orbit is nearly circular, decaying very little during one orbital period. If isolated from outside perturbations (e.g., gravitational encounters with other stars), it is expected that astrophysical binaries composed of compact stars (i.e., compact binaries) will ultimately evolve to a quasiequilibrium state following their formation at large separation.[1] The reason is that gravitational radiation loss drives orbital circularization as well as decay, as we will discuss in the next section. Only when the orbits become very close and highly relativistic, just prior to radial plunge and binary merger, do the deviations from circular motion become large.

The construction of quasiequilibrium binary initial data poses a number of conceptual challenges. Getting started, however, is fairly straightforward. To find solutions we shall follow the approaches outlined in Chapter 3. Specifically, we need to solve the constraint equations (2.132) and (2.133), which can be cast into a convenient form with the help of the conformal transformation (3.5) of the spatial metric γ_{ij},

$$\gamma_{ij} = \psi^4 \bar{\gamma}_{ij}. \tag{12.1}$$

Recall that ψ is the conformal factor and $\bar{\gamma}_{ij}$ the conformally related metric. Recall also that we can separate the trace of the extrinsic curvature from its traceless part and conformally rescale the latter according to equation (3.35), so that

$$K_{ij} = \psi^{-2} \bar{A}_{ij} + \frac{1}{3} \gamma_{ij} K. \tag{12.2}$$

---

[1] Exceptions include extreme-mass-ratio binaries consisting of stellar-mass companions (which may be compact stars) orbiting supermassive black holes in dense stellar systems. The orbits of the companions, which evolve via gravitational encounters (scattering) with other stars as well as by gravitational radiation reaction forces, may be highly eccentric. See Chapter 9.2.2 for discussion and references.

In terms of these quantities the Hamiltonian constraint then takes the form (3.37),[2]

$$8\bar{D}^2\psi - \psi\bar{R} - \frac{2}{3}\psi^5 K^2 + \psi^{-7}\bar{A}_{ij}\bar{A}^{ij} = 0, \qquad (12.3)$$

while the momentum constraint reduces to equation (3.38),

$$\bar{D}_j\bar{A}^{ij} - \frac{2}{3}\psi^6\bar{\gamma}^{ij}\bar{D}_j K = 0. \qquad (12.4)$$

Here we have specialized to the binary black hole problem and assumed that all matter sources vanish.

In Chapter 3.1 we saw that under the assumption of time-symmetry (where $K_{ij} = 0$) solutions containing multiple black holes can be constructed remarkably easily (see equation 3.23). Unfortunately, binary black holes in circular orbit are not time-symmetric, so that these simple solutions are not useful for purposes of obtaining quasiequilibrium solutions. In Chapter 3 we discussed two separate decompositions of the traceless part of the extrinsic curvature $\bar{A}_{ij}$, namely the transverse-traceless decomposition (CTT) in Chapter 3.2 and the conformal thin-sandwich decomposition (CTS) in Chapter 3.3. We will follow a very similar approach again in this chapter, adopting CTT in Section 12.2 and CTS in Section 12.3. Before constructing these binary black hole initial data, we first shall review some of the physical characteristics of compact binary inspiral, which will motivate some of the choices that we will make in later sections.

12.1 Binary inspiral: overview

The evolution of compact binaries – binary black holes, neutron stars, or black hole-neutron stars – proceeds in several distinct phases. By far the longest epoch is the initial *inspiral* phase, followed by the *plunge and merger* phase when the two objects coalesce, and finally the *ringdown* phase during which the merger remnant settles down to stationary equilibrium. These different phases are illustrated schematically in Figure 12.1.

As long as the binary is nearly Newtonian, with the orbital separation much larger than the total mass M, we may model the emitted gravitational waves as small perturbations propagating on an otherwise flat background spacetime. As reviewed in Chapters 1.1 and 9.1 the losses of orbital energy and angular momentum in the weak-field, slow-velocity regime appropriate to Newtonian binaries are then given to lowest order by the quadrupole expressions,

$$L_{\mathrm{GW}} \equiv -\frac{dE}{dt} = \frac{1}{5}\langle\dddot{\mathcal{I}}_{jk}\dddot{\mathcal{I}}_{jk}\rangle \qquad (12.5)$$

and

$$\frac{dJ_i}{dt} = -\frac{2}{5}\epsilon_{ijk}\langle\ddot{\mathcal{I}}_{jm}\dddot{\mathcal{I}}_{km}\rangle. \qquad (12.6)$$

---

[2] Remember that all barred quantities are associated with the conformally-related metric.

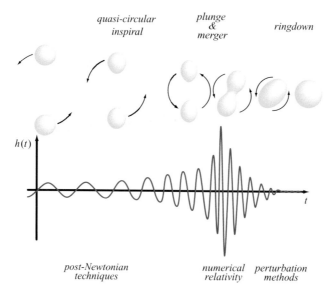

quasi-circular inspiral   *plunge & merger*   *ringdown*

post-Newtonian techniques   *numerical relativity*   *perturbation methods*

Figure 12.1 The different phases of compact binary inspiral and coalescence. The gravitational wave amplitude $h(t)$ is sketched schematically and the analysis technique is identified for each phase.

Here $\mathcal{I}_{ij}$ is the reduced quadrupole moment,

$$\mathcal{I}_{jk} \equiv \int \rho \left(x_j x_k - \frac{1}{3}\delta_{jk} r^2 \right) d^3 x = \sum_A m_A \left(x_j x_k - \frac{1}{3}\delta_{jk} r^2 \right), \qquad (12.7)$$

the bracket $\langle\,\rangle$ denotes an average over several orbital periods, and the triple dot denotes the third time derivative. For a binary at large separation we may treat the stars as point masses and insert their stellar masses m_A and Newtonian trajectories $x^i(t)$ into equation (12.7) to evaluate equations (12.5) and (12.6). For a binary orbit with eccentricity e the emission of gravitational radiation always leads to a decrease in the eccentricity, $\dot{e} < 0$.[3] Put differently, *gravitational radiation circularizes elliptical orbits*. This result implies that during the late stages of compact binary inspiral, we may approximate the orbit as circular.[4]

> **Exercise 12.1** Consider a Newtonian binary consisting of two point masses m_1 and m_2 at a binary separation r. Write the binary's Hamiltonian $H(r, \phi, P_r, P_\phi)$, which is equal to its conserved energy E, as
>
> $$E = H = \frac{1}{2}\frac{P_r^2}{\mu} + \frac{1}{2}\frac{P_\phi^2}{\mu r^2} - \frac{\mu M}{r}, \qquad (12.8)$$
>
> where $M = m_1 + m_2$ is the total mass and $\mu = m_1 m_2/M$ is the reduced mass. Define $\Omega_{\text{orb}} \equiv \dot{\phi}$ to be the orbital angular velocity, and $J \equiv P_\phi$ to be the orbital angular momentum.

---

[3] Peters (1964); see also Lightman *et al.* (1975), Problem 18.7.
[4] See also exercise 12.2, which shows that once they become circular, the orbits remain circular as they shrink in radius.

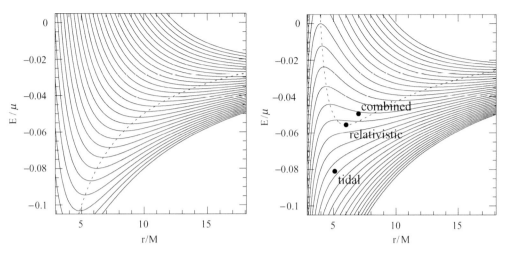

Figure 12.2 The energy E as a function of separation for a point-mass Newtonian binary (left panel), and a toy model problem that incorporates relativistic effects (right panel). The solid lines are contours of constant angular momentum J, with the values of J increasing from the bottom to the top. Extrema of these contours correspond to circular orbits (equation 12.9). The dashed line connects the circular orbits and represents the equilibrium energy E_{eq}. A turning point in this equilibrium energy represents the ISCO (equation 12.13) and is marked by dots for relativistic, tidal and combined relativistic and tidal effects. [From Baumgarte (2001).]

(a) Use Hamilton's equations to show the following: $P_r = \mu \dot{r}$, $J = \mu r^2 \Omega_{orb}$, and J is conserved.

(b) Apply Hamilton's equations to show that a circular orbit with $P_r = \dot{P}_r = 0$ satisfies

$$\boxed{\partial E / \partial r = 0.}$$
(12.9)

(c) Substitute equation (12.8) into equation (12.9) to obtain the virial relation between the kinetic energy T and potential energy W,

$$T = -\frac{1}{2} W,$$
(12.10)

which implies that the binary's *equilibrium* energy is

$$E_{eq}(r) = -\frac{1}{2} \frac{M\mu}{r}.$$
(12.11)

(d) Use the virial theorem to evaluate Ω_{orb} for circular equilibrium, obtaining Kepler's third law,

$$\Omega_{orb} = \left(\frac{M}{r^3}\right)^{1/2}.$$
(12.12)

The condition (12.9) shows that we can identify a circular orbit by finding an extremum of the energy at constant angular momentum, which we illustrate in Figure 12.2. A minimum of the energy corresponds to a stable circular orbit, while a maximum corresponds to an unstable orbit. The left panel in the figure plots the Newtonian energy (12.8), for which circular orbits can be found for arbitrarily small binary separations. The transition between stable and unstable circular orbits occurs at a turning point in the *equilibrium* energy curve, i.e., when

$$\boxed{\partial E_{\mathrm{eq}}/\partial r = 0.}$$

(12.13)

Equation (12.13) is the *turning-point criterion* for indentifying marginal stability in an equilibrium configuration, including a circular binary orbit.[5] For Newtonian point-mass binaries, the extrema of the energy are all minima, as seen in the figure; equivalently, there are no turning points in the equilibrium energy curve, as indicated by equation (12.11) and by the shape of the equilibrium energy curve plotted in the figure. Hence all of these circular orbits are stable. Condition (12.9), which identifies circular orbits, is equivalent to imposing the virial theorem (12.10) at every instant of time (and not simply as an orbit average, which applies for eccentric orbits also). These same general criteria for finding quasiequilibrium binaries obeying the virial theorem and identifying the onset of orbital instability also apply to relativistic systems (see, e.g., exercise 12.4), and they will be used in Sections 12.2.3 and 12.3.3.

> **Exercise 12.2**   Return to exercise (12.1) and show that changes in the equilibrium energy $E_{\mathrm{eq}}(r)$ and the angular momentum $J_{\mathrm{eq}}(r)$ of Newtonian point-mass binaries that reside along sequences of circular equilibrium are related according to
>
> $$\boxed{dE_{\mathrm{eq}} = \Omega_{\mathrm{orb}}\, dJ_{\mathrm{eq}}.}$$
>
> (12.14)
>
> Combined with the results of Exercise 9.3, which shows that the energy and angular momentum emitted by gravitational radiation satisfy the same relation, equation (12.14) implies that gravitational wave-driven binary inspiral will leave circular orbits circular.

As we have mentioned, an exactly circular orbit can only exist in Newtonian gravity, since in general relativity gravitational radiation makes the orbit shrink.[6] We can estimate the orbital decay rate dr/dt and the time dependence of the radial separation $r(t)$ from equation (12.5), as shown in exercise 12.3.

---

[5] We apply the same turning-point criterion to identify the onset of radial instability in equilibrium stars in Chapters 1.3 and 14.1.2.

[6] Unless, of course, the loss of energy and angular momentum by gravitational radiation emission is exactly cancelled by some ad hoc incoming gravitational radiation. In fact, several authors have suggested approaches for modeling equilibrium relativistic binaries by invoking such a "standing-wave" approximation to maintain exactly circular orbits (see, e.g., Detweiler 1989; Andrade *et al.* 2004).

Exercise 12.3   (a) Revisit the Newtonian binary of exercise 12.1 and use the results of exercise 9.3 to show that

$$L_{GW} = \frac{32}{5} \frac{M^3 \mu^2}{r^5} \qquad (12.15)$$

for a circular orbit.

(b) Combine equation (12.15) with the binary's equilibrium energy E_{eq} to find

$$\frac{dr}{dt} = \frac{dE_{eq}/dt}{dE_{eq}/dr} = -\frac{L_{GW}}{dE_{eq}/dr} = -\frac{64}{5} \frac{M^2 \mu}{r^3}. \qquad (12.16)$$

Now integrate equation (12.16) to find the binary separation r as a function of time,

$$r(t) = 4 \left(\frac{\mu M^2}{5} (T - t) \right)^{1/4}, \qquad (12.17)$$

where T is the time of the coalescence when $r = 0$.

(c) Combine equation (12.17) with Kepler's law (12.12) to find the gravitational wave quadrupole frequency as a function of time,

$$f_{GW}(t) = \frac{\Omega_{orb}}{\pi} = \frac{1}{8\pi} \left(\frac{5}{\mu M^{2/3} (T - t)} \right)^{3/8}. \qquad (12.18)$$

(d) Gravitational wave interferometers are sensitive to the number of wave cycles $d\mathcal{N}_{cyc}$ observed in a given wave frequency interval df_{GW}. Show that for the Newtonian binary, one obtains a "chirp signal"

$$\frac{d\mathcal{N}_{cyc}}{d \ln f_{GW}} \equiv \frac{f_{GW}^2}{df_{GW}/dt} = \frac{5}{96\pi} \frac{1}{(\pi \mathcal{M} f_{GW})^{5/3}}, \qquad (12.19)$$

where $\mathcal{M} \equiv \mu^{3/5} M^{2/5}$ is called the "chirp mass". Note that post-Newtonian corrections to equation (12.19) can be expressed as a series expansion in the parameter $\pi \mathcal{M} f_{GW} \ll 1$.

In exercise 12.3 we adopted circular orbits to estimate the deviation from circular motion. This procedure is justified as long as the deviation from circular motion is small. To assess the deviation from circularity, we first define the orbital decay time scale τ_{GW} due to gravitational wave emission according to

$$\tau_{GW} = \left| \frac{r}{dr/dt} \right|. \qquad (12.20)$$

Substituting equation (12.16) we find

$$\tau_{GW} = \frac{5}{64} \frac{r^4}{M^2 \mu}. \qquad (12.21)$$

We can now compare this orbital decay time scale with the orbital period,

$$P = \frac{2\pi}{\Omega_{orb}} = 2\pi \left(\frac{M}{r^3} \right)^{-1/2}, \qquad (12.22)$$

by forming the dimensionless ratio

$$\frac{\tau_{GW}}{P} = \frac{5}{128\pi} \frac{r^{5/2}}{M^{3/2}\mu}. \tag{12.23}$$

For a binary of mass ratio q we can express equation (12.23) as

$$\frac{\tau_{GW}}{P} = \frac{5}{128\pi} \frac{(q+1)^2}{q} \left(\frac{r}{M}\right)^{5/2}. \tag{12.24}$$

This ratio increases steeply with r and is greater than unity, thereby justifying the assumption of circular orbits, provided

$$\frac{r}{M} > \frac{r_{crit}}{M} \equiv \left(\frac{128\pi}{5}\right)^{2/5} \frac{q^{2/5}}{(1+q)^{4/5}} \leq 5.78. \tag{12.25}$$

The critical separation r_{crit} takes its maximum value of $5.78M$ for equal-mass binaries with $q = 1$. This value suggests that we may indeed approximate the binary orbit during the inspiral phase as quasicircular up to very small binary separations, on the order of a few Schwarzschild radii.

At these small binary separations deviations from the simple Newtonian point-mass potential in equation (12.8) affect the binary's evolution, both quantitatively and qualitatively. Deviations stem from relativistic corrections to the Newtonian interaction and from finite-size effects that lead to tidal interaction between the binary companions. Post-Newtonian techniques can handle the relativistic perturbations to the point-mass inspiral, furnishing the corrections as a power series in the parameter $\epsilon \equiv v^2/c^2 \sim GM/r$ (see Appendix E). Numerical simulations are required to handle tidal interactions, as well as the late inspiral, plunge and merger phases when relativistic and, in the case of neutron stars, hydrodynamic effects become increasingly strong.

Consider how these corrections to the Newtonian point-mass potential lead to qualitatively new features in the binary orbit. We can mimic the effect of the relativistic correction by borrowing a relativistic interaction potential

$$E_{rel} = -\frac{M}{\mu} \frac{J^2}{r^3} \tag{12.26}$$

from the effective potential for test particles in a Schwarzschild spacetime (see exercise 12.4). Adding this term to the Newtonian energy (12.8) and evaluating the circular equilibrium criterion (12.9) shows that circular orbits now have the equilibrium energy

$$E_{eq} = -\frac{1}{2} \frac{M\mu}{r} \frac{r - 4M}{r - 3M}. \tag{12.27}$$

This relativistic "toy model" is also illustrated in the right panel of Figure 12.2. Without the new term, the energy curves at constant angular momentum are dominated by the positive kinetic energy term $J^2/(2\mu r^2)$ at small binary separations r. With the addition of the relativistic correction, however, these curves are dominated by the new negative term (12.26) for small r. For sufficiently large J this correction adds a new maximum to the previous minimum in the $J = constant$ energy curves; for smaller values of J these curves no longer have extrema. This behavior means that circular orbits can only exist for sufficiently large values of J.

As a result of this correction, the equilibrium energy (12.27), which connects the extrema in the $J = constant$ energy curves and is represented by the dashed line in Figure 12.2, now exhibits a turning point at

$$r_{\text{ISCO}} = 6M, \qquad (12.28)$$

at which point equation (12.13) is satisfied. Beyond this turning point the extrema are minima and thus represent stable circular orbits, while inside they are maxima and represent unstable circular orbits. The turning point in the equilibrium energy therefore represents an *innermost stable circular orbit*, or ISCO for short. Binary companions on quasicircular inspiral trajectories start accelerating radially inward, plunging toward each other, after they reach the ISCO. This dynamical transition from the inspiral phase to the *plunge and merger* phase will induce a corresponding transition in the gravitational wave signal, yielding an observable diagnostic of the transition.

For a point particle orbiting a Schwarzschild black hole, from which we constructed the correction term (12.26), the location of the ISCO at $r_{\text{ISCO}} = 6M$ refers to the areal radius of the orbit. It is more useful to express the critical binary separation in terms of an invariant, like the orbital angular frequency Ω_{orb} measured by a distant observer. For a point particle orbiting a Schwarzschild black hole we have $M\Omega_{\text{ISCO}} = 1/6^{3/2} \approx 0.0680$. In Section 12.4 we will find similar values for binary black holes.

> **Exercise 12.4**   Consider the orbit of a test-particle m_{test} around a Schwarzschild black hole M. The exact relativistic analog of equation (12.8) in this case is[7]
>
> $$\tilde{E}^2 = \left(\frac{dr}{d\tau}\right)^2 + V(r), \qquad (12.29)$$
>
> where
>
> $$V(r) \equiv \left(1 - \frac{2M}{r}\right)\left(1 + \frac{\tilde{J}^2}{r^2}\right) \qquad (12.30)$$
>
> is the "effective potential", $\tilde{E} = E/m_{\text{test}}$, $\tilde{J} = J/m_{\text{test}}$, r is the areal radius, and τ is proper time.
> (a) Circular orbits occur when $dr/d\tau = 0$ and $\partial V/\partial r|_J = 0 = \partial \tilde{E}/\partial r|_J$, which is the relativistic counterpart of equation (12.9). Show that such orbits satisfy
>
> $$\tilde{E}^2_{\text{eq}} = \frac{(r - 2M)^2}{r(r - 3M)}, \qquad (12.31)$$
>
> $$\tilde{J}^2_{\text{eq}} = \frac{Mr^2}{r - 3M} \qquad (12.32)$$
>
> as quoted in Chapter 1.2.

---

[7] See, e.g., Shapiro and Teukolsky (1983), Chapter 12.3.

(b) The transition from orbital stability to instability occurs when $\partial^2 V/\partial r^2|_J = 0$, which gives $r_{\text{ISCO}} = 6M$. Show that this transition occurs when

$$\partial \tilde{E}_{\text{eq}}/\partial r = 0 = \partial \tilde{J}_{\text{eq}}/\partial r, \qquad (12.33)$$

which is the counterpart of equation (12.13). Thus the point of marginal stability corresponds to a *simultaneous* turning-point in the equilibrium energy and angular momentum curves. The existence of simultaneous turning-points in the two curves is a general result that follows from equation (12.14) in Newtonian gravitation and equation (12.112) in general relativity.

(c) Apply equation (12.14), together with equations (12.31) and (12.32), to show that Ω_{orb} is again given by equation (12.12) in these coordinates.

We emphasize, however, that for binaries with similar masses, $q \approx 1$, the transition from inspiral to plunge is not sharp. This can be seen from equation (12.24), which shows that τ_{orbit}/P approaches unity close to the ISCO if $q \approx 1$. For such a binary, the radial inspiral is already appreciable over an orbital period as it reaches the ISCO, so that the transition from inspiral to plunge and merger is gradual rather than sharp. Stated differently, the deviations from circular orbit are already large as binaries with companions of comparable mass approach our so-called ISCO. In Chapter 13 we will see that dynamical simulations of binary black hole merger indeed verify this expectation. The notion of the ISCO is nevertheless a useful concept for the comparison and characterization of initial data sets.

It is useful to note that *any* attractive interaction potential that falls off more steeply than $1/r^2$ results in an ISCO.

> **Exercise 12.5** (a) Show that any attractive interaction potential energy of the form $-\lambda a^{n+1}/r^n$, where $\lambda > 0$ is a dimensionless constant and $a > 0$ carries dimensions of length, leads to a positive contribution to the equilibrium energy E_{eq} provided that $n > 2$.
>
> (b) Find the minimum of E_{eq} to locate the ISCO at
>
> $$r_{\text{ISCO}}^{n-1} = \lambda n(n-2)\frac{a^{n+1}}{M\mu}, \qquad (12.34)$$
>
> which is again positive if $n > 2$.

The ISCO is therefore not only a relativistic phenomenon, but can also arise in Newtonian gravitation. For example, two fluid stars treated not as point masses but as hydrodynamic equilibria of finite size, experience a tidal interaction.[8] For irrotational, identical stars, for example, we can model these tidal effects by including a tidal interaction term

$$E_{\text{tidal}} = -\lambda \frac{\mu M R^5}{r^6}. \qquad (12.35)$$

Here R is the stellar radius, and the dimensionless parameter λ depends on the equation of state. For an incompressible fluid $\lambda = 3/2$, and from equation (12.34) we find the ISCO is

---

[8] See Lai *et al.* (1993b, 1994b) and references therein for a semi-analytic analysis and models.

located at

$$r_{\text{ISCO}} = (24\lambda)^{1/5} R \approx 2.05 R, \tag{12.36}$$

implying that such a binary would reach the ISCO just before the two stars touch. We have plotted this result in Figure 12.2 for identical stars of compaction $m/R = 0.2$, where $m = M/2$ is the mass of the individual stars; we also show the combined effects of relativistic corrections and tidal interactions in determining the ISCO. For compressible stars λ is smaller than $3/2$, so that in Newtonian gravitation these stars would merge before they could encounter any ISCO. However, in general relativity, for identical irrotational companions constructed with a moderately stiff equation of state, the entire equilibrium sequence actually terminates at a finite separation prior to merger, at which point the stellar density profiles form a cusp (see Chapter 15.3). Orbital decay beyond this point will then trigger some tidal disruption prior to eventual merger (see Chapter 16).

After reaching the ISCO, the binary companions plunge and merge on an orbital time scale. Depending on the nature and masses of the compact companions, their coalescence leads to the formation of a merged object, or remnant, that may either be a rotating neutron star or a black hole. During the final *ringdown* phase of the binary evolution this remnant settles down into an equilibrium state. In the case of a hypermassive neutron star, secular effects (viscosity or magnetic fields) can lead later to delayed collapse to a black hole (see Chapters 14 and 16.)

During the inspiral phase, up to reasonably small binary separations, the binary can be modeled very accurately by post-Newtonian expansions (see Appendix D). The ring-down phase can be described very well by strong-field perturbative techniques. Numerical relativity simulations are needed to connect these two regimes, beginning with the late inspiral phase, and continuing through the dynamical plunge and merger phase. Numerical relativity is also required to treat delayed collapse. Not surprisingly, considerable effort in numerical relativity has focused on compact binaries in close quasicircular orbits. In the rest of this chapter we will construct initial data for binary black holes in such orbits.

12.2 The conformal transverse-traceless approach: Bowen–York

12.2.1 Solving the momentum constraint

In the transverse-traceless approach we assume both conformal flatness, so that $\bar{\gamma}_{ij} = \eta_{ij}$, and maximal slicing, so that $K = 0$. The momentum constraint (12.4) then reduces to

$$\bar{D}_j \bar{A}^{ij} = 0, \tag{12.37}$$

where $\bar{D}_j$ is now a flat-space covariant derivative. In Cartesian coordinates this operator reduces to the partial derivative ∂_j. Our assumptions have simplified the momentum constraint to the point where the decomposition of $\bar{A}_{ij}$ into a transverse and a longitudinal part (see equation 3.48) is somewhat pointless, since both of them now have vanishing divergence. It is nevertheless useful to keep this decomposition in mind, since it makes the counting of freely specifiable variables more transparent. In particular we will choose $\bar{A}_{TT}^{ij} = 0$, which amounts to making two arbitrary choices for its two independent variables. We still have to find solutions $\bar{A}_{L}^{ij}$, but as we have seen in Chapter 3.2 we can construct these *analytically* given our assumptions of conformal flatness and maximal slicing. This, in fact, is the essence of this so-called *Bowen–York approach*: we can solve the momentum constraint analytically, leaving only the Hamiltonian constraint to be solved numerically.

In Chapter 3.2 we found two such analytical black hole solutions to (12.37); one that carries a linear momentum P^i,

$$\bar{A}_{\mathbf{P}}^{ij} = \frac{3}{2r^2}\left(P^i n^j + P^j n^i - (\eta^{ij} - n^i n^j)n^k P_k\right) \tag{12.38}$$

(see equation 3.80), and one that carries an angular momentum (spin) S^i,

$$\bar{A}_{\mathbf{S}}^{ij} = \frac{6}{r^3}n^{(i}\epsilon^{j)kl}S_k n_l \tag{12.39}$$

(see equation 3.66). Here we are assuming Cartesian coordinates, $r = \sqrt{x^2 + y^2 + z^2}$ is the coordinate distance to the center of the black hole (which is located at the origin), and $n^i = x^i/r$ is the normal vector pointing away from the black hole's center. In exercise 3.32 we evaluated the surface integral (3.195)

$$P^i = \frac{1}{8\pi}\oint_{\infty} K^{ij}dS_j \tag{12.40}$$

to show that the linear momentum associated with equation (12.38) is indeed P^i, and in exercise 3.29 we computed

$$J_i = \frac{\epsilon_{ijk}}{8\pi}\oint_{\infty} x^j K^{kl}dS_l \tag{12.41}$$

to verify that the angular momentum associated with equation (12.39) is S^i.

To allow for a black hole located at a point C^i we introduce a subscript (or occasionally superscript) **C**. We then have

$$\boxed{\bar{A}_{\mathbf{CP}}^{ij} = \frac{3}{2r_{\mathbf{C}}^2}\left(P^i n_{\mathbf{C}}^j + P^j n_{\mathbf{C}}^i - (\eta^{ij} - n_{\mathbf{C}}^i n_{\mathbf{C}}^j)n_{\mathbf{C}}^k P_k\right),} \tag{12.42}$$

and

$$\boxed{\bar{A}_{\mathbf{CS}}^{ij} = \frac{6}{r_{\mathbf{C}}^3}n_{\mathbf{C}}^{(i}\epsilon^{j)kl}S_k n_l^{\mathbf{C}},} \tag{12.43}$$

where $r_{\mathbf{C}} = ||x^i - C^i||$ is the coordinate distance to the center of the black hole located at $x^i = C^i$, and $n^i = (x^i - C^i)/r_{\mathbf{C}}$ is the normal vector.

Given that the momentum constraint (12.37) is linear, we can construct a *binary* black hole solution by superposition of single solutions

$$\boxed{\bar{A}^{ij} - \bar{A}^{ij}_{\mathbf{C}_1\mathbf{P}_1} + \bar{A}^{ij}_{\mathbf{C}_1\mathbf{S}_1} + \bar{A}^{ij}_{\mathbf{C}_2\mathbf{P}_2} + \bar{A}^{ij}_{\mathbf{C}_2\mathbf{S}_2}.}$$
(12.44)

This completes an analytic solution of the momentum constraint describing two black holes with arbitrary momenta and spins.

Exercise 12.6 Show that[9]

$$\mathbf{P} = \mathbf{P}_1 + \mathbf{P}_2$$
(12.45)

is the total linear momentum of the solution (12.44) and

$$\mathbf{J} = \mathbf{C}_1 \times \mathbf{P}_1 + \mathbf{C}_2 \times \mathbf{P}_2 + \mathbf{S}_1 + \mathbf{S}_2$$
(12.46)

its total angular momentum about the origin of the coordinate system.

For a binary system, $\mathbf{S}_1$ and $\mathbf{S}_2$ can be associated with the spin of the individual black holes only in the limit of *infinite* binary separation, but we nevertheless take the liberty to define the orbital angular momentum $\mathbf{L}$ as

$$\mathbf{L} \equiv \mathbf{J} - \mathbf{S}_1 - \mathbf{S}_2.$$
(12.47)

12.2.2 Solving the Hamiltonian constraint

With a solution to the momentum constraint at hand we can now proceed to solve the Hamiltonian constraint (12.3). Under the assumptions of conformal flatness and maximal slicing, this equation reduces to

$$\bar{D}^2\psi = -\frac{1}{8}\psi^{-7}\bar{A}_{ij}\bar{A}^{ij}.$$
(12.48)

We have reduced the construction of binary black hole initial data to solving a single nonlinear elliptic equation. On the right hand side, the term $\bar{A}_{ij}\bar{A}^{ij}$ can be computed analytically from (12.44). Unfortunately this term diverges at the black holes' centers $\mathbf{C}_i$. Dealing with this singularity requires some extra care.

Two different approaches have been adopted in the literature to solve this problem; they differ in the topology of the resulting solution. Recall our discussion in Chapter 3.1, which demonstrated that initial data sets representing multiple black holes are not unique. As a starting point, we can look for generalizations of the time-symmetric solution

$$\psi = 1 + \frac{\mathcal{M}_1}{r_{\mathbf{C}_1}} + \frac{\mathcal{M}_2}{r_{\mathbf{C}_2}},$$
(12.49)

---

[9] To represent the momenta P^i and spins S^i of these solutions it is convenient to use bold-face notation $\mathbf{P}$ and $\mathbf{S}$ instead of index notation.

which solves equation (12.48) for $\bar{A}_{ij} = 0$ (i.e., Laplace's equation). This solution represents the three-sheeted topology sketched in Figure 3.2. In this topology each black hole generates an Einstein–Rosen bridge to its own asymptotically flat universe, meaning that the solution is not symmetric across each throat. This symmetry can be recovered, however, by introducing "spherical inversion images". This approach leads to a two-sheeted topology, schematically represented in Figure 3.3, in which the two black holes connect two identical asymptotically flat universes.

In the *conformal imaging* approach[10] a two-sheeted topology is assumed. Establishing the isometry between the two sheets requires adding image terms to the extrinsic curvature (12.44).[11] Once this has been done, the two sheets, representing two asymptotically flat universes connected by two Einstein–Rosen bridges (see Figure 3.3), are exact mirror images of each other. We therefore need to solve the Hamiltonian constraint (12.48) only on one of these two sheets; we could always find the conformal factor on the other sheet from the symmetry. This means that we eliminate the black hole interior containing the black hole singularities from the computational domain and solve (12.48) only in the black hole exterior. This is an example of *black hole excision*, which we will encounter many more times.

Clearly we need a boundary condition for the conformal factor ψ on the black hole throats, the interior of which we excise. These boundary conditions follow immediately from the isometry that is imposed on the throats.

While the isometry of the resulting two-sheeted solutions is aesthetically appealing, the conformal-imaging approach requires complicated image terms for the extrinsic curvature as well as the imposition of boundary conditions on internal boundaries. The *puncture* approach,[12] which leads to a three-sheeted topology, avoids both of these difficulties.

The key idea of the puncture approach is to absorb the singularities arising in the solution to equation (12.48) in analytical terms. As we have said before, the expression (12.49) satisfies the Hamiltonian constraint (12.48) for $\bar{A}_{ij} = 0$. We may therefore write the general solution as equation (12.49), the homogeneous solution, plus a new term correcting for finite $\bar{A}_{ij}$. As it turns out, this new term satisfies an equation that is regular everywhere.

We begin by writing

$$\boxed{\psi = 1 + 1/\alpha + u,} \tag{12.50}$$

where α is defined by

$$\frac{1}{\alpha} \equiv \frac{\mathcal{M}_1}{r_{\mathbf{C}_1}} + \frac{\mathcal{M}_2}{r_{\mathbf{C}_2}}. \tag{12.51}$$

Here we may refer to the $\mathcal{M}_i$ as the *puncture masses*. It is important to realize, though, that they can be equated to the black hole's ADM mass or irreducible mass *only* in the limit of

---

[10] Cook (1991); Cook *et al.* (1993); Cook (1994).
[11] Kulkarni *et al.* (1983).
[12] Beig and Ó Murchadha (1994, 1996); Brandt and Brügmann (1997).

infinite separation. For any finite separation these parameters have *no* immediate physical meaning.

The Hamiltonian constraint now reduces to an equation for the correction term u,

$$\bar{D}^2 u = -\beta \left(\alpha(1 + u) + 1\right)^{-7},$$ (12.52)

where we have used the abbreviation

$$\beta \equiv \frac{1}{8}\alpha^7 \bar{A}_{ij} \bar{A}^{ij}.$$ (12.53)

Notice that α goes to zero at each one of the "punctures" $\mathbf{C}_i$. It also enters β with a sufficiently high power to suppress the divergence in $\bar{A}_{ij}\bar{A}^{ij}$. As a consequence the source term in equation (12.52) is regular everywhere. We have thus eliminated the need for both the image terms in $\bar{A}_{ij}$ and for black hole excision, and can, instead, solve for u with any standard method for nonlinear elliptic equations (see, e.g., our discussion in Chapter 6.2.2 following equation 6.54). To impose asymptotic flatness, we impose a Robin boundary condition

$$\frac{\partial(ru)}{\partial r} = 0$$ (12.54)

at large distances from the black holes.

Once we have found the conformal factor ψ numerically, we can compute the ADM mass M_{ADM} from equation (3.145),

$$M_{\mathrm{ADM}} = -\frac{1}{2\pi} \oint_{\infty} \bar{D}^i \psi\, d\bar{S}_i.$$ (12.55)

Within the puncture approach we can insert the ansatz (12.50) to find

$$M_{\mathrm{ADM}} = -\frac{1}{2\pi} \oint_{\infty} \bar{D}^i \left(\frac{1}{\alpha}\right) d\bar{S}_i - \frac{1}{2\pi} \int \bar{D}^2 u\, d^3x$$

$$= \mathcal{M}_1 + \mathcal{M}_2 + \frac{1}{2\pi} \int \beta \left(\alpha(1 + u) + 1\right)^{-7} d^3x.$$ (12.56)

Here the volume integral in the last term extends over all space.

12.2.3 Identifying circular orbits

Up to now we have discussed how, for particular choices of black hole positions $\mathbf{C}_i$, momenta $\mathbf{P}_i$, spins $\mathbf{S}_i$ and puncture masses $\mathcal{M}_i$, we can construct a solution to the constraint equations. We now ask how we can pick these parameters in such a way that the resulting black holes represent a binary in circular orbit.

In a zero-momentum frame we have $\mathbf{P} = 0$ and therefore

$$\mathbf{P}_1 = -\mathbf{P}_2.$$ (12.57)

For a circular orbit these momenta must also be perpendicular to $\mathbf{C} \equiv \mathbf{C}_1 - \mathbf{C}_2$,

$$\mathbf{C} \cdot \mathbf{P}_1 = 0. \tag{12.58}$$

This requirement corresponds to the condition $P_r = 0$ in exercise 12.1 and $dr/d\tau = 0$ in exercise 12.4. We will focus on black holes that have $\mathbf{S}_i = 0$, meaning that they are not spinning as seen in an inertial frame. We will later refer to such binaries as *nonspinning*. With this choice the problem has been reduced to a 4-dimensional parameter space. The four independent parameters may be taken to be the puncture masses $\mathcal{M}_1$ and $\mathcal{M}_2$, the momentum $P_1 = ||\mathbf{P}_1||$ and the coordinate separation $C = ||\mathbf{C}||$. For spinning black holes we also have to pick the black hole spins $\mathbf{S}_i$.[13]

Given these input parameters, we can construct a binary configuration by solving the constraint equations as outlined above in Sections 12.2.1 and 12.2.2. For each configuration we can then compute the significant physical quantities, including the total ADM mass M_{ADM} (equations 12.55 or 12.56) and the angular momentum J (equation 12.46). We then define the binding energy E_b as

$$\boxed{E_b \equiv M_{\text{ADM}} - m_1 - m_2.} \tag{12.59}$$

Here the black hole masses m_1 and m_2 are assigned to be their irreducible masses,[14]

$$\boxed{m_i = (\mathcal{A}_i/16\pi)^{1/2},} \tag{12.60}$$

where $\mathcal{A}_i$ is the proper area of the event horizon of the ith black hole (see equation 7.2). Sometimes the quantity E_b is referred to as the *effective potential* (recall exercise 12.4).

> **Exercise 12.7** Evaluate E_b for two black holes in circular orbit at infinite separation and determine the sign of E_b when the circular orbit is at finite separation.

As we discussed in Chapter 7.1, locating the black hole's event horizon requires knowledge of the entire future. We only construct data on one time slice Σ, so clearly we cannot compute the irreducible mass m_{irr} exactly. For most of the binary inspiral, however, the event horizons are very well approximated by the black hole's apparent horizons (see Chapter 7.3), and it is reasonable to compute m_{irr} from these.[15] In the conformal imaging approach these apparent horizons coincide with the throats on which the isometry is imposed. The horizons are therefore at a fixed coordinate location, and their proper area $\mathcal{A}$ can be computed quite easily. In the puncture method, however, we do not know the location

---

[13] Pfeiffer *et al.* (2000).

[14] Some authors define the binding energy as the difference between the ADM mass and the black holes' Kerr masses (7.3) (e.g., Cook 1994; Pfeiffer *et al.* 2000). For the nonspinning black holes that we focus on here this distinction clearly does not make any difference. In general, however, the definition (12.59) seems more natural. Since the irreducible mass M_{irr} is constant along quasistationary evolutionary sequences, the definition (12.59) leads to simultaneous turning points in the equilibrium binding energy and the ADM mass.

[15] But see Pfeiffer *et al.* (2000) for a discussion of rapidly spinning black holes in close binaries.

of the apparent horizons a priori, and instead we have to find them by implementing one of the methods discussed in Chapter 7.3 before their area can be computed.[16]

Lastly we compute the proper separation between the two horizons l. In practice it is much easier to compute the proper separation along $\mathbf{C}$, which is a very good approximation to the shortest proper, or geodesic, separation (within Σ) for many configurations.

The condition (12.58) insures that each black hole's momentum is perpendicular to the separation axis, at least momentarily. Imposing this condition is not sufficient for a circular orbit, however, since it also holds at the apocenter and pericenter of a noncircular orbit. To guarantee a circular orbit we must impose the additional condition that the binding energy (i.e., the effective potential) E_b be at a stationary point along a sequence of constant angular momentum J for fixed black hole masses m_1 and m_2,

$$\left. \frac{\partial E_b}{\partial l} \right|_{J,m_1,m_2} = 0. \tag{12.61}$$

We already employed this condition in exercises 12.1 and 12.4 to identify circular orbits. Once we identify the circular orbits, we can construct an equilibrium sequence of binaries of fixed black hole masses m_1 and m_2 at different separations. As in equation (12.14) (or equation 12.112) we can then find the binary's orbital angular velocity Ω_{orb} at each separation from

$$\Omega_{\text{orb}} = \left. \frac{\partial E_{\text{eq}}}{\partial J_{\text{eq}}} \right|_{m_1,m_2}. \tag{12.62}$$

To identify the ISCO, we can apply the turning-point criterion to the function $E_{\text{eq}}(l)$, as in equations (12.13) and (12.33). Alternatively, we can apply the turning-point criterion to the equilibrium angular momentum $J_{\text{eq}}(l)$, as noted in exercise 12.4. As pointed out in the same exercise, applying the turning-point criterion is equivalent to finding a point of inflection in the effective potential, $\partial^2 E_b / \partial l^2 = 0$.

Figure 12.3 illustrates the numerical construction of a quasiequilibrium binary consisting of equal-mass, nonspinning black holes in circular orbit using conformal imaging and the effective potential approach.[17] Convenient fitting formulae have been derived[18] for the same system, using the puncture method. These handy formulae specify the puncture masses $\mathcal{M}$, momenta P, and the quantities M_{ADM}, J_{ADM} and Ω_{orb} as functions of the puncture coordinate separation C, all normalized by the total ADM mass of the binary at infinite separation (which equals the total irreducible mass.)

We will review the resulting orbits from these numerical calculations in greater detail in Section 12.4. But first we will discuss an alternative approach to constructing binary black holes that probably provides a better approximation to binaries in quasiequilibrium.

---

[16] Baumgarte (2000).

[17] Cook (1994).

[18] Tichy and Brügmann (2004). Quasiequilibrium is achieved by imposing the relativistic virial relation $M_K = M_{\text{ADM}}$; see Chapter 3.5 and exercise 12.9. A lapse is required for this condition and is obtained by imposing $\partial K/\partial t = 0$ (equation 4.12).

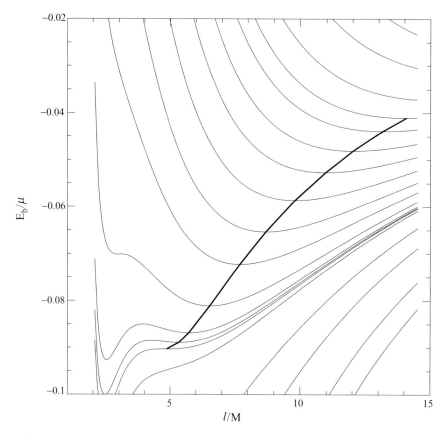

Figure 12.3 The binding energy E_b/μ as a function of separation l for various values of the angular momentum $J/\mu M$ (thin lines) as obtained by the Bowen–York conformal-imaging method for an equal mass, nonspinning black hole binary. The thick line connects a sequence of stable quasicircular orbits, which correspond to minima of the binding energy (equation 12.61), and yields E_{eq} as a function of l ("effective potential approach"). This sequence terminates at the ISCO, identified where an E_b curve at constant J exhibits a point of inflection. [From Cook (1994).]

12.3 The conformal thin-sandwich approach

12.3.1 The notion of quasiequilibrium

The appeal of the Bowen–York approach is that we can solve a large part of the problem (i.e., the momentum constraints) analytically, which furnishes useful insight into how black hole binary initial data can be constructed. However, in Section 12.4 we will find some evidence suggesting that the resulting binary solutions may not describe astrophysically realistic, circular-orbit, binary black holes as closely as we might like. The core of the problem is that the Bowen–York formalism does not provide any direct means of imposing quasiequilibrium conditions on the metric. We invoke global energy minimum arguments

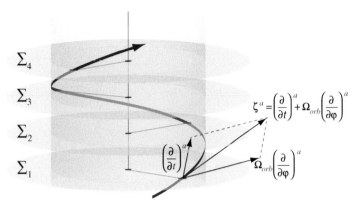

Figure 12.4 Illustration of a helical Killing vector generating a circular binary orbit.

to select models in the initial data set that presumably are close to quasiequilibrium, but there is no guarantee that the spacetime corresponding to these models will maintain quasiequilibrium once we evolve off the initial time slice.

Following our discussion in Section 12.1 we seek to construct black hole binaries at moderately close separations, at which point the orbits have become quasicircular. At these separations we assume that the individual black holes have settled into a quasiequilibrium state, whereby no matter and very little gravitational radiation fall into them to disturb this state. Now a stationary spacetime in *strict* equilibrium possesses two important Killing vectors: $\partial/\partial t$, associated with stationary symmetry and $\partial/\partial \phi$, associated with rotational symmetry. The spacetime for a *quasiequilibrium* binary system has no such Killing vectors, but it does possess an approximate *helical Killing vector* ξ^a_{hel} given by a linear combination of these two vectors. In particular, the approximate Killing vector ξ^a_{hel} generates a circular orbit with orbital angular velocity Ω_{orb}, as illustrated in Figure 12.4. In the coordinate system of an inertial observer at infinity we may write ξ^a_{hel} as

$$\xi^a_{\text{hel}} = (\partial/\partial t)^a + \Omega_{\text{orb}} (\partial/\partial \phi)^a. \tag{12.63}$$

If ξ^a_{hel} were an exact helical Killing vector we would have $\mathcal{L}_{\xi_{\text{hel}}} g_{ab} = 0$. In other words, the spacetime geometry of the binary system would be invariant if we were to move through an infinitesimal angle $d\phi = \Omega_{\text{orb}} dt$ in an infinitesimal time dt. Consequently, we could choose a corotating coordinate system in which the time axis $t^a = \alpha n^a + \beta^a$ remains aligned with ξ^a_{hel},[19] so that, in this coordinate system, all partial derivatives of the metric with respect to time would vanish. In quasiequilibrium, we can expect that, in such a corotating coordinate system, the corresponding time derivatives will vanish approximately.

The Bowen–York approach provides no mechanism to impose this notion of quasiequilibrium. We can construct a spatial metric and extrinsic curvature that solve the constraint

---

[19] Note that this choice leads to a spacelike time coordinate at large separations.

equations, but we have no control over their time evolution. In the conformal thin-sandwich decomposition of Chapter 3.3, on the other hand, we can explicitly set several of these time derivatives to zero. This approach therefore seems more promising for the construction of quasiequilibrium initial data.

In Chapter 3.3 we introduced the time derivative of the conformally related metric as

$$\bar{u}_{ij} \equiv \partial_t \bar{\gamma}_{ij} \tag{12.64}$$

(see equation 3.94) and then expressed the extrinsic curvature $\bar{A}_{ij}$ in terms of $\bar{u}_{ij}$ with the help of the evolution equation (2.134),

$$\bar{A}^{ij} = \frac{\psi^6}{2\alpha} \left((\bar{L}\beta)^{ij} - \bar{u}^{ij} \right) \tag{12.65}$$

(see equation 3.99). Inserting this into the momentum constraint (12.4) then yields

$$(\bar{\Delta}_L \beta)^i - (\bar{L}\beta)^{ij} \, \bar{D}_j \, \ln(\psi^{-6}\alpha) = \bar{\alpha} \, \bar{D}_j (\psi^6 \alpha^{-1} \bar{u}^{ij}) + \frac{4}{3} \alpha \bar{D}^i K, \tag{12.66}$$

while the Hamiltonian constraint (12.3) remains

$$8\bar{D}^2 \psi - \psi \bar{R} - \frac{2}{3} \psi^5 K^2 + \psi^{-7} \bar{A}_{ij} \bar{A}^{ij} = 0. \tag{12.67}$$

Here we have again assumed a vacuum spacetime.

For the construction of quasiequilibrium data it now seems natural to choose the requirements that

$$\bar{u}_{ij} = 0 \tag{12.68}$$

and

$$\partial_t K = 0 \tag{12.69}$$

in the corotating coordinate system where $t^a \propto \xi^a_{\text{hel}}$. The latter condition then yields an equation for the lapse α,

$$\bar{D}^2 (\alpha \psi) = \alpha \psi \left(\frac{7}{8} \psi^{-8} \bar{A}_{ij} \bar{A}^{ij} + \frac{5}{12} \psi^4 K^2 + \frac{1}{8} \bar{R} \right) + \psi^5 \beta^i \bar{D}_i K, \tag{12.70}$$

while the former reduces the momentum constraint (12.66) to

$$(\bar{\Delta}_L \beta)^i - (\bar{L}\beta)^{ij} \, \bar{D}_j \, \ln(\psi^{-6}\alpha) = \frac{4}{3} \alpha \bar{D}^i K. \tag{12.71}$$

The extrinsic curvature $\bar{A}^{ij}$ is given in terms of these unknowns by

$$\bar{A}^{ij} = \frac{\psi^6}{2\alpha} (\bar{L}\beta)^{ij}. \tag{12.72}$$

Equations (12.67), (12.70) and (12.71) now form a coupled system of elliptic equations for the lapse α, the conformal factor ψ and the shift β^i. Outer boundary conditions follow from asymptotic flatness and equation (12.63),

$$\lim_{r\to\infty} \psi = 1, \qquad \lim_{r\to\infty} \alpha = 1, \qquad \lim_{r\to\infty} \beta^i = \Omega_{\text{orb}} \left(\frac{\partial}{\partial\phi} \right)^i = (\mathbf{\Omega}_{\text{orb}} \times \mathbf{r})^i. \qquad (12.73)$$

Here we are assuming that the coordinate system is centered on the binary system's center of mass-energy at coordinate radius $r = 0$, where $r^2 = x^2 + y^2 + z^2$. The above asymptotic condition on the shift guarantees that the coordinate system corotates with the binary.

We have yet to choose a conformal background metric $\bar{\gamma}_{ij}$ and the extrinsic curvature K. The above equations simplify dramatically for conformally flat spaces $\bar{\gamma}_{ij} = \eta_{ij}$ and maximal slicing $K = 0$.[20] As an alternative, the background can be chosen to be a superposition of Kerr–Schild geometries.[21]

Once we have chosen a background solution we are almost ready to solve equations (12.67), (12.70) and (12.71) for α, ψ and β^i. The last issue that we need to address before attempting a numerical integration are the singularities that lurk in the interior of the black holes. In the context of the Bowen–York approach we discussed two different methods of eliminating these singularities. The first approach was the conformal-imaging method, which imposes an isometry and enables us to excise the black hole interior; the second approach was the puncture method which absorbs the singularity in analytical terms.

Unfortunately, neither one of these two methods works well in the context of the conformal thin-sandwich approach. The conformal-imaging approach leads to inconsistencies on the black hole throats. In a nutshell, the isometry requires both $\alpha = 0$ and $\beta^i = 0$ on the throats, which, according to (12.72), means that the extrinsic curvature $\bar{A}^{ij}$ can remain finite only if $(\bar{L}\beta)^{ij} = 0$. This requirement leads to boundary conditions on both the shift and its derivatives, making the system overdetermined.[22] While it would be desirable to avoid boundary conditions altogether with the puncture approach, it seems impossible to construct even single equilibrium black holes with a combination of the conformal thin-sandwich and puncture methods.[23] These considerations lead us to construct new black-hole boundary conditions that are based explicitly on quasiequilibrium requirements.

12.3.2 Quasiequilibrium black hole boundary conditions

Before discussing the quasiequilibrium black hole conditions[24] it is useful to review briefly some of the formalism that we introduced to describe apparent horizons in Chapter 7.3. In particular, we considered a closed 2-dimensional hypersurface of Σ and called it S.

---

[20] See Gourgoulhon *et al.* (2002); Grandclément *et al.* (2002) and Cook and Pfeiffer (2004).

[21] See Marronetti and Matzner (2000); Yo *et al.* (2004); Cook and Pfeiffer (2004).

[22] See Cook (2002) for a discussion. In Gourgoulhon *et al.* (2002) and Grandclément *et al.* (2002), who implemented this approach, the problem is avoided by adding a small and otherwise insignificant artificial term.

[23] At least if the background solution is based on an expression of the form (12.49); see Hannam *et al.* (2003), but compare with Hannam (2005).

[24] In this section we closely follow Cook and Pfeiffer (2004), but see Jaramillo *et al.* (2004) for an alternative approach.

We called its outward pointing unit normal s^a, and found that s^a induces a 2-dimensional metric,

$$m_{ab} = \gamma_{ab} - s_a s_b, \tag{12.74}$$

on S. We constructed a future-pointing, outgoing null vector k^a from

$$k^a = \frac{1}{\sqrt{2}}(n^a + s^a). \tag{12.75}$$

The geometry of all of these objects is illustrated in Figure 7.3. As before, we define the projection of the gradient of k^a into S as

$$\Theta_{ab} = m_a{}^c m_b{}^d \nabla_c k_d. \tag{12.76}$$

(see equation 7.78). We now have all the tools in hand to formulate the required conditions.

To begin with, we would like the surface S to be an apparent horizon. According to equations (7.15) and (7.17), this means that the trace of Θ_{ab}, i.e., the expansion of the k^a, has to vanish,

$$\boxed{\Theta = m^{ab}\Theta_{ab} = m^{ab}\nabla_a k_b = 0.} \tag{12.77}$$

In the context of the isolated horizon formalism in Chapter 7.4 we have also seen that for the black hole to be in equilibrium, the shear of Θ_{ab} also has to vanish,

$$\boxed{\sigma_{ab} = \Theta_{(ab)} - \frac{1}{2}m_{ab}\,\Theta = 0} \tag{12.78}$$

(see equation 7.84). Finally, we would like to construct a coordinate system that tracks the horizon, i.e., we want to require that in quasiequilibrium the coordinate location of the apparent horizon not change as the initial data begin to evolve. Consider a sequence of apparent horizons S residing in neighboring spatial slices Σ, as we did in Chapter 7.4. For isolated black holes, the resulting worldtube H is smooth and is generated by k^a. To obtain a coordinate system that tracks the horizon, the time vector of the evolution, t^a, must reside in the null surface generated by k^a. This requirement implies that t^a, on S, must be tangent to H. For that to be the case, t^a must be a linear combination of k^a, which generates H, and some spatial vector h^a tangent to S,

$$t^a = ak^a + bh^a, \tag{12.79}$$

where a and b are two arbitrary coefficients. Since h^a is spatial we have $n_a h^a = 0$, and since it is tangent to S we also have $s_a h^a = 0$. Combining equations (12.75) and (12.79), and using the fact that k^a is a null vector, it is then easy to see that we must have

$$\boxed{t^a k_a = 0.} \tag{12.80}$$

Conditions (12.77), (12.78) and (12.80) represent the geometric conditions that guarantee that the surface S is an apparent horizon, in quasiequilibrium, and will appear in quasiequilibrium in the coordinate system defined by the lapse and shift that we are constructing. Our next task is to translate these geometric conditions into horizon boundary conditions for the conformal factor ψ, the lapse α and the shift β^i.

We start with condition (12.80). Inserting both $t^a = \alpha n^a + \beta^a$ and equation (12.75) we immediately find

$$\beta^i s_i = \alpha. \tag{12.81}$$

The left-hand side of equation (12.81) suggests that we should split the shift into components that are normal and tangential to the surface S,

$$\beta^i = \beta^i_\parallel + \beta_\perp s^i, \tag{12.82}$$

where

$$\beta^i_\parallel \equiv m^i{}_j \beta^j \quad \text{and} \quad \beta_\perp \equiv \beta^i s_i. \tag{12.83}$$

With this notation the condition (12.81) becomes a condition on the normal component of the shift,

$$\boxed{\beta_\perp = \alpha.} \tag{12.84}$$

Exercise 12.8 Show that condition (12.84) holds for a Schwarzschild black hole in Kerr–Schild coordinates (see Table 2.1).

We next consider the apparent horizon condition (12.77). From equation (7.29) we may write the expansion as

$$\Theta = \frac{1}{\sqrt{2}} m^{ij}(D_i s_j - K_{ij}). \tag{12.85}$$

Since we would like to apply this condition to the conformally decomposed constraint equations, we also apply a conformal transformation to m_{ij} and s_i. Given the rescaling (12.1) for γ_{ij}, and given the relation (12.74) between γ_{ij}, m_{ij} and s_i, it is natural to choose

$$m_{ij} = \psi^4 \bar{m}_{ij} \quad \text{and} \quad s_i = \psi^2 \bar{s}_i. \tag{12.86}$$

Note that the normalization $\gamma_{ij} s^j s^i = 1$ then implies that the conformally rescaled normal vectors $\bar{s}^i$ are normalized with respect to the conformally related metric, $\bar{\gamma}_{ij} \bar{s}^j \bar{s}^i = 1$.

Exercise 12.9 Show that

$$D_i s_j = \psi^2 \left(\bar{D}_i \bar{s}_j - 2\bar{s}_i \bar{D}_j \ln \psi + 2\bar{\gamma}_{ij} \bar{s}^k \bar{D}_k \ln \psi \right). \tag{12.87}$$

Inserting equation (12.87) into the expansion (12.85), and using the abbreviation

$$J \equiv m^{ij} K_{ij}, \tag{12.88}$$

we find

$$\Theta = \frac{\psi^{-2}}{\sqrt{2}} \left(\bar{m}^{ij} \bar{D}_i \bar{s}_j + 4\bar{s}^k \bar{D}_k \ln \psi - \psi^2 J \right). \tag{12.89}$$

According to equation (12.77) the expansion Θ has to vanish for S to be an apparent horizon, which yields the condition

$$\boxed{\bar{s}^k \bar{D}_k \ln \psi = \frac{1}{4} \left(\psi^2 J - \bar{m}^{ij} \bar{D}_i \bar{s}_j \right).} \tag{12.90}$$

Equation (12.90) imposes a Neumann-type boundary condition on the normal gradient of the conformal factor ψ.

Before proceeding it is useful to illustrate this formalism by applying it to a Schwarzschild black hole in isotropic coordinates, for which

$$\bar{\gamma}_{ij} = \eta_{ij} \tag{12.91}$$

and

$$\psi = 1 + \frac{M}{2r}. \tag{12.92}$$

We also have $K_{ij} = 0$, which immediately gives us $J = 0$. To apply the boundary condition (12.90) on the horizon at $r = M/2$ we first have to construct the normal $\bar{s}^i$. They have to be radial, and they have to be normalized with respect to the conformally related metric, and therefore

$$\bar{s}^i = (1, 0, 0) \tag{12.93}$$

in spherical polar coordinates. We then have

$$\bar{m}^{ij} \bar{D}_i \bar{s}_j = -\bar{m}^{ij} \bar{\Gamma}^r_{ij} = \frac{2}{r}. \tag{12.94}$$

Evaluating the right hand side of equation (12.90) gives

$$\bar{s}^k \bar{D}_k \ln \psi = \psi^{-1} \partial_r \psi = -\psi^{-1} \frac{M}{2r^2}. \tag{12.95}$$

Inserting the last two equations into equation (12.90) we find

$$\frac{M}{2r^2} = \frac{1}{4} \left(1 + \frac{M}{2r} \right) \frac{2}{r}, \tag{12.96}$$

which holds on the horizon at $r = M/2$. This result is not surprising, of course, but it is reassuring.

Exercise 12.10 Show that condition (12.90) holds for a Schwarzschild black hole in Kerr–Schild coordinates (see Table 2.1).

We now return to our derivation of the quasiequilibrium boundary conditions and examine condition (12.78). For simplicity we start with the first term, $\Theta_{(ab)}$, and substitute equation (12.75) for k^a. Since in the end we are interested in an equation for the contravariant components of the shift vector β^i, it is also easier to consider the contravariant components of Θ^{ab}, obtaining

$$\Theta^{(ab)} = m^{ac} m^{bd} \nabla_{(c} k_{d)} = \frac{1}{\sqrt{2}} m^{ac} m^{bd} \left(\nabla_{(c} n_{d)} + \nabla_{(c} s_{d)} \right). \tag{12.97}$$

We now introduce the abbreviation

$$H^{ab} = m^{ac} m^{bd} \nabla_{(c} s_{d)} = \mathcal{D}^{(a} s^{b)}, \tag{12.98}$$

where we have also introduced $\mathcal{D}_i$ as the covariant derivative operator compatible with m_{ij}. In complete analogy with our construction of the 3-dimensional covariant derivative D_i from the 4-dimensional covariant derivative ∇_a by projecting all indices into Σ (see equation 2.40 and the following equations), we construct $\mathcal{D}_i$ by projecting all indices of the 4-dimensional covariant derivative into S. With the help of H^{ab}, as well as the definition of the extrinsic curvature (2.49), we may now rewrite equation (12.97) as

$$\Theta^{(ij)} = \frac{1}{\sqrt{2}} \left(H^{ij} - m^i{}_l m^j{}_m K^{lm} \right), \tag{12.99}$$

where, without loss of generality, we have also switched to spatial indices. It turns out to be convenient to write the extrinsic curvature as

$$K^{ij} = A^{ij} + \frac{1}{3} \gamma^{ij} K = \frac{1}{2\alpha} \left((L\beta)^{ij} - u^{ij} \right) + \frac{1}{3} \gamma^{ij} K, \tag{12.100}$$

where we have used equations (3.31) and (3.93). We also remind the reader that the longitudinal operator, i.e., the vector gradient or conformal Killing operator, is defined as

$$(L\beta)^{ij} = D^i \beta^j + D^j \beta^i - \frac{2}{3} \gamma^{ij} D_k \beta^k \tag{12.101}$$

(see equation 3.50). Before proceeding we decompose the shift term according to equation (12.82) and write

$$\frac{1}{2\alpha} m^i{}_l m^j{}_m (L\beta)^{lm} = \frac{1}{2\alpha} m^i{}_l m^j{}_m (L\beta_\parallel)^{lm} + \frac{\beta_\perp}{\alpha} H^{ij} - \frac{1}{3\alpha} m^{ij} D_k (\beta_\perp s^k), \tag{12.102}$$

where we have used $m_{ij} s^j = 0$ as well as the definition (12.98). Substituting the expressions (12.100) and (12.102) into equation (12.99) we now find

$$\sqrt{2}\, \Theta^{(ij)} = \frac{1}{2\alpha} \left(-m^i{}_l m^j{}_m (L\beta_\parallel)^{lm} + \frac{1}{3\alpha} m^{ij} D_k (\beta_\perp s^k) + m^i{}_l m^j{}_m u^{lm} \right)$$
$$- \frac{1}{3} m^{ij} K + H^{ij} \left(1 - \frac{\beta_\perp}{\alpha} \right). \tag{12.103}$$

Applying a conformal transformation to the first three terms casts this relation as

$$\sqrt{2}\, \Theta^{(ij)} = \frac{\psi^{-4}}{2\alpha} \left(-\bar{m}^i{}_l \bar{m}^j{}_m (\bar{L}\beta_\parallel)^{lm} + \frac{1}{3\alpha} \bar{m}^{ij} D_k (\beta_\perp s^k) + \bar{m}^i{}_l \bar{m}^j{}_m \bar{u}^{lm} \right)$$
$$- \frac{1}{3} m^{ij} K + H^{ij} \left(1 - \frac{\beta_\perp}{\alpha} \right), \tag{12.104}$$

where we have used the rescaling rule (3.98) derived in exercise 3.15 for $(L\beta_\parallel)^{ij}$. According to the condition (12.78) we can now compute the shear σ^{ij} by taking the trace-free part, with respect to m_{ij}, of $\Theta^{(ij)}$. In the process all terms in (12.104) that are proportional to m^{ij} drop out, and we find

$$\sigma^{ij} = -\frac{1}{\sqrt{2}}\frac{\psi^{-4}}{\alpha}\left(\left(\bar{\mathcal{D}}^{(i}\beta_\parallel^{j)} - \frac{1}{2}\bar{m}^{ij}\bar{\mathcal{D}}_c\beta_\parallel^c\right) - \frac{1}{2}\left(\bar{m}^i{}_l\bar{m}^j{}_m\bar{u}^{lm} - \frac{1}{2}\bar{m}^{ij}\bar{m}_{lm}\bar{u}^{lm}\right)\right)$$
$$+\frac{1}{\sqrt{2}}\left(H^{ij} - \frac{1}{2}m^{ij}H\right)\left(1 - \frac{\beta_\perp}{\alpha}\right). \tag{12.105}$$

So far this derivation has been completely general, and it is now time to specialize to the situation at hand. To begin with, we can insert the boundary condition (12.84), which immediately eliminates the last term. For the construction of quasiequilibrium data we also assume $\bar{u}_{ij} = 0$, which eliminates the middle term. For our purposes, the geometrical condition of vanishing shear (12.78) then reduces to

$$\boxed{\bar{\mathcal{D}}^{(i}\beta_\parallel^{j)} - \frac{1}{2}\bar{m}^{ij}\bar{\mathcal{D}}_c\beta_\parallel^c = 0.} \tag{12.106}$$

The trained eye will recognize this as the conformal Killing's equation (A.27) for a 2-dimensional surface S. This means that the shear σ^{ab} vanishes if the tangential shift $\beta_\parallel^i$ is a conformal Killing vector of $\bar{m}_{ij}$ on the black hole horizon S. This condition has a surprisingly transparent physical interpretation, as we discuss below.[25]

As it turns out, it is remarkably simple to construct a tangential shift $\beta_\parallel^i$ so that it is a conformal Killing vector of $\bar{m}_{ij}$ on S. We first observe that any closed, 2-dimensional surface S is conformally equivalent to the unit 2-sphere. By this we mean that we can construct any metric $\bar{m}_{ij}$ on S from the flat metric η_{ij} on the unit sphere through a combination of conformal and coordinate transformations.[26] In exercise A.7 of Appendix A we also show that a Killing vector of a metric g_{ab} is automatically a conformal Killing vector of the conformally related metric $\bar{g}_{ab}$. This means that a Killing vector of the flat metric η_{ij} on the unit sphere is automatically a conformal Killing vector of $\bar{m}_{ij}$ on S, and hence a solution to the boundary condition (12.106).

We have therefore reduced the problem to finding Killing vectors of the unit sphere. This is very simple – in spherical coordinates θ and ϕ, for example, we have $\xi^i = e^i_{(\phi)}$ (see equation A.22 in Appendix A). On the unit sphere, all azimuthal vectors associated with any rotation axis passing through its center are Killing vectors. Viewing the unit sphere as embedded in a flat 3-dimensional space we can express these vectors ξ^i in Cartesian coordinates as $\xi^i = \epsilon^{ijk}\hat{z}_j\hat{n}_k$, where $\hat{z}^i$ is a unit vector aligned with the axis of rotation and where $\hat{n}^i$ is the unit normal on the sphere. Since the product of a Killing vector ξ^i with a

---

[25] Cook and Pfeiffer (2004).

[26] The metric $\bar{m}_{ij}$ has at most three independent components, which we can account for by a conformal transformation and a transformation of the two coordinates.

constant is still a Killing vector, we see that

$$\boxed{\beta_{\parallel}^{i} = \Omega_{\text{spin}} \xi^{i}}$$

(12.107)

is a Killing vector of η_{ij} on the unit sphere, hence a conformal Killing vector of $\bar{m}_{ij}$ on the black hole horizon S, and thus the desired solution to (12.106).

It is remarkable that condition (12.107) leaves us exactly the freedom that we need to specify an arbitrary spin on the black hole. The parameter Ω_{spin} determines the angular speed of the black hole, and ξ^{i} the axis of rotation. Given that we implement this boundary condition in the corotating coordinate system associated with the helical Killing vector ξ_{hel}^{a} defined by equation (12.63), we can construct corotating black holes by setting $\Omega_{\text{spin}} = 0$. As seen by an inertial observer, these black holes are rotating at an angular velocity Ω_{orb}. To construct black holes that are not rotating as seen by an inertial observer it seems reasonable to set the magnitude of Ω_{spin} equal to Ω_{orb} and choose ξ^{i} perpendicular to the orbital plane. We will refer to this approach as the "leading-order" approximation. Evaluating the resulting black holes' quasi-local spin J_{S} with the help of equation (7.74) reveals that generally this approach does not lead to exactly zero spin. This suggests that we must vary Ω_{spin} until the J_{S} does indeed vanish.[27]

To summarize, we have now expressed the geometrical quasiequilibrium boundary conditions (12.77), (12.78) and (12.80) in terms of the boundary conditions (12.84) for $\beta_{\perp}$, (12.90) for ψ, and (12.107) for $\beta_{\parallel}^{i}$. Interestingly the geometric conditions do not provide a boundary condition for the lapse α, which apparently we can choose freely.[28] The lapse determines the rate at which proper time advances compared to coordinate time, and in fact it seems quite natural that we can freely choose this rate, even on the horizon of a black hole in quasiequilibrium.

We can now construct a black hole binary by solving the equations (12.67), (12.70) and (12.71) subject to the inner boundary conditions (12.84), (12.90), and (12.107). We impose these boundary conditions on two coordinate spheres chosen to represent the apparent horizons of the two black holes. The coordinate separation of these two spheres controls the binary separation, while the coordinate radius of the two spheres controls the black hole masses.

12.3.3 Identifying circular orbits

We now have almost all pieces in place to construct quasiequilibrium binary black holes in the conformal thin-sandwich approach. The one last missing piece is the value of the orbital angular velocity Ω_{orb}, which enters through the outer boundary for the shift β^{i} in

---

[27] Caudill *et al.* (2006).

[28] Cook (2002) derives a boundary condition for the lapse from an additional geometric condition on the expansion of the *ingoing* null vectors. However, Cook and Pfeiffer (2004) find the resulting system of boundary conditions to be degenerate, leading to an ill-posed elliptic system, and conclude that the boundary value of the lapse can be chosen freely; *cf.* Jaramillo *et al.* (2004); Matera *et al.* (2008).

equation (12.73). Finding the correct Ω_{orb} is equivalent to identifying a circular orbit. In the Bowen–York effective potential approach of Section 12.2 we identified circular orbits by locating extrema of the binding energy along curves of constant angular momentum (see exercises 12.1 and 12.4). Grandclément *et al.* (2002) suggest an alternative way of identifying circular orbits, namely by varying Ω_{orb} until the ADM mass M_{ADM} given by equation (3.128) equals the Komar mass M_K given by equation (3.179),

$$\boxed{M_{ADM} = M_K.}$$
(12.108)

For metrics that are sufficiently well-behaved asymptotically the ADM mass is always well defined, while the Komar mass is meaningful only for stationary spacetimes. Equating the two masses therefore singles out quasiequilibrium spacetimes, which, for binaries, identifies circular orbits.

We can also justify the condition (12.108) in terms of the virial relation. In exercises 3.23 and 3.28 we showed that, to first order, equation (12.108) imposes the virial theorem (3.184) in Newtonian gravity. Gourgoulhon and Bonazzola (1994) show that equation (12.108) in fact leads to a relativistic virial theorem, which we would expect to hold for circular orbits in binaries, at least approximately. In exercise 12.1 we have seen that finding energy extrema to identify circular orbits is equivalent to imposing the virial theorem for Newtonian point masses, and in Section 12.4 we will present evidence that a similar result holds for relativistic black hole binaries.[29]

Notice that computing the Komar mass (3.179) requires knowledge of a timelike Killing vector, which in a numerical relativity context is usually expressed in terms of the lapse α and the shift β^i. The conformal thin-sandwich approach allows us to impose the existence of at least an approximate timelike Killing vector, and provides us with the corresponding Killing lapse and shift. By contrast, the transverse-traceless approach of Section 12.2 only involves data intrinsic to one time slice Σ (as opposed to the slice and its neighborhood). It does not provide a true notion of stationary equilibrium, which requires vanishing time derivatives, it does not allow us to impose the existence of a timelike Killing vector, and it does not provide a lapse or shift. We therefore cannot compute the Komar mass from Bowen–York data alone.[30]

Before proceeding we point out one subtlety. The Komar mass (3.179) contains both lapse and shift terms. In many situations the latter fall off sufficiently rapidly so that they do not contribute to the mass – in fact, several authors define the Komar mass from the lapse term alone. In general, however, the shift term has to be included,[31] even though it leads to a complication in the current context.[32] A corotating observer moving along with

---

[29] Skoge and Baumgarte (2002) demonstrate this equivalence analytically for a relativistic spherical shell of collisionless particles.

[30] Given Bowen–York data, a lapse condition can be constructed that imposes some of the conditions for the existence of an approximate helical Killing vector and the computation of the Komar mass; see Tichy *et al.* (2003), Tichy and Brügmann (2004) and footnote 18.

[31] As exercise 3.27 demonstrates, the shift term carries the entire contribution to the Komar mass in some coordinate systems.

[32] See the discussion in Caudill *et al.* (2006).

the helical Killing vector (12.63) is not an inertial observer. For a true stationary spacetime we can split this helical Killing vector into separate rotational and timelike Killing vectors. For binaries this split cannot be done globally, but, assuming asymptotic flatness, the two can be separated at spatial infinity. There the timelike Killing vector defines an inertial observer, and we see from the boundary condition (12.73) that the shift approaches the rotational Killing vector

$$\lim_{r \to \infty} \beta^i = \beta^i_{\text{rot}} \equiv \Omega_{\text{orb}} \left(\frac{\partial}{\partial \phi} \right)^i = (\boldsymbol{\Omega}_{\text{orb}} \times \mathbf{r})^i. \tag{12.109}$$

In general the latter makes a nonvanishing contribution to the Komar mass.

Exercise 12.11 Show that substituting β^i_{rot} into the Komar mass integral (3.179) yields a contribution

$$M_{\text{rot}} = -2\Omega_{\text{orb}} J, \tag{12.110}$$

where J is the angular momentum (3.189).

Evaluating the Komar mass integral in the corotating frame, for which the shift behaves according to (12.109), yields a value that differs from what we would find in an inertial frame by M_{rot},

$$M_{\text{K}}^{\text{corot}} = M_{\text{K}}^{\text{inertial}} - 2\Omega_{\text{orb}} J. \tag{12.111}$$

This raises the question: which one of these values are we supposed to compare with the ADM mass M_{ADM} in equation (12.108)? The expression (3.128) for the ADM mass assumes that it is evaluated in an inertial frame, since otherwise the fall-off conditions (3.129) would not be satisfied. Moreover, the Komar mass requires a timelike Killing vector, and outside the "light cylinder" (on which ξ^i_{hel} is null) the helical Killing vector (12.63) becomes spacelike. This means that we have to evaluate the Komar mass in the asymptotic *inertial* frame. Our conformal thin-sandwich calculation is performed in the corotating frame, but we can account for that in one of three ways. If all "nonrotational" contributions to the shift fall off sufficiently fast, we can simply ignore the shift term in the integral (3.179) and compute the Komar mass from the lapse term alone (as many authors do). If the nonrotational contributions to the shift cannot be ignored, we can subtract β^i_{rot} before computing the Komar mass, or, equivalently, we can simply transform the Komar mass computed in the corotational frame to its value in the inertial frame with the help of equation (12.111).[33]

12.4 Quasiequilibrium sequences

As we discussed in Section 12.1, we can model the early (adiabatic) binary inspiral phase as a sequence of quasiequilibrium configurations. In the previous sections we have

---

[33] For a discussion of the invariance of the integrals defined in Chapter 3.5 for M_0, M_{ADM} and J_{ADM} when they are evaluated in the corotating *vs.* the inertial frame, see Duez *et al.* (2003), Appendix C.

developed techniques for constructing black hole binaries with certain black hole masses, spins and binary separations. Before we can stitch these together to construct evolutionary sequences we have to address one more question: what are the conserved quantities along these sequences?

For binary neutron stars it is quite evident that, in the absence of mass loss or accretion, the rest-mass (e.g., baryon number) of each neutron star is strictly conserved during binary inspiral (see Chapter 15.3). For binary black holes, identifying a conserved quantity is a little more subtle. As long as the individual black holes remain in quasiequilibrium as they evolve, it is reasonable to assume that their irreducible masses given by equation (12.60) remain constant.[34]

One approach to constructing an inspiral sequence is therefore to iterate, at a given coordinate separation of the binary pair, over the individual masses until the desired irreducible masses have been achieved. The procedure can then be repeated for different binary separations, until a sequence has been completed. A more elegant approach utilizes the relation[35]

$$\boxed{dM_{\mathrm{ADM}} = \Omega_{\mathrm{orb}}dJ,}$$ (12.112)

which strictly holds along stationary sequences of uniformly rotating configurations. We have effectively encountered this relation as it applies to binaries in exercise 12.2 and in the last part of exercise 12.4; it is the relativistic generalization of equation (12.14). In exercise 12.12 we illustrate how equation (12.112) can be enforced when connecting numerical models along a quasiequilibrium sequence.

Exercise 12.12 Let s denote some parameter along a sequence of binary black hole initial data, and let $\chi(s)$ denote an arbitrary length scale along this sequence. Also write the ADM mass, angular momentum and orbital frequency along the sequence as

$$M_{\mathrm{ADM}}(s) = \chi(s)\,e(s) \quad J(s) = \chi^2(s)\,j(s) \quad \Omega_{\mathrm{orb}}(s) = \chi^{-1}(s)\,\omega(s), \quad (12.113)$$

where $e(s)$, $j(s)$ and $\omega(s)$ are numerical values of these quantities in nondimensional units. Show that the identity (12.112) is automatically satisfied as long as χ changes according to

$$\frac{d\chi}{\chi} = -\frac{de - \omega dj}{e - 2\omega j}$$ (12.114)

along the sequence.[36]

---

[34] Black hole perturbation theory suggests that the fractional increase in the surface area of each hole in a binary consisting of equal-mass holes with spins aligned with the orbital angular momentum (and 99.8% of their maximal values) is at most 1% by the time the binary spirals down to a separation of $6M$, where M is the total (ADM) mass; Alvi (2001).

[35] See Friedman *et al.* (2002). See also equation (14.25) and related discussion.

[36] Cook and Pfeiffer (2004).

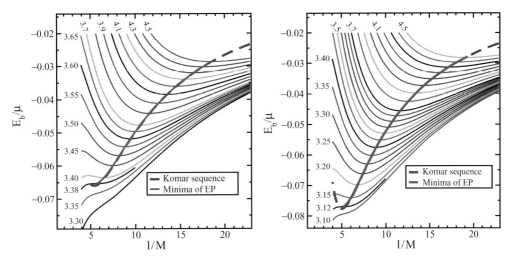

Figure 12.5 The binding energy E_b vs. separation l for corotational (left panel) and nonspinning (right panel) equal-mass binary black hole sequences constructed in the conformal thin-sandwich approach. The solid lines connecting the extrema of the curves of constant angular momentum, which are labeled by $J/\mu M$, represent sequences constructed with the effective potential method as in Figure 12.3; these lines terminate at the ISCO. The dashed lines represent sequences constructed from the mass criterion (12.108). These sequences agree extremely well up to very small separations, where the assumption of quasiequilibrium breaks down. [From Caudill *et al.* (2006).]

We have already seen an example of such a sequence in Figure 12.3 for Bowen–York binary data constructed with the transverse-traceless approach of Section 12.2. As with the point-mass models that we discussed in Section 12.1 this sequence does not extend to arbitrarily small binary separations, but instead terminates at the innermost stable circular orbit, or ISCO, marked by the turning point on the equilibrium energy curve.

In Figure 12.5 we show the equivalent figures for corotational and nonspinning binary black hole models constructed in the conformal thin-sandwich approach of Section 12.3. For most of the remainder of this section we will focus on the results of Cook and Pfeiffer (2004) and Caudill *et al.* (2006), who adopt spectral methods to integrate these equations, together with the quasiequilibrium black hole boundary conditions of Section 12.3.2. They also assume conformal flatness, $\bar{\gamma}_{ij} = \eta_{ij}$ and maximal slicing $K = 0$, as well as $\bar{u}_{ij} = 0$ and $\partial_t K = 0$.[37] Within this framework we can specify a corotational sequence by setting $\Omega_{\text{spin}} = 0$ in the solution (12.107) for the black hole boundary condition on the tangential piece of the shift.[38] Specifying an nonspinning sequence is slightly more involved. As we discussed below equation (12.107), we could attempt to construct such sequences by

---

[37] See Cook and Pfeiffer (2004) for some results that adopt "Kerr–Schild" slicing rather than maximal slicing.
[38] Recall that Ω_{spin} is a measure of the spin in the corotating frame.

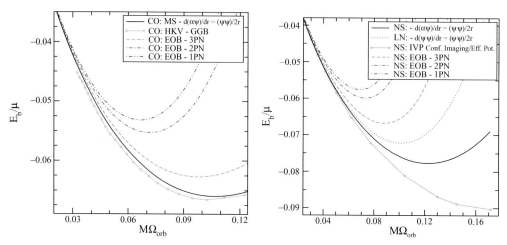

Figure 12.6 The binding energy E_b vs. the orbital angular velocity Ω_{orb} for corotational (left panel) and nonspinning (right panel) equal-mass binaries. The solid lines mark the results of Cook and Pfeiffer (2004) and Caudill *et al.* (2006), computed in the conformal thin-sandwich formalism with the equilibrium black hole boundary conditions of Section 12.3.2 for corotating and exactly nonspinning black holes. The dotted line labeled "LN" in the right panel was obtained similarly, except that the black holes are nonspinning to leading order only. The solid dotted line labeled "CO: HKV-GGB" in the left panel marks the results of Grandclément *et al.* (2002), who use an approximate isometry boundary condition on the horizon. The solid dotted line labeled "NS: IVP" marks the results from the Bowen–York effective potential approach, using the conformal imaging technique (Cook, 1994). Finally, the dashed and dashed-dotted lines denote post-Newtonian "effective one-body" results (Damour *et al.*, 2002). [From Cook and Pfeiffer (2004) and Caudill *et al.* (2006).]

setting the magnitude of Ω_{spin} equal to the magnitude of Ω_{orb} in equation (12.107). Evaluating the quasilocal spin (7.74) shows, however, that these black holes do not have exactly vanishing spin as seen in the inertial frame, meaning that this sequence is not a truly nonspinning sequence. We refer to this approach as the "leading-order approximation". More accurate results, representing exactly nonspinning black holes, can be obtained by iterating over the parameter Ω_{spin} until the quasilocal spin (7.74) does indeed vanish to a desired accuracy.

Figure 12.5 includes curves of constant angular momentum; the extrema of these lines mark quasicircular orbits as identified by the effective potential method. The solid line connecting these extrema forms the corresponding quasiequilibrium inspiral sequence. The dashed lines in Figure 12.5 mark the same sequence, but with circular orbits identified by the mass criterion (12.108). The two approaches agree extremely well up to very small binary separations, where the assumptions of quasiequilibrium break down.

In Figure 12.6 we graph the binding energy of these inspiral sequences as a function of the orbital angular speed Ω_{orb}, which is a gauge invariant quantity. Also included in these figures are results from other calculations. For the corotational sequence in the left panel of Figure 12.6 we include the numerical results of Grandclément *et al.* (2002), who also adopt the conformal thin-sandwich decomposition of Section 12.3, but, instead of using the black hole equilibrium boundary conditions of Section 12.3.2, adopt an approximate isometry condition.[39] It is reassuring that the numerical results differ only very little. For the nonspinning sequence in the right panel we also include results for black holes that are nonspinning to "leading-order" only, as well as the results from the Bowen–York effective potential approach highlighted in Figure 12.3. Post-Newtonian results are also plotted for comparison.

The turning points along the equilibrium binding energy curves in Figure 12.6 mark their respective ISCOs, as discussed in Section 12.1. We tabulate the ISCO parameters in Table 12.1 for corotational binaries and in Table 12.2 for nonspinning binaries. In these tables we include numerical results, post-Newtonian results, as well as the results for a test particle in circular orbit about a Schwarzschild black hole.

> **Exercise 12.13** Return to exercise 12.4 to reconsider a test particle of mass m_{test} in circular orbit about Schwarzschild black hole of mass M.
>
> (a) Evaluate the results of that exercise to show that at the ISCO (areal radius $r = 6M$) the test particle has an orbital angular velocity $M\Omega_{\text{orb}} = 1/6^{3/2} \approx 0.0680$, a binding energy $E_b/m_{\text{test}} \equiv -\tilde{E}_{\text{eq}} = \sqrt{8/9} - 1 \approx -0.0572$ and an angular momentum $J/m_{\text{test}} \equiv \tilde{J}_{\text{eq}} = 2\sqrt{3}M \approx 3.464M$.
>
> (b) Extrapolate the above results to estimate corresponding quantities for equal-mass binary black holes in circular orbit with *total* irreducible mass m. To do this, interpret the test mass m_{test} as the reduced mass, $m_{\text{test}} \to \mu = m/4$, and the black hole mass M as the total mass, $M \to m$ in the expressions found for part (a). Then derive the values quoted in the last rows of Tables 12.1 and 12.2.

Actually, Tables 12.1 and 12.2 include two different sets of post-Newtonian results. The first set is a "standard" post-Newtonian expansion as described in Appendix E, except that in the Appendix we focus on nonspinning objects only. For the nonspinning binaries of Table 12.2, the ISCO parameters can be obtained directly from the minima of the post-Newtonian expansion of the equilibrium binding energy (E.14). For the corotational binaries of Table 12.1, however, certain spin contributions must be added to the binding energy before the ISCO can be located.[40] The other set of post-Newtonian results is based on an alternative "effective one-body" treatment,[41] which in some cases may accelerate the convergence of the expansion.

It is instructive to graph the equilibrium binding energy E_b vs. the equilibrium angular momentum J as in Figure 12.7. It is quite noticeable that most curves form a cusp. These cusps are a consequence of equation (12.112), which implies that sequences of constant irreducible mass must have simultaneous turning points in the ADM mass (hence

---

[39] See the last paragraph in Section 12.3.1 and footnote 22.
[40] See, e.g., Blanchet (2002, 2006).
[41] Buonanno and Damour (1999); Damour *et al.* (2002).

Table 12.1 The orbital angular velocity Ω_{orb}, the equilibrium binding energy E_b, and the equilibrium angular momentum J at the ISCO of corotational, equal-mass, black hole binaries, as computed in different calculations. The mass m is the total irreducible mass of the binary. Grandclément *et al.* (2002) and Caudill *et al.* (2006) adopt the conformal thin-sandwich decomposition; different choices in the horizon boundary condition and the minimization procedure result in changes of at most a few percent. Two sets of post-Newtonian results are included; one based on a "standard" post-Newtonian expansion (Blanchet 2002) and another based on an "effective one-body" (EOB) approach (Damour *et al.* 2002). [After Caudill *et al.* (2006).]

Corotational binaries	$m\Omega_{\text{orb}}$	E_b/m	J/m^2
Grandclément *et al.* (2002)	0.103	−0.017	0.839
Caudill *et al.* (2006)	0.106	−0.0165	0.844
Blanchet (2002); 1PN – standard	0.5224	−0.0405	0.621
Blanchet (2002); 2PN – standard	0.0809	−0.0145	0.882
Blanchet (2002); 3PN – standard	0.0915	−0.0153	0.867
Damour *et al.* (2002); 1PN – EOB	0.0667	−0.0133	0.907
Damour *et al.* (2002); 2PN – EOB	0.0715	−0.0138	0.893
Damour *et al.* (2002); 3PN – EOB	0.0979	−0.0157	0.860
Test particle around Schwarzschild	0.068	−0.0143	0.866

Table 12.2 ISCO parameters for nonspinning, equal-mass black hole binaries for different calculations. In addition to calculations referenced in Table 12.1, we also list the Bowen–York effective potential results of Cook (1994) and Baumgarte (2000), who adopt the conformal imaging and puncture methods, respectively, to treat the black hole singularities (see Section 12.2.2). [After Caudill *et al.* (2006).]

Nonspinning binaries	$m\Omega_{\text{orb}}$	E_b/m	J/m^2
Caudill *et al.* (2006)	0.122	−0.0194	0.779
Cook (1994)	0.166	−0.0225	0.744
Baumgarte (2000)	0.18	−0.023	0.74
Blanchet (2002); 1PN – standard	0.5224	−0.0405	0.621
Blanchet (2002); 2PN – standard	0.1371	−0.0199	0.779
Blanchet (2002); 3PN – standard	0.1287	−0.0193	0.786
Damour *et al.* (2002); 1PN – EOB	0.0692	−0.0144	0.866
Damour *et al.* (2002); 2PN – EOB	0.0732	−0.0150	0.852
Damour *et al.* (2002); 3PN – EOB	0.0882	−0.0167	0.820
Test particle around Schwarzschild	0.068	−0.0143	0.866

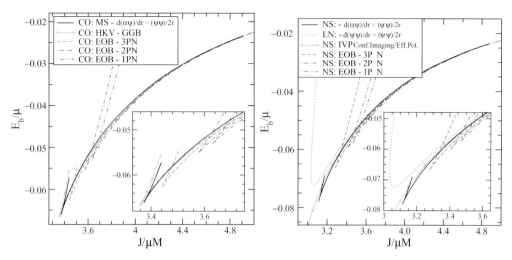

Figure 12.7 The equilibrium binding energy E_b vs. the total equilibrium angular momentum J for the same corotational (left panel) and nonspinning (right panel) binaries shown in Figure 12.6. [From Cook and Pfeiffer (2004) and Caudill et al. (2006).]

the binding energy) and the angular momentum.[42] The nonspinning sequence for which the black holes are nonspinning only to "leading-order", however, does not feature such a cusp – apparently this approximation is not sufficiently accurate at small binary separations to produce a simultaneous turning point.

Several other aspects of the above figures and tables are noteworthy. The numerical results of Grandclément et al. (2002) and Cook and Pfeiffer (2004) for corotational binaries agree very closely, which is reassuring. Both post-Newtonian expansions seem to converge to the conformal thin-sandwich data as the order of the expansion is increased. Since these results are based on completely different approaches, this convergence presumably reflects physical consistency in the solutions.

The ISCO parameters as obtained from the Bowen–York effective potential approach differ significantly from all other results. This discrepancy reconfirms our earlier concern that the transverse-traceless decomposition does not provide direct means to construct quasiequilibrium metrics, and suggests that the conformal thin-sandwich approach may be the more promising approach. For binary separations outside the ISCO, however, Bowen–York data agree reasonably well with other results (see, e.g., Figure 12.7). In fact, Bowen–York puncture data have been used as initial data in many recent dynamical simulations of binary black hole merger, partly because they are quite easy to construct, and partly because they form a natural starting-point for the robust "moving puncture" evolution technique that has been so widely adopted in many of these simulations (see Chapter 13).

---

[42] Provided the binding energy is defined as in equation (12.59); cf. footnote 14.

Given the increasing speed and dynamic range of these simulations, it is possible to use quasiequilibrium binaries of ever larger separation as initial data, whereby the distinctions between different initial data sets are minimized.

One way of improving the above numerical results might be to abandon the assumption of conformal flatness ($\bar{\gamma}_{ij} = \eta_{ij}$). We already know, for example, that this assumption limits the maximum allowed spin of a rotating black hole constructed in the conformal thin-sandwich approach (see the end of Section 3.3). By choosing the background metric for binary black holes to be the superposition of two Kerr–Schild metrics, quasiequilibrium binaries with quasilocal spin parmaters S_i / m_i^2 as large as 0.9998 have been constructed.[43] Most significant, when evolved, these rapidly-spinning holes remain rapidly spinning even after the initial relaxation, in contrast to the most rapidly spinning binaries constructed with conformally-flat metrics (either puncture or conformal thin-sandwich).

Another means of improvement might be to use post-Newtonian expressions for the background data in the constraint equations.[44] A promising alternative could be the "waveless approximation" described in Chapter 3.4. This approach allows a choice for the time-derivative of the extrinsic curvature, whereby the evolution equation for the extrinsic curvature turns into an equation for the conformally-related metric. Therefore, instead of making an ad hoc choice for the latter, one can determine the conformal geometry that results from some physical assumption. To date this approach has been implemented only for binary neutron stars,[45] but it could be interesting to study binary black holes in the future.

---

[43] Lovelace *et al.* (2008); see also Liu *et al.* (2009).
[44] Tichy *et al.* (2003).
[45] Uryū *et al.* (2006).

13 Binary black hole evolution

The dynamical simulation of the head-on collision of two black holes in axisymmetry was an early success of numerical relativity (see Chapter 10.2). Based on this success one might surmise that the subsequent simulation of the inspiral and merger of binary black holes initially in circular orbit represented a straightforward generalization of the head-on case. It turned out, however, that relaxing the assumption of axisymmetry to treat a binary in circular orbit, and then tracking the resulting evolution, presented several nontrivial challenges. As a consequence, dynamical simulations of these binaries were stalled for many years until these challenges were finally overcome. Today, the inspiral and merger of binary black holes is essentially a solved problem, constituting one of the major triumphs of numerial relativity.

One obvious complication that arises in numerical simulations when moving from two to three spatial dimensions is the burden of increased computational resources required to cover the added dimension. Even though computers have become considerably faster and can handle far more memory than the machines available when the first head-on black hole simulations were performed, the resources required to resolve inspiraling black holes in the strong-field, near-zone while simultaneously tracking and ultimately extracting gravitational waves in the weak-field, far-zone remain formidable. Different investigators have adopted different approaches to address this problem of "dynamic range", including the use of fixed or adaptive mesh refinement or the construction of novel coordinate systems that allocate gridpoints where they are most needed. In finite difference methods, it is also possible to use higher-order differencing schemes in both space and time, like fourth-order or higher, to increase the accuracy for a given number of grid points over the more traditional second-order schemes featured in Chapter 6.2.

Another complication in going from two to three spatial dimensions arises from the form of the evolution equations. In axisymmetry, it is not difficult to formulate evolution schemes that solve Einstein's field equations in a stable fashion when implemented numerically. One such scheme is presented in Appendix F. At their core, the earliest formulations in axisymmetry usually were based on the identification of good "radiation variables" (see, e.g., equations F.5 and F.6) and often employed one or more of the constraint equations to solve for one or more of the metric variables in lieu of integrating the corresponding evolution equation for the variable(s) ("constrained evolution"). In three spatial dimensions it has proven less obvious how to construct "radiation variables",[1] and, as a result, it is less clear how to arrive at a good choice for the form of the evolution equations. Numerical implementations of the standard $3 + 1$ or ADM equations as presented in Chapter 2 develop instabilities, even for small perturbations of flat space. As we discussed in Chapter 11, several reformulations of the evolution equations have been developed that *do* provide stable numerical implementations. These newer formulations, which include the generalized harmonic system (Chapter 11.4) and the BSSN system (Chapter 11.5), have led to dramatic breakthroughs in our ability to simulate the inspiral and merger of binary black holes.

Finally, evolving black holes requires handling the singularities in their interiors, which, if left unattended, could have dire consequences for a numerical simulation. For the head-on collisions in axisymmetry it proved sufficient to adopt singularity-avoiding coordinates to deal with this issue (see Chapters 4 and 10.2), but for the longer-term evolutions required for the simulations of binary black hole inspiral and merger, such coordinate systems can lead to coordinate pathologies. One alternative approach that has proven successful is the "excision" of the interior of the black hole in order to remove the singularity from the computational domain. Another successful approach is the "moving puncture" method, which is based on moving puncture gauge conditions. We will discuss both these approaches in Section 13.1, after which we will summarize some successful simulations of binary inspiral and merger in Section 13.2.

13.1 Handling the black hole singularity

Black hole interiors contain curvature singularities. If encountered during a numerical simulation, these singularities can cause the calculation to terminate prematurely (i.e., "crash"). Special care therefore needs to be taken to avoid encountering such a singularity.[2] In the following we will discuss three approaches that have been used to avoid singularities: singularity avoiding coordinates, black hole excision and the moving puncture method.

13.1.1 Singularity avoiding coordinates

To state the obvious, singularity avoiding coordinates are spacetime coordinates that avoid singularities. A singularity avoiding time coordinate, for example, is one in which the appearance of a spacetime singularity is postponed until $t = \infty$. A popular example of a such singularity avoiding slicing condition is maximal slicing (see Chapter 4.2 and, in greater detail, Chapter 8.1). Black hole simulations using maximal slicing and other singularity avoiding coordinate systems have been very successful in spherical symmetry

---

[1] But see Bonazzola *et al.* (2004).

[2] Unless, of course, the code is specifically designed to explore the properties of spacetime singularities; see, e.g., Hübner (1996); Berger (2002).

and axisymmetry (see Chapters 8 and 10). Ultimately, however, these coordinate systems tend to develop "grid stretching" along black hole throats (see Chapter 8.1), another cause of the premature demise of a numerical calculation.[3] For simulations of orbiting binaries in three spatial dimensions it is even harder to identify singularity avoiding coordinates that might lead to a long-term, stable evolution. As a consequence, most successful simulations have adopted other strategies for avoiding spacetime singularities.

13.1.2 Black hole excision

The idea of black hole excision goes back to a suggestion by Unruh.[4] As long as cosmic censorship holds, any spacetime singularity must be surrounded by an event horizon and hence must reside inside a black hole. By definition, no information can propagate from inside the black hole to the outside, so that none of the exterior spacetime can possibly be affected by the black hole's interior.[5] This suggests that it should be sufficient to simulate numerically only the exterior of any black holes, and to excise from the computational domain any region that lies inside an event horizon. This approach is referred to as black hole excision.[6]

While the underlying idea of black hole excision is very elegant and transparent, some of the details of its numerical implementation are more involved. To begin with, it is usually not possible to locate an event horizon during a numerical simulation. It is usually possible, however, to locate apparent horizons, which in general relativity are guaranteed to lie inside event horizons.[7] In practice, therefore, it is the region interior to an apparent horizon that is excised from the computational domain.[8]

Excision itself can be implemented in different ways. In many applications all gridpoints within a certain sphere (or ellipsoid) that lies within the apparent horizon are excised, leaving only a small buffer zone of gridpoints just outside the sphere but inside the apparent horizon on which valid data can reside.[9] Grid points within the excised region no longer contain valid data and therefore cannot be used in equations to determine the values of field quantities in the black hole exterior. This means that the boundary of the excision region has to be treated in a special way.

---

[3] We point out another unappealing feature of singularity avoiding coordinate systems. By slowing down the advance of proper time in areas close to a singularity, large amounts of computational resources are used to cover increasingly small regions of spacetime. Adopting singularity avoiding coordinates may therefore lead to an uneconomical use of the computational resources.

[4] Unruh (1984), as quoted in Thornburg (1987).

[5] See also the discussion in Chapter 7.1.

[6] Some early numerical implementations include Seidel and Suen (1992); Scheel *et al.* (1995a); Alcubierre and Brügmann (2001); Yo *et al.* (2002); Brügmann *et al.* (2004); Alcubierre *et al.* (2005); Pretorius (2005a,b).

[7] See Chapter 7, and in particular the discussion in Chapter 7.1.

[8] Recall, however, that in some slicings apparent horizons do not form, even though a black hole is present; see, e.g., exercise 8.10.

[9] In Cartesian coordinates, the resulting excised region represents an approximation to the smooth sphere or ellipsoid, and is often referred to as a "LEGO" sphere.

Mathematically, no boundary conditions for the evolution equations should be required on the excision surface, since all characteristics on that surface should be directed inward (see Figure 6.1 and the surrounding discussion in Chapter 6.1). Numerically, however, the excision surface may nevertheless require a special treatment. Many finite difference applications, for example, use centered differencing (see Chapter 6.2), which, when applied to a grid point on the surface, requires valid data on excised gridpoints. This problem can be avoided quite simply, either by using one-sided difference stencils on the excision surface,[10] or by extrapolating valid data to excised grid points from the outside before these points are used in a centered differencing scheme.

When black holes move through a numerical grid, grid points that had previously been excised may also re-emerge from a black hole, and have to be repopulated with valid data. In many applications this is done by extrapolating from adjacent exterior points. We will discuss some successful binary black hole simulations that have adopted black hole excision in Section 13.2.1.

13.1.3 The moving puncture method

We have already encountered the puncture method in Chapter 12.2.2 for the construction of black hole initial data. The key idea of the puncture method is to decompose the metric as a sum or product of one term that contains the singularity, but is analytic, and a correction term that must be obtained numerically, but is regular. In equation (12.50), for example, we wrote the conformal factor ψ as a sum of the analytical black hole "punctures" $1/\alpha$ and a correction term u. We then found that the Hamiltonian constraint (12.52) becomes a regular equation for u that can be solved everywhere, without any need for special boundary conditions at the black hole or excision surface. Clearly, it is tempting to try a similar approach for dynamical simulations containing black holes.

The first attempts to construct such a scheme employed a "fixed" puncture method. For a single Schwarzschild black hole, for example, we could write the conformal factor as the product

$$\psi = \left(1 + \frac{M}{2r}\right) f. \tag{13.1}$$

Starting with initial data on a slice of constant Schwarzschild time we have $f = 1$ initially. During the evolution, we could then leave the analytic singular part fixed, and evolve only the regular function f. In a binary evolution using this approach,[11] the singularities remain at fixed coordinate locations, given by the "punctures" in the analytical part of ψ. That means that suitable gauge conditions have to be found that are consistent with this prescription; i.e., the gauge conditions have to leave the singularities at their prescribed coordinate locations and have to preserve the nature of their singular behavior, in this

---

[10] See, e.g., exercise 6.5.
[11] Brügmann (1999).

case a $1/r$ divergence. An additional complication arises from the fact that, under these conditions, an evolution with a lapse that is positive everywhere cannot lead to a stationary solution, even for a single Schwarzschild black hole.[12] As a consequence, binary black hole simulations using the fixed puncture method have not proven very successful, since they have not led to long-term, stable evolution.[13]

The breakthrough came with the so-called "moving" puncture method, first suggested independently by Baker *et al.* (2006a) and Campanelli *et al.* (2006).[14] In some sense the moving puncture approach is even easier than the fixed puncture approach. The singular term in the metric is not factored out; instead, the metric is evolved in its entirety. The puncture is allowed to move freely in accord with the gauge conditions, except that care is taken that the singularity never hits an actual grid point. Usually this can be accomplished quite easily. In situations that feature equatorial symmetry, for example, the punctures always remain on the (orbital) plane of symmetry. Using a cell-centered finite difference scheme[15] no grid points are located on this plane, so that the punctures can never encounter a grid point. The moving puncture method is usually implemented in the BSSN formalism, or some close adaptation. It is sometimes advantageous, in the context of this formalism, to evolve a variable related to the inverse of the conformal factor, rather than the conformal factor itself, since the inverse will vanish at puncture singularities rather than diverge. With a suitable choice of gauge conditions, like the 1+log slicing condition (4.51) and the Gamma-driver condition (4.83), this prescription leads to remarkably stable evolutions.[16] Before we summarize some of the details of the solution, we are compelled to address one obvious question: how is it possible that the presence of puncture singularities in the computational domain does not spoil the numerical calculations?

Considerable insight and clarification of this question has been provided by Hannam *et al.* (2007, 2008) and Brown (2008). Consider the evolution of a single Schwarzschild black hole. A puncture evolution of this spacetime could start out with initial data expressing Schwarzschild geometry in isotropic coordinates at an instant of constant Schwarzschild time. The conformal factor at this instant is then given by

$$\psi = 1 + \frac{M}{2r}. \tag{13.2}$$

As we have seen in Chapter 3, these isotropic coordinates do not penetrate the black hole interior, but instead cover two copies of the black hole exterior. These two copies represent a "wormhole" solution, i.e., two sheets of asymptotically flat "Universes" that are connected by an Einstein–Rosen bridge at the black hole horizon, as illustrated in Figure 3.1. The singularity at isotropic radius $r = 0$, where the conformal factor diverges

---

[12] Hannam *et al.* (2003); Reimann and Brügmann (2004).

[13] Alcubierre *et al.* (2001); Brügmann *et al.* (2004).

[14] See Brügmann *et al.* (2008) and Hannam *et al.* (2008) for detailed discussions of this method, its numerical implementation, and geometrical interpretation.

[15] See, e.g., Figure 6.2.

[16] See van Meter *et al.* (2006) for a comparison of alternative gauge conditions for the evolution of moving punctures.

as $1/r$, corresponds to the asymptotically flat end of the "other" Universe, and is therefore a mere coordinate, and not a curvature, singularity.[17]

Let us now evolve these initial data with the 1+log slicing condition (4.51),

$$\partial_t \alpha - \beta^i \partial_i \alpha = -2\alpha K \tag{13.3}$$

and the Gamma-driver condition (4.83). These conditions are conditions on the time derivatives of the lapse and shift only; before we can start the evolution we therefore have to specify initial data for these variables. We could, for example, choose $\alpha = (1 - M/(2r))/(1 + M/(2r))$ and $\beta^i = \partial_t \beta^i = 0$, which we recognize from the familiar static form of the Schwarzschild line element in isotropic coordinates (equation 1.60). As we discussed in the context of equation (2.162), these choices represent the Killing lapse and shift; since they also satisfy the slicing conditions (4.51) and (4.83) with zero time-derivatives, the resulting time evolution would leave all metric quantities time-independent (see also equation 2.145 and the discussion that follows it). Consider instead the initial choices $\alpha = 1$ and $\beta^i = \partial_t \beta^i = 0$. It is easy to see that these choices lead to a nontrivial time evolution; specifically, the term $D_i D_j \alpha$ no longer cancels out the Ricci tensor αR_{ij} in equation (2.127), leading to a nonzero time-derivative of K_{ij} (see exercise 2.31).

Even though the spacetime geometry, which is still Schwarzschild, remains static, the metric quantities now vary with time and thereby constitute a dynamical coordinate system. Hannam *et al.* (2007) demonstrate that the solution behaves dynamically only for a brief period, after which it settles down to a new, time-independent solution that is different from the initial data. In particular, in the new, time-independent solution the conformal factor is characterized by a $1/\sqrt{r}$ singularity at $r = 0$ instead of the original $1/r$ singularity. As demonstrated in exercise 13.1, such a slice terminates on a (limit) surface of finite areal radius r_s.

> **Exercise 13.1** Assume a spherically symmetric spatial metric of the form $\gamma_{ij} = \psi^4 \eta_{ij}$ and assume that
>
> $$\psi \to \left(\frac{\mu M}{r} \right)^{1/2} \quad \text{as} \quad r \to 0. \tag{13.4}$$
>
> Show that the surface $r = 0$ has an areal radius of $r_s = \mu M$.

Consider, for a moment, evolving the data with a "nonadvective" version of the 1+log condition (13.3), i.e., drop the shift term $\beta^i \partial_i \alpha$ from that equation so that $\partial_t \alpha = -2\alpha K$. The late-time asymptotic solution, which is also time-independent, must satisfy $\partial_t \alpha = 0$, and must then be maximally sliced with $K = 0$. This solution must therefore be a member of the family of time-independent, maximal slicings (4.23)–(4.25), which are parametrized by C. Hannam *et al.* (2007) provide numerical evidence that this late-time solution corresponds to the particular member $C = 3\sqrt{3}M^2/4$, which has a limiting surface at $r_s = 3M/2$. In

---

[17] See the discussion following equation (3.20).

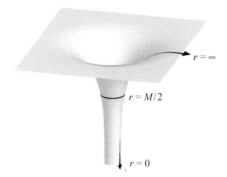

Figure 13.1 A schematic embedding diagram of a maximally sliced Schwarzschild geometry with $C = 3\sqrt{3}M^2/4$ (for $t = constant$ and $\theta = \pi/2$; see equations 4.23–4.25). In contrast to the "wormhole" solution shown in Figure 3.1, this "trumpet" solution has only one asymptotically flat end, while the other end, corresponding to $r \to 0$, approaches an infinitely long cylinder of areal radius $r_s = 3M/2$. [After Hannam *et al.* (2008).]

Appendix H we analyze this solution in more detail, and, in particular, we show that the conformal factor for this solution indeed features a $1/\sqrt{r}$ singularity at the isotropic radius $r = 0$, corresponding to an areal radius $r_s = 3M/2$ (see equation H.8). In Figure 13.1 we sketch an embedding diagram for this solution, the appearance of which suggests the name "trumpet" solution for this type of solution. For the "advective" version of the 1+log condition, only parts of the late-time solution can be constructed analytically (see Appendix H), but qualitatively the results are very similar. In particular, the asymptotic solution again features a "trumpet" geometry.[18]

The key point is that the singularity at isotropic radius $r = 0$ is again a mere coordinate singularity. Since the numerical grid terminates at $r = 0$ and $r_s > 0$, it does not include the spacetime singularity at $r_s = 0$. This result helps to explain in part the success of the moving puncture method. In a nut-shell, it can be described as "excision-without-excision"; instead of excising the spacetime explicitly as in the excision method, the moving puncture method excises the innermost part of the black hole interior containing the spacetime singularity with the help of a particular gauge choice.

Further insight into this behavior was provided by Brown (2008), who also performed moving puncture simulations of Schwarzschild. Starting again with "wormhole" initial data representing the Schwarzschild solution at an instant of constant Schwarzschild time, he also evolved the data with the 1+log slicing condition (4.51) and $\alpha = 1$ initially, but chose different shift conditions. Evolving with zero shift, Brown (2008) points out that the spatial slices, which start out connecting the two asymptotically flat ends of the "wormhole" solution, cannot possibly disconnect from either one of these two asymptotic regions. In fact, for zero shift the solution has to remain symmetric across the Kruskal–Szekeres coordinate $u = 0$ at all times, as shown in Figure 13.2. All spatial slices remain connected

---

[18] Hannam *et al.* (2008).

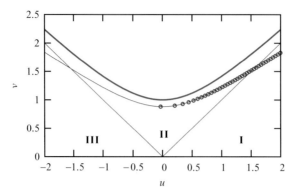

Figure 13.2 The spatial slice at $t = 3M$ in a moving puncture simulation of a single Schwarzschild black hole, plotted in a Kruskal–Szekeres diagram. The solid blue line shows the slice that results from an evolution with zero shift in equation (13.3), while the dots represent the location of individual grid points when evolved with the Gamma-driver condition (4.83). The upper red line represents the spacetime singularity at $r_s = 0$. [From Brown (2008).]

to both asymptotic ends. But since the slicing condition (4.51) is independent of the shift, the same result must hold for any other shift condition, including the Gamma-driver condition (4.83). The only difference is that individual spatial coordinate locations, or numerical grid points, now move on the spatial slices. For the Gamma-driver condition, the grid points, marked by dots in Figure 13.2, move "away" from the "other" asymptotic region very quickly, toward the right hand side in a Kruskal–Szekeres diagram. Even for a very fine grid resolution, and after only a very short time, no grid points remain in region III of the diagram. It is in this sense that a dynamical evolution approaches a "trumpet" solution like the one we discussed above. Even though the solutions have different asymptotic behavior, the Gamma-driver condition moves all grid points into a region where the two solutions converge towards each other.[19] Moreover, the grid points drift safely away from the coordinate singularity at large negative u.

Figure 13.2 also illustrates that the spatial slices again do not encounter the black hole spacetime singularity. In fact, these slices have properties similar to those of "singularity avoiding" coordinates, except that the gauge conditions used in moving puncture simulations avoid the coordinate pathologies (e.g., "grid stretching") often encountered when using other singularity avoiding coordinates.

13.2 Binary black hole inspiral and coalescence

In this section we will review some of the earliest successful simulations of binary black hole inspiral and coalescence. In Section 13.2.1 we focus on the first complete simulations, which marked an important breakthrough in the numerical evolution of strong-field,

---

[19] We refer to Hannam *et al.* (2008) for a more detailed discussion.

vacuum spacetimes. These simulations were performed for equal-mass binaries with zero spin. In Section 13.2.2 we will then discuss simulations of binaries containing black hole companions with mass ratios different from unity and with nonzero spin, as well as the calculation of black hole recoil. These results have important implications for black hole astrophysics and gravitational wave astronomy.

13.2.1 Equal-mass binaries

The simulation of binary black hole inspiral, plunge and merger had been a long-standing goal of numerical relativity. A vacuum problem in pure geometry, with no other complicating sources or physical interactions, this two-body scenario remained one of the most important unsolved problems in classical general relativity. In addition, theoretical templates of gravitational waveforms from binary black holes are of crucial importance for the identification and subsequent interpretation of such signals in the data of the new generation of gravitational wave interferometers. With the construction of these detectors, the solution of the binary black hole problem had become very urgent. Nevertheless, solving the problem, sometimes called the "holy grail" of numerical relativity, had resisted the efforts of numerous researchers for many years. It is not surprising, then, that Pretorius's announcement of his successful simulation of a binary black hole inspiral and merger at a workshop hosted by the Banff International Research Station in April 2005 was met with great excitement.

Pretorius (2005a) uses the generalized harmonic formalism described in Chapter 11.3, and eliminates black hole singularities with an excision method described in Section 13.1.2. In his initial calculations Pretorius created a black hole binary from nonsingular initial data by letting a scalar field (see Chapter 5.4) undergo gravitational collapse. By boosting the initial scalar field configuration, he generated a binary system in an approximate circular orbit. The companions each then collapsed to black holes prior to merger. While these calculations were not meant to simulate a realistic astrophysical scenario, they did provide the first calculation that could stably evolve a binary black hole from inspiral through merger to final ringdown, with the remnant settling down to a stationary Kerr black hole.

Figure 13.3 shows contours of the lapse function α in the orbital plane during one of these simulations. In these calculations, Pretorius adopts a compactified coordinate system in which the outer boundaries correspond to spatial infinity. He also uses adaptive mesh refinement (AMR, see Chapter 6.2.5) to achieve sufficient grid resolution close to the black holes. In his original calculations, the finest resolution is placed in the vicinity of each black hole and is approximately $M/20$.[20]

In later work, Pretorius replaced the scalar field initial data with conformal thin-sandwich, binary black hole initial data as described in Chapter 12.3.[21] This choice significantly reduces the eccentricity of the resulting orbit and allowed Pretorius to follow

---

[20] Here and below M is the total intial ADM mass of the binary system.
[21] Specifically, he used the initial data of Cook and Pfeiffer (2004) and Caudill *et al.* (2006).

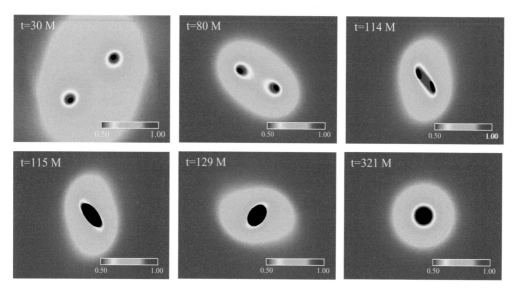

Figure 13.3 Evolution of the lapse function α in the orbital plane during the merger of an equal-mass, nonspinning binary black hole in the calculation of Pretorius (2005a). The black regions are the excised domains in the black hole interior. The apparent horizons merge between the times of $114M$ and $115M$, where M is the total initial ADM mass of the binary, and the initially distorted remnant subsequently settles down to an axisymmetric stationary Kerr black hole. [From Pretorius, private communication.]

the inspiral for several orbits prior to merger.[22] In Figure 13.4 we plot the orbital evolution of such an equal-mass binary, together with the location of the apparent horizons at several key instants of time.

Pretorius' success was quickly followed by another breakthrough that employed a completely independent approach. Both Campanelli *et al.* (2006) and Baker *et al.* (2006a) adopted versions of the BSSN formalism described in Chapter 11.5, together with the moving puncture approach of Section 13.1.3, and were also able to follow binary black hole coalescence through merger to ringdown.

The original implementations of Campanelli *et al.* (2006) and Baker *et al.* (2006a) differ in some of the details. Baker *et al.* (2006a) use the BSSN formalism and the moving puncture approach very similar to our descriptions in Chapter 11.5 and Section 13.1.3. As did Pretorius, they also use AMR to achieve sufficient grid resolution. Campanelli *et al.* (2006) implement a slight variation of the BSSN formalism. Instead of evolving the conformal exponent ϕ in the standard BSSN formulation (see equation 11.46), they evolve the inverse of the conformal factor, $\chi = \exp(-4\phi)$, which vanishes at the punctures. Given suitable gauge conditions it is possible, with this choice, to keep all quantities regular during the evolution. This approach has the advantage that the punctures may now encounter grid points without causing numerical difficulties. Also, Campanelli *et al.* (2006)

---

[22] See Buonanno *et al.* (2007).

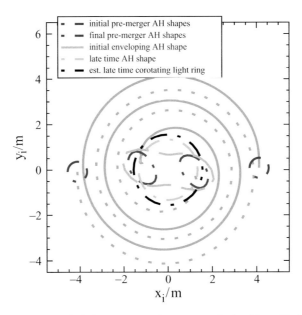

Figure 13.4 The evolution in the orbital plane of an equal-mass, nonspinning binary black hole, showing the coordinate locations of the apparent horizons (AH) at several key moments during the inspiral and merger. [From Buonanno *et al.* (2007).]

adopt a uniform numerical grid (as opposed to AMR), but introduce "multiple transition fisheye" spatial coordinates[23] as well as fourth-order finite differencing to provide sufficient resolution for the black holes. In many of the other aspects the two approaches are very similar. In particular, both Campanelli *et al.* (2006) and Baker *et al.* (2006a) adopt the puncture initial data described in Chapter 12.2 – a natural choice for evolution with the moving puncture method – to describe a quasiequilibrium binary configuration close to the ISCO.

Figures 13.5 and 13.6 illustrate some of the results of Campanelli *et al.* (2006) and Baker *et al.* (2006a). Figure 13.5 shows the gravitational waveform $Re(\psi_4)$ (i.e., the real part of the Newman–Penrose scalar ψ_4, see Chapter 9.4.2) of Campanelli *et al.* (2006). In this particular calculation the grid extends to an outer boundary of $60M$, and $Re(\psi_4)$ is extracted at $5M$.[24] The calculation is then performed with three different grid resolutions, $\Delta x = M/16$, $M/24$, and $M/36$ in the innermost region, where the resolution is the finest. The comparison of results suggests that the simulation is fourth-order accurate, as expected from a fourth-order finite difference implementation (see Chapter 6.4).

---

[23] See Chapter 14.2.3 for a brief description.

[24] To obtain waveforms measured by a distant observer, one must extract them at a much larger separation from the binary. In this implementation, however, the waveforms are affected by the outer boundaries. Consequently, for the purposes of the convergence test plotted in this figure, $Re(\psi_4)$ is determined at a smaller radius, so that it remains causally disconnected from the outer boundary until a later time.

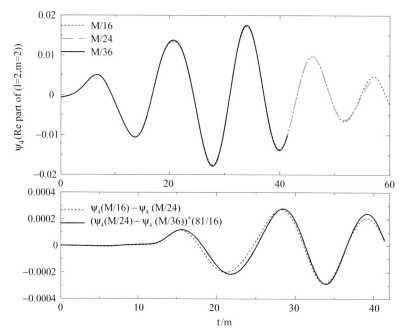

Figure 13.5 The gravitational waveform $\text{Re}(\psi_4)$ for an equal-mass, nonspinning binary black hole merger in the calculation of Campanelli *et al.* (2006). Results are plotted for different grid resolutions. This convergence test suggests fourth-order convergence, as expected from a fourth-order finite difference implementation. [From Campanelli *et al.* (2006)].

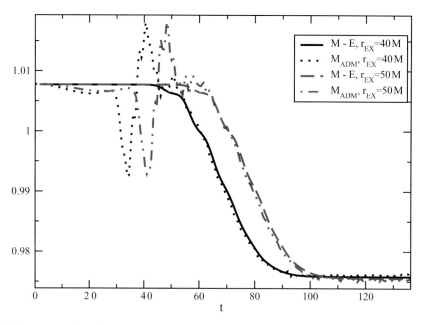

Figure 13.6 Demonstration of energy conservation in the calculation of Baker *et al.* (2006a). The initial total energy (the initial total ADM mass) M minus the energy E carried off in gravitational radiation agrees with the instantaneous total energy M_{ADM} to high accuracy. [From Baker *et al.* (2006a).]

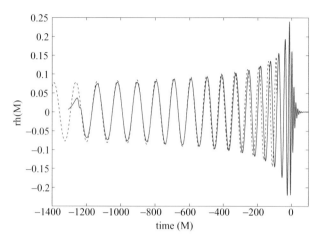

Figure 13.7 Comparison of the gravitational waveform $h(t)$ from an equal-mass, nonspinning black hole binary as obtained in the numerical simulations of Baker *et al.* (2007a) (solid line) with the corresponding post-Newtonian result to order 2.5PN in the amplitude and order 3.5PN in the phasing (dashed line; see Blanchet 2006). Here r is the distance to the source. The gravitational wave strain h is based on the dominant $l = 2, m = 2$ spin-weighted spherical harmonics of the radiation and represents an observation made in the system's equatorial plane, where only the h_+ polarization contributes to the strain. [From Baker *et al.* (2007a).]

Figure 13.6 demonstrates mass-energy conservation in the calculations of Baker *et al.* (2006a). This calculation used AMR, with a grid resolution of $\Delta x = M/32$ on the finest grid, and with the outer boundary at $128M$. The comparison is performed for gravitational wave extraction at both $40M$ and $50M$. Except for some fluctuations at early times, caused by a transient pulse in the gauge evolution that affects the measurement of the ADM mass, the agreement is remarkably good, and provides one measure of the accuracy of the simulation.

In a follow-up study, Baker *et al.* (2007a) compared their gravitational waveforms with those obtained from post-Newtonian approximations.[25] For such a comparison it is important to start the inspiral from a large binary separation, where one would expect the two approaches to agree well. Baker *et al.* (2007a) therefore use as initial data a binary with an orbital angular frequency of approximately $\Omega M = 0.0255$, which may be compared with the values at the ISCO listed in Table 12.2. This binary completes about seven orbits, and hence about 14 gravitational wave cycles, before coalescence. The agreement with post-Newtonian predictions well into the merger phase is remarkable. Figure 13.7 compares the gravitational waveform generated by the numerical simulations with the the post-Newtonian prediction to order 2.5PN in the amplitude and to order 3.5PN in the phase. The numerical simulation uses fourth-order finite differencing, AMR, and a resolution of

---

[25] See Appendix E as well as Blanchet (2006).

approximately $M/32$ on the finest grid. While the agreement is not perfect, it is consistent with the internal error estimates in either one of the two approaches.[26]

Given the different algorithms and implementations used in the simulation of binary black holes, it is of interest to compare how well predictions from these different codes agree. This has been addressed by Baker *et al.* (2007b), who provide a comparison of the results obtained with the codes of Pretorius (2005a), Campanelli *et al.* (2006) and Baker *et al.* (2006a). If the only differences between these simulations originate in the adopted formulation of Einstein's equations, gauge conditions and specifics of the numerical implementation, then clearly all of them should yield the same results for physical invariants in the limit of infinite resolution. However, the simulations differ in other aspects, too. One difference lies in the initial data. For the calculations presented in Buonanno *et al.* (2007), Pretorius adopts the corotational, conformal thin-sandwich binary black hole data of Cook and Pfeiffer (2004). The other two groups, on the other hand, adopt nonspinning puncture initial data. Even though the simulations start with quasiequilibrium initial data describing a binary at quite similar proper separations and angular velocities, they carry different total angular momenta.[27] Moreover, the different groups use different algorithms for the extraction of gravitational waves (see Chapter 9.4). Despite these differences, it is remarkable how well the predictions for the asymptotic gravitational wave signals agree, as shown in Figure 13.8. The agreement is particularly convincing for the waveforms emitted during the merger phase. At earlier times, shown in the inset, especially when the signal is dominated by noise associated with the initial data, the agreement is worse, as one would expect.

Figures 13.7 and 13.8 also demonstrate how "simple", in some sense, the gravitational wave signal from the merger of a binary black hole system appears. One can easily identify the familiar "chirp" signal, emitted during the late inspiral phase, during which both the wave amplitude and frequency increase. The signal terminates, as expected, with quasinormal ringing, as the merger remnant settles down to stationary equilibrium. The transition from the chirp to the ringdown part of the signal is rather smooth, and not at all abrupt. This feature is related to the relatively smooth transition of the binary orbit from inspiral to merger, as described in Chapter 12. While we have defined and located the ISCO as the last stable circular orbit for the quasiequilibrium models considered in Chapter 12, the ISCO does not mark a very sudden transition, but rather a gradual change from inspiral to plunge. This can be seen in Figure 13.4, which displays no sharp plunge prior to merger.

In fact, the inspiral is approximated remarkably well by the Newtonian quadrupole expressions of Exercise 12.3 until late times.[28] A dynamical instability in the binary orbit

---

[26] A similar analysis, based on the numerical simulations of Pretorius (2005a), has been presented in Buonanno *et al.* (2007).

[27] Even for identical angular momenta, the quasiequilibrium data constructed from the conformal thin-sandwich approach and the puncture data may not represent slices of identical spacetimes; see the discussion in Chapter 12.

[28] See Buonanno *et al.* (2007) for a detailed analysis.

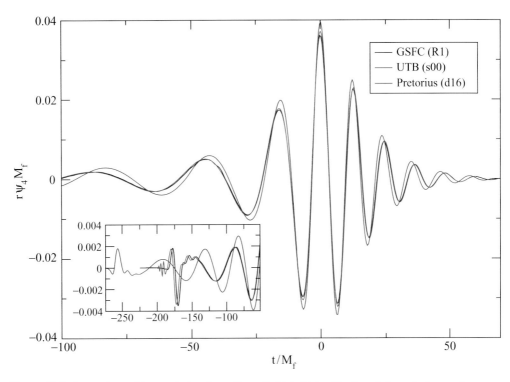

Figure 13.8 Comparison of the gravitational waveforms $\text{Re}(\psi_4)$ obtained by three independent research groups: Pretorius (Pretorius 2005a; Buonanno *et al.* 2007), Goddard Space Flight Center (GSFC, e.g., Baker *et al.* (2006a) and University of Texas at Brownsville (UTB, e.g., Campanelli *et al.* 2006). [From Baker *et al.* (2007b).]

at the ISCO should be reflected in a departure of the waveform from this expression. However, a common horizon forms sufficiently early to absorb much of the radiation generated during such an epoch and mask any such departure.

Simulations that can track binary black hole inspiral from large separation allow us to estimate the total energy emitted in gravitational radiation during the entire coalescence event. Baker *et al.* (2007b) find that approximately 3.5% of the total initial mass-energy M is radiated away during the late inspiral and merger, starting from a proper horizon separation of about $10M$; other investigations are in reasonable agreement. The amount of energy radiated during the early inspiral from infinite separation to the separation of $10M$ can be estimated from the binding energy of the initial data at $10M$. Using the values reported in Caudill *et al.* (2006) and Buonanno *et al.* (2007), the radiated energy is about 1.2% of M.[29] The total amount of energy radiated during inspiral from infinite separation is therefore between about 4.5 and 5% of the total initial mass-energy. This value is significantly less than the maximum possible value of 29% allowed by the area theorem

---

[29] See Tables 12.1 and 12.2, where we list the binding energies at the ISCO, i.e., at slightly smaller binary separation.

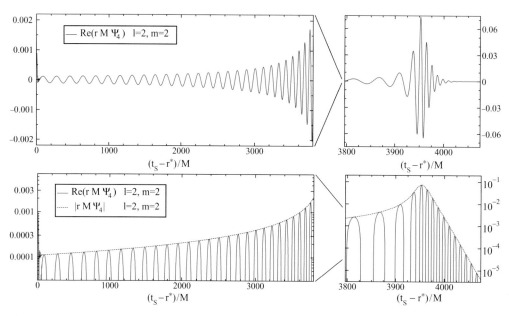

Figure 13.9 The gravitational waveform, extrapolated to spatial infinity, for an equal-mass, nonspinning, black hole binary. The top panel shows the real part of the dominant $l = 2$, $m = 2$ mode of ψ_4 on the linear scale, the bottom panel on a logarithmic scale. The panels on the right show an enlargement of the data for the merger and ringdown. [From Scheel *et al.* (2009).]

(see exercise 7.2), but considerably more than the energy radiated during a head-collision from rest at infinity ($\ll 1\%$; see Chapter 10.2.)

Building on these initial successes, many other investigators followed up with their own simulations of the inspiral and merger of equal-mass binary black holes. As an example we show results from a highly refined simulation, based on a carefully groomed, pseudo-spectral implementation of a first-order version of the generalized harmonic formalism, in Figure 13.9.[30]

In Appendix I we present results in greater detail for two simulations of the inspiral and merger of nonspinning binary black holes. One simulation describes an equal-mass binary and the other describes a binary with a mass ratio of 3:1 (see the following section). These particular simulations employ a finite difference implementation of the moving puncture method within the BSSN formalism and employ puncture initial data. In the Appendix we tabulate specific information about the adopted AMR grid set-up, the finite difference implementation, the initial data, and the diagnostics, as well as the emitted gravitational radiation and the black hole recoil. We also plot the orbital trajectories of the black holes and the emitted gravitational waveforms. The purpose of this Appendix is twofold: First, from a practical computational perspective, these two simulations provide handy

---

[30] See Scheel *et al.* (2009), as well as Lindblom *et al.* (2006), for a first-order version of the generalized harmonic formalism.

test-bed calculations for researchers seeking to perform similar simulations with their own algorithms and vacuum codes. Second, from a pedagogical perspective, the summary provided in the Appendix pinpoints where in the course of performing and evaluating a binary black hole simulation one is required to implement many of the basic concepts and tools that have been developed in this book. References to the places in the book where these concepts and tools are introduced are provided throughout Appendix I.

13.2.2 Asymmetric binaries, spin and black hole recoil

In the last section we focused on equal-mass, nonspinning binaries. While this is a very natural case to study first, it clearly is not a generic scenario: binary black holes in nature are likely to have unequal masses and nonzero spins. It therefore is of great interest to understand how the findings of the previous section change for mass ratios different from unity and how they are affected by black hole spin.

Allowing for unequal masses and black hole spin will certainly alter the gravitational signal emitted during the inspiral and merger. It is therefore crucial to understand these effects for the detection of gravitational waves from binaries. In fact, the accurate extraction of binary parameters from an observed gravitational wave signal requires a large catalog of templates of theoretical waveforms compiled for binaries with different mass ratios and spin parameters.

However, the detection and interpretation of gravitational wave signals is not the only astrophysical motivation for the study of asymmetric binaries. As it turns out, one important consequence – black hole recoil – is suppressed in the symmetric binary scenarios of Section 13.2.1, but plays an important role in several different astrophysical contexts, as we will sketch below.

Black hole recoil can be understood qualitatively as follows. Gravitational radiation generally carries both energy and momentum. In symmetric binaries, the total linear momentum radiated by the binary must be zero: the contributions to the radiated linear momentum from the two black holes cancel. This cancelation does not occur in an asymmetric binary, which therefore radiates a nonzero net linear momentum. Given this linear momentum loss, the center of mass of the system acquires a linear momentum in the opposite direction. In a frame in which the binary's center of mass was originally at rest prior to merger, the remnant will thus end up with a nonzero "kick" velocity.

Just how large this recoil or kick velocity V_{kick} is has important astrophysical and cosmological implications. For example, supermassive black holes with masses in the range $10^6 M_\odot - 10^9 M_\odot$ are believed to reside at the centers of many, if not most, bulge galaxies, including the Milky Way. One plausible scenario for their formation is through a combination of mergers with other black holes and gas accretion. The mass and age of the initial "seed" black hole is unknown, but it could be in the range $60-600 M_\odot$ if it is the collapsed remnant of a first-generation Population III star,[31] or much higher than $10^3 M_\odot$

---

[31] Madau and Rees (2001).

if it is the remant of a collapsed supermassive star.[32] Binary black holes can form during the merger of their host galaxies. This process is believed to be take place in the context of the cold dark matter (ΛCDM) model of structure formation in the early Universe, where dark matter halos merge hierachically and black holes are assumed to settle, merge and accrete in their gaseous centers.

Clearly, for the remnant of such mergers to remain within any host, the kick speed V_{kick} has to be less than the escape velocity of that host. In giant elliptical galaxies and spiral galaxies, the central escape velocities V_{esc} are roughly between 500 km/s and 2000 km/s.[33] If kick speeds routinely exceeded these escape speeds, this would clearly call into question whether supermassive black holes can form via hierarchical merger,[34] and might favor instead growth via pure accretion, or some other mechanism. On the other hand, even a modest kick speed V_{kick} would be sufficient to explain the apparent absence of massive black holes in dwarf galaxies and globular clusters, for which the central escape speed is significantly smaller than for giant galaxies.[35]

A typical kick speed just below the escape speed should result in a finite probability of finding remnant supermassive black holes displaced from the center of their host galaxies. Eventually, dynamical friction (gravitational scattering off other stars) will cause the orbit of the black hole to decay back to the galaxy center, as it transfers kinetic energy to the other stars in the galactic nucleus.

The ejection of merger remnants from globular clusters reduces the likelihood of further mergers and thus decreases the probability of observing binary black hole mergers in such clusters. For the same reason, black hole merger remnants have difficulty remaining in high redshift halos with relatively shallow potential wells.

Another important consequence of black hole mergers is its effect on the spin evolution of a supermassive black hole. The growth rate of a black hole by gas accretion is a function of its efficiency of conversion of accreted rest-mass into electromagnetic radiation, and this efficiency depends sensitively on the spin parameter a/M of the black hole. Black holes with smaller spin parameters have lower efficiency and thus grow more quickly for a given luminosity. It is thus a crucial question whether or not the combination of mergers and gas accretion at the Eddington limiting luminosity[36] is sufficiently rapid to build a supermassive black hole to power QSO SDSS $1148 + 5251$, the quasar with the highest known redshift ($z = 6.4$) at the time of the writing of this book. This quasar is believed to host a $10^9 M_\odot$ black hole, which therefore implies that a seed black hole must be able to grow to this size within 0.9 Gyr after the Big Bang in the standard ΛCDM cosmology. The

---

[32] See Rees (1984) and Shapiro (2004b) for discussions of supermassive star collapse and alternative scenarios for forming supermassive black hole seeds. See Chapter 14.2 for simulations of massive star collapse to black holes.

[33] See, e.g., Figure 2 of Merritt *et al.* (2004) for the central escape speeds of various types of galaxies and star clusters.

[34] We note that the kick speed is independent of the total mass of the binary, a result that is consistent with the fact that, when expressed in gravitational units, speed is dimensionless.

[35] See Volonteri (2007) and references therein for calculations of the effect of recoil on the formation of supermassive black holes.

[36] The Eddington limit L_{Edd} is the critical luminosity at which the outward force of radiation pressure equals the inward pull of gravity in an accreting plasma. See, e.g., Shapiro and Teukolsky (1983), Section 13.7 for a derivation.

existence of this black hole may constrain the spin evolution of its accreting progenitors and the viability of the merger-accretion scenario for supermassive black hole growth.[37]

We note that each time binary black holes of comparable mass merge (a "major merger"), the remnant acquires an appreciable spin, arising largely from the orbital angular momentum of the binary near the ISCO. For example, the merger of two equal-mass, nonspinning black holes results in a remnant with spin parameter $a/M \approx 0.7$ (see Table I.4). By contrast, the merger of a massive black hole with multiple small-mass companions ("minor mergers") that inspiral in isotropically-oriented orbits will spin down the massive black hole as it grows.[38] Gaseous disk accretion will drive up the spin of the black hole; spin equilibrium will be achieved for a value of a/M that depends on precise details of the accretion process. For example, ignoring radiation loss, accretion from a standard relativistic "thin disk" will drive the spin up to its maximal value, $a/M = 1$,[39] while accounting for photon emission shifts the equilibrium value back down to $a/M = 0.998$.[40] Simulations of relativistic MHD accretion onto Kerr black holes, however, suggest that the disk may not be so thin and the equilibrium spin value may be significantly lower, $a/m \sim 0.9$.[41] The spin of a supermassive black hole at any one time may thus be determined by the mechanism that has dominated its most recent growth.[42]

Finally, as we will see below, the merger of spinning black holes of comparable mass may cause the spin axis of the remnant to flip. It has been speculated that this phenomenon could explain the observation of X-shaped radio jets, in which the orientation of the emitted jets seems to have changed abruptly in the past.

Unequal masses

Historically, black hole recoil was first considered for unequal-mass binaries with non-spinning companions. In such a binary the less massive star or black hole resides at a larger separation from the center of mass than the more massive companion, and hence orbits with a larger orbital speed. The two objects therefore emit gravitational radiation at different rates, and, in a crude analogy to electromagnetism, the radiation from the faster object is more highly "beamed" in the forward direction than the radiation from the slower object. As a consequence the linear momentum emitted from the two companions no longer cancel, hence the center of mass acquires some linear momentum in the process.

In the absence of spin, the situation is symmetric across the orbital plane, which implies that the radiation reaction force must lie in this plane. Over the course of one orbit the direction of the force also completes a full circle – similar to a spinning lawn sprinkler that

---

[37] See Shapiro (2005) and Volonteri and Rees (2006) for discussion and references.

[38] $a/M \sim M^{-7/3}$; Hughes and Blandford (2003); Gammie *et al.* (2004). Spindown occurs because the orbital angular momenta of counter-rotating companions at their ISCOs are larger in magnitude than those of corotating companions.

[39] Bardeen (1970).

[40] Thorne (1974).

[41] De Villiers *et al.* (2003); Gammie *et al.* (2004).

[42] See Berti and Volonteri (2008) and references therein for studies of cosmological black hole spin evolution by mergers and accretion, with observational implications.

ejects water in a rotating beam. If the orbit were strictly circular, the motion of the center of mass would therefore also describe a circle, making the net effect vanish. Instead, however, the binary orbit inspirals slowly, so that the binary emits slightly more linear momentum at the end of one orbit than at the beginning. As a consequence, the center of motion does not follow a perfect circle, but instead describes an *outward* spiral – which can be pictured by imagining a spinning lawn sprinkler that emits water at an increasing rate. The process ends when the binary merges and ceases to emit linear momentum. At that point the center of mass will follow a rectilinear trajectory, having acquired a kick speed V_{kick} in a random direction in the orbital plane.

The first quantitative analysis of this process was performed by Fitchett (1983), who evaluated the lowest-order multipole moments for a Newtonian point-mass binary to obtain the emitted linear momentum. He finds that the kick speed is approximately[43]

$$V_{kick} \approx 1480 \, \text{km/s} \, \frac{f(q)}{f_{max}} \left(\frac{2M}{r_{term}}\right)^4 \tag{13.5}$$

where $q = m_1/m_2 \leq 1$ is the mass ratio, $M = m_1 + m_2$ is the total mass, and r_{term} is the binary separation at which the linear momentum emission terminates.[44] The function $f(q)$ is given by

$$f(q) = q^2 \frac{1-q}{(1+q)^5} \tag{13.6}$$

and assumes a maximum of $f_{max} \simeq 0.0179$ at $q = (3 - \sqrt{5})/2 \simeq 0.38$. The kick speed vanishes both for equal-mass binaries ($q = 1$) and in the test-particle limit, $q = 0$, as one would expect. Assuming that r_{term} may come close to the gravitational radius $2M$, the above expression predicts large kick speeds that easily exceed 1000 km/s. However, later analytical estimates, based on perturbation theory or post-Newtonian calculations, revised the maximum kick speed to smaller values of at most a few hundred km/s.[45]

With the availability of numerical codes that can simulate the inspiral and coalescence of binary black holes, it is possible to compute the recoil kick speed without approximation. The first such attempt was carried out by Herrmann *et al.* (2007a), who used very crude initial data. The first accurate calculation was presented by Baker *et al.* (2006b), who found a kick speed of 105 ± 10 km/s for a mass ratio of $q = 0.67$.[46]

A more comprehensive and very accurate study has been carried out by González *et al.* (2007). The computational methods used in these simulations are similar to those of Campanelli *et al.* (2006) and Baker *et al.* (2006a) described above; in particular they use the moving puncture method to model the black holes (see Section 13.1.3), the BSSN

---

[43] Favata *et al.* (2004).

[44] *cf.* exercise 13.3 below.

[45] See, e.g., Kidder (1995); Favata *et al.* (2004); Blanchet *et al.* (2005); Damour and Gopakumar (2006); Sopuerta *et al.* (2007).

[46] While Baker *et al.* (2006b) was published in a journal before Herrmann *et al.* (2007a), the latter appeared on the `xxx.arXiv.org` preprint server approximately two month before the former.

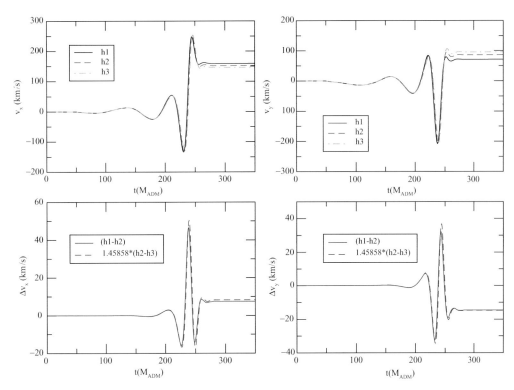

Figure 13.10 Components of the kick velocity as a function of time in the simulations of González *et al.* (2007) for a nonspinning binary of mass ratio $q = 0.33$. The graphs show results for three different resolutions, the finest of which are $h_1 = m_1/45$, $h_2 = m_1/51$, and $h_3 = m_1/58$, where m_1 is the smaller black hole's irreducible mass. The convergence test in the lower panels demonstrates second-order convergence. As expected, the kick velocity oscillates during the inspiral, and then remains constant once the emission of linear momentum ceases. [From González *et al.* (2007).]

formulation of Einstein's equations (Chapter 11.5), and an AMR grid structure.[47] González *et al.* (2007) construct puncture initial data (see Chapter 12.2) describing unequal mass binaries, fixing binary parameters as obtained in post-Newtonian calculations. These initial data contain some small amount of spurious gravitational radiation, which propagates off the numerical grid via an initial pulse. Once this pulse has passed, integration of equation (9.132), yields the kick momentum P_i. In the calculations of González *et al.* (2007), the Weyl scalar ψ_4 is extracted at $r_{\text{ext}} = 50M$.

In Figure 13.10 we show the x and y components of the kick velocity V_{kick} for a binary of mass ratio $q = 0.33$, calculated from the components of the momentum P_i by dividing by the final remnant black hole mass. The z component vanishes identically, since it would describe motion perpendicular to the orbital plane. As we would expect from our discussion

---

[47] Details of this numerical integration, together with many code tests and benchmarks, can be found in Brügmann *et al.* (2008).

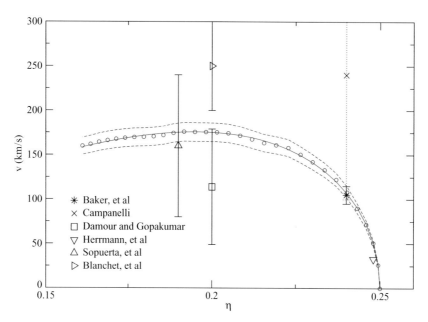

Figure 13.11 The kick velocity as a function of the symmetric mass-ratio parameter $\eta = q/(1+q)^2$, where q is the ratio of irreducible masses. The line connecting the open circles, together with the dashed line marking a 6% error, denote the results of González *et al.* (2007). Also plotted are the earlier numerical results of Baker *et al.* (2006b), Campanelli (2005), Herrmann *et al.* (2007a), as well as the analytical results of Blanchet *et al.* (2005); Damour and Gopakumar (2006); Sopuerta *et al.* (2007). [From González *et al.* (2007).]

above, the kick velocity initially oscillates with an increasing amplitude, until the remnant experiences a final kick during the black hole merger. Shortly after that the emission of linear momentum ceases, leaving the remnant to coast with a net kick speed V_{kick} in a random direction in the orbital plane.

Figure 13.11 shows the final kick speed V_{kick} for a number of different mass ratios. The maximum value of $V_{\text{kick}} = 175 \pm 11$ km/s is obtained for an irreducible mass ratio of $q = 0.36 \pm 0.03$.[48] Quite remarkably, this mass ratio is consistent with that of the maximum value of Fitchett's expression (13.5). In an extension of the work of Buonanno *et al.* (2007), Berti *et al.* (2007) synthesized numerical calculations of the inspiral, merger and ringdown of unequal mass binaries. In particular, they find that the total energy ΔE_{GW} emitted during the merger phase is approximately

$$\Delta E_{\text{GW}}/M = 0.032661 \left[\frac{4q}{(1+q)^2} \right]^2 + 0.004458 \left[\frac{4q}{(1+q)^2} \right]^4 \tag{13.7}$$

where M is the total initial ADM mass of the binary. Note that for equal-mass binaries with $q = 1$ we recover a value close to the 3.5% that we stated in Section 13.2.1. Berti *et al.* (2007) also find that the angular momentum of the remnant black hole is well approximated

---

[48] See also the simulation for $q = 1/3$ described in detail in Appendix I.

by

$$J_{\text{fin}}/M_{\text{fin}}^2 = 3.272\frac{q}{(1+q)^2} - 2.075\frac{q^2}{(1+q)^4}, \qquad (13.8)$$

where M_{fin} denotes the final ADM mass of the remnant black hole.

Exercise 13.2 Evaluate equations (13.7) and (13.8) for $q = 1$ and $q = 3$ and compare with the values recorded in Table I.4 for the puncture binary black hole simulations summarized in Appendix I.

Black hole spin

As we discussed above, the linear momentum emitted during one orbital period in the inspiral phase nearly cancels out for nonspinning black holes, leaving only a relatively small cumulative effect. Allowing for black hole spin, however, can enhance the effect. A post-Newtonian analysis shows that, to leading order, the spin-orbit contribution to the emitted linear momentum is[49]

$$\dot{\mathbf{P}}_{\text{SO}} = -\frac{8}{15}\frac{\mu^2 M}{r^5}\left(4\dot{r}\,\mathbf{v}\times\boldsymbol{\Delta} - 2v^2\,\hat{\mathbf{n}}\times\boldsymbol{\Delta} - (\hat{\mathbf{n}}\times\mathbf{v})(3\dot{r}\,\hat{\mathbf{n}}\cdot\boldsymbol{\Delta} + 2\,\mathbf{v}\cdot\boldsymbol{\Delta})\right). \quad (13.9)$$

Here and throughout the rest of this chapter boldfaced symbols represent 3-dimensional, spatial vectors. As before, $M = m_1 + m_1$ is the total mass, and $\mu = m_1 m_2/M$ is the reduced mass. We also define $\mathbf{x} = \mathbf{x}_1 - \mathbf{x}_2$ as the relative position vector, $r = |\mathbf{x}|$ as the coordinate distance, $\hat{\mathbf{n}} = \mathbf{x}/r$ as a unit vector pointing from one binary companion to the other, and $\mathbf{v} = \dot{\mathbf{x}}$ as the relative velocity, where the dot represents a derivative with respect to time. Finally, $\boldsymbol{\Delta} = M(\mathbf{S}_2/m_2 - \mathbf{S}_1/m_1)$ is a measure of the black hole spins.

Note that equation (13.9) provides an expression for the "instantaneous" emission of linear momentum. The final kick speed is again a cumulative quantity that results from the change in linear momentum integrated throughout the binary inspiral and merger. However, the rate equation (13.9) does have an immediate consequence for the direction of the resulting kick. To see this, first consider black hole spins $\mathbf{S}_1$ and $\mathbf{S}_2$ that are aligned with the orbital angular momentum and that are therefore perpendicular on $\mathbf{v}$ and $\hat{\mathbf{n}}$, both of which lie in the orbital plane. For such spins, the last two terms $\hat{\mathbf{n}}\cdot\boldsymbol{\Delta}$ and $\mathbf{v}\cdot\boldsymbol{\Delta}$ in equation (13.9) vanish, and the spin contribution to the kick originates from the first two terms. The resulting kick is therefore in the orbital plane, just like the contribution from unequal masses, though not necessarily in the same direction. We may call this contribution $\mathbf{V}^\perp$, with the $\perp$ symbol denoting a direction perpendicular to the orbital angular momentum. Then consider spins that are perpendicular to the orbital angular momentum and hence lie in the orbital plane. For such spins all four terms in (13.9) yield contributions that are

---

[49] See equation (3.31b) in Kidder (1995); spin-spin effects enter only at higher post-Newtonian order.

perpendicular on the orbital plane. We will call this contribution $\mathbf{V}^{\parallel}$, with the $\parallel$ symbol denoting a direction parallel to the orbital angular momentum.

To describe a general kick velocity we need to pick three independent unit vectors. We choose the first unit vector, $\mathbf{e}_1$, to be aligned with the direction of the kick velocity $\mathbf{V}^m$ that we would obtain in the absence of any black hole spins, i.e., the (unequal) mass contribution to the kick velocity. We choose the second unit vector $\mathbf{e}_2$ to lie in the orbital plane as well, but orthogonal to $\mathbf{e}_1$, and, finally, the third unit vector $\mathbf{e}_z$ to be orthogonal to the orbital plane. We may then write

$$\mathbf{V}_{\text{kick}} = V^m \mathbf{e}_1 + V^{\perp}(\cos\xi\,\mathbf{e}_1 + \sin\xi\,\mathbf{e}_2) + V^{\parallel}\mathbf{e}_z. \tag{13.10}$$

Here the angle ξ measures the angle between the mass and orbital plane spin contributions to the kick. The coefficients V^m, $V^{\perp}$ and $V^{\parallel}$ are the magnitudes of these kicks, which have to be determined numerically. However, we can already anticipate some of the properties of these coefficients. The coefficient of the mass contribution V^m should depend only on the mass ratio q; it should vanish for $q = 1$ and $q = 0$ and should presumably be similar to equation (13.5). At least to leading order, the spin contribution $V^{\perp}$ can depend only on components of the spins along the orbital angular momentum, $\mathbf{S}_i^{\parallel}$. Similarly, $V^{\parallel}$ should depend only on components of the spin in the orbital plane, $\mathbf{S}_i^{\perp}$.

Soon after accurate binary black hole simulations became feasible, a number of investigators leaped to explore the effect of black hole spin on black hole recoil.[50] Most of the initial calculations focused on black hole spins that are aligned or anti-aligned with the orbital angular momentum. The resulting kicks are several hundred km/s in magnitude, easily exceeding the maximum kick of approximately 175 km/s found for non-spinning black holes. Baker *et al.* (2007c) found that in this case, when the black hole spins are orthogonal to the orbital plane, the final, cumulative kick speed can be estimated from a fitting formula that combines the contributions from both the mass and spin asymmetries,[51]

$$V_{\text{kick}} = |\mathbf{V}^m + \mathbf{V}^{\perp}| \approx V_0 \frac{32q^2}{(1+q)^5} \left((1-q)^2 + 2(1-q)K\cos\xi + K^2 \right)^{1/2}. \tag{13.11}$$

Here the parameter V_0 determines the overall scaling of the kick speed and the quantity K is defined as $K = k(qS_1/m_1^2 - S_2/m_2^2)$. In the latter expression k is a dimensionless number representing the relative importance of the kick contributions from the spin and mass asymmetries, m_1 and m_2 are the irreducible black hole masses, and S_1 and S_2 are their spins. By choosing these parameters to be $V_0 = 276$ km/s, $\xi = 0.58$ rad and $k = 0.85$, Baker *et al.* (2007c) show that equation (13.11) reproduces all the numerical results to within about 10%.

---

[50] See, e.g, Herrmann *et al.* (2007b), Koppitz *et al.* (2007), Campanelli *et al.* (2007a) for some early studies.
[51] See also Favata *et al.* (2004); Koppitz *et al.* (2007).

Exercise 13.3   (a) Adopt equation (13.11) to estimate the maximum kick speed for nonspinning black holes.
(b) Show that equation (13.11) reduces to equation (13.5) for nonspinning black holes and use this identification to determine the terminal radius r_{term} in equation (13.5).

Campanelli *et al.* (2007a) realized that an even more dramatic effect can be achieved when the black hole spins are not aligned with the orbital angular momentum. As we have seen above, the kick velocity may then no longer lie in the orbital plane, but instead may pick up a component orthogonal to this plane that can be much larger than the component in the plane.

To synthesize the results of different simulations of the effect of black hole spin on the recoil we return to equation (13.10). This expression predicts the kick velocities rather well if the coefficients in equation (13.10) are chosen as follows.[52] Set the magnitude of the mass contribution V^m to be

$$V^m = V_0^m \frac{f(q)}{f_{\text{max}}} \left(1 + B \frac{q}{(1+q)^2} \right),$$ (13.12)

where $f(q)$ is given by equation (13.6), $B = -0.93$, and $V_0^m = 214$ km/s. Note that the second term in this expression does not appear in equation (13.5) or (13.11), which also explains why the overall coefficient differs. Now set the two spin contributions to be

$$V^\perp = V_0^\perp \frac{g(q)}{g_{\text{max}}} \left(\frac{S_2^\parallel}{m_2^2} - q \frac{S_1^\parallel}{m_1^2} \right)$$ (13.13)

and

$$V^\parallel = V_0^\parallel \frac{g(q)}{g_{\text{max}}} \cos(\Theta - \Theta_0) \left(\frac{S_2^\perp}{m_2^2} - q \frac{S_1^\perp}{m_1^2} \right).$$ (13.14)

Here we have defined

$$g(q) = \frac{f(q)}{1-q} = \frac{q^2}{(1+q)^5},$$ (13.15)

which assumes a maximum of $g_{\text{max}} = 2^2 3^3/5^5 \simeq 0.0346$ at $q = 2/3$. The quantity Θ measures the angle between the orbital plane component of the vector $\mathbf{\Delta}$ and the infall direction at merger. Finally, set the coefficients appearing above to be $V_0^\perp = 252$ km/s and $V_0^\parallel = 2,073$ km/s. Note that the coefficients V^m, $V^\perp$ and $V^\parallel$ have all the properties that we anticipated in the discussion below equation (13.10).

The size of the coefficient $V^\parallel$ suggests that binaries can pick up a much larger kick orthogonal to the orbital plane than in the orbital plane, if the individual black hole spins have a significant component in the plane. As an example, Figure 13.12 shows the trajectories of the black hole centers for an equal-mass binary.[53] This simulation starts

[52] Campanelli *et al.* (2007b).
[53] González *et al.* (2007).

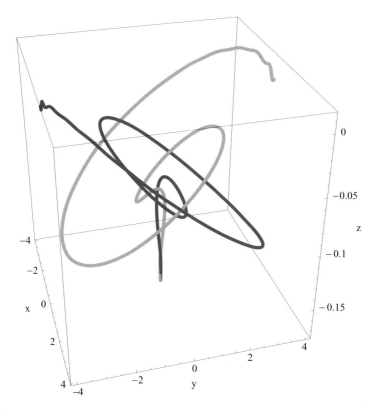

Figure 13.12 Coordinate trajectories of the black hole centers (punctures) for an equal-mass, spinning black hole binary. The initial spins are both perpendicular to the orbital angular momentum; each spin vector points away from the companion along the line joining the centers. Following merger, the remnant receives a kick in the negative z-axis, perpendicular to the orbital plane. [From González *et al.* (2007).]

with a binary in an approximately circular orbit. Both black holes have a spin of magnitude $S/m^2 = 0.8$, and both of the spin vectors initially point away from the companion along the line connecting the centers. As equation (13.10) suggests, this merger should lead to a kick orthogonal to the orbital plane, which can be seen in Figure 13.12. Given the symmetry of these particular initial data, the plane of the orbit is displaced along the z-axis during the inspiral and merger without being tilted. Ultimately, the remnant receives a kick of 2650 km/s in the negative z-direction.

This scenario has been explored further by Campanelli *et al.* (2007b), who evolve six different sets of initial data. All binaries have equal-mass black holes, and all have black hole spins of magnitude $S/m^2 = 0.5$ that lie in the orbital plane and are anti-aligned with each other. The different sets of initial data differ only in the direction of these black hole spins. It turns out that in many respects these particular initial data lead to a surprisingly similar evolution. The total radiated energy and angular momentum, for example, are all identical to within 3%. The projection of the black holes' trajectory into the orbital xy

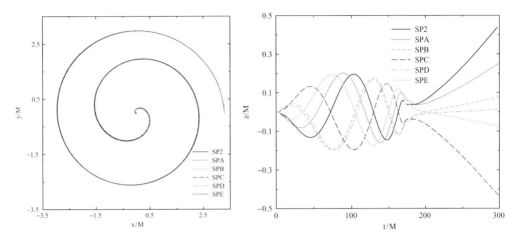

Figure 13.13 The left panel shows the projection into the xy (orbital) plane of the puncture trajectories for six different sets of initial data (with only one puncture shown per binary). All binaries have equal masses and black hole spins of magnitude $S/m^2 = 0.5$ that reside in the orbital plane and are anti-aligned with each other; however, the direction of the spins within the orbital plane are different for each set of initial data. In this case the spin hardly affects the trajectory of the black holes within the xy plane. The right panel shows the z-location of the orbital plane as a function of time. During the inspiral, the orbital plane oscillates along the z-axis, but does not tilt. Following merger, the remnant recoils along the z-axis. [From Campanelli *et al.* (2007b).]

plane is also hardly affected by the black hole spins; this is demonstrated in the left panel of Figure 13.13. What is affected, however, is the z location of the orbital plane. During the inspiral this plane containing the two punctures oscillates along the z-axis, i.e., orthogonal to the orbital plane. The two punctures always have identical z-coordinates throughout the inspiral, meaning that the orbital plane oscillates without tilting. This behavior can be understood in terms of the symmetry of the initial data: the two anti-aligned black hole spins "cancel" each other out and cannot affect the orbital angular momentum, which therefore has to keep its original orientation. The location of the orbital plane along the z-axis as a function of time for these different sets of initial data is shown in the right panel of Figure 13.13. Following merger, the emission of linear momentum again ceases, and the remnant recoils with the linear momentum it has at the time of merger. Campanelli *et al.* (2007b) find that the maximum kick that binary black holes can receive during merger is approximately 4000 km/s – significantly larger than the values for nonspinning, unequal-mass binaries.

Clearly these findings raise some very interesting astrophysical questions. Kick speeds of several thousand km/s easily exceed the escape speeds from the centers of even giant elliptical galaxies. However, we also have strong observational evidence that supermassive black holes reside at the centers of these galaxies, implying that these black holes have not undergone mergers resulting in such large kicks. Early analysis of these questions suggests that the likelihood of the black hole spins being aligned in such a way that merger leads to ejection is rather small.[54]

So far we have focused our discussion on the effect of black hole spin on the recoil of the binary remnant. Clearly, however, spin affects the inspiral and merger in other ways as well. For example, black hole spins that are aligned with the orbital angular momentum increase the binary's total angular momentum. If this total angular momentum exceeds the maximum angular momentum of a Kerr black hole, then the binary cannot merge until a sufficient amount of angular momentum has been radiated away. Quite generally, we expect binaries with black hole spins aligned with the orbital angular momentum to merge more slowly than binaries with spins that are anti-aligned. This effect, sometimes referred to as "orbital hang-up", has been explored with numerical simulations.[55]

Another very important effect is spin flip. In the discussion above, and in Figures 13.12 and 13.13, we have focused on binaries for which the two black hole spins are in the orbital plane and *anti-aligned* with each other. For these binaries we have found that the orbital plane gets shifted without being tilted, which we explained in terms of the two black hole spins canceling each other out. Clearly this situation changes if the two black hole spins are not anti-aligned. In this case a spin-orbit interaction affects both the black hole spins and the orbital angular momentum during the inspiral and may change their orientation.

As an example, we show results of a simulation by Campanelli *et al.* (2007) in Figure 13.14. Here the initial black hole spins, both of magnitude $S/m^2 = 0.5013$, again lie in the orbital plane, but they are now *aligned* with each other. At the initial time, they both point in the positive y-direction, perpendicular on the x-axis that initially connects the two black hole punctures. During the inspiral, the spins rotate by approximately $90°$ within the orbital plane and also pick up a nonzero z-component orthogonal to the orbital plane.[56] Simultaneously the orbital plane tilts, leaving a remnant with an angular momentum that has nonzero components in both the y and z directions. This is not too surprising, of course, since the total angular momentum of the initial data also had nonzero components in these two directions – the y-components from the black hole spins, and the z-component from the orbital angular momentum. Some of this angular momentum is radiated away during the inspiral, but part of it remains with the merger remnant. The individual black holes therefore experience a *spin flip*: originally, both of their spins were aligned with the y-axis, but after merger the remnant has a significant component in the z-direction.

These spin-flips may explain the so-called X-shaped radio jets. We show some examples of observations of such objects in Figure 13.15. While the details of jet emission are still

---

[54] Schnittman and Buonanno (2007); Bogdanovic *et al.* (2007).

[55] See, e.g., Campanelli *et al.* (2006b).

[56] The masses and spins of the individual black holes can be determined with the help of the isolated or dynamical horizon formalism introduced in Chapter 7.4; the spins, in particular, are given by (7.74), where ϕ^a is an approximate Killing vector on the horizon. Some subtleties in the determination of the direction of these spins are discussed in Campanelli *et al.* (2007).

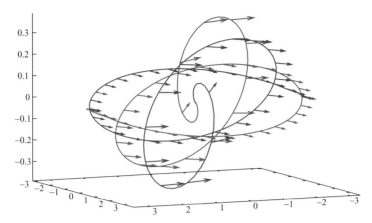

Figure 13.14 Puncture trajectories together with black hole spin directions, shown at time intervals of $4M$. The initial orbital plane is in the xy (horizontal) plane, and the initial black hole spins reside in this plane, both pointing in the positive y-direction (to the right). During the course of the inspiral the spins rotate by approximately $90°$ within the orbital plane, and also pick up a nonzero z-component orthogonal to the orbital plane. Simultaneously the orbital plane tilts, leaving a remnant with an angular momentum that has nonzero components in both the y and z directions. Note that the z-scale is 1/10th the x and y-scale. [From Campanelli *et al.* (2007).]

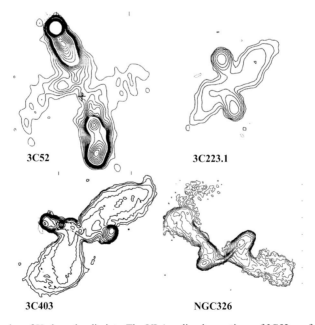

Figure 13.15 Examples of X-shaped radio jets. The VLA radio observations of 3C52 are from Leahy and Williams (1984), of 3C223.1 and 3C403 from Dennett-Thorpe *et al.* (1999), and of NGC326 from Murgia *et al.* (2001). [From Merritt and Ekers (2002).]

uncertain, it is believed that jets are emitted along the rotation axes of accreting Kerr black holes. While most jets point along well-defined directions, some jets appear as if the direction of the emission has changed very abruptly some time in the past: the more distant part of the jet points in one direction, while the closer part of the jet, emitted more recently, points along another direction. Together, these two axes form an X-shaped figure. The numerical simulations of binary black hole mergers described above provide a very plausibe explanation for this apparent spin flip of the spinning black hole, namely its merger with another black hole.

14 | **Rotating stars**

Rotating stars in general relativity are of long-standing theoretical interest. With the discovery of radio pulsars in 1967[1] and their identification as magnetized, rotating neutron stars, rotating relativistic stars have also become objects of intense observational scrutiny. Radio pulsars have become particularly useful for relativity, since they provide excellent cosmic clocks. The original idea of Baade and Zwicky (1934) that neutron stars could be formed during supernovae events, coupled with the discovery of radio pulsars in some supernova remnants like the Crab, firmly linked rotating neutron stars with stellar collapse and supernovae explosions. Following the detection of radio pulsars, X-ray pulsars were discovered in 1971.[2] X-ray pulsars proved the existence of rotating neutron stars that are accreting gas supplied by a companion star in a binary system. The discovery of the first binary pulsar by Hulse and Taylor[3] in 1974 proved that rotating, relativistic stars can reside in binary systems, making neutron stars even more interesting to relativists. General relativity is required to describe the gravitational field of neutron stars, since their compaction, M/R, where M is the mass and R is the characteristic radius of the star, is large (~ 0.1–0.2).[4] Numerical relativity is important for treating many dynamical processes involving neutron stars accurately. It provides a valuable tool for probing neutron star formation from stellar core collapse in a supernova and for tracking the dynamical evolution and assessing the final fate of neutron stars subject to instabilities. It also furnishes gravitational radiation waveforms generated by these events, as well as by binary inspirals and mergers. Numerical relativity is absolutely essential for following the catastrophic collapse of an unstable star to a black hole.

There are other types of rotating, relativistic stars besides neutron stars that are of astrophysical interest. These stars also require general relativity for an accurate description and numerical relativity for a reliable analysis of their dynamical evolution and final fate. For example, rotating supermassive stars, although they have not been observed as yet,

---

[1] Hewish *et al.* (1968). For their roles in the discovery of radio pulsars (with J. Bell), A. Hewish and M. Ryle shared the Nobel Prize in Physics in 1974.

[2] Schreier *et al.* (1972); Tananbaum *et al.* (1972). For his role in the discovery and interpretation of cosmic X-ray sources, R. Giacconi shared the Nobel Prize in Physics in 2002.

[3] Hulse and Taylor (1975). For their discovery of this binary, and for showing that its orbit decays at the rate predicted by general relativity via the emission of gravitational waves, R. Hulse and J. Taylor were awarded the Nobel Prize in Physics in 1993.

[4] For a detailed discussion of the physics of neutron stars, see, e.g., the texts by Shapiro and Teukolsky (1983), Glendenning (1996) and Haensel *et al.* (2007), as well as references therein.

might form in the early Universe and slowly evolve to the point of onset of collapse to black holes. Such stars might be the progenitors of supermassive black holes observed in the centers of most galaxies. Rotating white dwarfs are abundant in nature. While even massive white dwarfs in equilibrium are only marginally relativistic, they will undergo collapse once they accrete too much mass and exceed the Chandrasekhar limit.[5] Numerical relativity simulations are necessary to follow such collapse and determine the final outcome. Rotating boson stars, constructed from complex scalar fields, are also relativistic in nature and numerical relativity is again essential for analyzing their dynamical behavior.

Simulating rotating, relativistic stars using numerical relativity is a new enterprise. Its development is due partly to the construction of new formulations of numerical relativity that are able to integrate the multidimensional Einstein field equations in a stable fashion (see Chapter 11), and partly to the advent of parallel computer hardware with sufficient size and power to handle large-scale, multidimensional problems.

In this chapter we first will discuss the numerical construction of rotating, relativistic fluid stars in stationary equilibrium. Such models, interesting in their own right, comprise the initial data for many evolution calculations. We will then discuss some representative simulations of fluid stars in numerical relativity that probe the stability of these configurations. These simulations are capable of tracking the rapid dynamical and, in some cases, the slow secular evolution of rotating fluid stars, both in axisymmetry $(2 + 1)$ and in three spatial dimensions $(3 + 1)$.

14.1 Initial data: equilibrium models

Up until the 1970s, models of rotating stars in general relativity were restricted to slow rotation. Slowly rotating relativistic stars can be constructed by perturbing nonrotating, spherically symmetric models.[6] Such a treatment involves the integration of ordinary differential equations, which is straightforward. Since the 1970s, *rapidly* rotating relativistic stars have been constructed numerically by many authors.[7] We will focus on the method of Komatsu *et al.* (1989a,b, hereafter KEH), which allows for the construction of both uniformly and differentially rotating stellar configurations.

14.1.1 Field equations

We consider rotating equilibrium models that are stationary and axisymmetric. The metric can then be written, following KEH, in the form

$$ds^2 = -e^{\gamma + \rho} dt^2 + e^{2\sigma}(dr^2 + r^2 d\theta^2) + e^{\gamma - \rho} r^2 \sin^2 \theta (d\phi - \omega dt)^2, \qquad (14.1)$$

---

[5] The limit is $1.4 M_\odot$, the maximum mass of a nonrotating star supported by cold, degenerate electrons. For his derivation of this limit in 1931, Chandrasekhar shared the Nobel Prize in 1983.

[6] Hartle (1967); Hartle and Thorne (1968).

[7] See, e.g., Wilson (1972a); Bonazzola and Schneider (1974); Butterworth and Ipser (1975, 1976); Friedman *et al.* (1986), Komatsu *et al.* (1989a,b), Cook *et al.* (1992, 1994a,b), and Salgado *et al.* (1994a,b). For a comparison of methods, see Stergioulas and Friedman (1995); Nozawa *et al.* (1998).

where the metric potentials ρ, γ, ω, and σ are functions of r and θ only. We shall take the matter source to be described by a stress-energy tensor T^{ab}, which, for the moment, we shall leave unspecified, to allow for generalization to alternative matter sources.[8] The stellar model satisfies the Einstein field equations $G_{ab} = 8\pi T_{ab}$. These can be arranged to yield elliptic equations to determine ρ, γ and ω:

$$\nabla^2[\rho e^{\gamma/2}] = S_\rho(r, \mu), \tag{14.2}$$

$$\left(\nabla^2 + \frac{1}{r}\partial_r - \frac{\mu}{r^2}\partial_\mu \right)[\gamma e^{\gamma/2}] = S_\gamma(r, \mu), \tag{14.3}$$

$$\left(\nabla^2 + \frac{2}{r}\partial_r - \frac{2\mu}{r^2}\partial_\mu \right)[\omega e^{(\gamma-2\rho)/2}] = S_\omega(r, \mu), \tag{14.4}$$

where ∇^2 is the flat-space scalar Laplacian in spherical polar coordinates, $\mu = \cos\theta$, and S_ρ, S_γ and S_ω are effective source terms that include the nonlinear field and matter terms. The effective source terms are given in Appendix G. The fourth field equation for σ is slightly more complicated and is also listed in Appendix G.

The three elliptical field equations (14.2)–(14.4) can be solved by an integral Green function approach, as shown by KEH. Among the attractive features of this approach is that, for large r, the asymptotic flatness conditions $\rho \propto 1/r$, $\gamma \propto 1/r^2$ and $\omega \propto 1/r^3$ are imposed automatically. The integral equations for these functions are given in Appendix G. Auxiliary relations associated with the matter are discussed below.

14.1.2 Fluid stars

For perfect fluid stars the stress-energy tensor is given by equation (5.4),

$$T^{ab} = \rho_0 h u^a u^b + P g^{ab}. \tag{14.5}$$

The coordinate components of the 4-velocity of the matter can be written as

$$u^a = u^t[1, 0, 0, \Omega], \qquad u^t = \frac{e^{-(\rho+\gamma)/2}}{(1-v^2)^{1/2}}, \tag{14.6}$$

where Ω is the angular velocity of the matter as measured at infinity, and v is the proper velocity of the matter with respect to a normal observer, often referred to as a "zero angular momentum observer" or ZAMO[9] in the literature dealing with rotating equilibria,

$$v = (\Omega - \omega)r \sin\theta \, e^{-\rho}. \tag{14.7}$$

> **Exercise 14.1** Derive equation (14.7).
> **Hint:** Compute $\gamma_v \equiv (1 - v^2)^{-1/2}$ from $\gamma_v = -u^a n_a$.

---

[8] Shapiro and Teukolsky (1993a,b).
[9] Bardeen (1973).

The condition $\nabla_a T^a{}_b = 0$ yields the equation of hydrostatic equilibrium, which can be written in differential form as

$$dP - \rho_0 h[d \ln u^t - u^t u_\phi d\Omega] = 0. \tag{14.8}$$

Exercise 14.2 Derive equation (14.8).

The final relation required to specify an equilibrium configuration is the rotation law. For barotropic equations of state where P and h depend only on ρ_0, the integrability condition on equation (14.8) requires either that the rotation be uniform, with $d\Omega = 0$, or that

$$u^t u_\phi = F(\Omega) \tag{14.9}$$

for some function F of Ω. In the latter case the configuration is characterized by differential rotation. A simple choice adopted by KEH is

$$F(\Omega) = A^2(\Omega_c - \Omega), \tag{14.10}$$

where Ω_c is the angular velocity at the center of the configuration and A is a positive constant with dimensions of length. Combining equations (14.9) and (14.10) yields

$$A^2(\Omega_c - \Omega) = \frac{(\Omega - \omega)r^2 \sin^2\theta e^{2(\beta - \nu)}}{1 - (\Omega - \omega)^2 r^2 \sin^2\theta e^{2(\beta - \nu)}}, \tag{14.11}$$

where, following convention, we have introduced

$$\nu \equiv \frac{\gamma + \rho}{2} \quad \text{and} \quad \beta \equiv \frac{\gamma - \rho}{2}. \tag{14.12}$$

Equation (14.11) shows that the angular velocity on the rotation axis ($\theta = 0$) is equal to Ω_c. As $A \to \infty$, we have $\Omega \to \Omega_c$, hence for large A the rotation law approaches rigid rotation. Exercise 14.3 further illustrates how A parameterizes the characteristic length scale over which Ω changes.

Exercise 14.3 Show that in the Newtonian limit, rotation law (14.11) reduces to

$$\Omega/\Omega_c = 1/\left(1 + \varpi^2/A^2\right), \tag{14.13}$$

where $\varpi = r \sin\theta$ is the usual cylindrical radial coordinate.

Using the thermodynamic relation $dh = dP/\rho_0$ together with equation (14.9), allows us to integrate the equation of hydrostatic evolution (14.8) to obtain

$$\ln \frac{h}{u^t} = -\int d\Omega F(\Omega) + constant. \tag{14.14}$$

Adopting the rotation law (14.11) then gives

$$\ln \frac{h}{u^t} = \frac{1}{2} A^2 (\Omega - \Omega_c)^2 + constant. \tag{14.15}$$

For a polytropic equation of state (EOS) $P = K\rho_0^{(1+1/n)}$, as in equation (1.86),[10] equations (14.6) and (14.15) yield

$$\ln[1 + (1+n)K\rho_0^{1/n}] + \frac{1}{2}\ln(1 - v^2) + v - \frac{1}{2}A^2(\Omega - \Omega_c)^2 = constant. \quad (14.16)$$

Note that when using a polytropic EOS, it is always possible to scale out the constant K. We already noted this in Chapter 1.3 (see equation 1.87), but here we extend the analysis to nonspherical, rotating polytropes. In gravitational units $K^{n/2}$ has units of length, which leads to the following set of nondimensional quantities, denoted by a bar:

$$\bar{r} \equiv K^{-n/2}r, \quad \bar{t} \equiv K^{-n/2}t, \quad \bar{\omega} \equiv K^{n/2}\omega, \quad \bar{\Omega} \equiv K^{n/2}\Omega,$$

$$\bar{\rho}_0 \equiv K^n\rho_0, \quad \bar{P} \equiv K^n P, \quad \bar{M} \equiv K^{-n/2}M, \quad \bar{J} \equiv K^{-n}J, \quad (14.17)$$

and so forth. One can thus set $K = 1$ in numerical integrations and either use the above relations to scale the results to more physical values of K, or express answers in terms of nondimensional ratios (e.g., R/M, $M^2\rho_0$, $M\Omega$, J/M^2, etc.).[11]

The iterative algorithm of KEH to solve the coupled equations involves a further rescaling with respect to the equatorial radius $\bar{r}_e = K^{-n/2}r_e$. This rescaling simplifies the numerical implementation, since the unknown location of the stellar surface in the equatorial direction can now be set to unity. We can then parametrize the stellar rotation rate by setting the polar radius $\bar{r}_p/\bar{r}_e$ to a value equal or less than unity: for $\bar{r}_p/\bar{r}_e = 1$ the star is spherical and nonrotating, and decreasing this ratio (for oblate configurations) will increase the angular velocity $\bar{\Omega}$. Once all quantities have been expressed in terms of $\bar{r}_e$, this quantity appears as an eigenvalue in equation (14.16), together with Ω_c and the constant on the right-hand side. A self-consistent solution of Einstein's field equations coupled to the matter equation (14.16) then involves determining these three eigenvalues. This can be accomplished by evaluating equation (14.16), given a "guess" for the gravitational fields, at three different locations in the star, namely at the pole, on the equator, and at the stellar center, and then iterating until convergence. We will describe a similar algorithm for the construction of binary neutron star initial data in more detail in Chapter 15.2.

Diagnostics

Once the coupled system of equilibrium equations is solved self-consistently, a useful set of physical diagnostics can be computed for each configuration. For equilibrium configurations, the ADM mass must be equal to the Komar mass (see Chapter 3.5). We may therefore define the total mass-energy $M = M_{\text{ADM}}$ of these spacetimes as

$$M = M_{\text{K}} = -\int (2T^a{}_b - \delta^a{}_b T^c{}_c)\xi^b_{(t)}d^3\Sigma_a = \int (-2T^t{}_t + T^c{}_c)\sqrt{\gamma}\alpha d^3x \quad (14.18)$$

---

[10] Note that KEH adopt a different definition of a polytropic EOS: $P = K\rho^{*(1+1/n)}$, where $\rho^* = \rho_0(1 + \epsilon)$. Other authors (e.g., Cook *et al.* 1992, 1994a,b), use equation (5.18). As a result, equation (14.16) is modified for KEH.

[11] See exercises 14.4, 14.5 and 14.6 for examples.

(see exercise 3.31). Here $d^3\Sigma_a$ is defined below equation (3.124), and $\xi^b_{(t)}$ is the time Killing vector. The total rest-mass M_0 is given by equation (3.127),

$$M_0 = \int \rho_0 u^a d^3\Sigma_a = \int \rho_0 \sqrt{\gamma}\alpha u^t d^3 x. \qquad (14.19)$$

The total proper mass of the system M_p is defined as the rest-mass energy M_0 plus internal energy U of the star, i.e., the total energy stored in the configuration excluding gravitational potential and rotational energy,

$$M_p = M_0 + U, \qquad (14.20)$$

where

$$U = \int \rho_0 \epsilon u^a d^3\Sigma_a = \int \rho_0 \epsilon \sqrt{\gamma}\alpha u^t d^3 x. \qquad (14.21)$$

The total angular momentum of the system J is given by

$$J = \int T^a{}_b \xi^b_{(\phi)} d^3\Sigma_a = \int T^t{}_\phi \sqrt{\gamma}\alpha d^3 x, \qquad (14.22)$$

where $\xi^b_{(\phi)}$ is the angular Killing vector (see exercise 3.31). The total rotational kinetic energy of the system T is defined by

$$T = \frac{1}{2}\int \Omega dJ = \frac{1}{2}\int \Omega T^t{}_\phi \sqrt{\gamma}\alpha d^3 x. \qquad (14.23)$$

Finally, we can compute the gravitational potential energy of the star W as

$$W = M - M_p - T. \qquad (14.24)$$

All of the above integral diagnostics are gauge invariant quantities that reduce to their Newtonian counterparts in the weak-field limit.

Maximum masses, spins and stability

Numerical models of rotating stars obtained by solving the equilibrium equations have been constructed by many authors.[12] Equilibrium sequences constructed for a given barotropic EOS, along which either the rest mass or angular momentum is held constant, are particularly useful. Constant rest-mass sequences often can be used to represent quasistationary evolutionary sequences, along which some parameter, like the angular velocity, varies slowly, on a secular time scale. Along a uniformly rotating, equilibrium sequence of constant M_0, changes in M and J are related by[13]

$$dM = \Omega dJ. \qquad (14.25)$$

---

[12] See, e.g., footnote 7 for detailed model calculations in general relativity. For overviews and additional references concerning rotating equilibria and stability, see, e.g., Tassoul (1978) for Newtonian configurations and Shapiro and Teukolsky (1983) and Stergioulas (2003) for post-Newtonian and general relativistic configurations.

[13] See Ostriker and Gunn (1969) for a proof in the Newtonian case and Hartle (1970) for the relativistic case.

This relation provides a useful check on the numerical construction of such a sequence. Equilibrium sequences of constant J are required in order to apply the *turning-point criterion* of Friedman, Ipser and Sorkin[14] to test for quasiradial stability of a uniformly rotating configuration. This criterion states that along *uniformly* rotating sequences of constant J, members of which can be parametrized by their central mass-energy density $\rho_c^* = \rho_{0c}(1 + \epsilon_c)$, those configurations for which $\partial M/\partial \rho_c^* > 0$ are secularly stable against quasiradial perturbations, while those for which $\partial M/\partial \rho_c^* < 0$ are secularly unstable. The turning point criterion applied along such sequences can only identify the point of secular, but not dynamical, instability, since one is comparing neighboring, uniformly rotating configurations with the same angular momentum. Maintaining uniform rotation during perturbations tacitly assumes high viscosity, which acts on a secular time scale. In a dynamical perturbation, the star will preserve circulation, as well as angular momentum, but not uniform rotation. While a secular instability evolves on a dissipative (e.g., viscous) time scale, a dynamical instability evolves on a collapse (free-fall) time scale, $\tau_{ff} \sim 1/\sqrt{\rho^*}$, which is much shorter. It is thus possible that a secularly unstable star may be dynamically stable: for sufficiently small viscosity, the perturbed star may change to a differentially rotating, dynamically stable configuration. Ultimately, the presence of viscosity may bring the star back into rigid rotation, driving the star to an unstable state. Friedman, Ipser and Sorkin showed that along a sequence of uniformly rotating stars, a secular instability always occurs *before* a dynamical instability, implying that all secularly stable stars are also dynamically stable.

For spherical star sequences, the points of onset of secular and dynamical instability coincide, since for a nonrotating star a radial perturbation conserves both circulation and uniform rotation, and is located at the turning point on the M *vs.* ρ_c^* equilibrium curve. This result suggests that for uniformly rotating stars for which the rotational kinetic energy T is typically a small fraction of the gravitational binding energy $|W|$, the onset of dynamical instability is close to the onset of secular instability. Establishing the actual point of onset of dynamical instability for rotating stars is more complicated. Formalisms have been developed[15] to identify points of dynamical instability to axisymmetric perturbations along sequences of rotating stars. In such formalisms, however, a complicated functional for a set of trial functions has to be evaluated. Probably because of the complexity of this method, explicit calculations have never been performed. In practice, to identify the point of onset of a dynamical instability along an equilibrium sequence of uniformly or differentially rotating stars, and then to track the resulting evolution and determine the final fate of an unstable configuration, a full numerical simulation employing the equilibrium model as initial data is required. That is the role of numerical relativity (see Section 14.2).

Nonrotating polytropes with $n = 3$ or $\Gamma = 1 + 1/n = 4/3$ are marginally stable to radial perturbations in Newtonian gravitation, but dynamically unstable in general relativity. The

---

[14] Friedman *et al.* (1988).
[15] Chandrasekhar and Friedman (1972a,b,c); Schutz (1972).

critical Γ for radial stability for nonrotating stars is raised above $4/3$ in general relativity,[16] $\Gamma_{\text{crit}} = 4/3 + 1.125(M/R)$. Stars with $\Gamma > \Gamma_{\text{crit}}$ are stable, while those with $\Gamma < \Gamma_{\text{crit}}$ are unstable. Post-Newtonian analysis shows that unstable stars with Γ near $4/3$ can be stabilized by rotation, which lowers the critical value of Γ. Alternatively, there is a critical compaction $(M/R)_{\text{crit}}$ below which a rotating star with $n = 3$ is radially stable and above which it is unstable. The critical compaction depends on $T/|W|$, the ratio of rotational kinetic to gravitational binding energy. Determining the critical compaction for uniformly rotating stars with $n \approx 3$ requires an analysis correct to second post-Newtonian order.[17] Uniformly rotating stars at the mass-shedding limit (defined below) with n near 3 are useful to model rapidly rotating configurations supported either by the pressure of relativistic degenerate fermions (e.g., massive white dwarfs or cores of massive, evolved stars) or by thermal radiation pressure (e.g., very massive and supermassive stars).[18]

The maximum mass of a relativistic star, like a neutron star, is of great astrophysical interest. Configurations exceeding the maximum mass limit are likely to collapse to form black holes. For spherical (static) stars constructed from the same barotropic EOS, the maximum mass is located at the turning point along the M vs. ρ_c^* equilibrium curve and is thus marginally stable to quasiradial perturbations. Rotation can support stars with higher mass than the maximum static limit. Thus, uniform rotation can support these *supramassive* stars, but Cook *et al.* (1992, 1994a,b, hereafter CST) show that supramassive stars have masses which are typically at most $\lesssim 20\%$ larger than the static limit for the same EOS. In Table 14.1 we compare maximum mass models for nonrotating and uniformly rotating polytropes.[19] Beyond a critical spin rate, the gravitational attraction is insufficient to keep matter in uniform rotation from flying off the surface. At the so-called *mass-shedding limit*, matter at the equator has no outward support from pressure but is instead supported exclusively by centrifugal forces and therefore follows a circular geodesic. CST find that for a given EOS, the maximum Ω configuration (i.e., the configuration at the mass-shedding limit) does not coincide exactly with the maximum mass model, although they are quite close numerically.

Exercise 14.4   Consider modeling the EOS of a neutron star by a degenerate, nonrelativistic, neutron gas. Such a gas is described by a polytropic EOS with $n = 3/2$ and

$$K = \frac{3^{2/3}\pi^{4/3}}{5}\frac{\hbar^2}{m_n^{8/3}}, \tag{14.26}$$

where m_n is the neutron rest-mass.[20]

---

[16] Chandrasekhar (1964a,b); Feynmann, unpublished, as quoted in Fowler (1964).

[17] Zeldovich and Novikov (1971); Baumgarte and Shapiro (1999).

[18] See Baumgarte and Shapiro (1999); Shibata (2004) for numerical models.

[19] See, e.g., Cook *et al.* (1994a) for uniformly rotating models, including maximum-mass models, constructed for 14 realistic nuclear matter EOSs.

[20] See Shapiro and Teukolsky (1983), Section 2.3, for a derivation.

Table 14.1 Maximum mass models for nonrotating and uniformly rotating polytropes. For all stars we list the the polytropic index n, the total mass-energy $\bar{M}$, the rest-mass $\bar{M}_0$ and the central energy density $\bar{\rho}_c^*$. For nonrotating stars we also list the areal radius $\bar{R}$, while for rotating stars we give the equatorial areal radius $\bar{R}_e$, the eccentricity e, the angular velocity measured at infinity $\bar{\Omega}$, the total angular momentum $\bar{J}$ and the ratio of the rotational kinetic energy to the gravitational binding energy, $T/|W|$. All quantities are expressed in nondimensional units as defined in equation (14.17) (with $G = c = 1$). Here the notation $5.80(-3)$, for example, means 5.80×10^{-3}. [After Cook et al. (1994a).]

n	Nonrotating				Rotating							
	$\bar{M}$	$\bar{M}_0$	$\bar{R}$	$\bar{\rho}_c^*$	$\bar{M}$	$\bar{M}_0$	$\bar{R}_e$	e	$\bar{\rho}_c^*$	$\bar{\Omega}$	$\bar{J}$	$T/|W|$
0.5	0.125	0.151	0.395	1.29	0.153	0.182	0.536	0.784	1.04	0.976	0.0172	0.147
1.0	0.164	0.180	0.763	0.420	0.188	0.207	1.09	0.773	0.345	0.378	0.0202	0.0835
1.5	0.264	0.276	1.97	0.0718	0.290	0.304	2.88	0.760	0.0610	0.110	0.0387	0.0475
2.0	0.515	0.523	6.94	5.80(−3)	0.549	0.558	10.3	0.752	5.05(−3)	0.0224	0.120	0.0273
2.5	1.24	1.25	41.0	1.26(−4)	1.29	1.30	61.3	0.749	1.13(−4)	2.37(−3)	0.684	0.0157
2.9	3.23	3.23	620	1.54(−7)	3.32	3.32	947	0.749	1.38(−7)	6.36(−5)	7.82	0.0101

(a) Show that, in geometrized units, $K = 7.327\,\text{km}^{4/3}$. Note that $K^{n/2}$ has units of length, consistent with the scaling relations (14.17).

(b) Consult Table 14.1 and use equation (14.17) to find the maximum mass of nonrotating and uniformly rotating stars governed by this EOS. Given that many neutron stars are observed to have a mass close to $1.4M_{\odot}$, is this EOS realistic?

Exercise 14.5 The polytropic index n parametrizes the *stiffness* of the polytropic EOS. Smaller values of n (and hence larger values of $\Gamma = 1 + 1/n$) describe stiffer EOSs, while larger values of n (smaller Γ) describe softer EOSs. Stars governed by soft EOS are centrally condensed, i.e., most of the star's mass is concentrated in a small, high-density core surrounded by an extended, low-density envelope. Stiff EOSs, on the other hand, lead to stars that have a more uniform density profile.

(a) Compute the compaction M/R for the nonrotating maximum mass models listed in Table 14.1. Use your results to justify using a stiff polytropic EOS with $n \simeq 1$ to model typical neutron stars with masses $M \approx 1.4M_{\odot}$ and radii $R \approx 15$ km.

(b) Compute the fractional difference between M and M_0 for the nonrotating models of Table 14.1. Explain its dependence on n in terms of the results for M/R in part (a).

(c) Compute the fractional increase in the rest-mass M_0 between the nonrotating and uniformly rotating maximum mass models listed in Table 14.1. Explain its dependence on n in terms of the central concentration of these stars.

Exercise 14.6 PSR J1748−2446ad, with a spin frequency of $f = 716$ Hz, is the most rapidly spinning pulsar known to date.[21]

(a) Assume that the mass of this pulsar is at least $1.4M_{\odot}$ to find a lower limit on the dimensionless product $M\Omega$ for this pulsar.

(b) Now survey Table 14.1, assuming that the angular velocities Ω of the maximum mass models listed there are very close to the maximum spin models, to find which values of n can support a star like PSR J1748−2446ad.

By contrast with uniform rotation, differential rotation, can support *hypermassive stars*, i.e., equilibrium stars which exceed both the nonrotating static and uniformly rotating supramassive star mass limits. As an example, consider the rotation law given by equation (14.10) with $\hat{A} \equiv A/R_e$, where R_e is the equatorial coordinate radius, and apply it to an $n = 1$ polytrope.[22] Figure 14.1 shows the maximum M_0 *vs.* ρ_c^* relation along sequences of constant $\hat{A}$. It is clear from the figure that even modest differential rotation can support equilibrium stars with masses $\gtrsim 50\%$ larger than the static and supramassive limits.[23] Hypermassive neutron stars can be formed in nature, by, e.g., merging binary neutron stars or by stellar core collapse followed by accretion fall-back.

Exact criteria do not exist for determining the quasiradial stability of differentially rotating configurations, either to dynamical or secular perturbations. Such stability must be determined by numerical simulations. In Section 14.2 we discuss such simulations, which

---

[21] See Hessels *et al.* (2006).

[22] Baumgarte *et al.* (2000).

[23] For most other polytropic indices, as well as for realistic nuclear EOSs, the increases in maximum mass are typically somewhat smaller; see Lyford *et al.* (2003); Morrison *et al.* (2004).

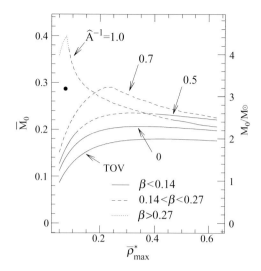

Figure 14.1 Maximum rest-mass configurations *vs.* maximum mass-energy density for differentially rotating $n = 1$ sequences of constant $\hat{A}^{-1}$. Values of $\beta = T/|W|$ for the models are indicated. The mass–density curve for static TOV equilibrium stars is shown for comparison. The supramassive limit for uniformaly rotating stars corresponds to the curve $\hat{A}^{-1} = 0$. Masses are given in nondimensional units on the left-hand side ($K = 1$) and in solar masses on the right-hand side; the later are calculated by assigning the maximum rest mass for nonrotating stars to $2M_\odot$. The dot marks one particular hypermassive configuration whose stability properties we will study below. [From Baumgarte *et al.* (2000).]

reveal, for example, that the hypermassive configuration marked by a dot in Figure 14.1 is dynamically stable. On the other hand, the presence of viscosity or magnetic fields will redistribute the angular momentum in a differentially rotating star on a secular time scale, presumably driving its core toward uniform rotation while depositing the excess angular momentum in the outermost layers. But uniform rotation alone cannot support such a core in equilibrium, if it is sufficiently massive. As a result, most, if not all, hypermassive stars are transient objects: configurations that are dynamically stable initially will evolve on a secular time scale and may ultimately undergo catastrophic collapse. Relativistic simulations that demonstrate this behavior are described in Section 14.2.

In addition to quasiradial instabilities, rotating stars are also subject to nonaxisymmetric instabilities. An exact treatment of these instabilities exists only for *incompressible* equilibrium fluids in Newtonian gravity.[24] For these configurations, global rotational instabilities arise from nonradial toroidal modes $e^{im\varphi}$ ($m = \pm1, \pm2, \ldots$) when $\beta \equiv T/|W|$ exceeds a certain critical value. Here φ is the azimuthal coordinate. In the following we will focus on the $m = \pm2$ bar mode, since it is the fastest growing mode when the rotation is sufficiently rapid.[25]

---

[24] See, e.g., Chandrasekhar (1969); Tassoul (1978); Shapiro and Teukolsky (1983).

[25] For a discussion of the *r*-mode instability, which may be important for slowly rotating neutron stars, see Stergioulas (2003) for review and references.

There exists two different mechanisms and corresponding timescales for bar-mode instabilities. Uniformly rotating, incompressible stars in Newtonian theory are *secularly* unstable to bar-mode formation when $\beta \geq \beta_s \geq 0.1375$. However, this instability can only grow in the presence of some dissipative mechanism, like viscosity or gravitational radiation reaction, and the growth time is determined by the dissipation time scale. Interestingly, for either viscosity or gravitational radiation, the point of onset of the bar-mode instability coincides for Newtonian stars. By contrast, a *dynamical* instability to bar-mode formation requires large spin rates and sets in when $\beta \geq \beta_d \geq 0.2738$. This instability is independent of any dissipation mechanism, and the growth time is determined by the hydrodynamical (collapse) time scale of the system.

In the case of *compressible* Newtonian stars, the secular bar-mode instability for both uniform and differential rotation has been analyzed numerically within linear perturbation theory by means of a variational principle and trial functions, and by other approximate means.[26] For uniformly rotating polytropes the $m = 2$ bar-mode instability is again found to set in at $\beta_s \simeq 0.14$.[27] However, this mode is reached only when the polytropic index of the star satisfies $n \leq 0.808$.[28] Stars with larger n (i.e., soft EOSs) are too centrally condensed to support high enough spin in uniform rotation without undergoing mass-shedding at the equator. This constraint does not apply to differentially rotating stars, which can support significantly more rotational energy in equilibrium, even when the degree of differential rotation is only moderate. The critical value for the onset of the secular $m = 2$ bar mode in Newtonian theory is again $\beta_s \simeq 0.14$ for a wide range of angular momentum distributions and barotropic equations of state[29] although for very strongly differentially rotating stars the critical value can be as small as $\beta_s < 0.1$.[30]

Similar approximate formalisms have also been applied to analyze the secular bar-mode instability in post-Newtonian theory[31] and in full general relativity.[32] For relativistic stars, the critical value of β_s depends on the compaction M/R of the star, the rotation law and the dissipative mechanism. The gravitational-radiation driven instability sets in for smaller rotation rates than in Newtonian theory, i.e., it is triggered for values of $\beta_s < 0.14$ as the compaction increases. By contrast, viscosity drives the instability to higher rotation rates $\beta_s > 0.14$ as the configurations become more compact.

Determining the onset of the dynamical bar-mode instability, as well as the subsequent evolution of an unstable star, generally requires a numerical simulation. Simulations performed in Newtonian theory[33] have shown that β_d depends only very weakly on the stiffness of the EOS. Once a bar has developed, the formation of spiral arms plays an

---

[26] Lynden-Bell and Ostriker (1967); Ostriker and Bodenheimer (1973); Friedman and Schutz (1975); Bardeen *et al.* (1977); Friedman and Schutz (1978a,b); Ipser and Lindblom (1989).

[27] See, e.g., Managan (1985); Imamura *et al.* (1985); Ipser and Lindblom (1990, 1991); Lai *et al.* (1993a).

[28] James (1964).

[29] See, e.g., Ostriker and Bodenheimer (1973); Bardeen *et al.* (1977); Tassoul (1978).

[30] Imamura *et al.* (1995).

[31] Cutler and Lindblom (1992); Shapiro and Zane (1998).

[32] Bonazzola *et al.* (1996); Stergioulas and Friedman (1998).

[33] See, e.g., Tohline *et al.* (1985); Smith *et al.* (1996); Pickett *et al.* (1996); New *et al.* (2000) and references therein.

important role in redistributing the angular momentum and forming a core-halo structure. It has been shown that, similar to the onset of secular instability, β_d can be smaller for stars with a higher degree of differential rotation.[34]

Until recently, almost nothing has been known about the dynamical bar-mode instability in relativistic gravitation. The reason was that stable numerical codes capable of performing reliable hydrodynamic simulations in three spatial dimensions plus time in full general relativity were not available. In Section 14.2 we shall highlight results from a few simulations which employ such codes to probe the dynamical stability of relativistic rotating stars for axisymmetric and nonaxisymmetric modes.

There are numerous evolutionary paths which may lead to the formation of rapidly rotating, relativistic stars, like neutron stars, with large values of β.

> **Exercise 14.7** Show that $\beta \sim 1/R$ during collapse, assuming conservation of J.

Exercise 14.7 suggests that during supernova collapse, as the core contracts from a radius of ~ 1000 km to ~ 10 km, β increases by about two orders of magnitude. Thus, even moderately rapidly rotating progenitor stars may yield rapidly rotating neutron stars that may reach the onset of dynamical instability.[35] These configurations will be differentially rotating at birth, even if they were uniformaly rotating prior to collapse, assuming that the collapse preserves angular momentum on cylindrical shells. Similar arguments hold for accretion-induced collapse of white dwarfs to neutron stars and for the merger of binary white dwarfs to neutron stars. In fact, X-ray and radio observations of supernova remnants have identified several young, isolated, rapidly rotating pulsars, suggesting that these stars may have been born with periods of several milliseconds.[36] These neutron stars could be the collapsed remnants of rapidly rotating progenitors. By the time they are observed as pulsars, they are presumably rotating uniformly, given that they are such good clocks.

Rapidly rotating neutron stars naturally arise in the merger of binary neutron stars. As we will discuss in Chapter 16, simulations reveal that such merger remnants could be hypermassive neutron stars. Hypermassive stars can also form during core collapse in massive stars. These configurations are transient objects that will likely undergo *delayed collapse*, producing a delayed burst of gravitational waves and, possibly, gamma-rays. Their fate may thus be important for the detection and identification of gravitational wave and gamma-ray burst sources.

14.1.3 Collisionless clusters

The field equations (14.2)–(14.4) and (G.8) can also be used to construct rotating, axisymmetric equilibrium configurations of collisionless matter.[37] The stress-energy tensor for the matter in this case is determined by the phase-space distribution function f, which

---

[34] Tohline and Hachisu (1990); Pickett *et al.* (1996); Shibata *et al.* (2003a).
[35] Rampp *et al.* (1998).
[36] Marshall *et al.* (1998); Kaspi *et al.* (1998).
[37] Shapiro and Teukolsky (1993a,b).

in turn is governed by the relativistic Vlasov equation.[38] The stress-energy tensor may be written in the form

$$T^{\hat{a}\hat{b}} = \int f p^{\hat{a}} p^{\hat{b}} \frac{d^3\hat{p}}{p^{\hat{t}}}, \tag{14.27}$$

where $p^{\hat{a}}$ are the orthonormal components of the particle 4-momentum (see equations 5.201, 5.213 and 5.215). To solve the field equations we need to evaluate the components of the stress-energy tensor in the ZAMO frame (see Appendix G). The simplest phase-space distribution functions that can generate nonspherical, axisymmetric equilibria are functions solely of the conserved particle energy E and conserved angular momentum J_z about the symmetry axis. Because E and J_z are integrals of the motion, choosing a distribution function of the form $f = f(E, J_z)$ guarantees that we have a solution of the Vlasov equation, provided the metric is determined self-consistently. No further dynamical equations need to be solved for the matter. By contrast, for equilibrium fluid systems, such as rotating stars, we need to integrate the equation of hydrostatic equilibrium, as discussed above.

For axisymmetric systems described by $f = f(E, J_z)$, the quantities E and J_z are the two constants of motion associated with the Killing vectors $\xi_{(t)}^a = (\partial/\partial t)^a$ and $\xi_{(\phi)}^a = (\partial/\partial\phi)^a$:

$$E \equiv -g_{ab} p^a \xi_{(t)}^b = e^\nu p^{\hat{t}} + \omega e^\beta r \sin\theta p^{\hat{\phi}}, \tag{14.28}$$

$$J_z \equiv g_{ab} p^a \xi_{(\phi)}^b = e^\beta r \sin\theta \, p^{\hat{\phi}}. \tag{14.29}$$

In evaluating the integrals (14.27) to obtain the matter source terms appearing in the field equations, we can write

$$p^{\hat{t}} = [(p^{\hat{r}})^2 + (p^{\hat{\theta}})^2 + (p^{\hat{\phi}})^2 + m^2]^{1/2} = [(p^\perp)^2 + (p^{\hat{\phi}})^2 + m^2]^{1/2}, \tag{14.30}$$

where

$$(p^\perp)^2 = (p^{\hat{r}})^2 + (p^{\hat{\theta}})^2. \tag{14.31}$$

The particle momentum distribution is isotropic in a plane containing the symmetry axis, perpendicular to the ϕ-direction. In other words,

$$p^{\hat{r}} = p^\perp \cos\psi, \qquad p^{\hat{\theta}} = p^\perp \sin\psi, \tag{14.32}$$

and

$$d^3\hat{p} = dp^{\hat{r}} dp^{\hat{\theta}} dp^{\hat{\phi}} = p^\perp dp^\perp dp^{\hat{\phi}} d\psi, \tag{14.33}$$

and f is now independent of the angle ψ in the symmetry plane. A detailed prescription for performing the necessary quadratures (14.27) for the source terms can be found in Shapiro and Teukolsky (1993a,b).

---

[38] See Chapter 5.3.

By suitable choice of the distribution in J_z, one can construct models that are either prolate or oblate. A depletion in the J_z-distribution produces a prolate configuration, while an enhancement produces an oblate one. The easiest way to implement this procedure is to write

$$f(E, J_z) = g(E)h(J_z) \qquad (14.34)$$

and vary h appropriately. When h is constant, we get spherical models with isotropic velocity distributions. If h depends only on the magnitude of J_z, the models are nonspherical with no net angular momentum. As h is varied, one can build entire sequences of equilibria with varying degrees of rotation. Bound systems of finite extent require that $g(E) = 0$ for $E > E_{\max}$, where $E_{\max} < m_0$ is the maximum particle energy and m_0 the particle rest mass. The construction of rotating clusters using this prescription has already been implemented in Chapter 10.4, where we focused on toroidal clusters and their collapse to black holes.

14.2 Evolution: instabilities and collapse

As suggested above, numerical relativity has proven to be an invaluable tool in probing the stability of relativistic rotating equilibria, particularly where no theorems exist to identify unstable configurations easily or unambiguously. Numerical relativity can also follow the evolution of unstable configurations and ascertain their final fate. We shall now illustrate these applications of numerical relativity by summarizing a few representative simulations.

14.2.1 Quasiradial stability and collapse

Uniformly rotating stars with stiff EOSs

Consider the point of onset of quasiradial dynamical stability along a sequence of uniformly rotating, fluid stars at the mass-shedding limit.[39] Adopt a polytropic EOS with index $n = 1$, $\Gamma = 2$, so as to crudely mimic a stiff nuclear EOS appropriate for a neutron star. Such a sequence is shown by the solid curve in Figure 14.2, together with a TOV sequence of nonrotating, spherical stars denoted by the dotted curve for comparison. We are particularly interested in the supramassive configurations residing along the solid curve.

The stability theorem of Friedman, Ipser and Sorkin can be applied to show that the onset of secular instability for these uniformly rotating models resides very close to the central rest-mass density $\rho_{0,c} = \rho_{\rm crit}$ at which M assumes its maximum value along the sequence.[40] According to our discussion in Section 14.1, stars with $\rho_{0,c} > \rho_{\rm crit}$ are likely candidates for dynamical instability, but there are no theorems to prove this expectation. The open circles on the curve identify 5 specific configurations chosen as initial data for hydrodynamical evolution. In some of the simulations, the star is subjected to an initial perturbation by

---

[39] Shibata *et al.* (2000a).
[40] See Figure 4 of Cook *et al.* (1994b).

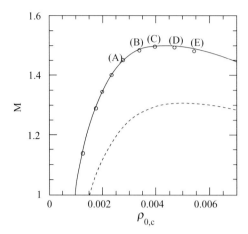

Figure 14.2 Total mass as a function of central rest-mass density for uniformly rotating polytropes with $\Gamma = 2$ and polytropic parameter $K = 200/\pi$. The solid line shows the exact sequence at the mass-shedding limit; the dashed line gives the TOV sequence of nonrotating, spherical stars. The open circles indicate configurations at the mass-shedding limit constructed from the conformal flatness approximation, which were used for the simulation. The supramassive configurations chosen as initial data for dynamical evolution calculations are labeled by (A)–(E). [From Shibata *et al.* (2000a).]

depleting the initial pressure everywhere a small, fixed, fractional amount below its equilibrium value to help induce an instability. In such cases, the Hamiltonian and momentum constraint equations are re-solved to restore valid initial data. The response of each star to such a perturbation then establishes where along the sequence a dynamical instability sets in.

The simulations are performed with a relativistic hydrodynamics code in $3 + 1$ dimensions.[41] The field solver is based on second-order finite-differencing of the BSSN equations, which were presented in Chapter 11.5. The adiabatic hydrodynamic equations are integrated by a Wilson scheme as described in Chapter 5.2.1, with the advection terms differenced by the second-order algorithm of van Leer (1977) and the use of artificial viscosity to handle shocks. The adiabatic index during the evolution is chosen to be $\Gamma = 2$, the same value that relates the pressure and rest-mass density in the equilibrium polytrope. An "approximate" maximal time slicing condition ($K_i{}^i \approx 0$) is employed to determine the lapse function α and an "approximate" minimum distortion (AMD) spatial gauge condition ($\tilde{D}_i(\partial_t \tilde{\gamma}^{ij}) \approx 0$) is used to determine the shift vector β^j.[42] Reflection symmetry about the equator at $z = 0$ and $\pi -$ rotation symmetry about the z-axis is assumed. The integrations are performed on a fixed, uniform grid, typically with $153 \times 77 \times 77$ zones in the x–y–z

---

[41] See Shibata (1999a) for details. We will discuss applications that employ more advanced codes later in this chapter and in other chapters.

[42] See Chapter 4; an extra contribution is added to the shift to increase the resolution near the origin when a black hole forms, as discussed at the end of Chapter 4.5.

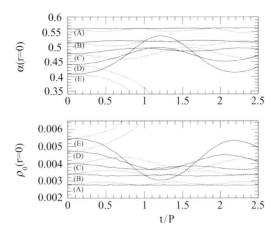

Figure 14.3 The central lapse α and rest-mass density $\rho_{0,c}$ as functions of time during the evolution of stars (A)–(E). Here time t is expressed in units of P, the initial rotation period of the star. The solid lines denote the results without an initial pressure perturbation, while dotted lines show results for a fractional pressure decrease of 1% everywhere. [From Shibata *et al.* (2000a).]

directions, respectively. With this grid the semi-major axes of the stars, along the x- and y-axes, are covered by 40 grid points while the semi-minor (rotation) axis along z is covered by 23 or 24 grid points.

The main results of the simulation are shown in Figure 14.3, which show the evolution of $\rho_{0,c}$ and α_c for each star. When $\rho_{0,c} < \rho_{crit}$ (i.e., stars A, B and C), the rotating stars oscillate independently of the initial pressure perturbation. Hence these stars are stable against gravitational collapse. Small amplitude oscillations at the fundamental quasiradial period arise for these stars even in the absence of initial pressure perturbations. These oscillations are caused by small deviations of the initial data from true equilibrium states, due partly to numerical truncation error and partly to approximations used in solving the equilibrium equations.[43] In the absence of a pressure perturbation ($\Delta P = 0$), star D does not collapse, but it does undergo a much larger amplitude oscillation than those of stars A–C. More significantly, when subjected to a perturbation $\Delta P/P = 1\%$, it collapses to a black hole, unlike stars A–C. The results for star E are similar to, but more pronounced, than those for star D. By studying these results and refining the initial models and the resolution, we conclude that the point of onset of dynamical instability along mass-shedding sequences nearly coincides with the onset of secular instability.

We also learn from these simulations that the unstable stars collapse to rotating black holes within about one rotation period. The appearance of a black hole is determined by

---

[43] The conformal flatness approximation for the spatial metric, $\gamma_{ij} = e^{4\phi}\eta_{ij}$, was adopted to construct these models. This simplification typically provides an excellent approximation to exact axisymmetric equilibrium models, as shown by Cook *et al.* (1996) and suggested by the proximity of the models in Figure 14.2 to the exact equilibrium curve.

finding an apparent horizon. All the rest mass and angular momentum, and almost all of the total mass-energy (apart from a small amount of gravitational radiation), wind up in the black holes. The spin parameters $J/M^2 \sim 0.6 < 1$ of all the stars are thus nearly conserved, and the resulting Kerr black holes have only moderate spin rates. In no case is there any formation of a massive disk or any ejecta around the newly formed Kerr holes, even though the progenitors are rapidly rotating. The qualitative reason for this outcome is clear: because they are slowly spinning, the spacetime outside each of the black holes is not far from Schwarzschild. The innermost stable circular orbit (ISCO) around each black hole is roughly $r_{\mathrm{ISCO}} \sim 5M$ in our adopted spatial gauge, which is comparable to isotropic coordinates in a spherical spacetime. However, the stellar equatorial radius R_e of the equilibrium star is less than $5M$, hence at the outset the star already resides inside the radius which becomes the ISCO of the final black hole. This same argument suggests that the formation of a disk around a black hole requires equilibrium progenitors constructed either with softer EOSs (higher n) if they are uniformly rotating, or with differential rotation, so that the initial equatorial radii will be larger. This suspicion is borne out by other simulations, which we shall summarize below.

During the collapse to a black hole, nonaxisymmetric perturbations do not have enough time to grow appreciably. But again this is not surprising, given that the progenitor is so compact, whereby the star can contract by at most a factor of three before a black hole forms. Hence $T/|W|$, which approximately scales with R_e^{-1}, can increase only by about a factor of three over its initial value of $(T/|W|)_{\mathrm{init}} \sim 0.09$, and only barely reach the critical value of dynamical instability for bar formation, $(T/|W|)_{\mathrm{dyn}} \sim 0.27$. We expect that these results hold for any uniformly rotating star constructed from a moderately stiff EOS, for which the corresponding critical configurations are similarly compact. For progenitors constructed with softer EOSs, or with differential rotation, the initial radii will be larger and $T/|W|$ may be amplified to larger values by the end of collapse. We shall return to the issue of bar instabilities in Section 14.2.2.

Uniformly rotating $n = 3$ polytropes

To study the fate of radially unstable, rotating stars constructed from softer EOSs, consider the adiabatic evolution of an $n = 3$ polytrope that is marginally unstable to quasiradial collapse and rotating uniformly at the mass-shedding limit.[44] Such a configuration can be used to model a spinning supermassive star, or the stellar core of an evolved, massive Population I star, or even a massive, "first generation", zero-metallicity Population III star, all at the onset of collapse.[45] The marginally unstable critical configuration is characterized by the nondimensional ratios $R_e/M \approx 640$, $R_p/M \approx 420$, $J/M^2 \approx 0.97$ and $T/|W| \approx 0.97$,

---

[44] Shibata and Shapiro (2002).

[45] Very massive ($M \gtrsim 10^3 M_\odot$) and supermassive stars supported by thermal radiation pressure can be modeled by $n \approx 3$ polytropes. The cores of young, high-mass Population I stars ($M \gtrsim 20 M_\odot$) are supported by degenerate relativistic electrons, for which the EOS satisfies $\Gamma = 4/3$. These cores can also be modeled by $n = 3$ polytropes. For an overview of the astrophysical and cosmological significance of this calculation, see Liu *et al.* (2007), and references therein.

independently of its mass,[46] and is very centrally condensed. Tracking the collapse of such a configuration is computationally challenging, since the equatorial radius will decrease by a factor of nearly 10^3 by the time a black hole forms, requiring a code with considerable dynamic range. The challenge can be met by (1) restricting the integrations to axisymmetry[47] and (2) moving the outer boundary inward at various stages during the collapse in order to rezone the spatial domain with finer grid spacing ("poor man's AMR").[48] Implementing both of these techniques improves the resolution of the strong-field, central regions where the black hole forms. Because the critical configuration is independent of M, a single calculation may be scaled to stars of *arbitrary* mass.

The simulation is carried out with the same basic $3 + 1$ code used for the previous problems, but now adapted to axisymmetry. The Einstein field equations are solved in Cartesian coordinates with a grid of size $(N, 3, N)$ in (x, y, z), covering a computational domain $0 \leq x, z \leq L$ and $-\Delta y \leq y \leq \Delta y$. Here the grid size N is a constant ($\gg 1$), L the location of the outer boundary, which is moved inwards at discrete moments during the simulation, and Δy is the grid spacing in the y-direction, with an axisymmetric boundary condition imposed at $y = \pm \Delta y$; the spin axis is along z. This way of "axisymmetrizing" a $3 + 1$ Cartesian code is referred to as the "cartoon method" in numerical relativity.[49] One can solve the relativistic hydrodynamics equations in axisymmetry with the same "cartoon" recipe. Instead, somewhat improved accuracy is achieved in this simulation if the equations are recast directly in cylindrical coordinates, using the same Cartesian variables in the expressions and the same grid points. Maximal time slicing and the AMD shift condition are adopted as gauge conditions. The appearance and growth of a black hole is again determined by finding an apparent horizon.

Violations of the constraints and conservation of mass and angular momentum are monitored as numerical accuracy checks during the simulation. Total angular momentum J and total baryon rest mass M_0 should be strictly conserved in axisymmetry. The gravitational mass M is not conserved, due to the emission of gravitational radiation, but the decrease in M is very small.

The collapse proceeds homologously early on and results in the appearance of an apparent horizon at the center at $t/M \sim 30\,630$. The final black hole contains about 90% of the total rest mass of the system and has a spin parameter $J/M^2 \sim 0.75$. The remaining gas forms a rotating disk about the black hole. In fact, these black hole and disk parameters can be calculated analytically to reasonable approximation from the initial stellar density and angular momentum distributions[50] using the fact that in axisymmetry, the specific angular momentum spectrum, i.e., the integrated rest mass of all fluid elements with

---

[46] Baumgarte and Shapiro (1999).
[47] A PN simulation in $3 + 1$ dimensions by Saijo *et al.* (2002) indicates that nonaxisymmetric instabilities are not excited.
[48] This trick is possible because for $n \approx 3$ and $T/|W|$ small, the collapse proceeds homologously in the central regions during the early, Newtonian stages; see Shapiro and Teukolsky (1979); Goldreich and Weber (1980); Saijo *et al.* (2002).
[49] Alcubierre *et al.* (2001). See Bardeen and Piran (1983); Evans (1986), Ruiz *et al.* (2008b) for procedures for regularizing the field equations to deal with the coordinate singularities that arise at the origin or along the axis when working in spherical symmetry or axisymmetry.
[50] Shapiro and Shibata (2002).

specific angular momentum j less than a fiducial value, is strictly conserved in the absence of viscosity.[51] The fact that a substantial disk forms containing about 10% of the initial rest mass may have important implications for the "collapsar" model of long-duration gamma-ray bursts[52] which posits that the central engines of such sources are rotating black holes surrounded by gaseous, magnetized disks that arise from the collapse of massive stars. To be confident that the evolution has reached stationary equilibrium, it is necessary to evolve the implosion well past the appearance of an apparent horizon. Accomplishing this requires the implementation of some technique that avoids the build up of numerical errors due to the presence of the black hole singularity. In Section 14.2.3 we will discuss simulations that adopt one such technique, namely "black hole excision", to repeat and extend this astrophysically important collapse calculation.

The mass fraction that forms an ambient disk about the rotating black hole is a sensitive function of the polytropic index of a marginally unstable progenitor star rotating uniformly at the mass-shedding limit. As the polytropic index decreases below $n \approx 3$, the mass of the disk decreases rapidly. This result has been demonstrated both numerically[53] and analytically.[54] For $2/3 < n < 2$, the disk mass fraction is very small ($< 10^{-3}$).

Differentially rotating hypermassive stars

Consider next the stability of a differentially rotating, hypermassive stars. Adopt the hypermassive equilibrium configuration marked by the dot in Figure 14.1 as initial data for the same $3 + 1$ code used to evolve the supramassive stars whose radially stability we studied above. This hypermassive model has $\hat{A}^{-1} = 1$, $R_e/M \sim 5$ and $\beta \sim 0.23$. The result of the simulation[55] is summarized in Figure 14.4. Little change is observed in the density variable $\rho_* \equiv \rho_0 u^t (-g)^{1/2}$ and velocity profiles of the configuration over several central rotation periods. This simulation thus proves that at least some hypermassive stars are dynamically stable, both to radial and nonradial modes. But their secular stability is another issue, to which we will return in Sections 14.2.4 and 14.2.5.

14.2.2 Bar-mode instability

Next consider the dynamical stability of differentially rotating, highly relativistic stars against bar-mode formation. Numerical simulations to study this instability have been performed in full $3 + 1$ general relativity, again with the same BSSN, relativistic hydrodynamics code discussed in the previous section.[56] Compact, radially stable, equilibrium stars with $M/R \gtrsim 0.1$ were constructed as initial data; here M is the total ADM mass and R the equatorial circumferential radius. A polytropic EOS with $n = 1$ and a differential

[51] Stark and Piran (1987).
[52] MacFadyen and Woosley (1999); MacFadyen et al. (2001).
[53] Shibata (2004).
[54] Shapiro (2004a).
[55] Baumgarte et al. (2000).
[56] Shibata et al. (2000b).

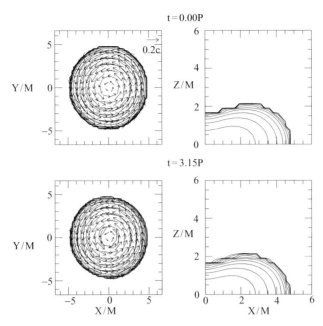

Figure 14.4 Snapshots of contours of the density ρ_* of the hypermassive star marked with a dot in Figure 14.1. We show contours of the initial data and after about 3 central rotation periods, both in the equatorial plane (left, including the velocity field $v^i = u^i/u^t$), and in a meridional plane containing the z-axis of rotation (right). The star has a rest-mass about 60% larger than the maximum nonrotating rest mass. The simulation shows that this model is dynamically stable. [From Baumgarte *et al.* (2000).]

angular velocity profile specified by equation (14.11) were employed, and the evolutions were performed adiabatically with adiabatic index $\Gamma = 2$.

To investigate the dynamical stability against bar-mode deformation, a nonaxisymmetric density perturbation was imposed of the form

$$\rho_0^{\text{new}} = \rho_0 \left(1 + \delta_b \frac{x^2 - y^2}{R_e^2} \right), \tag{14.35}$$

where ρ_0 denotes the original axisymmetric density distribution, and where δ_b was chosen to be 0.1 or 0.3. The 4-velocity components u_i were left unperturbed. The constraint equations were then re-solved to guarantee that, although the perturbed configuration is no longer in strict hydrostatic equilibrium, the Einstein equations are obeyed at $t = 0$. The growth of a bar mode can be followed by monitoring the stellar distortion parameter

$$\eta \equiv 2 \frac{x_{\text{rms}} - y_{\text{rms}}}{x_{\text{rms}} + y_{\text{rms}}}, \tag{14.36}$$

where x_{rms}^i denotes the mean square axial length

$$x_{\text{rms}}^i = \left[\frac{1}{M_0} \int (x^i)^2 \rho_* d^3 x \right]^{1/2}. \tag{14.37}$$

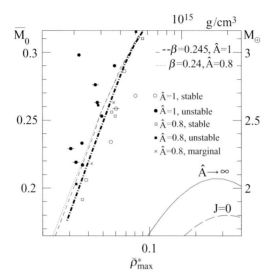

Figure 14.5 Models of differentially rotating stars in a $\bar{M}_0$ vs. $\bar{\rho}^*_{max}$ diagram. Circles denote stars with $\hat{A} \equiv A/R_e = 1$, squares with $\hat{A} = 0.8$. Solid (open) circles or squares represent stars that are unstable (stable). Marginally stable stars are denoted with a cross. The region for the stable stars is clearly separated from that of the unstable stars by the thick dashed-dotted line. This line is followed fairly closely by the dashed and dotted lines, which have been constructed for differentially rotating stars of $(\hat{A}, \beta) = (1, 0.245)$ and $(0.8, 0.24)$. The long-dashed and solid lines in the lower right hand corner are for nonrotating spherical stars and rigidly rotating stars at the mass-shedding limit. Scales for the top horizontal and right vertical axes are shown for polytropic constant $K = 100(G^3 M_\odot^2/c^4)$, for which the maximum rest-mass for spherical stars is about $1.8 M_\odot$. [From Shibata *et al.* (2000b).]

For dynamically unstable stars, η grows exponentially until reaching a saturation point, while for stable stars, it remains approximately constant for many rotational periods.

Simulations were performed using a fixed uniform grid with typical size $153 \times 77 \times 77$ in $x - y - z$ and assuming π-rotation symmetry around the z-axis, as well as a reflection symmetry about the equatorial ($z = 0$) plane. Test simulations with different grid resolutions were also performed to check that the results do not change significantly. By imposing π-rotation symmetry, one-armed spiral ($m = 1$) modes are ignored; these may be dominant for rotating stars in which Ω is a very steep function of axial radius ϖ.[57]

The simulations show that when plotted in a $\bar{M}_0$ vs. $\bar{\rho}^*_{max}$ diagram, a region of stable stars can be clearly distinguished from a region of unstable stars, with the onset of the bar-mode instability almost independent of the degree of differential rotation (see Figure 14.5).

The simulations also show that the parameter $\beta = T/|W|$ remains a good diagnostic of the onset point of instability in the relativistic domain as it did for Newtonian stars. The critical value for the instability onset depends only weakly on the degree of differential rotation for the models surveyed. In particular, the critical value of $\beta = \beta_d$ is $\sim 0.24 - 0.25$, only slightly smaller than the well-known Newtonian value of about 0.27 for incompressible

---

[57] Pickett *et al.* (1996); Centrella *et al.* (2001); Saijo *et al.* (2003); Ou and Tohline (2006); Baiotti *et al.* (2007).

Maclaurin spheroids. The critical value β_d decreases slightly for stars with a higher degree of differential rotation. When combined with other simulations that adopt the first post-Newtonian approximation of general relativity but are otherwise identical,[58] one finds that β_d decreases with increasing compaction as well. Thus relativistic gravitation *enhances* the dynamical bar-mode instability, i.e., the onset of the instability sets in for somewhat smaller values of β in relativistic gravitation than in Newtonian gravitation. To check the reliability of these simulations, the conservation of the relativistic circulation, defined by equation (5.59), was monitored, in addition to rest mass, total mass-energy, linear and angular momentum. Conservation of circulation indicates that the simulations are not seriously affected by any spurious numerical viscosity.

For selected models, the growth and saturation of bar-mode perturbations were followed up to late times. Stars with sufficiently large $\beta > \beta_d$ develop bars first and then form spiral arms, leading to some mass ejection; see Figure 14.6.

Stars with smaller values of $\beta \sim \beta_d$ also develop bars, but do not form spiral arms and eject only very little mass. In both cases, unstable stars appear to form differentially rotating, triaxial ellipsoids once the bar-mode perturbation saturates; see Figure 14.7.

Typically, these flattened ellipsoids appear to have $\beta \gtrsim 0.2$, so that they are secularly unstable to gravitational waves and viscosity. We expect that this secular instability will allow the stars to maintain a bar-like shape for many dynamical time scales, leading to quasiperiodic emission of gravitational waves. These expectations are borne out by simulations of merging binary neutron stars, which in some cases result in bar-like, hypermassive remnants that emit quasiperiodic gravitational waves prior to undergoing delayed collapse to black holes. We will return to this issue in Chapter 16.

As mentioned, relaxing the constraint of π-rotation symmetry can weaken the $m = 2$ mode via nonlinear coupling to the $m = 1$ mode. This effect was studied by Baiotti *et al.* (2007) with a $3 + 1$ general relativistic hydrodynamics HRSC code with a BSSN-like field solver. They found that such an effect can limit the persistence of bar-formation, and in some cases can even suppress the instability, when $\beta \sim \beta_d$, but has little influence when $\beta \gg \beta_d$. They also found, not surprisingly, that the nature of the initial perturbation had little effect on the growth time of the bar-mode instability, but could influence its duration when mode mixing is allowed.

14.2.3 Black hole excision and stellar collapse

The formation of a black hole during stellar collapse poses a serious challenge: the black hole singularity must be avoided to allow the exterior evolution to continue far into the future. In Chapter 13.1 we discussed several strategies for dealing with this issue in the context of binary black hole simulations in vacuum spacetimes. Here we will focus on one of the techniques, black hole excision, which has also proven very successfull in stellar collapse simulations.

---

[58] Saijo *et al.* (2001).

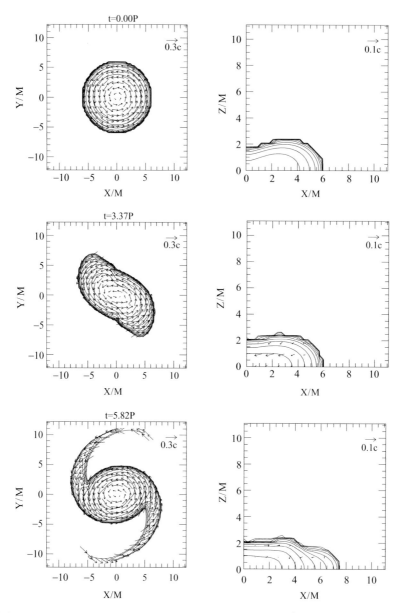

Figure 14.6 Snapshots of contours for the density ρ_* and the velocity field v^i in the equatorial plane (left) and in the meridional plane (right) for a bar-unstable model with $R/M \approx 7.1$ and $\beta \approx 0.26$. The contour lines are drawn for $\rho_*/\rho_{*\,max} = 10^{-0.3j}$ for $j = 0, 1, 2, \ldots, 10$ where $\bar{\rho}_{*\,max}$ is 0.126, 0.172 and 0.264 at the three different times. The lengths of arrows are normalized to $0.3c$ (left) and $0.1c$ (right). The time is shown in units of the initial central period $P \approx 15M$. [From Shibata *et al.* (2000b).]

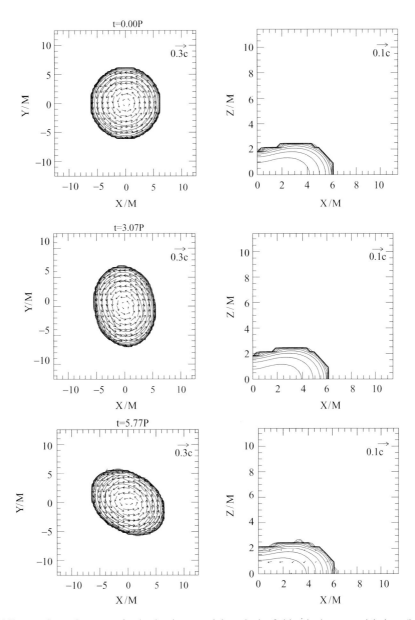

Figure 14.7 Snapshots of contours for the density ρ_* and the velocity field v^i in the equatorial plane (left) and in the meridional plane (right) for a bar-unstable model with $R/M \approx 7.2$ and $\beta \approx 0.25$. The contour lines are drawn for $\rho_*/\rho_{*\,\text{max}} = 10^{-0.3j}$ for $j = 0, 1, 2, \ldots, 10$ where $\bar{\rho}_{*\,\text{max}}$ is 0.128, 0.152 and 0.176 at the three different times. The lengths of arrows are normalized to $0.3c$ (left) and $0.1c$ (right). The time is shown in units of the initial central period $P \approx 15M$. [From Shibata *et al.* (2000b).]

As discussed in Chapter 13.1.2, black hole excision exploits the fact that the singularity resides inside an event horizon, in a region that is casually disconnected from the rest of the Universe (see Chapter 7.1). Since no physical information can propagate from inside the event horizon to the outside, we should be able to evolve the exterior independently of the interior spacetime. This entitles us to "excise" the black hole interior, or that part of the interior containing the singularity, from the computational domain and evolve the remaining region alone, without any adverse consequences.

Although it is guaranteed that no physical signal can propagate from inside the horizon to outside, unphysical signals often can propagate in evolution codes. Gauge modes can move acausally for many gauge conditions. Although they carry no physical content, such modes may destabilize the code. Thus, the choice of gauge is crucial to obtaining good excision evolutions. In addition, constraint-violating modes can, for some formulations of the field equations, propagate acausally, creating inaccuracies and instabilities. Thus, the choice of formulation is also crucial to obtaining good excision evolutions.

The feasibility of black hole excision was first demonstrated in spherically symmetric $1 + 1$ dimensional evolutions of a single black hole in the presence of a self-gravitating scalar field.[59] Excision was also implemented successfully to study the spherically symmetric collapse of collisionless matter to a black hole in Brans–Dicke theory.[60] Since that time, excision has been used successfully to evolve single and binary black holes in axisymmetry and full $3 + 1$ dimensions.[61] The first implementation of black hole excision in a $3 + 1$ relativistic hydrodynamics code was accomplished by Duez *et al.* (2004).[62] Their code evolves the gravitational field equations both in axisymmetry and full $3 + 1$ via the BSSN formulation. In their original version,[63] the hydrodynamic equations are evolved via a Wilson code, using the van Leer algorithm for the advection and artificial viscosity to handle shocks. Their subsequent version[64] evolves the fluid equations with a more sophisticated HRSC scheme along the lines discussed in Chapter 5; it also incorporates relativistic MHD. Duez *et al.* (2004) insert additional "constraint damping" terms into the BSSN field equations to help satisfy the constraints numerically and achieve improved accuracy, a technique we discussed in Chapter 11.5.[65] This addition consists of modifying equations (11.50), (11.51) and (11.52) according to

$$\partial_t \phi = \cdots + c_{H1} \Delta T \alpha \mathcal{H} \qquad (14.38)$$

$$\partial_t \bar{\gamma}_{ij} = \cdots + c_{H2} \Delta T \alpha \bar{\gamma}_{ij} \mathcal{H} \qquad (14.39)$$

$$\partial_t \tilde{A}_{ij} = \cdots - c_{H3} \Delta T \alpha \tilde{A}_{ij} \mathcal{H}, \qquad (14.40)$$

[59] Seidel and Suen (1992).
[60] Scheel *et al.* (1995a,b).
[61] See Baumgarte and Shapiro (2003c) and Chapter 13 for discussion and references.
[62] See also Baiotti *et al.* (2005).
[63] Duez *et al.* (2003).
[64] Duez *et al.* (2005b).
[65] Detweiler (1987); Frittelli (1997); Alcubierre *et al.* (2000); Yoneda and Shinkai (2001, 2002); Kelly *et al.* (2001); Yo *et al.* (2002).

where ΔT is the time step, $c_{H1} = 0.1$, $c_{H2} = 0.5$, and $c_{H3} = 1$ are dimensionless coefficients found empirically, and where $\mathcal{H}$ is given by the Hamiltonian constraint condition (11.48). In a similar spirit, the three BSSN "auxiliary constraint" conditions

$$0 = \mathcal{G}^i \equiv \bar{\Gamma}^i + \bar{\gamma}^{ij}_{,j} \tag{14.41}$$

$$0 = \mathcal{D} \equiv \det(\bar{\gamma}_{ij}) - 1 \tag{14.42}$$

$$0 = \mathcal{T} \equiv \mathrm{tr}(\tilde{A}_{ij}) \tag{14.43}$$

are incorporated into the right-hand sides of the evolution equation (11.55) for $\partial_t \bar{\Gamma}^i$, equation (11.51) for $\partial_t \bar{\gamma}_{ij}$ and equation (11.53) for $\partial_t \tilde{A}_{ij}$, respectively.[66]

The lapse and shift conditions are chosen so that, as the system settles into equilibrium, it appears stationary in the adopted coordinates. The lapse function is designed to be nonzero at the apparent horizon in order to maintain "horizon penetration", i.e., time evolution that proceeds smoothly across the horizon, without the steep build-up of gravitational waves and fluid that arises when the lapse and the advance of proper time plummet to zero there. Horizon penetration allows the location of an excision region *inside* the horizon; the evolution equations do not have to be solved inside the excision region. Duez *et al.* (2004) report great success with the hyperbolic "driver" conditions[67]

$$\partial_t^2 \beta^i = b_1(\alpha \partial_t \bar{\Gamma}^i - b_2 \partial_t \beta^i), \tag{14.44}$$

with $b_1 = 0.75$ and $b_2 = 0.27 M^{-1}$ and

$$\partial_t \alpha = \alpha \mathcal{A}$$
$$\partial_t \mathcal{A} = -a_1 \left(\alpha \partial_t K + a_2 \left[\partial_t \alpha + e^{-4\phi} \alpha (K - K_{\mathrm{drive}}) \right] \right). \tag{14.45}$$

In equation (14.45), the presence of α in front of $\mathcal{A}$ prevents the lapse from dropping to zero. With this feature, the lapse levels off at finite positive values everywhere on and outside the excision zone The term K_{drive} is the value to which the mean curvature K is "driven" at late times. Several choices for K_{drive} are useful. The simplest, and usually adequate choice, is zero. This drives K to zero (maximal slicing) and usually causes a very slow downward drift in the lapse near the horizon. For many astrophysical applications, where one only needs to evolve for several hundred M, this drift is usually unimportant. However, the effect can be removed and horzion penetration maintained by a better choice of K_{drive}. One possibility is K_{init}, the value of K at the time excision is introduced. Another choice is K_{KS}, a function which mimics the Kerr–Schild form for K for a Kerr black hole.[68]

The results of the simulations are insensitive to the precise boundary conditions applied at the edge of the excision zone, as they should be. For example, in their excision code, Duez *et al.* (2004) simply copy the time derivatives of field quantities onto the excision

---

[66] See also Alcubierre *et al.* (2000); Yo *et al.* (2002).
[67] Alcubierre *et al.* (2001); *cf.* the hyperbolic conditions discussed in Chapter 4.3 and 4.5.
[68] See Duez *et al.* (2004), equation (19)

boundary from the time derivatives at outer adjacent points. They use spherical excision regions inside the apparent horizon throughout, although this is not essential.

Tracking the collapse of rapidly rotating stars is one of the most important applications of black excision. As an example, consider the following issue: if the star collapses to a stationary black hole in vacuum, the "no-hair" theorem requires that it settle down to a Kerr black hole. In the Kerr spacetime, the singularity is covered by an event horizon only if $J/M^2 \leq 1$; otherwise the singularity is naked. Rotating stars, on the other hand, are not so restricted, and sufficiently rapidly rotating stars will have $J/M^2 > 1$. If the cosmic censorship hypothesis is true, then the collapse of the whole system must somehow be averted. How is this guaranteed? One way is if the star loses angular momentum as it collapses, either by gravitational wave emission[69] or by shedding matter with high specific angular momentum, so that the final black hole has $J/M^2 < 1$. A naked singularity can also be averted if the collapse of a $J/M^2 > 1$ star is always halted by centrifugal forces, so there will be no black hole and no singularity at all. It is easy to argue qualitatively[70] that a centrifugal barrier will arise to protect cosmic censorship in this way. Assuming no mass or angular momentum is shed during the collapse, the radius R_b at which the centrifugal force balances the gravitational force will be

$$\frac{M}{R_b^2} \sim \frac{J^2}{M^2 R_b^3},$$ (14.46)

so that

$$\frac{R_b}{M} \sim \left(\frac{J}{M^2}\right)^2.$$ (14.47)

Hence, if $J/M^2 < 1$ (i.e., the star is *sub-Kerr*), the star will already be inside a black hole before rotation can halt the collapse. For $J/M^2 > 1$ (i.e., the star is *supra-Kerr*), the collapse will be halted at a radius larger than M, and no black hole forms.

Simulating gravitational collapse of rotating objects and rigorously checking whether or not it complies with the Kerr limit for black hole spin has been done for representative cases of fluid stars[71] and collisionless clusters.[72] For some cases, the simulations did not (and sometimes could not) study in detail the final fate of the configuration or the outgoing gravitational radiation flux when a black hole forms. In hydrodynamic collapse, this is often the situation when the EOS is soft and the initial star is extended in size ($\gg M$) and centrally condensed, or when there is significant rotation to hold back the outer layers. Eventually, grid stretching, the collapse of the lapse, or some other complication in the vicinity of the black hole arises to terminate the integration before the outer layers have completed their evolution. Excision, which is designed to overcome this complication, enables one to address the issue of sub-Kerr vs. supra-Kerr collapse and identify the final

---

[69] But recall that in axisymmetry, no angular momentum is carried off by gravitational waves during collapse; see, e.g., Lightman *et al.* (1975), Problem 18.9.
[70] Nakamura (2002).
[71] Nakamura and Sato (1981); Stark and Piran (1985); Shibata (2000).
[72] Abrahams *et al.* (1994); see the discussion in Chapter 10.4.

state. Such an algorithm is useful not simply for determining whether or not a black hole forms, but also for determining how much rest mass and gravitational energy escapes collapse if it does.

To explore this cability with their excision algorithm, Duez *et al.* (2004) employed the routine of CST to construct three differentially rotating $n = 1$ polytropes of the same rest mass but varying spin to obtain both sub-Kerr and supra-Kerr models for initial data. Following pressure depletion to trigger their collapse, they tracked the adiabatic evolution of these configurations in axisymmetry on grids of 300^2 to 400^2 zones. Two of their models, stars A and B, are sub-Kerr ($J/M^2 = 0.57$ and 0.91, respectively) and collapse to Kerr black holes without disks. In the case of star B, a nearly spherical apparent horizon forms at $t/M = 28.4$ at a coordinate radius of $r_{AH}/M = 0.62$. Excision is introduced inside $r_{ex}/M = 0.08$ at $t = 29M$, at which time 22% of the rest mass is outside the excision zone, and 15% is outside the apparent horizon. All of the matter falls into the hole within $20M$ after excision is introduced, but the evolution is continued for an additional $20M$ after this. No instabilities arise and the system evolves to a stationary state. A third model, star C, is supra-Kerr ($J/M^2 = 1.2$). Following an initial implosion, the core of this star hits a centrifugal barrier, rebounds, and drives a shock into the infalling outer region. The star then expands into a torus and undergoes damped oscillations, settling down to a new stationary, nonsingular equilibrium state (a hot, rotating star), with nearly the same mass and angular momentum as the original star.

For the above class of initial data leading to collapse, not only is cosmic censorhip obeyed, but the spin parameter J/M^2 of the progenitor seems to provide a unique indicator of the final fate: if the spin is less than the Kerr limit, a black hole forms, otherwise there is no black hole but a new, nonsingular, equilibrium state. The situation is more complicated in general. Consider, for example, the fate of marginally unstable, rigidly rotating polytropes with n slightly above 3. Such stars crudely model the cores of massive stars in which thermal instabilities (e.g., iron photodissociation, or pair annihilation) are present to drive the adiabatic index slightly below 4/3 at the endpoint of stellar evolution. Consequently, tracking their collapse and determining their final fate is important astrophysically. Simulations of adiabatic collapse in axisymmetry by Shibata (2004) show that J/M^2 again proves to be a good predictor of the final outcome. But now a black hole may form even for $1 \lesssim J/M^2 \lesssim 2.5$; only for $J/M^2 \gtrsim 2.5$ does a centrifugal barrier prevent the direct formation of a black hole. For cases in which J/M^2 exceeds unity but a black hole forms, the effective value of the spin parameter in the central region of the progenitor is smaller than unity. The outcome is then a rapidly rotating black hole surrounded by a massive, hot, equilibrium disk.

Uniformly rotating $n = 3$ polytropes revisited

As we remarked in Section 14.2.1, the simulation of the collapse of a marginally unstable $n \approx 3$ rotating polytrope from the point of onset of radial instability to final stationary equilibrium is computationally challenging, even when restricting the analysis to axisymmetry. The reason, as we discussed, is dynamic range. In addition, even after the issue of dynamic range is resolved by a suitable spatial grid algorithm, black hole excision must be utilized in order that the simulation be able to reach the final, stationary, dynamical equilibrium state. In the presence of viscosity or magnetic fields, gas accretion from the disk onto the black hole will continue on a secular (i.e., viscous, or magnetic) time scale even after dynamical equilibrium is achieved. Reaching this dynamical equilibrium state, and then following the quasisteady accretion and secular growth of the black hole, clearly requires black hole excision.

The axisymmetric simulation of the adiabatic collapse of a marginally unstable $n = 3$ polytrope rotating uniformly at the mass-shedding limit performed by Shibata and Shapiro (2002) and discussed in Section 14.2.1 was repeated by Liu *et al.* (2007), who used black hole excision toward the end of the simulation to reach a final stationary equilibrium state. This simulation employed the basic BSSN, HRSC code of Duez *et al.* (2005b), using up to 1400^2 grid points and adopting the same rezoning technique as Shibata and Shapiro (2002) to handle the early Newtonian phase of the collapse. However, a transformation from the original set of coordinates $(\bar{x}, \bar{y}, \bar{z})$, with $\bar{r} = (\bar{x}^2 + \bar{y}^2 + \bar{z}^2)^{1/2}$ and uniform grid spacing $\Delta \bar{x}^i$, to new, radially "squeezed" coordinates (x, y, z) with $r = (x^2 + y^2 + z^2)^{1/2}$ and nonuniform grid spacing, was performed at late times to improve further the resolution near the black hole:[73] $\Delta r \approx (dr/d\bar{r})\Delta\bar{r}$, where $dr/d\bar{r}$ was constructed to equal $\approx 1/8$ near the hole, rising continuously to unity far away.[74] Most importantly, black hole excision was introduced soon after the appearance of an apparent horizon, and the simulation continued for an additional interval of time after this. The results are summarized in Figures 14.8, 14.9 and 14.10.

The pre-excision evolution from the onset of collapse to the appearance of an apparent horizon takes a total coordinate time $t/M \approx 28\,284$ and requires $32\,520$ Courant time steps. The post-excision evolution lasts only $\Delta t/M \approx 2220$, but requires about $18\,000$ additional time steps, due to the smaller grid spacing at late times. At $\Delta t/M \approx 219$ after the onset of the excision, gravitational field evolution is "turned off" and the metric is frozen while the hydrodynamic evolution of the residual matter continues until the end of the simulation. Holding the metric (i.e., gravitational field) fixed as the matter evolves is referred to as the "Cowling approximation". It is justified whenever the gravitational field settles into a quasistationary state, which is the case here. In the adopted gauge, the apparent horizon first appears at $r_{\rm AH}/M \approx 0.46$ and grows to $r_{\rm AH}/M \approx 1.6$ by the end of the post-excision evolution. The excision radius is initially set to $r_{\rm ex}/M \approx 0.77$ and is increased to $r_{\rm ex}/M \approx 1.1$ by $\Delta t/M \approx 100$ after the start of excision, after which it is held fixed. The outer boundary is moved inward to $L/M \approx 58$ at late times. About 1% of the rest-mass and 5% of the angular momentum is lost to regridding.

---

[73] This simple version of FMR is known as "multiple-transition fisheye" coordinates; see Campanelli *et al.* (2006a).

[74] The resulting radial grid spacing becomes $\Delta r \approx 0.025M$ for $r \lesssim 4M$ and rises to $\Delta r \approx 0.2M$ for $r \gtrsim 22M$; the outer boundary is near $60M$.

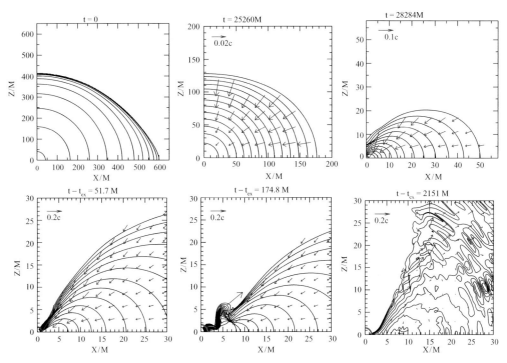

Figure 14.8 Snapshots of meridional rest-mass density contours and velocity vectors at selected times for $n = 3$ collapse. The initial model is rotating uniformly near the mass-shedding limit at the onset of gravitational instability to collapse. The density contour curves are drawn as follows: (a) Pre-excision top row, at $t = 0$, $\rho_0 = 10^{-j-0.1}\rho_{0c}(0)$ with $j = 0, 1, \ldots, 10$, where $\rho_{0c}(0)$ is the initial central density; at $t = 25\,260M$, $\rho_0 = \rho_{\text{scal}}10^{-0.3j}$ ($j = 0-12$), where $\rho_{\text{scal}} = 11\rho_{0c}(0)$ and at $t = 28\,284M$, $\rho_{\text{scal}} = 1000\rho_{0c}(0)$. (b) Post-excision bottom row, $\rho_0 = 100\rho_{0c}(0)10^{-0.3j}$ ($j = 0-10$). The thick solid (red) curve denotes the apparent horizon. For post-excision evolution, time is measured from the beginning of excision ($t_{\text{ex}} = 28\,284M$). Note that the coordinate scale decreases as time increases to show the central region in detail. [After Liu *et al.* (2007).]

Figure 14.8 shows the formation of a hot, thick disk about the black hole at late times. Figure 14.9 zooms into the region near the black hole at the end of the simulation and shows the relative scale of the excision zone, apparant horizon. and inner disk. Figure 14.10 plots the time evolution of several key parameters characterizing the hole. The irreducible mass M_{irr} is calculated from equation (7.2), using the area of the apparent horizon, which should coincide with the event horizon at late times. The black hole angular momentum is calculated from

$$J_{\text{h}} = J - J_{\text{matter}}(r > r_{AH})\qquad(14.48)$$

where the angular momentum of the matter outside the horizon is given by

$$J_{\text{matter}}(r > r_{AH}) = \int_{r > r_{AH}} S_\varphi \sqrt{\gamma}d^3x\,,\qquad(14.49)$$

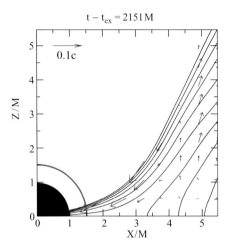

Figure 14.9 Blow-up of the meridional rest-mass density contours and velocity vectors at the end of the simulation, following excision and Cowling evolution. The density contours are drawn for $\rho_0 = 100\rho_{0c}(0)10^{-0.3j}$ ($j = 0-10$). The thick solid (red) curve denotes the apparent horizon, and the dark region denotes the excision zone. [After Liu *et al.* (2007).]

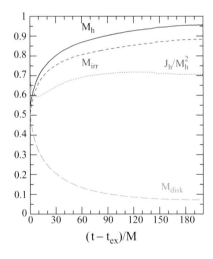

Figure 14.10 Post-excision evolution of the horizon mass M_h, spin parameter J_h/M_h^2, the irreducible mass M_{irr} of the central black hole, and the rest mass of the disk M_{disk} outside the apparent horizon. Time is measured from the beginning of excision ($t_{ex} = 28\,284M$). [From Liu *et al.* (2007).]

in accord with equation (14.22). Then, to estimate the black hole mass M_h, we can use equation (7.3). This relation is only approximate for this spacetime, which consists of a rotating black hole and an ambient disk, and is exact only for a vacuum Kerr spacetime. From this we estimate the black hole spin parameter, J_h/M_h^2. The figure shows that the loss of rest mass in the disk due to flow into the black hole reduces to a trickle by $\Delta t/M \gtrsim 150$

after the start of excision. As a result the mass and spin of the black hole settle down to their final stationary values, an indication that the adoption of the Cowling approximation to evolve the matter from this time forward is justified.[75] The final values are consistent with the earlier estimates of Shibata and Shapiro (2002): the final black hole contains about 95% of the total rest mass of the system and has a spin parameter $J_h/M_h^2 \approx 0.70$. This ideal gas simulation correctly finds that accretion of rest mass by the black hole eventually becomes quite small ($\dot{M}_0 \lesssim 10^{-6}$ in geometrized units[76]) due to the absence of a dissipative agent in the disk, like viscosity or magnetic fields, that can transport the angular momentum of the orbiting gas outwards. We shall return to this same problem in Section 14.2.5 where we consider the role of magnetic fields on the collapse and its aftermath.

An alternative determination of the mass and spin of a black hole, including one surrounded by fields and matter and possibly still undergoing growth, is provided by the "isolated-horizon" and "dynamical-horizon" framework discussed in Chapter 7.4. The approach assumes the existence of an axisymmetric Killing vector field residing on a marginally trapped surface like an apparent horizon. When there is some energy flux across the horizon, the isolated-horizon formalism must be replaced by the dynamical-horizon formalism.[77] Baiotti et al. (2005) have simulated the collapse of rotating, polytropic neutron stars to Kerr black holes to explore this framework and compare it to other measures of black hole masses and spins. They employ an HRSC general-relativistic code for the hydrodynamics, a BSSN-like scheme to integrate the gravitational field equations, a black hole excision algorithm and a dynamical-horizon routine.[78] The advantage of the dynamical-horizon approach is that it provides a measurement that can be computed locally, both in time and space, without having to determine the entire global spacetime. They also point out a limitation: the horizon itself, and its distortions, must be axisymmetric, although this is often not a major drawback in practical applications. Comparing different methods for measuring black hole mass and spin for cases leading to vacuum Kerr formation they conclude that the dynamical-horizon approach is simple to implement and can produce estimates that are accurate and more robust than those of other methods.

14.2.4 Viscous evolution

We have discussed the construction of equilibrium models for differentially rotating, hypermassive stars in Section 14.1.2 and pointed out in Section 14.2.1 how

---

[75] To adopt the Cowling approximation in a dynamical simulation, it is necessary that the spacetime be manifestly stationary in the adopted coordinates, so that time derivatives of the field variables vanish. The gauge choices of Liu et al. (2007) meet this criterion.

[76] In geometrized units the rate of rest-mass accretion, [mass/time], with [mass] ≡ M and [time] ≡ GM/c^3 ≡ M, is dimensionless.

[77] Note that for a given angular momentum J_h, equation (7.75) for the dynamical-horizon mass has the same form as equation (7.3) used by Liu et al. (2007) to estimate M_h. In the dynamical-horizon formalism, equation (7.74) is used to compute J_h.

[78] Their dynamical-horizon analysis is performed with the numerical algorithm of Dreyer et al. (2003).

relativistic simulations in full $3+1$ demonstrate that models exist that are stable on *dynamical* timescales. But viscosity and magnetic fields tend to drive differentially rotating objects toward uniform rotation on *secular* time scales, and such processes can have important consequences for hypermassive stars formed in nature. Consider, for example, the formation of a hypermassive neutron star following the inspiral and merger of a binary neutron star. The removal of the added centrifugal support provided by differential rotation can lead to the delayed collapse of the remnant to a black hole, accompanied by a delayed burst of gravitational radiation.[79] Both magnetic fields and viscosity alter the structure of differentially rotating stars on secular timescales, and tracking their lengthy secular evolution up to and beyond the onset of collapse presents a strenuous challenge for any numerical hydrodynamics code. Similarly, tracking the secular accretion of viscous or magnetized gas from a disk onto a central black hole is computationally taxing for a hydrodynamics code. Fortunately, numerical relativity has matured sufficiently that it is now up to the task in all of these cases. We will summarize in this section some representative simulations of dynamical spacetimes that treat viscosity in full general relativity; in the next section we will describe simulations that treat magnetic fields.

Duez *et al.* (2004) incorporated shear viscosity in their relativistic $2+1$ and $3+1$ Wilson–Van Leer hydrodynamics codes to solve the Navier–Stokes equations in general relativity for a viscous gas. They used their code to track the secular (viscous) evolution of differentially rotating stars, including hypermassive stars. They adopted equation (5.65) for the viscous stress-energy tensor, setting the bulk viscosity ζ to zero and the shear viscosity η according to $\eta = \nu_P P$, where ν_P is a constant parameter that can be varied from one simulation to the next. This form for the viscosity is sometimes chosen for modeling turbulence in fluids.[80] For sufficiently small coefficients ν_P secular effects will take many rotation periods to affect the velocity profile and structure of a differentially rotating star significantly. Accordingly, Duez *et al.* (2004) chose the coefficient large enough to make a numerical evolution feasible. At the same time, they were careful to keep it sufficiently small to guarantee that the viscous time scale $\tau_{\mathrm{vis}} \sim R^2/\nu$, where $\nu \equiv \eta/\rho_0$, remains much longer than the dynamical time scale $\tau_{\mathrm{dyn}} \approx (R^3/M)^{1/2}$ to mimic the physical behavior of any realistic fluid. Maintaining such an inequality usually guarantees that for a dynamically stable configuration, secular evolution proceeds in a quasistationary manner.

Duez *et al.* (2004) demonstrated that the secular evolution exhibits scaling behavior. Specifically, consider the quasistationary evolution of the same star, once with viscous time scale $\tau_{\mathrm{vis}} = \tau_1$ and once with $\tau_{\mathrm{vis}} = \tau_2$. Provided both τ_1 and τ_2 obey the inequality $\tau_{\mathrm{vis}} \gg \tau_{\mathrm{dyn}}$ then the configuration of the star with viscosity τ_1 at time t is the same as the configuration of the star with viscosity τ_2 at time $(\tau_2/\tau_1)t$. The results of any one secular simulation with a given choice of ν_P and τ_{vis} can then be scaled to any other choice without having to rerun the simulation.

---

[79] Baumgarte *et al.* (2000).
[80] This viscosity law is also used in some accretion disk models; see, e.g., Balbus and Hawley (1998).

Fluids with viscosity do not evolve isentropically; viscosity generates heat, as shown by equation (5.70). The gas may then cool, in principle, via neutrinos in the case of neutron stars. To survey the possible range of outcomes, Duez *et al.* treat cooling in two extreme opposite limits. In the *no-cooling* limit, which physically corresponds to $\tau_{\rm cool} \gg \tau_{\rm vis}$, all radiative cooling is ignored. In the *rapid-cooling* limit, where $\tau_{\rm cool} \ll \tau_{\rm vis}$, the viscous heating term is removed from the energy equation (5.69) on the assumption that all thermal energy generated by viscosity is radiated away immediately.

The simulations of hypermassive stars adopt $n = 1$ equilibrium polytropes for initial data, using the differential rotation law given by equations (14.9) and (14.10) with $\hat{A} = 1$. The value of ν_P is chosen so that the viscous time scale satisfies $\tau_{\rm vis} \approx 3 P_{\rm rot} \sim 10 \tau_{\rm dyn}$, where $P_{\rm rot}$ is the initial central rotation period. Even with viscosity of this magnitude, the stars need to be evolved for $100\text{–}200 P_{\rm rot}$ to complete their secular evolution and determine their final fate. The reason is that in most cases, viscosity generates a low-density envelope around the central core. Since the viscosity law has $\eta \propto P$, the viscosity in the low-density region is small. Hence the effective viscous time scale increases with time and it takes longer for the stars to reach a final state.

Snapshots during the evolution in axisymmetry of a representative configuration in the "no-cooling" case are shown in Figure 14.11. Here the initial configuration is a dynamically stable hypermassive star that is close to the model indicated by the solid dot in Figure 14.1. It has a mass $M_0 = 1.47 M_{0,\rm sup}$, where $M_{0,\rm sup}$ is the (supramassive) mass limit for uniformly rotating $n = 1$ polytropes (see Table 14.1). The spin parameter is $J/M^2 = 1.0$ and the central rotation period is $P_{\rm rot} = 38M$. As the evolution proceeds, viscosity brakes differential rotation and transfers angular momentum to the outer layers. The core then contracts and the outer layers expand in a quasistationary manner. As the core becomes more and more rigidly-rotating, it approaches instability because the star is hypermassive and cannot support a massive rigidly-rotating core. At time $t \approx 27 P_{\rm rot} \approx 11 \tau_{\rm vis}$, the star becomes dynamically unstable and collapses. An apparent horizon appears at time $t \approx 28.8 P_{\rm rot}$. Without black hole excision, the code begins to grow inaccurate about $10M$ after the horizon appears because of grid stretching. About 30% of rest mass remains outside the apparent horizon at this point. Using the excision technique described in Section 14.2.3, the evolution is extended for another $55M$, by which time the system settles down to a quasistationary, rotating black hole surrounded by a massive, hot ambient disk. The mass of the black hole at this point is $M_h \approx 0.82M$, while the rest mass and angular momentum of the ambient disk are $M_{0,\rm disk} \approx 0.23\ M_0$ and $J_{\rm disk} \approx 0.65J$. From conservation of angular momentum the angular momentum of the black hole is inferred to be $J_h \approx 0.35J$, giving a spin parameter $J_h/M_h^2 \approx 0.52(J/M^2) \approx 0.52$. Viscosity continues to drive slow, quasisteady accretion of material in the hot disk into the black hole.

The simulations summarized in the plots were performed in axisymmetry, using a grid of 128×128 zones and an outer boundary at $14M$. To check the reliability of the calculation, various diagnostics, like conservation of total mass and angular momentum, and the constraints, are monitored throughout the simulation. They all seemed to be

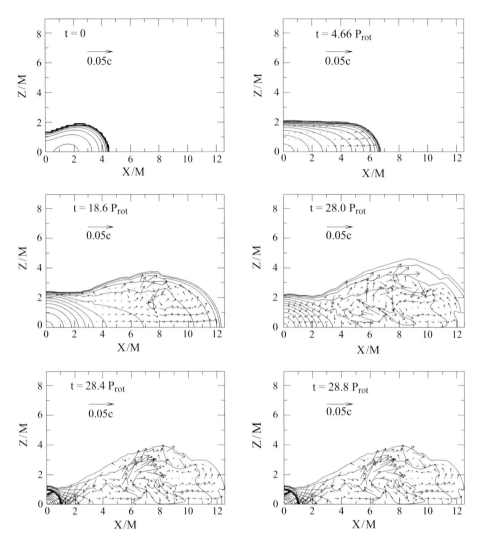

Figure 14.11 Meridional rest-mass density contours and velocity field at various times for a hypermassive star driven to catastrophic collapse by secular (viscous) redistribution of angular momentum. The simulation was performed by assuming that the system is axisymmetric and experiences no cooling. The levels of the contours (from inward to outward) are $\rho_0/\rho_{0,\max} = 10^{-0.15(2j+0.6)}$, where $j = 0, 1, \ldots, 12$. In the lower right panel ($t = 28.8 P_{\rm rot}$), the thick curve denotes the apparent horizon. [From Duez *et al.* (2004).]

well maintained throughout the simulation.[81] The circulation $\mathcal{C}$ on circular rings on the equatorial plane is also monitored. In the presence of viscosity, the flow is nonisentropic and circulation is not conserved, even in the absence of shocks (see equation 5.64). However,

---

[81] Gravitational radiation carries off some of the mass-energy, but the fractional amount is small. Angular momentum is automatically conserved in axisymmetry in a conservative code, but not in one based on the Wilson–van Leer scheme.

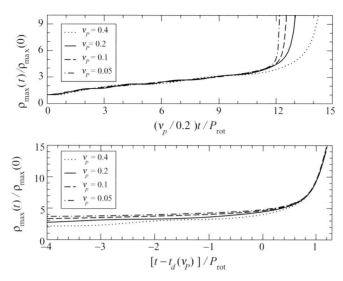

Figure 14.12 Scaling behavior exhibited by the evolution of the maximum rest-mass density of a hypermassive star for different strengths of viscosity, assuming rapid cooling. Upper panel: the curves coincide when plotted against the scaled time during secular evolution prior to dynamical collapse at $t_d(\nu_P)$. Lower panel: during dynamical collapse, it is possible to shift the time axes $[t \rightarrow t - t_d(\nu_P)]$ so that the curves again coincide, which indicates that viscosity plays an insignificant role during the dynamical collapse phase. [From Duez *et al.* (2004).]

the diagnostics show that the simulation properly acounts for the change in circulation due to viscosity.

In the "rapid-cooling" case, the behavior is qualitatively similar, but since there is no net viscous heating, the whole process occurs more quickly than in the "no-cooling" case. The collapse, in particular, now occurs at $t \approx 13.5 P_{\rm rot}$. A plot demonstrating that secular evolution exhibits scaling with viscosity, while dynamical collapse proceeds independently of the strength of viscosity, is shown in Figure 14.12. For rapid cooling, the total mass-energy is not conserved because the thermal energy generated by the viscous heating is removed, as described above. However, when the mass-energy carried away by thermal radiation, $M_{\rm cool}$, is accounted for via equation (5.70), the total mass-energy $M = M_{\rm ADM} + M_{\rm cool}$ is well-conserved.

14.2.5 MHD evolution

In any highly conducting astrophysical plasma, a frozen-in magnetic field can be amplified appreciably by gas compression or shear (e.g., differential rotation). Even when an initial seed magnetic field is weak, the field can grow to influence the system dynamics significantly.[82] Numerical codes have been constructed to treat such situations in relativistic

---

[82] See, e.g., Shapiro (2000) for a simple, but exact, Newtonian demonstration showing how an arbitrarily small magnetic seed field in a differentially rotating fluid can be amplified by winding on an Alfvén time scale to cause the magnetic braking of all differential motion.

dynamical spacetimes.[83] These codes solve the Einstein–Maxwell-MHD system of coupled equations along the lines discussed in Chapter 5.2.4. The two independent codes of Duez *et al.* (2005b) and Shibata and Sekiguchi (2005) have been used in tandem to tackle some of the same problems, and the two groups report good agreement. Both of these codes evolve the spacetime metric using the BSSN formulation and employ an HRSC scheme to integrate the general relativistic magnetohydrodynamics (GRMHD) equations. Multiple tests have been performed with these codes, including MHD wave propagation and shocks, magnetized Bondi accretion, MHD waves induced by linear gravitational waves, and magnetized accretion onto a neutron star.[84]

Hypermassive neutron stars

One of the early applications of the GRMHD codes has been to investigate the evolution of a hypermassive neutron star (HMNS). As we have already discussed, differentially rotating stars approach rigid rotation via transport of angular momentum on secular time scales. HMNSs, however, cannot settle down to rigidly rotating equilibria since their masses exceed the maximum allowed by uniform rotation. Thus, delayed collapse to a black hole, and possibly mass loss, can follow transport of angular momentum from the inner to the outer regions. In the previous section we have described relativistic calculations of HMNS collapse that have focused on angular momentum transport by viscosity; in Chapter 16 we will describe calculations of angular momentum loss due to gravitational radiation.[85] Here we focus on simulations of black hole formation triggered by seed magnetic fields in HMNSs.

There are at least two distinct effects which amplify the magnetic field in a shearing plasma like an HMNS: magnetic winding and the magnetorotational instability (MRI).[86] Magnetic winding will occur in a rotating star wherever $B^j \partial_j \Omega \neq 0$ and is a simple consequence of the "frozen-in" condition for a magnetic field in MHD. The MRI instability is a local shearing mode that is triggered wherever $\partial_\varpi \Omega < 0$, where ϖ is the cylindrical radius. When fully developed it will induce turbulence in the gas. Both effects combine to amplify a magnetic field, redistribute angular momentum and brake differential motion. The MRI poses a challenge for a numerical simulation in that the wavelength of the fastest-growing MRI mode must be well resolved on the computational grid in order for the instability to be observed. Since this wavelength is proportional to the magnetic field

---

[83] Duez *et al.* (2005b); Shibata and Sekiguchi (2005); Antón *et al.* (2006).

[84] For early computations involving GRMHD in *dynamical* spacetimes, see, e.g., Wilson (1975), who adopts the conformal flatness approximation to general relativity (see Chapter 16.2), and Sloan and Smarr (1985) and Baumgarte and Shapiro (2003a,b), who work in full general relativity, and references therein. For computations involving GRMHD in *stationary* (e.g., fixed Kerr metric) spacetimes, with emphasis on accretion, see, e.g., Yokosawa (1993); Koide *et al.* (1999, 2000); Komissarov (2004); De Villiers and Hawley (2003); Gammie *et al.* (2003); Anninos *et al.* (2005) and references therein.

[85] Shibata *et al.* (2003b).

[86] Velikhov (1959); Chandrasekhar (1960, 1961); Balbus and Hawley (1991, 1998).

strength, it becomes very difficult to resolve for small seed fields. However, we shall describe simulations below which succeed in resolving the MRI.

Duez *et al.* and Shibata used their two independent codes to collaborate on the study of the effects of magnetic fields on HMNSs.[87] In one of their simulations they employed as initial data essentially the same hypermassive $n = 1$ polytrope used in the study of viscous evolution described in Section 14.2.4, with a mass of $M_0 = 1.46 M_{0,\text{sup}}$ and angular momentum $J/M^2 = 1.0$. The matter has the same initial differential rotation profile and is evolved as a $\Gamma = 2$ ideal gas, as before. Before it is evolved, however, the configuration is threaded by a weak, purely poloidal magnetic field. Cases are then treated in which the maximum value of the ratio of the magnetic energy density to gas pressure, $C \equiv \max(b^2/P)$, is chosen between 10^{-3} and 10^{-2}. Such small initial magnetic fields introduce negligible violations of the Hamiltonian and momentum constraints, and leave the configurations in near dynamical equilibrium. When scaled to an actual HMNS that may arise from a binary neutron star merger, the value of the adopted initial maximum magnetic field strength is about $10^{16}(2.8 M_\odot/M)$ G. Though numerically small, such a field is probably too strong to model a typical HMNS, but is similar to the fields believed to form in "magnetars".[88] Regardless, it becomes increasingly difficult to evolve fields that are appreciably smaller, due to the need to resolve the MRI.

The calculations were performed in axisymmetry on a uniform grid of size (N, N), adopting cylindrical coordinates for the MHD. To check the convergence of numerical results, the simulations were repeated with four different grid resolutions: $N = 250, 300,$ 400 and 500. Figure 14.13 shows the evolution of the central density ρ_{0c}, central lapse α_c, and the maximum values of $|B^x|(\equiv |B^\varpi|)$ and $|B^y|(\equiv \varpi |B^\varphi|)$ as functions of t/P_c. Here $P_c \approx 39M = 0.54(M/2.8 M_\odot)$ ms denotes the central rotation period at $t = 0$. The central density increases monotonically with time up to the formation of a black hole. Evolutions with various grid resolutions demonstrate that the results begin to converge when $N \gtrsim 400$. On the other hand, results seem far from convergent for $N \lesssim 300$. For example, for low resolutions the maximum values of $|B^x|$ prior to saturation are much smaller than those with higher resolutions, and the growth rate of $|B^x|$ is underestimated. Hence, the effect of MRI, which is responsible for the exponential growth of $|B^x|$ on a dynamical time scale, is not computed accurately for low resolutions. This is because the wavelength of the fastest growing MRI mode is not well-resolved for low resolutions (see below).

For the chosen initial seed magnetic field, the early evolution is dominated by magnetic winding of the poloidal field, which generates a toroidal field B^y. When the poloidal seed field is weak and does not influence the flow, the induction equation (5.140) shows that B^y grows approximately linearly:

$$B^y(t; \varpi, z) \approx t \varpi B^i(0; \varpi, z) \partial_i \Omega(0; \varpi, z). \tag{14.50}$$

---

[87] Duez *et al.* (2006a,b), Shibata *et al.* (2006a).
[88] Duncan and Thompson (1992).

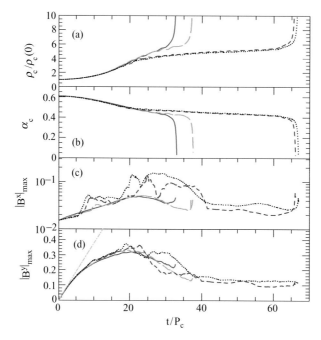

Figure 14.13 Evolution of the central rest-mass density, central lapse, and maximum values of $|B^x|$ and $|B^y|$ (the behavior of $|B^z|_{max}$ is similar to the behavior of $|B^x|_{max}$ and is therefore not shown). $|B^x|_{max}$ and $|B^y|_{max}$ are plotted in units of $\sqrt{\rho_{max,0}}$ where $\rho_{max,0}$ is the maximum rest-mass density at $t = 0$. The solid, long-dashed, dashed, and dotted curves denote the results with $N = 250$, 300, 400, and 500 respectively. The dot-dashed line in the last panel represents the predicted linear growth of B^y at early times. [From Duez *et al.* (2006a).]

Exercise 14.8   Adopt cylindrical coordinates in axisymmetry to verify equation (14.50). Assume that the star is rotating with an angular velocity $\Omega(\varpi, z)$ and remains nearly stationary as the field amplifies.

Indeed, the early growth rate agrees with the predicted one, as shown in Figure 14.13. When the energy stored in the toroidal field becomes comparable to the energy in differential rotation, $|B^y|$ grows more slowly and the degree of differential rotation is reduced. Eventually $|B^y|$ reaches a maximum and starts to decrease. This should happen on the Alfvén time scale t_A, where the Alfvén speed is $v_A = \sqrt{b^2/(\rho_0 h + b^2)}$. For the model considered here, the minimum value of t_A in the star is $t_A = 15.8 P_c$. So it is not surprising that the figure shows the maximum value of $|B^y|$ to begin decreasing when $t \gtrsim 20 P_c$,

The MRI is evident at times $t \gtrsim 6 P_c$ as shown in Figure 14.13, where the maximum value of $|B^x|$ suddenly increases rapidly. The wavelength for the fastest growing mode is $\lambda_{MRI} \approx 2\pi v_A/\Omega$ and the e-folding time of the growth is $\tau_{MRI} = 2 (\partial\Omega/\partial \ln \varpi)^{-1}$.[89] For the adopted initial magnetic field strength and rotation profile, this gives $\lambda_{MRI} \sim R/10$

---

[89] The estimates are based on a local, linear Newtonian analysis as in Balbus and Hawley (1991, 1998).

and $\tau_{\mathrm{MRI}} \sim 1 P_c$. The figure shows the growth of MRI when $N \gtrsim 400$. For this resolution the grid spacing is $\Delta \lesssim \lambda_{\mathrm{MRI}}/10$, the required scale to study the effect of MRI accurately. The central density begins to grow more slowly once $|B^x|$ saturates, presumably due to MRI-induced turbulence redistributing some of the angular momentum to slow down the contraction of the core.

The evolution sequence shown in Figure 14.14 reveals how magnetic braking due to winding and MRI turbulence combine to trigger gravitational collapse to a black hole. An apparent horizon forms at $t \approx 66 P_c \approx 36 (M/2.8 M_\odot)$ ms. As we will discuss in Chapter 16, simulations of binary neutron star mergers show that for a sufficiently stiff EOS and typical observed binary masses, hypermassive star formation is a very possible outcome. Such remnants become triaxial and strong emitters of gravitational waves in these simulations. The dissipation time scale of angular momentum due to gravitational radiation is ~ 100 ms. Therefore, a hypermassive remnant with an initially large magnetic field ($B \gtrsim 10^{16}$G) will be subject to delayed collapse due to MHD effects (magnetic braking and MRI) rather than by the emission of gravitational waves. For seed magnetic fields which are much weaker than the cases summarized here, gravitational radiation may be the trigger of collapse. However, it is possible that the MRI may dominate the evolution even in this situation, since the e-folding time of MRI is independent of the initial field strength. A more careful study of this scenario has to be carried out. However, since any dissipative agent (viscosity, magnetic fields, gravitational radiation) serves to redistribute and/or carry off angular momentum, the final fate of a hypermassive star – collapse to a black hole, accompanied by a gravitational wave burst – is assured.

To follow the evolution after the formation of the apparent horizon, black hole excision is implemented. The evolution of the irreducible mass of the black hole and the total rest mass outside the apparent horizon are shown in Figure 14.15.

Soon after formation, the black hole grows rapidly, swallowing the surrounding matter. However, the accretion rate $\dot{M}_0$ gradually decreases and the black hole settles down to a quasiequilibrium state. At the end of the simulation, $\dot{M}_0$ decreases to a steady value of about $0.01 M_0/P_c$. The black hole angular momentum is computed from equations (14.48) and (14.49), and the mass of the black hole M_h is then estimated from equation (7.3). The estimated value of the black hole spin parameter is then $J_h/M_h^2 \sim 0.8$.

The value of $\dot{M}_0$ indicates that the accretion time scale is approximately 10–$20 P_c \approx 5$–10 ms$(M/2.8 M_\odot)$. Also, the specific internal thermal energy in the ambient gaseous torus near the surface is substantial because of shock heating, indicating that the torus can be a strong emitter of neutrinos. The final system formed after such a delayed collapse of a magnetized HMNS – a rotating black hole + hot accretion torus + collimated magnetic field – has all the attributes proposed for a gamma-ray burst central engine.[90] Given the timescales and energetics associated with such a delayed collapse, this scenario was suggested by Duez *et al.* (2006a) as a particularly good candidate for a short, hard

---

[90] See, e.g., Piran (1999, 2005); Aloy *et al.* (2005).

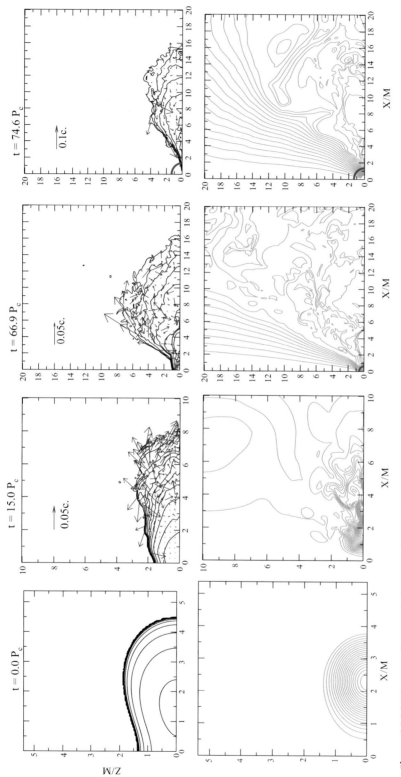

Figure 14.14 The upper four panels show snapshots of the rest-mass density contours and velocity vectors on the meridional plane. The lower panels show the field lines (lines of constant A_ϕ) for the poloidal magnetic field at the same times as the upper panels. The density contours are drawn for $\rho^*/\rho^*_{\mathrm{max},0} = 10^{-0.3i-0.09}$ ($i = 0-12$). The field lines are drawn for $A_\phi = A_{\phi,\mathrm{min}} + (A_{\phi,\mathrm{max}} - A_{\phi,\mathrm{min}})i/20$ ($i = 1-19$), where $A_{\phi,\mathrm{max}}$ and $A_{\phi,\mathrm{min}}$ are the maximum and minimum value of A_ϕ respectively at the given time. The thick solid curves in the lower left corners denote the apparent horizon. [From Duez et al. (2006a).]

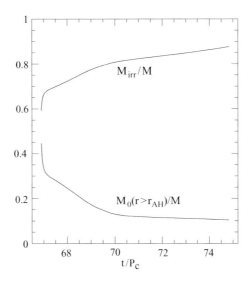

Figure 14.15 Evolution of the irreducible mass and the total rest mass outside the apparent horizon. [From Duez *et al.* (2006a).]

gamma-ray burst source.[91] They have followed up this suggestion by performing further simulations with a more realistic nuclear EOS, and these appear to strengthen the argument.[92]

It is interesting to compare the evolution and final fate of a magnetized hypermassive star with dynamically stable, differentially rotating, magnetized stars that are not hypermassive. Consider two distinct categories of differentially rotating, nonhypermassive configurations: (1) "normal stars", which have rest masses below the maximum achievable with uniform rotation, and angular momentum below the maximum for uniform rotation at the same rest mass, and (2) "ultraspinning stars", which have angular momentum exceeding the maximum for uniform rotation at the same rest mass. Suppose they are again threaded by a weak, initially poloidal magnetic field with $C = \max(b^2/P) = 2.5 \times 10^{-3}$. Simulations in axisymmetry[93] show that a normal star evolves to a uniformly rotating equilibrium configuration. By contrast, an ultraspinning star cannot settle into a uniformly rotating equilibrium state. Instead, it evolves to an equilibrium configuration consisting of a nearly uniformly rotating central core, surrounded by a differentially rotating torus characterized by an angular velocity that is constant along magnetic field lines ($B^j \partial_j \Omega \to 0$). Hence, although the final state exhibits differential rotation, it ceases to wind the magnetic field. In addition, the final state appears to be stable against the MRI. It will be interesting to explore whether the fates of these configurations are altered when the constraint of axisymmetry is relaxed and high resolution simulations in full $3 + 1$ are performed.

---

[91] For discussion and references to the observational literature, see Berger (2006).
[92] Shibata *et al.* (2006a), Stephens *et al.* (2008).
[93] Duez *et al.* (2006b).

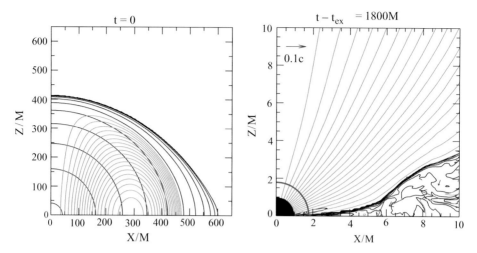

Figure 14.16 Snapshots of meridional rest-mass density contours (black), velocity vectors and magnetic field lines (green) for the initial and endpoint configurations for $n = 3$ magnetorotational collapse. The initial model is identical to the configuration in Figure 14.8 except that it is threaded by a poloidal magnetic field. Field lines coincide with contours of vector potential A_φ and are drawn for $A_\varphi = A_{\varphi,\text{max}}(j/20)$ with $j = 1, 2, \ldots, 19$ where $A_{\varphi,\text{max}}$ is the maximum value of A_φ. In the final, post-excision model ($t_{\text{ex}} = 29\,150M$), the density levels are drawn for $\rho_0 = 100\rho_{0c}(0)10^{-0.3j}$ ($j = 0$–10). The thick (red) line near the lower left corner denotes the apparent horizon. [After Liu *et al.* (2007).]

Magnetorotational collapse of $n = 3$ polytropes

The astrophysical and cosmological importance of massive star collapse motivates fully relativistic simulations of the magnetorotational collapse of $n = 3$ polytropes. Recall that such polytropes are useful to model very massive and supermassive stars, including "first generation" Population III stars, as well as the degenerate cores of massive Population I stars. It is likely that all stars are rotating and contain magnetic fields, so it is important to extend the calculations described in Sections 14.2.1 and 14.2.3 to include small seed magnetic fields in the initial configurations and explore the consequences. Liu *et al.* (2007) performed such simulations in axisymmetry and we will briefly summarize their findings below.

The HRSC GRMHD code of Liu *et al.* (2007) is closely adapted from the code of Duez *et al.* (2005b) described previously. The basic setup of the simulation, including the adjustable grid algorithm and the gauge conditions, are those used for magnetic-free $n = 3$ simulation summarized in Section 14.2.3. Now, however, a small poloidal magnetic field is inserted into the initial configuration (see Figure 14.16). Once again the initial configuration is uniformly rotating near the mass-shedding limit and at the onset of radial instability to collapse. The ratio of magnetic to rotational kinetic energy is chosen to be small (10% for the case shown here) to study whether the amplification of a small seed magnetic field can have large consequences. The simulation reveals that such magnetic fields do not affect the *initial* collapse significantly. The core collapses to a black hole, after which black hole excision is employed to continue the evolution long enough for the hole to reach a quasistationary state. As in the magnetic-free case, the black hole mass is $M_h = 0.95M$ and its spin parameter is $J_h/M_h^2 = 0.7$, with the remaining matter forming a disk or torus around the black hole. Once quasistationarity is reached, the spacetime metric is again frozen ("Cowling approximation") and the evolution of the torus continued. Interestingly, the subsequent evolution of the torus *does* depend on the strength of the magnetic field. In the absence of strong magnetic fields, the torus settles down, following ejection of a small amount of matter due to shock heating. When magnetic fields are present, the field lines gradually collimate along the hole's rotation axis. MHD shocks and the magnetorotational instability (MRI) combine to generate MHD turbulence in the the torus and stochastic accretion onto the central black hole (see Figure 14.16 again). When the magnetic field in the initial configuration is at least 10% of the rotational energy, a wind is generated in the torus, and the torus undergoes radial oscillations that drive episodic accretion onto the hole. At late times, the accretion rate has an amplitude of order $\dot{M} \lesssim 10^{-4}$ in geometrized units, significantly higher than the amplitude found in the magnetic-free case.

Both a collimated magnetic field and a massive, accretion torus surrounding a central black hole are essential ingredients for launching ultrarelativistic jets.[94] The black hole-torus system observed in these relativistic simulations comprises a plausible central engine for producing jets and long-duration, soft-spectrum ("long-soft") gamma-ray bursts. The simulation represents the starting point of a viable "collapsar" scenario for generating such bursts.[95]

The collapse of a very massive star could result in the simultaneous detection of a gamma-ray burst and gravitational waves. In addition to a gravitational wave burst arising from the initial collapse, the radial oscillations of the disk produce long-wavelength, quasi-periodic gravitational waves. The oscillation period of about $500M$ corresponds to an observed gravitational wave frequency $f \sim 1/[500M(1+z)] \sim 0.04(10^4 M_\odot/M)/(1+z)$ Hz for a source at redshift z. For a very massive star with $M \gtrsim 10^4 M_\odot$, the signal is in the LISA frequency band. To estimate its amplitude h, Liu *et al.* (2007) apply the quadrupole formula to an oscillating disk of mass $M_{\text{disk}} \sim 0.05M$ to obtain

$$h \sim 4 \times 10^{-23} \left(\frac{M}{10^4 M_\odot} \right) \left(\frac{48\text{Gpc}}{D_L} \right), \tag{14.51}$$

where $D_L = 48$ Gpc is the luminosity distance of a source at redshift $z = 5$ in the concordance ΛCDM cosmology model, where $H_0 = 71$ km s^{-1} Mpc^{-1}, $\Omega_M = 0.27$ and $\Omega_\Lambda = 0.73$.[96] If the quasiperiodic signal can be tracked for n cycles, where n is expected to be a few, the effective wave strength will be increased by a factor of $\sqrt{n}$. Such a gravitational wave signal may be strong enough to be detectable by LISA.[97]

---

[94] De Villiers *et al.* (2005).
[95] MacFadyen and Woosley (1999); MacFadyen *et al.* (2001).
[96] Spergel *et al.* (2007).
[97] See LISA's projected sensitivity curve in Figure 9.6.

Magnetorotational core collapse to neutron stars

In previous sections we described stellar collapse to black holes using an $n = 3$ polytropic profile to construct the initial model and an adiabatic $\Gamma = 4/3$ EOS to follow the evolution. Recall that this EOS is a good approximation for a very massive Population III star or a supermassive star, where the pressure is dominated by thermal radiation. For a Population I star, which has smaller mass, the pressure of the pre-collapse core is dominated by the relativistic degenerate electron pressure, which is also well-approximated by an $n = 3$ polytrope initially and a $\Gamma = 4/3$ EOS during the early phases of collapse. But during the final phases, the EOS stiffens when the density exceeds nuclear density $\rho_{\text{nuc}} \approx 2 \times 10^{14}$ g cm^{-3}. Only if the mass of the collapsing core exceeds a critical value M_{crit} will a black hole horizon appear before the star reaches nuclear density. Otherwise, the stiffening of the EOS can significantly effect the collapse, causing the core to bounce and possibly producing a neutron star remnant rather than a black hole.

> **Exercise 14.9** Estimate M_{crit} by considering the collapse of a uniform density sphere. A horizon appears when the areal radius of the sphere reaches $R \sim 2M$. At this time, the density is $\rho_0 \sim 3M/(4\pi R^3) \approx 1.7 \times 10^{16}(M_\odot/M)^2$. Set $\rho_0 = \rho_{\text{nuc}}$ to find M_{crit}.
> (**Ans.** $\sim 10 M_\odot$)

For a supermassive star, the mass, which collapses homologously, is much larger than M_{crit}. For a Population III star of mass $M = 300 M_\odot$, the mass of the collapsing core is $180 M_\odot$,[98] which is still much larger than M_{crit}. Hence the $\Gamma = 4/3$ EOS is also a reasonable approximation during the entire collapse phase for such stars.[99] For Population I stars, on the other hand, the core mass is less than $2 M_\odot$ and a more realistic EOS is required during the late stages. In addition, neutrino emission and transport are also dynamically important.

To simulate crudely the collapse of a magnetized, rotating stellar core of a Population I star to a neutron star in general relativity, Shibata *et al.* (2006b) adopt a hybrid EOS that consists of the sum of a cold component and a thermal component:[100]

$$P(\rho_0, \epsilon) = P_{\text{P}}(\rho_0) + P_{\text{th}}(\rho_0, \epsilon). \tag{14.52}$$

The "cold" component, $P_{\text{P}}(\rho_0)$, is chosen according to the prescription

$$P_{\text{P}}(\rho_0) = \begin{cases} K_1(\rho_0)^{\Gamma_1}, & \rho_0 \leq \rho_{\text{nuc}}, \\ K_2(\rho_0)^{\Gamma_2}, & \rho_0 \geq \rho_{\text{nuc}}, \end{cases} \tag{14.53}$$

where K_1, K_2, Γ_1 and Γ_2 are constants. Setting $K_2 = K_1(\rho_{\text{nuc}})^{\Gamma_1 - \Gamma_2}$ makes P_{P} continuous at $\rho_0 = \rho_{\text{nuc}}$. Typical choices are $\Gamma_1 \approx 4/3$ and $\Gamma_2 = 2.5 - 2.7$. Setting $K_1 = 5 \times 10^{14}$

---

[98] Fryer *et al.* (2001).
[99] Neutrino generation and transport also play a role in the collapse of massive stars, but are probably not dynamically important for the most massive Population III progenitors or for supermassive stars because of their low temperatures and densities.
[100] This form follows from the work of Obergaulinger *et al.* (2006), who simulate magnetorotational core collapse in Newtonian gravitation.

(cgs), ensures that the cold component represents relativistic degenerate electron pressure for $\rho_0 < \rho_{nuc}$. This simplified cold EOS is designed to mimic a more complicated, cold, stiff nuclear EOS. The specific internal energy density, ε_P, associated with the cold component of the pressure P_P is obtained by integrating the first law of thermodynamics using equation (14.53). The thermal part of the pressure P_{th} plays an important role when shocks occur, but is absent otherwise. It is given by the familiar ideal gas law

$$P_{th} = (\Gamma_{th} - 1)\rho_0 \varepsilon_{th}, \tag{14.54}$$

where $\varepsilon_{th} \equiv \varepsilon - \varepsilon_P$ is defined as the thermal specific internal energy. The value of Γ_{th} determines the efficiency of converting kinetic energy to thermal energy at shocks; the rule $\Gamma_{th} = \Gamma_1$ is adopted to conservatively account for shock heating.

Shibata *et al.* (2006b) use their two GRMHD codes described earlier in this section to perform simulations of core collapse in axisymmetry with a grid of 2500×2500 zones. The grid is large, and adjusted during the collapse as in the $n = 3$ simulations described previously, in order to cover the full dynamic range of the imploding core and to resolve the fastest growing MRI wavelength. Not surprisingly, they find that significant differential rotation results even when the rotation of the progenitor is initially uniform. Consequently, a seed magnetic field is amplified both by magnetic winding and the MRI. Even if the ratio of magnetic energy to rotational kinetic energy is quite small at the time of proto-neutron star (PNS) formation, the ratio increases to 0.1–0.2 by magnetic winding. Following PNS formation, MHD outflows lead to losses of rest mass, energy, and angular momentum from the system. The earliest outflow is produced primarily by the increasing magnetic stress caused by magnetic winding. The MRI amplifies the poloidal field and increases the magnetic stress, causing further angular momentum transport and helping to drive the outflow. After the magnetic field saturates, a nearly stationary, collimated magnetic field forms near the rotation axis and a Blandford–Payne type outflow[101] develops along the field lines. These outflows remove angular momentum from the PNS and, as a result, the rotation period quickly increases for a strongly magnetized PNS until the degree of differential rotation decreases. The simulations thus suggest that rapidly rotating, magnetized PNSs may not give rise to rapidly rotating neutron stars.

These simulations should be regarded as preliminary, prototype calculations that illustrate how magnetorotional collapse can be tackled in numerical relativity. A number of simplifying assumptions have been incorporated in these simulations, including the adoption of equatorial and axisymmetry, a crude hybrid EOS, and the neglect of all neutrino transport and emission. These restrictions will have to be relaxed and the calculations repeated with greater sophistication in the future. But the tools are in place and rapid progress is now possible.

---

[101] Blandford and Payne (1982).

Binary neutron star initial data

We now turn to binary neutron star systems. In this chapter we will discuss strategies for constructing initial data that describe two neutron stars in quasiequilibrium and quasicircular orbit; in Chapter 16 we will review some simulations of their dynamical evolution, including their coalescence and merger.

As we have already discussed in Chapter 12.1, relativistic binaries cannot possibly be in exact equilibrium, since the emission of gravitational radiation leads to a slow inspiral. For sufficiently large separations, however, the time scale for this inspiral is much longer than the orbital period, so that we can approximate the inspiral as a sequence of binary configurations in circular orbit. We refer to these binaries as being in "quasiequilibrium", and to their orbit as "quasicircular".

In Chapter 12.3 we saw how binary black hole circular orbits are generated by a *helical Killing vector* ξ^a_{hel} (see equation 12.63 and Figure 12.4), and how this helical Killing vector can be used to construct solutions to the constraint equations that describe gravitational fields that are in approximate equilibrium. We will follow a very similar approach in this chapter to construct quasiequilibrium binary neutron stars in quasicircular orbit. The added complication here is that we also have to find a matter distribution that is in equilibrium with the gravitational field.

15.1 Stationary fluid solutions

To construct quasiequilibrium binary neutron stars we need to find quasistationary solutions to the relativistic equations of hydrodynamics. For starters, we need to find a rest-mass density ρ_0 and a fluid 4-velocity u^a that, given an equation of state, satisfy the conservation of energy-momentum (5.6),

$$\nabla_b T^{ab} = 0, \tag{15.1}$$

and the conservation of rest mass (5.7),

$$\nabla_a(\rho_0 u^a) = 0. \tag{15.2}$$

Given a one-parameter (e.g., cold nuclear) equation of state (EOS) we can express the specific internal energy density ϵ and the pressure P, both of which appear in the stress-energy tensor T^{ab} defined by equation (5.4), entirely in terms of ρ_0.

A solution to equations (15.1) and (15.2) automatically satisfies the relativistic equations of hydrodynamics, but need not be in equilibrium. Enforcing equilibrium requires an additional condition. What we mean by stationary equilibrium is that an observer corotating with the binary would not notice any change in the binary's structure with time. In other words, the binary configuration at any time could be generated from its configuration at any other time by a simple coordinate transformation consisting of a rotation around the axis of rotation. This rotation is generated by a helical Killing vector ξ_{hel}^a (see equation 12.63) that, like a corkscrew, is tangent to the binary's orbit through spacetime. Such a helical Killing vector is illustrated in Figure 12.4.

As explored in Appendix A, the Lie derivative $\mathcal{L}_X T$ of a tensor field T along a vector field X^a measures the difference between the actual change in the tensor field T and the change that would arise under a coordinate transformation generated by the vector field X^a. To insure, then, that the fluid quantities ρ_0 and u^a describe an equilibrium configuration, we require these quantities to be Lie-dragged along the helical Killing vector ξ_{hel}^a,

$$\mathcal{L}_{\xi_{hel}} \rho_0 = 0 \qquad \mathcal{L}_{\xi_{hel}} u^a = 0. \tag{15.3}$$

It is clear that the above equations still cannot determine the solution uniquely, since the individual stars in the binary may or may not be spinning. In the case of binary black holes we determined the spin of the individual black holes through the horizon boundary condition (12.107); here we will determine the spin of the individual neutron stars by making an assumption about the fluid velocity. We will see that the equations simplify for two special, but astrophysically relevant, cases: *corotational* and *irrotational* binaries.

The equations become especially simple for corotational binaries, whereby the spin angular velocity of each star is equal to the orbital angular velocity of the binary (when measured in the inertial frame of a distant observer). In such systems each star always shows the same side to its companion. A familiar example is the Moon, which corotates with the Earth. Since the equations simplify dramatically for corotational binaries, this case has usually been studied first. Unfortunately, this case is not physically realistic for neutron stars. Maintaining synchronization (corotation) during the inspiral requires a viscosity that acts on a time scale that is short compared to the inspiral time. It turns out that such a large viscosity is very unphysical for binary neutron stars.[1] Instead, the assumption of an irrotational fluid flow is more realistic for neutron star binaries, as we shall now explain. Isolated neutron stars, like radio pulsars, are observed to have spin. However, their spin frequencies, with notable exceptions, are small compared to the high orbital frequencies ($\sim$kHz) characterizing neutron star binaries at the small separations they reach prior to merger. These orbital frequencies approach the high frequencies that isolated neutron stars would have at break-up due to centrifugal forces (i.e., the mass-shedding limit). It is thus reasonable to expect that typical neutron stars have much smaller frequencies when they

---

[1] See Kochanek (1992a); Bildsten and Cutler (1992).

form a binary. Viscous interactions cannot spin up such a star significantly during the inspiral. Tidal and gravitational radiation reaction forces conserve circulation; in fact, circulation is strictly conserved for isentropic flow (this is the Kelvin–Helmholtz theorem, proven for relativistic flow in Chapter 5.2.1). We therefore expect binary neutron stars to maintain their low spin frequencies during inspiral. Therefore, at small separations, the fluid motion in neutron star binaries can be well approximated as irrotational.

Most studies of binary neutron stars assume the opposite extremes of corotational or irrotational fluid flow,[2] and in the following we also will focus on these two limiting cases. We will show how the Euler equation (15.1) reduces to an algebraic equation in both cases, and will see how the continuity equation (15.2) is satisfied identically for corotational binaries and can be recast as an elliptic equation for irrotational binaries. However, before going through this formal derivation in general relativity, it is instructive to derive the corresponding equations in a Newtonian framework.

15.1.1 Newtonian equations of stationary equilibrium

The spatial components of equation (15.1) yield the Euler equations (5.14), which in the Newtonian limit reduce to

$$\partial_t v_i + v^j D_j v_i = -\frac{1}{\rho_0} D_i P - D_i \Phi_N. \tag{15.4}$$

Here Φ_N is the Newtonian gravitational potential and is a solution to Poisson's equation,

$$D^2 \Phi_N = 4\pi \rho_0. \tag{15.5}$$

In the Newtonian limit, the continuity equation (15.2) becomes

$$\partial_t \rho_0 + D_i(\rho_0 v^i) = 0. \tag{15.6}$$

The first law of thermodynamics is given by equation (5.16) for adiabatic changes and remains the same in the Newtonian limit. This law relates $P(\rho_0)$ to $\epsilon(\rho_0)$ in a one-parameter EOS.

Before proceeding, we use the thermodynamic relation (5.63) for isentropic configurations,

$$\frac{1}{\rho_0} D_i P = D_i h, \tag{15.7}$$

to rewrite the Euler equation (15.4) as

$$\partial_t v_i + v^j D_j v_i = -D_i h - D_i \Phi_N. \tag{15.8}$$

Here h is the specific enthalpy defined by equation (5.5),

$$h = 1 + \epsilon + P/\rho_0. \tag{15.9}$$

[2] See Marronetti and Shapiro (2003) for an exception.

In a Newtonian context, the enthalpy is usually defined without the first term, which accounts for the contribution of the rest mass-energy,

$$h = 1 + h_{\text{Newt}}. \tag{15.10}$$

In general we will use the relativistic convention, but when considering Newtonian limits it is sometimes useful to use h_{Newt}.

In anticipation of our derivation of the corresponding relativistic equations we follow an approach that is somewhat unconventional in a Newtonian context and express the time derivatives of v_i and ρ_0 in equations (15.8) and (15.6) in terms of a Lie derivative along the helical Killing vector ξ_{hel}^a (see equation 12.63). We adopt this approach as preparation for our relativistic analysis in Section 15.1.2, which is greatly simplified by the formalism introduced here. We start by writing ξ_{hel}^a as

$$\xi_{\text{hel}}^a = t^a + k^a. \tag{15.11}$$

Here t^a is a timelike vector that points "up" in a spacetime diagram, while k^a is purely spatial and describes rotations around the axis of rotation. We can write k^a as

$$k^a = \Omega \left(\frac{\partial}{\partial \phi} \right)^a, \tag{15.12}$$

or, in Cartesian coordinates, as

$$k^i = \epsilon^{ijk} \Omega_j x_k, \tag{15.13}$$

where Ω^i is aligned with the axis of rotation and measures the orbital angular velocity Ω. Together, t^a and k^a form the "corkscrew" shape of the helical Killing vector ξ_{hel}^a.

We can now express the Lie derivative of v^i as

$$\mathcal{L}_{\xi_{\text{hel}}} v_i = \partial_t v_i + \mathcal{L}_{\mathbf{k}} v_i = \partial_t v_i + k^j D_j v_i - v_j D_i k^j, \tag{15.14}$$

and that of ρ_0 as

$$\mathcal{L}_{\xi_{\text{hel}}} \rho_0 = \partial_t \rho_0 + \mathcal{L}_{\mathbf{k}} \rho_0 = \partial_t \rho_0 + k^i D_i \rho_0. \tag{15.15}$$

We also write the fluid velocity as

$$v^i = k^i + V^i, \tag{15.16}$$

where V^i now measures the fluid velocity relative to k^i. For a corotating fluid, for example, we must have $V^i = 0$.

Exercise 15.1   Show that the Euler equation (15.8) can be written as

$$\mathcal{L}_{\xi_{\text{hel}}} v_i - \frac{1}{2} D_i v^2 + D_i(v_j V^j) + V^j(D_j v_i - D_i v_j) = -D_i h - D_i \Phi_{\text{N}}, \tag{15.17}$$

and the continuity equation (15.6) as

$$\mathcal{L}_{\xi_{\text{hel}}} \rho_0 + V^i D_i \rho_0 + \rho_0 D_i v^i = 0. \tag{15.18}$$

Equations (15.17) and (15.18) are still completely general, since we have not specialized yet to stationary solutions. However, at this point it is very easy to do so, since can simply set the Lie derivatives of v^i and ρ_0 along ξ^a_{hel} to zero, in which case equation (15.17) reduces to

$$-\frac{1}{2}D_i v^2 + D_i(v_j V^j) + V^j(D_j v_i - D_i v_j) = -D_i h - D_i \Phi_{\text{N}}. \qquad (15.19)$$

Since we also have $D_i k^i = 0$ (which follows from equation 15.13), we can write equation (15.18) in the compact form

$$D_i(\rho_0 V^i) = 0. \qquad (15.20)$$

It is now clear that these equations simplify for two special types of fluid flow: corotational flow, where $V^i = 0$, and irrotational flow, where $D_j v_i - D_i v_j = 0$ (i.e., **curl v** = 0). We will discuss these two cases separately.

Corotational binaries

For corotational, or synchronized, binaries we set $V^i = 0$, and hence

$$v^i = k^i \qquad \text{(corotational flow)} \qquad (15.21)$$

in equations (15.19) and (15.20). The continuity equation (15.20) is satisfied identically, and does not provide any information. The Euler equations (15.19) reduce to

$$-\frac{1}{2}D_i k^2 = -D_i h - D_i \Phi_{\text{N}}, \qquad (15.22)$$

which we can integrate once to find the integrated Euler equation,

$$h + \Phi_{\text{N}} - \frac{1}{2}k^2 = C. \qquad (15.23)$$

Here C is a constant of integration. Squaring equation (15.13) to obtain

$$k^2 = \Omega^2 \varpi^2, \qquad (15.24)$$

where ϖ measures the distance from the axis of rotation, we can write equation (15.23) as

$$h_{\text{Newt}} + \Phi_{\text{N}} - \frac{1}{2}\Omega^2 \varpi^2 = C_{\text{Newt}}, \qquad (15.25)$$

where we have introduced $C_{\text{Newt}} = C - 1$. This is a remarkable result: assuming corotational fluid flow (15.21) we have reduced the equations of hydrodynamics, which in general are a reasonably complicated set of coupled partial differential equations, to a single algebraic equation for the enthalpy h. Given a one-parameter EOS we can express the enthalpy in terms of the density ρ_0. Any density distribution that satisfies equation (15.25) and is Lie dragged along $v^a = k^a$ automatically satisfies the Euler equation (15.4) and the continuity equation (15.6), and is therefore a stationary solution to the Newtonian equations of hydrodynamics.

Incidently, we could have found equation (15.25) directly by evaluating equation (15.8) in a corotating frame. In the corotating frame the velocity must vanish; that means that the continuity equation (15.6) is satisfied identically if $\partial_t \rho_0 = 0$ in that frame. The left hand side of (15.8) also vanishes with $\partial_t v_i = 0$. Since the corotating frame is not inertial, we have to include a fictitious centrifugal force in the Euler equation (15.8). We can account for this term by replacing the Newtonian potential Φ_N with the effective potential

$$\Phi_{\text{eff}} = \Phi_N - \frac{1}{2}\Omega^2\varpi^2. \tag{15.26}$$

Substituting this relation into equation (15.8) yields equation (15.25) as before.

In summary, we can construct Newtonian models of corotational binaries by solving the (algebraic) integrated Euler equation (15.25) for the enthalpy h together with Poisson's equation (15.5) for the Newtonian potential Φ_N. In these equations the orbital angular velocity Ω and the constant of integration C appear as eigenvalues that must be determined along with solving the equations. We will provide an example of how this can be done with the help of an iterative algorithm in Section 15.2 below.

Irrotational binaries

For irrotational binaries the curl of the velocity field v^i has to vanish,

$$\epsilon_{ijk}D^j v^k = 0. \tag{15.27}$$

We can enforce this by choosing v^i to be the gradient of a velocity potential Φ,

$$v_i = D_i\Phi \qquad \text{(irrotational flow)}. \tag{15.28}$$

The Euler equations (15.19) now become

$$-\frac{1}{2}D_i(D_j\Phi D^j\Phi) + D_i(V^j D_j\Phi) = -D_i h - D_i\Phi_N, \tag{15.29}$$

which we can again integrate once to find the integrated Euler equation

$$-\frac{1}{2}D_j\Phi D^j\Phi + V^j D_j\Phi + h + \Phi_N = C, \tag{15.30}$$

where C is again a contant of integration. We now eliminate V^j with the help of equation (15.16) to find

$$\frac{1}{2}D_j\Phi D^j\Phi - k^l D_l\Phi + h_{\text{Newt}} + \Phi_N = C_{\text{Newt}}. \tag{15.31}$$

For a given velocity field, expressed in terms of Φ, we can intepret this result as an algebraic equation for the enthalpy h, as we did before for corotating flow.

In contrast to the corotating case, the continuity equation (15.20) is not satisfied identically for irrotational flow. Substituting equation (15.28) we find that equation (15.20) now becomes an elliptic equation for the velocity potential Φ,

$$D_i D^i \Phi = \left(k^i - D^i \Phi \right) D_i \ln \rho_0. \tag{15.32}$$

Since the fluid velocity is only defined in the stellar interior, this equation holds in the stellar interior only, and we need to supply a boundary condition on the stellar surface. At the surface the density vanishes, so regularity of the right-hand side of equation (15.32) demands that

$$\left(k^i - D^i \Phi \right) D_i \rho_0 \Big|_{\text{surface}} = 0. \tag{15.33}$$

Since $D_i \rho_0$ is normal to the surface, this relation represents a Neumann boundary condition for Φ.

> **Exercise 15.2** Show that the boundary condition (15.33) can also be derived from demanding that, in the corotating frame, the fluid velocity must be tangent to the stellar surface.

One challenging conceptional issue is already evident: we need to solve equation (15.32) subject to the boundary condition (15.33), but a priori we don't know where the stellar surface, and hence the boundary, is located. One common approach to solving this problem is to introduce "surface-fitting" coordinates, in which the stellar surface always corresponds to a fixed coordinate surface.[3]

In summary, then, we can construct Newtonian models of irrotational binary neutron stars by solving the integrated Euler equation (15.31) for the enthalpy h, the continuity equation (15.32), subject to the boundary conditions (15.33), for the velocity potential Φ, and Poisson's equation (15.5) for the Newtonian potential Φ_N. As in the corotational case, the orbital angular velocity Ω and the constant of integration C appear as eigenvalues, and have to be solved for together with the equations.

15.1.2 Relativistic equations of stationary equilibrium

We now follow a very similar approach to derive the corresponding equations for relativistic fluids, starting with the conservation of energy-momentum (15.1) and the continuity equation (15.2). We first note that we can rewrite (15.1) as

$$u^b \nabla_b (h u_a) + \nabla_a h = 0. \tag{15.34}$$

for isentropic configurations (see equation 5.63). This result is now the relativistic equivalent of the Newtonian Euler equations (15.8).

As in the Newtonian case we proceed by expressing the derivative operators in equations (15.2) and (15.34) in terms of a Lie derivative along ξ^a_{hel}. Towards that end, we write

---

[3] See Uryū and Eriguchi (1999) for a numerical implemention.

the fluid 4-velocity u^a as[4]

$$u^a = u^t(\xi^a_{\mathrm{hel}} + V^a). \tag{15.35}$$

Here we assume that ξ^a_{hel} is timelike inside the stars, and we normalize ξ^a_{hel} so that its time component is unity, $\xi^t_{\mathrm{hel}} = 1$. We also choose V^a to be purely spatial, $n_a V^a = 0$, so that it plays a role that is very similar to that of its Newtonian counterpart in equation (15.16). In a coordinate system that is comoving with ξ^a_{hel}, i.e., a corotating coordinate system, the spatial components of the fluid velocity reduce to $u^t V^a$. For a corotating fluid we again have $V^a = 0$. For convenience of notation we will also define the spatial projection of hu_a as

$$\hat{u}_i = \gamma_i{}^a hu_a. \tag{15.36}$$

We now ask the reader to prove two useful identities in exercise 15.3.

Exercise 15.3 Derive the two relations

$$\gamma_i{}^a \xi^b_{\mathrm{hel}} \nabla_b(hu_a) = \gamma_i{}^a \mathcal{L}_{\xi_{\mathrm{hel}}}(hu_a) + hD_i\left(\frac{1}{u^t}\right) + \hat{u}_b D_i V^b - hu_b n^b V^a K_{ia}$$

$$\tag{15.37}$$

and

$$\gamma_i{}^a V^b \nabla_b(hu_a) = V^c D_c \hat{u}_i + hu_b n^b V^a K_{ai}. \tag{15.38}$$

Hint: Choose a coordinate basis as in Chapter 2.7 so that $n_i = 0$ and use equation (2.62).

We can now relate the spatial projection of the covariant derivative along u^a in equation (15.34) to the Lie derivative along ξ^a_{hel} and spatial derivatives by adding equations (15.37) and (15.38),

$$\gamma_i{}^a u^b \nabla_b(hu_a) = u^t\left(\gamma_i{}^a \xi^b_{\mathrm{hel}} \nabla_b(hu_a) + \gamma_i{}^a V^b \nabla_b(hu_a)\right)$$

$$= u^t\left(\gamma_i{}^a \mathcal{L}_{\xi_{\mathrm{hel}}}(hu_a) + hD_i\left(\frac{1}{u^t}\right) + \hat{u}_b D_i V^b + V^c D_c \hat{u}_i\right). \tag{15.39}$$

Combining this result with equation (15.34) we find

$$\gamma_i{}^a \mathcal{L}_{\xi_{\mathrm{hel}}}(hu_a) + D_i\left(\frac{h}{u^t} + \hat{u}_j V^j\right) + V^j(D_j \hat{u}_i - D_i \hat{u}_j) = 0. \tag{15.40}$$

With the help of equation (2.63) we can also express equation (15.2) as

$$\alpha\left(\mathcal{L}_{\xi_{\mathrm{hel}}}(\rho_0 u^t) + \rho_0 u^t \nabla_a \xi^a_{\mathrm{hel}}\right) + D_i(\alpha u^t \rho_0 V^i) = 0. \tag{15.41}$$

Since we have not yet used the fact that ξ^a_{hel} is a Killing vector, equations (15.40) and (15.41) are still completely general. As in the Newtonian derivation we can now specialize

---

[4] In this section we follow the notation and approach of Shibata (1998).

to equilibrium configurations by invoking that ξ_{hel}^a be a helical Killing vector.[5] In this case the Lie derivatives along ξ_{hel}^a, as well as the divergence $\nabla_a \xi_{\text{hel}}^a$, must vanish, so that equations (15.40) and (15.41) reduce to

$$D_i \left(\frac{h}{u^t} + \hat{u}_j V^j \right) + V^j (D_j \hat{u}_i - D_i \hat{u}_j) = 0 \qquad (15.42)$$

and

$$D_i(\alpha u^t \rho_0 V^i) = 0. \qquad (15.43)$$

Exercise 15.4   Show that in the Newtonian limit equations (15.42) and (15.43) reduce to the Newtonian equations (15.19) and (15.20).
Hint: Use exercise 2.28 and equation (5.22) to find the Newtonian limits of α and αu^t.

As in the Newtonian case, equations (15.42) and (15.43) simplify further for either corotational or irrotational fluid flow. We will discuss these two cases separately in the following two sections.

15.2 Corotational binaries

We can construct corotational binaries by requiring that the fluid flow vanish in the frame corotating with the binary,

$$V^a = 0. \qquad (15.44)$$

With this choice, the continuity equation (15.43) is satisfied identically, as in our Newtonian analysis, and the Euler equations (15.42) reduce to

$$D_i \left(\frac{h}{u^t} \right) = 0. \qquad (15.45)$$

We can again integrate these equations immediately to obtain the integrated Euler equation[6]

$$\frac{h}{u^t} = C, \qquad (15.46)$$

where C is a constant of integration. In fact, this result proves relation (5.58).
From the normalization $u_a u^a = -1$ we also have

$$\alpha u^t = \left(1 + \gamma^{ij} u_i u_j \right)^{1/2} \qquad (15.47)$$

---

[5] Recall that for relativistic binaries, we seek a *quasiequilibrium* state, for which ξ_{hel}^a is really only an *approximate* Killing vector; see the discussion leading up to equation (12.63).
[6] In some of the literature this expression is incorrectly refered to as the Bernoulli equation; compare equations (5.57) and (5.58) and the related discussion in Chapter 5.

(see equation 5.11). In a typical application, the spatial metric will be rescaled conformally, $\gamma_{ij} = \psi^4 \bar{\gamma}_{ij}$. Given a choice for the conformal background metric $\bar{\gamma}_{ij}$, and given values for the lapse α, the conformal factor ψ, and the shift β^i, we can use equations (15.35) and (15.11) (adopting the form 15.13 when working in Cartesian coordinates) to obtain u^t. Substituting this result into equation (15.46) yields an algebraic expression for h.

> **Exercise 15.5** Show that, for a conformally flat spatial metric, the enthalpy h satisfies
>
> $$h \left\{ \alpha^2 - \psi^4 \left((\Omega y - \beta^x)^2 + (\Omega x + \beta^y)^2 + (\beta^z)^2 \right) \right\}^{1/2} = C \qquad (15.48)$$
>
> if the axis of rotation is aligned with the z-axis. Verify that this expression reduces to equation (15.25) in the Newtonian limit.
> **Hint:** Use exercise 2.28.

In typical applications equation (15.48) – or an equivalent equation if the background is not conformally flat – is solved in conjunction with the conformal thin-sandwich equations (3.109)–(3.112) listed in Box 3.3. The latter provide a set of equations for the conformal factor ψ, the lapse α, and the shift β^i in terms of matter sources that depend on h. Furthermore, we can use equation (15.48) to compute h algebraically in terms of ψ, α, and β^i. As before, the constants Ω and C appear as eigenvalues in this equation, and have to be determined in the course of solving for the field and matter variables.

It is worth noting why it is the conformal thin-sandwich formalism of Chapter 3.3, rather than the conformal transverse-traceless approach of Chapter 3.2, that is relevant in this context. A quick glance at equation (15.48) reveals that we require a lapse α and a shift β^i to solve the quasiequilibrium problem. The conformal thin-sandwich formalism provides these functions, while the conformal transverse-traceless formalism does not. More fundamentally, we wish to construct fluid configurations that are in quasiequilibrium, so that Lie derivatives along a timelike Killing vector vanish. To impose this condition, we need to constrain the behavior of the spacetime in a *neighborhood* of a spatial slice Σ and not merely *on* Σ. The conformal thin-sandwich formalism is an approach that allows us to impose quasiequilibrium in just this fashion.

A number of different numerical algorithms have been implemented to construct simultaneous solutions to the integrated Euler equation (15.48) and the field equations (3.109)–(3.112).[7] We will describe one such scheme, namely that of Baumgarte *et al.* (1997, 1998a), which is based on a similar scheme for constructing rotating stars developed by Hachisu (1986). We will also focus on the Newtonian problem, which is simpler than the relativistic case, and yet captures all the key ingredients of the numerical scheme. Instead of the relativistic integrated Euler equation (15.48) we will therefore solve its Newtonian counterpart (15.25), and instead of the field equations (3.109)–(3.112) we will solve Poisson's equation (15.5).

[7] See, e.g., Baumgarte *et al.* (1998a); Marronetti *et al.* (1998); Gourgoulhon *et al.* (2001); Taniguchi and Gourgoulhon (2002).

We start by choosing an equation of state. For simplicity, we adopt a Γ-law equation of state

$$P = (\Gamma - 1)\rho_0 \epsilon \qquad (15.49)$$

(see equation 5.17). For isentropic fluids, this equation of state is equivalent to a polytropic equation of state of the form

$$P = K\rho_0^\Gamma, \quad \Gamma = 1 + 1/n, \qquad (15.50)$$

where n is the polytropic index and K is the gas constant (see exercise 5.5). Choosing a polytropic equation of state allows for several simplifications in our scheme, but very similar algorithms can be used for any other equation of state, even if it exists only in tabulated form.

The Newtonian enthalpy h_{Newt} that appears in equation (15.25) can then be written as

$$h_{\text{Newt}} = \epsilon + \frac{P}{\rho_0} = (n+1)\frac{P}{\rho_0}. \qquad (15.51)$$

If we also assume that the binary rotates about the z-axis, the integrated Euler equation (15.25) takes the form

$$(n+1)\frac{P}{\rho_0} + \Phi_{\text{N}} - \frac{1}{2}\Omega^2(x^2 + y^2) = C_{\text{Newt}}. \qquad (15.52)$$

For polytropes it is convenient to introduce the dimensionless density parameter

$$q \equiv \frac{P}{\rho_0}, \qquad (15.53)$$

in terms of which we have

$$\rho_0 = K^{-n}q^n, \qquad P = K^{-n}q^{n+1}. \qquad (15.54)$$

Since physical units enter the problem only through the constant K, we can introduce dimensionless coordinates $\bar{x} = K^{-n/2}x$ and similar for y and z, as in equation (14.17). The angular velocity scales as $\bar{\Omega} = K^{n/2}\Omega$ and the Laplace operator as[8] $\bar{D}^2 = K^n D^2$. In terms of these quantities, equation (15.52) becomes

$$(n+1)q + \Phi_{\text{N}} - \frac{1}{2}\bar{\Omega}^2(\bar{x}^2 + \bar{y}^2) = C_{\text{Newt}}, \qquad (15.55)$$

and Poisson's equation becomes

$$\bar{D}^2 \Phi_{\text{N}} = 4\pi q^n. \qquad (15.56)$$

We simplify the problem further by treating equal-mass binaries. These binaries form a two-parameter family of solutions that can be parametrized, for example, by the maximum

---

[8] Here we depart from our previous notation convention: the bar denotes the nondimensional operator, and not the conformally-related operator as defined elsewhere in the book.

density of each star and their separation. Instead of using the absolute separation of the binary stars, we will specify their *relative* separation as follows: Assume that the binary lies along the x-axis, i.e., the x-axis connects the points of maximum density in the two stars. Let $x = 0$ reside at the center of mass, midway between the stars. Denote the coordinate along x at the nearest point to the origin on the surface of one star by x_A and at the farthest point by x_B. Also label the point of maximum matter density in the star by x_C. We can then parametrize the binary separation in terms of the relative separation $s = x_A/x_B = \bar{x}_A/\bar{x}_B$. An infinite separation corresponds to $s = 1$, while a contact binary has a separation $s = 0$.[9]

In fact, it proves convenient to rescale not only $\bar{x}_A$ with respect to $\bar{x}_B$ (to define the relative separation s), but all coordinates as well. Denoting these rescaled coordinates with a caret, we define $\hat{x} = \bar{x}/\bar{x}_B$ and similarly for $\bar{y}$ and $\bar{z}$. The Laplace operator now rescales as $\hat{D}^2 = \bar{x}_B^2 \bar{D}^2$, and we also have $\hat{\Omega} = \bar{x}_B \bar{\Omega}$. Poisson's equation now takes on the form

$$\hat{D}^2 \Phi_N = 4\pi \bar{x}_B^2 q^n. \tag{15.57}$$

Equation (15.57) motivates the scaling relation

$$\Phi_N = \bar{x}_B^2 \hat{\Phi}_N \tag{15.58}$$

for the Newtonian potential, whereby equation (15.57) reduces to

$$\hat{D}^2 \hat{\Phi}_N = 4\pi q^n. \tag{15.59}$$

The integrated Euler equation (15.55) then takes the form

$$(n+1)q + \bar{x}_B^2 \hat{\Phi}_N - \frac{1}{2}\hat{\Omega}^2(\hat{x}^2 + \hat{y}^2) = C_{\text{Newt}}. \tag{15.60}$$

We now have to find simultaneous solutions to equations (15.59) and (15.60) for q and $\hat{\Phi}_N$, together with the eigenvalues $\hat{\Omega}$, C_{Newt} and $\bar{x}_B$. This can be accomplished by the following iteration scheme:

Choose the polytropic index, then select a particular binary model by specifying the maximum density, q_{max}, and the relative binary separation s.

1. As a first step in the iteration, provide an initial guess for the density profile q. For example, this can be a spherical density profile with maximum density q_{max} and properly rescaled so that it is confined between $\hat{x}_A$ and $\hat{x}_B$.
2. Next, solve Poisson's equation (15.59), which requires solving an elliptic equation. This can be accomplished, for example, by employing the algorithms described in Chapter 6.2.2 for finite difference methods, or in Chapter 6.3.4 for spectral methods. Solving equation (15.59) provides the rescaled Newtonian potential $\hat{\Phi}_N$.

---

[9] This parametrization is not unlike that for a family of isolated, uniformly rotating stars, members of which can be parametrized by their maximum density and the ratio between their polar and equatorial radii. It is therefore possible to construct binary neutron stars with an iterative algorithm similar to one used for isolated rotating stars; see Chapter 14.1.2.

3. Then, determine the eigenvalues $\hat{\Omega}$, C_{Newt} and $\bar{x}_B$ by evaluating the integrated Euler equation (15.60) at the three points $\hat{x}_A = s$, $\hat{x}_B = 1$ and $\hat{x}_C$. To find $\hat{x}_C$, locate the (current) point of maximum density along the x-axis. At all of these three points we know the value of q: at $\hat{x}_A$ and $\hat{x}_B$ we must have $q = 0$, since both points lie on the stellar surface, and at $\hat{x}_C$ we must have $q = q_{\text{max}}$. Evaluating (15.60) at these three points yields the three equations

$$\bar{x}_B^2 \hat{\Phi}_{\text{N}} - \frac{1}{2} \hat{\Omega}^2 \hat{x}_A^2 = C_{\text{Newt}}$$

$$\bar{x}_B^2 \hat{\Phi}_{\text{N}} - \frac{1}{2} \hat{\Omega}^2 \hat{x}_B^2 = C_{\text{Newt}} \qquad (15.61)$$

$$(n+1)\, q_{\text{max}} + \bar{x}_B^2 \hat{\Phi}_{\text{N}} - \frac{1}{2} \hat{\Omega}^2 \hat{x}_C^2 = C_{\text{Newt}}.$$

This set of equations can be solved iteratively for the eigenvalues $\hat{\Omega}$, C_{Newt} and $\bar{x}_B$.
4. With the values of the eigenvalues $\hat{\Omega}$, C_{Newt} and $\bar{x}_B$ find the new density distribution q by solving equation (15.60) in the stellar interior.
5. Given the new density distribution q, now evaluate the residual of Poisson's equation (15.59). If this residual is larger than some predetermined tolerance, then the iteration continues, beginning with step 2 above. Otherwise the solution has been obtained to within the desired accurcay, and the iteration terminates.

The above iterative scheme describes the method to solve the Newtonian problem, but a very similar algorithm can be used to solve the corresponding relativistic problem. The equations are more complicated, of course, and instead of solving one equation for the Newtonian potential we now have to solve the five equations (3.109)–(3.112) for α, ψ and β^i, but all of these changes can be accommodated by the above iteration scheme. One complication arises from the scaling of the gravitational fields, which are less obvious in the relativistic case. However, for the purposes of this iteration, we can rescale α and ψ in such a way that in the Newtonian limit we recover the scaling relation (15.58).

Solving the above equations then allows us to construct relativistic, quasiequilibrium binary neutron star models. Figure 15.1 shows a typical neutron star binary at a small separation. Subjected to a small tidal distortion, the profile of each star assumes a characteristic prolate shape.

As a first application, we can now generalize Oppenheimer–Volkoff sequences (see Chapter 1.3). In an Oppenheimer–Volkoff sequence we fix the equation of state and construct spherically symmetric, static neutron stars with different central densities. Along such a sequence, the configuration with the maximum total mass-energy $M(= M_{\text{ADM}})$ is also the configuration with the maximum rest-mass M_0 and locates the maximum central density beyond which stars in isolation are radially unstable. Generalizing such sequences to binaries, we can fix the separation s and construct sequences of binaries with identical

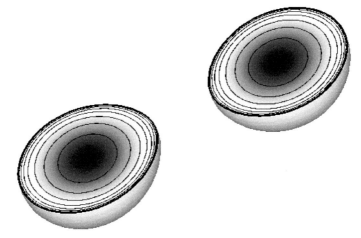

Figure 15.1 Rest-mass density contours in the equatorial plane for a neutron star binary orbiting close to the ISCO. These stars satisfy a polytropic equation of state with $n = 1$, and each star has a rest mass of $\bar{M}_0 = 0.169$, corresponding to a compaction in isolation of $(M/R)_\infty = 0.175$. (The maximum-mass configuration in isolation has $\bar{M}_0 = 0.180$ and $(M/R)_\infty = 0.215$.) The lines correspond to contours of constant rest-mass density in decreasing factors of 0.556. [From Baumgarte *et al.* (1998a).]

companions, parametrized by the maximum density inside each star.[10] Clearly, for infinite separation ($s = 1$) this sequence should reduce to an Oppenheimer–Volkoff sequence for each companion, at least up to numerical error.[11]

In Figure 15.2 we show some results for $n = 1$ polytropic binaries constructed by finite differencing on a very coarse, 3-dimensional grid. These particular models were computed on a uniform grid of $(64)^3$ grid points, with the outer boundary placed at such a distance that the stellar interior along the x-axis is always covered by 17 grid points. The dashed line represents the Oppenheimer–Volkoff sequence, while the solid lines represent binaries of successively decreasing binary separation s, moving from the bottom to top curve. The finite difference error in the binary calculations systematically underestimates the mass, which explains why these curves do not converge to the Oppenheimer–Volkoff result as one would expect. We nevertheless notice that the curves for smaller binary separations lie above those for larger binary separation in the graph. This result implies that the maximum allowed mass increases with decreasing binary separation and reflects the fact that tidal forces serve to stabilize neutron stars against radial instability.[12]

Figure 15.2 also has an important consequence for evolutionary sequences. For sufficiently large binary separations, at which the binary evolves very slowly, we can construct

---

[10] We allow for the possibility that the maximum density does not reside at the coordinate "center" of the star.

[11] Recall that Oppenheimer–Volkoff solutions satisfy *ordinary* differential equations, which can be solved to essentially arbitrary precision.

[12] See Baumgarte *et al.* (1998b) for a stability analysis of corotating, relativistic binary stars.

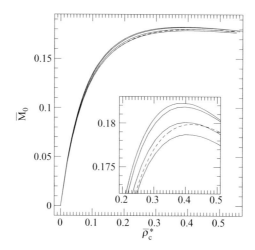

Figure 15.2 Rest mass $\bar{M}_0$ *vs.* central density $\bar{\rho}_c^*$ along sequences of equal-mass, corotational binary companions obeying an $n = 1$ polytropic EOS. Successive curves are for binary separations ranging from $s = 0.3$ (bottom solid line), through 0.2 and 0.1, and extending to 0.0 (top line). The dashed line is the Oppenheimer–Volkoff limiting curve for isolated static stars ($s = 1$). Due to finite difference errors, the numerical values systematically underestimate the mass, hence some of the curves creep below the Oppenheimer–Volkoff curve. The insert is a blow-up of the region around the maximum masses. [From Baumgarte *et al.* (1997).]

a quasiequilibrium inspiral sequence by "gluing" together quasiequilibrium models at different binary separations. For binary neutron stars the rest mass M_0 (e.g., the baryon number) must be constant along such an evolutionary sequence.[13] In Figure 15.2, such an evolutionary sequence must start at very large separation with configurations lying close to some point on the stable branch of the Oppenheimer–Volkoff sequence. As the binary emits gravitational radiation, loses energy and angular momentum and spirals inward, it must evolve along a horizontal line of constant rest mass M_0 in Figure 15.2. This implies that the maximum density *decreases* as the binary separation decreases and the stars are tidally elongated.

Whether the maximum density in binary neutron stars increases or decreases with decreasing binary separation was once the subject of an interesting controversy.[14] Originally, a "star-crushing" effect was reported by Wilson and Mathews (1995),[15] whereby the maximum density in binary neutron stars increases as the two stars approach each other, ultimately triggering a radial instability and a "binary-induced" collapse to two individual black holes prior to binary merger. This effect runs counter to Newtonian intuition and calculations, which argue that a tidal elongation in a quasiequilibrium fluid binary star

---

[13] Compare the discussion in Chapter 12.4 for binary black holes.
[14] See Kennefick (2000) for an account of the sociological aspects of this controversy.
[15] See also Wilson *et al.* (1996).

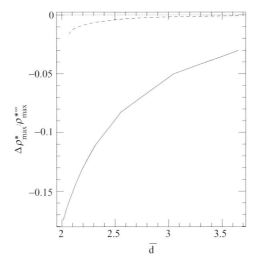

Figure 15.3 The relative change in the maximum density $\Delta\rho^*_{\max}/\rho^{*\infty}_{\max}$ as a function of the separation $\bar{d}$ between the locations of the maximum densities in an equal-mass, $n = 1$ polytropic binary with companions of rest-mass $\bar{M}_0 = 0.18$ (close to the maximum allowed mass for static stars). The solid line marks the changes in the density for corotational binaries, and the dashed line for irrotational binaries. The dotted lines marks $\Delta\rho^*_{\max} = 0$. This graph shows that the central density decreases with decreasing binaries for both corotational and irrotational binaries. Similar results hold for binaries of different masses and polytropic indices, and also for unequal-mass binaries. [After Taniguchi and Gourgoulhon (2002).]

should reduce the density inside a star.[16] As a consequence, the "star-crushing" effect was met with great skepticism in the community.

Our argument above, based on Figure 15.2, also suggests that the central density should decrease, and not increase, with decreasing binary separation. However, this argument only holds for corotational binaries. At small binary separations corotational binaries have to spin very fast to keep up with the orbit. This spin by itself leads to a flattening of the star and hence may lead to a reduction in the central density, even in isolated stars. Moreover, corotation is not maintained during binary inspiral for realistic neutron stars, as we discussed in Section 15.1. The more relevant question is therefore how the central density behaves in irrotational binaries. Anticipating findings from Section 15.3 we show in Figure 15.3 results for irrotational quasiequilibrium binaries, as well as more accurate results for corotational, binaries. This plot demonstrates that, while the effect is indeed much smaller for irrotational binaries than for corotational binaries, the central density decreases with decreasing binary separation in both cases.

Ultimately, the original controversy was resolved when a mistake was found in the calculations of Wilson and Mathews (1995) and Wilson *et al.* (1996) (see Flanagan 1999). Correcting this mistake reduced the size of the reported effect dramatically.[17] More significantly, no sign of a "crushing instability" has been observed in any other post-Newtonian

---

[16] See Rasio and Shapiro (1999) for a review and references.
[17] Mathews and Wilson (2000).

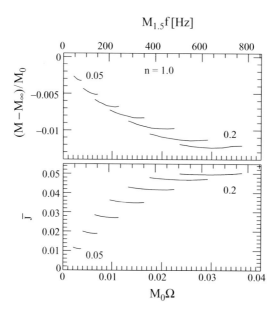

Figure 15.4 The binding energy and angular momentum as functions of orbital angular velocity for several different values of the rest mass $\bar{M}_0$. The curves are labeled by the compaction $(M/R)_\infty$ that the stars would have at infinite binary separation, starting with 0.05 and increasing in steps of 0.0025 up to 0.2. The maximum compaction of a stable, isolated, nonrotating $n = 1$ polytrope is 0.217. The upper label gives the orbital frequency for stars with a rest mass of $1.5M_\odot$. [From Baumgarte *et al.* (1998a).]

or fully relativistic simulation of neutron star inspiral and merger[18] as we will discuss in Chapter 16, although no theorem has been proven to completely rule out the possibility.[19]

As we discussed in Chapter 12.1, the innermost stable circular orbit (ISCO) marks a transition from a slow, quasistationary inspiral to a rapid plunge and merger. While this transition may not be particularly abrupt, crossing the the ISCO will mark a transition in the emitted gravitational waveform away from a quasiperiodic chirp signal (see Chapter 16). Locating the ISCO and the orbital frequency to which it corresponds are therefore important goals when constructing quasiequilibrium sequences of binary neutron stars.

We can identify the ISCO by locating the minimum of the total ADM energy M_{ADM} along an evolutionary sequence of constant rest-mass M_0 (see Chapter 12.1). As discussed in Chapter 12 in conjunction with equation (12.112),

$$dM_{\text{ADM}} = \Omega dJ, \tag{15.62}$$

a minimum in the energy should coincide with a minimum in the angular momentum. In Figure 15.4 we show some numerical results obtained on a coarse grid for the fractional binding energy $(M_{\text{ADM}} - M_{\text{ADM}}^\infty)/M_0$ and the angular momentum $\bar{J}$ for sequences of

---

[18] Such an effect *can* occur in binaries consisting of *collisionless* clusters; Shapiro (1998); Duez *et al.* (1999).
[19] See, e.g., Favata (2006).

Table 15.1 The orbital angular velocity Ω, angular momentum J and ADM mass $\bar{M}_{\mathrm{ISCO}}$ at the ISCO, for corotating, equal mass, binary neutron stars with polytropic index $n = 1$. We tabulate the rest-mass $\bar{M}_0$, the ADM mass $\bar{M}_\infty$ and the compaction $(M/R)_\infty$ each star would have in isolation (where R is the areal radius), as well as the angular velocity $M_0\Omega$, angular momentum $J_{\mathrm{tot}}/M_{\mathrm{tot}}^2$ (where "tot" means total ADM value) and ADM mass $\bar{M}_{\mathrm{ISCO}} \equiv M_{\mathrm{tot}}/2$ at the ISCO. For this EOS the maximum rest mass in isolation is $\bar{M}_0^{\mathrm{max}} = 0.180$. [After Taniguchi and Gourgoulhon (2002).]

$\bar{M}_0$	$\bar{M}_\infty$	$(M/R)_\infty$	$M_0\Omega_{\mathrm{ISCO}}$	$(J_{\mathrm{tot}}/M_{\mathrm{tot}}^2)_{\mathrm{ISCO}}$	$\bar{M}_{\mathrm{ISCO}}$
0.130	0.122	0.12	0.0139	1.136	0.1213
0.146	0.136	0.14	0.0175	1.075	0.1350
0.160	0.148	0.16	0.0217	1.028	0.1464
0.171	0.157	0.18	0.0253	0.994	0.1550

constant rest mass $\bar{M}_0$. In the definition of the binding energy, M_{ADM}^∞ denotes the ADM mass that this star would have at infinite binary separation. We plot these quantities *vs.* the orbital frequency instead of binary separation, because the former is a gauge-invariant quantity while the latter is not.[20] Figure 15.4 suggests that the minima in the binding energy (and hence the ADM mass) do indeed coincide with those of the angular momentum, and thereby identify the ISCO.

In Table 15.1 we list the angular velocity $M_0\Omega$, the total angular momentum $J_{\mathrm{tot}}/M_{\mathrm{tot}}^2$ (where M_{tot} is the total ADM mass of the binary system), and the individual ADM mass $M_{\mathrm{ISCO}} \equiv M_{\mathrm{tot}}/2$ at the ISCO for different values of the stellar rest mass M_0 for equal-mass, $n = 1$ polytropes. As it turns out, only stars with sufficiently stiff equations of state ($n \lesssim 1.0$) encounter an ISCO before they touch, which is in qualitative agreement with Newtonian results.[21] Stars with softer equations of state are more centrally condensed and have more extended envelopes, so that they come into contact and merge before reaching an ISCO.

As we discussed in Section 15.1 above, the assumption of corotation is very unrealistic, since it requires an unrealistically large viscosity. To obtain results that are of greater astrophysical relevance than those found in this section, we must consider irrotational binaries.

15.3 Irrotational binaries

We can find a set of equations that governs irrotational binaries in close analogy to our Newtonian treatment in Section 15.1.1.[22] In general relativity, the vorticity tensor ω_{ab} can

---

[20] Also, the observed gravitational wave spectrum is dominated by mass quadrupole emission at frequency $f_{GW} = \Omega/\pi$.

[21] See, e.g., Rasio and Shapiro (1999) for a review and references.

[22] The first relativistic formalism for irrotational binaries in quasiequilibrium was presented in Bonazzola *et al.* (1997). Subsequently, other approaches were developed by Asada (1998), Teukolsky (1998) and Shibata (1998). Gourgoulhon (1998) demonstrated that all of these formulations are equivalent. The derivation here follows that of Shibata (1998).

be defined as

$$\omega_{ab} \equiv P_a{}^c P_b{}^d \left(\nabla_d (h u_c) - \nabla_c (h u_d) \right), \tag{15.63}$$

where $P_a{}^b = g_a{}^b + u_a u^b$ is the projection operator with respect to the fluids's 4-velocity u^a (equation 5.67). For irrotational binaries the vorticity must vanish,

$$\omega_{ab} = 0, \tag{15.64}$$

which is equivalent to the Newtonian condition (15.27). For this to be the case, the quantity $h u_a$ must be the gradient of a velocity (scalar) potential Φ,

$$h u_a = \nabla_a \Phi; \tag{15.65}$$

cf. (15.28).

> **Exercise 15.6**   Show that with equation (15.65) the continuity equation (15.2) reduces to
>
> $$\nabla_a \left(\frac{\rho_0}{h} \nabla^a \Phi \right) = 0, \tag{15.66}$$
>
> while the Euler equation (15.34) is satisfied identically.

> **Exercise 15.7**   (a) Use the Euler equation (15.34) to show that ω_{ab} can be written
>
> $$\omega_{ab} = \nabla_b (h u_a) - \nabla_a (h u_b). \tag{15.67}$$
>
> (b) Evaluate equation (5.59) to show that for irrotational flow in an isentropic fluid the circulation satisfies $\mathcal{C} = 0$.

With the definition (15.36), the spatial projection of equation (15.65) becomes

$$\hat{u}_i = D_i \Phi. \tag{15.68}$$

In the Euler equation (15.42) this result now eliminates the second term, so that we can integrate the first term once to find

$$\frac{h}{u^t} + \hat{u}_i V^i = C, \tag{15.69}$$

where C is again a constant of integration.

It is now convenient to introduce a rotational shift vector

$$B^a = \beta^a + k^a, \tag{15.70}$$

where k^a is given by equation (15.11) as $k^a = \xi_{\mathrm{hel}}^a - t^a$. From equation (2.98) we have $t^a = \alpha n^a + \beta^a$, so that we can write B^a as

$$B^a = \xi_{\mathrm{hel}}^a - \alpha n^a. \tag{15.71}$$

We now solve equation (15.35) for V^a, substitute equation (15.71) for ξ_{hel}^a, take the spatial projection (to eliminate the αn^a term), substitute equation (15.68) for u^i, and obtain finally

$$V^i = \frac{1}{u^t h} D^i \Phi - B^i. \tag{15.72}$$

As we did for corotational binaries, we now use the normalization $u_a u^a = -1$ to express u^t in terms of the spatial components of the 4-velocity. Substituting equation (15.68) into (15.47) yields

$$\alpha u^t = \left(1 + h^{-2} D_i \Phi D^i \Phi\right)^{1/2}. \tag{15.73}$$

We next combine equations (15.69), (15.72) and (15.73) to find

$$\alpha u^t = \frac{1}{\alpha h} \left(C + B^i D_i \Phi\right). \tag{15.74}$$

We digress to note an interesting result derived in exercise 15.8.

Exercise 15.8 Use $\mathcal{L}_{\xi_{hel}}(hu_a) = 0$ to show that

$$\mathcal{L}_{\xi_{hel}} \Phi = \xi^a \nabla_a \Phi = -C' \tag{15.75}$$

where C' is a constant.

Naively, we might have expected that the Lie derivative of Φ along the Killing vector ξ_{hel} vanishes. However, Φ is only a potential and not a measurable (physical) quantity. Only the gradient of Φ, $\nabla_a \Phi = hu_a$, is measurable, and therefore only for the gradient can we impose the Lie derivative along ξ_{hel} to vanish.[23] As we have found in exercise 15.8, this introduces a constant of integration when we evaluate the Lie derivative of Φ itself. We can now use the result (15.75) to provide an alternative derivation of equation (15.74) (see exercise 15.9).

Exercise 15.9 Replace n^a in $h\alpha u^t = -hn_a u^a = -n^a \nabla_a \Phi$ with the help of equation (15.71) to rederive equation (15.74).

We can finally combine equations (15.73) and (15.74) to find

$$h^2 = \frac{1}{\alpha^2} (C + B^i D_i \Phi)^2 - D_i \Phi D^i \Phi. \tag{15.76}$$

As in the Newtonian derivation, the integrated Euler equation (15.69) furnishes an algebraic expression for the enthalpy h. We could have left this equation in the form (15.69), but since h also appears in V^i we have taken a few more steps to isolate h in equation (15.76).

We still need an equation for the velocity potential Φ, which we can find from the continuity equation (15.43). Inserting (15.72) into (15.43) we obtain

$$D_i(\alpha \rho_0 h^{-1} D^i \Phi) - D_i(\alpha u^t \rho_0 B^i) = 0. \tag{15.77}$$

---

[23] See also Teukolsky (1998), who shows formally that $\mathcal{L}_{\xi_{hel}} hu_a = 0 = \mathcal{L}_{\xi_{hel}} \widetilde{\mathbf{d}\Phi} = \widetilde{\mathbf{d}}\mathcal{L}_{\xi_{hel}} \Phi$ (see exercise A.3), which implies equation (15.75).

To eliminate u^t we can now insert equation (15.74) and find, just like in the Newtonian case, an elliptic equation for the velocity potential Φ,

$$\boxed{D_i D^i \Phi - D_i \left(\frac{C + B^j D_j \Phi}{\alpha^2} B^i \right) = \left(\frac{C + B^j D_j \Phi}{\alpha^2} B^i - D^i \Phi \right) D_i \ln \frac{\alpha \rho_0}{h}.} \qquad (15.78)$$

Equations (15.76) and (15.78) now provide two equations for the two fluid variables h and Φ. Equation (15.78) is an elliptic equation and hence needs to be supplemented with boundary conditions for Φ on the stellar surface. Since ρ_0 vanishes there, regularity requires

$$\left(\frac{C + B^j D_j \Phi}{\alpha^2} B^i - D^i \Phi \right) D_i \rho_0 \Bigg|_{\text{surface}} = 0. \qquad (15.79)$$

This is again a Neumann boundary condition. Just like its Newtonian equivalent (15.33) this condition enforces the fluid flow to be tangent to the surface.

> **Exercise 15.10** Show that in the Newtonian limit equations (15.76), (15.78) and (15.79) reduce to the Newtonian equations (15.31), (15.32) and (15.33).

In most applications to date, the matter equations (15.76) and (15.78) have been solved together with the conformal thin-sandwich equations (3.109)–(3.112) for the lapse α, the shift β^i and the conformal factor ψ. Constructing irrotational binaries is much more involved than constructing corotational binaries. By contrast with corotational binaries, where only one algebraic equation (15.46) has to be solved for the enthalpy, we now have to solve the enthalpy equation (15.76) together with the elliptic equation (15.78) for the velocity potential Φ, subject to the boundary condition (15.79) on the surface of the star. The latter adds another complication, since the location of the surface is not known a priori.

Bonazzola *et al.* (1999a) and Gourgoulhon *et al.* (2001) solved this problem with the help of a multidomain, spectral method that covers the entire computational domain with several patches of coordinate systems. In particular, the interiors of the stars are covered with spherical-type coordinate systems, which are constructed so that the surface of the star lies at a constant value of the radial coordinate. Such coordinate systems are called "surface-fitting" coordinates, and are very well suited for imposing the boundary condition (15.79). A similar algorithm, based on Newtonian simulations of irrotational neutron star binaries,[24] has been used by Uryū *et al.* (2000).[25]

As a brief, technical detour, we point out why it is natural for these calculations in spherical polar coordinate systems to use two computational domains, each one centered on one of the binary companions. Consider an equation of the form

$$\bar{D}^2 \phi = \text{RHS}, \qquad (15.80)$$

---

[24] Uryū and Eriguchi (1999).
[25] See also Uryū and Eriguchi (2000).

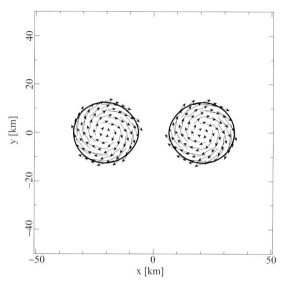

Figure 15.5 The internal velocity field with respect to the corotating frame in the orbital plane for identical stars of rest mass $M_0 = 1.625 M_\odot$ at a coordinate separation of 41 km. The binary stars are irrotational and constructed from an $n = 1$ polytropic EOS with $K = 1.8 \times 10^{-2} \, \mathrm{J \, m^3/kg^2}$. The thick lines mark the stellar surfaces. [From Gourgoulhon *et al.* (2001).]

where RHS symbolizes some potentially nonlinear source terms on the right hand side. We can choose to solve this equation by splitting it into the two equations

$$\bar{D}^2 \phi_1 = \mathrm{RHS}_1,$$
$$\bar{D}^2 \phi_2 = \mathrm{RHS}_2, \qquad (15.81)$$

where $\phi = \phi_1 + \phi_2$ and $\mathrm{RHS} = \mathrm{RHS}_1 + \mathrm{RHS}_2$, and where each equation is now associated with one of the two stars. The two equations in (15.81) can then be solved on two separate computational domains, each one centered on one star. Clearly, the separation of the source terms RHS into the two parts RHS_1 and RHS_2 is far from unique. One guiding principle is to move those parts of RHS that are large in the neighborhood of star 1 into RHS_1, and similarly for the other star. Another principle is that each of the source terms should asymptotically coincide with those for the corresponding isolated star when the binary separation is large.

We show a typical binary configuration and its internal velocity field in Figure 15.5. As we did for corotational binaries, we can determine the maximum allowed mass of neutron stars in irrotational binaries by first finding the mass as a function of central density for fixed separation, and then varying the separation. In Figure 15.6 we show results of Uryū *et al.* (2000), which demonstrate that, as in corotational binaries, the maximum mass increases with decreasing separation. However, by comparing with Figure 15.2, we note that the increase in maximum mass is smaller for irrotational binaries than for corotational

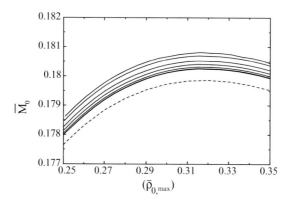

Figure 15.6 The rest mass $\bar{M}_0$ as a function of maximum density $\bar{\rho}_0$ for separations $\bar{d} = 1.3125, 1.375,$ 1.5, 1.625, 1.75, 1.875 and 2 (thick lines running from top to bottom) of irrotational, binary neutron stars constructed from an $n = 1$ polytropic EOS. The dashed line is the Oppenheimer–Volkoff result. [From Uryū *et al.* (2000).]

binaries.[26] This result is not surprising, since neutron stars in corotational binaries are spinning with respect to the inertial frame at rest with respect to the binary center. This spin by itself increases the maximum mass of neutron stars.[27] It is also evident from Figures 15.6 and 15.2 that, while the density along evolutionary sequences of irrotational binaries of constant rest mass $\bar{M}_0$ decreases with decreasing separation, the decrease is less than that for corotational binaries, which we had already anticipated in Figure 15.3.

As we have seen in Section 15.2, evolutionary sequences of corotational binaries typically end either at the ISCO or at contact. Irrotational sequences, on the other hand, typically end when a *cusp* forms on the stellar surface, prior to reaching the ISCO or contact.[28] Such a cusp forms when the stellar surface reaches an inner Lagrange point, so that matter can start flowing from one star to its companion.

A potential disadvantage of using spectral methods for constructing these binaries is that the appearance of such a cusp on the stellar surface leads to Gibbs phenomena and hence decreasing accuracy. Typically, sequences constructed with spectral methods must therefore terminate shortly before the appearance of a cusp. To identify cusp formation, many authors[29] introduce a "mass-shedding indicator" χ, defined as the ratio between the radial derivative of the specific enthalpy h on the neutron star equator at the point nearest to the companion to the radial derivative at the pole,

$$\chi \equiv \frac{\partial_r(\ln h)|_{\text{eq}}}{\partial_r(\ln h)|_{\text{pole}}}. \tag{15.82}$$

---

[26] This result is especially evident taking into account the coarse resolution used by Baumgarte *et al.* (1998a), which underestimate the masses in Figure 15.2.

[27] See, e.g., Cook *et al.* (1994b) as well as Chapter 14.

[28] Bonazzola *et al.* (1999a); Uryū and Eriguchi (2000); Uryū *et al.* (2000).

[29] See, e.g., Gourgoulhon *et al.* (2001).

Table 15.2 The orbital angular velocity Ω, angular momentum J and the ADM mass $\bar{M}$ at cusp formation, for irrotational, equal-mass, binary neutron stars with polytropic index $n = 1$. We tabulate the individual rest mass $\bar{M}_0$, the ADM mass $\bar{M}_\infty$ and the compaction $(M/R)_\infty$ each star would have in isolation (where R is the areal radius), as well as the angular velocity $M_0\Omega$, the angular momentum $J_{\text{tot}}/M_{\text{tot}}^2$ (where "tot" means total ADM value) and the ADM mass $\bar{M}_{\text{CUSP}} = M_{\text{tot}}/2$ at cusp formation. For this EOS, the maximum rest mass in isolation is $\bar{M}_0^{\max} = 0.180$. [After Taniguchi and Gourgoulhon (2002).]

$\bar{M}_0$	$\bar{M}_\infty$	$(M/R)_\infty$	$M_0\Omega_{\text{CUSP}}$	$(J_{\text{tot}}/M_{\text{tot}}^2)_{\text{CUSP}}$	$\bar{M}_{\text{CUSP}}$
0.130	0.122	0.12	0.0142	0.997	0.1211
0.146	0.136	0.14	0.0181	0.947	0.1347
0.160	0.148	0.16	0.0222	0.910	0.1460
0.171	0.157	0.18	0.0265	0.881	0.1546

This indicator is identically one for a spherical star and falls to zero at the appearance of a cusp. We can then tabulate χ as a function of binary separation, or orbital angular velocity, and extrapolate to $\chi = 0$ to identify the onset of cusp formation.

Uryū *et al.* (2000) find that equal-mass, irrotational binaries reach an ISCO before a cusp appears only for very stiff equations of state ($n \lesssim 2/3$), while binaries with softer equations of state form a cusp first. In Table 15.2 we list some parameters for irrotational, equal-mass $n = 1$ binaries at cusp formation. Comparing with the corresponding ISCO parameters of corotational binaries in Table 15.1, we note that the cusp and ISCO occur at quite similar frequencies. The corotational binaries have more angular momentum, because the individual stars carry a spin in addition to the orbital angular momentum of the binary. We also find that the binding energy $(\bar{M} - \bar{M}_\infty)/\bar{M}_0$ of corotational binaries is slightly larger than for irrotational binaries. This is because the ADM mass $\bar{M}$ of the former include the additional spin kinetic energies of the individual stars.[30] For additional numerical results, including models for different equations of state and unequal-mass binaries, we refer the reader to the literature.[31]

In all of the above calculations the matter equations (15.76), (15.78) and (15.79) have been solved together with the conformal thin-sandwich formalism of Chapter 3.3 for the gravitational fields, as summarized in Box 3.3. However, it is not clear that this formalism is the best possible approach for constructing quasiequilibrium binaries. In particular, this approach makes an ad hoc choice for the conformally-related background metric $\bar{\gamma}_{ij}$, which may or may not lead to the best approximation of quasiequilibrium. Several authors have therefore suggested alternative methods that may produce more accurate models of compact binaries in quasiequilibrium.[32]

---

[30] See also the discussion in Duez *et al.* (2002).
[31] See, e.g., Taniguchi and Gourgoulhon (2002, 2003).
[32] See, e.g., Blackburn and Detweiler (1992); Andrade *et al.* (2004); Shibata *et al.* (2004); Friedman and Uryū (2006) and references therein.

One of these alternative approaches is the "waveless" approximation of Shibata *et al.* (2004) that we discussed briefly in Chapter 3.4. Uryū *et al.* (2006) implemented this scheme for binary neutron stars. As expected, for stars with small compaction they obtain results that are very similar to those found with the conformal thin-sandwich approach assuming conformal flatness. For stars with a larger compaction, however, their results show some deviation from both the conformally flat models as well as third post-Newtonian (3PN) point-mass calculations. For stars with a compaction $(M/R)_\infty = 0.17$, at a binary separation at which the coordinate distance from the orbital center to the geometric center of each star is $1.75 R_\infty$, the conformally related metric, for example, deviates from a flat metric by about 1%. The deviations in the binding energy are comparable. To establish whether or not these models are closer to quasiequilibrium than the conformally flat models presumably requires a fully dynamical hydrodynamic simulation. We describe such simulations for binary neutron stars in Chapter 16.

15.4 Quasiadiabatic inspiral sequences

In the previous sections we have discussed how we can construct individual models of binary neutron stars in circular orbit at arbitrary binary separations. We have also mentioned that we can "stitch" together models of constant rest mass M_0 to build evolutionary inspiral sequences. In this section we will describe how such a "quasiadiabatic" approach can be used to calculate the gravitational wave signal from the late inspiral phase, prior to plunge and merger.[33]

We start by approximating the binary orbit outside the ISCO as circular, and by treating the orbital decay as a small correction. For each binary separation we can then find the matter distribution using the quasiequilibrium methods of the previous sections. Next we need to find the gravitational wave signal and luminosity at a given binary separation. As a crude approximation we could simply use the quadrupole formula, equation (9.37), but we can do better than that. We can substitute the quasiequilibrium matter profiles that we obtained above as source terms in the Einstein field equations, and then evolve these equations. In effect, we are performing a relativistic hydrodynamics simulation without having to resolve the hydrodynamic equations, given that the matter is in near equilibrium. This approach is sometimes referred to as "hydro-without-hydro".[34]

Evolving the gravitational fields for the given binary matter distribution, we can then read off the gravitational waveform and luminosity, and hence the rate at which the binary loses energy at a given separation r. Combining this luminosity $dM_{\mathrm{ADM}}/dt = -L_{\mathrm{GW}}$ with the derivative of the ADM mass with respect to separation r along a quasiequilibrium

---

[33] Similar ideas have been suggested as possible solutions to the "intermediate compact binary problem" that may arise if (point-mass) post-Newtonian techniques, which are favored for the modeling of the adiabatic inspiral phase, break down at too large a binary separation for fully dynamical numerical relativity simulations to track the entire remaining inspiral. Bridging the resulting gap between post-Newtonian and numerical relativity methods might then require some "quasiadiabatic" technique similar to the one sketched in this section to treat the full inspiral phase (see Brady *et al.* 1998 and Duez *et al.* 2002).

[34] See Baumgarte *et al.* (1999).

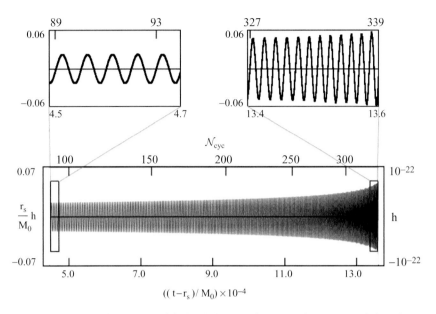

Figure 15.7 The final hundreds of cycles of the inspiral gravitational waveform measured along the axis of rotation by a distant observer as a function of retarded time and cycle number for an irrotational binary neutron star system. The numbers give the strain h for a binary of total rest mass $M_0 = 2 \times 1.5 M_\odot$ at a distance $r_s = 100$ Mpc. [From Duez *et al.* (2002).]

sequence of fixed rest-mass yields the inspiral rate

$$\frac{dr}{dt} = \frac{dM_{\mathrm{ADM}}/dt}{dM_{\mathrm{ADM}}/dr}. \tag{15.83}$$

Integrating equation (15.83) then yields the separation as a function of time, $r(t)$. For example, equation (12.17) quotes the result for point-mass, Newtonian binaries radiating in the quadrupole approximation. Given that we already know the gravitational waveform as a function of r, we can then construct the entire gravitational wave train as a function of t.

> **Exercise 15.11** Use the results of exercise 12.3, together with equation (9.27), to calculate the quasiadiabatic inspiral gravitational wave train for a Newtonian binary. In particular, determine as functions of time the amplitude $A(t)$ and the phase $\Phi(t)$ of the wave amplitude along the binary axis of rotation in the quadrupole approximation,
>
> $$r_s h_+ = r_s h_\times = A(t) \cos\Phi(t), \tag{15.84}$$
>
> where r_s is the distance to the source.

Duez *et al.* (2002) implemented a crude but illustrative version of such a scheme for both corotational and irrotational relativistic binary inspiral.[35] In Figure 15.7 we show the resulting quasiadiabatic gravitational wave train for irrotational binary inspiral.

---

[35] See also Duez *et al.* (2001); Shibata and Uryū (2001).

For a given separation, the gravitational wave luminosity dM_{ADM}/dt is very similar for corotational and irrotational models.[36] However, the ADM energy M_{ADM} of corotational models includes the spin kinetic energy of the individual stars, and as a consequence M_{ADM} decreases less for corotational binaries than for irrotational binaries as the binary separation decreases. For corotational binaries $|dM_{\mathrm{ADM}}/dr|$ is therefore smaller than for irrotational binaries, so that, according to (15.83), the inspiral of corotational binaries proceeds faster than that of irrotational binaries.

Finally, we can construct the entire gravitational wave train by matching the quasiadiabatic gravitational wave train to the waveform arising from plunge and merger phases.[37] Calculating the waveform emitted during the highly dynamical plunge and merger phases is the focus of Chapter 16.

---

[36] This result is not surprising, since the gravitational wave emission is dominated by the matter mass density, which is fairly similar for corotational and irrotational binaries, while the matter current density is less important.
[37] Figure 6 in Duez *et al.* (2002) exhibits such a match.

16 Binary neutron star evolution

Binary neutron stars have always been of great interest to relativists and astrophysicists. Binary neutron stars are known to exist. Approximately a half-dozen have been identified to date in our own galaxy, and, for some of these, general relativistic effects in the binary orbit have been measured to high precision.[1] The discovery of the first binary pulsar, PRS 1913 + 16, by Hulse and Taylor (1975), led to the observational confirmation of Einstein's quadrupole formula for gravitational wave emission in the slow-motion, weak-field regime of general relativity. The inspiral and coalescence of binary neutron stars is one of the most promising scenarios for the generation of gravitational waves detectable by laser interferometers. With the construction of the first of these interferometers completed, and planned upgrades already scheduled, it is of growing urgency that theorists be able to predict the gravitational waveform emitted during the merger of the two stars. The low-frequency *inspiral* waveform is emitted early on, before tidal distortions of the stars become important, and it can be calculated fairly accurately by performing high-order post-Newtonian expansions of the equations of motion for two *point* masses.[2] The high-frequency *coalescence* waveform is emitted at the end, during the epoch of tidal distortion, disruption and merger, and it requires the combined machinery of relativistic hydrodynamics (or MHD) and numerical relativity. These tools are necessary to determine not only the waveform in the strong-field regime but also the final fate of the merged remnant. One of the key issues is determining whether a merged remnant collapses to a black hole immediately after coalescence ("prompt collapse") or instead forms a transient, dynamically stable, differentially rotating, hypermassive star that only later undergoes collapse due to dissipative secular effects ("delayed collapse"). These different outcomes will leave distinguishing imprints on the late-epoch gravitational waveform.

Gravitational wave generation may be only one of several observable consequences of binary neutron star merger. Gamma-ray bursts (GRBs) may be another, as there are many theoretical models for which the coalescence of a binary neutron star provides the energy source for a GRB.[3] Binary neutron stars, as well as black hole-neutron star binaries, are currently invoked to explain one class of GRBs, namely the short, hard GRBs that are characterized by their short duration and hard radiation spectrum. Currently favored are scenarios in which the merger leads to the formation of a rapidly rotating black

---

[1] Taylor and Weisberg (1989); Stairs *et al.* (1998).
[2] See Chapter 9.4 and Appendix E for discussion and references; see also Chapter 12.1 for an overview of binary inspiral.
[3] Paczynski (1986); Eichler *et al.* (1989); Narayan *et al.* (1992); Ruffert and Janka (1998).

hole surrounded by a torus of debris, wherein the energy of the burst originates from either $\nu\bar{\nu}$ annihilation or from the rotational energy of the black hole.[4] In addition, decompressed nuclear matter ejected during binary neutron star coalescence may provide an explanation for the observed abundance of r-process nuclei.[5]

Relativistic binary systems, like binary black holes, neutron stars and black hole-neutron stars, pose a fundamental challenge to theorists. Indeed, the two-body problem has been one of the most outstanding problems in classical general relativity. Tackling this problem in its many incarnations represents one of the most important applications of numerical relativity, and providing solutions constitutes one of numerical relativity's greatest triumphs.[6]

16.1 Peliminary studies

The earliest computational work on binary neutron stars consisted of simulations of head-on collisions of identical stars.[7] The restriction to head-on collisions enables the calculation to be performed in axisymmetry (i.e., $2 + 1$ dimensions), which requires far fewer computational resources than simulations in full $3 + 1$ dimensions. However, astrophysical scenarios that lead to binary neutron star collisions most likely involve inspiral in a quasi-circular orbit, since gravitational radiation will circularize a bound, eccentric binary orbit well before stellar contact.[8]

Newtonian studies of instabilities in binary systems in close circular orbit, and the nonlinear evolution of unstable binaries all the way to complete coalescence, have been the subject of many early investigations. The classical work by Chandrasekhar[9] for equilibrium binaries composed of incompressible fluids has been extended to compressible fluids.[10] These analytic studies identify the existence of dynamical and secular instabilities in sufficiently close systems. They even give rise to an approximate set of hydrodynamical evolution equations, incorporating viscosity and gravitational radiation reaction, that can be integrated to model binary inspiral and the corresponding gravitational radiation waveforms.[11] Although these simplified analytic studies can give appreciable physical insight into tidal effects and global fluid instabilities, fully numerical calculations are essential for establishing rigorous stability limits for close binaries and for following the nonlinear evolution of unstable systems all the way to complete coalescence, even in Newtonian theory.

---

[4] See Piran (2005) for review and references.
[5] Symbalisty and Schramm (1982); Eichler *et al.* (1989); Rosswog *et al.* (1998).
[6] For early overviews and references, see, e.g., Rasio and Shapiro (1999); Font (2000); Baumgarte and Shapiro (2003c).
[7] For Newtonian simulations, see Gilden and Shapiro (1984); for relativistic simulations, see Wilson (1979); Abrahams and Evans (1992); Jin and Suen (2007), and references therein.
[8] Peters (1964); see Chapter 12.1.
[9] Chandrasekhar (1969).
[10] Lai *et al.* (1993a, 1994a,c); Taniguchi and Nakamura (2000a,b).
[11] Lai *et al.* (1994c); Lai and Shapiro (1995b), and references therein. In the triaxial ellipsoid treatment adopted here, the hydrodynamical equations in $3 + 1$ reduce to *ordinary* differential equations. The treatment is exact for incompressible fluids orbiting in tidal gravitational fields. See also Carter and Luminet (1983, 1985); Luminet and Carter (1986); Kochanek (1992a,b); Kosovichev and Novikov (1992).

Given the absence of any underlying spatial symmetry in the problem, these calculations must be done in $3 + 1$ dimensions.

Newtonian simulations of coalescing neutron stars have been performed by numerous investigators employing a variety of numerical methods and emphasizing different aspects of the problem. Nakamura and collaborators[12] were the first to perform hydrodynamic simulations of binary neutron star (sometimes abbreviated NSNS) coalescence from circular orbits. They used an Eulerian finite-difference code for the hydrodynamics and focused on gravitational wave generation. Rasio and Shapiro[13] employed the Lagrangian SPH method (see Chapter 5.2). They focused on determining the stability properties of initial binary models in strict hydrostatic equilibrium and calculating the emission of gravitational waves from the coalescence of unstable binaries. Many of their results were confirmed by New and Tohline,[14] who used completely different numerical methods but also focused on stability questions, and by Zhuge *et al.*,[15] who also used SPH. The later group also explored the dependence of the gravitational wave signals on the initial neutron star spins. Davies *et al.*,[16] who used SPH, and Ruffert *et al.*[17] who employed a high-resolution shock-capturing (HRSC) hydrodynamics scheme, incorporated a simple treatment of the nuclear physics in their hydrodynamic calculations, motivated by models of gamma-ray bursts and the ejection of r-process nuclei.

16.2 The conformal flatness approximation

One of the first approaches used to simulate binary neutron star coalescence in general relativity was the conformal flatness approximation, pioneered by Wilson and Mathews.[18] In this approach, one assumes that the dynamical degrees of freedom of the gravitational field, i.e., the gravitational radiation, play a negligible role in determining the dynamical behavior or structure of the neutron stars. Simplifying the spacetime to reduce radiative influences, one sets the initial spatial metric to be conformally flat, $\gamma_{ij} = \psi^4 \eta_{ij}$, so that the spacetime metric takes the form

$$ds^2 = -\alpha^2 dt^2 + \psi^4 \eta_{ij}(dx^i + \beta^i dt)(dx^j + \beta^j dt). \tag{16.1}$$

One further assumes that the spatial metric remains conformally flat *at all times*, an approximation that greatly simplifies the resulting field equations. For example, in the ADM scheme, the traceless part of equation (2.134) then has to vanish, which, according to equations (3.92) and (3.93), yields

$$A^{ij} = \frac{1}{2\alpha}(L\beta)^{ij}. \tag{16.2}$$

---

[12] See Nakamura and Oohara (1991) and references therein.
[13] Rasio and Shapiro (1992, 1994, 1995).
[14] New and Tohline (1997).
[15] Zhuge *et al.* (1994, 1996).
[16] Davies *et al.* (1994); Rosswog *et al.* (1999).
[17] Ruffert *et al.* (1996); Ruffert and Janka (1998).
[18] Wilson and Mathews (1989, 1995). See also Isenberg (1978); Wilson *et al.* (1996).

Here A^{ij} is the traceless part of the extrinsic curvature, and the vector gradient L is defined in (3.50). We will discuss the validity of this approximation below.

The resulting equations can be derived in complete analogy to our treatment of the conformal thin-sandwich decomposition in Chapter 3.3. In particular, we can substitute equation (16.2) into the momentum constraint (2.133) to yield an equation for the shift vector β^i. The conformal factor ψ can be found from the Hamiltonian constraint (2.132), and, adopting maximal slicing with $K = 0 = \partial_t K$ (see Chapter 4.2), we can also obtain an equation for the lapse function α. With the conformal rescaling relation (3.35) for A^{ij}, these equations then reduce to the thin-sandwich equations (3.116)–(3.118) with $\bar{R} = 0$ and flat-space differential operators,[19]

$$\Delta^{\text{flat}}\psi = -\frac{1}{8}\psi^{-7}\bar{A}_{ij}\bar{A}^{ij} - 2\pi\psi^5\rho \tag{16.3}$$

$$(\Delta_L^{\text{flat}}\beta)^i = 2\bar{A}^{ij}\bar{D}_j(\alpha\psi^{-6}) + 16\pi\alpha\psi^4 j^i \tag{16.4}$$

$$\Delta^{\text{flat}}(\alpha\psi) = \alpha\psi\left(\frac{7}{8}\psi^{-8}\bar{A}_{ij}\bar{A}^{ij} + 2\pi\psi^4(\rho + 2S)\right). \tag{16.5}$$

This coupled system now completely determines the metric (16.1).

All unknowns in the metric (16.1) are determined by elliptic equations, and in this sense all dynamical degrees of freedom have been removed from the gravitational fields. In this approach, one solves an initial value problem at each instant of time, as opposed to dynamically evolving the gravitational fields. While one may be concerned about the accuracy of this approximation (see below), it greatly simplifies the field equations and allowed Wilson and Mathews (1995) to perform some of the first relativistic simulations of binary neutron stars. In this approach, the time step is limited by the matter sound speed and not the light speed, since there are no dynamical field equations. Hence the time step can be much larger than in fully self-consistent algorithms. In their simulations, Wilson and Mathews (1995) solved equations (3.116)–(3.118) for the metric (16.1) simultaneously with Wilson's formulation of the relativistic equations of hydrodynamics, equations (5.12)–(5.14). In their approach, at each timestep one first evolves the matter variables, and then solves the field equations for the metric, with the new matter distribution used to compute the source terms. If desired, the gravitational wave emission can then be estimated a posteriori using the quadrupole formula or some other low-order expansion scheme. Alternatively, the matter profile computed via the conformal flatness approximation can be used to calculate the source terms in the full dynamical field equations to determine the gravitational radiation as a perturbation. The later approach has already been discussed in Chapter 15.4.

A useful computational check of a code that adopts the conformal flatness approxima- tion is provided by the fact that the conformally flat evolution equations maintain *strict* stationary equilibrium for initial data satisfying the conformal thin-sandwich equations

---

[19] See exercise 3.17 for an expansion of the differential operators into partial derivatives.

for the metric coupled to the stationary equilibrium equations for the matter. These are the initial data we obtained for relativistic binaries in circular equilibrium in Chapter 15.1.1.

> **Exercise 16.1**   Consider a corotating, circular equilibrium binary constructed as in Chapter 15.1.1 using the conformal thin-sandwich approximation coupled to the integrated Euler equation (15.46). Confirm that these initial data are strictly conserved in the the conformal flatness approximation. Specifically, show that the conformal thin-sandwich initial value solution provides a stationary solution to the hydrodynamic matter evolution equations (5.12), (5.13) and (5.14), coupled to the metric equations (16.3)–(16.5). Use the fact that for corotating binaries, the fluid velocity satisfies $v^j = 0$ in the corotating frame.
> **Hint:** It also may be helpful to use the relativistic Gibbs–Duhem relation $dh = dP/\rho_0$.

Of course, in full general relativity, binaries constructed from such initial data are only in *quasi*equilibrium, as they undergo inspiral due to the emission of gravitational radiation. To mimic inspiral in the conformal flatness approximation, a post-Newtonian gravitational radiation-reaction potential[20] is sometimes added in the Euler equation.

Validity of the conformal flatness approximation

In Chapter 3 we found that conformal flatness greatly simplifies the initial value equations. Here we have seen that it greatly simplifies the evolution treatment as well. It is important, however, to appreciate that the two treatments are very different in nature. For the construction of initial data, the assumption of conformal flatness still leads to exact solutions to Einstein's constraint equations, and therefore does not represent an approximation in this sense.[21] The true dynamical evolution of such initial data, however, will generally lead to a spatial geometry that does not remain conformally flat. Assuming conformal flatness during a dynamical simulation therefore yields solutions that, in general, are only approximate spacetime solutions to Einstein's equations.

The conformal flatness approximation has been used frequently to model relativistic systems in which gravitational radiation plays a minimal role in the structure and evolution of a system on the time scales of interest. This is certainly the case for spherical spacetimes (e.g., as in spherical stellar collapse) for which the formalism is exact. It is also applicable, albeit approximately, to quasiequilibrium binary systems in circular orbit over a few orbital periods.

In Chapter 3 we found that the dynamical degrees of freedom of the gravitational fields can be identified with parts of the conformally-related spatial metric and the

---

[20] See equation (1.50) and Chapter 9.1.1, as well as, e.g., Burke (1971) or Misner *et al.* (1973), Chapter 36.8, for a discussion of the radiation-reaction potential.

[21] Conformally flat initial data, even though they constitute exact solutions to Einstein's constraint equations, may not represent configurations that are astrophysically realistic. It is in this sense they are sometimes referred to as approximations.

transverse-traceless part of the extrinsic curvature. This suggests that the assumptions of conformal flatness and the vanishing of $\bar{A}_{ij}^{TT}$ may indeed "minimize the gravitational radiation content" of a spatial slice Σ. This argument cannot be strictly true, however; it does not even hold for single rotating black holes. Rotating Kerr black holes, which are stationary and do not emit any gravitational radiation, are not conformally flat.[22] Similarly, conformally-flat models of rotating black holes that are constructed in the Bowen–York formalism do contain gravitational radiation.[23]

For rapidly rotating, isolated neutron stars in stationary equilibrium, the restriction to conformal flatness introduces an error of at most a few percent, and the error is this large only for the most relativistic and rapidly rotating configurations.[24] Similarly, small differences exist between conformally flat binary neutron star models and binary models constructed under different assumptions.[25] These small deviations are not surprising, since differences between a conformally flat metric and the "correct" metric already appear at second post-Newtonian order,[26] and thus are on the order of a few percent for neutron stars. To gauge the importance such an error, it should be compared with other approximations and errors made in the calculations, including finite resolution error, the treatment of outer boundaries, uncertainties in the equation of state, and the effect of neglecting other physical processes like neutrino transport or magnetic fields in the simulation.

To calibrate the conformal flatness approximation it is useful to compare how well it performs in comparison to fully relativistic calculations. Shibata and Sekiguchi (2004) have performed axisymmetric simulations of rotating stellar core collapse to a neutron star in full general relativity, using the BSSN scheme to integrate the gravitational field equations. They find that the evolution of the central density during the collapse, bounce and formation of the protoneutron star agrees well with the evolution found by Dimmelmeier *et al.* (2002a,b), who use the conformal flatness approximation to simulate the same problem. Both groups employ an HRSC scheme to integrate the relativistic fluid equations for the matter. Both groups computed gravitational waves using the quadrupole approximation, although they adopted slightly different forms for the quadrupole formula. Their waveforms are in good qualitative agreement, but exhibit some quantitative differences.[27] The differences in their adopted quadrupole formulae are likely responsible for most of

---

[22] At least slices of constant Boyer–Lindquist time are not conformally flat, nor are axisymmetric foliations that smoothly reduce to slices of constant Schwarzschild time in the Schwarzschild limit; Garat and Price (2000).

[23] Brandt and Seidel (1995a,b, 1996); Gleiser *et al.* (1998); Jansen *et al.* (2003).

[24] Cook *et al.* (1996).

[25] Usui *et al.* (2000); Usui and Eriguchi (2002).

[26] Rieth and Schäfer (1996).

[27] Shibata and Sekiguchi (2004) had to use the quadrupole approximation to compute waveforms since the wave amplitudes were too small ($< 10^5$) to be extracted accurately from the metric data with their uniform spatial grid of $2500 \times 3 \times 2500$ zones. The formula for the quadrupole moment is not defined uniquely in strong-field general relativity; different forms of the integrand all have the same Newtonian limit. The formula adopted by Shibata and Sekiguchi (2004) was calibrated against fully relativistic waveforms calculated for highly oscillating and rapidly rotating neutron stars of high compaction. The quadrupole formula wave amplitudes were found to be reliable to within 10% error, with waveform phase errors considerably smaller.

the discrepancy, suggesting that the conformal flatness approximation is quite adequate for handling many features of stellar collapse to neutron stars.

> **Exercise 16.2** (a) Show from the quadrupole formula that the gravitational wave amplitude of the + mode in axisymmetric spacetimes may be obtained from
>
> $$h_+ = \frac{\ddot{I}_{xx}(t_{\text{ret}}) - \ddot{I}_{zz}(t_{\text{ret}})}{r}\sin^2\theta, \qquad (16.6)$$
>
> where the symmetry axis is along z, I_{ij} is the quadrupole moment, r is the distance to the observer, θ is the polar angle of the observer and t_{ret} is retarded time. (What can you say about the $\times$ mode in axisymmetry in the quadrupole approximation?) (b) Define the quadrupole moment as in Shibata and Sekiguchi (2004):
>
> $$I_{ij} \equiv \int \rho_* x^i x^j d^3 x, \qquad (16.7)$$
>
> where $\rho_* \equiv \gamma^{1/2} D = \alpha u^t \gamma^{1/2} \rho_0$. Use the continuity equation (5.12) to show that the first time derivative of I_{ij} can be obtained from
>
> $$\dot{I}_{ij} = \int \rho_* (v^i v^j + x^i v^j) d^3 x. \qquad (16.8)$$
>
> Hence argue that to compute $\ddot{I}_{ij}$ in equation (16.6), only *one* numerical time derivative of the data is required provided equation (16.8) is used, which is much more reliable than using equation (16.7) and taking two numerical derivatives.[28]

This suggesion is strengthened considerably by the relativistic simulations of rotating stellar core collapse in both $3 + 1$ and $2 + 1$ (axisymmetry) by Ott *et al.* (2006), who also incorporated more detailed microphysics in their modeling.[29] They compared fully general relativistic evolution using a BSSN scheme with a treatment adopting conformal flatness and focused on gravitational wave emission from the rotating collapse, core bounce and the early post-bounce phases. The waveforms computed by extracting the metric data directly in the fully relativistic simulation closely match those computed from the quadrupole formula in the simulation that adopts conformal flatness. This suggests that the conformal flatness approximation is sufficiently accurate to handle the core-collapse supernova problem, at least when the outcome is a neutron star.

As we emphasized above, the conformal flatness approximation is, in general, inconsistent with Einstein's evolution equations. Miller and Suen (2003) have attempted to calibrate this error for the inspiral of binary neutron stars. To do so, they compute the Bach tensor B_{ij} (see equation 3.15), which vanishes if and only if the spatial metric γ_{ij} is conformally flat.[30] To construct a diagnostic that measures departure from conformal

---

[28] This trick was first pointed out by Finn and Evans (1990).

[29] Ott *et al.* (2006) employ a finite-temperature equation of state (EOS) and an approximate treatment of deleptonization during collapse.

[30] See the related discussion in Chapter 3.1.2; note that Miller and Suen (2003) use a conformal rescaling that is different from that in equation (3.15).

flatness they compute the matrix norm $|B_{ij}|$,[31] and normalize this quantity to the magnitude of the covariant derivative of the 3-Ricci tensor, $B \equiv |B_{mn}|/(D_i R_{jk} D^i R^{jk})^{1/2}$. They then integrate this quantity across a star and obtain a rest-mass weighted average of B, denoted by $\langle B \rangle$. Miller and Suen (2003) evaluate the diagnostic $\langle B \rangle$ for a binary system of identical, corotating neutron stars, modeled initially as relativistic polytropes with index $n = 1$ and evolved adiabatically with adiabatic index $\Gamma = 2$. They normalize their polytropic equation of state so that the maximum ADM mass of a spherical, static star is $1.79 M_\odot$, whereby the maximum rest mass M_0 is $1.97 M_\odot$. They treat binaries with stellar rest masses of $1.49 M_\odot$, approximately 75% of the maximum mass; in isolation, static stars with this rest mass have an ADM mass of about $1.4 M_\odot$. They begin their simulations with quasiequilibrium binaries constructed in the conformal thin-sandwich formalism, assuming conformal flatness so that $\langle B \rangle$ vanishes initially. They treat close binaries orbiting at several different initial separations, including an initial separation close to the ISCO. In all cases $\langle B \rangle$ increases from zero to a maximum value in a fraction of an orbital period, and then begins to decay. The maximum value is larger for tighter binaries, but is never more than a few percent, in agreement with our rough estimate above. Even for the tightest orbit, where the spatial geodesic separation between the density maxima in the two stars is $l/M_0 = 23.44$ and the orbital frequency is $\Omega M_0 = 0.01547$, the value never rises above 5% before it begins to decay. This finding helps quantify the degree to which the conformal flatness approximation provides a reasonable model for the spacetime of a relativistic binary neutron star, at least during the inspiral phase.

Numerical results

The pioneering simulations of binary inspiral by Wilson and his collaborators introducing the conformal flatness approximation have been followed by several other treatments of increasing sophistication.[32] One of the earliest was by Oechslin *et al.* (2002), who employed a Lagrangian SPH hydrodynamics code with a multigrid elliptic solver to handle the metric equations, and adopted corotating initial configurations. While these configurations are believed to be unphysical for neutron stars,[33] these simulations confirmed earlier SPH simulations in post-Newtonian (PN) gravitation[34] that suggested that relativistic effects suppress mass loss during merger.

---

[31] The matrix norm of the tensor B_{ij} is defined as the square root of the largest eigenvalue of $B_{ij} B^j{}_k$.

[32] The original version of Wilson and Mathews (1995) contained a mathematical error, pointed out by Flanagan (1999), which is now believed to be mainly responsible for the spurious finding of a "crushing" instability that triggers the collapse of the neutron stars prior to merger; see our discussion in Chapter 15.2. Numerous earlier analyses, including the post-Newtonian dynamical simulations of Shibata *et al.* (1998), had cast considerable doubt on the existence of a "crushing" instability.

[33] Viscous stresses are thought to be too weak to maintain tidal synchronization in binary neutron stars; Bildsten and Cutler (1992); Kochanek (1992a); see Chapter 15.1.

[34] Faber and Rasio (2002).

Faber *et al.* (2004) subsequently performed SPH simulations in the conformal flatness approximation using a spectral elliptic solver in spherical coordinates[35] for the metric equations. They considered the adiabatic evolution and merger of equal-mass relativistic binary polytropes with $n = 1$ and $\Gamma = 2$. For initial data they employed the quasiequilibrium, irrotational binary models of Taniguchi and Gourgoulhon (2002), which are constructed using the conformal thin-sandwich formalism. Faber *et al.* (2004) consider binaries whose members each have rest mass $M_0 = 0.146$ in units where the polytropic gas parameter $K = 1$ (i.e., $\bar{M}_0 = 0.146$). At infinite separation, such a star has a total mass-energy (i.e., ADM mass) equal to $M = 0.136$ and a compaction $M/R = 0.14$. (For comparison, the maximum-mass configuration for a static $n = 1$ polytrope has $M_0^{\mathrm{max}} = 0.180$ and $M^{\mathrm{max}} = 0.164$.) The binary profiles form a cusp slightly within an orbital radius $r_0/M_{\mathrm{ch}} = 19.9$, inside of which quasiequilibrium, irrotational, circular-orbit solutions do not exist for this EOS. Here $M_{\mathrm{ch}} \equiv \mu^{3/5} M_t^{2/5} = M/2^{1/5}$ is the "chirp mass" at large separation, where M_t is the total (ADM) mass-energy of the system at large separation and $\mu \equiv M_1 M_2 / M_t$ is the reduced (ADM) mass.

Typical runs employ approximately 10^5 SPH particles to solve the hydrodynamic equations and three spheroidal computational domains around each star to evaluate the field equations. As a test, Faber *et al.* (2004) demonstrate that the binaries remain dynamically stable and maintain circular equilibrium for all separations up to cusp formation. By inserting an approximate PN radiation-reaction potential term in the Euler equation they trigger a small inward spiral motion and follow the complete coalescence of a binary from just outside the cusp radius through merger, remnant formation and ringdown. They find that mass loss is highly suppressed, but that the massive remnant is dynamically stable against gravitational collapse because of its strong differential rotation. In other words, the remnant settles into an equilibrium hypermassive neutron star like those discussed in Chapter 14. Snapshots of the merger scenario are depicted in Figures 16.1 and 16.2. The rotation profile of the hypermassive merger remnant is plotted in Figure 16.3. The gravitational radiation waveforms, calculated from the quadrupole formula, are shown in Figure 16.4.

> **Exercise 16.3** Assume that each neutron star in the binary system shown in Figures 16.1–16.4 has a rest mass of $M_0 = 1.5 M_\odot$. Compute the ADM mass, chirp mass M_{ch} and stellar radius R at large separation. Convert the length, time, and angular velocity scales plotted in the figures to physical units. Evaluate the maximum amplitude of the waveform shown in Figure 16.4 if the binary were located in the Virgo cluster ($d \approx 20$ Mpc, where 1 Mpc $= 3.09 \times 10^{24}$ cm).

One of the most extensive sets of binary neutron star merger calculations based on the conformal flatness approximation to Einstein's field equations is that of Oechslin *et al.* (2007). Like Faber *et al.* (2004), they employ a relativistic SPH code for the hydrodynamics,

---

[35] The LORENE numerical libraries, developed by Grandclément *et al.* (2001) and publically available online at `http://www.lorene.obspm.fr` were used in these simulations.

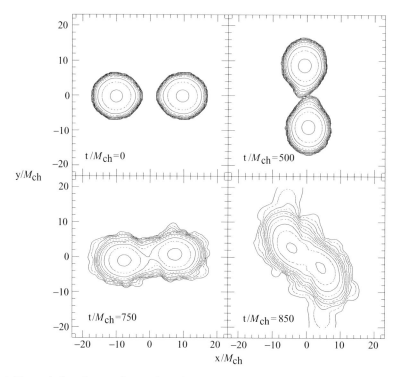

Figure 16.1 The evolution of an equal-mass, irrotational binary neutron star system in the conformal flatness approximation, as viewed in the orbital plane. The simulation begins just outside the orbital separation where a cusp develops and continues through the merger and formation of a remnant. Density contours shown here are logarithmically spaced, two per decade, ranging from $M_{\rm ch}^2 \rho_* = 10^{-6.5}$ to 10^{-1}. A significant tidal lag angle develops at $t/M_{\rm ch} = 500$, followed by an "off-center" collision. This process leads to the formation of a vortex sheet and a small amount of matter ejection by $t/M_{\rm ch} = 850$. [From Faber *et al.* (2004).]

but unlike Faber *et al.* (2004), they include artificial viscosity to handle shocks (see Chapter 5.2.1). They solve the metric equations with the help of a multigrid solver and employ approximately 120 000 SPH particles in their simulations. Most significantly, they consider the effect of varying the nuclear EOS, the neutron star masses and the neutron star spins.[36] Oechslin *et al.* (2007) supplement their treatment by adding a small, nonconformally flat gravitational radiation back-reaction contribution to the metric,[37] which is only important during the inspiral phase before the plunge.[38] They ignore neutrino transport by arguing that any back-reaction of the neutrino emission on the matter is

---

[36] Oechslin *et al.* (2007) impose the irrotational spin condition, as well as their other choices for the initial spin, only approximately. However, when they compare their initial models to corresponding models obtained via the conformal thin-sandwich scheme, they find deviations in global quantities like the total angular momentum J and orbital angular velocity Ω to be small, of order 1%.

[37] They adopt the 3.5 PN prescription of Faye and Schäfer (2003).

[38] The merger and postmerger phases are found to depend very weakly on the inspiral dynamics.

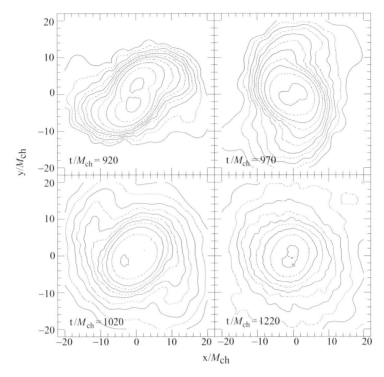

Figure 16.2 The advanced evolution of the merger depicted in Figure 16.1; labels are the same as in that figure. A dense remnant forms in the center of the system, surrounded by a thin halo. Some ellipticity is seen shortly after the merger, but the system quickly relaxes toward a spheroidal, hypermassive equilibrium configuration, with maximum density in the center of the system. [From Faber *et al.* (2004).]

negligible on the 10 ms (dynamical) time scales of interest here. They consider several models for the finite-temperature nuclear EOS, including the "hard" EOS of Shen *et al.* (1998a,b), which gives a cold, static, maximum neutron star mass of about $2.2 M_\odot$, and the "soft" EOS of Lattimer and Swesty (1991), for which the maximum mass is about $1.8 M_\odot$. The simulations show that the dynamics and the final outcome of the merger depend sensitively on the EOS and the binary parameters. For example, Oechslin *et al.* (2007) find that with the soft EOS the remnant collapses to a black hole either immediately or within a few dynamical timescales after merger, while for all other EOSs, the remnant does not collapse but instead forms a hypermassive equilibrium configuration. The mass of the low-density torus that forms around the black hole is larger for unequal-mass systems (reaching $0.3 M_\odot$ for a mass ratio of 0.55), large initial neutron stars and for neutron star spin states that result in a larger total angular momentum. The characteristic temperatures achieved in the torus are 3–10 MeV, depending on the degree of shear at the collision interface; the shear, in turn, depends on the initial spins. Only about 10^{-3} to 10^{-2} of the total rest mass escapes from the system.

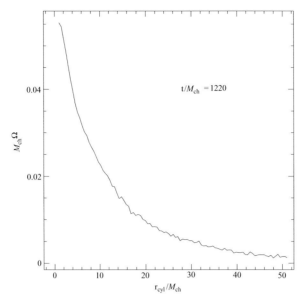

Figure 16.3 Angular velocity of the hypermassive merger remnant in Figure 16.2 at $t/M_{ch} = 1220$, shown as a function of cylindrical radius, $r_{cyl} \equiv \sqrt{x^2 + y^2}$. The remnant exhibits strong differential rotation, with the highest angular velocity in the center, decreasing monotonically with radius. It is this differential rotation that supports the hypermassive remnant against collapse. [From Faber *et al.* (2004).]

From these simulations Oechslin and Janka (2006) suggest that binary neutron star mergers may be the origin of short-hard gamma-ray bursts (GRBs). Previous simulations[39] indicated that insufficient energy is liberated during the highly dynamical phase of the merger event to account for the observed energies. Moreover, the possibility that a GRB is triggered during the neutrino emitting, pre-collapse lifetime of the hypermassive remnant is also doubtful, because the high mass-loss rates in the neutrino-driven wind will likely quench the propagation of gamma-rays (the "baryon loading" problem). Instead, Oechslin and Janka (2006) propose that the bursts could be powered by neutrino-antineutrino annihilation in the hot gas in the torus surrounding the black hole remnant. In particular, the measured GRB energies and durations lead to estimates for the accreted mass and accretion rate that are compatible with their theoretical estimates derived from their simulation data. Their model is not very different from the scenario proposed by Shibata *et al.* (2006a), based on their general relativistic MHD simulations of the evolution and collapse of a magnetized hypermassive star and the black hole-magnetized torus

---

[39] Ruffert *et al.* (1996).

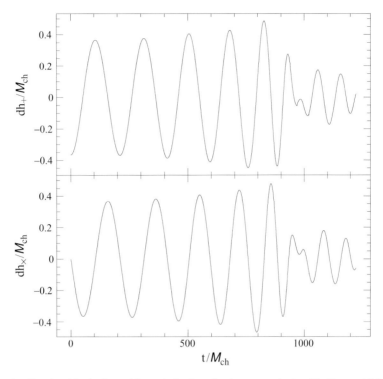

Figure 16.4 The GW signal in the h_+ and $h_\times$ polarizations for the merger depicted in Figures 16.1 and 16.2 as seen by an observer situated along the vertical axis. The amplitude is scaled with the distance of the observer, d/M_{ch}. A chirp signal is followed by a modulated ringdown spike. The modulation is caused by the alignment between quadrupole deformations in the inner regions of the remnant core and and those at larger radius. [From Faber *et al.* (2004).]

configuration that results.[40] While the origin of GRBs is far from resolved, the key point is that numerical relativity is now sufficiently mature to address the issue.

16.3 Fully relativistic simulations

Several groups have launched efforts over the years to solve the full set of Einstein's field equations self-consistently with the equations of relativistic hydrodynamics to model the merger of binary neutron stars.[41] Many of the most advanced and detailed simulations to date have been performed by Shibata and his collaborators.[42] These simulations represent a major achievement of numerical relativity and demonstrate the degree to which fully

[40] See Chapter 14.2.5.
[41] See, e.g., Oohara and Nakamura (1999); Baumgarte *et al.* (1999); Font *et al.* (2000, 2002).
[42] Shibata (1999a); Shibata and Uryū (2000, 2002); Shibata *et al.* (2003b); Shibata *et al.* (2005); Shibata (2005); Shibata and Taniguchi (2006).

relativistic calculations of binary neutron stars have advanced. We will summarize some of the highlights of these calculations below.

Equal-mass binaries

The first truly successful simulations of binary neutron star mergers in full general relativity were those of Shibata and Uryū (2000), which used a computational scheme developed and tested earlier by Shibata (1999a). The original scheme employed the BSSN formulation of Einstein's equations to evolve the field equations (see Chapter 11.5) and the Wilson form of the equations of relativistic hydrodynamics, equations (5.12), (5.14) and (5.19). The stars are constructed using an $n = 1$ polytropic EOS; the matter is assumed to evolve adiabatically according to a Γ-law EOS with $\Gamma = 2$. The transport terms in the hydrodynamics equations are handled by a second-order van Leer algorithm.[43] Artificial viscosity is used to capture shocks, as in equation (5.24). Shibata and Uryū (2000) employ "approximate maximal slicing" (Chapter 4.2) to specify the lapse α and "approximate minimal distortion" (Chapter 4.5) to determine the shift β^i. They also add a radial component to the shift vector to avoid grid stretching in collapse situations (see the discussion at the very end of Chapter 4.5). Most of their simulations in this study used a fixed uniform grid with $233 \times 233 \times 117$ grid points in the x–y–z directions, respectively, and assumed reflection symmetry across the equatorial (orbital) plane at $z = 0$.

As initial data, Shibata and Uryū (2000) prepared equal-mass polytropic ($n = 1$) models of binary neutron stars in quasiequilibrium with both corotational (see Chapter 15.2) and irrotational (see Chapter 15.3) velocity profiles. For both velocity profiles they generated three different models with individual stellar masses ranging from about 70% to 100% of the maximum allowed mass of nonrotating stars in isolation. For corotational models, they adopted contact models ($z_A = 0$ in the parametrization of Chapter 15.2) as initial data, which are fairly close to the ISCO. As we discussed in Chapter 15.3, irrotational sequences terminate at cusp formation, which is still outside of the ISCO. Shibata and Uryū (2000) therefore adopted the cusp model as initial data, and induced collapse by artificially reducing the angular momentum by about 2.5%. Since irrotational velocity profiles are probably more realistic, we will discuss their models (I1), (I2) and (I3) for irrotational binaries. The initial data for these three models are summarized in Table 16.1.[44]

The simulation for model I1 is for a binary of total rest mass $\bar{M}_0 = 0.261$.[45] The angular momentum of the initial data is $J/M^2 = 0.98$, where M is the total ADM mass, and is hence smaller than the Kerr limit $J/M^2 = 1$. We show snapshots of density contours in Figure 16.5.

---

[43] van Leer (1977).
[44] More information can be found in Table 1 of Shibata and Uryū (2000).
[45] We have converted the units of Shibata and Uryū (2000) into the same dimensionless ("barred") units adopted earlier in this book, corresponding to setting the polytropic gas constant $K = 1$.

Table 16.1 Summary of the initial data for the coalescence simulations of irrotational binary neutron stars obeying an $n = 1$ polytropic EOS. Here $\bar{M}_0$, $\bar{M}$ and J are the total rest mass, ADM mass and angular momentum of the binary. In these dimensionless units, the maximum allowed rest mass of an isolated, nonrotating star is $\bar{M}_0^{\mathrm{max}} = 0.180$. [Table adapted from Shibata and Uryū (2000).]

Model	$\bar{\rho}_{\mathrm{max}}$	$\bar{M}_0$	$\bar{M}$	J/M^2	remnant
I1	0.0726	0.261	0.242	0.98	neutron star
I2	0.120	0.294	0.270	0.93	black hole
I3	0.178	0.332	0.301	0.88	black hole

In contrast to the coalescence of corotational binaries, no significant spiral arms form during the merger of irrotational stars, and hardly any matter is ejected. Nevertheless, the remnant settles down to an equilibrium neutron star and does not collapse to a black hole, at least not on a dynamical time scale, even though its rest mass exceeds the maximum allowed rest mass of a spherical, nonrotating star by about 45%. There is only a small amount of shock heating, which rules out thermal pressure as the origin of the extra support against collapse. As noted, the angular momentum J/M^2 is smaller than the Kerr limit and therefore cannot prevent black hole formation. Uniform rotation can increase the maximum allowed rest mass for $\Gamma = 2$ ($n = 1$) polytropes only by about 20%.[46] However, the remnant core is differentially rotating as opposed to uniformly rotating, as shown in Figure 16.6. The remnant is thus a "hypermassive neutron star" of the type discussed in Chapter 14.1.2. The formation of hypermassive neutron stars following binary neutron mergers were foreshadowed by Newtonian simulations;[47] relativistic equilibrium models for such stars were then constructed numerically and tested for dynamical stability by performing simulations in general relativity.[48] But the calculations of Shibata and Uryū (2000) were the first simulations in full general relativity to demonstrate that hypermassive stars can actually form during binary neutron star mergers.

Both models I2 and I3, for which the total rest mass exceeds the maximum allowed spherical mass by 63% and 85%, respectively, form a black hole promptly upon merger. In Figure 16.7 we show snapshots of the density contours and velocity profiles for the more massive model I3. In the last frame the thick solid line denotes the location of an apparent horizon, indicating the formation of a black hole. By means of these simulations, Shibata and Uryū (2000) conclude that hypermassive remnants will form whenever the rest mass of each star is less than about 70–80% of the maximum allowed mass of the spherical star, otherwise a black hole will form. As demonstrated by simulations discussed

---

[46] Cook *et al.* (1994b); see Table 14.1.
[47] See, e.g., Rasio and Shapiro (1999).
[48] Baumgarte *et al.* (2000); see Chapter 14.2.

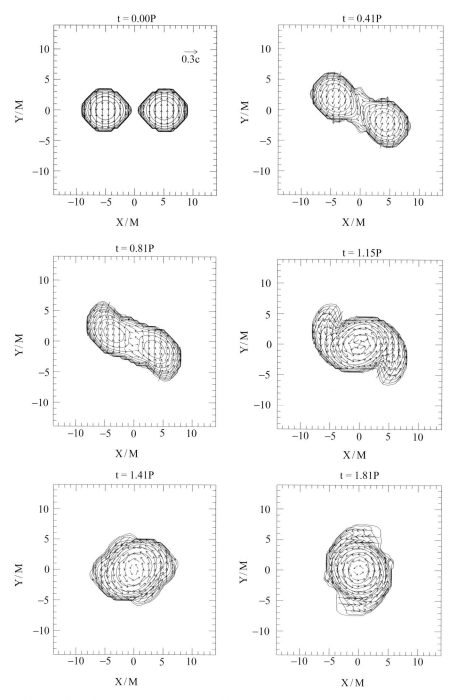

Figure 16.5 Snapshots of density contours for $\rho_* = \gamma^{1/2}\rho_0 W$ (see equation 5.12) and the velocity field (v^x, v^y) in the equatorial plane for the coalescence of an irrotational binary of total rest mass $\bar{M}_0 = 0.261$ (Model I1 in Table 16.1). Time is measured in terms of the initial orbital period P. The contour lines denote densities $\rho_*/\rho_{*\,\mathrm{max}} = 10^{-0.3j}$ with $\rho_{*\,\mathrm{max}} = 0.255$ and $j = 0, 1, 2, \ldots, 10$. [From Shibata and Uryū (2000).]

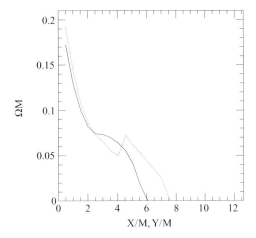

Figure 16.6 The angular velocity Ω along the x-axis (solid line) and the y-axis (dotted line) at $t = 1.81\,P$ for model I1. [From Shibata and Uryū (2000).]

in Chapter 14.2, hypermassive stars prove to be unstable on secular timescales: dissipative mechanisms such as shear viscosity or magnetic fields tend to bring the stars into more uniform rotation, resulting in a "delayed collapse" to a black hole.

Shibata and Uryū (2000) find fairly similar results for corotational binary models. Probably the most significant difference is that corotational binaries have more angular momentum prior to merger, which leads to the formation of spiral arms during the coalescence. The spiral arms contain a few percent of the total mass, and may ultimately form a disk around the central object. For corotational binaries that promptly form black holes, a disk of mass ~ 0.05–$0.1 M_0$ forms, where M_0 is the total rest-mass of the system. For irrotational binaries, the disk mass is much smaller, less than $0.01 M_0$.

In these early simulations one of the largest limitations on the accuracy involved is the outer boundaries. Because of limited computational resources, these boundaries had to be imposed well within one gravitational radiation wavelength from the binary (at about $\lambda_{\rm GW}/3$). This means that the waves were not being extracted in the radiation zone, which necessarily introduces error unless a special near-zone wave extraction algorithm is implemented. After gaining access to a more powerful supercomputer, Shibata and Uryū (2002) decided to repeat their calculations on computational grids that extend further, to about a gravitational wavelength. Their improved calculation, performed on uniform grids with typical size $505 \times 505 \times 253$, gave results in good qualitative agreement with their earlier ones, although the onset of black hole formation shifts to slightly smaller masses. Most strongly affected by this improvement are the gravitational waveforms. In Figure 16.8, we show examples from these improved simulations for models that are similar to model I1, leading to a hypermassive neutron star remnant, and model I2, leading to a black hole. The quantities are plotted for an observer along the z-axis and are defined

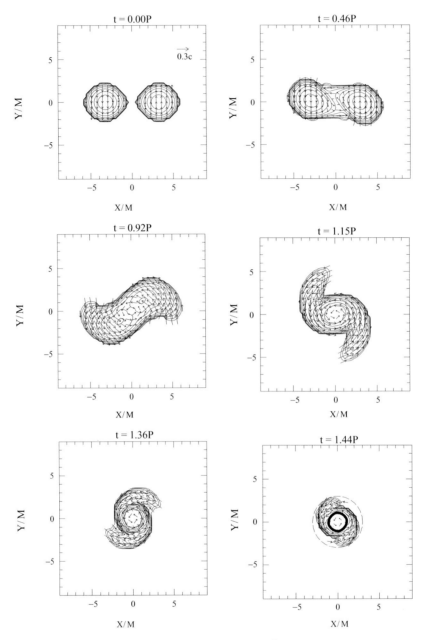

Figure 16.7 Same as Figure 16.5, but for a binary of total rest mass $\bar{M}_0 = 0.332$ (Model I3 in Table 16.1). Contour lines denote densities $\rho_*/\rho_{* \, max} = 10^{-0.3j}$ with $\rho_{* \, max} = 0.866$ and $j = 0, 1, 2, \ldots, 10$. The dashed line in the last snapshot is the circle at $r = 3M$, which encloses over 99% of the total rest mass. The thick solid line at $r/M \approx 1$ denotes the location of the apparent horizon. [From Shibata and Uryū (2000).]

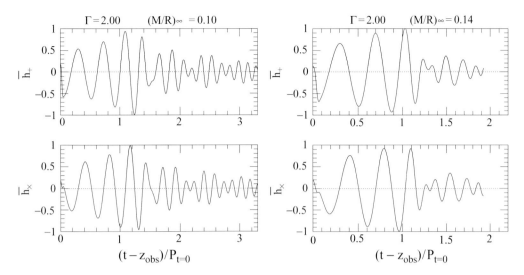

Figure 16.8 Gravitational wave amplitudes $\bar{h}_+$ and $\bar{h}_\times$ as functions of retarded time for the $n = 1$ irrotational binary neutron star models E and H of Shibata and Uryū (2002). Model E (left panel) corresponds to a slightly smaller mass then I1 and results in a differentially rotating hypermassive neutron star, while model H (right panel) is similar to model I2 and results in a black hole. [From Shibata and Uryū (2002).]

according to

$$\bar{h}_+ \equiv \frac{z_{\text{obs}}}{2M_{\text{ADM}}(M/R)_\infty}\left(\bar{\gamma}_{xx} - \bar{\gamma}_{yy}\right), \qquad \bar{h}_\times \equiv \frac{z_{\text{obs}}}{2M_{\text{ADM}}(M/R)_\infty}\left(\bar{\gamma}_{xy}\right). \qquad (16.9)$$

Here z_{obs} is the location of the observer. The quantities $\bar{h}_+$ and $\bar{h}_\times$ calculated here are expected to provide a reasonable approximation to the asymptotic gravitational waves, since the adopted gauge condition is approximately transverse-traceless in the wave zone.[49] Note that the wave amplitude attains its maximum along the z-axis, along which it is composed only of $|m| = 2$ modes.

> **Exercise 16.4** Show that at the point of contact of two equal-mass, spherical, Newtonian stars in a circular binary, the quadrupole formula gives the amplitude of gravitational waves along the z-axis to be
>
> $$h_{\text{GW}} = \frac{2M(M/R)_\infty}{z_{\text{obs}}}. \qquad (16.10)$$
>
> This result explains the adopted normalization in equation (16.9).

The difference between the formation of a hypermassive remnant versus a black hole immediately upon merger is reflected in their waveforms, as seen in Figure 16.8. Quasiperiodic oscillations are observed to persist for a long duration for the hypermassive remnant,

---

[49] In the wave zone, the conformally related spatial metric used in equation (16.9) and the physical spatial metric are nearly equal.

while such oscillations last only for a short time scale in the case of a prompt black hole formation.

Unequal-mass binaries

Following up these simulations, Shibata *et al.* (2003b) next considered the merger of binary neutron stars of *unequal* mass. Here again they adopted a simple Γ-law EOS with $\Gamma = 2$ to explore the effect of varying the mass ratio q between 0.85 and 1. This is an appropriate range to consider, since the mass ratio for all the observed galactic binary neutron stars for which each mass is determined accurately resides in the this range.[50] The numerical implementation has been improved in these simulations, beginning with the hydrodynamics, which is performed by means of an HRSC scheme. Also, the configurations are followed beginning from the late inspiral phase through the merger phase; this "early start" allows the transition from inspiral to plunge to be triggered more properly by gravitational radiation reaction, rather than by an artificial initial angular momentum depletion. A hyperbolic shift equation similar in spirit to the one given by equation (14.44) is used to replace the time-consuming elliptic AMD gauge condition. A typical spatial grid size of $633 \times 633 \times 317$ zones is employed in these simulations.

A number of interesting results emerge from this study. If the total rest mass of the system exceeds approximately 1.7 times the maximum allowed rest mass of a spherical neutron star, a black hole forms promptly upon merger, independently of the mass ratio. Otherwise a hypermassive neutron star forms. The disk mass around the black holes increases with decreasing rest-mass ratios and with increasing neutron star compactness. The merger process and the gravitational waveforms are sensitive to the rest-mass ratios, even in the restricted range $q = 0.85$–1. For example, the maximum amplitude is smaller for models with smaller mass ratios. This behavior is due to the enhanced role of tidal effects for smaller mass ratios, causing tidal disruption at larger orbital separation.[51] Also, emission in modes of odd values of m, which is absent in the case of equal-mass binaries due to π-rotation symmetry, is not negligible for the merger of unequal-mass binaries, although the amplitude of this radiation never exceeds 5% of the $l = 2$, $m = 2$ mode.

Binaries with realistic EOSs

Having succeeded in simulating binary neutron star mergers for simple polytropes and Γ-law EOSs, Shibata *et al.* (2005) proceeded to explore mergers with more realistic EOSs. They constructed a convenient "hybrid" hot nuclear EOS consisting of two parts, $P = P_{\text{cold}} + P_{\text{th}}$. For the cold nuclear matter contribution P_{cold} they used both the SLy[52] and

---

[50] Stairs (2004).

[51] This effect was reported earlier in Newtonian and post-Newtonian studies; see Rasio and Shapiro (1994); Faber and Rasio (2000, 2002).

[52] Douchin and Haensel (2001).

FPS[53] nuclear EOSs, for which the maximum allowed ADM mass of an isolated spherical neutron star is $2.04 M_\odot$ and $1.80 M_\odot$, respectively.[54] For the thermal contribution P_{th}, they adopted the law $P_{th} = (\Gamma_{th} - 1)\rho_0 \epsilon_{th}$, where ρ_0 is the rest-mass density, and $\epsilon_{th} = \epsilon - \epsilon_{cold}$ is the specific thermal energy density. To match the stiff behavior of the cold contribution, they chose $\Gamma_{th} = 2$, but also experimented with other values.

> **Exercise 16.5**   Assuming that the stars are completely cold when they begin their plunge towards each other, what physical effect causes them to acquire nonzero thermal energy? Estimate the resulting characteristic gas temperature in the gas.

Simulations were performed for binary systems with total ADM mass in the range between $2.4 M_\odot$ and $2.8 M_\odot$ and with rest-mass ratios q is the range $0.9 \lesssim q \lesssim 1$. Uniform grids with as many as $633 \times 633 \times 317$ zones were employed. They found that when the total ADM mass exceeds a threshold M_{thr}, a black hole forms promptly after merger, independently of mass ratio. Otherwise a differentially rotating hypermassive neutron star remnant forms. The value of M_{thr} is found to be approximately $2.7 M_\odot$ for SLy and $2.5 M_\odot$ for FPS, which is larger than the maximum spherical or uniformly rotating mass in each case.

For binaries with total masses exceeding M_{thr}, over 99% of the rest-mass forms a black hole promptly on merger. The spin of the black hole falls in the range $J/M^2 \approx 0.7$–0.8, not very different from the spins typically arising from the collapse of rapidly spinning stars.[55] The quasinormal mode ringdown radiation of the hole will then be at a frequency $f \approx 6.5$–$7(2.8 M_\odot/M)$ kHz, which, unfortunately, exceeds the frequency range of optimal sensitivity of all current laser interferometers.[56]

For $M < M_{thr}$ the binary results in a hypermassive remnant. In contrast to the remnants governed by a pure $\Gamma = 2$ EOS, the hypermassive remnants formed here are characterized by a large *ellipticity*, a consequence of their high spin and high effective adiabatic index.[57] A typical relativistic ellipsoidal remnant is shown in Figure 16.9.[58]

The remnant shown here results from the merger of an irrotational binary consisting of identical neutron stars, each of which has an ADM mass of $1.3 M_\odot$ in isolation. The matter is governed by the SLy hybrid EOS. The simulation begins from a quasiequilibrium circular orbit just beyond the ISCO; the initial orbital period is about 2 ms and the merger occurs after about one orbit. As a result of its spin and high ellipticity, the hypermassive remnant

---

[53] Pandharipande and Ravenhall (1989).
[54] For comparison, a Γ-law EOS of the form $P = K\rho_0^\Gamma$ with $\Gamma = 2$ has a maximum ADM mass of $1.72(K/1.6 \times 10^5)^{1/2} M_\odot$, where K is given in cgs units. However, the radii of neutron stars with this Γ-law EOS are considerably larger than the radii given by the adopted realistic EOSs; see Shibata *et al.* (2005), Figure 2b.
[55] See Chapter 14.2.
[56] See Chapter 9.3.
[57] Rapidly rotating, nonaxisymmetric quasiequilibria do not exist unless the adiabatic index is sufficiently high. In Newtonian theory, a uniform rotating Jacobi-like ellipsoid exists only if $\Gamma \gtrsim 2.25$ and $T/|W| \gtrsim 0.14$; James (1964); Tassoul (1978).
[58] Such triaxial ellipsoids were first seen in the Newtonian coalescence calculations of Rasio and Shapiro (1994).

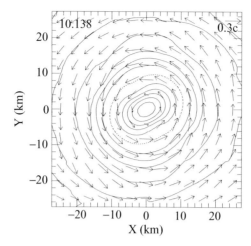

Figure 16.9 Contours of rest-mass density ρ_0 in the equatorial plane of an ellipsoidal hypermassive remnant at time $t = 10.138$ ms. The simulation begins at $t = 0$ and merger occurs at $t \sim 2$ ms. The solid curves are drawn for contours defined by $\rho_0 = 2 \times 10^{14} \times i$ g/cm^3 ($i = 2, \dots, 10$) and for $2 \times 10^{14} \times 10^{-0.5i}$ g/cm^3 ($i = 1, \dots, 7$). The dotted curve denotes the contour near nuclear density, $\rho_0 = 2 \times 10^{14}$ g/cm^3. Vectors indicate the local velocity field (v^x, v^y), where the scale is shown in the upper right-hand corner. [From Shibata (2005).]

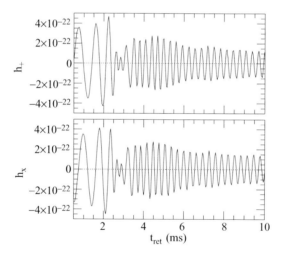

Figure 16.10 Waveforms for the binary merger that produced the ellipsoidal hypermassive neutron star remnant depicted in Figure 16.9. The observer is located along the rotational axis of the binary at a distance of 50 Mpc and t_{ret} is retarded time. [From Shibata (2005).]

is a strong emitter of quasiperiodic gravitational waves. The waveforms for the simulation depicted in Figure 16.9 are shown in Figure 16.10. Their characteristic frequency is about 3 kHz and their amplitude is roughly 1.5×10^{-22} (assuming a distance of 50 Mpc). These features hold constant over many rotation periods since the emission damping time, about 50 ms, is very long.

Exercise 16.6 Consider gravitational radiation emission in the quadrupole approximation from a uniformly rotating, homogeneous, Newtonian ellipsoid.
(a) Show that the energy and angular momentum loss rates are given by

$$\left(\frac{dE}{dt}\right)_{\text{GW}} = \Omega \left(\frac{dJ}{dt}\right)_{\text{GW}} = -\frac{32}{5}\Omega^6 (I_{11} - I_{22})^2, \qquad (16.11)$$

where $I_{ii} = Ma_i^2/5$ are the components of the star's quadrupole moment along the principle axes in its equatorial plane, a_i are the semimajor axes in the equatorial plane, M is the mass, and Ω is the angular velocity, which points in the polar direction.
(b) Show that the waveform amplitudes of the two polarization states are

$$h_+ = \frac{2}{D}\Omega^2 (I_{11} - I_{22})\cos\Phi(1 + \cos^2\theta), \qquad (16.12)$$

$$h_\times = \frac{4}{D}\Omega^2 (I_{11} - I_{22})\sin\Phi\cos\theta, \qquad (16.13)$$

where D is the distance to the source, θ is the angle between the rotation axis of the star and line of sight from the Earth, and $\Phi \equiv 2\int^t \Omega dt$ is twice the orbital phase.

Gravitational radiation emitted by the nonaxisymmetric hypermassive remnant carries off angular momentum. Since angular momentum is crucial in holding up the hypermassive remnant against collapse, its dissipation by wave emission will eventually lead to a "delayed" collapse to a rotating black hole. Unfortunately, the simulations performed here were forced to terminate long before the collapse. The presence and amplification of a seed magnetic field can accelerate the process, as demonstrated by simulations presented in Chapter 14.2.5. Which mechanism dominates the dissipation of angular momentum probably depends on the initial field strength, the degree of differential rotation, the EOS, the importance of other dissipative agents, like neutrino emission, and other factors. But the final fate – delayed collapse – is almost certain.

Shibata (2005) has pointed out that the quasiperiodic radiation from an ellipsoidal remnant may be sufficiently strong for detection by advanced laser interferometers, even though the frequency is near the high-end of their current dectability limit. The reason is that a long-term integration over the many cycles of the wavetrain may increase the effective peak amplitude above the noise level of detectors like Advanced LIGO. Also, identifying the chirp signal from the binary during the inspiral phase prior to merger should improve the search for a quasiperiodic signal from a hypermassive remnant. Detecting such a quasiperiodic signal will then provide a lower limit to M_{thr}, and will therefore constrain the neutron star EOS. In fact, only a single detection of such a signal is necessary to furnish a useful constraint. Since the theoretical values for M_{thr} found from these binary merger simulations are quite close to the total masses of the observed binary neutron stars with accurate mass determinations, binary neutron star mergers leading to hypermassive remnants might occur frequently.

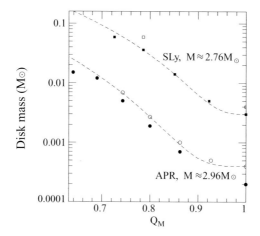

Figure 16.11 Baryon rest mass of the disk around the black hole as a function of the mass ratio Q_M in the case of prompt black hole formation. The disk mass is evaluated 0.5 ms after the appearance of an apparent horizon. The open circles and squares denote the results with $633 \times 633 \times 317$ zones for the APR and SLy EOSs, respectively; the filled circles and squares denote the results with $377 \times 377 \times 189$ zones. The results for the APR and SLy EOSs are obtained for a total initial binary ADM mass of $M \approx 2.96 M_\odot$ and $2.76 M_\odot$, respectively. [From Shibata and Taniguchi (2006).]

Shibata and Taniguchi (2006) have extended some of these calculations in two areas. First, they use a stiff cold nuclear EOS by Akmal *et al.* (1998) (APR) in place of FPS. This choice was motivated by reports of the discovery of a heavy neutron star (pulsar) with a mass of $2.1 \pm 0.2 M_\odot$.[59] Such measurement would establish a lower limit on the maximum ADM mass of a spherical neutron star, and it is consistent with the APR maximum mass $(2.18 M_\odot)$, but not the FPS value. Their other motivation was to apply black hole excision in cases of prompt black hole formation to evolve longer and thereby better determine the final state of the black hole system. In particular, they wanted to determine the final mass of the quasistationary disk surrounding the black hole. The disk is an important ingredient in black hole models of the central engine of short-hard GRB sources, as we have noted in Chapter 14.2.5. Finally, they wanted to simulate a wider range of mass ratios, $0.65 \lesssim q \lesssim 1$.

Collecting their results, Shibata and Taniguchi (2006) conclude that the the value of M_{thr} depends on the adopted cold nuclear EOS, but can be approximated by 1.3–$1.35 M_{sph}$, where M_{sph} is the maximum mass of a spherical neutron star constructed with the same EOS. Since uniform rotation can account for mass increases of at most about 20%, differential rotation is again required to support hypermassive neutron star remnants.[60] Shibata and Taniguchi

---

[59] PSR J0751+1807; Nice *et al.* (2005). But note that more recent timing observations have led to a substantially lower mass for this pulsar, $1.26 \pm 0.14 M_\odot$; Nice *et al.* (2008).

[60] The value of M_{thr}/M_{sph} is thus smaller for stars constructed from realistic nuclear EOSs than for polytropes. This result is consistent with the finding of Morrison *et al.* (2004) that, while the maximum mass of a differentially rotating, hypermassive star is typically 50% larger than the maximum mass of a nonrotating star constructed from the same realistic nuclear EOS, the increase is less than that for a polytrope.

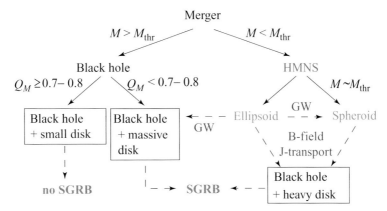

Figure 16.12 Plausible routes for the formation of a short-hard gamma-ray burst (SGRB) central engine following the merger of a binary neutron star. Here "HMNS" stands for hypermassive neutron star remnant and Q_M represents the mass ratio of the initial stars. For HMNS remnants, "GW" *vs.* "B-field J-transport" denotes the outward transport of angular momentum dominated by gravitational wave emission *vs.* magnetic torques. [From Shibata and Taniguchi (2006).]

(2006) also find that in the case of prompt black hole formation the disk mass increases sharply with decreasing q for a fixed ADM mass and EOS, as shown in Figure 16.11.[61]

By contrast, in the case of hypermassive remnant formation, followed by delayed collapse to a black hole, the disk mass is likely to exceed $0.01 M_\odot$, independently of the mass ratio. If the dissipation of angular momentum responsible for inducing the collapse is due solely to the emission of gravitational waves from the ellipsoidal remnant, the disk mass can only be estimated, and amounts to 0.01–$0.03 M_\odot$.[62] If, however, the dissipation is accelerated by the amplification of a seed magnetic field in the remnant, the evolution of the system following collapse can be followed using black hole excision until a quasistationary state is reached. At this point the disk mass is found to be roughly $0.1 M_\odot$ (see Figure 16.11).

Several scenarios now suggest themselves by which the merger of a binary neutron star can trigger a short-hard GRB. Some plausible routes are summarized in Figure 16.12. Another scenario for forming the central engine of a short-hard GRB is the merger of a black hole-neutron star binary, which we will discuss in Chapter 17. While all of these

[61] Shibata and Taniguchi (2006) therefore suggest that a merger with a smaller q may be a better candidate for producing a short-hard gamma-ray burst central engine.

[62] The duration of this nonaxisymmetric secular dissipation process, which typically lasts 50 ms, is too long to be tracked reliably using a fully relativistic $3 + 1$ hydrodynamic code like the one of Shibata and Taniguchi (2006). However, it has been possible to follow this dissipation epoch for incompressible (Miller 1974; Detweiler and Lindblom 1977) and compressible (Ipser and Managan 1984; Lai and Shapiro 1995a) triaxial ellipsoids in Newtonian gravitation. These treatments allow for differential rotation (vorticity) and viscosity and incorporate a gravitational radiation-reaction potential. It is found that in the absence of viscosity, radiation-reaction forces typically drive a Jacobi-like ellipsoid to a nonradiating triaxial state (a Dedekind ellipsoid), for which $\Omega = 0$ in the inertial frame (a "stationary football"). Other possibilites exist, however, depending on the vorticity and (conserved) circulation of the configuration. For every scenario, the emitted quasiperiodic waves exhibit a unique signature that may be detectable by a gravitational wave laser interferometer; see Lai and Shapiro (1995a) and references therein.

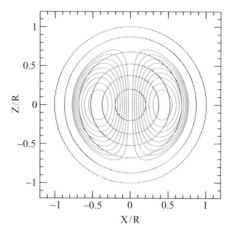

Figure 16.13 Magnetic field lines in a widely separated, spherical neutron star companion as viewed in the meridional plane. The star is an $n = 1$ polytrope with a rest mass $\bar{M}_0 = 0.146$ (the maximum being 0.180 for this EOS). Dotted (black) concentric circles are rest-mass density contours drawn for $\rho_0/\rho_0^{\max} = 0.9, 0.7, 0.5, 0.3, 0.1$, and 0.001. Solid (green) lines show representative magnetic field lines; the field is purely poloidal. Coordinates are in units of the stellar radius R. [From Liu *et al.* (2008).]

scenarios still await detailed analyses, it seems clear that numerical relativity is capable of supplying the computational firepower necessary to construct viable theoretical models. Perhaps one of the most important roles of numerical relativity will prove to be laying the theoretical groundwork for the *simultaneous* detection and identification of a gravitational wave signal and gamma-ray burst from the same cosmic source.[63]

Magnetized binary neutron star mergers

Since neutron stars are threaded by magnetic fields, it is important to assess the effect of such fields on binary mergers. The *exterior* fields of the two neutron stars will interact during the inspiral phase, and in principle this could affect the dynamics and gravitational waveforms. However, calculations show[64] that an exterior dipole field has a negligible influence provided the surface field strength is below 10^{16} G. Typical neutron stars are expected to have fields much smaller than this, and only for extreme "magnetars" do the inferred strengths begin to approach this limit. Any appreciable dynamical effect from magnetic fields must thus originate from *interior* magnetic fields and can occur only during and/or after the merger phase.

The effects of an interior field on binary neutron star merger have been investigated by Liu *et al.* (2008), who adopted the same basic HRSC GRMHD scheme of

---

[63] Also, simulations coupled with broadband gravitational wave observations at frequencies between 500 and 1000 Hz may constrain the neutron star EOS and measure its radius to an accuracy of $\delta R \sim 1$ km at 100 Mpc; Read *et al.* (2009).

[64] Ioka and Taniguchi (2000).

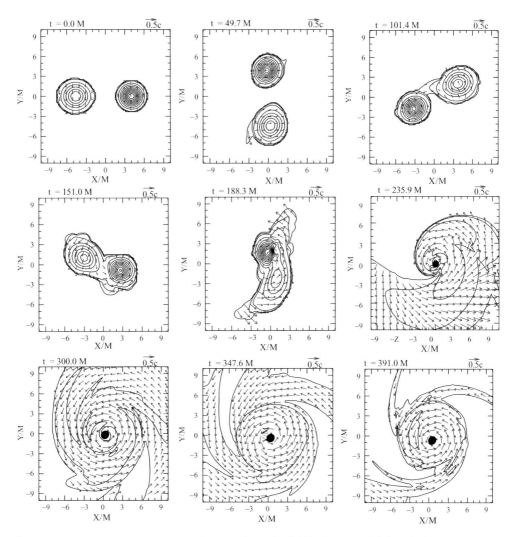

Figure 16.14 Snapshots of the rest-mass density and velocity field in the equatorial plane during the merger of a magnetized, neutron star binary. The initial binary is irrotational with $n = 1$ polytropic companions of unequal-mass (the ratio of rest masses is $q = 0.855$). The binary has a total ADM mass $\bar{M} = 0.290$, a rest mass $\bar{M}_0 = 0.317$ and an angular momentum $J/M^2 = 0.933$. Density contours are drawn for $\rho_0/\rho_0^{\max}(0) = 0.9, 0.8, \ldots, 0.1, 0.01, 0.001$, and 0.0001. The initial magnetic field profile in each star is depicted in Figure 16.13. The 3-velocity vectors are normalized as indicated above each frame. The black circle locates the apparent horizon. [From Liu *et al.* (2008).]

Duez *et al.* (2005b) described earlier in the book and used to evolve rotating, magnetized stars in Chapter 14.2.5. Liu *et al.* (2008) evolve binaries constructed via the conformal thin-sandwich approach for $n = 1$ irrotational polytropes. Adopting equatorial symmetry, they employ spatial grids with as many as $400 \times 400 \times 200$ zones in x–y–z directions

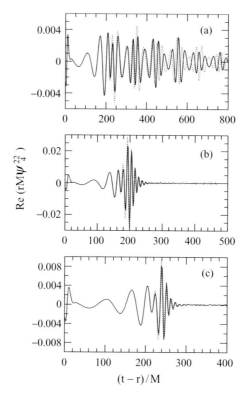

Figure 16.15 A comparison of gravitational waveforms for neutron star mergers in three different binary systems; each system is treated with and without magnetic fields. Binary system (a) has low, equal-mass companions and a total ADM mass $\bar{M} = 0.269$ and rest mass $\bar{M}_0 = 0.292$; it produces a hypermassive remnant. Binary system (b) has high, equal-mass companions and a total ADM mass $\bar{M} = 0.292$ and rest mass $\bar{M}_0 = 0.320$; it leads to prompt collapse to a black hole. Binary system (c) has high, unequal-mass companions and a total ADM mass $\bar{M} = 0.290$ and rest mass $\bar{M}_0 = 0.317$; it also leads to prompt collapse to a black hole. The merger of system (c) is highlighted in Figure 16.14. The solid (dotted) lines show the waveforms for unmagnetized (magnetized) binaries. [After Liu *et al.* (2008).]

(with the binary rotation axis along z) and use a simple FMR scheme[65] to enhance the resolution and dynamic range. As a check, they first repeated the simulations of Shibata *et al.* (2003b) for several unmagnetized cases, using the same initial data as in Shibata *et al.* (2003b), and found good agreement. For cases in which the mergers result in a prompt collapse to a black hole, Liu *et al.* (2008) employ moving puncture gauge conditions (Chapter 4.3) to extend the simulations in time to better determine the final mass of any debris that remains in an ambient disk about the remnant hole. They find that the disk mass is less than 2% of the total rest mass in all the cases studied. Liu *et al.* (2008) then add a poloidal magnetic field with a volume-averaged strength of about 10^{16} G to the initial configurations and explore the subsequent evolution; Figure 16.13 shows the initial

---

[65] "Multiple-transition fisheye" coordinates; see Chapter 14.2.3

magnetic field profile. Such a field is small enough to be dynamically unimportant initially, but large enough to reveal any effects of a field once the stars merge. For low-mass cases in which the remnant is a hypermassive neutron star, Liu *et al.* (2008) find small, but measurable differences in both the amplitude and phase of the gravitational waveforms following the merger when compared to the unmagnetized cases. For these cases they again find no appreciable disk. For high-mass cases in which the remnant is a rotating black hole, they find that the hole is surrounded by a disk, and that the disk mass and the gravitational waveforms are about the same as in the corresponding unmagnetized cases.

The simulation of a magnetized, unequal-mass binary with a total rest mass equal to 1.76 times the Oppenheimer–Volkoff limit is shown in Figure 16.14. The merger leads to prompt collapse to a black hole surrounded by a disk containing about 2% of the initial rest mass at the end of the simulation, as in the unmagnetized case.

A comparison of gravitational waveforms is plotted in Figure 16.15 for the merger of binaries with three different pairs of neutron star companions. Each binary is evolved both with and without interior magnetic fields. The quantities plotted are the $l = 2, m = 2, s = -2$ spin-weighted spherical harmonics of the Weyl tensor ψ_4, which may be related to the wave amplitudes h_+ and $h_\times$ using equations (9.126) and (9.133). In particular, $\text{Re}\,(\psi_4^{22}) = \ddot{h}_+^{22}$. The upper panel corresponds to a merger leading to a hypermassive remnant. The difference in waveforms between the magnetized and unmagnetized mergers is more pronounced for this case than for the other two, which involve prompt collapse to black holes. The simulation involving a hypermassive remnant was terminated after the remnant reached a quasistationary state, but before "delayed collapse" occurs.

Following binary merger and relaxation to a quasistationary state, the magnetic field in a hypermassive remnant can grow substantially by winding and instabilities like MRI. As a consequence, magnetic fields can drastically affect the long-term, secular evolution of a hypermassive remnant. As demonstrated in Chapter 14.2.5, such magnetic field amplification and turbulence can trigger delayed collapse to a black hole and subsequently drive gas accretion from the ambient disk onto the hole. The preliminary simulations summarized here suggest that, for coalescing binary neutron stars with astrophysically realistic magnetic field strengths, it is during this secular evolution phase in hypermassive remants that magnetic fields may play their most important role.

17 Binary black hole–neutron stars: initial data and evolution

Binary black hole–neutron stars have received significantly less attention than binary black holes or binary neutron stars. No black hole–neutron star binary has been identified to date. However, stellar population synthesis models suggest that such systems represent a significant fraction of all compact binary mergers ultimately visible in gravitational waves by the LIGO detector.[1] In addition, the study of black hole–neutron star mergers is important in light of the localizations of short-hard gamma-ray bursts.[2] These GRB sources are found in galactic regions of low star-formation devoid of supernovae associations, ruling out massive stars as progenitors: massive stars have very short lifetimes and would need to be replenished more rapidly than is possible in low star-formation regions to account for these bursts. A more plausible progenitor for a short-hard GRB is a compact binary containing a neutron star, i.e., either a binary neutron star or binary black hole-neutron star.[3] The short-hard burst time scales and energetics are consistent with GRB models based on the coalescence of such compact binaries, and the evolution time scale of over 1 Gyr between formation and merger is consistent with the low star-formation rate.[4]

Black hole-neutron star binaries can merge in two distinct ways. The neutron star may either be tidally disrupted by the black hole companion before being consumed, or it may be swallowed by the black hole more or less intact. Which one of these two scenarios is realized depends on whether the tidal disruption occurs sufficiently far outside the ISCO, which, in turn, depends largely on the binary mass ratio[5] $q \equiv M_{BH}/M_{NS}$ and the neutron star compaction $\mathcal{C} \equiv M_{NS}/R_{NS}$. To understand the qualitative dependence we can invoke a very simple Newtonian argument. Consider a particle of mass m on the surface of the neutron star. The inward gravitational force F_{grav} exerted by the neutron star on this mass is

$$F_{grav} = \frac{m M_{NS}}{R_{NS}^2}.$$ (17.1)

---

[1] Kalogera *et al.* (2007).
[2] See, e.g., Berger (2006) and references therein for the physical parameters measured or inferred from *Swift* and *HETE-2* gamma-ray satellite observations.
[3] Paczynski (1986); Narayan *et al.* (1992).
[4] Belczynski *et al.* (2002); but note that other GRB models have been proposed.
[5] Note that some authors define the mass ratio q as the inverse of our convention.

The outward tidal force on m caused by the presence of the black hole companion is approximately

$$F_{\text{tid}} \approx \frac{m M_{\text{BH}} R_{\text{NS}}}{r^3}, \tag{17.2}$$

where r is the binary separation. We can now estimate the tidal separation r_{tid} at which the star is tidally disrupted by equating these two forces, which yields

$$\frac{r_{\text{tid}}}{M_{\text{BH}}} \simeq q^{-2/3} \frac{R_{\text{NS}}}{M_{\text{NS}}} \simeq q^{-2/3} \mathcal{C}^{-1}. \tag{17.3}$$

For a sufficiently massive black hole (and thus large q) r_{tid} is smaller than the ISCO of the black hole, so that the neutron star plunges through the ISCO without being disrupted. For a typical neutron star of compaction $\mathcal{C} \approx 0.15$, the critical mass ratio at which tidal disruption occurs at the ISCO is $q_{\text{crit}} \approx 4$ (see equation 17.20 below).

> **Exercise 17.1**   (a) Consider a binary black hole–white dwarf. Use equation (17.3) to estimate the critical mass of the black hole above which a typical white dwarf can be swallowed by the hole without disruption. Take $M_{\text{WD}} = M_\odot$ and $R_{\text{WD}} = 10^{-2} R_\odot$.
> (b) Repeat the above calculation for a binary consisting of a black hole and a main sequence star like the Sun.

Neutron stars with *stellar-mass* black hole companions are likely to be formed through normal stellar binary evolution at a rate that depends on the distribution of binary mass ratios, common-envelope dynamics, the magnitude of imparted supernovae kicks, and other effects, many of which remain uncertain.[6] For sufficiently tight binaries, merger will occur within a Hubble time and the orbit will circularize by the time the binary enters the LIGO gravitational wave detector band. As we can see from the estimate (17.3), the neutron star will be tidally disrupted before being swallowed if the black hole mass is sufficiently small.

Binaries consisting of neutron stars orbiting *very massive* black holes form quite differently from those orbiting stellar-mass black holes. As we have already noted, there is strong evidence from astrometric observations that massive, compact, "dark" objects reside in the cores of every bulge galaxy, including the Milky Way. These objects are believed to be supermassive black holes with masses in the range 10^6–$10^9 M_\odot$. Such objects are also believed to be the engines that power quasars and active galactic nuclei.[7] Conservative, N-body gravitational encounters (small-angle, Coulomb scattering) between stars in the dense cores of galaxies can inject stars, including neutron stars and stellar-mass black holes, into orbits that move close to the central supermassive black hole. If they pass sufficiently close, these stars may be captured by the hole and subsequently spiral inward due to the emission of gravitational waves.[8] The estimate (17.3) shows that neutron stars

---

[6] See, e.g., Kalogera *et al.* (2007) and references therein for a discussion.
[7] Rees (1998).
[8] Shapiro (1985); Sigurdsson (2003); Merritt and Poon (2004); Hopman and Alexander (2005).

captured in this way will be consumed by supermassive black holes before undergoing tidal disruption. In fact, the internal structure of the neutron star has very little effect on the orbital evolution of these extreme-mass-ratio inspiral binaries ("EMRIs"). They are, however, subject to a host of relativistic effects, including periastron and Lense–Thirring precession, and these strong-field effects will leave an imprint on the gravitational wave signals. For central black holes in the mass range $10^5 \lesssim M/M_\odot \lesssim 10^7$ the emitted waves will peak in the low frequency band near $\sim 10^{-3}$Hz, close to the lower limit anticipated for the LISA gravitational wave interferometer.[9] Rough estimates suggest that LISA might be able to detect EMRIs out to a redshift $z \sim 1$ and that as many as 10^3 EMRIs might be observed during its lifetime. The measured waveforms will be able to chart the spacetime of the black hole, trace its multipolar structure and confirm that it obeys the Kerr solution.[10] The orbital dynamics and the expected gravitational waveforms of EMRIs can be treated by black hole perturbation techniques in the stationary Kerr field of the central hole.[11]

However, in this chapter we shall be primarily interested in tight stellar-mass BH-NS binaries. These systems ultimately inspiral in quasistationary, nearly circular orbits as the neutron stars approach the horizon. The neutron stars are subject to appreciable tidal distortion and, in some cases, disruption prior to merger. The strong-field spacetime is dynamical, nonstationary and nonperturbative. For these systems the full machinery of numerical relativity is necessary to follow the evolution.

As in the case of a binary neutron star merger, a black hole-neutron merger, if observed, could potentially provide insight into the physics of matter at nuclear densities. For example, the onset of mass transfer from the neutron star to the black hole depends on the neutron star radius, given the stellar masses. The stability and nature of the mass transfer provides information about the stiffness of the EOS.[12] Whereas for binary neutron stars the characteristic frequencies of gravitational wave emission during the merger and formation of a remnant (either a black hole or hypermassive neutron star) fall beyond the peak sensitivity of an advanced LIGO detector (100–500 Hz), the characteristic frequencies of the onset of neutron star mass transfer and tidal disruption occur at a lower values, closer to LIGO's most sensitive band. Should a gravitational wave signal from a merger be observed in coincidence with a short-duration GRB, one can determine its distance, luminosity and characteristic beaming angle.[13]

Black hole–neutron star merger calculations have in common with binary neutron star calculations the challenge of solving relativistic hydrodynamics in a strong, dynamical, gravitational field. They have in common with binary black hole calculations the additional complications associated with the presence of a spacetime singularity inside the black hole from the onset of the simulation. The combination of these two effects helps explain why

---

[9] See Chapter 9.2.2.
[10] Collins and Hughes (2004).
[11] See, e.g., Babak *et al.* (2007) and references therein.
[12] Faber *et al.* (2006a).
[13] See Kobayashi and Mészáros (2003) for a discussion.

progress on simulating black hole–neutron star mergers has come later than the progress achieved on the other two problems. But the situation is advancing rapidly, as the techniques that proved successful in simulating binary neutron stars and black hole binaries have been adapted successfully to handle black hole–neutron star binaries.

17.1 Initial data

As we found for the other binary systems considered in this book, constructing initial data serves two independent purposes. Clearly, we need such solutions as initial data for dynamical simulations, as discussed in Section 17.2 below. In addition we can construct sequences of constant mass, parametrized by the binary separation, to mimic evolutionary sequences, and to locate the ISCO or the onset of tidal disruption. The approximate scaling of the tidal separation $r_{\rm tid}$ with the mass ratio and the neutron star compaction is given by (17.3), but clearly a fully relativistic calculation is required to obtain a more reliable result.

Constructing initial data that model black hole–neutron star binaries basically requires three ingredients: a solution for the gravitational fields, a solution for the relativistic fluid profiles inside the neutron star, and a suitable description of the black hole. Fortunately, we can assemble these ingredients from previous sections in this book: we have developed decompositions of Einstein's initial value equations in Chapter 3, derived the equations governing relativistic, stationary fluid solutions in Chapter 15, and modeled binaries containing black holes in Chapter 12. As in the case of binary black hole initial data, black hole–neutron star initial data have also been constructed using both the conformal transverse-traceless decomposition of Einstein's constraint equations (Chapter 3.2) and the conformal thin-sandwich decomposition (Chapter 3.3). We shall begin with the latter approach in Section 17.1.1, and will then briefly review the former in Section 17.1.2.

17.1.1 The conformal thin-sandwich approach

A systematic study of quasiequilibrium, black hole–neutron star binary initial data has been carried out by the group at the University of Illinois.[14] Their hierarchical treatment, which is based on the conformal thin-sandwich approach, started with several simplifying assumptions, including extreme mass ratios and corotating neutron stars, and then relaxed these assumptions one at a time. Here we will focus on their models of irrotational binaries, by which we mean irrotational neutron stars orbiting nonspinning black holes, with companions of comparable mass. They have also generated initial data by this method to describe binaries of comparable mass containing irrotational neutron stars and spinning black holes. They have used these initial data sets to evolve representative binary configurations,[15] and these simulations will be discussed in Section 17.2.2.

[14] Baumgarte *et al.* (2004); Taniguchi *et al.* (2005, 2006, 2007, 2008); see also Grandclément (2006) for a similar approach.
[15] See, e.g., Etienne *et al.* (2008, 2009) and references to earlier work.

The gravitational field equations

In Chapter 3.3 we introduced the conformal thin-sandwich decomposition of Einstein's constraint equations and saw how it is well suited for the construction of equilibrium initial data (see Box 3.3 for a summary). To describe an equilibrium binary we assume the existence of an approximate, helical Killing vector. In a corotating coordinate system we may then set to zero the time derivative of both the conformally related metric and the mean curvature, i.e., $\bar{u}_{ij} = \partial_t \bar{\gamma}_{ij} = 0$ and $\partial_t K = 0$. If we further adopt maximal slicing $K = 0$ and conformal flatness $\bar{\gamma}_{ij} = \eta_{ij}$, the equations reduce to the Hamiltonian constraint

$$\bar{D}^2 \psi = -\frac{1}{8}\psi^{-7}\bar{A}_{ij}\bar{A}^{ij} - 2\pi\psi^5\rho, \tag{17.4}$$

the momentum constraint

$$(\bar{\Delta}_L \beta)^i = 2\bar{A}^{ij}\bar{D}_j(\alpha\psi^{-6}) + 16\pi\alpha\psi^4 j^i, \tag{17.5}$$

and the condition $\partial_t K = 0$, which, from equation (2.137), yields

$$\bar{D}^2(\alpha\psi) = \alpha\psi\left(\frac{7}{8}\psi^{-8}\bar{A}_{ij}\bar{A}^{ij} + 2\pi\psi^4(\rho + 2S)\right). \tag{17.6}$$

Here $\bar{D}^2 = \nabla^2$ is the flat Laplace operator, $\bar{\Delta}_L$ the flat vector Laplacian defined in equation (3.51),[16] and $\bar{A}^{ij}$ is given by

$$\bar{A}^{ij} = \frac{\psi^6}{2\alpha}(\bar{L}\beta)^{ij}, \tag{17.7}$$

where the vector gradient $\bar{L}$ is defined in equation (3.50). For given matter sources and boundary conditions these equations can be solved as in Chapters 12 and 15.

Relativistic equations of stationary equilibrium

We can determine the matter sources appearing in equations (17.4)–(17.6) by solving the relativistic hydrodynamic equations in stationary equilibrium, which we derived in Chapter 15. In particular, we need to find stationary solutions of the Euler equation (15.34) and the continuity equation (5.7).

For a *corotational* star, which is static as viewed in a corotating coordinate system, the continuity equation is satisfied identically, and the Euler equation can be integrated once to yield the algebraic equation (15.46),

$$\frac{h}{u^t} = C. \tag{17.8}$$

Here h is the fluid's enthalpy, u^t is the time-component of the fluid's 4-velocity, and C is a constant of integration that has to be determined as part of the iteration. Using

---

[16] See exercise 3.8 for an example.

the normalization of the 4-velocity $u_a u^a = -1$, we can express u^t in terms of its spatial components (see equation 15.47) and, hence, in terms of the binary's orbital angular velocity Ω. The solution for h, which can expressed in terms of the matter source profiles once we specify an equation of state, depends on the gravitational fields through this normalization condition. The field and fluid equations therefore have to be solved simultaneously, as outlined in Chapter 15.2.

For more realistic *irrotational* stars,[17] we can express the 4-velocity in terms of a gradient of a velocity potential Φ as in equation (15.65),

$$h u_a = \nabla_a \Phi. \tag{17.9}$$

Defining B^i as the shift vector in the corotating coordinate system (15.70), we can again integrate the Euler equation to find an algebraic equation for the enthalpy h

$$h^2 = \frac{1}{\alpha^2}(C + B^i D_i \Phi)^2 - D_i \Phi D^i \Phi, \tag{17.10}$$

where C is again a constant of integration (see equation 15.76). The continuity equation, however, is no longer satisfied identically, and now results in an elliptic equation for the velocity potential Φ,

$$D_i D^i \Phi - D_i \left(\frac{C + B^j D_j \Phi}{\alpha^2} B^i \right) = \left(\frac{C + B^j D_j \Phi}{\alpha^2} B^i - D^i \Phi \right) D_i \ln \frac{\alpha \rho_0}{h} \tag{17.11}$$

(see equation 15.78). As in the case of corotational stars, these fluid equations have to be solved simultaneously with the constraint equations (17.4)–(17.6). Before we can do that, though, we have to impose appropriate boundary conditions on the gravitational field variables.

Quasi-equilibrium black hole boundary conditions

Assuming asymptotic flatness, we may impose the asymptotic fall-off conditions (12.73) at infinity, or at a large separation from the binary. To avoid the black hole singularity, we excise the black hole interior, usually taken to be a coordinate sphere, and impose black hole equilibrium boundary conditions on the surface of this excised region. We derived these boundary conditions in Chapter 12.3.1 from the notions of apparent and isolated horizons (see Chapters 7.3 and 7.4).

To formulate these boundary conditions it proved convenient to split the shift vector β^i into parts that are normal and tangential to the excision surface, $\beta_\perp$ and $\beta_\parallel^i$ (see equation 12.82). The normal component must then satisfy condition (12.84)

$$\beta_\perp = \alpha, \tag{17.12}$$

---

[17] See Faber *et al.* (2006a), who show why binary black hole-neutron stars are likely to be irrotational, or footnote 27 for a brief summary.

while the tangential components $\beta_\parallel^i$ must form a conformal Killing vector of the the induced metric on the excision surface (see equation 12.106), which can be constructed from

$$\beta_\parallel^i = \Omega_{spin}\xi^i \tag{17.13}$$

(see equation 12.107). Here $\xi^i = \epsilon^{ijk}\hat{z}_j\hat{n}_k$ is the Killing vector of a unit sphere, where $\hat{z}^i$ is a unit vector aligned with the axis of rotation, $\hat{n}^i$ the unit normal on the sphere's surface, and Ω_{spin} is a yet undetermined parameter associated with the black hole's spin. We furthermore found that the conformal factor ψ has to satisfy the Neumann boundary condition (12.90), while the lapse α can be chosen arbitrarily.

We can now construct an iterative algorithm to solve the above equations self-consistently. In Chapter 15.2 we outlined such an algorithm for an equal-mass, corotating neutron star binary, and we have discussed the construction of irrotational binaries in Chapter 15.3. Here the situation is more complicated, because we have replaced one of the two neutron stars with a black hole, and we can no longer assume the two masses to be equal. Allowing for these differences therefore entails several additional nested iterations. One of these new iterations adjusts the location of the binary's axis of rotation – assumed to intersect the coordinate axis connecting the centers of the black hole and the neutron star – until the binary's total linear momentum vanishes. Another iteration adjusts the coordinate radius of the excised sphere until the black hole's irreducible mass (7.2) equals its desired value. This still leaves undetermined the spin parameter Ω_{spin} in the boundary condition (17.13). A corotating black hole would be nonspinning in a corotating coordinate system, which corresponds to $\Omega_{spin} = 0$. More interesting are nonspinning black holes. We might attempt to construct such binaries by setting Ω_{spin} equal to the orbital angular speed, which we called the "leading-order spin approximation" in Chapter 12.4. As we discussed there, however, it is more accurate to iterate over Ω_{spin} until the black hole's quasi-local spin (7.74) vanishes. To treat spinning black holes, we iterate over Ω_{spin} until the black hole spin equals the desired value.

Numerical results

Taniguchi *et al.* (2008) solve the above equations with the spectral methods of the LORENE[18] package to construct irrotational black hole–neutron star binaries. As we discussed in the context of binary neutron stars in Chapter 15.3, it is again useful to split all gravitational field variables into parts that are associated with one or the other companion star (see equation 15.81). The resulting two equations can then be solved on two separate computational domains, one centered on the black hole, and one on the neutron star. Only the equation associated with the black hole needs to be excised, if all source terms affected by those excised quantities are moved into the black hole equation.[19]

---

[18] See http://www.lorene/obspm.fr/.
[19] See the discussion following equation (15.81); also see Appendix A in Taniguchi *et al.* (2007) for a detailed treatment.

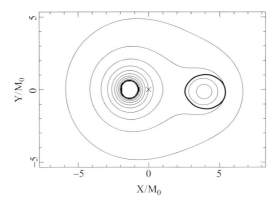

Figure 17.1 Contours of the conformal factor ψ in the orbital plane for an irrotational binary black hole–neutron star close to tidal break-up. This binary has a mass ratio of $q = 3$, and the neutron star, governed by a polytropic equation of state with polytropic index $n = 1$, has a compaction of $M_{NS}/R_0 = 0.1452$, where M_{NS} is the ADM mass that the star would have in isolation and R_0 its areal radius. The circle on the left marks the black hole's apparent horizon. The interior of this coordinate circle is excised from the computational grid. The other thick line on the right marks the neutron star's deformed surface. The cross "$\times$" indicates the position of the rotation axis. [From Taniguchi *et al.* (2008).]

Taniguchi *et al.* (2008) adopt a polytropic equation of state (5.18) with $\Gamma = 2$ (i.e., polytropic index $n = 1$) to describe the neutron star matter. They consider mass ratios $1 \leq q \leq 10$, where $q \equiv M_{BH}/M_{NS}$ is the ratio of the irreducible mass of the black hole to the ADM mass of a spherical, isolated neutron star. The neutron stars in isolation would have a compaction in the range $0.1088 \leq M_{NS}/R_0 \leq 0.1780$, where R_0 is the areal radius. The most compact neutron star model has a rest mass that corresponds to 94% of the maximum allowed rest mass for a nonrotating, $\Gamma = 2$ polytrope in isolation. A typical configuration, close to the formation of a cusp in the neutron star's surface that marks the onset of tidal disruption,[20] is shown in Figure 17.1.

We can mimic inspiral sequences by constructing constant-mass sequences of black hole-neutron stars, parametrized by the binary separation. In this case, the conserved masses are the rest mass M_0 of the neutron star and the irreducible mass M_{irr} of the black hole. For these sequences we can measure the total angular momentum J as well as the binding energy E_b, defined as

$$E_b \equiv M_{ADM} - M_0. \tag{17.14}$$

Here M_0 is the total ADM mass of the system at infinite binary separation, i.e., the sum of the black hole's irreducible mass and the neutron star's ADM mass in isolation,

$$M_0 = M_{BH} + M_{NS}. \tag{17.15}$$

---

[20] The spectral methods used in these calculations break down when the stellar surface forms a cusp – see our discussion in Chapter 15.3. The authors therefore introduce a "mass-shedding indicator" χ as in equation (15.82) and use extrapolation to locate the onset of tidal disruption.

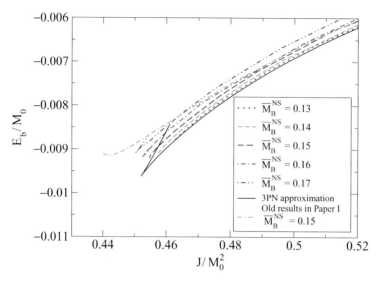

Figure 17.2 Binding energy E_b vs. the total angular momentum J for a polytropic $n = 1$ binary with mass ratio $q = 5$. The "old results in Paper I" refer to results obtained with the "leading-order spin approximation" for the parameter Ω_{spin} in the black hole boundary conditions. The quantity $\bar{M}_{\mathrm{B}}^{\mathrm{NS}}$ denotes the dimensionless (baryon) rest mass of the neutron star and is larger for higher compactions; the maximum isolated mass corresponds to $\bar{M}_{\mathrm{B}}^{\mathrm{NS}} = 0.18$. [From Taniguchi *et al.* (2008).]

In Figure 17.2 we show the binding energy as a function of angular momentum for binaries of mass ratio $q = 5$. The graph includes results for several different neutron star compactions, the post-Newtonian point-mass result (which is independent of neutron star compaction; see Appendix E), as well as a "leading-order" spin result that we return to below. We first observe that the numerical results agree with the the post-Newtonian result to very high precision for low neutron star compactions; for larger compactions, which imply stronger gravitational fields, the deviation increases slightly.

A simultaneous turning-point in the binding energy E_b and the angular momentum J marks the onset of an orbital instability; we therefore identify the corresponding orbit with the ISCO. When we plot E_b vs. J, as in Figure 17.2, such a simultaneous turning point results in a cusp. The post-Newtonian curve, for example, displays such a cusp[21] very clearly. For the numerical results, however, only those with the larger compaction feature a cusp. This is because the numerical results terminate just before the onset of tidal disruption. For small neutron star compactions, the tidal separation is larger than the ISCO separation, so that the corresponding curves never reach an ISCO. We anticipated this finding qualitatively from equation (17.3), which demonstrates that, for a given binary mass ratio, a smaller compaction $\mathcal{C} = M_{\mathrm{NS}}/R_0$ leads to a larger tidal separation $r_{\mathrm{tid}}/M_{\mathrm{BH}}$. If

---

[21] The cusp in the E_b vs. J curve, identifying an ISCO, should not be confused with the cusp in the stellar surface at the onset of tidal disruption.

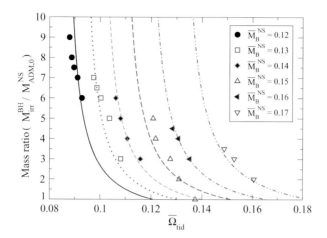

Figure 17.3 The tidal break-up limit. The lines represent the estimate (17.17) with $\mathcal{A} = 0.27$ for different neutron star compactions, while the other symbols show the corresponding numerical results. The quantity $\bar{M}_{\mathrm{B}}^{\mathrm{NS}}$ denotes the dimensionless (baryon) rest mass of the neutron star and is larger for higher compactions; the maximum isolated mass corresponds to $\bar{M}_{\mathrm{B}}^{\mathrm{NS}} = 0.18$. [From Taniguchi *et al.* (2008).]

the tidal separation is larger than the ISCO separation, i.e., if the compaction is sufficiently small, the neutron star breaks apart before encountering an ISCO. The post-Newtonian curve cannot capture this behavior, since it ignores the neutron star's internal structure.

Finally, we note that the "leading-order spin approximation", which amounts to setting Ω_{spin} to the orbital angular speed in the boundary condition (17.13), evidently introduces some numerical error. This curve shows a larger deviation from the post-Newtonian result than the corresponding curve for which the black hole's quasilocal spin is set to zero, and the minimum in the binding energy no longer coincides with a minimum in the angular momentum – exactly as for the nonspinning black hole binaries of Chapter 12.4.

It is of interest to see how well the estimate (17.3) predicts the actual numerical results. Instead of expressing the latter in terms of the binary separation, which is a gauge-dependent quantity, it is more natural to parametrize the separation in terms of the orbital angular speed Ω. In equation (17.3) we can eliminate r_{tid} with the help of Kepler's law

$$\Omega^2 \simeq \frac{M_{\mathrm{BH}} + M_{\mathrm{NS}}}{r^3} \tag{17.16}$$

(where equality holds for Newtonian point masses). Combining equations (17.3) and (17.16) we find

$$\Omega_{\mathrm{tid}} M_{\mathrm{NS}} = \mathcal{A} \left(\frac{M_{\mathrm{NS}}}{R_{\mathrm{NS}}} \right)^{3/2} \left(\frac{1+q}{q} \right)^{1/2}, \tag{17.17}$$

where $\mathcal{A}$ is a yet-to-be-determined constant of order unity.

In Figure 17.3 we show a comparison between the estimate (17.17) and the actual numerical results for a number of different neutron star compactions $M_{\mathrm{NS}}/R_{\mathrm{NS}}$. Choosing

$\mathcal{A} = 0.27$ in the estimate (17.17) captures the numerical tidal break-up results rather well, although we caution that these results cover only a very small part of parameter space – and, in particular, only $n = 1$ polytropes. Using equation (17.17) to estimate the characteristic gravitational wave frequency $f_{GW} = \Omega/\pi$ at tidal break-up yields

$$f_{GW} \simeq 730 \text{ Hz} \left(\frac{M_{NS}}{1.4 M_\odot} \right)^{1/2} \left(\frac{R_{NS}}{15 \text{ km}} \right)^{-3/2} \left(\frac{(1+q)/q}{4/3} \right)^{1/2}. \qquad (17.18)$$

As mentioned earlier, this frequency resides within the LIGO band.

Having identified $\mathcal{A} = 0.27$ in equation (17.17), we can now refine equation (17.3) to obtain

$$\frac{r_{tid}}{M_{BH}} \simeq 2.4 \, q^{-2/3} \frac{R_{NS}}{M_{NS}}. \qquad (17.19)$$

Assuming a typical neutron star compaction of $M_{NS}/R_{NS} \simeq 0.15$ we can now estimate the critical mass ratio at which tidal separation occurs at the ISCO by setting the left-hand side of equation (17.19) equal to $r_{ISCO}/M_{BH} \simeq 6$, which yields

$$q_{crit} \simeq 4. \qquad (17.20)$$

Evidently, neutron stars are tidally disrupted only in binaries with black holes of comparable mass. The tidal disruption of the neutron star and the subsequent formation of a gaseous disk may be necessary ingredients for a GRB model based on the merger of a black hole–neutron star binary. As these estimates suggest, and as we discuss in greater detail in Section 17.2, it is difficult to form such a disk for nonspinning black holes. For rapidly rotating black holes, however, the ISCO occurs at a smaller binary separation, enhancing the possibility of tidal break-up and the formation of a disk.

17.1.2 The conformal transverse-traceless approach

Shibata and Uryū (2006, 2007) adopt an alternative approach to construct black hole–neutron star binaries. Instead of using the conformal thin-sandwich decomposition of the Einstein's constraint equations, as in the previous section, they employ the conformal transverse-traceless decomposition to construct the extrinsic curvature.

Following Chapter 3.2, we can split $\bar{A}^{ij}$ into a transverse-traceless piece and a longitudinal piece. We may assume the former to vanish, and write the latter as the vector gradient of a vector W^i,

$$\bar{A}_L^{ij} = (\bar{L}W)^{ij} \qquad (17.21)$$

(see equation 3.50). Assuming maximal slicing $K = 0$, the momentum constraint then reduces to

$$(\bar{\Delta}_L W)^i = \bar{D}_j \bar{A}^{ij} = 8\pi \, \psi^{10} S^i \qquad (17.22)$$

(see equation 3.53). Given the linear nature of this equation, we can write its solution as the sum of a particular and a homogeneous solution. We may choose the latter, which satisfies the homogeneous equation $\bar{D}_j \bar{A}^{ij} = 0$, to describe the black hole, in which case the former accounts for the neutron star only.

As in Chapter 12.2, where we constructed binary black holes in exactly the same way, we may adopt conformal flatness $\bar{\gamma}_{ij} = \eta_{ij}$ and take the boosted Bowen–York black hole solution (3.80) as the homogeneous solution

$$\bar{A}^{ij}_{\text{BH}} = \frac{3}{2r^2} \left(P^i n^j + P^j n^i - (\eta^{ij} - n^i n^j) n_k P^k \right). \tag{17.23}$$

The coordinate distance r from the black hole's center and the normal vector n^i are defined in the context of equation (3.80) (see also Chapter 12.2.1), and for circular orbits, the momentum P^i of this solution has to point in a direction perpendicular on the axis connecting the black hole and the center of the neutron star (see Chapter 12.2.3). This analytical homogeneous solution to equation (17.22) then accounts for the black hole of the black hole–neutron star binary.

Given a matter source S^i, we can now solve equation (17.22) with appropriate asymptotic fall-off boundary conditions to find the neutron star contribution $\bar{A}^{ij}_{\text{NS}}$. This can be done by solving equation (17.22) for W^i, using one of the decompositions of W^i described in Appendix B, and then computing $\bar{A}^{ij}_{\text{NS}}$ from equation (17.21). Adding this result to the black hole contribution yields

$$\bar{A}^{ij} = \bar{A}^{ij}_{\text{BH}} + \bar{A}^{ij}_{\text{NS}}. \tag{17.24}$$

We can now insert this extrinsic curvature into the Hamiltonian constraint (17.4), which, given the matter source ρ, we can solve via the "puncture" method (12.50) described in Chapter 12.2.2. This completes the construction of the initial data, assuming we know the matter sources ρ and S^i.

Of course, we do *not* know these matter sources a priori. Instead, we need to determine them self-consistently with the gravitational field variables, so that they describe the neutron star in quasiequilibrium circular orbit about its black hole companion. This determination entails solving the fluid equations above, either equation (17.8) for corotating stars, or equations (17.10) and (17.11) for irrotational stars. Since the notion of an equilibrium fluid flow involves a condition on the fluid 4-velocity u^a, i.e., a 4-dimensional object, it is clear that we need information about the spacetime not only on the initial slice, but in a neighborhood of this slice. This observation is one of the reasons why the conformal thin-sandwich decomposition, which automatically introduces the lapse and shift that are needed in the above fluid equations, is more commonly used for the construction of quasiequilibrium fluid initial data (see also the discussion in Chapter 15.2).

The conformal transverse-traceless decomposition, by contrast, only provides the metric and extrinsic curvature of the initial slice, meaning that we now need to construct a lapse and shift independently. This cannot be just any lapse and shift, of course – instead they must describe the coordinate system in which the binary appears momentarily stationary, as assumed in the derivation of the fluid equations. In other words, we require the 4-velocity of the resulting coordinate-observer to be aligned with the approximate helical Killing vector describing the binary orbit. We can construct such an approximate "Killing" lapse and shift by solving equations (17.5) and (17.6) of the conformal thin-sandwich decomposition.[22] Unlike we did in the previous section, however, we do not express $\bar{A}^{ij}$ in these equations in terms of the shift as in equation (17.7), but instead use the transverse-traceless expression (17.24). The latter is based on equation (17.21) instead of equation (17.7), and does not result from, and therefore does not self-consistently embody, the assumption of a helical Killing vector. In this case, the solutions of equations (17.5) and (17.6) do not require excision: all terms in equation (17.5) are regular, and the singular terms in equation (17.6) can be treated analytically by writing

$$\alpha\psi = 1 + \frac{C}{r} + v \tag{17.25}$$

in complete analogy to the puncture approach (12.50) for the conformal factor.[23] The resulting equation for the new variable v is then regular, and the constant C can be determined by imposing a virial relation, namely by equating the Komar mass (3.179) with the ADM mass (3.145) as in equation (12.108).

Given the lapse and shift, we can finally solve the fluid equations, which in turn provide the matter sources ρ and S^i in equations (17.4) and (17.22). A self-consistent solution can then be found by solving all of the above equations iteratively until convergence has been achieved. These solutions have been used as initial data for dynamical simulations,[24] which we will describe in Section 17.2.2 below.

17.2 Dynamical simulations

17.2.1 The conformal flatness approximation

The merger of a black hole-neutron star binary, focusing on the tidal disruption of the neutron star, has been investigated in a number of Newtonian[25] and semirelativistic simulations.[26] Some of the first dynamical simulations of merging binaries that attempted to evolve the black hole–neutron star spacetime self-consistently in a general relativistic framework were performed by Faber *et al.* (2006a,b) in the conformal flatness approximation. We have already discussed in Chapter 16.2 the application of this approximation to track binary neutron star mergers. However, the conformal flatness approximation is not

---

[22] *cf.* Tichy *et al.* (2003).
[23] See also Hannam *et al.* (2003); Tichy *et al.* (2003); Hannam and Cook (2005); Hannam (2005).
[24] Shibata and Uryū (2006, 2007); Shibata and Taniguchi (2007); Yamamoto *et al.* (2008).
[25] See, e.g., Lee and Kluźniak (1999); Janka *et al.* (1999); Rosswog (2005).
[26] See, e.g., Rantsiou *et al.* (2007), who studied merging black hole–neutron star systems with high mass ratios $q \sim 10$ by evolving neutron stars in a fixed Kerr background metric, treating the self-gravity of the matter by a Newtonian potential in the hydrodynamic equations.

well suited to handle an arbitrary black hole–neutron star spacetime since it contains a moving, vacuum black hole containing an interior singularity. The metric field equations in the conformal approximation are not evolution equations but spatial elliptic equations, identical to the conformal thin-sandwich equations (17.4)–(17.6) used to solve the initial value problem. Consequently they do not involve time and cannot be used to determine the motion of the black hole or track the changing position of its interior spacetime singularity. Nor can these field equations encounter the singularity without blowing up. The one limit in which the conformal approximation can be applied to the binary black hole–neutron star problem is the case in which the black hole mass is considerably larger than the mass of the neutron star. In this situation, the black hole remains nearly fixed in space and the spacetime near the black hole is dominated by the massive hole. This limit enables one to determine the metric by avoiding the black hole singularity while still accounting for the strong-field contributions from *both* the black hole and neutron star matter.

Consider the Hamiltonian equation (17.4) used to solve for the conformal factor ψ. Embed the neutron star in a computational spatial domain whose characteristic size d satisfies $M_{NS} \ll d \ll M_{BH}$. Exploiting this inequality makes it possible to construct a computational grid that encompasses the neutron star matter but avoids the black hole horizon. Decompose ψ according to $\psi = \psi_{BH} + \psi_{NS}$, where $\psi_{BH} = 1 + M_{BH}/2r$ may be taken to be the exact solution for a stationary, isolated, nonspinning Schwarzschild black hole in isotropic coordinates as a function of the radius from the black hole. Since $\nabla^2 \psi_{BH} = 0$, equation (17.4) results in a nonlinear elliptic equation for $\nabla^2 \psi_{NS}$. This equation can be solved on the computational grid by considering all the source terms inside the grid interior and imposing suitable asymptotic conditions on ψ_{NS} at the outer boundary of the grid, where ψ_{NS} falls off to zero. The resulting metric solution for ψ then accounts self-consistently for the field that arises from the combined nonlinear contributions of the black hole and the neutron star matter sources inside the grid. The (small) contribution from ψ_{NS} in the region outside the computational grid can be obtained analytically by a suitable multipole expansion involving multipole moments of the interior source. As long as the metric outside the computational grid is dominated by the black hole, the solution remains roughly valid even if some of the matter flows out into this region. The solutions to the lapse and shift equations are obtained by similar considerations.

The approach described above was developed by Faber *et al.* (2006a) to handle binaries with large black hole-to-neutron star mass ratios in conformal gravitation. They used a spheroidal spectral routine based on the LORENE code to solve the field equations and a relativistic SPH code to evolve the hydrodynamics.[27] Subsequently, Faber *et al.* (2006b) switched to an FFT convolution-based scheme to solve the field equations, whereby they linearized the nonlinear terms in the Hamiltonian about a previous guess for ψ_{NS} and iterated until convergence. The SPH approach for the hydrodynamics has the advantage

---

[27] See Chapter 16.2, where similar tools were used for the treatment of binary neutron star mergers in the conformal approximation.

that, given the metric field, SPH does not require a grid on which to evolve the matter, hence the matter can be followed everywhere, even in the region in exterior to the computational grid constructed to solve the field equations.

For initial data Faber *et al.* (2006b) consider both corotational and irrotational neutron star companions, modeled as polytropes.[28] One of the goals of Faber *et al.* (2006b) is to delineate regimes of stable *vs.* unstable mass transfer and to identify the point of onset of tidal disruption for different neutron star compactions, EOS stiffness, and binary mass ratios. Their simulations reveal that whenever mass transfer begins while the neutron star orbit is still outside the ISCO the transfer is more unstable than predicted by earlier analytic formalisms. The reason is that mass transfer causes the orbital separation to increase and the neutron star radius to expand, both on a dynamical time scale. Models of mass transfer that assume the orbit remains quasicircular are thus not applicable. The unstable mass loss drives the evolution of the binary orbit and leads to the disruption of the neutron star in a few orbital periods. Most of the mass is accreted promptly by the black hole. Some matter is shed outward, becomes gravitationally unbound and is ejected from the system. The remaining matter forms an accretion disk around the black hole.

As a representative example, Faber *et al.* (2006b) consider the evolution of an irrotational binary in which the mass ratio is $q = M_{BH}/M_{NS} = 10$, the neutron star compaction is $M_{NS}/R_{NS} = 0.09$ and the polytropic index is $n = 1$. The initial separation is chosen to be $r/M_{BH} = 5.5$, just outside the ISCO at $r_{ISCO}/M_{BH} \approx 5$ in the adopted (isotropic) coordinates. The initial data for this simulation describe a relativistic, irrotational BHNS model calculated in the conformal thin-sandwich formalism by Taniguchi *et al.* (2005).[29] According to equation (17.19) the tidal break-up radius is at $r_{tid}/M_{BH} \approx 5.8$. Not surprisingly, the simulation shows that tidal disruption occurs. Rapid angular momentum transfer during the disruption ejects a significant fraction of the matter back outside the ISCO, part of which forms an accretion disk.

The simulation is depicted in Figures 17.4 and 17.5. The rapid redistribution of angular momentum during the tidal disruption of the neutron star near the ISCO causes an outwardly directed spiral arm to form, sending some of the matter outside the ISCO. By the end of the simulation, the black hole accretes $\sim 75\%$ of the total neutron star rest mass, while $\sim 12\%$ of the matter forms a disk and $\sim 13\%$ is ejected completely from the system. Here bound and unbound matter trajectories are distinguished by the sign of $u_0 + 1$, where u_0 is the time component of the matter 4-velocity. Note that the quantity u_0 remains nearly constant in time for the (low pressure) outflowing gas streaming along quasigeodesic worldlines in the nearly static spacetime that forms following the accretion of the bulk of the disrupted star. The bound matter in the disk, while cold at first and ejected into the

---

[28] Neutron star binaries are likely to be irrotational since the magnitude of the viscosity required to synchronize the stars is unphysically large (Bildsten and Cutler, 1992; Kochanek, 1992a). By the same reasoning, typical black hole–neutron star binaries are also likely to be irrotational, since the required viscosity to synchronize a neutron star increases as the primary mass increases.

[29] see Chapter 17.1 for discussion.

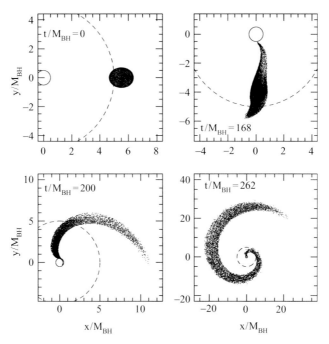

Figure 17.4 Snapshots of the disruption of a neutron star at selected times, projected into the orbital plane of the black hole–neutron star binary, as computed assuming conformally flat gravitation. The neutron star is an $n = 1$ polytrope with a compaction at infinite separation of $M_{NS}/R_{NS} = 0.09$; the binary mass ratio is $M_{BH}/M_{NS} = 10$. The neutron star, represented by approximately 60 000 SPH particles, disrupts near the ISCO (dashed curve) to produce a mass-transfer stream that eventually wraps around the black hole horizon (solid curve) to form a torus. The initial orbital period is $P/M_{BH} = 105$. [After Faber *et al.* (2006b).]

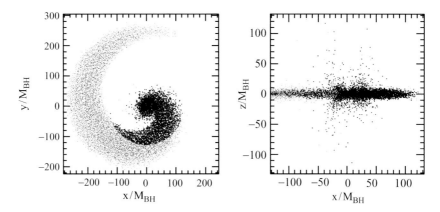

Figure 17.5 Snapshots of the same simulation depicted in Figure 17.4 at late time, $t/M_{BH} = 990$, projected onto the orbital (left panel) and meridional (right panel) planes. A hot torus at $r/M_{BH} \lesssim 50$ is evident. Bound fluid elements satisfying $u_0 + 1 < 0$ are shown as heavy dots, unbound elements as light dots. [From Faber *et al.* (2006b).]

spiral arm without strong shock heating, eventually undergoes shock heating as it falls back and wraps around the black hole forming a torus. The torus extends to a radius $r/M_{\rm BH} \sim 50$ within a time $t/M_{\rm BH} \sim 1000$ (corresponding to ~ 0.07s for typical neutron star masses). The characteristic temperature in the inner part of the torus is estimated to be $T \sim 3 - 10$ MeV $\sim (2\text{–}7) \times 10^{10}$ K. Faber *et al.* (2006b) point out that this system is a plausible model for the central engine of a short-hard GRB. Among its other attributes, a funnel is cleared out along the polar axis along which a GRB jet can presumably propagate (no "baryon pollution"). Also, the high temperatures can produce an appreciable thermal neutrino luminosity, $L_\nu \sim 10^{54}$ ergs^{-1}, which in turn can produce an appreciable electron pair annihilation luminosity required by some GRB models to generate the observed gamma-ray flux.[30] However, there are other scenarios involving neutron stars and compact binary mergers that are possible progenitors of short-hard GRBs.[31]

We have thus seen that the main virtue of binary black hole–neutron star simulations in conformal gravitation is that, to first approximation, they can track the bulk motion of matter in a relativistic gravitational field, at least in the case when the black hole-to-neutron star mass ratio is large. In particular, such simulations can follow the tidal-break up of the neutron star by a black hole when break-up occurs, and they can trace the subsequent dynamical flow of the matter, including possible disk formation about the black hole. But since the spatial metric is restricted to be conformally flat, the simulations described here are only qualitatively reliable at best. In particular, the predictions of disk masses and ejected mass fractions are not very trustworthy, especially when the black hole is rapidly spinning. Moreover, because the metric is not dynamical, these simulations do not convey any direct information about the emitted gravitational wave content of the spacetime. At best, information about gravitational waves can be estimated by evaluating the wave components of the metric perturbatively, after the simulation is completed. We did this in Chapter 16.2 for binary neutron stars when we adopted the quadrupole approximation to calculate gravitational waveforms. Here we shall postpone a discussion of the gravitational waves generated by black hole–neutron star mergers to the next section, where we will describe fully relativistic treatments of the evolution.

17.2.2 Fully relativistic simulations

As we mentioned earlier, simulations of binary black hole–neutron star mergers provide the ultimate challenge of evolving compact binaries in full general relativity: they involve all of the complications of relativistic hydrodynamics, including shocks, in a strong, dynamical

---

[30] See, e.g., Aloy *et al.* (2005); Piran (2005) and references therein.

[31] For high-mass binary neutron stars containing unequal mass companions with $q \sim 0.7$, prompt collapse to a black hole following merger also leads to the formation of a substantial, hot, neutrino-radiating accretion disk for which the polar axes are also free of intervening matter (Shibata and Taniguchi, 2006). However it is not clear whether such unequal mass binaries exist, since all observed systems containing pulsars that yield reliable mass measurements have $q \gtrsim 0.9$. Lower-mass binary neutron stars produce hypermassive neutron stars upon merger which undergo delayed collapse to a black hole. Simulations show that substantial, hot accretion disks with collimated magnetic fields are also formed in this case, so hypermassive neutron stars are also possible progenitors of short-hard GRB central engines. See Chapters 14 and 16 and Figure 16.12 for more extensive discussion.

field, together with all of the hurdles of tracking moving black holes without encountering their interior spacetime singularities. It is not surprising that progress in this area has been achieved by adapting and extending many of the tools and techniques used to evolve binary black holes and binary neutron stars. In this section we will summarize some of the earliest simulations that have tackled this problem successfully.

Corotational neutron stars and nonspinning black holes

Shibata and Uryū (2006, 2007) performed some of the earliest simulations of black hole–neutron star mergers in full general relativity. Their initial data consist of corotational neutron stars in quasiequilibrium circular orbit about nonspinning black holes and are determined using the conformal transverse-traceless approach discussed in Section 17.1.2.[32] The black hole is modeled by a moving puncture with no spin and the neutron star by a Γ-law EOS, with $\Gamma = 2$ (i.e., polytropic index $n = 1$), and a corotating velocity field. The approach of treating the black hole by moving punctures, which has worked so well in the case of vacuum binary black hole mergers (see Chapter 13.1.3), again proves to be a robust technique for tracking the motion and determining the evolution of the black hole without encountering singularities. Handling a moving black hole puncture in the presence of relativistic matter turns out to be straightforward. In fact, the ability of the moving puncture method to incorporate matter without imposing black hole excision has been bolstered by independent test-bed simulations.[33]

The evolution code employed by Shibata and Uryū (2006, 2007) is based on the BSSN scheme for the gravitational field, with a few refinements, and an HRSC scheme for the hydrodynamic matter. One of the refinements concerns the function ϕ appearing in the BSSN conformal factor (recall equation 11.46). They choose to evolve ϕ^{-6} instead of ϕ, since ϕ diverges at the puncture.[34] Note, however, that for the binaries considered here, the location of the puncture remains in the orbital plane and, with a cell-centered Cartesian grid, never encounters a grid point.

> **Exercise 17.2** Show that the evolution equation for ϕ^{-6} is
>
> $$\partial_t \phi^{-6} = \partial_i(\phi^{-6}\beta^i) + (\alpha K - 2\partial_i\beta^i)\phi^{-6}. \qquad (17.26)$$

Another refinement deals with the handling of the advective terms in the field evolution equations, which have the form $(\partial_t - \beta^i\partial_i)Q$, where Q is one of the field variables. Shibata and Uryū (2006, 2007) recast these terms in "conservative" form as

---

[32] See Löffler *et al.* (2006) for simulations in general relativity of head-on collisions between neutron stars and black holes of comparable mass.

[33] See, e.g., Faber *et al.* (2007), who treat relativistic Bondi accretion onto a moving black hole puncture to test their algorithm.

[34] Campanelli *et al.* (2006) and Baker *et al.* (2006a) employed a similar substitution for handling puncture binary black hole simulations. Marronetti *et al.* (2007) advocate evolving the variable $W \equiv \exp(-2\phi)$: not only is W regular everywhere, varying linearly with radius near black hole punctures, but it enters other field equations only as W^2. Hence even if numerical errors drive W slightly negative this will not cause other field variables to change sign and cause codes to crash.

$\partial_t Q - \partial_i(Q\beta^i) + Q\partial_i\beta^i$ and then adopt the same high-resolution, conservative scheme to evolving the field equations as they use for the hydrodynamic equations. They surmise that this technique proves important because, like many of the other field variables, the term $\beta^i \partial_i Q$ varies sharply near the puncture. As in other puncture calculations, they choose an advective "1+log" time slicing condition for the lapse,

$$\partial_t \ln\alpha = \beta^i \partial_i \ln\alpha - 2K. \tag{17.27}$$

Their shift condition is a modified parabolic "Gamma-driver" condition,

$$\partial_t \beta^i = 0.75\bar{\gamma}^{ij}(F_j + \Delta t \partial_t F_j), \tag{17.28}$$

where $\bar{\gamma}^{ij}$ is the inverse conformal 3-metric and $F_i \equiv \delta^{jk}\partial_j \bar{\gamma}_{ik}$. Here Δt is the time step in the simulation and the second term on the right-hand side of equation (17.28) is employed to stablize the integration.

> **Exercise 17.3** Discuss the role of the advective term in equation (17.27).
> **Hint:** $\partial_t x^i_{\text{punct}} = -\beta^i_{\text{punct}}$.

> **Exercise 17.4** Recast equation (17.28) in terms of the conformal connection function $\bar{\Gamma}^i = -\partial_j \bar{\gamma}^{ij}$ and compare with the usual Gamma-driver condition (4.82).

In their simulations, Shibata and Uryū (2006, 2007) adopt equatorial symmetry and a nonuniform cell-centered Cartesian grid (x, y, z) with a total grid size $(2N, 2N, N)$. The inner grid domain has a size $(2N_0, 2N_0, N_0)$ and is uniform, while the outer domain is nonuniform with gradually increasing grid separation. To test convergence they vary the resolution, choosing $N = 160 - 220$, $N_0 = 105 - 150$, with a grid spacing in the inner domain of size $\Delta x/\mathcal{M} = 1/8 - 7/120$, where $\mathcal{M} \approx M_{\text{irr}}$ is the puncture mass parameter. The outer boundary falls at $L/\lambda = 0.46$–0.83, where λ is the wavelength of the initial gravitational waveform ($\lambda/M \approx 50$, where $M \approx 5$ is the total ADM mass).

They study three cases for low-mass mergers ($q \lesssim 3$) in which the initial black hole irreducible masses are in the range $M_{\text{BH}} = 3.2$–$4.0M_\odot$ and the neutron stars at infinite separation have ADM masses $M_{\text{NS}} = 1.3M_\odot$ and radii $R = 13$–14 km, corresponding to compactions $M_{\text{NS}}/R_{\text{NS}} = 0.14$–$0.15$. The total ADM masses of the systems are $M = 4.5$–$5.3M_\odot$ and the total angular momenta are $J/M^2 = 0.65$–0.73, corresponding to close, circular orbits with periods $P/M = 110$–120.

In all the cases studied, the neutron star is tidally disrupted near the ISCO. Most of the matter (80–90%) is swallowed by the black hole, with the remainder going into a disk. The final state is a rotating black hole of spin $J_{\text{BH}}/M_{\text{BH}}^2 \approx 0.4$–$0.6$ surrounded by a disk of mass ~ 0.1–$0.3M_\odot$. An appreciable fraction of the initial angular momentum of the system is deposited in the disk. As we will discuss below, these disk masses are unrealistically large, in part because of the assumption of corotation and in part because of the (nonadaptive) grid structure used in the calculation. Nevertheless, they strengthen the argument that, since a rotating black hole with an ambient disk could serve as the central engine for a

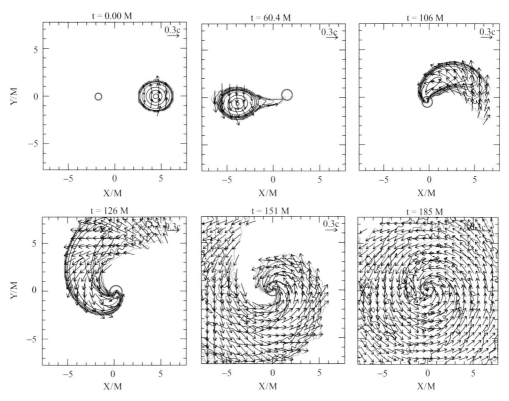

Figure 17.6 Snapshots of rest-density contours and 3-velocity vectors at selected times during the binary merger of a $1.3 M_\odot$ neutron star corotating about a $3.2 M_\odot$ nonspinning black hole. The total initial ADM mass of the system is $M = 4.5 M_\odot$. Contours are drawn in the orbital plane at $\rho_0 = 10^{14}$ (blue), 10^{12} (cyan), 10^{11} (magenta), and 10^{10} g cm^{-3} (green). The maximum density at $t = 0$ is $\approx 7.2 \times 10^{14}$ g cm^{-3}. The thick red circle denotes the apparent horizon. [From Shibata and Uryū (2007).]

short-hard GRB, the merger and disruption of a neutron star with a low-mass black hole might be the progenitor of such a source. A representative tidal disruption is illustrated in Figure 17.6.

Gravitational waveforms for a typical merger are shown in Figure 17.7. The waves are extracted from the metric near the outer boundaries of the grid, using the gauge-invariant Moncrief formalism discussed in Chapter 9.4.1. In particular, Shibata and Uryū (2006, 2007) focus on the dominant even-parity, $l = 2$ mass quadrupole modes and define the quantities

$$R_+ = \sqrt{\frac{5}{16\pi}} R_{22+}, \quad R_\times = \sqrt{\frac{5}{16\pi}} R_{22-}, \tag{17.29}$$

where $R_{lm\pm}$ are the even-parity functions defined in equations (9.117).

Exercise 17.5 Why do the functions $R_{21\pm}$ vanish for these simulations?

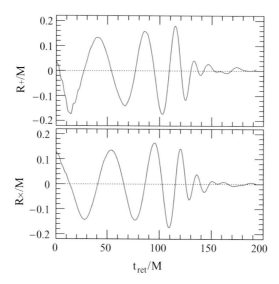

Figure 17.7 Gravitational waveforms for the binary merger of a $1.3 M_\odot$ neutron star corotating about a $4.0 M_\odot$ black hole. The total initial ADM mass of the system is $M = 5.3 M_\odot$ (see text). [From Shibata and Uryū (2007).]

The wave amplitude assumes its maximum for observers viewing along the binary axis, perpendicular to the orbital plane, for which we may write

$$h \equiv \sqrt{h_+^2 + h_\times^2} = \frac{\sqrt{R_+^2 + R_\times^2}}{r} \approx 1.0 \times 10^{-22} \left(\frac{\sqrt{R_+^2 + R_\times^2}}{0.31\ \mathrm{km}} \right) \left(\frac{100\ \mathrm{Mpc}}{D} \right),$$

$$(17.30)$$

where D is the distance of the observer to the source.

> **Exercise 17.6** Verify equation (17.30).
> **Hint:** Use equations (9.115) and (9.116).

From Figure 17.7, which applies to a system with total mass $M \approx 5.3 M_\odot$, this maximum amplitude is $h \approx 5 \times 10^{-22}$ for a characteristic distance of $D = 100$ Mpc. Such a waveform should be detectable by a gravitational wave laser interferometer like advanced LIGO. The waveform exhibits the characteristic increase in amplitude and frequency during the late inspiral phase, $t_{\mathrm{ret}} \equiv t - D \lesssim 120M$. At the end of the inspiral the wavelength of the radiation is $\sim 25M$, appropriate for orbital motion near the ISCO. For $120 \lesssim t_{\mathrm{ret}}/M \lesssim 150$, black hole ringdown radiation is evident, with an expected quasinormal wavelength of $\sim 15 M_{\mathrm{BH}}^f$, where the final black hole mass M_{BH}^f is comparable to the total initial mass of the binary.

Irrotational binaries

The merger of irrotational binaries (irrotational neutron stars orbiting nonspinning black holes) has been treated by several investigators in full general relativity.[35] The emphasis has been on low-mass cases ($q \lesssim 5$) for which the neutron star undergoes tidal disruption prior to capture. In most of these investigations, the gravitational fields are evolved by the BSSN scheme, and the matter is evolved by a HRSC algorithm on the same spacetime lattice. By contrast, Duez *et al.* (2008) evolve Einstein's equations by a first-order representation of the generalized harmonic formulation using pseudo-spectral methods. The equations of hydrodynamics are evolved on a separate grid using shock-capturing finite difference or finite volume techniques. The consensus of all the studies is that a toroidal disk forms around the black hole in most cases but that the typical disk masses are much lower than the values quoted in the previous sections. In the following discussion we shall highlight the calculations of Etienne *et al.* (2009), in part to facilitate a comparison in the next section with spinning black holes, which they also treat. For irrotational binaries, Etienne *et al.* (2009) find that, though indeed lower, the disk masses of 0.01–$0.05 M_\odot$ that form around the remnant black hole following neutron star disruption are still astrophysically significant.

The code developed by Etienne *et al.* (2009) is a refinement of the BSSN HRSC relativistic MHD scheme of Duez *et al.* (2005b) discussed earlier in the book in several other contexts.[36] Their most significant improvement consists of the implementation of AMR[37] to track the evolution of the strong-field, inner regions at high resolution while simultaneously allowing the grid to extend far out into the wave zone for more accurate wave extraction and more reliable boundary conditions. Etienne *et al.* (2009) launched a new suite of tests to check the AMR version of their code. These tests involved both vacuum spacetimes and spacetimes with hydrodynamic matter sources and included shock-tube tests and simulations of linear gravitational waves, single stationary and boosted puncture black holes, puncture black hole binary mergers,[38] rapidly and differentially rotating equilibrium stars, and relativistic Bondi accretion onto a Schwarzschild black hole. Convergence tests performed in all cases confirmed the reliability of the AMR version.

The BSSN equations are integrated with fourth-order accurate, centered, finite-differencing stencils, except on shift advection terms, where fourth-order accurate upwind stencils are used. Outgoing wave boundary conditions are adopted for all BSSN fields. Fourth-order Runge–Kutta time-stepping is managed by a MoL (Method of Lines) routine, with a Courant–Friedrichs–Lewy factor set to 0.45.[39] Etienne *et al.* (2009) find that in the presence of hydrodynamic matter, the evolution of the conformal variable ϕ rather than

[35] See, e.g., Shibata and Taniguchi (2007); Etienne *et al.* (2008); Yamamoto *et al.* (2008); Duez *et al.* (2008); Etienne *et al.* (2009).

[36] See, e.g., Chapters 5.2.4, 14.2.3, and 16.3.

[37] The implementation uses the moving-box AMR infrastructure provided by the "Carpet" algorithm; see Schnetter *et al.* (2004) for a description of the FMR version of the AMR algorithm and http://www.carpetcode.org for publically available modules of the AMR version.

[38] See Appendix I, where these simulations of binary black hole mergers are summarized, together with some of the features of the vacuum sector of this AMR code.

[39] See Chapter 6 for a discussion of these concepts.

$\chi = e^{-4\phi}$ or $W = e^{-2\phi}$ yields a more stable evolution near the black hole puncture and better rest-mass conservation.[40] The equations are evolved using standard puncture gauge conditions: an advective "1+log" slicing condition for the lapse and a "Gamma-driver" condition for the shift.[41]

The HRSC scheme for the hydrodynamics is second-order accurate for laminar flow, but first-order accurate when discontinuities like shocks arise. To stabilize the scheme in vacuum regions, a tenuous atmosphere is maintained, with a density floor ρ_{atm} set equal to 10^{-10} times the initial maximum density on the grid.

Etienne *et al.* (2009) consider irrotational binaries with mass ratios $q = 1, 3$ and 5. For initial data they use the quasiequilibrium models constructed using the conformal thin-sandwich method in Taniguchi *et al.* (2008). All three cases are tracked from approximately the same starting value of $M\Omega$, where Ω is the orbital angular velocity and M is the total ADM mass. The black hole interiors are excised in this method, but with suitable care these regions can be filled with smooth, but otherwise arbitrary, constraint-violating "junk" initial data that does not affect the black hole exterior when evolved.[42] The neutron stars obey an $n = 1$ polytropic EOS initially and evolve accordingly to a $\Gamma = 2$ adiabatic EOS. All neutron stars in this study have the same nondimensional rest mass, $\bar{M}_0 = 0.15$, which is 83% of the maximum rest mass of a nonrotating star built with the same polytropic EOS; the compaction of the adopted neutron star is $\mathcal{C} = 0.145$ in isolation.

The $q = 3$ case (case A) represents a generic tidal disruption scenario. Evolution starts from a binary coordinate separation of $D_0/M = 8.81$, corresponding to $M\Omega = 0.0333$. The AMR grid used in this case has nine levels of refinement; there are two sets of nested refinement boxes, one centered on the black hole and the other on the neutron star. The maximum resolution (minimum spacing) is $M/32.5$ in the innermost box centered on the black hole. Approximately 41 grid points cover the diameter of the (spherical) apparent horizon initially and 85 grid points cover the (smallest) diameter of the neutron star. The outer boundary of the grid is located at $r/M = 197$.

Figure 17.8 shows the inspiral trajectories of the black hole and neutron star while Figure 17.9 shows snapshots of the density and velocity vectors at selected times during the evolution for the $q = 3$ case. Note that the equilibrium shape of the neutron star is hardly disturbed for the first two orbits (upper middle). The lower panels show how the neutron star tail deforms into a quasistationary disk, as the bulk of the matter is accreted onto the black hole. A time history of the rest-mass consumption by the black hole is plotted in Figure 17.10 for this case, as well as for the irrotational $q = 1$ (case E) and 5 (case D) mergers. Astrophysically, we expect that the formation of a binary black hole–neutron star will typically occur with $q > 1$.

Figure 17.10 suggests that for $q = 3$ and $q = 1$ there are two distinct phases during which matter falls into the black hole following tidal disruption of the neutron star: an

---

[40] In contrast to evolution on a fixed grid using a conservative HRSC scheme, rest-mass conservation is not guaranteed on an AMR grid, where spatial and temporal prolongation (interpolation) is performed at the grid boundaries.
[41] See equations (4.51) and (4.83), implementing the "shifting shift" version in the later case.
[42] Etienne *et al.* (2007); see also Brown *et al.* (2007).

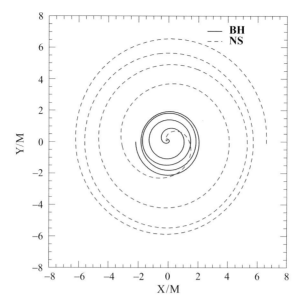

Figure 17.8 Trajectories of the black hole and neutron star coordinate centroids for the merger of an irrotational binary with mass ratio $q = 3$ (case A). The black hole (BH) centroid corresponds to the centroid of the apparent horizon, and the neutron star (NS) centroid is computed via the integral $\tilde{X}_c^i \equiv (\int_V x^i \rho_* d^3x)/(\int_V \rho_* d^3x)$, where V is the total simulation volume. [From Etienne *et al.* (2009).]

initial plunge phase and a secondary accretion phase. The plunge phase occurs early in the merger as part of the disrupted star streams onto the black hole and the remainder deforms into a tidal tail. The plunge phase ends after 70–90% of the matter falls into the BH, resulting in a sudden drop in the slope of the exterior M_0 vs. time plot in the figure. The debris in the tail, having larger specific angular momentum, spreads out and forms a disk. Material with lower specific angular momentum in the disk accretes onto the BH. Since there is neither viscosity nor magnetic fields in this simulation, the accretion should eventually cease as the evolution continues. However, in realistic astrophysical environments, viscosity and magnetic fields frozen into the initial gas will drive further accretion on secular time scales.[43]

For the $q = 5$ case plotted in Figure 17.10 the neutron star essentially plunges into hole, leaving $< 1\%$ of its rest mass outside the horizen at the end of the simulation. This result is not surprising since, at the ISCO, the tidal effect of the black hole is smaller for larger q, resulting in tidal disruption occurring closer to the ISCO. Moreover, for a given neutron star, the horizon size of the black hole is larger for larger q, making it easier to capture neutron star matter during the plunge. More surprising is the result that the disk mass in the $q = 1$ case ($M_{\text{disk}}/M_0 \approx 2.3\%$) is less than the disk mass in the $q = 3$ case ($M_{\text{disk}}/M_0 \approx 3.9\%$). This result is presumably due to the complicated interaction of the disrupted matter in the

---

[43] Recall the examples of viscous and MHD disk accretion onto black holes discussed in Chapter 14.2.4 and 14.2.5.

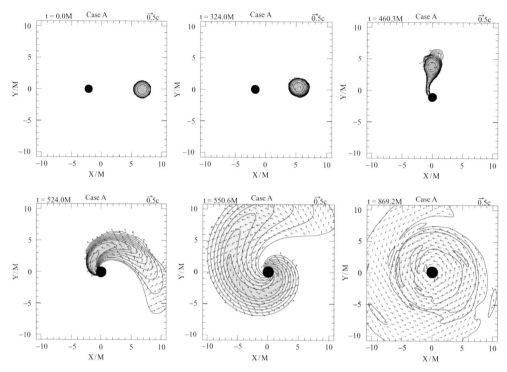

Figure 17.9 Snapshots of rest-density contours and 3-velocity vectors at selected times during the merger of the irrotational black hole-neutron star binary with mass ratio $q = 3$ shown in Figure 17.8 (case A). The contours represent the density in the orbital plane, plotted according to $\rho_0 = \rho_{0,\text{max}} 10^{-0.38j-0.04}$, $(j = 0, 1, \ldots, 12)$, with darker gray scaling for higher density. The maximum initial neutron star density is $K\rho_{0,\text{max}} = 0.126$, where K is the initial polytropic gas constant, or $\rho_{0,\text{max}} = 9 \times 10^{14}$ g cm$^{-3}(1.4M_\odot/M_0)^2$. Arrows represent the velocity field in the orbital plane. The apparent horizon interior is marked by a filled black circle. In cgs units, the initial ADM mass for this case is $M = 2.5 \times 10^{-5}(M_0/1.4M_\odot)$ s $= 7.6(M_0/1.4M_\odot)$ km. [From Etienne *et al.* (2009).]

merging process. A low-density, hot spiral region of disrupted matter winds around the black hole and smashes into the tidal tail of the disrupted star, causing significant shock heating. As shown in exercise 17.7 the degree of shock heating can be considerable in a supersonic, low-density stream. Some of the heated matter transfers angular momentum to the other part of the tail and then falls into the black hole. The remaining matter in the tail deforms into an inhomogeneous disk before settling into a quasistationary state in which a high density, relatively low temperature torus of matter surrounds the black hole.

> **Exercise 17.7** Shock heating can be measured by the degree to which the polytropic gas constant $K \equiv P/\rho_0^\Gamma$ increases in a gas that passes through a shock front. Recall that this quantity remains constant for adiabatic flow in the absence of shocks. Assume Newtonian hydrodynamics in doing this exercise.

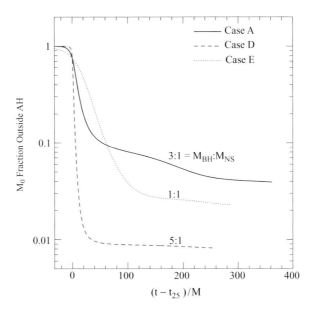

Figure 17.10 Rest-mass fraction outside the black hole as a function of time for irrotational binaries with different mass ratios. Here, the time coordinate is shifted by t_{25}, the time at which 25% of the neutron star rest-mass has fallen into the apparent horizon. [From Etienne *et al.* (2009).]

(a) Assume that matter evolves adiabatically with the same adiabatic constant $\Gamma > 1$ both before and after entering a planar shock front. Let the upstream Mach number in a fluid entering a shock be defined by $\mathcal{M} \equiv v_1/c_1$, where v_1 is the upstream flow as viewed in the frame of the shock front and c_1 is the upstream sound speed. Use the fact that the flow always enters a shock front supersonically (i.e., $\mathcal{M} > 1$) to show that the downstream-to-upstream ratio of the gas constants is given by

$$K_2/K_1 = \frac{2\Gamma\mathcal{M}^2 - (\Gamma - 1)}{(\Gamma + 1)^{\Gamma+1}} \left(\Gamma - 1 + \frac{2}{\mathcal{M}^2} \right)^{\Gamma}. \qquad (17.31)$$

Prove that this ratio is always larger than unity.
(b) Evaluate $\mathcal{M}$ in terms of v_1 and ρ_1 and argue that for a given upstream velocity, $\mathcal{M}$ is larger for a lower density fluid.
(c) Assume highly supersonic flow $\mathcal{M} \gg 1$ and show that

$$K_2/K_1 \approx \frac{2v_1^2}{(\Gamma + 1)K_1\rho_1^{\Gamma-1}} \left(\frac{\Gamma - 1}{\Gamma + 1} \right)^{\Gamma}. \qquad (17.32)$$

Equation (17.32) shows that when the density ρ_1 of the upstream matter is low, the degree of shock heating downstream can be substantial, i.e., $K_2/K_1 \gg 1$.

Etienne *et al.* (2009) find that the disk is smallest and most dense for the $q = 1$ binary. Specifying the length scale in km via the relation $M = 1.9(q + 1)(M_0/1.4M_\odot)$km, and setting the neutron star rest mass M_0 equal to $1.4M_\odot$, they obtain a characteristic radius of $r_{disk} \approx 20$ km and a maximum density of $\rho_{0,max} \approx 6 \times 10^{12}$ g cm^{-3} for this disk. The corresponding values for the disk in the $q = 3$ case are $r_{disk} \approx 50$ km and $\rho_{0,max} \approx 4 \times 10^{11}$g cm^{-3}. In all cases the disk forms a hot, thick torus whose height is about 15–20% of the characteristic radius and whose density plummets at the black hole ISCO. The low-density regions are hottest, due to shock heating (see exercise 17.7). However, the gas constant $K \equiv P/\rho_0^\Gamma$ increases by more than a factor of 6 everywhere in the $q = 1$ disk and by more than 85 everywhere in the $q = 3$ disk.

Etienne *et al.* (2009) determine the thermal temperature T in the gas from the specific energy density ϵ, which can determined from the evolved hydrodynamic variables. For a polytropic equation of state, the "cold" contribution ϵ_{cold} is defined as

$$\epsilon_{cold} = -\int P_{cold} d(1/\rho_0) = \frac{K_{cold}}{\Gamma - 1}\rho_0^{\Gamma-1}, \tag{17.33}$$

where $P_{cold} \equiv K_{cold}\rho_0^\Gamma$ governs the initial (cold) neutron star. Now define the thermal contribution to the specific energy density generated by shock heating as $\epsilon_{th} = \epsilon - \epsilon_{cold}$ and compute the thermal contribution according to

$$\epsilon_{th} = \epsilon - \epsilon_{cold} = \frac{1}{\Gamma - 1}\frac{P}{\rho_0} - \frac{K_{cold}}{\Gamma - 1}\rho_0^{\Gamma-1}$$
$$= (K/K_{cold} - 1)\epsilon_{cold}. \tag{17.34}$$

To estimate T, Etienne *et al.* (2009) model ϵ_{th} according to

$$\epsilon_{th} = \frac{3kT}{2m_n} + f_s\frac{aT^4}{\rho_0}, \tag{17.35}$$

where m_n is the mass of a nucleon, k is the Boltzmann constant, and a is the radiation constant. The first term represents the approximate thermal energy of the nucleons, and the second term accounts for the thermal energy due to radiation and (thermal) relativistic particles. The factor f_s reflects the number of species of ultrarelativistic particles that contribute to thermal energy. When $T \ll 2m_e/k \sim 10^{10}$ K, where m_e is the mass of electron, thermal energy is dominated by photons and $f_s = 1$. When $T \gg 2m_e/k$, electrons and positrons become ultrarelativistic and also contribute to the energy, and $f_s = 1 + 2 \times (7/8) = 11/4$. At sufficiently high temperatures and densities ($T \gtrsim 10^{11}$ K, $\rho_0 \gtrsim 10^{12}$ g cm^{-3}), thermal neutrinos are copiously generated and become trapped, so, taking into account three flavors of neutrinos and antineutrinos, $f_s = 11/4 + 6 \times (7/8) = 8$.

Using equation (17.35), the characteristic temperature is $T \sim 5 \times 10^{10}$ K (or ~4 MeV) in the disk formed following the $q = 3$ binary merger, with comparable values for the other disks. Numerical models of rotating black holes with ambient disks with similar temperatures and densities suggest that such systems can produce a total gamma-ray energy $E \sim 10^{47}$–10^{50}erg from neutrino-antineutrino annihilation.[44] This result is promising for generating a short-hard GRB from a merging black hole–neutron star binary.

---

[44] Setiawan *et al.* (2006).

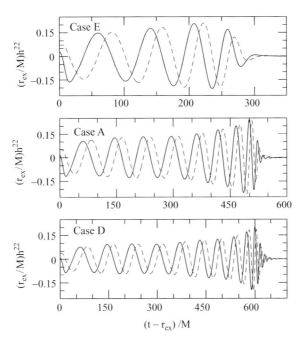

Figure 17.11 Gravitational wave signal for irrotational binaries with mass ratios (top to bottom) $q = 1, 3$, and 5. The black solid (blue dash) line denotes the (2,2) mode of $h_+ (h_\times)$ extracted at $r_{ex} = 43.4M$. [From Etienne *et al.* (2009).]

Gravitational waveforms are computed for the three cases described above and the results are compared in Figure 17.11. Waveforms extracted at different radii are found to overlap well provided the extraction radius is sufficiently large: $r_{ex} \gtrsim 40M$. With approximately the same initial $M\Omega$, we observe more wave cycles are emitted as the mass ratio q is decreased. The amplitudes are attenuated at frequencies roughly equal to double the orbital frequency at which tidal disruption begins.

Exercise 17.8 The gravitational waveforms h_+ and $h_\times$ can be decomposed into $s = -2$ spin-weighted spherical harmonics $_{-2}Y_{lm}$ according to

$$h_+ - ih_\times = \sum_{l,m} (h_+^{lm} - ih_\times^{lm})_{-2}Y_{lm}, \qquad (17.36)$$

where h_+^{lm} and $h_\times^{lm}$ are real functions. Each (l, m) mode is a function of radius and time only.
(a) Use the results of Chapter 9.4 to express h_+^{lm} and $h_\times^{lm}$ in terms of gauge-invariant Moncrief functions.
(b) Use the results of Chapter 9.4 to express h_+^{lm} and $h_\times^{lm}$ in terms of the Weyl scalar ψ_4.

One significant measure of the accuracy of the simulations is provided by the degree to which energy and angular momentum are conserved. Etienne *et al.* (2009) calculate the diagnostic quantities

$$\delta E \equiv M - M_f - \Delta E_{\mathrm{GW}}, \tag{17.37}$$

$$\delta J \equiv J - J_f - \Delta J_{\mathrm{GW}}, \tag{17.38}$$

where M and J are the ADM mass and angular momentum of the initial binary, M_f and J_f are the ADM mass and angular momentum of the final system, and ΔE_{GW} and ΔJ_{GW} are the energy and angular momentum carried off by gravitational waves. Assuming that no matter or other form of radiation leaves the computational domain, which is the case here, strict conservation of energy and angular momentum demands that $\delta E = 0 = \delta J$. Etienne *et al.* (2009) find that $(\delta E/M, \delta J/J) = (-2 \times 10^{-4}, 2.2 \times 10^{-2})$ for the $q = 3$ case, which takes approximately 4.5 orbits before merger, with comparable errors for the other cases. Maintaining conservation of angular momentum is typically more difficult than energy conservation, due to spurious numerical viscosity (shear) effects. Moreover, the determination of the final disk mass is particularly sensitive to errors in angular momentum conservation, so they must be kept to a minimum ($\lesssim$ a couple percent) for a reliable measure.

The radiated fractions $(\Delta E_{GW}/M, \Delta J_{GW}/J)$ are found to be $(0.35\%, 7.2\%)$ for the $q = 1$ system, $(0.93\%, 17.4\%)$ for $q = 3$, and $(0.98\%, 19.2\%)$ for $q = 5$. The spins J_f/M_f^2 of the black hole remnants are 0.85 for $q = 1$, 0.56 for $q = 3$ and 0.42 for $q = 5$. Binary black hole–neutron star mergers typically impart kick velocities to the remnant black holes due to recoil.[45] The kick velocity v_{kick} is calculated to be 17 km/s for $q = 1$, 33 km/s for $q = 3$ and 73 km/s for $q = 5$.

By evolving initial binaries along the same quasiequilibrium sequences as those chosen here but with larger initial separations (smaller $M\Omega$) Etienne *et al.* (2009) find that the outcomes (apart from a phase shift in the case of gravitational waves) are essentially unchanged. This indicates that choosing the initial separation corresponding to $M\Omega = 0.0333$ is sufficiently large to predict the onset of tidal disruption, merger evolution, gravitational waveforms and disk masses reliably, although this result may only be marginally true in the case of $q = 1$ (see exercise 17.9).

> **Exercise 17.9** Explain why fixing the initial value of $M\Omega$ leads to closer initial separations as the binary mass ratio q is decreased. Specifically, estimate the separation in a circular binary as a function of $M\Omega$ and q.

Figure 17.12 plots the gravitational wave power spectra (i.e., the effective wave strain $h_{\mathrm{eff}}(f) \equiv (\sqrt{2}/\pi D)(\sqrt{dE/df})$ vs. f, where $E(f)$ is the Fourier transform of the energy, f the frequency and D the source distance) for the three cases, using only the dominant (2,2) and (2,−2) wave modes, and compares them to the Advanced LIGO sensitivity curve, or effective strain, $h_{\mathrm{LIGO}}(f) \equiv \sqrt{f S_h(f)}$.[46] To set the scale in the figure, the neutron

---

[45] See Chapter 13.2.2 for a discussion of recoil in the context of binary black hole mergers.
[46] The quantity $h_{\mathrm{LIGO}}(f)$ is related to the LIGO "strain per root hertz", or noise, $\tilde{h}(f)$ plotted in Figure 9.4 by $h_{\mathrm{LIGO}}(f) = \tilde{h}(f)\sqrt{f}$.

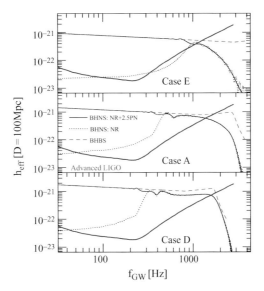

Figure 17.12 Gravitational wave power spectra for the $q = 1, 3$, and 5 cases (cases E, A, and D, respectively). The solid curve in each panel shows the combined waveform found by attaching the restricted 2.5 order PN waveform to the dominant modes computed by the numerical relativity simulation (NR), while the dotted curve shows the contribution from the latter alone. The dashed curve shows the power spectrum for a nonspinning black hole binary merger (BHBH) for black holes with the same mass ratios as the black hole–neutron star binaries (BHNS); the later exhibit significantly less power at higher frequencies, due to tidal disruption. The heavier solid curve is the effective strain of the Advanced LIGO detector, defined by $h_{\rm LIGO}(f)$ (see text). To set physical units, we assume a neutron star rest mass of $M_0 = 1.4 M_\odot$ and a source distance of $D = 100$ Mpc. [From Etienne *et al.* (2009).]

star is assumed to have a rest-mass of $1.4 M_\odot$ and the binary is placed at a distance of 100 Mpc. This is the distance required to reach one merger per year in one estimate,[47] assuming an overall rate of 10 mergers per Myr per Milky Way-equivalent galaxy (and a density of these of 0.1 gal/Mpc^3). This distance is roughly that of the Coma cluster, and approximately five times the distance to the Virgo cluster. The gravitational wave spectra of nonspinning binary black hole mergers with the same mass ratios as in the three black hole–neutron star cases are also plotted for comparison. The figure shows that the wave signal drops appreciably near the frequency corresponding to the onset of neutron star tidal disruption. The difference in wave signals between binary black hole and black hole mergers is seen to be marginally observable in the Advanced LIGO frequency band in most cases. Distinguishing binary black hole from binary black hole–neutron star inspirals and mergers may thus require narrow-band wave detection techniques with advanced detectors. The observation of an accompanying GRB would also serve to distinguish the two types of events (and would be a dramatic discovery!). Should the chirp mass determination, combined with higher order PN waveform phase effects, allow for an

---

[47] Belczynski *et al.* (2002).

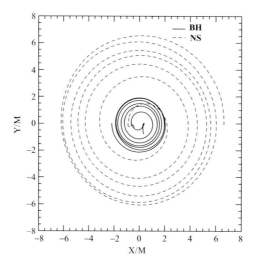

Figure 17.13 Trajectories of the black hole and neutron star coordinate centroids for the merger of an irrotational neutron star orbiting a spinning black hole with spin $J_{\mathrm{BH}}/M_{\mathrm{BH}}^2 = 0.75$ (case B). Compare with Figure 17.8 for the merger of an identical neutron star with a nonspinning black hole, assuming the same binary mass ratio $q = 3$ and initial orbital angular velocity $M\Omega \approx 0.033$. [From Etienne *et al.* (2009).]

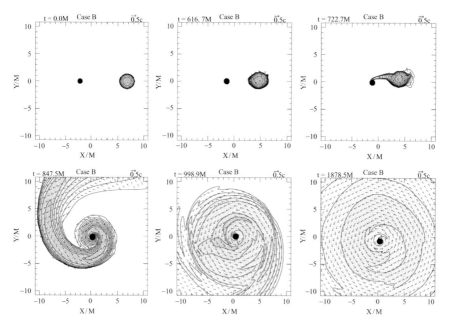

Figure 17.14 Snapshots of rest-density contours and 3-velocity vectors at selected times during the merger of an irrotational neutron star orbiting a black hole with spin $a/M_{\mathrm{BH}} = J_{\mathrm{BH}}/M_{\mathrm{BH}}^2 = 0.75$, as shown in Figure 17.13 (case B). The contours represent the density in the orbital plane, plotted according to $\rho_0 = \rho_{0,\mathrm{max}} 10^{-0.38j-0.04}$, ($j = 0, 1, \ldots, 12$), with darker gray scaling for higher density. The maximum initial neutron star density is $K\rho_{0,\mathrm{max}} = 0.126$, where K is the initial polytropic gas constant, or $\rho_{0,\mathrm{max}} = 9 \times 10^{14}\,\mathrm{g/cm^3}(1.4M_\odot/M_0)^2$. Arrows represent the velocity field in the orbital plane. The apparent horizon interior is marked by a filled black circle. In cgs units, the initial ADM mass for this case is $M = 2.5 \times 10^{-5}(M_0/1.4M_\odot)\,\mathrm{s} = 7.6(M_0/1.4M_\odot)\,\mathrm{km}$. [From Etienne *et al.* (2009).]

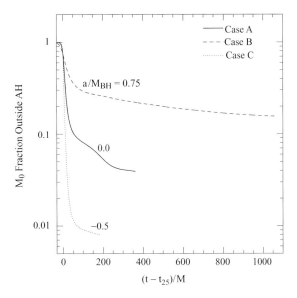

Figure 17.15 Rest-mass fraction outside the black hole as a function of time for the merger of an irrotational neutron star orbiting a black hole with different initial black hole spins a/M_{BH}. Here the time coordinate is shifted by t_{25}, the time at which 25% of the neutron star rest-mass has fallen into the apparent horizon. [From Etienne *et al.* (2009).]

independent determination of the individual masses and spins of the binary companions, the measurement of the black hole-neutron star tidal disruption frequency should give a good estimate of the neutron star radius and, hence, insight into the nuclear EOS.

Irrotational neutron stars with spinning black holes

Etienne *et al.* (2009) also explore the effects of black hole spin on the merger of black hole–neutron star binaries. They consider irrotational neutron stars orbiting spinning black holes, taking the black hole spins to be either aligned or anti-aligned with the orbital angular momentum. They again use conformal thin-sandwich quasiequilibrium initial data for the binaries and evolve models with the initial spin parameter of the black hole between $J_{BH}/M_{BH}^2 = -0.5$ (anti-aligned) and 0.75, fixing the mass ratio at $q = 3$ and the initial orbital angular velocity at $M\Omega \approx 0.033$. As the initial spin parameter increases, the total initial angular momentum increases, requiring more gravitational wave cycles to emit sufficient angular momentum to bring the companions close enough to merge. Not surprisingly, the binary undergoes more orbits before merger as J_{BH}/M_{BH}^2 increases: for spins -0.5, 0.0, and 0.75, the binary inspiral phase lasts for 3.25, 4.5, and 6.5 orbits, respectively. The coordinate trajectories of the apparent horizon and neutron star centroids are shown in Figure 17.13 for the case with $J_{BH}/M_{BH}^2 = 0.75$ (case B).

Figure 17.14 shows snapshots of the density and velocity vectors at selected times during the evolution for the $J_{BH}/M_{BH}^2 = 0.75$ case. The upper plots in the figure demonstrate that the neutron star retains its shape for four orbits (upper middle), and begins shedding

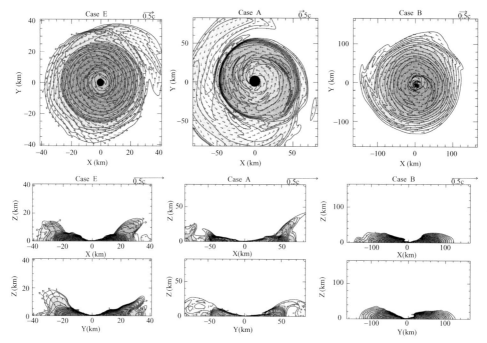

Figure 17.16 Snapshots of rest-density contours and 3-velocity vectors at the end of the simulations for three binary black hole–neutron star merger scenarios (see text). The remnant disks are shown in the orbital plane (upper row) and meridional plane (bottom rows). The contours represent the density plotted according to $\rho_0 = \rho_{0,\mathrm{max}} 10^{-0.1j-3.3} (j = 0, 1, \ldots, 13)$ for cases A and B and $\rho_0 = \rho_{0,\mathrm{max}} 10^{-0.19j-2.32}$ $(j = 0, 1, \ldots, 12)$ for case E, with darker gray scaling for higher density. In all cases the maximum initial neutron star density is $K\rho_{0,\mathrm{max}} = 0.126$, where K is the initial polytropic gas constant, or $\rho_{0,\mathrm{max}} = 9 \times 10^{14}$ g cm$^{-3}(1.4M_\odot/M_0)^2$. Arrows represent the velocity field. The apparent horizon interior in the orbital plane is marked by a filled black circle. Length scales are specified in km, assuming the neutron star has a rest-mass of $1.4M_\odot$; it can be converted to units of M via the formula $M = 1.9(q + 1)$ km. [After Etienne *et al.* (2009).]

its outer layers due to tidal disruption only after about five orbits (upper right). In this case the tail is quite massive (lower left), and by the time 75% of the mass has been accreted (lower center) it is much larger than the tail measured at the same point for a nonspinning black hole (lower left plot in Figure 17.9). The lower right plot is a snapshot taken near the end of the simulation, when a quasistationary disk resides outside the BH. The disk is massive: $M_{\mathrm{disk}}/M_0 \approx 15\%$, corresponding to $0.2M_\odot$ for a $1.4M_\odot$ neutron star. The maximum density of the disk is $\rho_{0,\mathrm{max}} \approx 5 \times 10^{11}$g cm^{-3} and the characteristic temperature is $T \sim 5 \times 10^{10}$ K (or $\sim$4 MeV), values which are similar to the nonspinning case. However, the disk is about twice as large in size, with a characteristic radius of $r_{\mathrm{disk}} = 100$ km, in addition to being much more massive.

Etienne *et al.* (2009) conclude from their simulations that the black hole-neutron star inspiral phase lasts longer the larger the spin $J_{\mathrm{BH}}/M_{\mathrm{BH}}^2$. Also, larger aligned spins form more extensive and more massive disks: the final disk mass grows from $< 1\%$ of the initial neutron star rest mass when the black hole has a spin $J_{\mathrm{BH}}/M_{\mathrm{BH}}^2 = -0.5$ (case C) to $\approx 4\%$

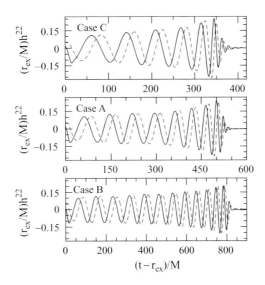

Figure 17.17 Gravitational wave signal for three binaries with mass ratio $q = 3$ and initial black hole spins (top to bottom) $J_{BH}/M_{BH}^2 = -0.5, 0,$ and 0.75. The black solid (blue dash) line denotes the (2,2) mode of $h_+(h_\times)$ extracted at $r_{ex} = 43.4M$. [From Etienne *et al.* (2009).]

for a spin of 0 (case A) to $\approx 15\%$ for a spin of 0.75 (case B). Figure 17.15 shows the rest mass outside the black hole as a function of time for these three cases, illustrating once again the two phases of mass consumption by the black hole. If the formation of a massive disk about a rotating black hole is a prerequisite for a short-hard GRB, then it appears that the more rapidly spinning and closely aligned the initial black hole, the more likely it is to power such a source by binary black hole–neutron star merger.

A comparison between the final disks formed in the three merger scenarios is shown in Figure 17.16. Both the mass ratios and initial black hole spins are varied for a cross-the-board comparison: recall that $(q, J_{BH}/M_{BH}^2) = (1, 0)$ for case E, $(3, 0)$ for case A and $(3, 0.75)$ for case B.

Finally, we show a comparison between gravitational waveforms emitted for three merger scenarios in Figure 17.17. The mass ratios are all identical ($q = 3$) but the initial black hole spins are varied: $J_{BH}/M_{BH}^2 = -0.5$ for case C, 0 for case A and 0.75 for case B.

Though tentalizing, all of these results are quite preliminary. Simulations that take into account the detailed microphysics, including the correct hot, nuclear EOS, magnetic fields and neutrino and photon transport, are necessary to fully assess the viability of and mechanism for generating short-hard GRBs from binary black hole–neutron star mergers. But much of the groundwork, especially the tool of numerical relativity, has been prepared to carry out such simulations.

18 Epilogue

This is not the end.
It is not even the beginning of the end.
But it is, perhaps, the end of the beginning.
Winston Churchill (1942)

This brings us to the end of our introduction to numerical relativity. A quick glance at the table of contents shows that we have covered a wide range of subjects, starting with the foundations of numerical relativity and continuing with applications to different areas of gravitational physics and astrophysics. Despite the breadth of our survey, we have had to be selective in our choice of topics. Our focus has been on solving the Cauchy problem in general relativity for dynamical, asymptotically flat spacetimes, with applications to compact objects and compact binaries. There are a number of alternative approaches for solving Einstein's equations that we did not touch on at all, such as the characteristic approach and the Regge calculus. We also did not discuss in any detail applications involving strictly stationary spacetimes, such as gas accretion onto Kerr black holes, although some of the same schemes we described for matter evolution have been used successfully to treat problems with fixed background metrics. We trust that interested readers will find discussions of the subjects we omitted elsewhere in the literature.

We hope that our treatment laying out the foundations of numerical relativity will remain relevant for the foreseeable future. However, we suspect that some of the large-scale simulations we have chosen to illustrate different implementations will be superseded by more sophisticated calculations. Such calculations will invariably incorporate more detailed microphysics and will take advantage of more advanced computational algorithms and hardware. Yet we would like to believe that, at the very least, the examples we have selected to highlight will help preview and elucidate these more sophisticated, future simulations. We invite our readers to assess this situation for themselves.

Exercise 18.1 Pick one of the applications of numerical relativity treated in this book, such as binary neutron star mergers, or magnetorotational stellar collapse, or one of the other topics. Perform a search of the recent computational literature to determine what new physical input or methods of solution have been implemented and what new results have emerged for this application since the writing of this book.

One exciting prospect is that numerical relativity will be applied to areas of gravitational physics and astrophysics that have yet to take full advantage of this powerful computational tool. It is possible, for example, that numerical relativity will prove useful to probe certain aspects of cosmology where homogeneity is not applicable and where neither perturbation theory nor Newtonian gravitation is adequate. It is also possible that numerical relativity will play a greater role in investigating the higher-dimensional spacetimes that arise in quantum field theories, such as string theory. More than likely, there are areas that we cannot even fathom now that will turn to numerical relativity in the future for computational insight. We again leave it to future readers to identify such areas, or, even better, to lead the charge!

> **Exercise 18.2** Do a casual search of the research literature and see if there are any new applications of numerical relativity in physics or astrophysics that have not been mentioned at all in this book.

A | Lie derivatives, Killing vectors, and tensor densities

In this appendix we introduce a few of the mathematical concepts used in the book with which some readers may be less familiar, namely, the Lie derivative, Killing vectors and tensor densities. The aim of our presentation is to make these concepts transparent and easy to use in applications; for a more complete and rigorous treatment we refer the reader to the discussions in, for example, Schutz (1980) and Wald (1984).

A.1 The Lie derivative

Consider a (nonzero) vector field X^a in a manifold M.[1] We can find the integral curves $x^a(\lambda)$ (or orbits, or trajectories) of X^a by integrating the ordinary differential equations

$$\frac{dx^a}{d\lambda} = X^a(\mathbf{x}(\lambda)). \tag{A.1}$$

Here λ is some affine parameter, and we use bold-face notation $\mathbf{x}$ instead of index notation x^a for the coordinate location in the argument to make the expressions more transparent. For a sufficiently well-behaved vector field X^a a solution is guaranteed, at least locally, by the existence and uniqueness theorem for ordinary differential equations. This procedure is completely equivalent to finding the paths of fluid particles, given their fluid velocity. The integral curves $x^a(\lambda)$ are now a family of curves in M, so that exactly one curve passes through each point in M (see Figure A.1). Such a family is known as a *congruence* of curves. Obviously, at each point the tangent to the integral curves is given by X^a.

We would now like to define a derivative of a tensor field, say T^a_b, using X^a. This involves comparing the tensor field at two different points along X^a, say P and Q, and taking the limit as Q tends to P. This is where we encounter a conceptual problem: what do we mean by comparing two tensors at two different locations in the manifold M?

We could, of course, simply compare components of the tensor field T^a_b at P, $T^a_b(P)$ and at Q, $T^a_b(Q)$. This leads to the definition of the partial derivative. Note, however, what happens under a coordinate transformation. The tensors $T^a_b(P)$ and $T^a_b(Q)$ have to be transformed with the transformation matrix evaluated at the two points P and Q. In general the two matrices will be different, and, accordingly, the result of this differentiation cannot be a tensor. This is one way to understand why the partial derivative of a tensor is not a tensor.

In order to differentiate a tensor in a tensorial manner, we therefore have to evaluate the two tensors at the same point. To do so, we have to *drag* one tensor to the other point before we can compare the two tensors. For example, we can drag $T^a_b(P)$ along X^a to the point Q. At Q, we

---

[1] This section closely follows the discussion in d'Inverno (1992).

598

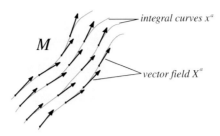

Figure A.1 A vector field X^a generates a congruence of curves x^a.

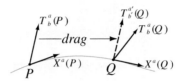

Figure A.2 Dragging a tensor $T^a{}_b$ from P to Q.

can then compare the dragged tensor, which we will denote with primes, $T^{a'}_{b'}(Q)$, with the tensor already present at Q, $T^a_b(Q)$ (see Figure A.2).

However, this recipe still leaves open *how* we drag T^a_b along X^a. One approach would be to parallel-transport the tensor T^a_b from P to Q. This idea leads to the definition of the covariant derivative. Put into words, the covariant derivative measures by how much the changes in a tensor field along X^a differ from being parallel-transported.

Parallel-transporting is not the only way of dragging T^a_b along X^a. In some sense an even more straightforward approach is to view the dragging as a simple coordinate transformation from P to Q. This, in fact, defines the Lie derivative. In other words, *the Lie derivative along a vector field X^a measures by how much the changes in a tensor field along X^a differ from a mere infinitesimal coordinate transformation generated by X^a.* Unlike the covariant derivative, the Lie derivative does not require an affine connection and hence requires less structure.

Consider now the infinitesimal coordinate transformation

$$x^{a'} = x^a + \delta\lambda\, X^a(\mathbf{x}), \tag{A.2}$$

which maps the point P, with coordinates x^a, into the point Q, with coordinates $x^{a'}$. We regard this as an *active* coordinate transformation, which maps points (and tensors) to new locations in the old coordinate system. Completely equivalently, a *passive* coordinate transformation assigns new coordinate values to the old point.

Assuming a coordinate basis we can differentiate (A.2) to find

$$\frac{\partial x^{a'}}{\partial x^b} = \delta^a{}_b + \delta\lambda\, \partial_b X^a, \tag{A.3}$$

and, to first order in $\delta\lambda$,

$$\frac{\partial x^a}{\partial x^{b'}} = \delta^a{}_b - \delta\lambda\, \partial_b X^a. \tag{A.4}$$

We now start at point P, where the components of the tensor field T^a_b are $T^a_b(\mathbf{x})$. We map this tensor into the primed tensor $T^{a'}_{b'}(\mathbf{x}')$ at Q with the help of the coordinate transformation (A.2)

$$
\begin{aligned}
T^{a'}_{b'}(\mathbf{x}') &= \frac{\partial x^{a'}}{\partial x^c}\frac{\partial x^d}{\partial x^{b'}}T^c{}_d(\mathbf{x}) \\
&= (\delta^a{}_c + \delta\lambda\,\partial_c X^a)(\delta^d{}_b - \delta\lambda\,\partial_b X^d)T^c{}_d(\mathbf{x}) \\
&= T^a_b(\mathbf{x}) + \delta\lambda\,(\partial_c X^a\, T^c{}_b(\mathbf{x}) - \partial_b X^c\, T^a{}_c(\mathbf{x})) + \mathcal{O}(\delta\lambda^2).
\end{aligned}
\tag{A.5}
$$

For the purpose of defining the Lie derivative this is the result of dragging T^a_b along X^a from P to Q. The components of the unprimed tensor already present at Q, $T^a_b(\mathbf{x}')$, can be related to $T^a_b(\mathbf{x})$ by Taylor expanding

$$T^a_b(\mathbf{x}') = T^a_b(x^{c'}) = T^a_b(x^c + \delta\lambda\, X^c) = T^a_b(\mathbf{x}) + \delta\lambda\, X^c \partial_c T^a_b + \mathcal{O}(\delta\lambda^2). \tag{A.6}$$

We now denote the *Lie derivative* of T^a_b with respect to X^a as $\mathcal{L}_\mathbf{X} T^a_b$ and define

$$\boxed{\; \mathcal{L}_\mathbf{X} T^a_b \equiv \lim_{\delta\lambda\to 0} \frac{T^a_b(\mathbf{x}') - T^{a'}_{b'}(\mathbf{x}')}{\delta\lambda}. \;} \tag{A.7}$$

This definition holds for any tensor of arbitrary rank and type (i.e., covariant and contravariant). Note that we evaluate both tensors at the same point, so that the Lie derivative of a tensor is again a tensor, and moreover a tensor of the same rank. Note also a subtlety of our abstract tensor notation: The expression $\mathcal{L}_\mathbf{X} T^a_b = (\mathcal{L}_\mathbf{X} T)^a_b$ implies that the Lie derivative of the tensor T^a_b is again a tensor of rank $\binom{1}{1}$; it does not denote the Lie derivative of the a-b component of T^a_b.[2]

For our tensor T^a_b of rank $\binom{1}{1}$ we can insert (A.5) and (A.6) into (A.7) to find

$$\boxed{\; \mathcal{L}_\mathbf{X} T^a_b = X^c \partial_c T^a_b - T^c{}_b \partial_c X^a + T^a{}_c \partial_b X^c. \;} \tag{A.8}$$

The Lie derivative of a general tensor field can be found by first taking a partial derivative of the tensor and contracting it with X^a, and then adding additional terms involving derivatives of X^a as in (A.8) for each index, with a negative sign for contravariant indices and a positive sign for covariant indices.

We can always introduce an adapted coordinate system, in which, for example, the coordinate basis vector $e^a_{(0)}$ is aligned with X^a, and all the other coordinates are constant along X^a. Setting $X^a = e^a_{(0)}$ then yields

$$X^a = \delta^a{}_0 = (1, 0, \ldots, 0). \tag{A.9}$$

We have used Greek indices as opposed to the Latin indices of our abstract index notation, since this relationship for components only holds in these adapted coordinates. Writing equation (A.8) in this coordinate system, we immediately find

$$\mathcal{L}_\mathbf{X} T^\alpha_\beta = \frac{\partial}{\partial x^0} T^\alpha_\beta. \tag{A.10}$$

---

[2] See also exercise (8.13) in Lightman *et al.* (1975).

In this sense the Lie derivative is a tensorial generalization of the partial derivative. In fact, this relationship is sometimes used as a starting point for defining the Lie derivative.

It is an important property of the Lie derivative that, for any symmetric affine connection (i.e., $\Gamma^a{}_{bc} = \Gamma^a{}_{(bc)}$) we can replace all the partial derivatives with covariant derivatives.

Exercise A.1 Show that the expression

$$\mathcal{L}_{\mathbf{X}} T^a_b = X^c \nabla_c T^a_b - T^c_b \nabla_c X^a + T^a_c \nabla_b X^c, \qquad (A.11)$$

where ∇_a denotes a covariant derivative[3] with a symmetric connection, is equivalent to equation (A.8).

As we have emphasized, our derivation holds only for coordinate or holonomic bases, for which the affine connection associated with the metric is symmetric. But for these bases we can replace all the partial derivatives with covariant derivatives, obtaining covariant (coordinate-free) expressions for the Lie derivative. Accordingly, these covariant expressions for the Lie derivative, like equation (A.11), hold even in noncoordinate bases, and are hence more general. In the following we will therefore use covariant derivatives when writing out a Lie derivative.

For a scalar f, the Lie derivative naturally reduces to the partial derivative

$$\mathcal{L}_{\mathbf{X}} f = X^b \nabla_b f = X^b \partial_b f, \qquad (A.12)$$

since there are no free indices on f to take care of. For a vector field v^a, we find from (A.8)

$$\mathcal{L}_{\mathbf{X}} v^a = X^b \nabla_b v^a - v^b \nabla_b X^a = [X, v]^a, \qquad (A.13)$$

which is the commutator of the two vector fields, while for a 1-form ω_a we have

$$\mathcal{L}_{\mathbf{X}} \omega_a = X^b \nabla_b \omega_a + \omega_b \nabla_a X^b. \qquad (A.14)$$

Note that the Lie derivative satisfies the Leibnitz rule for outer products.

Exercise A.2 Show that

$$\mathcal{L}_{\mathbf{X}}(v^a \omega_b) = v^a \mathcal{L}_{\mathbf{X}} \omega_b + \omega_b \mathcal{L}_{\mathbf{X}} v^a. \qquad (A.15)$$

Also, the Lie derivative commutes with the exterior derivative of a p-form $\widetilde{\Omega}$.

Exercise A.3 Show that for a p-form $\widetilde{\Omega}$

$$\mathcal{L}_{\mathbf{X}} \widetilde{\mathbf{d}\Omega} = \widetilde{\mathbf{d}} \mathcal{L}_{\mathbf{X}} \widetilde{\Omega}. \qquad (A.16)$$

Exercise A.4 Let $x^a(\lambda)$ be the integral curves of a vector field X^a, and let Y^a be a second vector field. Show that if Y^a is *Lie dragged* along X^a,

$$\mathcal{L}_{\mathbf{X}} Y^a = 0, \qquad (A.17)$$

then it will connect points of equal λ along the congruence $x^a(\lambda)$.

The last exercise provides a geometrical interpretation of a property of the Lie derivative that is very important in constructing $3 + 1$ evolution equations in Chapter 2. If a vector Y^a resides in

---

[3] We depart from our convention in the rest of the book and denote by ∇_a a covariant derivative in *any* number of dimensions.

a spacelike hypersurface, then it connects, in the language of Chapter 2, points in spacetime that have the same coordinate time t. Exercise (A.4) shows that if we Lie drag Y^a along the vector αn^a, where α is the lapse function and n^a the normal vector, then it will still connect points of equal t, and hence will continue to reside in our spacelike hypersurfaces.

A.2 Killing vectors

As an important application, consider the Lie derivative along X^a of a metric g_{ab}

$$\mathcal{L}_{\mathbf{X}} g_{ab} = X^c \nabla_c g_{ab} + g_{cb} \nabla_a X^c + g_{ca} \nabla_b X^c. \tag{A.18}$$

If ∇_a is compatible with the metric, the first term vanishes, and we find

$$\mathcal{L}_{\mathbf{X}} g_{ab} = \nabla_a X_b + \nabla_b X_a. \tag{A.19}$$

A *Killing* vector field ξ^a can now be defined by

$$\boxed{\mathcal{L}_{\xi} g_{ab} = 0.} \tag{A.20}$$

In other words, a Killing field ξ^a generates an isometry of the spacetime, and a displacement along ξ^a leaves the metric invariant. From (A.19), we immediately recover *Killing's equation*

$$\boxed{\nabla^a \xi^b + \nabla^b \xi^a = 0.} \tag{A.21}$$

In some cases it is very easy to identify a Killing vector. If the metric components are independent of a coordinate x^i, then it follows from the property (A.10) that the coordinate basis vector $e^a_{(i)}$ is a Killing vector. An example is the flat metric of a unit sphere in spherical polar coordinates θ and ϕ,

$$\eta_{ij} = \mathrm{diag}\left(1, \sin^2\theta\right). \tag{A.22}$$

The above metric is independent of ϕ, making $e^i_{(\phi)}$ a Killing vector.

In general it is much less trivial to identify Killing vectors. If we happen to know a Killing vector ξ^a and its derivatives at one point of a manifold, the following relation can help to find ξ^a everywhere.

Exercise A.5 Show that a Killing vector field ξ^a satisfies

$$\nabla_a \nabla_b \xi_c = -R_{bca}{}^d \xi_d, \tag{A.23}$$

where $R_{bca}{}^d$ is the Riemann tensor.

We can now define the quantity $L_{ab} \equiv \nabla_a \xi_b$. If both ξ_a and L_{ab} are given at a point P of a manifold, then we can find ξ^a and L_{ab} at another point Q by integrating

$$v^a \nabla_a \xi_b = v^a L_{ab} \tag{A.24}$$

$$v^a \nabla_a L_{bc} = -R_{bca}{}^d \xi_d v^a, \tag{A.25}$$

along a curve connecting P and Q, where v^a denotes the tangent to this curve.

Among other useful properties, Killing vectors can be used to identify conserved quantities. For example, given a stress energy tensor T^{ab} and a Killing vector field ξ^a, we can show that the flux $J^a = T^{ab} \xi_b$ is conserved,

$$\nabla_a J^a = \nabla_a (T^{ab} \xi_b) = \xi_b \nabla_a T^{ab} + T^{ab} \nabla_a \xi_b = 0. \tag{A.26}$$

Here the first term vanishes because the stress energy tensor is divergence free (conservation of stress-energy), and the second term vanishes as a consequence of Killing's equation (A.21): the term $\nabla_a \xi_b$ is antisymmetric, so that the contraction with T^{ab}, which is symmetric, vanishes.

> **Exercise A.6** Let ξ^a be a Killing vector field and let p^a be the 4-momentum vector of a particle (or photon) moving on a geodesic curve. Show that $\xi^a p_a$ is conserved along the geodesic.

Since freely falling particles (or photons) follow geodesics, we see that each symmetry of the spacetime, i.e., each linearly independent Killing vector field, gives rise to a conserved quantity for such particles. Specifically, if the metric is independent of some coordinate x^b, then that coordinate basis vector $e^a_{(b)}$ is a Killing vector and the particle momentum conjugate to that coordinate, p_b, is conserved.

Before leaving this section we briefly discuss conformal Killing vectors.

> **Exercise A.7** Use the conformal transformation laws for the connection coefficients (3.7) to show that a Killing vector ξ^a of the metric g_{ab}, i.e., a vector that satisfies Killing's equation (A.21), also satisfies the *conformal Killing's equation*
>
> $$\bar{\nabla}^a \xi^b + \bar{\nabla}^b \xi^a - \frac{2}{n}\bar{\gamma}^{ab}\bar{\nabla}_c \xi^c = 0, \qquad (A.27)$$
>
> where $\bar{\nabla}_a$ is the covariant derivative associated with the conformally related metric $\bar{g}_{ab} = \psi^{-4} g_{ab}$, and where $n = \delta^a_a$ is the number of dimensions.

A vector field ξ^a that satisfies the conformal Killing's equation (A.27) is called a *conformal Killing vector*. Exercise (A.7) demonstrates that a Killing vector of the metric g_{ab} is automatically a conformal Killing vector of the conformally related metric $\bar{g}_{ab}$.

We point out that in three dimensions the differential operator in (A.27) is identical to the longitudinal operator, or vector gradient,

$$(\bar{L}W)^{ij} = \bar{\nabla}^i W^j + \bar{\nabla}^j W^i - \frac{2}{3}\bar{\gamma}^{ij}\bar{\nabla}_k W^k \qquad (A.28)$$

defined in (3.50), which is therefore also known as the *conformal Killing operator*.

A.3 Tensor densities

So far we have restricted our attention to *tensors*, which are defined by their transformation properties from one coordinate system x^a to another $x^{a'}$. A rank $\binom{1}{1}$ tensor, for example, is defined by transforming as

$$T^{a'}_{b'} = \frac{\partial x^{a'}}{\partial x^c}\frac{\partial x^d}{\partial x^{b'}} T^c_d \qquad (A.29)$$

under a coordinate transformation. There is a related class of objects that behaves similar to tensors, except that they pick up a certain power of the Jacobian

$$J \equiv \det\left|\frac{\partial x^a}{\partial x^{b'}}\right| \qquad (A.30)$$

during the transformation. An example of such an object is the determinant of the metric. The spacetime metric transforms as

$$g_{a'b'} = \frac{\partial x^c}{\partial x^{a'}} \frac{\partial x^d}{\partial x^{b'}} g_{cd}, \tag{A.31}$$

so taking the determinant of this equation yields

$$g' = J^2 g. \tag{A.32}$$

We now define a *tensor density of weight* W and rank $\binom{1}{1}$ as an object that transforms as

$$\boxed{T^{a'}_{b'} = J^W \frac{\partial x^{a'}}{\partial x^c} \frac{\partial x^d}{\partial x^{b'}} T^c_d.} \tag{A.33}$$

Tensors of different rank transform in an analogous way, namely as the normal tensor transformation times J to the power W. Normal "plain-old" tensors are therefore tensor densities of weight zero. The determinant of the spacetime metric, on the other hand, is a scalar density of weight $W = +2$. Another example of a tensor density is the "natural" conformal 3-metric (3.6)

$$\bar{\gamma}_{ij} = \gamma^{-1/3} \gamma_{ij} \tag{A.34}$$

that we encounter in several places throughout this book.[4] Given that $\gamma = \det(\gamma_{ij})$ is a scalar density of weight $+2$, $\bar{\gamma}_{ij}$ is then a tensor density of weight $-2/3$.

We now construct the Lie derivative of a tensor density. Following the procedure of Section A.1 we drag T^a_b from a point P to a point Q, assuming that the tensor transforms as if under an infinitesimal coordinate transformation generated by X^a. That means that we simply have to replace the transformation (A.5) for a tensor with (A.33) for a tensor density. To first order in $\delta\lambda$ the Jacobian of the coordinate transformation (A.4) is

$$J = 1 - \delta\lambda\, \partial_c X^c + \mathcal{O}(\delta\lambda^2), \tag{A.35}$$

so that (A.5) now becomes

$$\begin{aligned} T^{a'}_{b'}(\mathbf{x}') &= J^W \frac{\partial x^{a'}}{\partial x^c} \frac{\partial x^d}{\partial x^{b'}} T^c_d(\mathbf{x}) \\ &= (1 - W\delta\lambda\, \partial_c X^c)(\delta^a_c + \delta\lambda\, \partial_c X^a)(\delta^d_b - \delta\lambda\, \partial_b X^d) T^c_d(\mathbf{x}) + \mathcal{O}(\delta\lambda^2) \\ &= T^a_b(\mathbf{x}) + \delta\lambda \left(\partial_c X^a T^c_b(\mathbf{x}) - \partial_b X^c T^a_c(\mathbf{x}) - W \partial_c X^c T^a_b(\mathbf{x}) \right) + \mathcal{O}(\delta\lambda^2). \end{aligned} \tag{A.36}$$

The Taylor expansion (A.6) is unaffected by T^a_b now being a tensor density, so inserting (A.36) together with (A.6) into the definition of the Lie derivative (A.7) we find[5]

$$\boxed{\mathcal{L}_X T^a_b = X^c \partial_c T^a_b - T^c_b \partial_c X^a + T^a_c \partial_b X^c + W T^a_b \partial_c X^c.} \tag{A.37}$$

This expression can be generalized for a tensor of any other rank. The rule is that the Lie derivative of a tensor density of weight W is the same as that of a tensor, except that we have to add a term

---

[4] e.g., in the context of the minimal distortion gauge condition in Chapter 4.5 and the BSSN formulation in Chapter 11.5.
[5] See exercise 11.3 for an example of how the Lie derivative of an even more general object can be constructed.

$WT\partial_c X^c$, where $\mathcal{T}$ is the tensor density of whatever rank we are considering. From the definition (A.7) we note that the Lie derivative of a tensor density of weight W is again a tensor density of the same weight.

As in exercise A.1 it might be desirable to express the Lie derivative (A.37) in terms of covariant derivatives instead of partial derivatives. The attentive reader may object that we do not yet know what the covariant derivative of a tensor density is. We may counter that, by construction, the Lie derivative is independent of the connection, so that we can express it in terms of *any* connection (as we have seen in Section A.1). In fact, *requiring* that the Lie derivative of a tensor density take the same form as (A.37) when expressed in terms of covariant derivatives will guide us in defining the covariant derivative of a tensor density.

Since X^a is a vector, we have

$$\nabla_a X^b = \partial_a X^b + X^d \Gamma^b_{ad}. \tag{A.38}$$

Using equation (A.38) to substitute for all the partial derivatives of X^a in equation (A.37) gives

$$\mathcal{L}_{\mathbf{X}} \mathcal{T}^a_b = X^c \left(\partial_c \mathcal{T}^a_b + \mathcal{T}^d_b \Gamma^a_{cd} - \mathcal{T}^a_d \Gamma^d_{cb} - W \mathcal{T}^a_b \Gamma^d_{dc} \right)$$
$$- \mathcal{T}^c_b \nabla_c X^a + \mathcal{T}^a_c \nabla_b X^c + W \mathcal{T}^a_b \nabla_c X^c. \tag{A.39}$$

We now identify the first line in (A.39) with the covariant derivative of a tensor density of weight W and rank $\binom{1}{1}$ as

$$\boxed{\nabla_c \mathcal{T}^a_b = \partial_c \mathcal{T}^a_b + \mathcal{T}^d_b \Gamma^a_{cd} - \mathcal{T}^a_d \Gamma^d_{cb} - W \mathcal{T}^a_b \Gamma^d_{dc}.} \tag{A.40}$$

The generalization to other tensors of arbitrary rank is straightforward following this rule: the covariant derivative of a tensor density $\mathcal{T}^{a\cdots}{}_{b\cdots}$ of weight W is the same as that of a tensor $T^{a\cdots}{}_{b\cdots}$ of the same rank, except that we have to subtract a term $W \mathcal{T}^{a\cdots}{}_{b\cdots} \Gamma^d_{dc}$ in the case of the tensor density.

Exercise A.8 Use the identity

$$\partial_a g = g g^{bc} \partial_a g_{bc} \tag{A.41}$$

to show that

$$\nabla_a g = 0. \tag{A.42}$$

Exercise A.9 Show that for a vector X^a we have

$$\nabla_a(|g|^{1/2} X^a) = \partial_a(|g|^{1/2} X^a), \tag{A.43}$$

and hence

$$\nabla_a X^a = \frac{1}{|g|^{1/2}} \partial_a(|g|^{1/2} X^a). \tag{A.44}$$

With the definition (A.40) we can now, by construction, write the Lie derivative of a tensor density in terms of any covariant derivative

$$\mathcal{L}_{\mathbf{X}} \mathcal{T}^a_b = X^c \nabla_c \mathcal{T}^a_b - \mathcal{T}^c_b \nabla_c X^a + \mathcal{T}^a_c \nabla_b X^c + W \mathcal{T}^a_b \nabla_c X^c. \tag{A.45}$$

Exercise A.10 Show that the Lie derivative of the conformal metric (A.34) is

$$\mathcal{L}_{\mathbf{X}}\bar{\gamma}_{ij} = \bar{\nabla}_i X_j + \bar{\nabla}_j X_i - \frac{2}{3}\bar{\gamma}_{ij}\bar{\nabla}_k X^k = (\bar{L}X)_{ij}, \qquad (A.46)$$

where we have lowered the indices of X^a with $\bar{\gamma}_{ij}$.

Exercise A.10 provides an alternative definition of a conformal Killing vector. The definition (A.20) states that the Lie derivative of the metric along a Killing vector vanishes, while (A.46) together with (A.27) states that the Lie derivative of the conformal metric along a conformal Killing vector vanishes.

B | Solving the vector Laplacian

In this appendix we discuss two approaches to solving the vector Poisson equation in flat space,

$$(\Delta_L^{\text{flat}} W)^i = S^i, \tag{B.1}$$

which in Cartesian coordinates becomes

$$\partial^j \partial_j W_i + \frac{1}{3} \partial_i \partial^j W_j = S_i. \tag{B.2}$$

We have encountered this operator in several places in this book, starting with the momentum constraint (3.60) in the conformal transverse-traceless decomposition in Chapter 3.2.

The first approach[1] to solving this equation entails writing the vector field W_i as a sum of a vector field V_i and the gradient of a scalar field U,

$$W_i = V_i + \partial_i U. \tag{B.3}$$

Inserting this decomposition into equation (B.2) yields

$$\partial^j \partial_j V_i + \frac{1}{3} \partial_i \partial^j V_j + \partial^j \partial_j \partial_i U + \frac{1}{3} \partial_i \partial^j \partial_j U = S_i. \tag{B.4}$$

We can now choose U in such a way that the two terms involving U in equation (B.4) cancel the second term,

$$\partial^j \partial_j U = -\frac{1}{4} \partial_j V^j, \tag{B.5}$$

so that equation (B.4) reduces to

$$\partial^j \partial_j V_i = S_i. \tag{B.6}$$

We have thereby rewritten the vector Poisson equations (B.2) as a set of four coupled scalar Poisson equations for U and the three components of V_i. We use this decomposition to construct the solution for a spinning black hole in Chapter 3.2.

A second approach[2] adopts the ansatz

$$W_i = \frac{7}{8} V_i - \frac{1}{8} \left(\partial_i U + x^k \partial_i V_k \right). \tag{B.7}$$

Inserting this decomposition into equation (B.2) yields

$$\frac{5}{6} \partial^j \partial_j V_i - \frac{1}{6} \partial_i \partial^j \partial_j U - \frac{1}{6} x^k \partial_i \partial^j \partial_j V_k = S_i. \tag{B.8}$$

If we now choose U so that it satisfies

$$\partial^j \partial_j U = -S_j x^j, \tag{B.9}$$

---

[1] Bowen and York, Jr. (1980).
[2] Shibata (1999b); see also Oohara *et al.* (1997).

then equation (B.8) reduces to

$$\frac{5}{6}\partial^j\partial_j V_i + \frac{1}{6}x^j\partial_i S_j - \frac{1}{6}x^j\partial_i\partial^k\partial_k V_j = \frac{5}{6}S_i,$$ (B.10)

which is solved by

$$\partial^j\partial_j V_i = S_i.$$ (B.11)

The vector equation (B.2) now has been reduced to a set of four decoupled scalar Poisson equations (B.9) and (B.11). In Chapter 3.2 we use this decomposition to construct boosted black holes.

While the second approach (B.7) seems a little more complicated, it has the advantage that the equations are decoupled and the source terms in equations (B.9) and (B.11) are nonzero only where S_i is nonzero. In some cases this later feature results in compact sources,[3] which can have advantages for numerical implementations. In the approach (B.3), on the other hand, the source term of equation (B.5) is never compact. Other approaches have been suggested,[4] and we refer the reader to the literature[5] for a comparison of the numerical performance of these different methods.

---

[3] By "compact" we mean that the sources are nonzero only on a bounded subset of the coordinate space.
[4] See, e.g., Oohara and Nakamura (1999).
[5] See Grandclément *et al.* (2001).

C   The surface element on the apparent horizon

In this appendix we outline a method for computing the surface element on a closed 2-surface S embedded in some 3-dimensional spatial hypersurface Σ. The approach is useful for diverse numerical applications that require areas of closed 2-surfaces on spatial slices, but our prime motivation here is to compute the proper area of an apparent horizon of a black hole or a system of black holes. Recall that for stationary spacetimes, the apparent horizon coincides with the event horizon and its area then determines the irreducible mass (7.2) of a black hole.

One approach to computing this surface element is to construct the induced metric m_{ij} on the surface S (see Chapter 7). Typically. however, the resulting line element will not be expressed in terms of coordinates that are convenient for setting up the surface element. One way of constructing such a coordinate system is the following:[1] We start with the line element for the spatial hypersurface Σ,

$$dl^2 = \gamma_{ij} dx^i dx^j, \tag{C.1}$$

where the x^i denote the spatial coordinates that have been used to label points on Σ in the numerical calculation. We can now transform to spherical polar coordinates, centered on some fiducial point C^i. The coordinate separation between nearby points as expressed in the two coordinate systems will then be related by the usual transformation,

$$dx^i = \frac{\partial x^i}{\partial r_C} dr_C + \frac{\partial x^i}{\partial \theta} d\theta + \frac{\partial x^i}{\partial \phi} d\phi. \tag{C.2}$$

Using the notation of Chapter 7, we define the closed 2-surface S around C^i as the level surface

$$\tau(x^i) = r_C(x^i) - h(\theta, \phi) = 0, \tag{C.3}$$

where r_C is the coordinate separation between the point x^i and the point C^i. For example, in Cartesian coordinates, we have $r_C^2 = x^2 + y^2 + z^2$, provided the centers of the polar and Cartesian coordinate systems coincide. The function $h(\theta, \phi)$ then measures the coordinate distance from C^i to the 2-surface S in the (θ, ϕ) direction.[2] In the following we assume that we have already determined $h(\theta, \phi)$ by means, for example, of the techniques described in Chapter 7 in the case where S is an apparent horizon.

On S, where $r_C = h(\theta, \phi)$, we must have

$$dr_C = \frac{\partial h}{\partial \theta} d\theta + \frac{\partial h}{\partial \phi} d\phi . \tag{C.4}$$

---

[1]   See Appendix D in Baumgarte *et al.* (1996).
[2]   By our notation $h(\theta, \phi)$ we mean that h can be expressed in terms of the coordinates θ and ϕ via, for example, a superposition of spherical harmonics (*cf.* Chapter 7.3.3). In Cartesian coordinates, it may be more convenient to express h in terms of combinations of x, y and z via, for example, a superpositions of symmetric, tracefree "STF" tensors.

Inserting this into equation (C.2) we obtain

$$dx^i = \left(\frac{\partial x^i}{\partial r_C}\frac{\partial h}{\partial \theta} + \frac{\partial x^i}{\partial \theta}\right)d\theta + \left(\frac{\partial x^i}{\partial r_C}\frac{\partial h}{\partial \phi} + \frac{\partial x^i}{\partial \phi}\right)d\phi. \tag{C.5}$$

We now define the vectors

$$\Theta^i = \frac{1}{r_C}\left(\frac{\partial x^i}{\partial r_C}\frac{\partial h}{\partial \theta} + \frac{\partial x^i}{\partial \theta}\right) \tag{C.6}$$

and

$$\Phi^i = \frac{1}{r_C \sin\theta}\left(\frac{\partial x^i}{\partial r_C}\frac{\partial h}{\partial \phi} + \frac{\partial x^i}{\partial \phi}\right), \tag{C.7}$$

both of which are tangent to S.

> **Exercise C.1** Show that both Θ^i and Φ^i are orthogonal to $m^i = D^i\tau$ (i.e., $m_i = \partial_i\tau$), which demonstrates that both are tangent to S.

With the help of Θ^i and Φ^i we can now write the line element on S as

$$
\begin{aligned}
dl^2 &= \gamma_{ij}(r_C\Theta^i\,d\theta + r_C\sin\theta\,\Phi^i\,d\phi)(r_C\Theta^j\,d\theta + r_C\sin\theta\,\Phi^j\,d\phi)\\
&= r_C^2(\Theta^i\Theta_i d\theta^2 + 2\sin\theta\,\Theta^i\Phi_i d\theta d\phi + \Phi^i\Phi_i \sin^2\theta\phi^2),
\end{aligned} \tag{C.8}
$$

where we have lowered the indices on Θ^i and Φ^i with γ_{ij}. In the above equation we may, of course, replace r_C with h. We have now expressed the line element on S in terms of the polar coordinates θ and ϕ. The outward-oriented surface element on the apparent horizon is then given by $dS_i = s_i\sqrt{^{(2)}\gamma}\,d\theta d\phi$ (cf. Chapter 3.5), which yields

$$dS_i = s_i\left((\Theta^i\Theta_i)(\Phi^j\Phi_j) - (\Theta^k\Phi_k)^2\right)^{1/2} h^2 \sin\theta d\theta d\phi . \tag{C.9}$$

Here $s^i = \lambda D^i\tau$ is the unit normal on S (see equation 7.33), λ is a normalization factor, and $^{(2)}\gamma$ is the determinant of the 2-metric appearing in equation (C.8).

What remains to be done is computing the vectors Θ^i and Φ^i, whose form depends on the original coordinate system x^i. If these coordinates were already spherical polar coordinates centered on C^i, then they take the particularly simple form

$$
\begin{aligned}
\Theta^i &= \frac{1}{r_C}(\partial_\theta h, 1, 0)\\
\Phi^i &= \frac{1}{r_C\sin\theta}(\partial_\phi h, 0, 1)
\end{aligned}
\qquad \text{(spherical polar coordinates)}. \tag{C.10}
$$

If the original coordinates x^i are Cartesian, we find the more complicated expressions

$$\Theta^x = \frac{l^x l^z}{\sqrt{1-(l^z)^2}}(1 + l^x \partial_x h + l^y \partial_y h) - l^x \sqrt{1-(l^z)^2}\,\partial_z h$$

$$\Theta^y = \frac{l^y l^z}{\sqrt{1-(l^z)^2}}(1 + l^x \partial_x h + l^y \partial_y h) - l^y \sqrt{1-(l^z)^2}\,\partial_z h$$

$$\Theta^z = \frac{(l^z)^2}{\sqrt{1-(l^z)^2}}(l^x \partial_x h + l^y \partial_y h) - \sqrt{1-(l^z)^2}\,(1 + l^z \partial_z h)$$

$$\Phi^x = \frac{1}{\sqrt{1-(l^z)^2}}\left(l^x(l^x \partial_y h - l^y \partial_x h) - l^y\right)$$ (Cartesian coordinates),

$$\Phi^y = \frac{1}{\sqrt{1-(l^z)^2}}\left(l^y(l^x \partial_y h - l^y \partial_x h) + l^x\right)$$

$$\Phi^z = \frac{l^z}{\sqrt{1-(l^z)^2}}(l^x \partial_y h - l^y \partial_x h) \qquad\qquad\qquad\text{(C.11)}$$

where $l^i = \partial x^i / \partial r_C = x^i / r_C = (\sin\theta\cos\phi,\, \sin\theta\sin\phi,\, \cos\theta)$.

Scalar, vector and tensor spherical harmonics

In this appendix we provide a partial list of useful properties of scalar, vector and tensor spherical harmonics. We focus on those properties that we use in our treatment of gravitational wave extraction in Chapter 9.4.1. We adopt the notation of Nagar and Rezzolla (2005) and refer to Thorne and Campolattaro (1967), Sandberg (1978) and Thorne (1980) for additional details and references.

Here it is useful to depart from our general notation convention. Instead of separating spatial indices from spacetime indices, it is more convenient, in the context of the spherical polar coordinates used here, to separate the angular coordinates θ and ϕ from the remaining spacetime coordinates r and t. Following this convention, we label the former (θ and ϕ) with lower-case letters $a, b, \ldots$, and the latter (t and r) with upper-case letters $A, B, \ldots$ We also write the two-dimensional metric on the unit sphere S^2 as

$$\sigma_{ab}dx^a dx^b = d\theta^2 + \sin^2\theta d\phi^2. \tag{D.1}$$

Scalar spherical harmonics

The *scalar spherical harmonics* $Y_{lm}(\theta, \phi)$ satisfy the equations

$$\sigma^{ab}\nabla_a\nabla_b Y_{lm} = -l(l+1)Y_{lm} \tag{D.2}$$

and

$$\partial_\phi Y_{lm} = im Y_{lm} \tag{D.3}$$

and can be expressed in terms of associated Legendre polynomials $P_l^m(\cos\theta)$ as

$$Y_{lm}(\theta, \phi) = \sqrt{\frac{2l+1}{4\pi}}\sqrt{\frac{(l-m)!}{(l+m)!}}\, P_l^m(\cos\theta)e^{im\phi}. \tag{D.4}$$

Under a space inversion, spherical harmonics have even (or polar) parity (see our discussion in Chapter 9.1.2)

$$Y_{lm}(\pi-\theta, \pi+\phi) = (-1)^l\, Y_{lm}(\theta, \phi). \tag{D.5}$$

We also note the property

$$Y_{l-m}(\theta, \phi) = (-1)^m\, Y_{lm}^*(\theta, \phi). \tag{D.6}$$

The first few scalar spherical harmonics, up to $l = 2$, are listed in Table D.1.

The derivative $\partial_\theta Y_{lm}$ can be computed in terms of the associated Legendre polynomials as

$$\partial_\theta Y_{lm}(\theta, \phi) = -\sqrt{\frac{2l+1}{4\pi}}\sqrt{\frac{(l-m)!}{(l+m)!}}\Big((l+m)(l-m+1)P_l^{m-1}(\cos\theta) + m\cot\theta\, P_l^m(\cos\theta)\Big)e^{im\phi}. \tag{D.7}$$

Table D.1 The scalar spherical harmonics Y_{lm} up to $l = 2$.

	$l = 0$	$l = 1$	$l = 2$
$m = -2$			$\frac{1}{4}\sqrt{\frac{15}{2\pi}}\sin^2\theta e^{-2i\phi}$
$m = -1$		$\frac{1}{2}\sqrt{\frac{3}{2\pi}}\sin\theta e^{-i\phi}$	$\frac{1}{2}\sqrt{\frac{15}{2\pi}}\sin\theta\cos\theta e^{-i\phi}$
$m = 0$	$\frac{1}{2}\sqrt{\frac{1}{\pi}}$	$\frac{1}{2}\sqrt{\frac{3}{\pi}}\cos\theta$	$\frac{1}{4}\sqrt{\frac{5}{\pi}}(3\cos^2\theta - 1)$
$m = 1$		$-\frac{1}{2}\sqrt{\frac{3}{2\pi}}\sin\theta e^{i\phi}$	$-\frac{1}{2}\sqrt{\frac{15}{2\pi}}\sin\theta\cos\theta e^{i\phi}$
$m = 2$			$\frac{1}{4}\sqrt{\frac{15}{2\pi}}\sin^2\theta e^{2i\phi}$

The scalar spherical harmonics satisfy the orthogonality relation

$$\int Y_{lm}^* Y_{l'm'} d\Omega = \delta_{ll'}\delta_{mm'}. \tag{D.8}$$

A generalization of the scalar spherical harmonics are the *spin-weighted spherical harmonics*, which can be computed from the regular spherical harmonics by applying certain spin-raising or lowering operators. Other than the $s = 0$ harmonics, which are the usual scalar spherical harmonics, the $s = -2$ spin-weighted harmonics,

$$_{-2}Y_{lm}(\theta, \phi) = \sqrt{\frac{(l-2)!}{(l+2)!}}\left(W_{lm}(\theta, \phi) - i\frac{X_{lm}(\theta, \phi)}{\sin\theta}\right), \tag{D.9}$$

are of special interest for our purposes in Chapter 9.4.1. Here the functions W_{lm} and X_{lm} are defined in terms of derivatives of the scalar spherical harmonics by

$$W_{lm} = \left(\partial_\theta^2 - \cot\theta\,\partial_\theta - \frac{1}{\sin^2\theta}\partial_\phi^2\right)Y_{lm} \tag{D.10}$$

and

$$X_{lm} = 2\partial_\phi(\partial_\theta - \cot\theta)Y_{lm}. \tag{D.11}$$

The function W_{lm} can also be expressed as

$$W_{lm} = l(l+1)Y_{lm} + 2\partial_\theta^2 Y_{lm} = -l(l+1)Y_{lm} - 2\cot\theta\,\partial_\theta Y_{lm} + \frac{2m^2}{\sin^2\theta}Y_{lm}. \tag{D.12}$$

We will see below that the functions X_{lm} and W_{lm} are also very useful in the context of the tensor spherical harmonics (see equation D.25 below). The spin-weighted spherical harmonics satisfy the same orthogonality relation as the spherical harmonics,

$$\int {}_sY_{lm}^* \, {}_sY_{l'm'} d\Omega = \delta_{ll'}\delta_{mm'}. \tag{D.13}$$

Exercise D.1 Compute the $s = -2$ spin-weighted spherical harmonic $_{-2}Y_{20}$ and show that it has the same angular behavior as the h_+ polarization of a $l = 2$, $m = 0$ linearized wave at large radii r (see equation 9.57; the $h_\times$ polarization vanishes identically for this mode).

Vector spherical harmonics

In addition to the familiar scalar spherical harmonics, we can also define *vector* and *tensor spherical harmonics*. Vector spherical harmonics form a basis for vectors on S^2. Evidently, we need two linearly independent vectors to form a basis on S^2, and we can create such a basis by taking two different derivatives of the Y_{lm}. The electric vector spherical harmonics E_a^{lm} are defined as gradients of the scalar spherical harmonics

$$E_a^{lm} \equiv \nabla_a Y_{lm}, \tag{D.14}$$

and have even (or polar) parity under space inversion, while the magnetic vector spherical harmonics S_a^{lm} are defined as the duals[1] of the gradients,

$$S_a^{lm} \equiv \epsilon^b{}_a \nabla_b Y_{lm}, \tag{D.15}$$

and have odd (or axial) parity. Here ϵ_{ab} is the 2-dimensional Levi-Civita tensor on S^2, which has

$$\epsilon_{\theta\phi} = -\epsilon_{\phi\theta} = \sin\theta \tag{D.16}$$

as the only nonvanishing components.

> **Exercise D.2** Show that that the electric and magnetic vector spherical harmonics are orthogonal,
> $$E_{lm}^a S_a^{lm} = \sigma^{ab} E_a^{lm} S_b^{lm} = 0. \tag{D.17}$$

Since the scalar spherical harmonics Y_{lm} are scalars, the covariant derivatives in the above expressions can be expressed in terms of partial derivatives, and can be evaluated with the help of equations (D.3) and (D.7).

The vector spherical harmonics satisfy the orthogonality relations

$$\int (E_a^{lm})^* E_{l'm'}^a d\Omega = l(l+1)\delta_{ll'}\delta_{mm'} \tag{D.18}$$

and

$$\int (S_a^{lm})^* S_{l'm'}^a d\Omega = l(l+1)\delta_{ll'}\delta_{mm'}. \tag{D.19}$$

> **Exercise D.3** Show that the electric and magnetic $l=2$, $m=0$ vector spherical harmonics are
> $$E_a^{lm} = \begin{pmatrix} -\frac{3}{2}\sqrt{\frac{5}{\pi}}\cos\theta\sin\theta \\ 0 \end{pmatrix} \quad \text{and} \quad S_a^{lm} = \begin{pmatrix} 0 \\ -\frac{3}{2}\sqrt{\frac{5}{\pi}}\cos\theta\sin^2\theta \end{pmatrix}. \tag{D.20}$$

Tensor spherical harmonics

It will also be useful to have a basis for symmetric, traceless, rank-2 tensors on S^2. In two dimensions, these objects again have only two independent components, and we can therefore

---

[1] In two dimensions the dual of a vector is again a vector.

create a basis, in complete analogy to the vector spherical harmonics, by taking two different second derivatives of the scalar spherical harmonics. In particular, we define

$$Z_{ab}^{lm} \equiv \nabla_a \nabla_b Y_{lm} + \frac{l(l+1)}{2} \sigma_{ab} Y_{lm}, \qquad (D.21)$$

which have even (or polar) parity under space inversion, and

$$S_{ab}^{lm} \equiv \nabla_{(a} S_{b)}^{lm}, \qquad (D.22)$$

which have odd (or axial) parity. By construction, both Z_{ab}^{lm} and S_{ab}^{lm} are symmetric.

> **Exercise D.4**   (a) Show that the two tensor spherical harmonics S_{ab}^{lm} and Z_{ab}^{lm} are orthonormal,
>
> $$S_{lm}^{ab} Z_{ab}^{lm} = 0. \qquad (D.23)$$
>
> (b) Show that the tensor spherical harmonics S_{ab}^{lm} and Z_{ab}^{lm} are traceless,
>
> $$\sigma^{ab} S_{ab}^{lm} = 0 \quad \text{and} \quad \sigma^{ab} Z_{ab}^{lm} = 0. \qquad (D.24)$$

Both Z_{ab}^{lm} and S_{ab}^{lm} are symmetric, traceless, two-dimensional rank-2 tensors, meaning that each one has only two independent components. The orthogonality relation (D.23) reduces the combined number of independent components to three. One of these three can be accounted for by an arbitrary normalization, so that we are left with only two independent components for the Z_{ab}^{lm} and S_{ab}^{lm}. That is not surprising, of course, since we have constructed the Z_{ab}^{lm} and S_{ab}^{lm} as a basis for symmetric, traceless, 2-dimensional rank-2 tensors, which only have two independent components. As it turns out, we can conveniently express these two independent components in terms of the functions X_{lm} and W_{lm} defined by equations (D.10) and (D.11).

> **Exercise D.5**   (a) Show that
>
> $$Z_{ab}^{lm} = \frac{1}{2}\begin{pmatrix} W_{lm} & X_{lm} \\ X_{lm} & -\sin^2\theta \, W_{lm} \end{pmatrix} \quad \text{and} \quad S_{ab}^{lm} = \frac{1}{2\sin\theta}\begin{pmatrix} -X_{lm} & \sin^2\theta \, W_{lm} \\ \sin^2\theta \, W_{lm} & \sin^2\theta \, X_{lm} \end{pmatrix}. \qquad (D.25)$$
>
> (b) Show that for $l = 2$, $m = 0$, the tensor spherical harmonics are
>
> $$Z_{ab}^{20} = \begin{pmatrix} \frac{3}{4}\sqrt{\frac{5}{\pi}} \sin^2\theta & 0 \\ 0 & -\frac{3}{4}\sqrt{\frac{5}{\pi}} \sin^4\theta \end{pmatrix} \quad \text{and} \quad S_{ab}^{20} = \begin{pmatrix} 0 & \frac{3}{4}\sqrt{\frac{5}{\pi}} \sin^3\theta \\ \frac{3}{4}\sqrt{\frac{5}{\pi}} \sin^3\theta & 0 \end{pmatrix}. \qquad (D.26)$$

Finally, the tensor spherical harmonics satisfy the orthogonality relations

$$\int (Z_{cd}^{lm})^* Z_{l'm'}^{cd} d\Omega = \frac{(l-1)l(l+1)(l+2)}{2} \delta_{ll'} \delta_{mm'} \qquad (D.27)$$

and

$$\int (S_{ab}^{lm})^* S_{l'm'}^{ab} d\Omega = \frac{(l-1)l(l+1)(l+2)}{2} \delta_{ll'} \delta_{mm'}. \qquad (D.28)$$

E ☐ Post-Newtonian results

The central premise of this book is that, for many situations of greatest physical interest, Einstein's field equations (1.32),

$$G_{ab} = 8\pi T_{ab}, \tag{E.1}$$

do not admit exact, analytic solutions. In order to treat these situations we therefore need to construct approximate solutions. This book is an introduction to numerical relativity, which aims at constructing such solutions by numerical means. The intrinsic accuracy of this approach depends on the reliability of the numerical algorithms and the adopted computational resources, but it can handle gravitational fields that are arbitrarily strong and speeds v of the sources that are arbitrarily close to the speed of light.

Post-Newtonian methods provide an alternative approach to constructing approximate solutions to Einstein's equations (E.1). In this approach a solution is constructed iteratively, starting with the corresponding Newtonian solution. In each step of the iteration a new correction term of order v^2 times the previous one is added so that the resulting solution is effectively a Taylor expansion in the parameter v^2 about the Newtonian solution. Given that this expansion is carried out analytically, this approach does not introduce any numerical error *per se*. Instead, its accuracy is limited by the number of terms retained and the rate of convergence of the expansion. Practically speaking, then, the method is limited both by the strength of the gravitational field and the speed of the sources, which are required to be small for any finite expansion to be accurate.

For compact binaries, numerical relativity and post-Newtonian approaches complement each other very well. For moderately large binary separations the binary inspirals very slowly, and the orbital speeds and tidal deformations are small (see the discussion of binary inspiral in Chapter 12.1). Post-Newtonian approximations based on point-mass companions are very accurate in this regime, while numerical models are severely challenged by the vast dynamic range of the problem that stretches computational resources. For close binary separations, on the other hand, the different time and length scales characterizing binary inspiral all become comparable. The orbital speeds become large and approach the speed of light, nonlinear gravitation, including the tidal interaction between the configurations, becomes significant, and the evolution drives the sources far from equilibrium. In this regime, post-Newtonian models become inaccurate, while numerical methods become both crucial and well-suited to tracking the dynamical behavior. In the intermediate regimes, both approaches are reliable and comparisons between post-Newtonian and numerical calculations provide very valuable checks (see Chapter 13.2 for some examples). Furthermore, realistic numerical binary initial data, which by practical necessity consist of binaries at relatively small separation, can incorporate post-Newtonian results that encode effects of the prior inspiral.[1]

In this appendix we sketch how post-Newtonian expansions are constructed, and list some of the most important results for compact binaries.[2]

Post-Newtonian expansions

Einstein's equations (E.1) form a set of differential equations for the spacetime metric g_{ab}. We may express this metric in terms of new variables[3]

$$h^{ab} \equiv \sqrt{-g}g^{ab} - \eta^{ab}, \tag{E.2}$$

which are sometimes referred to as the *gravitational field perturbation amplitudes*. Here g is the determinant of the 4-metric, $g = \det(g_{ab})$, and η_{ab} is the Minkowski metric. Evidently, for a flat spacetime we have $h^{ab} = 0$, so that h^{ab} measures deviations from Minkowski spacetime. In general, however, h^{ab} is neither zero, nor small; in strong-field regions we might have $|h^{ab}| \gtrsim 1$.

Einstein's equations cannot determine a solution for the metric uniquely, since they allow for a choice of four coordinate conditions. If we choose *harmonic* coordinates, also known as *de Donder* coordinates,[4]

$$\partial_b h^{ab} = 0 \tag{E.3}$$

(see Chapter 4.3), we can bring Einstein's equations (E.1) into the form

$$\Box h^{ab} = 16\pi \tau^{ab}. \tag{E.4}$$

Here $\Box \equiv \eta^{cd}\partial_c\partial_d$ is the flat d'Alembertian operator, and the source term

$$\tau^{ab} \equiv -gT^{ab} + \frac{1}{16\pi}\Lambda^{ab} \tag{E.5}$$

contains contributions from both the stress-energy tensor T^{ab} as well as terms that are nonlinear in h^{ab}, which we have absorbed in a new quantity Λ^{ab}. For our purposes it is sufficient to state that Λ^{ab}, the *stress-energy pseudo-tensor*, contains terms that are at least quadratic in h^{ab} as well as its first two derivatives.[5]

We can now write the formal solution to Einstein's equations (E.4) in terms of the Green function for the flat d'Alembertian operator,

$$h^{ab}(t, \mathbf{x}) = -4 \int \frac{\tau^{ab}(\mathbf{x}', t - |\mathbf{x} - \mathbf{x}'|)d^3\mathbf{x}'}{|\mathbf{x} - \mathbf{x}'|}, \tag{E.6}$$

---

[1] We briefly discuss some of these approaches in Chapter 12.

[2] Our brief summary of post-Newtonian methods and results is based on the review by Blanchet (2006), to which we refer the reader for a more detailed treatment, as well as a list of references. We also refer the reader to Blanchet *et al.* (2008) for further results on 3PN gravitational wave polarizations.

[3] Note that some authors introduce these quantities with the opposite sign, which leads to sign differences in many of the following equations.

[4] Compare with the Lorentz gauge condition (9.3), which we introduced in the context of perturbation theory and linearized waves in Chapter 9.1.1.

[5] For a precise definition, see Blanchet (2006).

where, for convenience, we have denoted the three spatial coordinates with a bold-face $\mathbf{x}$. Unfortunately, this solution is not of very much immediate help, since the source term τ^{ab} in the integrand depends on the integral h^{ab}. It forms, however, the basis for the iteration that is involved in constructing post-Newtonian approximations.

We start the iteration with some known Newtonian solution. For example, for a binary this solution could describe two point-masses in a circular orbit. Using this Newtonian solution we can compute a Newtonian stress-energy tensor $T_0^{ab} = T_{\mathrm{Newt}}^{ab}$ and set the corresponding Newtonian values of the gravitational field perturbation amplitudes to zero, $h_0^{ab} = 0$. This determines the Newtonian source term τ_0^{ab}, which we may insert into equation (E.6). Solving the integral then yields the first correction to the gravitational field perturbations, h_1^{ab}. Given these, we can re-evaluate the equations of motion, compute τ_1^{ab}, insert these into the integrand of equation (E.6), and compute the next correction, h_2^{ab}. In general, we can obtain the $(n + 1)$th correction by inserting the previous one into the integrand of equation (E.6),

$$h_{n+1}^{ab}(t, \mathbf{x}) = -4 \int \frac{\tau_n^{ab}(\mathbf{x}', t - |\mathbf{x} - \mathbf{x}'|)d^3\mathbf{x}'}{|\mathbf{x} - \mathbf{x}'|}. \tag{E.7}$$

If all goes well, this iteration converges to give the correct solution $h^{ab}(t, \mathbf{x})$.

While our crude recipe grossly over-simplifies the problem, it does lay out a starting point for constructing a post-Newtonian expansion. Each iteration in the above procedure adds a new correction that improves the previous one by an order $1/c^2$, where, in a rare reappearance in this volume, c is the speed of light. An nth-order post-Newtonian expansion therefore includes terms up to order $1/c^{2n}$; a "3PN" expansion, for example, goes to order $1/c^6$.

Starting out with a purely Newtonian expression we obtain correction terms that are even powers in $1/c$. These terms capture only the so-called conservative effects, but not the radiation-reaction effects. The latter first appear at the odd order $1/c^5$, or in a 2.5PN expansion. We have seen this, for example, in exercise 9.3, where we computed the leading-order gravitational wave luminosity dE/dt for a binary in circular orbit using the weak-field, slow-velocity quadrupole formula. There we expressed the result (9.41) in geometrized units with $c = G = 1$, where G is the gravitational constant; if we now restore these constants we find

$$\frac{dE}{dt} = -\frac{32}{5} \frac{G}{c^5} \mu^2 R^4 \Omega^6, \tag{E.8}$$

demonstrating that the leading-order radiation-reaction effects indeed appear in a 2.5PN expansion. When referring to post-Newtonian expansions of the gravitational radiation, an nPN expansion usually means a correction of order $1/c^{2n}$ *beyond* the leading order quadrupole formula above.

Results for compact binaries

In lieu of pursuing the construction of post-Newtonian expansions in greater detail here, we shall simply summarize some of the most important post-Newtonian results for compact binary inspiral. Binary inspiral has been the focus of considerable attention, given its importance as a promising source of gravitational radiadion.[6]

---

[6] We refer the reader to the review article by Blanchet (2006) for a much more detailed discussion and references.

Label the masses of the two stars m_1 and m_2, and define the total mass as $M = m_1 + m_2$ and the reduced mass as $\mu = m_1 m_2 / M$. It will also be useful to introduce the *symmetric mass ratio*

$$\nu \equiv \frac{\mu}{M} = \frac{m_1 m_2}{(m_1 + m_2)^2}, \tag{E.9}$$

which takes its maximum value of $\nu_{\max} = 1/4$ for equal-mass binaries and approaches zero for extreme mass ratios. We will focus on *nonspinning binaries*[7] in quasicircular orbit, a problem which, at the time of this writing, has been solved up to order 3.5PN for the orbital phase evolution and to order 3PN for the gravitational waveform.[8]

We first write the binary's orbital angular velocity ω as a function of the binary separation R, where R is the coordinate separation in the harmonic gauge (equation E.3). Following Blanchet (2006), we introduce a dimensionless measure of the binary separation

$$\gamma \equiv \frac{G\,M}{c^2\,R} = \mathcal{O}\left(\frac{1}{c^2}\right). \tag{E.10}$$

In terms of this ratio we then obtain

$$\omega^2 = \frac{GM}{R^3}\left\{ 1 + \left(-3 + \nu\right)\gamma + \left(6 + \frac{41}{4}\nu + \nu^2\right)\gamma^2 \right.$$
$$\left. + \left(-10 + \left(-\frac{75707}{840} + \frac{41}{64}\pi^2 + 22\ln\left(\frac{R}{R_0}\right)\right)\nu + \frac{19}{2}\nu^2 + \nu^3\right)\gamma^3 \right\} + \mathcal{O}\left(\frac{1}{c^8}\right). \tag{E.11}$$

This is the 3.5PN version of Kepler's third law in harmonic coordinates; the leading-order term is recognizable as the Newtonian expression.[9] Upon further thought, we realize that the harmonic coordinate conditions (E.3) do not fully specify the coordinates uniquely.[10] This gauge ambiguity is encoded in the term $\ln(R/R_0)$ in the above expression, which includes as yet an undetermined length scale R_0.

The energy E of a circular orbit to the same order is given by

$$E = -\frac{\mu c^2 \gamma}{2}\left\{ 1 + \left(-\frac{7}{4} + \frac{1}{4}\nu\right)\gamma + \left(-\frac{7}{8} + \frac{49}{8}\nu + \frac{1}{8}\nu^2\right)\gamma^2 \right.$$
$$\left. + \left(-\frac{235}{64} + \left(\frac{46031}{2240} - \frac{123}{64}\pi^2 + \frac{22}{3}\ln\left(\frac{R}{R_0}\right)\right)\nu + \frac{27}{32}\nu^2 + \frac{5}{64}\nu^3\right)\gamma^3 \right\} + \mathcal{O}\left(\frac{1}{c^8}\right). \tag{E.12}$$

We again recognize the Newtonian result as the leading-order term.

Both the angular velocity ω and the energy E are gauge-invariant quantities. In equations (E.11) and (E.12), however, both of these quantities still appear to depend on the gauge-dependent length

---

[7] Effects of spin have been included up to order 2.5PN, see Faye *et al.* (2006); Blanchet *et al.* (2006) (and errata).

[8] See Blanchet *et al.* (2008); recall that a 3PN expansion of the gravitational waveforms includes corrections up to order $1/c^6$ *beyond* the weak-field, slow-velocity quadrupole formula (E.8).

[9] Recall that in terms of the areal radius R_s, the angular velocity of a *test* particle in circular orbit about a Schwarzschild black hole is given by $\omega^2 = GM/R_s^3$ exactly, to all orders of γ.

[10] Compare the discussion following equation (9.4) in the context of the Lorentz gauge for linearized waves.

scale R_0. The dependence is an artifact that results from having expressed ω and E in terms of the harmonic coordinate binary separation R, which itself is a gauge-dependent quantity. To eliminate this gauge dependence we express E in terms of ω. For this purpose it is convenient to introduce a dimensionless angular velocity according to

$$x \equiv \left(\frac{GM\omega}{c^3} \right)^{2/3} = \mathcal{O}\left(\frac{1}{c^2} \right).$$ (E.13)

We can now invert equation (E.11) to find γ in terms of x. Substituting this result into equation (E.12) then yields

$$E = -\frac{\mu c^2 x}{2}\left\{ 1 + \left(-\frac{3}{4} - \frac{1}{12}v \right)x + \left(-\frac{27}{8} + \frac{19}{8}v - \frac{1}{24}v^2 \right)x^2 \right.$$
$$\left. + \left(-\frac{675}{64} + \left(\frac{34445}{576} - \frac{205}{96}\pi^2 \right)v - \frac{155}{96}v^2 - \frac{35}{5184}v^3 \right)x^3 \right\} + \mathcal{O}\left(\frac{1}{c^8} \right),$$ (E.14)

which, as expected, is gauge-invariant, and independent of the length scale R_0.

The above expressions hold for binaries in quasicircular orbit. To model the slow, adiabatic inspiral, caused by the emission of gravitational radiation, we also need to compute the gravitational radiation luminosity

$$L_{\text{GW}} = -\frac{dE}{dt}.$$ (E.15)

In order to keep this expression consistent with the 3.5PN expressions above, we need to compute this luminosity beyond the weak-field, slow-velocity expression (E.8) to order 3.5PN. The result, when expressed in terms of x, is

$$L_{\text{GW}} = \frac{32}{5}\frac{c^5}{G}v^2 x^5 \left\{ 1 + \left(-\frac{1247}{336} - \frac{35}{12}v \right)x + 4\pi x^{3/2} + \left(-\frac{44711}{9072} + \frac{9271}{504}v + \frac{65}{18}v^2 \right)x^2 \right.$$
$$+ \left(-\frac{8191}{672} - \frac{583}{24}v \right)\pi x^{5/2}$$
$$+ \left(\frac{6643739519}{69854400} + \frac{16}{3}\pi^2 - \frac{1712}{105}C - \frac{856}{105}\ln(16x) \right)$$
$$+ \left(\frac{41}{48}\pi^2 - \frac{134543}{7776} \right)v - \frac{94403}{3024}v^2 - \frac{775}{324}v^3 \right)x^3$$
$$\left. + \left(-\frac{16285}{504} + \frac{214745}{1728}v + \frac{193385}{3024}v^2 \right)\pi x^{7/2} + \mathcal{O}\left(\frac{1}{c^8} \right) \right\}.$$ (E.16)

Here $C = 0.57722\ldots$ is the Euler constant. It is easy to verify that the leading-order term is the familiar quadrupole formula (E.8).

We can now use the above results to obtain the binary's orbital phase Φ as a function of time. The idea is to integrate

$$\frac{dx}{dt} = \frac{dE/dt}{dE/dx} = -\frac{L_{\text{GW}}}{dE/dx}$$ (E.17)

to find x, and hence ω, as a function of time. Since

$$\frac{d\Phi}{dt} = \omega, \tag{E.18}$$

a further integration then yields the phase Φ. It is convenient to express these results in terms of a dimensionless measure of time,

$$\Theta \equiv \frac{\nu c^3}{5GM}(T - t), \tag{E.19}$$

where T denotes the time of coalescence, at which point the angular speed ω diverges. Clearly, we should abandon post-Newtonian methods well before this time. The first integration now yields x in terms of Θ,

$$
\begin{aligned}
x = \frac{1}{4}\Theta^{-1/4}\Bigg\{ & 1 + \left(\frac{743}{4032} + \frac{11}{48}\nu\right)\Theta^{-1/4} - \frac{1}{5}\pi\Theta^{-3/8} + \left(\frac{19583}{254016} + \frac{24401}{193536}\nu + \frac{31}{288}\nu^2\right)\Theta^{-1/2} \\
& + \left(-\frac{11891}{53760} + \frac{109}{1920}\nu\right)\pi\Theta^{-5/8} + \left(-\frac{10052469856691}{6008596070400} + \frac{1}{6}\pi^2 + \frac{107}{420}C\right. \\
& \left. - \frac{107}{3360}\ln\left(\frac{\Theta}{256}\right) + \left(\frac{3147553127}{780337152} - \frac{451}{3072}\pi^2\right)\nu - \frac{15211}{442368}\nu^2 + \frac{25565}{331776}\nu^3\right)\Theta^{-3/4} \\
& + \left(-\frac{113868647}{433520640} - \frac{31821}{143360}\nu + \frac{294941}{3870720}\nu^2\right)\pi\Theta^{-7/8} + \mathcal{O}\left(\frac{1}{c^8}\right)\Bigg\}. \tag{E.20}
\end{aligned}
$$

It is easy to verify that the leading-order term again agrees with the weak-field, slow-velocity result, which we derived in exercise 12.3. Integrating this expression again then yields the phase Φ in terms of Θ,

$$
\begin{aligned}
\Phi = -\frac{1}{\nu}\Theta^{5/8}\Bigg\{ & 1 + \left(\frac{3715}{8064} + \frac{55}{96}\nu\right)\Theta^{-1/4} - \frac{3}{4}\pi\Theta^{-3/4} + \left(\frac{9275495}{14450688} + \frac{284875}{258048}\nu + \frac{1855}{2048}\nu^2\right)\Theta^{-1/2} \\
& + \left(-\frac{38645}{172032} + \frac{65}{2048}\nu\right)\pi\Theta^{-5/8}\ln\left(\frac{\Theta}{\Theta_0}\right) + \left(\frac{831032450749357}{57682522275840} - \frac{53}{40}\pi^2 - \frac{107}{56}C\right. \\
& \left. + \frac{107}{448}\ln\left(\frac{\Theta}{256}\right) + \left(-\frac{126510089885}{4161798144} + \frac{2255}{2048}\pi^2\right)\nu + \frac{154565}{1835008}\nu^2 - \frac{1179625}{1769472}\nu^3\right)\Theta^{-3/4} \\
& + \left(\frac{188516689}{173408256} + \frac{488825}{516096}\nu - \frac{141769}{516096}\nu^2\right)\pi\Theta^{-7/8} + \mathcal{O}\left(\frac{1}{c^8}\right)\Bigg\}. \tag{E.21}
\end{aligned}
$$

Here Θ_0 is a constant of integration that can be fixed by the initial conditions. It is also possible to combine the last two results and obtain the phase Φ in terms of the angular speed x,

$$
\begin{aligned}
\Phi = -\frac{x^{-5/2}}{32\nu}\Bigg\{ & 1 + \left(\frac{3715}{1008} + \frac{55}{12}\nu\right)x - 10\pi x^{3/2} + \left(\frac{15293365}{1016064} + \frac{27145}{1008}\nu + \frac{3085}{144}\nu^2\right)x^2 \\
& + \left(\frac{38645}{1344} - \frac{65}{16}\nu\right)\pi x^{5/2}\ln\left(\frac{x}{x_0}\right) \\
& + \left(\frac{12348611926451}{18776862720} - \frac{160}{3}\pi^2 - \frac{1712}{21}C - \frac{856}{21}\ln(16x)\right)
\end{aligned}
$$

$$+\left(-\frac{15737765635}{12192768}+\frac{2255}{48}\pi^2\right)v+\frac{76055}{6912}v^2-\frac{127825}{5184}v^3\right)x^3$$

$$+\left(\frac{77096675}{2032128}+\frac{378515}{12096}v-\frac{74045}{6048}v^2\right)\pi x^{7/2}+\mathcal{O}\left(\frac{1}{c^8}\right)\right\},$$ (E.22)

where x_0 is another constant of integration.

Finally we list the two gravitational wave polarization amplitudes h_+ and $h_\times$. Following Blanchet *et al.* (2008) we write these as

$$h_{+,\times}=\frac{2G\mu x}{c^2 r}H_{+,\times}+\mathcal{O}\left(\frac{1}{r^2}\right),$$ (E.23)

where r is the distance to the binary, and expand

$$H_{+,\times}=\sum_{n=0}^{\infty}x^{n/2}H_{+,\times}^{(n/2)}.$$ (E.24)

To order 3PN, the coefficients of the "+" polarization are given by[11]

$$H_+^{(0)}=-(1+c_i^2)\cos 2\psi-\frac{1}{96}s_i^2(17+c_i^2),$$ (E.25)

$$H_+^{(0.5)}=-s_i\,\Delta\left[\cos\psi\left(\frac{5}{8}+\frac{1}{8}c_i^2\right)-\cos 3\psi\left(\frac{9}{8}+\frac{9}{8}c_i^2\right)\right],$$ (E.26)

$$H_+^{(1)}=\cos 2\psi\left[\frac{19}{6}+\frac{3}{2}c_i^2-\frac{1}{3}c_i^4+v\left(-\frac{19}{6}+\frac{11}{6}c_i^2+c_i^4\right)\right]$$
$$-\cos 4\psi\left[\frac{4}{3}s_i^2(1+c_i^2)(1-3v)\right],$$ (E.27)

$$H_+^{(1.5)}=s_i\,\Delta\cos\psi\left[\frac{19}{64}+\frac{5}{16}c_i^2-\frac{1}{192}c_i^4+v\left(-\frac{49}{96}+\frac{1}{8}c_i^2+\frac{1}{96}c_i^4\right)\right]$$
$$+\cos 2\psi\left[-2\pi(1+c_i^2)\right]$$
$$+s_i\,\Delta\cos 3\psi\left[-\frac{657}{128}-\frac{45}{16}c_i^2+\frac{81}{128}c_i^4+v\left(\frac{225}{64}-\frac{9}{8}c_i^2-\frac{81}{64}c_i^4\right)\right]$$
$$+s_i\,\Delta\cos 5\psi\left[\frac{625}{384}s_i^2(1+c_i^2)(1-2v)\right],$$ (E.28)

$$H_+^{(2)}=\pi s_i\,\Delta\cos\psi\left[-\frac{5}{8}-\frac{1}{8}c_i^2\right]$$
$$+\cos 2\psi\left[\frac{11}{60}+\frac{33}{10}c_i^2+\frac{29}{24}c_i^4-\frac{1}{24}c_i^6+v\left(\frac{353}{36}-3c_i^2-\frac{251}{72}c_i^4+\frac{5}{24}c_i^6\right)\right.$$
$$\left.+v^2\left(-\frac{49}{12}+\frac{9}{2}c_i^2-\frac{7}{24}c_i^4-\frac{5}{24}c_i^6\right)\right]$$

---

[11] Note that the nonlinear memory (DC) term is included in the lowest order ("Newtonian") $H_+^{(0)}$ term. For a discussion see Arun *et al.* (2004).

$$+ \pi\, s_i\, \Delta \cos 3\psi \left[\frac{27}{8}(1 + c_i^2) \right]$$

$$+ \cos 4\psi \left[\frac{118}{15} - \frac{16}{5} c_i^2 - \frac{86}{15} c_i^4 + \frac{16}{15} c_i^6 + \nu \left(-\frac{262}{9} + 16\, c_i^2 + \frac{166}{9} c_i^4 - \frac{16}{3} c_i^6 \right) \right.$$

$$\left. + \nu^2 \left(14 - 16\, c_i^2 - \frac{10}{3} c_i^4 + \frac{16}{3} c_i^6 \right) \right]$$

$$+ \cos 6\psi \left[-\frac{81}{40} s_i^4 (1 + c_i^2)\left(1 - 5\nu + 5\nu^2 \right) \right]$$

$$+ s_i\, \Delta \sin \psi \left[\frac{11}{40} + \frac{5 \ln 2}{4} + c_i^2 \left(\frac{7}{40} + \frac{\ln 2}{4} \right) \right]$$

$$+ s_i\, \Delta \sin 3\psi \left[\left(-\frac{189}{40} + \frac{27}{4} \ln(3/2) \right)(1 + c_i^2) \right] , \qquad (E.29)$$

$$H_+^{(2.5)} = s_i\, \Delta \cos \psi \left[\frac{1771}{5120} - \frac{1667}{5120} c_i^2 + \frac{217}{9216} c_i^4 - \frac{1}{9216} c_i^6 \right.$$

$$+ \nu \left(\frac{681}{256} + \frac{13}{768} c_i^2 - \frac{35}{768} c_i^4 + \frac{1}{2304} c_i^6 \right)$$

$$\left. + \nu^2 \left(-\frac{3451}{9216} + \frac{673}{3072} c_i^2 - \frac{5}{9216} c_i^4 - \frac{1}{3072} c_i^6 \right) \right]$$

$$+ \pi \cos 2\psi \left[\frac{19}{3} + 3\, c_i^2 - \frac{2}{3} c_i^4 + \nu \left(-\frac{16}{3} + \frac{14}{3} c_i^2 + 2\, c_i^4 \right) \right]$$

$$+ s_i\, \Delta \cos 3\psi \left[\frac{3537}{1024} - \frac{22977}{5120} c_i^2 - \frac{15309}{5120} c_i^4 + \frac{729}{5120} c_i^6 \right.$$

$$+ \nu \left(-\frac{23829}{1280} + \frac{5529}{1280} c_i^2 + \frac{7749}{1280} c_i^4 - \frac{729}{1280} c_i^6 \right)$$

$$\left. + \nu^2 \left(\frac{29127}{5120} - \frac{27267}{5120} c_i^2 - \frac{1647}{5120} c_i^4 + \frac{2187}{5120} c_i^6 \right) \right]$$

$$+ \cos 4\psi \left[-\frac{16\pi}{3} (1 + c_i^2)\, s_i^2 (1 - 3\nu) \right]$$

$$+ s_i\, \Delta \cos 5\psi \left[-\frac{108125}{9216} + \frac{40625}{9216} c_i^2 + \frac{83125}{9216} c_i^4 - \frac{15625}{9216} c_i^6 \right.$$

$$+ \nu \left(\frac{8125}{256} - \frac{40625}{2304} c_i^2 - \frac{48125}{2304} c_i^4 + \frac{15625}{2304} c_i^6 \right)$$

$$\left. + \nu^2 \left(-\frac{119375}{9216} + \frac{40625}{3072} c_i^2 + \frac{44375}{9216} c_i^4 - \frac{15625}{3072} c_i^6 \right) \right]$$

$$+ \Delta \cos 7\psi \left[\frac{117649}{46080} s_i^5 (1 + c_i^2)(1 - 4\nu + 3\nu^2) \right]$$

$$+ \sin 2\psi \left[-\frac{9}{5} + \frac{14}{5} c_i^2 + \frac{7}{5} c_i^4 + \nu \left(32 + \frac{56}{5} c_i^2 - \frac{28}{5} c_i^4 \right) \right]$$

$$+ s_i^2 (1 + c_i^2) \sin 4\psi \left[\frac{56}{5} - \frac{32 \ln 2}{3} + \nu \left(-\frac{1193}{30} + 32 \ln 2 \right) \right] , \qquad \text{(E.30)}$$

$$H_+^{(3)} = \pi \, \Delta \, s_i \cos \psi \left[\frac{19}{64} + \frac{5}{16} c_i^2 - \frac{1}{192} c_i^4 + \nu \left(-\frac{19}{96} + \frac{3}{16} c_i^2 + \frac{1}{96} c_i^4 \right) \right]$$

$$+ \cos 2\psi \left[-\frac{465497}{11025} + \left(\frac{856 \, C}{105} - \frac{2\pi^2}{3} + \frac{428}{105} \ln(16 \, x) \right) (1 + c_i^2) \right.$$

$$- \frac{3561541}{88200} c_i^2 - \frac{943}{720} c_i^4 + \frac{169}{720} c_i^6 - \frac{1}{360} c_i^8$$

$$+ \nu \left(\frac{2209}{360} - \frac{41\pi^2}{96} (1 + c_i^2) + \frac{2039}{180} c_i^2 + \frac{3311}{720} c_i^4 - \frac{853}{720} c_i^6 + \frac{7}{360} c_i^8 \right)$$

$$+ \nu^2 \left(\frac{12871}{540} - \frac{1583}{60} c_i^2 - \frac{145}{108} c_i^4 + \frac{56}{45} c_i^6 - \frac{7}{180} c_i^8 \right)$$

$$\left. + \nu^3 \left(-\frac{3277}{810} + \frac{19661}{3240} c_i^2 - \frac{281}{144} c_i^4 - \frac{73}{720} c_i^6 + \frac{7}{360} c_i^8 \right) \right]$$

$$+ \pi \, \Delta \, s_i \cos 3\psi \left[-\frac{1971}{128} - \frac{135}{16} c_i^2 + \frac{243}{128} c_i^4 + \nu \left(\frac{567}{64} - \frac{81}{16} c_i^2 - \frac{243}{64} c_i^4 \right) \right]$$

$$+ s_i^2 \cos 4\psi \left[-\frac{2189}{210} + \frac{1123}{210} c_i^2 + \frac{56}{9} c_i^4 - \frac{16}{45} c_i^6 \right.$$

$$+ \nu \left(\frac{6271}{90} - \frac{1969}{90} c_i^2 - \frac{1432}{45} c_i^4 + \frac{112}{45} c_i^6 \right)$$

$$+ \nu^2 \left(-\frac{3007}{27} + \frac{3493}{135} c_i^2 + \frac{1568}{45} c_i^4 - \frac{224}{45} c_i^6 \right)$$

$$\left. + \nu^3 \left(\frac{161}{6} - \frac{1921}{90} c_i^2 - \frac{184}{45} c_i^4 + \frac{112}{45} c_i^6 \right) \right]$$

$$+ \Delta \cos 5\psi \left[\frac{3125 \, \pi}{384} s_i^3 (1 + c_i^2)(1 - 2\nu) \right]$$

$$+ s_i^4 \cos 6\psi \left[\frac{1377}{80} + \frac{891}{80} c_i^2 - \frac{729}{280} c_i^4 + \nu \left(-\frac{7857}{80} - \frac{891}{16} c_i^2 + \frac{729}{40} c_i^4 \right) \right.$$

$$\left. + \nu^2 \left(\frac{567}{4} + \frac{567}{10} c_i^2 - \frac{729}{20} c_i^4 \right) + \nu^3 \left(-\frac{729}{16} - \frac{243}{80} c_i^2 + \frac{729}{40} c_i^4 \right) \right]$$

$$+ \cos 8\psi \left[-\frac{1024}{315} s_i^6 (1 + c_i^2)(1 - 7\nu + 14\nu^2 - 7\nu^3) \right]$$

$$+ \Delta \, s_i \sin \psi \left[-\frac{2159}{40320} - \frac{19 \ln 2}{32} + \left(-\frac{95}{224} - \frac{5 \ln 2}{8} \right) c_i^2 + \left(\frac{181}{13440} + \frac{\ln 2}{96} \right) c_i^4 \right.$$

$$+ v \left(\frac{81127}{10080} + \frac{19 \ln 2}{48} + \left(-\frac{41}{48} - \frac{3 \ln 2}{8} \right) c_i^2 + \left(-\frac{313}{480} - \frac{\ln 2}{48} \right) c_i^4 \right) \Bigg]$$

$$+ \sin 2\psi \left[-\frac{428 \pi}{105} (1 + c_i^2) \right]$$

$$+ \Delta s_i \sin 3\psi \left[\frac{205119}{8960} - \frac{1971}{64} \ln(3/2) + \left(\frac{1917}{224} - \frac{135}{8} \ln(3/2) \right) c_i^2 \right.$$

$$+ \left(-\frac{43983}{8960} + \frac{243}{64} \ln(3/2) \right) c_i^4$$

$$+ v \left(-\frac{54869}{960} + \frac{567}{32} \ln(3/2) + \left(-\frac{923}{80} - \frac{81}{8} \ln(3/2) \right) c_i^2 \right.$$

$$\left. + \left(\frac{41851}{2880} - \frac{243}{32} \ln(3/2) \right) c_i^4 \right) \Bigg]$$

(E.31)

$$+ \Delta s_i^3 (1 + c_i^2) \sin 5\psi \left[-\frac{113125}{5376} + \frac{3125}{192} \ln(5/2) + v \left(\frac{17639}{320} - \frac{3125}{96} \ln(5/2) \right) \right] ,$$

while the coefficients for the cross-polarization are

$$H_\times^{(0)} = -2c_i \sin 2\psi ,$$ (E.32)

$$H_\times^{(0.5)} = s_i c_i \Delta \left[-\frac{3}{4} \sin \psi + \frac{9}{4} \sin 3\psi \right] ,$$ (E.33)

$$H_\times^{(1)} = c_i \sin 2\psi \left[\frac{17}{3} - \frac{4}{3} c_i^2 + v \left(-\frac{13}{3} + 4 c_i^2 \right) \right]$$

$$+ c_i s_i^2 \sin 4\psi \left[-\frac{8}{3} (1 - 3v) \right] ,$$ (E.34)

$$H_\times^{(1.5)} = s_i c_i \Delta \sin \psi \left[\frac{21}{32} - \frac{5}{96} c_i^2 + v \left(-\frac{23}{48} + \frac{5}{48} c_i^2 \right) \right]$$

$$- 4\pi c_i \sin 2\psi$$

$$+ s_i c_i \Delta \sin 3\psi \left[-\frac{603}{64} + \frac{135}{64} c_i^2 + v \left(\frac{171}{32} - \frac{135}{32} c_i^2 \right) \right]$$

$$+ s_i c_i \Delta \sin 5\psi \left[\frac{625}{192} (1 - 2v) s_i^2 \right] ,$$ (E.35)

$$H_\times^{(2)} = s_i c_i \Delta \cos \psi \left[-\frac{9}{20} - \frac{3}{2} \ln 2 \right]$$

$$+ s_i c_i \Delta \cos 3\psi \left[\frac{189}{20} - \frac{27}{2} \ln(3/2) \right]$$

$$- s_i c_i \Delta \left[\frac{3\pi}{4} \right] \sin \psi$$

$$+ c_i \sin 2\psi \left[\frac{17}{15} + \frac{113}{30} c_i^2 - \frac{1}{4} c_i^4 + v \left(\frac{143}{9} - \frac{245}{18} c_i^2 + \frac{5}{4} c_i^4 \right) \right.$$

$$+ v^2 \left(-\frac{14}{3} + \frac{35}{6} c_i^2 - \frac{5}{4} c_i^4 \right) \Bigg]$$

$$+ s_i c_i \, \Delta \sin 3\psi \left[\frac{27\pi}{4} \right]$$

$$+ c_i \sin 4\psi \left[\frac{44}{3} - \frac{268}{15} c_i^2 + \frac{16}{5} c_i^4 + v \left(-\frac{476}{9} + \frac{620}{9} c_i^2 - 16 c_i^4 \right) \right.$$

$$\left. + v^2 \left(\frac{68}{3} - \frac{116}{3} c_i^2 + 16 c_i^4 \right) \right]$$

$$+ c_i \sin 6\psi \left[-\frac{81}{20} s_i^4 (1 - 5v + 5v^2) \right] , \tag{E.36}$$

$$H_\times^{(2.5)} = \frac{6}{5} s_i^2 c_i \, v$$

$$+ c_i \cos 2\psi \left[2 - \frac{22}{5} c_i^2 + v \left(-\frac{282}{5} + \frac{94}{5} c_i^2 \right) \right]$$

$$+ c_i s_i^2 \cos 4\psi \left[-\frac{112}{5} + \frac{64}{3} \ln 2 + v \left(\frac{1193}{15} - 64 \ln 2 \right) \right]$$

$$+ s_i c_i \, \Delta \sin \psi \left[-\frac{913}{7680} + \frac{1891}{11520} c_i^2 - \frac{7}{4608} c_i^4 \right.$$

$$+ v \left(\frac{1165}{384} - \frac{235}{576} c_i^2 + \frac{7}{1152} c_i^4 \right)$$

$$\left. + v^2 \left(-\frac{1301}{4608} + \frac{301}{2304} c_i^2 - \frac{7}{1536} c_i^4 \right) \right]$$

$$+ \pi c_i \sin 2\psi \left[\frac{34}{3} - \frac{8}{3} c_i^2 + v \left(-\frac{20}{3} + 8 c_i^2 \right) \right]$$

$$+ s_i c_i \, \Delta \sin 3\psi \left[\frac{12501}{2560} - \frac{12069}{1280} c_i^2 + \frac{1701}{2560} c_i^4 \right.$$

$$+ v \left(-\frac{19581}{640} + \frac{7821}{320} c_i^2 - \frac{1701}{640} c_i^4 \right)$$

$$\left. + v^2 \left(\frac{18903}{2560} - \frac{11403}{1280} c_i^2 + \frac{5103}{2560} c_i^4 \right) \right]$$

$$+ s_i^2 c_i \sin 4\psi \left[-\frac{32\pi}{3} (1 - 3v) \right]$$

$$+ \Delta s_i c_i \sin 5\psi \left[-\frac{101875}{4608} + \frac{6875}{256} c_i^2 - \frac{21875}{4608} c_i^4 \right.$$

$$+ v \left(\frac{66875}{1152} - \frac{44375}{576} c_i^2 + \frac{21875}{1152} c_i^4 \right)$$

$$\left. + v^2 \left(-\frac{100625}{4608} + \frac{83125}{2304} c_i^2 - \frac{21875}{1536} c_i^4 \right) \right]$$

$$+ \Delta\, s_i^5\, c_i \sin 7\psi \left[\frac{117649}{23040} \left(1 - 4\nu + 3\nu^2 \right) \right], \tag{E.37}$$

$$
\begin{aligned}
H_\times^{(3)} =\ & \Delta\, s_i\, c_i \cos\psi \left[\frac{11617}{20160} + \frac{21}{16} \ln 2 + \left(-\frac{251}{2240} - \frac{5}{48} \ln 2 \right) c_i^2 \right. \\
& \left. + \nu \left(-\frac{48239}{5040} - \frac{5}{24} \ln 2 + \left(\frac{727}{240} + \frac{5}{24} \ln 2 \right) c_i^2 \right) \right]. \\
& + c_i \cos 2\psi \left[\frac{856\,\pi}{105} \right] \\
& + \Delta\, s_i\, c_i \cos 3\psi \left[-\frac{36801}{896} + \frac{1809}{32} \ln(3/2) + \left(\frac{65097}{4480} - \frac{405}{32} \ln(3/2) \right) c_i^2 \right. \\
& \left. + \nu \left(\frac{28445}{288} - \frac{405}{16} \ln(3/2) + \left(-\frac{7137}{160} + \frac{405}{16} \ln(3/2) \right) c_i^2 \right) \right] \\
& + \Delta\, s_i^3\, c_i \cos 5\psi \left[\frac{113125}{2688} - \frac{3125}{96} \ln(5/2) + \nu \left(-\frac{17639}{160} + \frac{3125}{48} \ln(5/2) \right) \right] \\
& + \pi\, \Delta\, s_i\, c_i \sin\psi \left[\frac{21}{32} - \frac{5}{96} c_i^2 + \nu \left(-\frac{5}{48} + \frac{5}{48} c_i^2 \right) \right] \\
& + c_i \sin 2\psi \left[-\frac{3620761}{44100} + \frac{1712\,C}{105} - \frac{4\,\pi^2}{3} + \frac{856}{105} \ln(16\,x) \right. \\
& \qquad - \frac{3413}{1260} c_i^2 + \frac{2909}{2520} c_i^4 - \frac{1}{45} c_i^6 \\
& \qquad + \nu \left(\frac{743}{90} - \frac{41\,\pi^2}{48} + \frac{3391}{180} c_i^2 - \frac{2287}{360} c_i^4 + \frac{7}{45} c_i^6 \right) \\
& \qquad + \nu^2 \left(\frac{7919}{270} - \frac{5426}{135} c_i^2 + \frac{382}{45} c_i^4 - \frac{14}{45} c_i^6 \right) \\
& \qquad \left. + \nu^3 \left(-\frac{6457}{1620} + \frac{1109}{180} c_i^2 - \frac{281}{120} c_i^4 + \frac{7}{45} c_i^6 \right) \right] \\
& + \pi\, \Delta\, s_i\, c_i \sin 3\psi \left[-\frac{1809}{64} + \frac{405}{64} c_i^2 + \nu \left(\frac{405}{32} - \frac{405}{32} c_i^2 \right) \right] \\
& + s_i^2\, c_i \sin 4\psi \left[-\frac{1781}{105} + \frac{1208}{63} c_i^2 - \frac{64}{45} c_i^4 \right. \\
& \qquad + \nu \left(\frac{5207}{45} - \frac{536}{5} c_i^2 + \frac{448}{45} c_i^4 \right) \\
& \qquad + \nu^2 \left(-\frac{24838}{135} + \frac{2224}{15} c_i^2 - \frac{896}{45} c_i^4 \right) \\
& \qquad \left. + \nu^3 \left(\frac{1703}{45} - \frac{1976}{45} c_i^2 + \frac{448}{45} c_i^4 \right) \right] \\
& + \Delta \sin 5\psi \left[\frac{3125\,\pi}{192} s_i^3\, c_i (1 - 2\nu) \right]
\end{aligned}
$$

$$+ s_i^4 c_i \sin 6\psi \left[\frac{9153}{280} - \frac{243}{35} c_i^2 + \nu \left(-\frac{7371}{40} + \frac{243}{5} c_i^2 \right) \right.$$
$$+ \nu^2 \left(\frac{1296}{5} - \frac{486}{5} c_i^2 \right) + \nu^3 \left(-\frac{3159}{40} + \frac{243}{5} c_i^2 \right) \right]$$
$$+ \sin 8\psi \left[-\frac{2048}{315} s_i^6 c_i (1 - 7\nu + 14\nu^2 - 7\nu^3) \right] . \tag{E.38}$$

We have used several new quantities and abbreviations in the above expressions. The phase variable ψ is related to the orbital phase Φ by

$$\psi = \Phi - \frac{2G M_{\mathrm{ADM}}\omega}{c^3} \ln \left(\frac{\omega}{\omega_0} \right) , \tag{E.39}$$

where

$$M_{\mathrm{ADM}} = m \left(1 - \frac{\nu\gamma}{2} \right) + \mathcal{O}\left(\frac{1}{c^4} \right) \tag{E.40}$$

is the ADM mass, and where ω_0 is some arbitrary constant frequency. It could be chosen, for example, to be the entry frequency of a gravitational wave detector. We also use the dimensionless mass difference $\Delta = (m_1 - m_2)/M$, and the abbreviations $c_i = \cos i$ and $s_i = \sin i$, where i is the inclination angle between the binary's axis of rotation and the direction to the observer. The waveforms can also be decomposed into spin-weighted tensor spherical harmonics, but we shall refer the reader elsewhere for the expressions and further details.[12]

Before closing this appendix we mention two alternative "resummation" techniques that have been suggested to enhance the convergence of post-Newtonian approximations. Our description above outlines the construction of a post-Newtonian approximation as a Taylor expansion in the parameters γ or x (see equations E.10 or E.13). Such a polynomial expansion can converge well only in the absence of poles in the functions that we aim to approximate. For small binary separations, however, poles may appear. It has therefore been suggested that one construct, from the information contained in the polynomial expansion, an expansion in terms of rational functions.[13] The resulting expansion is an example of a so-called *Padé* approximant, which sometimes converges faster than the corresponding Taylor expansion.[14]

In yet another approach, the general relativistic binary problem is mapped onto that of a test particle moving in an effective external metric, thereby reducing a two-body problem to an effective one-body (EOB) problem.[15] Starting with a post-Newtonian expansion for the relativistic two-body equations of motion, this approach may be viewed as a nonperturbative means of resumming the post-Newtonian expressions. In Chapter 12.4 we have included EOB results in our comparison with post-Newtonian predictions for locating the innermost stable circular orbit for binary black holes.

---

[12] See Blanchet *et al.* (2008).
[13] See Damour *et al.* (1998).
[14] See, e.g., Press *et al.* (2007) for a brief introduction.
[15] See Buonanno and Damour (1999).

F | Collisionless matter evolution in axisymmetry: basic equations

Here we list the key equations describing a mean-field, particle simulation scheme that can treat the evolution of collisionless matter in axisymmetry according to general relativity.[1] The scheme is a generalization of the one described in Chapter 8.2 for spherical systems and employs the standard ADM form of the field equations as listed in Box 2.1. We adopt spherical polar spacetime coordinates (t, r, θ, ϕ), assume axisymmetry and specialize to the case where there is no net angular momentum.[2] In axisymmetry all quantities are functions only of (t, r, θ). We also impose maximal slicing and quasi-isotropic spatial coordinates as our gauge conditions.[3] This spatial gauge condition reduces to isotropic coordinates for Schwarzschild geometry.[4] The field equations listed below constitute a fully constrained approach to solving the Einstein field equations for this problem, i.e., one which solves all of the constraint equations in lieu of integrating evolution equations for some of the variables. Fully constraint schemes have the advantage over unconstrained schemes that the constraints are guaranteed to be satisfied at all times, which may in some cases also eliminate some instabilities associated with the evolution equations. Their disadvantage is that the constraints constitute elliptic equations, which typically require more computational resources to solve than explicit time evolution equations. This disadvantage is not so severe, however, in $1 + 1$ or $2 + 1$ spacetimes. A similar set of variables and field equations to the ones summarized below has been used to simulate the gravitational collapse of hydrodynamic fluids[5] and vacuum gravitational waves[6] in nonrotating, axisymmetric spacetimes, as well as the head-on collision of neutron stars.[7]

Gravitational field equations

The metric is written as

$$ds^2 = -\alpha^2 dt^2 + A^2(dr + \beta^r dt)^2 + A^2 r^2(d\theta + \beta^\theta dt)^2 + B^2 r^2 \sin^2\theta \, d\phi^2. \tag{F.1}$$

Here we used the fact that the absence of axial rotation demands invariance with respect to all changes $\phi \to -\phi$, which implies $\beta^\phi = 0 = \gamma_{r\phi} = \gamma_{\theta\phi}$. The simplified quasi-isotropic form of

---

[1] Shapiro and Teukolsky (1991b, 1992a,b).
[2] Evans (1984). For extension to rotating spacetimes, see Abrahams *et al.* (1994).
[3] See Chapter 4 for a discussion.
[4] For an alternative radial gauge choice, which reduces to Schwarzschild (areal) coordinates, see Bardeen and Piran (1983) and Chapter 4.
[5] See, e.g., Evans (1986).
[6] Abrahams and Evans (1993).
[7] Abrahams and Evans (1992).

the 3-metric $\gamma_{ij} = \mathrm{diag}(A^2, A^2 r^2, B^2 r^2 \sin^2\theta)$ results from imposing the gauge conditions

$$\partial_t \gamma_{r\theta} = 0, \qquad \partial_t (r^2 \gamma_{rr} - \gamma_{\theta\theta}) = 0, \tag{F.2}$$

with initial conditions

$$\gamma_{r\theta} = 0, \qquad r^2 \gamma_{rr} - \gamma_{\theta\theta} = 0. \tag{F.3}$$

In addition to the metric coefficients α, β^r, β^θ, A and B, we need to determine the components of the extrinsic curvature tensor, $K^i{}_j$. It is convenient to introduce the related quantities

$$\hat{K}^i{}_j \equiv A^2 B K^i{}_j, \tag{F.4}$$

and similarly for all other variables with a caret (i.e., all quantities with a caret are related to "uncareted" quantities by a factor $A^2 B$).

The evolution equations are used to determine the "radiation variables" $\eta \equiv \ln(A/B)$ and $\hat{K}^r{}_\theta$ according to

$$\partial_t \eta = \frac{\alpha}{A^2 B} \hat{\lambda} + \beta^r \partial_r \eta + \beta^\theta \partial_\theta \eta + \partial_\theta \beta^\theta - \beta^\theta \cot\theta, \tag{F.5}$$

$$\begin{aligned}
\partial_t \hat{K}^r{}_\theta =\ & \frac{1}{r^2} \partial_r (r^2 \beta^r \hat{K}^r{}_\theta) + \frac{1}{\sin\theta} \partial_\theta (\sin\theta\, \beta^\theta\, \hat{K}^r{}_\theta) \\
& - \frac{\alpha A}{r \sin\theta} \left\{ \partial_\theta \left[\frac{\sin\theta}{A} \partial_r (Br) \right] - \frac{1}{A^2} \partial_r (Ar) \partial_\theta (B \sin\theta) \right\} \\
& + \hat{K}^r{}_\theta (\partial_\theta \beta^\theta - \partial_r \beta^r) + (2\hat{\lambda} - 3\hat{K}^\phi{}_\phi) \partial_\theta \beta^r - AB \partial_\theta \left(\frac{1}{A} \partial_r \alpha \right) \\
& + \frac{B}{Ar} \partial_r (Ar) \partial_\theta \alpha - \frac{\alpha}{A^2} \hat{S}_{r\theta}, \tag{F.6}
\end{aligned}$$

where

$$\hat{\lambda} \equiv \hat{K}^r{}_r + 2\hat{K}^\phi{}_\phi, \tag{F.7}$$

and where matter source terms like $\hat{S}_{r\theta}$ will be defined below.

The momentum constraint equations, which we use to solve for $\hat{K}^r{}_r$, are

$$\frac{1}{\sin\theta} \partial_\theta (\sin\theta\, \hat{K}^r{}_r) + \frac{T}{\sin^2\theta} \partial_\theta \left(\frac{\sin^2\theta}{T} \hat{K}^\phi{}_\phi \right) = -\hat{S}_\theta + \frac{1}{r^2} \partial_r (r^2 \hat{K}^r{}_\theta), \tag{F.8}$$

$$\frac{1}{r^3} \partial_r (r^3 \hat{K}^r{}_r) + \hat{K}^\phi{}_\phi \partial_r \eta = \hat{S}_r - \frac{1}{r^2 \sin\theta} \partial_\theta (\sin\theta\, \hat{K}^r{}_\theta), \tag{F.9}$$

where $T \equiv A/B$. The Hamiltonian constraint yields an elliptic equation for $\psi \equiv B^{1/2}$,

$$\begin{aligned}
\frac{1}{r^2} \partial_r (r^2 \partial_r \psi) + \frac{1}{r^2 \sin\theta} \partial_\theta (\sin\theta\, \partial_\theta \psi) =\ & -\frac{1}{4} \psi \left[\frac{1}{r} \partial_r (r \partial_r \eta) + \frac{1}{r^2} \partial_\theta^2 \eta \right. \\
& \left. + \frac{1}{T^2 \psi^8} \left\{ \hat{\lambda}^2 - 3\hat{\lambda}\hat{K}^\phi{}_\phi + 3(\hat{K}^\phi{}_\phi)^2 + \left(\hat{K}^r{}_\theta / r \right)^2 \right\} \right] - \frac{\hat{\rho}}{4\psi}. \tag{F.10}
\end{aligned}$$

Maximal slicing $\partial_t K = 0 = K$ yields an elliptic equation for the lapse, which we combine with the Hamiltonian constraint (F.10) to obtain an elliptic equation for $\alpha\psi$,

$$\begin{aligned}
\frac{1}{r^2} \partial_r [r^2 \partial_r (\alpha\psi)] + \frac{1}{r^2 \sin\theta} \partial_\theta [\sin\theta\, \partial_\theta (\alpha\psi)] =\ & \frac{1}{4} \alpha\psi \left[-\frac{1}{r} \partial_r (r \partial_r \eta) - \frac{1}{r^2} \partial_\theta^2 \eta \right. \\
& + \frac{7}{A^2 B^2} \left\{ \hat{\lambda}^2 - 3\hat{\lambda}\hat{K}^\phi{}_\phi + 3(\hat{K}^\phi{}_\phi)^2 + \left(\hat{K}^r{}_\theta / r \right)^2 \right\} \\
& \left. + \frac{1}{B} (\hat{\rho} + 2\hat{S}) \right]. \tag{F.11}
\end{aligned}$$

An attractive feature of this equation for the lapse is that the simple flat space form of the 3-dimensional Laplacian appears on the left hand side. Imposing the quasi-isotropic spatial gauge conditions (F.2) leads to a system of coupled first-order equations for the two shift components,[8] β^r and β^θ:

$$r\partial_r\left(\frac{\beta^r}{r}\right) - \partial_\theta\beta^\theta = \frac{\alpha}{A^2 B}(2\hat{\lambda} - 3\hat{K}^\phi_{\ \phi}), \tag{F.12}$$

$$r\partial_r\beta^\theta + \partial_\theta\left(\frac{\beta^r}{r}\right) = \frac{2\alpha}{A^2 Br}\hat{K}^r_{\ \theta}. \tag{F.13}$$

In spherical symmetry, the above equations reduce to those presented in Chapter 8.2. The variable η is a measure of the deviation from spherical symmetry since $\eta = 0$ in spherical spacetimes. In an asymptotically flat, axisymmetric spacetime, η also measures the even-parity gravitational wave amplitude at large distances according to

$$\eta = h^{TT}_+ + \mathcal{O}(r^{-2}). \tag{F.14}$$

According to equation (F.5), we also have $\partial_t\eta \to \lambda$ as $r \to \infty$. With $h^{TT}_\times = 0$ in the absence of rotation, we can then use equation (9.30) to compute the energy loss due to gravitational radiation,

$$\frac{dE}{dt} = -\frac{1}{16\pi}\lim_{r\to\infty}r^2\int d\Omega\lambda^2. \tag{F.15}$$

No angular momentum is carried off by gravitational waves in axisymmetry.

Matter equations

The geodesic equations of motion for the collisionless particles are:

$$\frac{dr}{dt} = \frac{\alpha u_r}{\Gamma A^2} - \beta^r, \tag{F.16}$$

$$\frac{d\theta}{dt} = \frac{\alpha u_\theta}{\Gamma A^2 r^2} - \beta^\theta, \tag{F.17}$$

$$\frac{d\phi}{dt} = \frac{\alpha}{B^2 r^2\sin^2\theta}\frac{u_\phi}{\Gamma}, \tag{F.18}$$

$$\frac{du_r}{dt} = -\Gamma\partial_r\alpha + u_r\partial_r\beta^r + u_\theta\partial_r\beta^\theta$$
$$- \frac{\alpha}{2\Gamma}\left[\partial_r\left(\frac{1}{A^2}\right)u_r^2 + \partial_r\left(\frac{1}{r^2 A^2}\right)u_\theta^2 + \partial_r\left(\frac{1}{B^2 r^2}\right)\frac{u_\phi^2}{\sin^2\theta}\right], \tag{F.19}$$

$$\frac{du_\theta}{dt} = -\Gamma\partial_\theta\alpha + u_r\partial_\theta\beta^r + u_\theta\partial_\theta\beta^\theta$$
$$- \frac{\alpha}{2\Gamma}\left[\partial_\theta\left(\frac{1}{A^2}\right)u_r^2 + \partial_\theta\left(\frac{1}{r^2 A^2}\right)u_\theta^2 + \partial_\theta\left(\frac{1}{B^2 r^2\sin^2\theta}\right)u_\phi^2\right], \tag{F.20}$$

$$\frac{du_\phi}{dt} = 0, \tag{F.21}$$

---

[8] See Evans (1984), p. 121, for a proof that this is an elliptic system.

where the normalization condition $u^\mu u_\mu = -1$ gives

$$\Gamma \equiv \alpha u^0 = \left[1 + \frac{u_r^2}{A^2} + \frac{u_\theta^2}{r^2 A^2} + \frac{u_\phi^2}{B^2 r^2 \sin^2\theta} \right]^{1/2}. \tag{F.22}$$

Assuming that all N particles have the same rest mass $m = M_0/N$, where M_0 is the total rest mass of the system, they may be binned to determine the source terms for the field equations as follows:

$$\hat{\rho} = \sum_j \frac{m\Gamma_j}{(r^2 \sin\theta \, \Delta r \, \Delta\theta \, \Delta\phi)_j}, \tag{F.23}$$

$$\hat{S}_r = \sum_j \frac{m u_r^j}{(r^2 \sin\theta \, \Delta r \, \Delta\theta \, \Delta\phi)_j}, \tag{F.24}$$

$$\hat{S}_\theta = \sum_j \frac{m u_\theta^j}{(r^2 \sin\theta \, \Delta r \, \Delta\theta \, \Delta\phi)_j}, \tag{F.25}$$

$$\hat{S}_{r\theta} = \sum_j \frac{m u_r^j u_\theta^j}{\Gamma_j (r^2 \sin\theta \, \Delta r \, \Delta\theta \, \Delta\phi)_j}, \tag{F.26}$$

$$\hat{S} = \hat{\rho} - \sum_j \frac{m}{\Gamma_j (r^2 \sin\theta \, \Delta r \, \Delta\theta \, \Delta\phi)_j}. \tag{F.27}$$

Since the particles are distributed axisymmetrically in rings the bin width $\Delta\phi$ can be set to 2π.

Disks

The above scheme is easily adapted to treat infinitely thin, axisymmetric disks of matter residing in the equatorial plane.[9] In this case, "jump" conditions satisfied by the field variables across the equator replace the usual matter source terms that appear in the field equations. As an example of how the jump conditions are obtained, consider equation (F.9). Functions like $\hat{K}_r^r$, $\hat{K}_\phi^\phi$ and η are symmetric across the equatorial plane and continuous there. Multiplying the equation by $r\sin\theta$ and integrating across the equator therefore yields

$$0 = \int_-^+ \hat{S}_r r \sin\theta d\theta - \frac{1}{r} \hat{K}_\theta^r |_-^+, \tag{F.28}$$

where $\pm$ denotes $\theta = \pi/2 \pm \epsilon, \epsilon \to 0$. Since $\hat{K}_\theta^r$ is antisymmetric across the equator, equation (F.28) gives

$$\hat{K}_\theta^r |^+ = -\hat{K}_\theta^r |^- = \frac{r}{2} \int_-^+ \hat{S}_r r \sin\theta d\theta. \tag{F.29}$$

The jump condition (F.29) is used to set the value of $\hat{K}_\theta^r$ at the boundary of an integration along the equatorial plane. In the vacuum outside of the equatorial plane, $\hat{K}_\theta^r$ is determined by evolving equation (F.6) in the usual way. Since particles are confined to the equatorial plane where $\beta^\theta = 0$,

[9] Abrahams *et al.* (1994, 1995).

the particle 4-velocity satisfies $u^\theta = u_\theta = 0$, hence $S_\theta = 0$. Integrating equation (F.8) across the equator then reduces to $0 = 0$.

In a similar fashion, integration of the Hamiltonian constraint equation (F.10) gives

$$\frac{1}{r}\sin\theta \, \partial_\theta \psi\,|^+_- = -\frac{1}{4r}\psi \, \partial_\theta \eta\,|^+_- - \frac{1}{8\psi}\int_-^+ \hat{\rho}r \, \sin\theta d\theta. \tag{F.30}$$

Integrating the lapse equation (F.11) yields

$$\frac{1}{r}\sin\theta \, \partial_\theta(\alpha\psi)\,|^+_- = -\frac{1}{4r}\alpha\psi \, \partial_\theta \eta\,|^+_- + \frac{1}{8}\frac{\alpha\psi}{B}\int_-^+ (\hat{\rho} + 2\hat{S})r \, \sin\theta d\theta. \tag{F.31}$$

In finite differencing, say, the Hamiltonian constraint, the derivative terms $\partial_\theta \psi$ and $\partial_\theta \eta$ appear in exactly the same combination as in the jump condition (F.30). Thus the only place where the matter source term appears is through the boundary conditions. The same result holds for the lapse equation.

The evolution equation (F.5) for η and the shift equations (F.12) and (F.13) for β^r and β^ϕ do not contain any matter sources and remain unchanged. We recall that η, β^r and β^ϕ are metric coefficients and must be continuous across the equator.

The geodesic equations for the particles are unchanged, except for the simplification that $\theta = \pi/2$ and $u_\theta = 0$. Accordingly, only the radial motion of the particles is dynamical in an infinitely thin, axisymmetric disk, as in spherical symmetry.

The matter source terms appearing in the jump conditions can be determined by binning the particles in radial annuli, yielding

$$\hat{\sigma} \equiv \int_-^+ \hat{\rho} \, \sin\theta d\theta = \sum_j \frac{m\Gamma_j}{(2\pi r \Delta r)_j}, \tag{F.32}$$

$$\hat{\Sigma}_r \equiv \int_-^+ \hat{S}_r r \, \sin\theta d\theta = \sum_j \frac{m(u_r)_j}{(2\pi r \Delta r)_j}, \tag{F.33}$$

$$\hat{\Sigma} \equiv \int_-^+ \hat{S}r \, \sin\theta d\theta = \hat{\sigma} - \sum_j \frac{m}{\Gamma_j(2\pi r \Delta r)_j}. \tag{F.34}$$

Once again, the particle rest mass m is related to the total rest mass M_0 by $m = M_0/N$, where N is the total particle number. The rest mass can be found from

$$M_0 = \int \hat{\sigma}_0 2\pi r dr, \tag{F.35}$$

where

$$\hat{\sigma}_0 \equiv \int_-^+ \left(\frac{\hat{\rho}}{\Gamma}\right)r \, \sin\theta d\theta = \sum_j \frac{m}{(2\pi r \Delta r)_j}. \tag{F.36}$$

G   Rotating equilibria: gravitational field equations

In this appendix we assemble the gravitational field equations that are required to construct numerical models of rotating, relativistic, equilibrium configurations, such as the rotating fluid stars discussed in Chapter 14 and the rotating collisionless clusters discussed in Chapters 10 and 14.

Following Komatsu *et al.* (1989a) we assume stationary equilibrium and write the spacetime metric in the form (14.1),

$$ds^2 = -e^{\gamma+\rho}dt^2 + e^{2\sigma}(dr^2 + r^2 d\theta^2) + e^{\gamma-\rho}r^2 \sin^2\theta(d\phi - \omega dt)^2, \tag{G.1}$$

where the metric potentials ρ, γ, ω, and σ are functions of r and θ only. Einstein's equations $G_{ab} = 8\pi T_{ab}$ then provide three elliptic equations (14.2)–(14.4) for the metric potentials ρ, γ and ω,

$$\nabla^2[\rho e^{\gamma/2}] = S_\rho(r, \mu), \tag{G.2}$$

$$\left(\nabla^2 + \frac{1}{r}\partial_r - \frac{\mu}{r^2}\partial_\mu\right)[\gamma e^{\gamma/2}] = S_\gamma(r, \mu), \tag{G.3}$$

$$\left(\nabla^2 + \frac{2}{r}\partial_r - \frac{2\mu}{r^2}\partial_\mu\right)[\omega e^{(\gamma-2\rho)/2}] = S_\omega(r, \mu), \tag{G.4}$$

where ∇^2 is the flat-space Laplace operator in spherical polar coordinates and $\mu = \cos\theta$. The effective sources terms S_ρ, S_γ and S_ω are given by

$$S_\rho(r, \mu) = e^{\gamma/2}\left\{\frac{1}{r}\partial_r\gamma - \frac{\mu}{r^2}\partial_\mu\gamma + \frac{\rho}{2}\left[-\partial_r\gamma\left(\frac{1}{2}\partial_r\gamma + \frac{1}{r}\right) - \frac{1}{r^2}\partial_\mu\gamma\left(\frac{1-\mu^2}{2}\partial_\mu\gamma - \mu\right)\right]\right.$$

$$+ r^2(1-\mu^2)e^{-2\rho}\left((\partial_r\omega)^2 + \frac{1-\mu^2}{r^2}(\partial_\mu\omega)^2\right)$$

$$\left. + 8\pi e^{2\sigma}\left[T^{\hat{\phi}}{}_{\hat{\phi}} - T^{\hat{t}}{}_{\hat{t}} + \frac{\rho}{2}(T^{\hat{r}}{}_{\hat{r}} + T^{\hat{\theta}}{}_{\hat{\theta}})\right]\right\} \tag{G.5}$$

$$S_\gamma(r, \mu) = e^{\gamma/2}\left\{\frac{\gamma}{2}\left[-\frac{1}{2}(\partial_r\gamma)^2 - \frac{1-\mu^2}{2r^2}(\partial_\mu\gamma)^2\right] + 8\pi e^{2\sigma}\left(1 + \frac{\gamma}{2}\right)(T^{\hat{r}}{}_{\hat{r}} + T^{\hat{\theta}}{}_{\hat{\theta}})\right\}, \tag{G.6}$$

$$S_\omega(r, \mu) = e^{(\gamma-2\rho)/2}\left\{\omega\left[-\frac{1}{r}\left(2\partial_r\rho + \frac{1}{2}\partial_r\gamma\right) + \frac{\mu}{r^2}\left(2\partial_\mu\rho + \frac{1}{2}\partial_\mu\gamma\right) + \frac{1}{4}\left(4(\partial_r\rho)^2 - (\partial_r\gamma)^2\right)\right.\right.$$

$$\left. + \frac{1-\mu^2}{4r^2}\left(4(\partial_\mu\rho)^2 - (\partial_\mu\gamma)^2\right) - r^2(1-\mu^2)e^{-2\rho}\left((\partial_r\omega)^2 + \frac{1-\mu^2}{r^2}(\partial_\mu\omega)^2\right)\right]$$

$$\left. + 8\pi e^{2\sigma}\left[-\omega(T^{\hat{\phi}}{}_{\hat{\phi}} - T^{\hat{t}}{}_{\hat{t}}) + \frac{\omega}{2}(T^{\hat{r}}{}_{\hat{r}} + T^{\hat{\theta}}{}_{\hat{\theta}}) - \frac{2e^\rho T^{\hat{t}}{}_{\hat{\phi}}}{r(1-\mu^2)^{1/2}}\right]\right\}. \tag{G.7}$$

Here $T^{\hat{a}}{}_{\hat{b}}$ are the orthonormal components of the stress-energy tensor for matter as observed by a normal observer, n^a. In the literature dealing with stationary, rotating equilibria, normal observers are often referred to as *zero angular momentum observers* (ZAMOs).[1] We purposely leave the stress-energy tensor unspecified to allow for different matter models[2] (see Chapter 5 for some astrophysically relevant examples).

The fourth field equation determines σ and is given by

$$\partial_\mu \sigma = -\frac{1}{2}(\partial_\mu \rho + \partial_\mu \gamma) - \{(1-\mu^2)(1+r\partial_r\gamma)^2 + [\mu - (1-\mu^2)\partial_\mu\gamma]^2\}^{-1}$$

$$\times \left\{ \frac{1}{2}[r^2(\partial_r^2\gamma + (\partial_r\gamma)^2) - (1-\mu^2)(\partial_\mu^2\gamma + (\partial_\mu\gamma)^2)][-\mu + (1-\mu^2)\partial_\mu\gamma] \right.$$

$$+ r\partial_r\gamma\left[\frac{1}{2}\mu + \mu r\partial_r\gamma + \frac{1}{2}(1-\mu^2)\partial_\mu\gamma\right] + \frac{3}{2}\partial_\mu\gamma[-\mu^2 + \mu(1-\mu^2)\partial_\mu\gamma]$$

$$- r(1-\mu^2)\left(\partial_r\partial_\mu\gamma + (\partial_r\gamma)(\partial_\mu\gamma)\right)(1+r\partial_r\gamma) - \frac{1}{4}\mu r^2(\partial_r\rho + \partial_r\gamma)^2$$

$$- \frac{r}{2}(1-\mu^2)(\partial_r\rho + \partial_r\gamma)(\partial_\mu\rho + \partial_\mu\gamma) + \frac{1}{4}\mu(1-\mu^2)(\partial_\mu\rho + \partial_\mu\gamma)^2$$

$$- \frac{r^2}{2}(1-\mu^2)\partial_r\gamma(\partial_r\rho + \partial_r\gamma)(\partial_\mu\rho + \partial_\mu\gamma)$$

$$+ \frac{1}{4}(1-\mu^2)\partial_\mu\gamma[r^2(\partial_r\rho + \partial_r\gamma)^2 - (1-\mu^2)(\partial_\mu\rho + \partial_\mu\gamma)^2]$$

$$+ (1-\mu^2)e^{-2\rho}\left[\frac{1}{4}r^4\mu(\partial_r\omega)^2 + \frac{1}{2}r^3(1-\mu^2)(\partial_r\omega)(\partial_\mu\omega) - \frac{1}{4}r^2\mu(1-\mu^2)(\partial_\mu\omega)^2\right.$$

$$\left.+ \frac{1}{2}r^4(1-\mu^2)(\partial_r\gamma)(\partial_r\omega)(\partial_\mu\omega) - \frac{1}{4}r^2(1-\mu^2)\partial_\mu\gamma[r^2(\partial_r\omega)^2 - (1-\mu^2)(\partial_\mu\omega)^2]\right]$$

$$\left.- r^2[\mu - (1-\mu^2)\partial_\mu\gamma]e^{2\sigma}4\pi(T^{\hat{r}}{}_{\hat{r}} - T^{\hat{\theta}}{}_{\hat{\theta}}) + r^2(1-\mu^2)^{1/2}(1+r\partial_r\gamma)e^{2\sigma}8\pi T^{\hat{r}}{}_{\hat{\theta}} \right\}.$$

$$(G.8)$$

The equations for the metric components assembled by Komatsu *et al.* (1989a) and listed above are most easily derived by projecting the Einstein field equations onto the orthonormal tetrad of a ZAMO.

Following Komatsu *et al.* (1989a), the integral Green function solutions to the three elliptical field equations (G.2)–(G.4) are

$$\rho = -e^{-\gamma/2}\sum_{n=0}^{\infty}\int_0^\infty dr' \int_0^1 d\mu' r'^2 f_{2n}^2(r,r')P_{2n}(\mu)P_{2n}(\mu')S_\rho(r',\mu'),\qquad (G.9)$$

$$\gamma = -\frac{2}{\pi}\frac{e^{-\gamma/2}}{r\sin\theta}\sum_{n=1}^{\infty}\int_0^\infty dr'\int_0^1 d\mu' r'^2 \frac{f_{2n-1}^1(r,r')}{2n-1}\sin((2n-1)\theta)\sin\left((2n-1)\theta'\right)S_\gamma(r',\mu'),$$

$$(G.10)$$

---

[1] Bardeen (1973).

[2] Shapiro and Teukolsky (1993a).

$$\omega = -\frac{e^{(2\rho-\gamma)/2}}{r\sin\theta}\sum_{n=1}^{\infty}\int_0^\infty dr'\int_0^1 d\mu' r'^3\sin\theta'\frac{f_{2n-1}^2(r,r')}{2n(2n-1)}P_{2n-1}^1(\mu)P_{2n-1}^1(\mu')S_\omega(r',\mu'),$$

(G.11)

where

$$f_n^1(r,r') = \begin{cases} (r'/r)^n, & \text{for } r'/r \le 1, \\ (r/r')^n, & \text{for } r'/r > 1, \end{cases}$$

(G.12)

and

$$f_n^2(r,r') = \begin{cases} (1/r)(r'/r)^n, & \text{for } r'/r \le 1, \\ (1/r')(r/r')^n, & \text{for } r'/r > 1. \end{cases}$$

(G.13)

Here the P_n are Legendre polynomials and the P_n^m are associated Legendre functions. Equation (G.8) for σ is solved by integrating the linear ordinary differential equation from the pole ($\mu = 1$) to the equator with the initial condition that

$$\sigma = \frac{\gamma - \rho}{2} \quad \text{at} \quad \mu = 1,$$

(G.14)

which arises from the requirement of local flatness on the coordinate axis.

The coupled equations for the gravitational and matter fields can be solved by a stable iteration scheme like the one described in Komatsu *et al.* (1989a) for fluid stars.[3] It is modeled after the approach devised by Hachisu (1986) for Newtonian systems. The integral equations can be solved on a discrete 2-dimensional spatial grid covering the computational domain over a large but finite radius, $0 \le r \le r_{\max}$ and $0 \le \mu \le 1$. In the implementation by Cook *et al.* (1992), a coordinate transformation to a new radial coordinate s, defined by

$$r = r_e\frac{s}{1-s}, \quad 0 \le s \le 1,$$

(G.15)

allows us to extend the computational domain to spatial infinity ($r_{\max} = \infty$). Here r_e is the radius of the surface of the matter at the equator. Derivatives of quantities appearing in the effective source terms are handled by finite differencing.

---

[3] See Chapter 14.1.2 for a brief sketch.

Moving puncture representions of Schwarzschild: analytical results

As we discussed in Chapter 13.1.3, moving puncture simulations typically employ the 1+log slicing condition (4.51),

$$\partial_t \alpha - \beta^i \partial_i \alpha = -2\alpha K. \tag{H.1}$$

We also pointed out that moving puncture simulations that start with Schwarzschild initial data – specifically, the $v = 0$ time slice of Schwarzschild in a Kruskal–Szekeres diagram (Figure 1.1) – settle down asymptotically to a stationary solution. In this appendix we derive some analytical representations of these asymptotic solutions. These solutions are very valuable for two reasons: they provide very useful code tests,[1] and help delineate the geometrical properties of moving puncture solutions.

Maximal slicing

Let us first consider a "nonadvective" version of the 1+log slicing condition (H.1), i.e.,

$$\partial_t \alpha = -2\alpha K. \tag{H.2}$$

If at late times the solution settles down and becomes time-independent, we must have $\partial_t \alpha = 0$, implying that the late-time solution must be maximally sliced with $K = 0$. The solution must therefore be a member of the family of time-independent, maximal slicings of Schwarzschild described by equations (4.23)–(4.25) and parametrized by the constant C. Hannam *et al.* (2007) show that evolving with the Gamma-driver shift condition (4.83) yields the late-time solution corresponding to the particular member $C = 3\sqrt{3}M^2/4$, which has a limiting surface at the areal radius $r_s = 3M/2$. The family (4.23)–(4.25) is expressed in terms of an areal radius r_s, but for numerical purposes it is often more convenient to express this solution in terms of an isotropic radius r. As it turns out, the cases $C = 0$ (see exercise 3.4) and $C = 3\sqrt{3}M^2/4$ are the only cases for which these solutions can be expressed in isotropic coordinates in terms of elementary functions.[2]

> **Exercise H.1** (a) To transform from the areal radius r_s in (4.23) to an isotropic radius r, identify the spatial metric (4.23) with its counterpart in isotropic, polar

---

[1] Employing moving puncture gauge conditions, and using these solutions as initial data, should result in a time-independent solution. Any nonzero time evolution is therefore a measure of the numerical error.

[2] The remainder of this section follows Baumgarte and Naculich (2007).

coordinates, $\gamma_{ij} = \psi^4 \eta_{ij}$. Show that this yields

$$\psi = \left(\frac{r_s}{r}\right)^{1/2} \tag{H.3}$$

as well as the differential equation

$$\pm \int \frac{dr}{r} = \int \frac{r_s \, dr_s}{\sqrt{r_s^4 - 2Mr_s^3 + C^2}}. \tag{H.4}$$

(b) In the general case the right-hand side of equation (H.4) may be expressed in terms of elliptic integrals. The integral simplifies, however, for $C = 0$ (in which case we recover the situation of exercise 3.4), and for $C = 3\sqrt{3}M^2/4$. Show that for $C = 3\sqrt{3}M^2/4$ the quartic polynomial in equation (H.4) has a double root at $r_s = 3M/2$, and show that integration yields

$$r = \frac{2r_s + M + (4r_s^2 + 4Mr_s + 3M^2)^{1/2}}{4}$$

$$\times \left(\frac{(4 + 3\sqrt{2})(2r_s - 3M)}{8r_s + 6M + 3(8r_s^2 + 8Mr_s + 6M^2)^{1/2}}\right)^{1/\sqrt{2}}, \tag{H.5}$$

where a constant of integration has been fixed so that $r \to r_s$ as $r_s \to \infty$. Also show that $r \to 0$ as $r_s \to 3M/2$.

Given equation (H.5) we can now construct the entire spacetime metric

$$ds^2 = -\alpha^2 dt^2 + \psi^4 \eta_{ij}(dx^i + \beta^i dt)(dx^j + \beta^j dt) \tag{H.6}$$

as follows. From equation (H.3) we find the conformal factor

$$\psi = \left(\frac{4r_s}{2r_s + M + (4r_s^2 + 4Mr_s + 3M^2)^{1/2}}\right)^{1/2} \left(\frac{8r_s + 6M + 3(8r_s^2 + 8Mr_s + 6M^2)^{1/2}}{(4 + 3\sqrt{2})(2r_s - 3M)}\right)^{1/2\sqrt{2}}. \tag{H.7}$$

To express ψ in terms of the isotropic radius r would require inverting equation (H.5); clearly that is not possible. Instead, equations (H.5) and (H.7) together provide a parametric representation of $\psi(r)$. Exercise H.2 demonstrates that the conformal factor features a $1/\sqrt{r}$ coordinate singularity at $r = 0$, so that this point corresponds to a surface of finite areal radius r_s (*cf.* exercise 13.1). We show $\psi(r)$ together with its asymptotic limits in Figure H.1.

> **Exercise H.2** Show that the conformal factor (H.7) has the expected asymptotic limits
>
> $$\psi \to \begin{cases} \left(\dfrac{3M}{2r}\right)^{1/2} & r \to 0 \\ 1 + \dfrac{M}{2r} & r \to \infty. \end{cases} \tag{H.8}$$

The lapse α in equation (4.24) transforms like a scalar under spatial coordinate transformations and hence remains

$$\alpha = \left(1 - \frac{2M}{r_s} + \frac{27M^4}{16r_s^4}\right)^{1/2}, \tag{H.9}$$

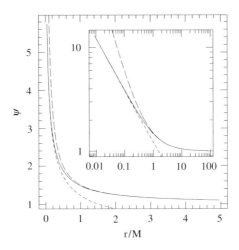

Figure H.1 The conformal factor ψ (equation H.7) for the maximally sliced asymptotic solution (solid lines), together with its asymptotic limits (equation H.8) (dashed lines), as functions of the isotropic radius r. [From Baumgarte and Naculich (2007).]

while the shift has to be transformed from the areal radius r_s in equation (4.25) to the isotropic radius r,

$$\beta^r = \frac{dr}{dr_s}\beta^{r_s} = \frac{r}{r_s}\frac{1}{f}\frac{Cf}{r_s^2} = \frac{3\sqrt{3}M^2}{4}\frac{r}{r_s^3}. \tag{H.10}$$

Both the lapse and shift can again be represented parametrically in terms of r.

Stationary 1+log slicing

Now let us return to the "advective" 1+log slicing condition (H.1). The time-independent, asymptotic solution must then satisfy[3]

$$K = \frac{\beta^i \partial_i \alpha}{2\alpha}. \tag{H.11}$$

We can construct these slices using the same "height-function" approach that we used in Chapter 4.2 to find the family of maximal slices (4.23)–(4.25). Recall that, starting with the standard Schwarzschild coordinates, we introduced a new time coordinate $\bar{t} = t + h(r_s)$, where t is the original Schwarzschild time and $h(r_s)$ the height function. We then identified the lapse and the shift on $\bar{t} = constant$ surfaces and found the expressions (4.18)

$$\beta^{r_s} = \frac{f_0^2 h'}{1 - f_0^2 h'^2}, \qquad \alpha^2 = \frac{f_0}{1 - f_0^2 h'^2}, \tag{H.12}$$

where $h' \equiv dh/dr_s$ and $f_0(r_s) = 1 - 2M/r_s$. It is now convenient to express h' in terms of the lapse,

$$h' = \frac{1}{\alpha f_0}\left(\alpha^2 - f_0\right)^{1/2}, \tag{H.13}$$

---

[3] This section follows Hannam *et al.* (2008).

so that the shift becomes

$$\beta^{r_s} = \alpha \left(\alpha^2 - f_0 \right)^{1/2} . \tag{H.14}$$

From equation (4.19) we can also show that

$$K = \frac{1}{r_s^2} \frac{d}{dr_s} \left(r_s^2 f_0 \alpha h' \right) = \frac{1}{r_s^2} \frac{d}{dr_s} \left(r_s^2 \left(\alpha^2 - f_0 \right)^{1/2} \right) . \tag{H.15}$$

Inserting the shift (H.14) and the mean curvature (H.15) into the slicing condition (H.11) then yields a differential equation for the lapse

$$\frac{d\alpha}{dr_s} = -\frac{2}{r_s} \frac{3M - 2r_s + 2r_s \alpha^2}{r_s - 2M + 2r_s \alpha - r_s \alpha^2} . \tag{H.16}$$

Exercise H.3 Show that the differential equation (H.16) is solved by the implicit equation

$$\alpha^2 = 1 - \frac{2M}{r_s} + \frac{C^2 e^\alpha}{r_s^4} \tag{H.17}$$

for any value of the constant C.

Remarkably, the solution (H.17) for the lapse differs from the maximal slicing result (4.24) only by the exponential term e^α. This term complicates matters considerably, of course, since now the lapse is given only implicitly. As for the maximal slicing case, again we have discovered an entire family of solutions that satisfies the slicing condition (H.11), and again we can parametrize this family in terms of the constant C.

We notice, however, that the solutions may become singular if the denominator in equation (H.16) vanishes. To ensure that the solution remains regular we demand that the numerator of the equation vanish simultaneously with the denominator; this condition determines the constant C. The numerator of equation (H.16) vanishes when

$$\alpha = \left(1 - \frac{3M}{2r_s} \right)^{1/2} \tag{H.18}$$

(where we have chosen the positive root). Given this value of the lapse, the denominator of equation (H.16) vanishes at a critical radius r_c given by

$$r_c = \frac{3 + \sqrt{10}}{4} M. \tag{H.19}$$

At r_c, the lapse is

$$\alpha_c = \sqrt{10} - 3. \tag{H.20}$$

We can now insert both equations (H.19) and (H.20) into equation (H.17), which yields

$$C = \frac{\sqrt{2}}{16} (\sqrt{10} + 3)^{3/2} e^{(3 - \sqrt{10})/2} M^2 \simeq 1.24672 M^2. \tag{H.21}$$

This value of C identifies the member of the family in which we are interested. Substituting C into equation (H.17) we can then find, at least implicitly, α for all values of the areal radius r_s. Knowing α we can find h' from equation (H.13), which then determines all other metric quantities.

Exercise H.4 (a) Recall that the 1+log slicing condition (H.1) is a member of a larger family of slicing conditions (4.49)

$$\partial_t \alpha - \beta^i \partial_i \alpha = -\alpha^2 f(\alpha) K \tag{H.22}$$

with $f(\alpha) = 2/\alpha$. Retrace our derivation above for the more general form $f(\alpha) = n/\alpha$, where n is some arbitrary constant.

Hint: Try $\alpha^2 = 1 - 2M/r_s + C^2 e^{2\alpha/n}/r_s^4$ as the generalization of (H.17). You should find $C^2 = (3n + \sqrt{4 + 9n^2})^3/(128n^3)\, e^{-2\alpha_c/n} M^4$.

(b) In Chapter 4.3 we argued that maximal slicing can be recovered in the limit $n \to \infty$. Demonstrate that the solution of part (a) does indeed reduce to the maximal slicing solution of the previous section (with $C = 3\sqrt{3}M^2/4$) in this limit.

The above steps complete the solution in terms of the areal radius r_s. For numerical purposes it would again be desirable to express this solution in terms of an isotropic radius; while that does not seem to be possible analytically, the transformation can be carried out numerically. We refer the reader to Hannam *et al.* (2008) for a more detailed discussion of these solutions, as well as their geometrical properties.

Binary black hole puncture simulations as test problems

Simulating the late inspiral, merger and ringdown of a binary black hole was for many years the "holy grail" of numerical relativity. Dozens of researchers have spent many years attempting to formulate a stable algorithm capable of solving this important problem. As discussed in Chapter 13, several different techniques and corresponding codes now exist that can evolve coalescing binary black holes successfully. In this appendix we summarize two simulations in sufficient detail to enable the reader to assess the requirements and caliber of a typical computation. We employ one widely adopted method for the simulations, based on the BSSN evolution scheme and binary puncture initial data. The examples chosen here, and the quoted results, can be used as test problems for the construction of new algorithms and new modules.

We emphasize that performing a simulation of binary black hole coalescence provides a highly nontrivial and comprehensive laboratory for testing a $3 + 1$ code capable of solving Einstein's field equations in vacuum in the strong-field regime of general relativity. It tests the caliber of not only the basic evolution scheme, but also the all-important diagnostic routines. These routines measure globally conserved quantities like the total mass and angular momentum of the system, the masses and spins of the individual black holes, the location and area of the black hole horizons, the asymptotic gravitational waveforms, the recoil kick imparted to the black hole remnant, and other "vital statistics" characterizing a dynamical spacetime containing black holes. Appreciable effort invariably goes into implementing and testing these diagnostic modules in order to extract meaningful and reliable physical information from the numerical output.

We consider two cases of merging, nonspinning black holes: one is an equal-mass system, and the other has a 3:1 mass ratio. Both cases employ initial data constructed using the binary puncture technique to obtain quasiequilibrium binaries in circular orbit, as discussed in Chapter 12.2. The initial data for the two cases are specified by the puncture parameters listed in Table I.1. The code employs AMR with equatorial symmetry; the AMR grid setup is described in Table I.2. The basic evolution scheme is BSSN with moving puncture gauge conditions. Details concerning the implementation of the evolution scheme are summarized in Table I.3. A number of the global diagnostic parameters, and several key physical results, are listed in Table I.4.

In reading through the summaries of these simulations as provided by the tables and figures below it should become apparent that performing such calculations requires an understanding of much of the formalism and computational machinery reviewed in this book. This point is implicit in the many references cited below to earlier equations and discussions throughout the main text where many of the quantities that appear in tables and figures are introduced. The simulation of the inspiral and merger of a binary black hole thus represents the culmination of our training and ability to evolve a pure vacuum, dynamical spacetime in $3 + 1$ dimensions. It is also the logical

Table I.1 Binary black hole puncture initial data. The freely chosen parameters are each puncture's coordinate position $\mathbf{x}$, "bare" mass $\mathcal{M}$, momentum $\mathbf{P}$ and spin $\mathbf{S}$. The resulting initial puncture geometry is specified by the 3-metric γ_{ij} and extrinsic curvature K_{ij}, as constructed in Chapter 12.2 for nonspinning black holes in quasiequilibrium circular orbit. In the last five lines we list diagnostic parameters computed from these initial data, including the total ADM mass M (equation 3.140), the total angular momentum J (equation 3.191), and the irreducible mass M_{irr} of each black hole (equation 7.2). Quantities like $\mathbf{x}$, $\mathbf{P}$, etc. are listed above in "code units", (where 1 code unit $= 1.011\,M$) and can be scaled to arbitrary M. [After Etienne *et al.* (2009).]

	Equal mass ($q = 1$)	Unequal mass ($q = 3$)
Location of punctures	$\mathbf{x}_+ = (0, 4.891, 0)$,	$\mathbf{x}_+ = (5.25, 0, 0)$,
	$\mathbf{x}_- = (0, -4.891, 0)$	$\mathbf{x}_- = (-1.75, 0, 0)$
"Bare" mass	$\mathcal{M}_+ = \mathcal{M}_- = 0.4856$	$\mathcal{M}_+ = 0.234$, $\mathcal{M}_- = 0.735$
Spin	$\mathbf{S}_+ = \mathbf{S}_- = 0$	$\mathbf{S}_+ = \mathbf{S}_- = 0$
Momentum	$\mathbf{P}_+ = (0.0969, 0, 0)$,	$\mathbf{P}_+ = (0, 0.09407, 0)$,
	$\mathbf{P}_- = (-0.0969, 0, 0)$	$\mathbf{P}_- = (0, -0.09407, 0)$
ADM mass	$M = 0.9894$	$M = 0.9895$
Angular momentum	$J = 0.9479$	$J = 0.6587$
Irreducible mass	$M_{\text{irr}}^+ = M_{\text{irr}}^- = 0.5000$	$M_{\text{irr}}^+ = 0.2498$, $M_{\text{irr}}^- = 0.7494$
Mean coord. radius of BHs' horizon	$\mathcal{R}_+ = \mathcal{R}_- = 0.2359$	$\mathcal{R}_+ = 0.1097$, $\mathcal{R}_- = 0.3605$
Binary coord. separation ($|\mathbf{x}_+ - \mathbf{x}_-|$)	$9.887\,M$	$7.074\,M$

Table I.2 Grid setup. The parameter L is the location of the outer boundary, i.e., $x_{\max} = y_{\max} = z_{\max} = L$, $x_{\min} = y_{\min} = -L$, $z_{\min} = 0$ (equatorial symmetry). $N_+(N_-)$ is the number of AMR refinement levels (Chapter 6.2.5) centered at the "+" ("−") puncture. R_+ (R_-) are the radii of the refinement levels centered at the "+" ("−") puncture in code units (where 1 code unit $= 1.011\,M$). $\Delta_+(\Delta_-)$ is the grid spacing in each refinement level centered at the "+" ("−") puncture (in code units). Note that the grid spacing doubles in each successive refinement level. The coarsest grid spacing is twice that of the outermost refinement level, which is 8 for the equal-mass case and 10.24 for the unequal-mass case. $N_{\text{AH}}^+(N_{\text{AH}}^-)$ is the number of grid points across the mean diameter of the apparent horizon of the "+" ("−") black hole at $t = 0$. [After Etienne *et al.* (2009).]

	Equal-mass ($q = 1$)	Unequal mass ($q = 3$)
L	$320\ (= 323.4M)$	$409.6\ (= 413.9M)$
N_+	8	10
N_-	8	9
R_+	$(160, 80, 140/3, 20, 5, 2.5, 1.25, 0.625)$	$(256, 128, 64, 24, 12, 6, 3, 1.5, 0.75, 0.375)$
Δ_+	$(4, 2, 1, \ldots, 0.0625, 0.03125)$	$(5.12, 2.56, 1.28, \ldots, 0.02, 0.01)$
R_-	$(160, 80, 140/3, 20, 5, 2.5, 1.25, 0.625)$	$(256, 128, 64, 32, 16, 8, 4, 2, 1)$
Δ_-	$(4, 2, 1, \ldots, 0.0625, 0.03125)$	$(5.12, 2.56, 1.28, \ldots, 0.04, 0.02)$
N_{AH}^+	≈ 15	≈ 22
N_{AH}^-	≈ 15	≈ 36

Table I.3 Evolution scheme. The basic BSSN scheme is summarized in Chapter 11.5 (see Box 11.1). "Moving puncture" gauge conditions are adopted for the lapse and shift (see equation 4.51 and discussion following equation 4.83). To enforce the auxiliary constraints in (1) the technique described in Etienne *et al.* (2008) is implemented; to enforce the auxiliary constraint in (2) the technique suggested by Marronetti *et al.* (2007) is adopted (see also equations 14.41–14.43 and related discussion). Kreiss and Oliger (1973) dissipation (Chapter 6.2.3) is employed here to reduce any high frequency noise arising at the moving refinement boundaries. Carpet is a Cactus-based module that implements AMR (see http://www.carpetcode.org for full documentation). Upwind differencing *vs.* centered spatial differencing, as well as the method of lines (MOL) algorithm for time-integrations, are discussed in Chapter 6.2.3. The Courant factor is defined below equation (6.72). Prolongation refers to interpolations in space and time at the AMR refinement boundaries. A range of extraction radii is chosen to confirm convergence of waveforms. [After Etienne *et al.* (2009).]

Formalism	BSSN
Lapse	1+log slicing: $\partial_0 \alpha = -2\alpha K$, where $\partial_0 \equiv \partial_t - \beta^j \partial_j$
Shift	Γ-freezing: $\partial_0 \beta^i = (3/4)B^i$, $\partial_0 B^i = \partial_0 \tilde{\Gamma}^i - \eta B^i$, $\eta = 0.25 = 0.2474/M$
Conformal variable	Evolve $e^{-2\phi}$ (instead of ϕ), where $\det \gamma_{ij} \equiv e^{12\phi}$
Symmetry	Equatorial
Other Numerical techniques	(1) Enforce auxiliary constraints $\det(\tilde{\gamma}_{ij}) = 1$ and $\mathrm{Tr}(\tilde{A}_{ij}) = 0$ (2) Enforce auxiliary constraint $\tilde{\Gamma}^i = -\partial_j \tilde{\gamma}^{ij}$ (3) Add 5th order Kreiss–Oliger dissipation of the form $(\epsilon/64)(\Delta x^5 \partial_x^6 + \Delta y^5 \partial_y^6 + \Delta z^5 \partial_z^6)f$ in all lapse, shift and BSSN evolution equations, $\partial_t f$. Here Δx, Δy and Δz are grid spacings and $\epsilon = 0.1$.
Grid-driver code	Carpet
Spatial differencing	4th order upwind on shift advection terms, 4th order centered otherwise
Temporal differencing	Method of lines (MOL) with 4th order Runge–Kutta (RK4)
Courant (CFL) factor	0.25
Prolongation	5th order spatial prolongation, 2nd order temporal prolongation
Wave extraction radii	$10.11M$–$54.58M$ for equal-mass case, $30.32M$–$70.75M$ for unequal-mass case

starting point of many nonvacuum dynamical scenarios involving strong gravitational fields and requiring the tools of numerical relativity for solution.

Of special signficance is the degree to which total energy and angular momentum are conserved when properly accounting for losses due to gravitational wave emission. The fractional errors are seen in Table I.4 to be a few times 10^{-4} for the energy and 10^{-3} for the angular momentum. Also, different measures of the mass and spin of the final (stationary) remnant Kerr black hole agree closely, given the adopted computational resources.

Table I.4 Simulation results: the black hole remnant and emitted radiation. M_f and J_f are the final ADM mass and angular momentum measured at radius $43.90M$ in the equal-mass case, and at $50.35M$ in the unequal-mass case. Here M is the initial ADM mass as listed in Table I.1. The value of J_f fluctuates slightly; the quoted error is the fluctuation amplitude at late times. $M_{\rm BH}$ is the mass of the final black hole as determined by its irreducible mass (equations 7.2 and 7.4); $J_{\rm BH}/M_{\rm BH}^2$ is the spin parameter of the final hole as determined by the ratio of polar and equatorial circumferences, assuming the Kerr relation. $\Delta E_{\rm GW}$ and $\Delta J_{\rm GW}$ are the energy and angular momentum carried off by gravitational radiation (equations 9.130 and 9.131.) The waves are extracted at radius $50.53M$, where the Weyl scalar ψ_4 is evaluated (equation 9.121). The parameter $v_{\rm kick}$ is the kick velocity imparted to the black hole in the unequal-mass case (equation 9.132). (Wave data for $t - r < 50M$, arising from gravitational radiation present in the initial data, are removed before computing the kick velocity.) The parameter $\delta E \equiv (M - M_{\rm BH} - \Delta E_{\rm GW})/M$ measures the fractional violation of energy conservation; $\delta J \equiv (J - J_{\rm BH} - \Delta J_{\rm GW})/J$ measures the fractional violation of angular momentum conservation. [After Etienne *et al.* (2009).]

	Equal mass ($q = 1$)	Unequal mass ($q = 3$)
M_f/M	0.9615	0.9808
J_f/M^2	0.636 ± 0.008	0.52 ± 0.04
$M_{\rm BH}/M$	0.9617	0.9809
$J_{\rm BH}/M_{\rm BH}^2$	0.6852	0.5413
$\Delta E_{\rm GW}/M$	0.03794	0.01934
$\Delta J_{\rm GW}/M^2$	0.3306	0.1513
$v_{\rm kick}$	—	174 km s^{-1}
δE	4×10^{-4}	-2×10^{-4}
δJ	4×10^{-3}	9×10^{-4}

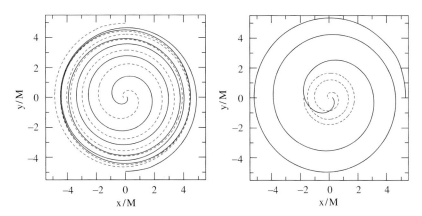

Figure I.1 Trajectory of the coordinate centroid of the "+" (black solid line) and "−" (blue dash line) black hole during nonspinning binary black hole inspiral and merger. The left-hand panel is the equal-mass case, the right-hand panel in the unequal mass case. [From Etienne *et al.* (2009).]

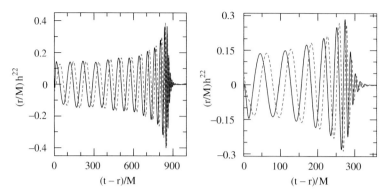

Figure I.2 Gravitational waveforms for the inspiral and merger of nonspinning binary black holes. Shown here are the dominant $l = 2$, $m = 2$ wave amplitudtes $(r/M)h_+^{22}$ (black solid line) and $(r/M)h_\times^{22}$ (blue dash line) as functions of retarded time $t - r$. The left-hand panel is the equal-mass case while the right-hand panel is the unequal-mass case. The waveforms are extracted at radius $r = 50.53\,M$ and computed by integrating the $s = -2$ spin-weighted spherical harmonics of the Weyl scalar ψ_4 twice with time (equations 9.126 and 9.134). [From Etienne *et al.* (2009).]

The coordinate trajectories of the two black holes are plotted in Figure I.1 for the two cases; the emitted gravitational waveforms are shown in Figure I.2.

The cases analyzed here have been considered by many different investigators. The actual numerical results reported here were obtained by Etienne *et al.* (2009), who performed these simulations to test the vacuum sector of their relativistic hydrodynamics AMR code. Usually, vacuum codes that can simulate these two cases successfully can automatically evolve more complicated scenarios, such as spinning black hole binaries. Such codes can also treat black hole binaries for which approaches other than the puncture method are adopted to construct the initial data (e.g., the conformal thin-sandwich approach is one such alternative; see Chapter 12.3).

References

Abrahams, A., A. Anderson, Y. Choquet-Bruhat, and J. W. York, Jr. (1995). Einstein and Yang-Mills theories in hyperbolic form without gauge fixing. *Phys. Rev. Lett.* **75**, 3377–3381.

Abrahams, A. M. and G. B. Cook (1994). Collisions of boosted black holes: Perturbation theory prediction of gravitational radiation. *Phys. Rev. D* **50**, R2364–R2367.

Abrahams, A. M., G. B. Cook, S. L. Shapiro, and S. A. Teukolsky (1994). Solving Einstein's equations for rotating spacetimes: Evolution of relativistic star clusters. *Phys. Rev. D* **49**, 5153–5164.

Abrahams, A. M. and C. R. Evans (1988). Reading off gravitational radiation waveforms in numerical relativity calculations: Matching to linearized gravity. *Phys. Rev. D* **37**, 318–332.

Abrahams, A. M. and C. R. Evans (1990). Gauge-invariant treatment of gravitational radiation near the source: Analysis and numerical simulations. *Phys. Rev. D* **42**, 2585–2594.

Abrahams, A. M. and C. R. Evans (1992). Gravitational waveforms from the head-on collision of relativistic stars. In F. Sato and T. Nakamura (Eds.), *Proceedings of the 6th Marcel Grossmann Meeting on General Relativity*, pp. 1331–1333. World Scientific, Singapore.

Abrahams, A. M. and C. R. Evans (1993). Critical behavior and scaling in vacuum axisymmetric gravitational collapse. *Phys. Rev. Lett.* **70**, 2980–2983.

Abrahams, A. M., K. R. Heiderich, S. L. Shapiro, and S. A. Teukolsky (1992). Vacuum initial data, singularities, and cosmic censorship. *Phys. Rev. D* **46**, 2452–2463.

Abrahams, A. M., S. L. Shapiro, and S. A. Teukolsky (1994). Disk collapse in general relativity. *Phys. Rev. D* **50**, 7282–7291.

Abrahams, A. M., S. L. Shapiro, and S. A. Teukolsky (1995). Calculation of gravitational waveforms from black hole collisions and disk collapse: Applying perturbation theory to numerical spacetimes. *Phys. Rev. D* **51**, 4295–4301.

Abrahams, A. M. and J. W. York, Jr., (1997). 3+1 general relativity in hyperbolic form. In *Relativistic Gravitation and Gravitational Radiation*. Cambridge University Press, Cambridge.

Abramovici, A., W. E. Althouse, R. W. P. Drever, Y. Gursel, S. Kawamura, F. J. Raab, D. Shoemaker, L. Sievers, R. E. Spero, and K. S. Thorne (1992). LIGO – The Laser Interferometer Gravitational-Wave Observatory. *Science* **256**, 325–333.

Akmal, A., V. R. Pandharipande, and D. G. Ravenhall (1998). Equation of state of nucleon matter and neutron star structure. *Phys. Rev. C* **58**, 1804–1828.

Alcubierre, M. (1997). The appearance of coordinate shocks in hyperbolic formalisms of general relativity. *Phys. Rev. D* **55**, 5981–5991.

Alcubierre, M. (2008a). Formulations of the 3+1 evolution equations for numerical relativity. Presentation at the Conference *Numerical modeling of astrophysical sources of gravitational radiation*, Valencia, Spain.

Alcubierre, M. (2008b). *Introduction to 3+1 Numerical Relativity*. Oxford University Press, New York.

Alcubierre, M., G. Allen, B. Brügmann, E. Seidel, and W.-M. Suen (2000). Towards an understanding of the stability properties of the 3+1 evolution equations in general relativity. *Phys. Rev. D* **62**, 124011/1–15.

Alcubierre, M., W. Benger, B. Brügmann, G. Lanfermann, L. Nerger, E. Seidel, and R. Takahashi (2001). 3D Grazing Collision of Two Black Holes. *Phys. Rev. Lett.* **87**, 271103/1–4.

Alcubierre, M. and B. Brügmann (2001). Simple excision of a black hole in 3+1 numerical relativity. *Phys. Rev. D* **63**, 104006/1–6.

Alcubierre, M., B. Brügmann, P. Diener, F. S. Guzmán, I. Hawke, S. Hawley, F. Herrmann, M. Koppitz, D. Pollney, E. Seidel, and J. Thornburg (2005). Dynamical evolution of quasicircular binary black hole data. *Phys. Rev. D* **72**, 044004/1–14.

Alcubierre, M., B. Brügmann, T. Dramlitsch, J. A. Font, P. Papadopoulos, E. Seidel, N. Stergioulas, and R. Takahashi (2000). Towards a stable numerical evolution of strongly gravitating systems in general relativity: The conformal treatments. *Phys. Rev. D* **62**, 044034/1–16.

Alcubierre, M., B. Brügmann, D. Holz, R. Takahashi, S. Brandt, E. Seidel, J. Thornburg, and A. Ashtekar (2001). Symmetry Without Symmetry. *International Journal of Modern Physics D* **10**, 273–289.

Alcubierre, M., B. Brügmann, D. Pollney, E. Seidel, and R. Takahashi (2001). Black hole excision for dynamic black holes. *Phys. Rev. D* **64**, 061501/1–5.

Alcubierre, M. and J. Massó (1998). Pathologies of hyperbolic gauges in general relativity and other field theories. *Phys. Rev. D* **57**, R4511–R4515.

Aloy, M. A., H.-T. Janka, and E. Müller (2005). Relativistic outflows from remnants of compact object mergers and their viability for short gamma-ray bursts. *Astron. Astrophys.* **436**, 273–311.

Alvi, K. (2001). Energy and angular momentum flow into a black hole in a binary. *Phys. Rev. D* **64**, 104020/1–9.

Anderson, A., Y. Choquet-Bruhat, and J. W. York, Jr. (1997). Einstein-Bianchi hyperbolic system for general relativity. *Topol. Methods Nonlinear Anal.* **10**, 353–377.

Anderson, A. and J. W. York, Jr. (1998). Hamiltonian time evolution for general relativity. *Phys. Rev. Lett.* **81**, 1154–1157.

Anderson, A. and J. W. York, Jr. (1999). Fixing Einstein's equations. *Phys. Rev. Lett.* **82**, 4384–4387.

Andrade, Z., C. Beetle, A. Blinov, B. Bromley, L. M. Burko, M. Cranor, R. Owen, and R. H. Price (2004). The periodic standing-wave approximation: Overview and three dimensional scalar models. *Phys. Rev. D* **70**, 064001/1–14.

Anninos, P., D. Bernstein, S. Brandt, J. Libson, E. Seidel, L. Smarr, W.-M. Suen, and P. Walker (1995). Dynamics of apparent and event horizons. *Phys. Rev. Lett.* **74**, 630–633.

Anninos, P., D. Bernstein, S. R. Brandt, D. Hobill, E. Seidel, and L. Smarr (1994). Dynamics of black hole apparent horizons. *Phys. Rev. D* **50**, 3801–3815.

Anninos, P., K. Camarda, J. Libson, J. Massó, E. Seidel, and W.-M. Suen (1998). Finding apparent horizons in dynamic 3d numerical spacetimes. *Phys. Rev. D* **58**, 024003/1–12.

Anninos, P., J. Centrella, and R. Matzner (1989). Nonlinear solutions for initial data in the vacuum Einstein equations in plane symmetry. *Phys. Rev. D* **39**, 2155–2171.

Anninos, P., P. C. Fragile, and J. D. Salmonson (2005). Cosmos++: Relativistic Magnetohydrodynamics on Unstructured Grids with Local Adaptive Refinement. *Astrophys. J.* **635**, 723–740.

Antón, L., O. Zanotti, J. A. Miralles, J. M. Martí, J. M. Ibáñez, J. A. Font, and J. A. Pons (2006). Numerical 3+1 General Relativistic Magnetohydrodynamics: A Local Characteristic Approach. *Astrophys. J.* **637**, 296–312.

Arnett, D. (1996). *Supernovae and nucleosynthesis. an investigation of the history of matter, from the Big Bang to the present.* Princeton University Press, Princeton.

Arnowitt, R., S. Deser, and C. W. Misner (1962). The dynamics of general relativity. In L. Witten (Ed.), *Gravitation: an Introduction to Current Research*, pp. 227–265. Wiley, New York. Also available at gr-qc/0405109.

Arun, K. G., L. Blanchet, B. R. Iyer, and M. S. S. Qusailah (2004). The 2.5PN gravitational wave polarizations from inspiralling compact binaries in circular orbits. *Class. Quantum Grav.* **21**, 3771–3801.

Asada, H. (1998). Formulation for the internal motion of quasi-equilibrium configurations in general relativity. *Phys. Rev. D* **57**, 7292–7298.

Ashtekar, A., C. Beetle, and S. Fairhurst (2000). Mechanics of isolated horizons. *Class. Quantum Grav.* **17**, 253–298.

Ashtekar, A. and B. Krishnan (2003). Dynamical horizons and their properties. *Phys. Rev. D* **68**, 104030/1–25.

Ashtekar, A. and B. Krishnan (2004). Isolated and dynamical horizons and their applications. *Living Rev. Relativity* **7**, 1–91.

Baade, W. and F. Zwicky (1934). Cosmic Rays from Super-novae. *Proc. Nat. Acad. Sci. U.S.* **20**, 259–263.

Babak, S., H. Fang, J. R. Gair, K. Glampedakis, and S. A. Hughes (2007). "Kludge" gravitational waveforms for a test-body orbiting a Kerr black hole. *Phys. Rev. D* **75**, 024005/1–25.

Baiotti, L., I. Hawke, P. J. Montero, F. Löffler, L. Rezzolla, N. Stergioulas, J. A. Font, and E. Seidel (2005). Three-dimensional relativistic simulations of rotating neutron-star collapse to a Kerr black hole. *Phys. Rev. D* **71**, 024035/1–30.

Baiotti, L., R. D. Pietri, G. M. Manca, and L. Rezzolla (2007). Accurate simulations of the dynamical bar-mode instability in full general relativity. *Phys. Rev. D* **75**, 044023/1–24.

Baker, J. G., W. D. Boggs, J. Centrella, B. J. Kelly, S. T. McWilliams, M. C. Miller, and J. van Meter (2007c). Modeling kicks from the merger of non-precessing black-hole binaries. *Astrophys. J.* **668**, 1140–1144.

Baker, J. G., M. Campanelli, F. Pretorius, and Y. Zlochower (2007b). Comparisons of binary black hole merger waveforms. *Class. Quantum Grav.* **24**, S25–S31.

Baker, J. G., J. Centrella, D.-I. Choi, M. Koppitz, and J. van Meter (2006a). Gravitational-wave extraction from an inspiraling configuration of merging black holes. *Phys. Rev. Lett.* **96**, 111102/1–4.

Baker, J. G., J. Centrella, D.-I. Choi, M. Koppitz, J. R. van Meter, and M. C. Miller (2006b). Getting a Kick Out of Numerical Relativity. *Astrophys. J. Lett.* **653**, L93–L96.

Baker, J. G., J. R. van Meter, S. T. McWilliams, J. Centrella, and B. J. Kelly (2007a). Consistency of post-Newtonian waveforms with numerical relativity. *Phys. Rev. Lett.* **99**, 181101/1–4.

Balakrishna, J., G. Daues, E. Seidel, W.-M. Suen, M. Tobias, and E. Wang (1996). Coordinate conditions in three-dimensional numerical relativity. *Class. Quantum Grav.* **13**, L135–L142.

Balberg, S. and S. L. Shapiro (2002). Gravothermal Collapse of Self-Interacting Dark Matter Halos and the Origin of Massive Black Holes. *Phys. Rev. Lett.* **88**, 101301/1–4.

Balberg, S., L. Zampieri, and S. L. Shapiro (2000). Black Hole Emergence in Supernovae. *Astrophys. J.* **541**, 860–882.

Balbus, S. A. and J. F. Hawley (1991). A powerful local shear instability in weakly magnetized disks. I – Linear analysis. II – Nonlinear evolution. *Astrophys. J.* **376**, 214–233.

Balbus, S. A. and J. F. Hawley (1998). Instability, turbulence, and enhanced transport in accretion disks. *Reviews of Modern Physics* **70**, 1–53.

Bardeen, J. M. (1970). Kerr Metric Black Holes. *Nature* **226**, 64–65.

Bardeen, J. M. (1973). Rapidly rotating stars, disks, and black holes. In C. DeWitt and B. S. DeWitt (Eds.), *Black holes*, pp. 241–289. Gordon and Breach, New York.

Bardeen, J. M., J. L. Friedman, B. F. Schutz, and R. Sorkin (1977). A new criterion for secular instability of rapidly rotating stars. *Astrophys. J. Lett.* **217**, L49–L53.

Bardeen, J. M. and T. Piran (1983). General relativistic axisymmetric rotating systems: Coordinates and equations. *Phys. Rept.* **96**, 205–250.

Bardeen, J. M., W. H. Press, and S. A. Teukolsky (1972). Rotating Black Holes: Locally Nonrotating Frames, Energy Extraction, and Scalar Synchrotron Radiation. *Astrophys. J.* **178**, 347–370.

Baumgarte, T. W. (2000). Innermost stable circular orbit of binary black holes. *Phys. Rev. D* **62**, 024018/1–8.

Baumgarte, T. W. (2001). The innermost stable circular orbit in compact binaries. In J. M. Centrella (Ed.), *Astrophysical sources for ground-based gravitational wave detectors*, pp. 176–188. AIP Conference Prodeedings 575, Melville, New York.

Baumgarte, T. W., G. B. Cook, M. A. Scheel, S. L. Shapiro, and S. A. Teukolsky (1996). Implementing an apparent-horizon finder in three dimensions. *Phys. Rev. D* **54**, 4849–4857.

Baumgarte, T. W., G. B. Cook, M. A. Scheel, S. L. Shapiro, and S. A. Teukolsky (1997). Binary Neutron Stars in General Relativity: Quasiequilibrium Models. *Phys. Rev. Lett.* **79**, 1182–1185.

Baumgarte, T. W., G. B. Cook, M. A. Scheel, S. L. Shapiro, and S. A. Teukolsky (1998a). General relativistic models of binary neutron stars in quasiequilibrium. *Phys. Rev. D* **57**, 7299–7311.

Baumgarte, T. W., G. B. Cook, M. A. Scheel, S. L. Shapiro, and S. A. Teukolsky (1998b). Stability of relativistic neutron stars in binary orbit. *Phys. Rev. D* **57**, 6181–6184.

Baumgarte, T. W., S. A. Hughes, L. Rezzolla, S. L. Shapiro, and M. Shibata (1999). Implementing Fully Relativistic Hydrodynamics in Three Dimensions. In C. P. Burgess and R. C. Myers (Eds.), *General Relativity and Relativistic Astrophysics*, Volume 493 of *American Institute of Physics Conference Series*, pp. 53–59. AIP, Melville, New York.

Baumgarte, T. W., S. A. Hughes, and S. L. Shapiro (1999). Evolving Einstein's field equations with matter: The "hydro without hydro" test. *Phys. Rev. D* **60**, 087501/1–4.

Baumgarte, T. W., N. Ó. Murchadha, and H. P. Pfeiffer (2007). Einstein constraints: uniqueness and nonuniqueness in the conformal thin sandwich approach. *Phys. Rev. D* **75**, 044009/1–9.

Baumgarte, T. W. and S. G. Naculich (2007). Analytical representation of a black hole puncture solution. *Phys. Rev. D* **75**, 067502/1–4.

Baumgarte, T. W. and S. L. Shapiro (1998). Numerical integration of Einstein's field equations. *Phys. Rev. D* **59**, 024007/1–7.

Baumgarte, T. W. and S. L. Shapiro (1999). Evolution of Rotating Supermassive Stars to the Onset of Collapse. *Astrophys. J.* **526**, 941–952.

Baumgarte, T. W. and S. L. Shapiro (2003a). Collapse of a Magnetized Star to a Black Hole. *Astrophys. J.* **585**, 930–947.

Baumgarte, T. W. and S. L. Shapiro (2003b). General Relativistic Magnetohydrodynamics for the Numerical Construction of Dynamical Spacetimes. *Astrophys. J.* **585**, 921–929.

Baumgarte, T. W. and S. L. Shapiro (2003c). Numerical relativity and compact binaries. *Phys. Rept.* **376**, 41–131.

Baumgarte, T. W., S. L. Shapiro, and M. Shibata (2000). On the Maximum Mass of Differentially Rotating Neutron Stars. *Astrophys. J. Lett.* **528**, L29–L32.

Baumgarte, T. W., S. L. Shapiro, and S. A. Teukolsky (1995). Computing supernova collapse to neutron stars and black holes. *Astrophys. J.* **443**, 717–734.

Baumgarte, T. W., S. L. Shapiro, and S. A. Teukolsky (1996). Computing the Delayed Collapse of Hot Neutron Stars to Black Holes. *Astrophys. J.* **458**, 680–691.

Baumgarte, T. W., M. L. Skoge, and S. L. Shapiro (2004). Black hole-neutron star binaries in general relativity: quasiequilibrium formulation. *Phys. Rev. D* **70**, 064040/1–18.

Beig, R. and N. Ó Murchadha (1994). Trapped Surfaces in Vacuum Spacetimes. *Class. Quantum Grav.* **11**, 419–430.

Beig, R. and N. Ó Murchadha (1996). Vacuum spacetimes with future trapped surfaces. *Class. Quantum Grav.* **13**, 739–751.

Beig, R. and N. Ó Murchadha (1998). Late time behavior of the maximal slicing of the Schwarzschild black hole. *Phys. Rev. D* **57**, 4728–4737.

Bekenstein, J. D. (1973). Black Holes and Entropy. *Phys. Rev. D* **7**, 2333–2346.

Bekenstein, J. D. (1975). Statistical black-hole thermodynamics. *Phys. Rev. D* **12**, 3077–3085.

Belczynski, K., T. Bulik, and B. Rudak (2002). Study of Gamma-Ray Burst Binary Progenitors. *Astrophys. J.* **571**, 394–412.

Berger, B. K. (2002). Numerical Approaches to Spacetime Singularities. *Living Rev. Relativity* **5**, 1–59.

Berger, E. (2006). The Afterglows and Host Galaxies of Short GRBs: An Overview. In S. S. Holt, N. Gehrels, and J. A. Nousek (Eds.), *Gamma-Ray Bursts in the Swift Era*, Volume 836, pp. 33–42. AIP, Melville, New York.

Berger, M. J. and J. Oliger (1984). Adaptive mesh refinement for hyperbolic partial differential equations. *J. Comput. Phys.* **53**, 484–512.

Berti, E., V. Cardoso, J. A. Gonzalez, U. Sperhake, M. Hannam, S. Husa, and B. Brügmann (2007). Inspiral, merger and ringdown of unequal mass black hole binaries: a multipolar analysis. *Phys. Rev. D* **76**, 064034/1–40.

Berti, E. and M. Volonteri (2008). Cosmological Black Hole Spin Evolution by Mergers and Accretion. *Astrophys. J.* **684**, 822–828.

Bildsten, L. and C. Cutler (1992). Tidal interactions of inspiraling compact binaries. *Astrophys. J.* **400**, 175–180.

Binney, J. and S. Tremaine (1987). *Galactic dynamics*. Princeton University Press, Princeton.

Bishop, N. T. (1982). The closed trapped region and the apparent horizon of two Schwarzschild black holes. *Gen. Rel. Grav.* **14**, 717–723.

Blackburn, J. K. and S. Detweiler (1992). Close black-hole binary systems. *Phys. Rev. D* **46**, 2318–2333.

Blanchet, L. (2002). Innermost circular orbit of binary black holes at the third post-Newtonian approximation. *Phys. Rev. D* **65**, 124009/1–7.

Blanchet, L. (2006). Gravitational Radiation from Post-Newtonian Sources and Inspiralling Compact Binaries. *Living Rev. Relativity* **9**, 1–114.

Blanchet, L., A. Buonanno, and G. Faye (2006). Higher-order spin effects in the dynamics of compact binaries. ii: Radiation field. *Phys. Rev. D* **74**, 104034/1–17.

Blanchet, L., G. Faye, B. R. Iyer, and S. Sinha (2008). The third post-Newtonian gravitational wave polarisations and associated spherical harmonic modes for inspiralling compact binaries in quasi-circular orbits. *Class. Quantum Grav.* **25**, 125003/1–12.

Blanchet, L., M. S. S. Qusailah, and C. M. Will (2005). Gravitational Recoil of Inspiraling Black Hole Binaries to Second Post-Newtonian Order. *Astrophys. J.* **635**, 508–515.

Blandford, R. D. and D. G. Payne (1982). Hydromagnetic flows from accretion discs and the production of radio jets. *Mon. Not. R. Astron. Soc.* **199**, 883–903.

Bogdanovic, T., C. S. Reynolds, and M. C. Miller (2007). Alignment of the spins of supermassive black holes prior to merger. *Astrophys. J. Lett.* **661**, L147–L150.

Bona, C., T. Ledvinka, C. Palenzuela, and M. Zacek (2003). General-covariant evolution formalism for Numerical Relativity. *Phys. Rev. D* **67**, 104005/1–5.

Bona, C. and J. Massó (1988). Harmonic synchronizations of spacetime. *Phys. Rev. D* **38**, 2419–2422.

Bona, C. and J. Massó (1992). Hyperbolic evolution system for numerical relativity. *Phys. Rev. Lett.* **68**, 1097–1099.

Bona, C., J. Massó, E. Seidel, and J. Stela (1995). New Formalism for Numerical Relativity. *Phys. Rev. Lett.* **75**, 600–603.

Bonazzola, S., J. Frieben, and E. Gourgoulhon (1996). Spontaneous Symmetry Breaking of Rapidly Rotating Stars in General Relativity. *Astrophys. J.* **460**, 379–389.

Bonazzola, S., E. Gourgoulhon, P. Grandclément, and J. Novak (2004). A constrained scheme for einstein equations based on dirac gauge and spherical coordinates. *Phys. Rev. D* **70**, 104007/1–24.

Bonazzola, S., E. Gourgoulhon, and J.-A. Marck (1997). A relativistic formalism to compute quasi-equilibrium configurations of non-synchronized neutron star binaries. *Phys. Rev. D* **56**, 7740–7749.

Bonazzola, S., E. Gourgoulhon, and J.-A. Marck (1999a). Numerical models of irrotational binary neutron stars in general relativity. *Phys. Rev. Lett.* **82**, 892–895.

Bonazzola, S., E. Gourgoulhon, and J.-A. Marck (1999b). Spectral methods in general relativistic astrophysics. *J. Comput. Appl. Math.* **109**, 433–473.

Bonazzola, S. and J. Schneider (1974). An Exact Study of Rigidly and Rapidly Rotating Stars in General Relativity with Application to the Crab Pulsar. *Astrophys. J.* **191**, 273–290.

Bondi, H., F. A. E. Pirani, and I. Robinson (1959). Gravitational Waves in General Relativity. III. Exact Plane Waves. *Royal Society of London Proceedings Series A* **251**, 519–533.

Bowen, J. M. and J. W. York, Jr. (1980). Time-asymmetric initial data for black holes and black hole collisions. *Phys. Rev. D* **21**, 2047–2056.

Boyd, J. P. (2001). *Chebyshev and Fourier Spectral Methods*. Dover, New York. Also available at http://www-personal.engin.umich.edu/~jpboyd.

Boyer, R. H. and R. W. Lindquist (1967). Maximal analytic extension of the Kerr metric. *J. Math. Phys.* **8**, 265–281.

Brady, P. R., J. D. E. Creighton, and K. S. Thorne (1998). Computing the merger of black-hole binaries: The IBBH problem. *Phys. Rev. D* **58**, 061501/1–5.

Brandt, S. and B. Brügmann (1997). A simple construction of initial data for multiple black holes. *Phys. Rev. Lett.* **78**, 3606–3609.

Brandt, S. R. and E. Seidel (1995a). Evolution of distorted rotating black holes. I. Methods and tests. *Phys. Rev. D* **52**, 856–869.

Brandt, S. R. and E. Seidel (1995b). Evolution of distorted rotating black holes. II. Dynamics and analysis. *Phys. Rev. D* **52**, 870–886.

Brandt, S. R. and E. Seidel (1996). Evolution of distorted rotating black holes. III. Initial data. *Phys. Rev. D* **54**, 1403–1416.

Brill, D. R. (1959). On the positive definite mass of the Bondi-Weber-Wheeler time-symmetric gravitational waves. *Ann. Phys.* 7, 466–483.

Brill, D. R. and R. W. Lindquist (1963). Interaction energy in geometrostatics. *Phys. Rev.* **131**, 471–476.

Brodbeck, O., S. Frittelli, P. Hübner, and O. Reula (1999). Einstein's equations with asymptotically stable constraint propagation. *J. Math. Phys.* **40**, 909–923.

Browdy, S. F. and G. J. Galloway (1995). Topological censorship and the topology of black holes. *J. Math. Phys.* **36**, 4952–4961.

Brown, D., O. Sarbach, E. Schnetter, M. Tiglio, P. Diener, I. Hawke, and D. Pollney (2007). Excision without excision. *Phys. Rev. D* **76**, 081503/1–5.

Brown, J. D. (2008). Puncture Evolution of Schwarzschild Black Holes. *Phys. Rev. D* **77**, 044018/1–5.

Brügmann, B. (1999). Binary black hole mergers in 3d numerical relativity. *Int. J. Mod. Phys. D* **8**, 85–100.

Brügmann, B., J. A. Gonzalez, M. Hannam, S. Husa, U. Sperhake, and W. Tichy (2008). Calibration of Moving Puncture Simulations. *Phys. Rev. D* **77**, 024027/1–25.

Brügmann, B., W. Tichy, and N. Jansen (2004). Numerical Simulation of Orbiting Black Holes. *Phys. Rev. Lett.* **92**, 211101/1–4.

Buchdahl, H. A. (1959). General relativistic fluid spheres. *Phys. Rev.* **116**, 1027–1034.

Buonanno, A., G. B. Cook, and F. Pretorius (2007). Inspiral, merger and ring-down of equal-mass black-hole binaries. *Phys. Rev. D* **75**, 124018/1–42.

Buonanno, A. and T. Damour (1999). Effective one-body approach to general relativistic two-body dynamics. *Phys. Rev. D* **59**, 084006/1–24.

Burgay, M., N. D'Amico, A. Possenti, R. N. Manchester, A. G. Lyne, B. C. Joshi, M. A. McLaughlin, M. Kramer, J. M. Sarkissian, F. Camilo, V. Kalogera, C. Kim, and D. R. Lorimer (2003). An increased estimate of the merger rate of double neutron stars from observations of a highly relativistic system. *Nature* **426**, 531–533.

Burke, W. L. (1971). Gravitational Radiation Damping of Slowly Moving Systems Calculated Using Mathched Assymptotic Expansions. *J. Math. Phys.* **12**, 402–418.

Burko, L. M., T. W. Baumgarte, and C. Beetle (2006). Towards a wave-extraction method for numerical relativity. iii. analytical examples for the beetle-burko radiation scalar. *Phys. Rev. D* **73**, 024002/1–19.

Burrows, A., L. Dessart, C. D. Ott, and E. Livne (2007). Multi-dimensional explorations in supernova theory. *Phys. Rept.* **442**, 23–37.

Butterworth, E. M. and J. R. Ipser (1975). Rapidly rotating fluid bodies in general relativity. *Astrophys. J. Lett.* **200**, L103–L106.

Butterworth, E. M. and J. R. Ipser (1976). On the structure and stability of rapidly rotating fluid bodies in general relativity. I – The numerical method for computing structure and its application to uniformly rotating homogeneous bodies. *Astrophys. J.* **204**, 200–223.

Cabannes, H. (1970). *Theoretical magnetofluiddynamics*. Academic Press, New York.

Čadež, A. (1974). Apparent horizons in the two-black-hole problem. *Ann. Phys.* **83**, 449–457.

Caldwell, R. R., R. Dave, and P. J. Steinhardt (1998). Quintessential Cosmology Novel Models of Cosmological Structure Formation. *Ap. Sp. Sci.* **261**, 303–310.

Campanelli, M. (2005). Understanding the fate of merging supermassive black holes. *Class. Quantum Grav.* **22**, S387–S393.

Campanelli, M., C. O. Lousto, P. Marronetti, and Y. Zlochower (2006). Accurate evolutions of orbiting black-hole binaries without excision. *Phys. Rev. Lett.* **96**, 111101/1–4.

Campanelli, M., C. O. Lousto, and Y. Zlochower (2006a). Spin-orbit interactions in black-hole binaries. *Phys. Rev. D* **74**, 084023/1–12.

Campanelli, M., C. O. Lousto, and Y. Zlochower (2006b). Spinning black holes: The orbital hang-up. *Phys. Rev. D* **74**, 041501/1–5.

Campanelli, M., C. O. Lousto, Y. Zlochower, B. Krishnan, and D. Merritt (2007). Spin flips and precession in black-hole-binary mergers. *Phys. Rev. D* **75**, 064030/1–17.

Campanelli, M., C. O. Lousto, Y. Zlochower, and D. Merritt (2007a). Large merger recoils and spin flips from generic black-hole binaries. *Astrophys. J. Lett.* **659**, L5–L8.

Campanelli, M., C. O. Lousto, Y. Zlochower, and D. Merritt (2007b). Maximum gravitational recoil. *Phys. Rev. Lett.* **98**, 231102/1–4.

Cao, Z., H.-J. Yo, and J.-P. Yu (2008). A Reinvestigation of Moving Punctured Black Holes with a New Code. *Phys. Rev. D* **78**, 124011/1–17.

Carroll, S. (2004). *Spacetime and Geometry: An Introduction to General Relativity*. Addison-Wesley, San Francisco.

Carter, B. (1968). Global Structure of the Kerr Family of Gravitational Fields. *Phys. Rev.* **174**, 1559–1571.

Carter, B. and J.-P. Luminet (1983). Tidal compression of a star by a large black hole. I Mechanical evolution and nuclear energy release by proton capture. *Astron. Astrophys.* **121**, 97–113.

Carter, B. and J. P. Luminet (1985). Mechanics of the affine star model. *Mon. Not. R. Astron. Soc.* **212**, 23–55.

Castor, J. I. (1972). Radiative Transfer in Spherically Symmetric Flows. *Astrophys. J.* **178**, 779–792.

Caudill, M., G. B. Cook, J. D. Grigsby, and H. P. Pfeiffer (2006). Circular orbits and spin in black-hole initial data. *Phys. Rev. D* **74**, 064011/1–24.

Centrella, J. and J. R. Wilson (1983). Planar numerical cosmology. I – The differential equations. *Astrophys. J.* **273**, 428–435.

Centrella, J. and J. R. Wilson (1984). Planar numerical cosmology. II – The difference equations and numerical tests. *Astrophys. J. Suppl.* **54**, 229–249.

Centrella, J. M., K. C. B. New, L. L. Lowe, and J. D. Brown (2001). Dynamical Rotational Instability at Low T/W. *Astrophys. J. Lett.* **550**, L193–L196.

Chandrasekhar, S. (1931). The Maximum Mass of Ideal White Dwarfs. *Astrophys. J.* **74**, 81–82.

Chandrasekhar, S. (1960). The Stability of Non-Dissipative Couette Flow in Hydromagnetics. *Proceedings of the National Academy of Science* **46**, 253–257.

Chandrasekhar, S. (1961). *Hydrodynamic and hydromagnetic stability*. Oxford University Press, London.

Chandrasekhar, S. (1964a). Dynamical Instability of Gaseous Masses Approaching the Schwarzschild Limit in General Relativity. *Phys. Rev. Lett.* **12**, 114–116.

Chandrasekhar, S. (1964b). The Dynamical Instability of Gaseous Masses Approaching the Schwarzschild Limit in General Relativity. *Astrophys. J.* **140**, 417–433.

Chandrasekhar, S. (1969). *Ellipsoidal figures of equilibrium*. Yale University Press, New Haven.

Chandrasekhar, S. and J. L. Friedman (1972a). On the Stability of Axisymmetric Systems to Axisymmetric Perturbations in General Relativity, I. The Equations Governing Nonstationary, and Perturbed Systems. *Astrophys. J.* **175**, 379–406.

Chandrasekhar, S. and J. L. Friedman (1972b). On the Stability of Axisymmetric Systems to Axisymmetric Perturbations in General Relativity. II. a Criterion for the Onset of Instability in Uniformly Rotating Configurations and the Frequency of the Fundamental Mode in Case of Slow Rotation. *Astrophys. J.* **176**, 745–768.

Chandrasekhar, S. and J. L. Friedman (1972c). On the Stability of Axisymmetric Systems to Axisymmetric Perturbations in General Relativity. III. Vacuum Metrics and Carter's Theorem. *Astrophys. J.* **177**, 745–756.

Choptuik, M. (1998). The (Unstable) Threshold of Black Hole Formation. In N. Dadhich and J. Narlikar (Eds.), *Gravitation and Relativity: At the Turn of the Millennium*, pp. 67–86. Inter-University Centre for Astronomy and Astrophysics, Ganeshkhind, Pune.

Choptuik, M. W. (1993). Universality and scaling in gravitational collapse of a massless scalar field. *Phys. Rev. Lett.* **70**, 9–12.

Choquet-Bruhat, Y. (1952). Théorème d'existence pour certains systèmes d'équations aux dérivées partielles non linéaires. *Acta. Math.* **88**, 141–225.

Choquet-Bruhat, Y. (1962). The cauchy problem. In L. Witten (Ed.), *Gravitation: an Introduction to Current Research*, pp. 130–168. Wiley, New York.

Choquet-Bruhat, Y. and T. Ruggeri (1983). Hyperbolicity of the 3+1 system system of Einstein equations. *Commun. Math. Phys.* **89**, 269–275.

Christodoulou, D. (1970). Reversible and irreversible transformations in black-hole physics. *Phys. Rev. Lett.* **25**, 1596–1597.

Christodoulou, D. (1986). The problem of a self-gravitating scalar field. *Commun. Math. Phys.* **105**, 337–361.

Christodoulou, D. (1991). The formation of black holes and singularities in spherically symmetric gravitational collapse. *Comm. Pure and Appl. Math.* **44**, 339–373.

Christodoulou, D. (1993). Bounded variation solutions of the spherically symmetric Einstein-scalar field equations. *Comm. Pure and Appl. Math.* **46**, 1131–1220.

Christodoulou, D. and R. Ruffini (1971). Reversible Transformations of a Charged Black Hole. *Phys. Rev. D* **4**, 3552–3555.

Colella, P. and P. R. Woodward (1984). The Piecewise Parabolic Method (PPM) for Gas-Dynamical Simulations. *Journal of Computational Physics* **54**, 174–201.

Collins, N. A. and S. A. Hughes (2004). Towards a formalism for mapping the spacetimes of massive compact objects: Bumpy black holes and their orbits. *Phys. Rev. D* **69**, 124022/1–16.

Colpi, M., S. L. Shapiro, and I. Wasserman (1986). Boson stars – Gravitational equilibria of self-interacting scalar fields. *Phys. Rev. Lett.* **57**, 2485–2488.

Cook, G. B. (1991). Initial data for axisymmetric black-hole collisions. *Phys. Rev. D* **44**, 2983–3000.

Cook, G. B. (1994). Three-dimensional initail data for the collision of two black holes. II. Quasicircular orbits for equal-mass black holes. *Phys. Rev. D* **50**, 5025–5032.

Cook, G. B. (2000). Initial data for numerical relativity. *Living Rev. Relativity* **5**, 1–53.

Cook, G. B. (2002). Corotating and irrotational binary black holes in quasi-circular orbits. *Phys. Rev. D* **65**, 084003/1–13.

Cook, G. B., M. W. Choptuik, M. R. Dubal, S. Klasky, R. A. Matzner, and S. R. Oliveira (1993). Three-dimensional initial data for the collision of two black holes. *Phys. Rev. D* **47**, 1471–1490.

Cook, G. B. and H. P. Pfeiffer (2004). Excision boundary conditions for black-hole initial data. *Phys. Rev. D* **70**, 104016/1–24.

Cook, G. B. and M. A. Scheel (1997). Well-behaved harmonic time slices of a charged, rotating, boosted black hole. *Phys. Rev. D* **56**, 4775–4781.

Cook, G. B., S. L. Shapiro, and S. A. Teukolsky (1992). Spin-up of a rapidly rotating star by angular momentum loss – Effects of general relativity. *Astrophys. J.* **398**, 203–223.

Cook, G. B., S. L. Shapiro, and S. A. Teukolsky (1994a). Rapidly rotating neutron stars in general relativity: Realistic equations of state. *Astrophys. J.* **424**, 823–845.

Cook, G. B., S. L. Shapiro, and S. A. Teukolsky (1994b). Rapidly rotating polytropes in general relativity. *Astrophys. J.* **422**, 227–242.

Cook, G. B., S. L. Shapiro, and S. A. Teukolsky (1996). Testing a simplified version of einstein's equations for numerical relativity. *Phys. Rev. D* **53**, 5533–5540.

Cook, G. B. and B. F. Whiting (2007). Approximate Killing vectors on S^2. *Phys. Rev. D* **76**, 041501/1–5.

Cook, G. B. and J. W. York, Jr. (1990). Apparent horizons for boosted or spinning black holes. *Phys. Rev. D* **41**, 1077–1085.

Courant, R. and K. O. Friedrichs (1948). *Supersonic flow and shock waves*. Pure and Applied Mathematics, Interscience, New York.

Cutler, C. and É. E. Flanagan (1994). Gravitational waves from merging compact binaries: How accurately can one extract the binary's parameters from the inspiral waveform? *Phys. Rev. D* **49**, 2658–2697.

Cutler, C. and L. Lindblom (1992). Post-Newtonian frequencies for the pulsations of rapidly rotating neutron stars. *Astrophys. J.* **385**, 630–641.

Cutler, C. and K. S. Thorne (2002). An Overview of Gravitational-Wave Sources. In *Proceedings of the 16th International Conference on General Relativity and Gravitation*, pp. 72–111. World Scientific, Singapore.

Damour, T. and A. Gopakumar (2006). Gravitational recoil during binary black hole coalescence using the effective one body approach. *Phys. Rev. D* **73**, 124006/1–23.

Damour, T., E. Gourgoulhon, and P. Grandclément (2002). Circular orbits of corotating binary black holes: comparison between analytical and numerical results. *Phys. Rev. D* **66**, 024007/1–15.

Damour, T., B. R. Iyer, and B. S. Sathyaprakash (1998). Improved filters for gravitational waves from inspiraling compact binaries. *Phys. Rev. D* **57**, 885–907.

Darmois, G. (1927). Les équations de la gravitation einsteinienne. *Mém. Sc. Math.* **25**, 1–47.

Davies, M. B., W. Benz, T. Piran, and F. K. Thielemann (1994). Merging neutron stars. 1. Initial results for coalescence of noncorotating systems. *Astrophys. J.* **431**, 742–753.

De Donder, T. (1921). *La gravifique einsteinienne*. Gauthier-Villars, Paris.

De Villiers, J.-P. and J. F. Hawley (2003). A Numerical Method for General Relativistic Magnetohydrodynamics. *Astrophys. J.* **589**, 458–480.

De Villiers, J.-P., J. F. Hawley, and J. H. Krolik (2003). Magnetically Driven Accretion Flows in the Kerr Metric. I. Models and Overall Structure. *Astrophys. J.* **599**, 1238–1253.

De Villiers, J.-P., J. Staff, and R. Ouyed (2005). GRMHD Simulations of Disk/Jet Systems: Application to the Inner Engines of Collapsars. pp. 1–16.

Dedner, A., F. Kemm, D. Kröner, C.-D. Munz, T. Schnitzer, and M. Wesenberg (2002). Hyperbolic Divergence Cleaning for the MHD Equations. *Journal of Computational Physics* **175**, 645–673.

Del Zanna, L., O. Zanotti, N. Bucciantini, and P. Londrillo (2007). ECHO: an Eulerian Conservative High Order scheme for general relativistic magnetohydrodynamics and magnetodynamics. *Astron. Ap.* **473**, 11–30.

Dennett-Thorpe, J., A. H. Bridle, R. A. Laing, and P. A. G. Scheuer (1999). Asymmetry of jets, lobe size and spectral index in radio galaxies and quasars. *Mon. Not. R. Astron. Soc.* **304**, 271–280.

Dennison, K. A., T. W. Baumgarte, and H. P. Pfeiffer (2006). Approximate initial data for binary black holes. *Phys. Rev. D* **74**, 064016/1–13.

Detweiler, S. (1987). Evolution of the contraint equations in general relativity. *Phys. Rev. D* **35**, 1095–1099.

Detweiler, S. (1989). Kepler's third law in general relativity. In C. R. Evans, L. S. Finn, and D. W. Hobill (Eds.), *Frontiers in numerical relativity*, pp. 43–56. Cambridge University Press, Cambridge.

Detweiler, S. L. and L. Lindblom (1977). On the evolution of the homogeneous ellipsoidal figures. *Astrophys. J.* **213**, 193–199.

Dimmelmeier, H., J. A. Font, and E. Müller (2002a). Relativistic simulations of rotational core collapse I. Methods, initial models, and code tests. *Astron. Ap.* **388**, 917–935.

Dimmelmeier, H., J. A. Font, and E. Müller (2002b). Relativistic simulations of rotational core collapse II. Collapse dynamics and gravitational radiation. *Astron. Ap.* **393**, 523–542.

d'Inverno, R. (1992). *Introducing Einstein's Relativity*. Oxford University Press, Oxford.

Douchin, F. and P. Haensel (2001). A unified equation of state of dense matter and neutron star structure. *Astron. Ap.* **380**, 151–167.

Dreyer, O., B. Krishnan, D. Shoemaker, and E. Schnetter (2003). Introduction to isolated horizons in numerical relativity. *Phys. Rev. D* **67**, 024018/1–14.

Duez, M. D., T. W. Baumgarte, and S. L. Shapiro (2001). Computing the complete gravitational wavetrain from relativistic binary inspiral. *Phys. Rev. D* **63**, 084030/1–6.

Duez, M. D., T. W. Baumgarte, S. L. Shapiro, M. Shibata, and K. Uryū (2002). Comparing the inspiral of irrotational and corotational binary neutron stars. *Phys. Rev. D* **65**, 024016/1–8.

Duez, M. D., E. T. Engelhard, J. M. Fregeau, K. M. Huffenberger, and S. L. Shapiro (1999). Binary-induced collapse of a compact, collisionless cluster. *Phys. Rev. D* **60**, 104024/1–8.

Duez, M. D., F. Foucart, L. E. Kidder, H. P. Pfeiffer, M. A. Scheel, and S. A. Teukolsky (2008). Evolving black hole-neutron star binaries in general relativity using pseudospectral and finite difference methods. *Phys. Rev. D* **78**, 104015/1–16.

Duez, M. D., Y. T. Liu, S. L. Shapiro, M. Shibata, and B. C. Stephens (2006a). Collapse of Magnetized Hypermassive Neutron Stars in General Relativity. *Phys. Rev. Lett.* **96**, 031101/1–4.

Duez, M. D., Y. T. Liu, S. L. Shapiro, M. Shibata, and B. C. Stephens (2006b). Evolution of magnetized, differentially rotating neutron stars: Simulations in full general relativity. *Phys. Rev. D* **73**, 104015/1–25.

Duez, M. D., Y. T. Liu, S. L. Shapiro, and B. C. Stephens (2004). General relativistic hydrodynamics with viscosity: Contraction, catastrophic collapse, and disk formation in hypermassive neutron stars. *Phys. Rev. D* **69**, 104030/1–24.

Duez, M. D., Y. T. Liu, S. L. Shapiro, and B. C. Stephens (2005a). Excitation of magnetohydrodynamic modes with gravitational waves: A testbed for numerical codes. *Phys. Rev. D* **72**, 024029/1–10.

Duez, M. D., Y. T. Liu, S. L. Shapiro, and B. C. Stephens (2005b). Relativistic magnetohydrodynamics in dynamical spacetimes: Numerical methods and tests. *Phys. Rev. D* **72**, 024028/1–21.

Duez, M. D., P. Marronetti, S. L. Shapiro, and T. W. Baumgarte (2003). Hydrodynamic simulations in 3+1 general relativity. *Phys. Rev. D* **67**, 024004/1–22.

Duez, M. D., S. L. Shapiro, and H.-J. Yo (2004). Relativistic hydrodynamic evolutions with black hole excision. *Phys. Rev. D* **69**(10), 104016/1–16.

Duncan, R. C. and C. Thompson (1992). Formation of very strongly magnetized neutron stars – Implications for gamma-ray bursts. *Astrophys. J. Lett.* **392**, L9–L13.

Eardley, D. M. and L. Smarr (1979). Time functions in numerical relativity: Marginally bound dust collapse. *Phys. Rev. D* **19**, 2239–2259.

Ehlers, J. and W. Kundt (1962). Exact solutions of the gravitational field equations. In L. Witten (Ed.), *Gravitation: an Introduction to Current Research*, pp. 49–101. Wiley, New York.

Eichler, D., M. Livio, T. Piran, and D. N. Schramm (1989). Nucleosynthesis, neutrino bursts and gamma-rays from coalescing neutron stars. *Nature* **340**, 126–128.

Eisenhart, L. P. (1926). *Riemannian Geometry*. Princeton University Press, Princeton.

Eppley, K. (1977). Evolution of time-symmetric gravitational waves: Initial data and apparent horizons. *Phys. Rev. D* **16**, 1609–1614.

Estabrook, F., H. Wahlquist, S. Christensen, B. Dewitt, L. Smarr, and E. Tsiang (1973). Maximally slicing a black hole. *Phys. Rev. D* **7**, 2814–2817.

Etienne, Z. B., J. A. Faber, Y. T. Liu, S. L. Shapiro, and T. W. Baumgarte (2007). Filling the holes: Evolving excised binary black hole initial data with puncture techniques. *Phys. Rev. D* **76**, 101503/1–5.

Etienne, Z. B., J. A. Faber, Y. T. Liu, S. L. Shapiro, K. Taniguchi, and T. W. Baumgarte (2008). Fully general relativistic simulations of black hole-neutron star mergers. *Phys. Rev. D* **77**, 084002/1–22.

Etienne, Z. B., Y. T. Liu, S. L. Shapiro, and T. W. Baumgarte (2009). General relativistic simulations of black hole-neutron star mergers: Effects of black-hole spin. *Phys. Rev. D* **79**, 044024/1–26.

Evans, C. R. (1986). An approach for calculating axisymmetric gravitational collapse. In J. M. Centrella (Ed.), *Dynamical Spacetimes and Numerical Relativity*, pp. 3–39. Cambridge University Press, Cambridge.

Evans, C. R., L. S. Finn, and D. W. Hobill (1989). *Frontiers in Numerical Relativity*. Cambridge University Press, Cambridge.

Evans, C. R. and J. F. Hawley (1988). Simulation of magnetohydrodynamic flows – A constrained transport method. *Astrophys. J.* **332**, 659–677.

Evans, II, C. R. (1984). *A method for numerical relativity: Simulation of axisymmetric gravitational collapse and gravitational radiation generation*. Ph. D. thesis, (Univ. Texas, Austin.).

Evans, E., S. Iyer, W.-M. Suen, J. Tao, R. Wolfmeyer, and H.-M. Zhang (2005). Computational relativistic astrophysics with adaptive mesh refinement: testbeds. *Phys. Rev. D* **71**, 081301/1–5.

Faber, J. A., T. W. Baumgarte, Z. B. Etienne, S. L. Shapiro, and K. Taniguchi (2007). Relativistic hydrodynamics in the presence of puncture black holes. *Phys. Rev. D* **76**, 104021/1–21.

Faber, J. A., T. W. Baumgarte, S. L. Shapiro, and K. Taniguchi (2006b). General Relativistic Binary Merger Simulations and Short Gamma-Ray Bursts. *Astrophys. J. Lett.* **641**, L93–L96.

Faber, J. A., T. W. Baumgarte, S. L. Shapiro, K. Taniguchi, and F. A. Rasio (2006). Dynamical evolution of black hole-neutron star binaries in general relativity: Simulations of tidal disruption. *Phys. Rev. D* **73**, 024012/1–31.

Faber, J. A., P. Grandclément, and F. A. Rasio (2004). Mergers of irrotational neutron star binaries in conformally flat gravity. *Phys. Rev. D* **69**, 124036/1–26.

Faber, J. A. and F. A. Rasio (2000). Post-Newtonian SPH calculations of binary neutron star coalescence: Method and first results. *Phys. Rev. D* **62**, 064012/1–23.

Faber, J. A. and F. A. Rasio (2002). Post-Newtonian SPH calculations of binary neutron star coalescence. III. Irrotational systems and gravitational wave spectra. *Phys. Rev. D* **65**, 084042/1–18.

Fackerell, E. D. (1970). Relativistic, Spherically Symmetric Star Clusters.IV. a Sufficient Condition for Instability of Isotropic Clusters against Radial Perturbations. *Astrophys. J.* **160**, 859–874.

Fackerell, E. D., J. R. Ipser, and K. S. Thorne (1969). Relativistic Star Clusters. *Comments on Astrophysics and Space Physics* **1**, 134–139.

Fan, X. *et al.* (2003). A Survey of z greater than 5.7 Quasars in the Sloan Digital Sky Survey. II. Discovery of Three Additional Quasars at z greater than 6. *Astron. J.* **125**, 1649–1659.

Farris, B. D., T. K. Li, Y. T. Liu, and S. L. Shapiro (2008). Relativistic Radiation Magnetohydrodynamics in Dynamical Spacetimes: Numerical Methods and Tests. *Phys. Rev. D* **78**, 024023/1–20.

Favata, M. (2006). Are neutron stars crushed? gravitomagnetic tidal fields as a mechanism for binary-induced collapse. *Phys. Rev. D* **73**, 104005/1–26.

Favata, M., S. A. Hughes, and D. E. Holz (2004). How Black Holes Get Their Kicks: Gravitational Radiation Recoil Revisited. *Astrophys. J. Lett.* **607**, L5–L8.

Faye, G., L. Blanchet, and A. Buonanno (2006). Higher-order spin effects in the dynamics of compact binaries. i: Equations of motion. *Phys. Rev. D* **74**, 104033/1–19.

Faye, G. and G. Schäfer (2003). Optimizing the third-and-a-half post-Newtonian gravitational radiation-reaction force for numerical simulations. *Phys. Rev. D* **68**, 084001/1–25.

Finn, L. S. and C. R. Evans (1990). Determining gravitational radiation from Newtonian self-gravitating systems. *Astrophys. J.* **351**, 588–600.

Fischer, A. and J. Marsden (1972). The Einstein evolution equations as a first-order quasi-linear symmetric hyperbolic system I. *Commun. Math. Phys.* **28**, 1–38.

Fitchett, M. J. (1983). The influence of gravitational wave momentum losses on the centre of mass motion of a Newtonian binary system. *Mon. Not. R. Astron. Soc.* **203**, 1049–1062.

Flanagan, É. É. (1999). Possible explanation for star-crushing effect in binary neutron star simulations. *Phys. Rev. Lett.* **82**, 1354–1357.

Flanagan, É. É. and S. A. Hughes (2005). The basics of gravitational wave theory. *New Journal of Physics* **7**, 204/1–52.

Font, J. A. (2000). Numerical Hydrodynamics in General Relativity. *Living Rev. Relativity* **3**, 1–81.

Font, J. A., T. Goodale, S. Iyer, M. Miller, L. Rezzolla, E. Seidel, N. Stergioulas, W.-M. Suen, and M. Tobias (2002). Three-dimensional numerical general relativistic hydrodynamics. II. Long-term dynamics of single relativistic stars. *Phys. Rev. D* **65**, 084024/1–18.

Font, J. A., M. Miller, W.-M. Suen, and M. Tobias (2000). Three-dimensional numerical general relativistic hydrodynamics: Formulations, methods, and code tests. *Phys. Rev. D* **61**, 044011/1–26.

Fourès-Bruhat, Y. (1956). Sur líntégration des équations de la relativité générale. *J. Rational Mech. Anal.* **5**, 951–966.

Fowler, W. A. (1964). Massive Stars, Relativistic Polytropes, and Gravitational Radiation. *Reviews of Modern Physics* **36**, 545–554.

Freedman, W. L., B. F. Madore, B. K. Gibson, L. Ferrarese, D. D. Kelson, S. Sakai, J. R. Mould, R. C. Kennicutt, Jr., H. C. Ford, J. A. Graham, J. P. Huchra, S. M. G. Hughes, G. D. Illingworth, L. M. Macri, and P. B. Stetson (2001). Final Results from the Hubble Space Telescope Key Project to Measure the Hubble Constant. *Astrophys. J.* **553**, 47–72.

Fridman, A. M. and V. L. Polyachenko (1984). *Physics of Gravitating Systems*. Springer-Verlag, New York.

Friedman, J. L., J. R. Ipser, and R. D. Sorkin (1988). Turning-point method for axisymmetric stability of rotating relativistic stars. *Astrophys. J.* **325**, 722–724.

Friedman, J. L., L. Parker, and J. R. Ipser (1986). Rapidly rotating neutron star models. *Astrophys. J.* **304**, 115–139.

Friedman, J. L., K. Schleich, and D. M. Witt (1993). Topological censorship. *Phys. Rev. Lett.* **71**, 1486–1489.

Friedman, J. L. and B. F. Schutz (1975). Gravitational radiation instability in rotating stars. *Astrophys. J. Lett.* **199**, L157–L159.

Friedman, J. L. and B. F. Schutz (1978a). Lagrangian perturbation theory of nonrelativistic fluids. *Astrophys. J.* **221**, 937–957.

Friedman, J. L. and B. F. Schutz (1978b). Secular instability of rotating Newtonian stars. *Astrophys. J.* **222**, 281–296.

Friedman, J. L. and K. Uryū (2006). Post-Minkowski action for point particles and a helically symmetric binary solution. *Phys. Rev. D* **73**, 104039/1–28.

Friedman, J. L., K. Uryū, and M. Shibata (2002). Thermodynamics of binary black holes and neutron stars. *Phys. Rev. D* **65**, 064035/1–20.

Friedrich, H. (1981). The asymptotic characteristic initial value problem for Einstein's vacuum field equations as an initial value problem for a first-order quasilinear symmetric hyperbolic system. *Proc. R. Soc. London* **A378**, 401–421.

Friedrich, H. (1985). On the hyperbolicity of Einstein's and other gauge field equations. *Commun. Math. Phys.* **100**, 525–543.

Friedrich, H. (1996). Hyperbolic reductions for Einstein's equations. *Class. Quantum Grav.* **13**, 1451–1469.

Frittelli, S. (1997). Note on the propagation of the constraints in standard 3+1 general relativity. *Phys. Rev. D* **55**, 5992–5996.

Frittelli, S. and O. Reula (1994). On the Newtonian limit of general relativity. *Commun. Math. Phys.* **166**, 221–235.

Fryer, C. L., S. E. Woosley, and A. Heger (2001). Pair-Instability Supernovae, Gravity Waves, and Gamma-Ray Transients. *Astrophys. J.* **550**, 372–382.

Fujiwara, T. (1981). Vlasov Simulations of Stellar Systems – Infinite Homogeneous Case. *Pub. Astr. Soc. Jap.* **33**, 531–540.

Fujiwara, T. (1983). Integration of the collisionless Boltzmann equation for spherical stellar systems. *Pub. Astr. Soc. Jap.* **35**, 547–558.

Gammie, C. F., J. C. McKinney, and G. Tóth (2003). HARM: A Numerical Scheme for General Relativistic Magnetohydrodynamics. *Astrophys. J.* **589**, 444–457.

Gammie, C. F., S. L. Shapiro, and J. C. McKinney (2004). Black Hole Spin Evolution. *Astrophys. J.* **602**, 312–319.

Gannon, D. (1976). On the topology of spacelike hypersurfaces, singularities, and black holes. *Gen. Rel. Grav.* **7**, 219–232.

Garat, A. and R. H. Price (2000). Nonexistence of conformally flat slices of the Kerr spacetime. *Phys. Rev. D* **61**, 124011/1–4.

Garfinkle, D. (2002). Harmonic coordinate method for simulating generic singularities. *Phys. Rev. D* **65**, 044029/1–6.

Garfinkle, D. and G. C. Duncan (2001). Numerical evolution of Brill waves. *Phys. Rev. D* **63**, 044011/1–8.

Gilden, D. L. and S. L. Shapiro (1984). Gravitational radiation from colliding compact stars Hydrodynamical calculations in two dimensions. *Astrophys. J.* **287**, 728–744.

Gingold, R. A. and J. J. Monaghan (1977). Smoothed particle hydrodynamics – Theory and application to non-spherical stars. *Mon. Not. R. Astron. Soc.* **181**, 375–389.

Gleiser, R. J., G. Khanna, and J. Pullin (2002). Evolving the bowen-york initial data for boosted black holes. *Phys. Rev. D* **66**, 024035/1–9.

Gleiser, R. J., C. O. Nicasio, R. H. Price, and J. Pullin (1998). Evolving the Bowen-York initial data for spinning black holes. *Phys. Rev. D* **57**, 3401–3407.

Glendenning, N. K. (1996). *Compact Stars: Nuclear Physics, Particle Physics, and General Relativity*. Springer-Verlag, New York.

Goldreich, P. and S. V. Weber (1980). Homologously collapsing stellar cores. *Astrophys. J.* **238**, 991–997.

González, J. A., M. D. Hannam, U. Sperhake, B. Brugmann, and S. Husa (2007). Supermassive recoil velocities for binary black-hole mergers with antialigned spins. *Phys. Rev. Lett.* **98**, 231101/1–4.

González, J. A., U. Sperhake, B. Bruegmann, M. Hannam, and S. Husa (2007). Total recoil: the maximum kick from nonspinning black-hole binary inspiral. *Phys. Rev. Lett.* **98**, 091101/1–4.

Gottlieb, D. and S. A. Orszag (1977). *Numerical analysis of spectral methods: theory and applications.* Society for Industrial and Applied Mathematics, Philadelphia.

Gourgoulhon, E. (1998). Relations between three formalisms for irrotational binary neutron stars in general relativity. pp. 1–5.

Gourgoulhon, E. (2002). Introduction to spectral methods. Presentation at the 4th EU Network Meeting, Palma de Mallorca; available at http://www.lorene.obspm.fr/.

Gourgoulhon, E. and S. Bonazzola (1994). A formulation of the virial theorem in general relativity. *Class. Quantum Grav.* **11**, 443–452.

Gourgoulhon, E., P. Grandclément, and S. Bonazzola (2002). Binary black holes in circular orbits. I. A global spacetime approach. *Phys. Rev. D* **65**, 044020/1–19.

Gourgoulhon, E., P. Grandclément, K. Taniguchi, J.-A. Marck, and S. Bonazzola (2001). Quasiequilibrium sequences of synchronized and irrotational binary neutron stars in general relativity: Method and tests. *Phys. Rev. D* **63**, 064029/1–27.

Grandclément, P. (2006). Accurate and realistic initial data for black hole neutron star binaries. *Phys. Rev. D* **74**, 124002/1–6.

Grandclément, P., S. Bonazzola, E. Gourgoulhon, and J.-A. Marck (2001). A Multidomain Spectral Method for Scalar and Vectorial Poisson Equations with Noncompact Sources. *Journal of Computational Physics* **170**, 231–260.

Grandclément, P., E. Gourgoulhon, and S. Bonazzola (2002). Binary black holes in circular orbits. II. Numerical methods and first results. *Phys. Rev. D* **65**, 044021/1–18.

Gundlach, C. (1998). Pseudospectral apparent horizon finders: An efficient new algorithm. *Phys. Rev. D* **57**, 863–875.

Gundlach, C. (2000). Critical phenomena in gravitational collapse. *Living Rev. Relativity* **2**, 1–58.

Gundlach, C. (2003). Critical phenomena in gravitational collapse. *Phys. Repts.* **376**, 339–405.

Gundlach, C. and J. M. Martin-Garcia (2006a). Hyperbolicity of second-order in space systems of evolution equations. *Class. Quantum Grav.* **23**, S387–S393.

Gundlach, C. and J. M. Martin-Garcia (2006b). Well-posedness of formulations of the Einstein equations with dynamical lapse and shift conditions. *Phys. Rev. D* **74**, 024016/1–19.

Gundlach, C., J. M. Martin-Garcia, G. Calabrese, and I. Hinder (2005). Constraint damping in the Z4 formulation and harmonic gauge. *Class. Quantum Grav.* **22**, 3767–3773.

Hachisu, I. (1986). A versatile method for obtaining structures of rapidly rotating stars. *Astrophys. J. Suppl.* **61**, 479–507.

Haensel, P., A. Y. Potekhin, and D. G. Yakovlev (2007). *Neutron Stars: Equation of State and Structure.* Springer-Verlag, New York.

Hannam, M., S. Husa, N. Ó. Murchadha, B. Brügmann, J. A. González, and U. Sperhake (2007). Where do moving punctures go? *J. Phys. Conf. Series* **66**, 012047/1–9.

Hannam, M., S. Husa, F. Ohme, B. Brügmann, and N. Ó. Murchadha (2008). Wormholes and trumpets: the Schwarzschild spacetime for the moving-puncture generation. *Phys. Rev. D* **78**, 064020/1–19.

Hannam, M., S. Husa, D. Pollney, B. Bruegmann, and N. O'Murchadha (2007). Geometry and Regularity of Moving Punctures. *Phys. Rev. Lett.* **99**, 241102/1–4.

Hannam, M. D. (2005). Quasi-circular orbits of conformal thin-sandwich puncture binary black holes. *Phys. Rev. D* **72**, 044025/1–8.

Hannam, M. D. and G. B. Cook (2005). Conformal thin-sandwich puncture initial data for boosted black holes. *Phys. Rev. D* **71**, 084023/1–12.

Hannam, M. D., C. R. Evans, G. B. Cook, and T. W. Baumgarte (2003). Can a combination of the conformal thin-sandwich and puncture methods yield binary black hole solutions in quasi-equilibrium? *Phys. Rev. D* **68**, 064003/1–7.

Harten, A., P. D. Lax, and B. van Leer (1983). On upstream differencing and Godunov-type schemes for hyperbolic conservation laws. *SIAM Review* **25**, 35–61.

Hartle, J. B. (1967). Slowly Rotating Relativistic Stars. I. Equations of Structure. *Astrophys. J.* **150**, 1005–1030.

Hartle, J. B. (1970). Slowly-Rotating Relativistic Stars.IV. Rotational Energy and Moment of Inertia for Stars in Differential Rotation. *Astrophys. J.* **161**, 111–118.

Hartle, J. B. and K. S. Thorne (1968). Slowly Rotating Relativistic Stars. II. Models for Neutron Stars and Supermassive Stars. *Astrophys. J.* **153**, 807–834.

Hawking, S. W. (1971). Gravitational Radiation from Colliding Black Holes. *Phys. Rev. Lett.* **26**, 1344–1346.

Hawking, S. W. (1972). Black holes in general relativity. *Commun. Math. Phys.* **25**, 152–166.

Hawking, S. W. (1973). The Event Horizon. In C. DeWitt and B. S. DeWitt (Eds.), *Black Holes*, pp. 1–55. Gordon and Breach, New York.

Hawking, S. W. (1974). Black hole explosions? *Nature* **248**, 30–31.

Hawking, S. W. (1975). Particle Creation by Black Holes. *Commun. Math. Phys.* **43**, 199–220.

Hawking, S. W. and G. F. R. Ellis (1973). *The large scale structure of space-time*. Cambridge University Press, Cambridge.

Hawley, J. F., L. L. Smarr, and J. R. Wilson (1984). A numerical study of nonspherical black hole accretion. I Equations and test problems. *Astrophys. J.* **277**, 296–311.

Hernandez, Jr., W. C. and C. W. Misner (1966). Observer Time as a Coordinate in Relativistic Spherical Hydrodynamics. *Astrophys. J.* **143**, 452–464.

Hernquist, L. (1993). Some cautionary remarks about smoothed particle hydrodynamics. *Astrophys. J.* **404**, 717–722.

Herrmann, F., I. Hinder, D. Shoemaker, and P. Laguna (2007a). Unequal-Mass Binary Black Hole Plunges and Gravitational Recoil. *Class. Quantum Grav.* **24**, S33–S42.

Herrmann, F., I. Hinder, D. Shoemaker, P. Laguna, and R. A. Matzner (2007b). Gravitational recoil from spinning binary black hole mergers. *Astrophys. J.* **661**, 430–436.

Hessels, J. W. T., S. M. Ransom, I. H. Stairs, P. C. C. Freire, V. M. Kaspi, and F. Camilo (2006). A Radio Pulsar Spinning at 716 Hz. *Science* **311**, 1901–1904.

Hewish, A., S. J. Bell, J. D. Pilkington, P. F. Scott, and R. A. Collins (1968). Observation of a Rapidly Pulsating Radio Source. *Nature* **217**, 709–713.

Hinshaw, G., J. L. Weiland, R. S. Hill, N. Odegard, D. Larson, C. L. Bennett, J. Dunkley, B. Gold, M. R. Greason, N. Jarosik, E. Komatsu, M. R. Nolta, L. Page, D. N. Spergel, E. Wollack, M. Halpern, A. Kogut, M. Limon, S. S. Meyer, G. S. Tucker, and E. L. Wright (2009). Five-Year Wilkinson Microwave Anisotropy Probe (WMAP) Observations: Data Processing, Sky Maps, and Basic Results. *Astrophys. J. Suppl.* **180**, 225–245.

Ho, L. C. (Ed.) (2004). *Coevolution of Black Holes and Galaxies*. Cambridge Univ. Press, Cambridge.

Hockney, R. W. and J. W. Eastwood (1981). *Computer Simulation Using Particles*. McGraw-Hill, New York.

Hopman, C. and T. Alexander (2005). The Orbital Statistics of Stellar Inspiral and Relaxation near a Massive Black Hole: Characterizing Gravitational Wave Sources. *Astrophys. J.* **629**, 362–372.

Hübner, P. (1996). A method for calculating the structure of (singular) spacetimes in the large. *Phys. Rev. D* **53**, 701–721.

Hughes, S. A. and R. D. Blandford (2003). Black Hole Mass and Spin Coevolution by Mergers. *Astrophys. J. Lett.* **585**, L101–L104.

Hughes, S. A., C. R. Keeton, II, P. Walker, K. T. Walsh, S. L. Shapiro, and S. A. Teukolsky (1994). Finding black holes in numerical spacetimes. *Phys. Rev. D* **49**, 4004–4015.

Hulse, R. A. and J. H. Taylor (1975). Discovery of a pulsar in a binary system. *Astrophys. J. Lett.* **195**, L51–L53.

Huq, M. F., M. W. Choptuik, and R. A. Matzner (2002). Locating boosted Kerr and Schwarzschild apparent horizons. *Phys. Rev. D* **66**, 084024/1–15.

Imamura, J. N., J. L. Friedman, and R. H. Durisen (1985). Secular stability limits for rotating polytropic stars. *Astrophys. J.* **294**, 474–478.

Imamura, J. N., J. Toman, R. H. Durisen, B. K. Pickett, and S. Yang (1995). Nonaxisymmetric secular instabilities driven by star/disk coupling. *Astrophys. J.* **444**, 363–375.

Inagaki, S., M. T. Nishida, and J. A. Sellwood (1984). A check on the non-linear behaviour of galaxy simulations. *Mon. Not. R. Astron. Soc.* **210**, 589–596.

Ioka, K. and K. Taniguchi (2000). Gravitational Waves from Inspiraling Compact Binaries with Magnetic Dipole Moments. *Astrophys. J.* **537**, 327–333.

Ipser, J. R. (1969a). Relativistic, Spherically Symmetric Star Clusters. II. Sufficient Conditions for Stability against Radial Perturbations. *Astrophys. J.* **156**, 509–528.

Ipser, J. R. (1969b). Relativistic, Spherically Symmetric Star Clusters. III. Stability of Compact Isotropic Models. *Astrophys. J.* **158**, 17–44.

Ipser, J. R. (1980). A binding-energy criterion for the dynamical stability of spherical stellar systems in general relativity. *Astrophys. J.* **238**, 1101–1110.

Ipser, J. R. and L. Lindblom (1989). Oscillations and stability of rapidly rotating neutron stars. *Phys. Rev. Lett.* **62**, 2777–2780.

Ipser, J. R. and L. Lindblom (1990). The oscillations of rapidly rotating Newtonian stellar models. *Astrophys. J.* **355**, 226–240.

Ipser, J. R. and L. Lindblom (1991). The oscillations of rapidly rotating Newtonian stellar models. II - Dissipative effects. *Astrophys. J.* **373**, 213–221.

Ipser, J. R. and R. A. Managan (1984). On the emission of gravitational radiation from inhomogeneous Jacobi configurations. *Astrophys. J.* **282**, 287–290.

Ipser, J. R. and K. S. Thorne (1968). Relativistic, Spherically Symmetric Star Clusters. I. Stability Theory for Radial Perturbations. *Astrophys. J.* **154**, 251–270.

Isenberg, J. (1978). Unpublished.

Jacobson, T. and S. Venkataramani (1995). Topology of event horizons and topological censorship. *Class. Quantum Grav.* **12**, 1055–1061.

James, R. A. (1964). The Structure and Stability of Rotating Gas Masses. *Astrophys. J.* **140**, 552–582.

Janka, H.-T., T. Eberl, M. Ruffert, and C. L. Fryer (1999). Black Hole-Neutron Star Mergers as Central Engines of Gamma-Ray Bursts. *Astrophys. J. Lett.* **527**, L39–L42.

Janka, H.-T., K. Langanke, A. Marek, G. Martínez-Pinedo, and B. Müller (2007). Theory of core-collapse supernovae. *Phys. Rept.* **442**, 38–74.

Jansen, N., P. Diener, J. Hansen, A. Khokhlov, and I. Novikov (2003). Local and global properties of conformal initial data for black-hole collisions . *Class. Quantum Grav.* **20**, 51–73.

Jaramillo, J. L., E. Gourgoulhon, and G. A. Mena Marugan (2004). Inner boundary conditions for black hole initial data derived from isolated horizons. *Phys. Rev. D* **70**, 124036/1–13.

Jin, K.-J. and W.-M. Suen (2007). Critical Phenomena in Head-On Collisions of Neutron Stars. *Phys. Rev. Lett.* **98**, 131101/1–4.

Källberg, A., G. Brodin, and M. Bradley (2004). Nonlinear coupled Alfvén and gravitational waves. *Phys. Rev. D* **70**, 044014/1–10.

Kalogera, V., K. Belczynski, C. Kim, R. O'Shaughnessy, and B. Willems (2007). Formation of double compact objects. *Phys. Reports* **442**, 75–108.

Kaspi, V. M., F. Crawford, R. N. Manchester, A. G. Lyne, F. Camilo, N. D'Amico, and B. M. Gaensler (1998). The 69 Millisecond Radio Pulsar near the Supernova Remnant RCW 103. *Astrophys. J. Lett.* **503**, L161–L165.

Katz, J., G. Horwitz, and M. Klapisch (1975). Thermodynamic stability of relativistic stellar clusters. *Astrophys. J.* **199**, 307–321.

Kaup, D. J. (1968). Klein-Gordon Geon. *Phys. Rev.* **172**, 1331–1342.

Kelly, B., P. Laguna, K. Lockitch, J. Pullin, E. Schnetter, D. Shoemaker, and M. Tiglio (2001). A cure for unstable numerical evolutions of single black holes: adjusting the standard adm equations. *Phys. Rev. D* **64**, 084013/1–14.

Kemball, A. J. and N. T. Bishop (1991). The numerical determination of apparent horizons. *Class. Quantum Grav.* **8**, 1361–1367.

Kennefick, D. (2000). Star crushing: Theoretical practice and the theoreticians' regress. *Social Studies of Science* **30**, 5–40.

Kerr, R. P. (1963). Gravitational Field of a Spinning Mass as an Example of Algebraically Special Metrics. *Phys. Rev. Lett.* **11**, 237–238.

Khokhlov, A. M. and I. D. Novikov (2002). Gauge stability of 3 + 1 formulations of general relativity. *Class. Quantum Grav.* **19**, 827–846.

Kidder, L. E. (1995). Coalescing binary systems of compact objects to (post)$^{5/2}$-Newtonian order. V. Spin effects. *Phys. Rev. D* **52**, 821–847.

Kidder, L. E., M. A. Scheel, and S. A. Teukolsky (2001). Extending the lifetime of 3d black hole computations with a new hyperbolic system of evolution equations. *Phys. Rev. D* **64**, 064017/1–13.

Kidder, L. E., M. A. Scheel, S. A. Teukolsky, E. D. Carlson, and G. B. Cook (2000). Black hole evolution by spectral methods. *Phys. Rev. D* **62**, 084032/1–20.

Knapp, A. M., E. J. Walker, and T. W. Baumgarte (2002). Illustrating stability properties of numerical relativity in electrodynamics. *Phys. Rev. D* **65**, 064031/1–5.

Kobayashi, S. and P. Mészáros (2003). Gravitational Radiation from Gamma-Ray Burst Progenitors. *Astrophys. J.* **589**, 861–870.

Kochanek, C. S. (1992a). Coalescing binary neutron stars. *Astrophys. J.* **398**, 234–247.

Kochanek, C. S. (1992b). The dynamical evolution of tidal capture binaries. *Astrophys. J.* **385**, 604–620.

Koide, S., D. L. Meier, K. Shibata, and T. Kudoh (2000). General Relativistic Simulations of Early Jet Formation in a Rapidly Rotating Black Hole Magnetosphere. *Astrophys. J.* **536**, 668–674.

Koide, S., K. Shibata, and T. Kudoh (1999). Relativistic Jet Formation from Black Hole Magnetized Accretion Disks: Method, Tests, and Applications of a General RelativisticMagnetohydrodynamic Numerical Code. *Astrophys. J.* **522**, 727–752.

Kokkotas, K. and B. Schmidt (1999). Quasi-Normal Modes of Stars and Black Holes. *Living Rev. Relativity* **2**, 1–72.

Komar, A. (1959). Covariant conservation laws in general relativity. *Phys. Rev.* **113**, 934–936.

Komatsu, H., Y. Eriguchi, and I. Hachisu (1989a). Rapidly rotating general relativistic stars. I – Numerical method and its application to uniformly rotating polytropes. *Mon. Not. R. Astron. Soc.* **237**, 355–379.

Komatsu, H., Y. Eriguchi, and I. Hachisu (1989b). Rapidly rotating general relativistic stars. II – Differentially rotating polytropes. *Mon. Not. R. Astron. Soc.* **239**, 153–171.

Komissarov, S. S. (1997). On the properties of Alfvén waves in relativistic magnetohydrodynamics. *Physics Letters A* **232**, 435–442.

Komissarov, S. S. (1999). A Godunov-type scheme for relativistic magnetohydrodynamics. *Mon. Not. R. Astron. Soc.* **303**, 343–366.

Komissarov, S. S. (2004). Electrodynamics of black hole magnetospheres. *Mon. Not. R. Astron. Soc.* **350**, 427–448.

Koppitz, M., D. Pollney, C. Reisswig, L. Rezzolla, J. Thornburg, P. Diener, and E. Schnetter (2007). Recoil velocities from equal-mass binary-black-hole mergers. *Phys. Rev. Lett.* **99**, 041102/1–4.

Kosovichev, A. G. and I. D. Novikov (1992). Non-linear effects at tidal capture of stars by a massive black hole. I – Incompressible affine model. *Mon. Not. R. Astron. Soc.* **258**, 715–724.

Kreiss, H.-O. and J. Lorenz (1989). *Initial-boundary value problems and the Navier-Stokes equations*. Academic Press, San Diego.

Kreiss, H. O. and J. Oliger (1973). *Methods for the Approximate Solution of Time Dependent Problems, GARP Publication Series No. 10*. World Meterological Organization, Geneva.

Kruskal, M. D. (1960). Maximal Extension of Schwarzschild Metric. *Phys. Rev.* **119**, 1743–1745.

Kulkarni, A. D., L. C. Shepley, and J. W. York, Jr. (1983). Initial data for N black holes. *Phys. Lett.* **96A**, 228–230.

Lai, D., F. A. Rasio, and S. L. Shapiro (1993a). Ellipsoidal figures of equilibrium – Compressible models. *Astrophys. J. Suppl.* **88**, 205–252.

Lai, D., F. A. Rasio, and S. L. Shapiro (1993b). Hydrodynamic instability and coalescence of close binary systems. *Astrophys. J. Lett.* **406**, L63–L66.

Lai, D., F. A. Rasio, and S. L. Shapiro (1994a). Equilibrium, stability, and orbital evolution of close binary systems. *Astrophys. J.* **423**, 344–370.

Lai, D., F. A. Rasio, and S. L. Shapiro (1994b). Hydrodynamic instability and coalescence of binary neutron stars. *Astrophys. J.* **420**, 811–829.

Lai, D., F. A. Rasio, and S. L. Shapiro (1994c). Hydrodynamics of rotating stars and close binary interactions: Compressible ellipsoid models. *Astrophys. J.* **437**, 742–769.

Lai, D. and S. L. Shapiro (1995a). Gravitational radiation from rapidly rotating nascent neutron stars. *Astrophys. J.* **442**, 259–272.

Lai, D. and S. L. Shapiro (1995b). Hydrodynamics of coalescing binary neutron stars: Ellipsoidal treatment. *Astrophys. J.* **443**, 705–716.

Lanczos, C. (1922). Ein vereinfachtes Koordinatensystem für die Einsteinschen Gravitations-gleichungen. *Phys. Z.* **23**, 537–539.

Landau, L. D. and E. M. Lifshitz (1959). *Fluid Mechanics*. Pergamon Press, London.

Landry, W. and S. A. Teukolsky (1999). An efficient method for fully relativistic simulations of coalescing binary neutron stars. pp. 1–10.

Lattimer, J. M. and F. D. Swesty (1991). A generalized equation of state for hot, dense matter. *Nuclear Physics A* **535**, 331–376.

Leahy, J. P. and A. G. Williams (1984). The bridges of classical double radio sources. *Mon. Not. R. Astron. Soc.* **210**, 929–951.

Lee, T. D. and Y. Pang (1992). Nontopological solitons. *Phys. Rept.* **221**, 251–350.

Lee, W. H. and W. Ł. Kluźniak (1999). Newtonian Hydrodynamics of the Coalescence of Black Holes with Neutron Stars. I. Tidally Locked Binaries with a Stiff Equation of State. *Astrophys. J.* **526**, 178–199.

Lehner, L. and O. M. Moreschi (2007). Dealing with delicate issues in waveforms calculations. *Phys. Rev. D* **76**, 124040/1–12.

Libson, J., J. Massó, E. Seidel, and W.-M. Suen (1996). A 3d apparent horizon finder. In R. T. Jantzen, G. M. Keiser, and R. Ruffini (Eds.), *The Seventh Marcel Grossmann Meeting*, pp. 631–633. World Scientific, Singapore.

Libson, J., J. Massó, E. Seidel, W.-M. Suen, and P. Walker (1996). Event horizons in numerical relativity: Methods and tests. *Phys. Rev. D* **53**, 4335–4350.

Lichnerowicz, A. (1944). L'intégration des équations de la gravitation relativiste et la problème des *n* corps. *J. Math. Pure Appl* **23**, 37–63.

Lichnerowicz, A. (1967). *Relativistic Hydrodynamics and Magnetohydrodynamics*. Benjamin, New York.

Lightman, A. P., W. H. Press, R. H. Price, and S. A. Teukolsky (1975). *Problem book in relativity and gravitation*. Princeton University Press, Princeton.

Lightman, A. P. and S. L. Shapiro (1978). The dynamical evolution of globular clusters. *Reviews of Modern Physics* **50**, 437–481.

Lin, C. C., L. Mestel, and F. H. Shu (1965). The Gravitational Collapse of a Uniform Spheroid. *Astrophys. J.* **142**, 1431–1446.

Lindblom, L. and M. A. Scheel (2002). Energy norms and the stability of the Einstein evolution equations. *Phys. Rev. D* **66**, 084014/1–16.

Lindblom, L., M. A. Scheel, L. E. Kidder, R. Owen, and O. Rinne (2006). A new generalized harmonic evolution system. *Class. Quantum Grav.* **23**, S447–S462.

Lindquist, R. W. (1963). Initial-value problem on Einstein-Rosen manifolds. *J. Math. Phys.* **4**, 938–950.

Lindquist, R. W. (1966). Relativistic transport theory. *Annals of Physics* **37**, 487–518.

Liu, Y. T., S. L. Shapiro, Z. B. Etienne, and K. Taniguchi (2008). General relativistic simulations of magnetized binary neutron star mergers. *Phys. Rev. D* **78**, 024012/1–20.

Liu, Y. T., S. L. Shapiro, and B. C. Stephens (2007). Magnetorotational collapse of very massive stars to black holes in full general relativity. *Phys. Rev. D* **76**, 084017/1–17.

Löffler, F., L. Rezzolla, and M. Ansorg (2006). Numerical evolutions of a black hole-neutron star system in full general relativity: Head-on collision. *Phys. Rev. D* **74**, 104018/1–16.

Lombardi, J. C., A. Sills, F. A. Rasio, and S. L. Shapiro (1999). Tests of Spurious Transport in Smoothed Particle Hydrodynamics. *J. Comp. Phys.* **152**, 687–735.

Lovelace, G., R. Owen, H. P. Pfeiffer, and T. Chu (2008). Binary-black-hole initial data with nearly-extremal spins. *Phys. Rev. D* **78**, 084017/1–28.

Lucy, L. B. (1977). A numerical approach to the testing of the fission hypothesis. *Astron. J.* **82**, 1013–1024.

Luminet, J.-P. and B. Carter (1986). Dynamics of an affine star model in a black hole tidal field. *Astrophys. J. Suppl.* **61**, 219–248.

Lyford, N. D., T. W. Baumgarte, and S. L. Shapiro (2003). Effects of Differential Rotation on the Maximum Mass of Neutron Stars. *Astrophys. J.* **583**, 410–415.

Lynden-Bell, D. (1967). Statistical mechanics of violent relaxation in stellar systems. *Mon. Not. R. Astron. Soc.* **136**, 101–121.

Lynden-Bell, D. and J. P. Ostriker (1967). On the stability of differentially rotating bodies. *Mon. Not. R. Astron. Soc.* **136**, 293–310.

MacFadyen, A. I. and S. E. Woosley (1999). Collapsars: Gamma-Ray Bursts and Explosions in "Failed Supernovae". *Astrophys. J.* **524**, 262–289.

MacFadyen, A. I., S. E. Woosley, and A. Heger (2001). Supernovae, Jets, and Collapsars. *Astrophys. J.* **550**, 410–425.

Madau, P. and M. J. Rees (2001). Massive Black Holes as Population III Remnants. *Astrophys. J. Lett.* **551**, L27–L30.

Managan, R. A. (1985). On the secular instability of axisymmetric rotating stars to gravitational radiation reaction. *Astrophys. J.* **294**, 463–473.

Marronetti, P. (2005). Hamiltonian relaxation. *Class. Quantum Grav.* **22**, 2433–2451.

Marronetti, P. (2006). Momentum constraint relaxation. *Class. Quantum Grav.* **23**, 2681–2695.

Marronetti, P., G. J. Mathews, and J. R. Wilson (1998). Binary neutron-star systems: From the Newtonian regime to the last stable orbit. *Phys. Rev. D* **58**, 107503/1–4.

Marronetti, P. and R. A. Matzner (2000). Solving the initial value problem of two black holes. *Phys. Rev. Lett.* **85**, 5500–5503.

Marronetti, P. and S. L. Shapiro (2003). Relativistic models for binary neutron stars with arbitrary spins. *Phys. Rev. D* **68**, 104024/1–19.

Marronetti, P., W. Tichy, B. Brügmann, J. González, M. Hannam, S. Husa, and U. Sperhake (2007). Binary black holes on a budget: simulations using workstations. *Class. Quantum Grav.* **24**, S43–S48.

Marronetti, P., W. Tichy, B. Brügmann, J. González, and U. Sperhake (2007). High-spin binary black hole mergers. *Phys. Rev. D* **77**, 064010/1–8.

Marshall, F. E., E. V. Gotthelf, W. Zhang, J. Middleditch, and Q. D. Wang (1998). Discovery of an Ultrafast X-Ray Pulsar in the Supernova Remnant N157B. *Astrophys. J. Lett.* **499**, L179–L182.

Martí, J. M. and E. Müller (1999). Numerical Hydrodynamics in Special Relativity. *Living Rev. Relativity* **2**, 1–101.

Matera, K., T. W. Baumgarte, and E. Gourgoulhon (2008). Shells around black holes: The effect of freely specifiable quantities in Einstein's constraint equations. *Phys. Rev. D* **77**, 024049/1–9.

Mathews, G. J. and J. R. Wilson (2000). Revised relativistic hydrodynamical model for neutron-star binaries. *Phys. Rev. D* **61**, 127304/1–4.

Mathews, J. and R. L. Walker (1970). *Mathematical Methods of Physics.* Benjamin/Cummings, Menlo Park.

Matzner, R. A., H. E. Seidel, S. L. Shapiro, L. Smarr, W.-M. Suen, S. A. Teukolsky, and J. Winicour (1995). Geometry of a black hole collision. *Science* **270**, 941–947.

May, M. M. and R. H. White (1966). Hydrodynamic Calculations of General-Relativistic Collapse. *Phys. Rev.* **141**, 1232–1241.

May, M. M. and R. H. White (1967). Stellar Dynamics and Gravitational Collapse. *Meth. Computat. Phys.* **7**, 219–258.

Merritt, D. and R. D. Ekers (2002). Tracing Black Hole Mergers Through Radio Lobe Morphology. *Science* **297**, 1310–1313.

Merritt, D., M. Milosavljević, M. Favata, S. A. Hughes, and D. E. Holz (2004). Consequences of Gravitational Radiation Recoil. *Astrophys. J. Lett.* **607**, L9–L12.

Merritt, D. and M. Y. Poon (2004). Chaotic Loss Cones and Black Hole Fueling. *Astrophys. J.* **606**, 788–798.

Mezzacappa, A. and R. A. Matzner (1989). Computer simulation of time-dependent, spherically symmetric spacetimes containing radiating fluids – Formalism and code tests. *Astrophys. J.* **343**, 853–873.

Mihalas, D. and B. W. Mihalas (1984). *Foundations of radiation hydrodynamics.* Oxford University Press, New York.

Miller, B. D. (1974). The Effect of Gravitational Radiation-Reaction on the Evolution of the Riemann S-Type Ellipsoids. *Astrophys. J.* **187**, 609–620.

Miller, M. and W.-M. Suen (2003). Towards a Realistic Neutron Star Binary Inspiral. pp. 1–4.

Misner, C. W. (1960). Wormhole initial conditions. *Phys. Rev.* **118**, 1110–1111.

Misner, C. W. (1963). The method of images in geometrostatics. *Ann. Phys.* **24**, 102–117.

Misner, C. W. and D. H. Sharp (1964). Relativistic Equations for Adiabatic, Spherically Symmetric Gravitational Collapse. *Phys. Rev.* **136**, 571–576.

Misner, C. W., K. S. Thorne, and J. A. Wheeler (1973). *Gravitation.* Freeman, New York.

Monaghan, J. J. and J. C. Lattanzio (1985). A refined particle method for astrophysical problems. *Astron. ApJ.* **149**, 135–143.

Moncrief, V. (1974). Gravitational perturbations of spherically symmetric systems. I. The exterior problem. *Annals of Physics* **88**, 323–342.

Moortgat, J. and J. Kuijpers (2003). Gravitational and magnetosonic waves in gamma-ray bursts. *Astron. Ap.* **402**, 905–911.

Moortgat, J. and J. Kuijpers (2004). Gravitational waves in magnetized relativistic plasmas. *Phys. Rev. D* **70**, 023001/1–12.

Morrison, I. A., T. W. Baumgarte, and S. L. Shapiro (2004). Effect of Differential Rotation on the Maximum Mass of Neutron Stars: Realistic Nuclear Equations of State. *Astrophys. J.* **610**, 941–947.

Murgia, M., P. Parma, H. R. de Ruiter, M. Bondi, R. D. Ekers, R. Fanti, and E. B. Fomalont (2001). A multi-frequency study of the radio galaxy NGC 326. I. The data. *Astron. Astrophys.* **380**, 102–116.

Nagar, A. and L. Rezzolla (2005). Gauge-invariant non-spherical metric perturbations of schwarzschild black-hole spacetimes. *Class. Quantum Grav.* **22**, R167–R192.

Nakamura, T. (2002). Possible formation scenario of the quark star of maximum mass around 0.7 solar masses. In M. Shibata, Y. Eriguchi, and K. Taniguchi (Eds.), *Proceedings of the 12th Workshop on General Relativity and Gravitation in Japan*, pp. 81–86. University of Tokyo, Tokyo.

Nakamura, T., Y. Kojima, and K. Oohara (1984). A method of determining apparent horizons in three-dimensional numerical relativity. *Phys. Lett.* **106A**, 235–238.

Nakamura, T., Y. Kojima, and K. Oohara (1985). Apparent horizons of time-symmetric initial value for three black holes. *Phys. Lett.* **107A**, 452–455.

Nakamura, T. and K. Oohara (1991). Gravitational Radiation from Coalescing Binary Neutron Stars. IV —Tidal Disruption—. *Progress of Theoretical Physics* **86**, 73–88.

Nakamura, T., K. Oohara, and Y. Kojima (1987). General relativistic collapse to black holes and gravitational waves from black holes. *Prog. Theor. Phys. Suppl.* **90**, 1–218.

Nakamura, T. and H. Sato (1981). General Relativistic Collapse of Rotating Supermassive Stars. *Progress of Theoretical Physics* **66**, 2038–2051.

Nakamura, T., S. L. Shapiro, and S. A. Teukolsky (1988). Naked singularities and the Hoop conjecture: An analytic exploration. *Phys. Rev. D* **38**, 2972–2978.

Narayan, R., B. Paczynski, and T. Piran (1992). Gamma-ray bursts as the death throes of massive binary stars. *Astrophys. J. Lett.* **395**, L83–L86.

New, K. C. B., J. M. Centrella, and J. E. Tohline (2000). Gravitational waves from long-duration simulations of the dynamical bar instability. *Phys. Rev. D* **62**, 064019/1–16.

New, K. C. B. and J. E. Tohline (1997). The Relative Stability against Merger of Close, Compact Binaries. *Astrophys. J.* **490**, 311–327.

Newman, E. T., E. Couch, K. Chinnapared, A. Exton, A. Prakash, and R. Torrence (1965). Metric of a Rotating Charged Mass. *J. Math. Phys.* **6**, 918–919.

Newman, E. T. and R. Penrose (1962). An approach to gravitational radiation by a method of spin coefficients. *J. Math. Phys.* **3**, 566–578.

Newman, E. T. and R. Penrose (1963). Errata: An approach to gravitational radiation by a method of spin coefficients. *J. Math. Phys.* **4**, 998–998.

Nice, D. J., E. M. Splaver, I. H. Stairs, O. Löhmer, A. Jessner, M. Kramer, and J. M. Cordes (2005). A $2.1 M_\odot$ Pulsar Measured by Relativistic Orbital Decay. *Astrophys. J.* **634**, 1242–1249.

Nice, D. J., I. H. Stairs, and L. E. Kasian (2008). Masses of Neutron Stars in Binary Pulsar Systems. In C. Bassa, Z. Wang, A. Cumming, and V. M. Kaspi (Eds.), *40 Years of Pulsars: Millisecond Pulsars, Magnetars and More*, Volume 983 of *American Institute of Physics Conference Series*, pp. 453–458.

Nishida, M. T. (1986). Collisionless Boltzmann simulation of stellar disks having a small bulge. *Astrophys. J.* **302**, 611–616.

Norman, M. L. and K. H. Winkler (1983). Why ultrarelativistic numerical hydrodynamics is difficult. In M. L. Norman and K. H. Winkler (Eds.), *Radiation Hydrodynamics*. D. Reidel Publishing, Dordrecht.

Novikov, I. D. and K. S. Thorne (1973). Astrophysics of black holes. In C. DeWitt and B. S. DeWitt (Eds.), *Black holes*, pp. 343–450. Gordon and Breach, New York.

Nozawa, T., N. Stergioulas, E. Gourgoulhon, and Y. Eriguchi (1998). Construction of highly accurate models of rotating neutron stars – comparison of three different numerical schemes. *Astron. Astrophys. Suppl.* **132**, 431–454.

Ó Murchadha, N. and J. W. York (1974). Gravitational energy. *Phys. Rev. D* **10**, 2345–2357.

Obergaulinger, M., M. A. Aloy, and E. Müller (2006). Axisymmetric simulations of magneto-rotational core collapse: dynamics and gravitational wave signal. *Astron. Astrophys.* **450**, 1107–1134.

Oechslin, R. and H.-T. Janka (2006). Torus formation in neutron star mergers and well-localized short gamma-ray bursts. *Mon. Not. R. Astron. Soc.* **368**, 1489–1499.

Oechslin, R., H.-T. Janka, and A. Marek (2007). Relativistic neutron star merger simulations with non-zero temperature equations of state. I. Variation of binary parameters and equation of state. *Astron. Astrophys.* **467**, 395–409.

Oechslin, R., S. Rosswog, and F.-K. Thielemann (2002). Conformally flat smoothed particle hydrodynamics application to neutron star mergers. *Phys. Rev. D* **65**, 103005/1–15.

Olabarrieta, I. and M. W. Choptuik (2002). Critical phenomena at the threshold of black hole formation for collisionless matter in spherical symmetry. *Phys. Rev. D* **65**(2), 024007/1–10.

Oohara, K. and T. Nakamura (1999). 3D General Relativistic Simulations of Coalescing Binary Neutron Stars. *Progress of Theoretical Physics Supplement* **136**, 270–286.

Oohara, K., T. Nakamura, and M. Shibata (1997). A way to 3D numerical relativity. *Prog. Theor. Phys. Suppl.* **128**, 183–249.

Oppenheimer, J. R. and H. Snyder (1939). On Continued Gravitational Contraction. *Phys. Rev.* **56**, 455–459.

Oppenheimer, J. R. and G. M. Volkoff (1939). On Massive Neutron Cores. *Phys. Rev.* **55**, 374–381.

Ostriker, J. P. and P. Bodenheimer (1973). On the Oscillations and Stability of Rapidly Rotating Stellar Models. 111. Zero-Viscosity Polytropic Sequences. *Astrophys. J.* **180**, 171–180.

Ostriker, J. P. and J. E. Gunn (1969). On the Nature of Pulsars. I. Theory. *Astrophys. J.* **157**, 1395–1418.

Ott, C. D., H. Dimmelmeier, A. Marek, H. . Janka, I. Hawke, B. Zink, and E. Schnetter (2006). 3D Collapse of Rotating Stellar Iron Cores in General Relativity including Deleptonization and a Nuclear Equation of State. *Phys. Rev. Lett.* **98**, 261101/1–4.

Ou, S. and J. E. Tohline (2006). Unexpected Dynamical Instabilities in Differentially Rotating Neutron Stars. *Astrophys. J.* **651**, 1068–1078.

Paczynski, B. (1986). Gamma-ray bursters at cosmological distances. *Astrophys. J. Lett.* **308**, L43–L46.

Pandharipande, V. R. and D. G. Ravenhall (1989). Hot Nuclear Matter. In M. Soyeur, H. Flocard, B. Tamain, and M. Porneuf (Eds.), *NATO ASIB Proc. 205: Nuclear Matter and Heavy Ion Collisions*, pp. 103–132. D. Reidel Publishing, Dordrecht.

Pavlidou, V., K. Tassis, T. W. Baumgarte, and S. L. Shapiro (2000). Radiative falloff in neutron star spacetimes. *Phys. Rev. D* **62**, 084020/1–8.

Peacock, J. A. (1999). *Cosmological Physics*. Cambridge University Press, Cambridge.

Peebles, P. J. E. and B. Ratra (1988). Cosmology with a time-variable cosmological 'constant'. *Astrophys. J. Lett.* **325**, L17–L20.

Penrose, R. (1969). Gravitational collapse: The role of general relativity. *Nuovo Cimento* **1**, 252–276.

Peskin, M. E. and D. V. Schroeder (1995). *An Introduction to Quantum Field Theory*. Addison-Wesley, Reading, Mass.

Peters, P. C. (1964). Gravitational Radiation and the Motion of Two Point Masses. *Phys. Rev.* **136**, 1224–1232.

Petrich, L. I., S. L. Shapiro, and S. A. Teukolsky (1985). Oppenheimer-Snyder collapse with maximal time slicing and isotropic coordinates. *Phys. Rev. D* **31**, 2459–2469.

Petrich, L. I., S. L. Shapiro, and S. A. Teukolsky (1986). Oppenheimer-Snyder collapse in polar time slicing. *Phys. Rev. D* **33**, 2100–2110.

Pfeiffer, H. P., L. E. Kidder, M. A. Scheel, and S. A. Teukolsky (2003b). A multidomain spectral method for solving elliptic equations. *Comput. Phys. Commun.* **152**, 253–273.

Pfeiffer, H. P., S. A. Teukolsky, and G. B. Cook (2000). Quasicircular orbits for spinning black holes. *Phys. Rev. D* **62**, 104018/1–11.

Pfeiffer, H. P. and J. W. York, Jr. (2003a). Extrinsic curvature and the Einstein constraints. *Phys. Rev. D* **67**, 044022/1–8.

Pfeiffer, H. P. and J. W. York, Jr. (2005). Uniqueness and non-uniqueness in the Einstein constraints. *Phys. Rev. Lett.* **95**, 091101/1–4.

Pickett, B. K., R. H. Durisen, and G. A. Davis (1996). The Dynamic Stability of Rotating Protostars and Protostellar Disks. I. The Effects of the Angular Momentum Distribution. *Astrophys. J.* **458**, 714–738.

Piran, T. (1979). Gravitational waves and gravitational collapse in cylindrical systems. In L. L. Smarr (Ed.), *Sources of Gravitational Radiation*, pp. 409–422. Cambridge University Press, Cambridge.

Piran, T. (1999). Gamma-ray bursts and the fireball model. *Phys. Rept.* **314**, 575–667.

Piran, T. (2005). The physics of gamma-ray bursts. *Reviews of Modern Physics* **76**, 1143–1210.

Poisson, E. (2004). *A relativist's toolkit – the mathematics of black-hole mechanics*. Cambridge University Press, Cambridge.

Press, W. H., S. A. Teukolsky, W. T. Vetterling, and B. P. Flannery (2007). *Numerical recipes in C++: The art of scientific computing, Third edition*. Cambridge University Press, Cambridge.

Pretorius, F. (2005a). Evolution of binary black-hole spacetimes. *Phys. Rev. Lett.* **95**, 121101/1–4.

Pretorius, F. (2005b). Numerical relativity using a generalized harmonic decomposition. *Class. Quantum Grav.* **22**, 425–451.

Price, R. H. (1972a). Nonspherical perturbations of relativistic gravitational collapse. I. Scalar and gravitational perturbations. *Phys. Rev. D* **5**, 2419–2438.

Price, R. H. (1972b). Nonspherical perturbations of relativistic gravitational collapse. II. Integer-spin, zero-rest-mass fields. *Phys. Rev. D* **5**, 2439–2454.

Price, R. H. and J. Pullin (1994). Colliding black holes: The close limit. *Phys. Rev. Lett.* **72**, 3297–3300.

Rampp, M., E. Mueller, and M. Ruffert (1998). Simulations of non-axisymmetric rotational core collapse. *Astron. Astrophys.* **332**, 969–983.

Rantsiou, E., S. Kobayashi, P. Laguna, and F. Rasio (2007). Mergers of Black Hole – Neutron Star binaries. I. Methods and First Results. *Astrophys. J.* **680**, 1326–1349.

Rasio, F. A. and S. L. Shapiro (1992). Hydrodynamical evolution of coalescing binary neutron stars. *Astrophys. J.* **401**, 226–245.

Rasio, F. A. and S. L. Shapiro (1994). Hydrodynamics of binary coalescence. 1: Polytropes with stiff equations of state. *Astrophys. J.* **432**, 242–261.

Rasio, F. A. and S. L. Shapiro (1995). Hydrodynamics of binary coalescence. 2: Polytropes with gamma = 5/3. *Astrophys. J.* **438**, 887–903.

Rasio, F. A. and S. L. Shapiro (1999). Coalescing binary neutron stars. *Class. Quantum Grav.* **16**, R1–R29.

Rasio, F. A., S. L. Shapiro, and S. A. Teukolsky (1989a). On the existence of stable relativistic star clusters with arbitrarily large central redshifts. *Astrophys. J. Lett.* **336**, L63–L66.

Rasio, F. A., S. L. Shapiro, and S. A. Teukolsky (1989b). Solving the Vlasov equation in general relativity. *Astrophys. J.* **344**, 146–157.

Read, J. S., C. Markakis, M. Shibata, K. Uryū, J. D. E. Creighton, and J. L. Friedman (2009). Measuring the neutron star equation of state with gravitational wave observations. *Phys. Rev. D 79*(12), 124033/1–12.

Rees, M. J. (1984). Black Hole Models for Active Galactic Nuclei. *Ann. Rev. Astro. ApJ.* **22**, 471–506.

Rees, M. J. (1998). Astrophysical Evidence for Black Holes. In R. M. Wald (Ed.), *Black Holes and Relativistic Stars*, pp. 79–101. University of Chicago Press, Chicago.

Regge, T. and J. A. Wheeler (1957). Stability of a Schwarzschild Singularity. *Phys. Rev.* **108**, 1063–1069.

Reimann, B. and B. Brügmann (2004). Maximal slicing for puncture evolutions of Schwarzschild and Reissner-Nordström black holes. *Phys. Rev. D* **69**, 044006/1–21.

Reinhart, B. (1973). Maximal foliations of extended Schwarzschild space. *J. Math. Phys.* **14**, 719–719.

Reula, O. (1998). Hyperbolic methods of einstein's equations. *Living Rev. Relativity* **1**, 1–40.

Rezzolla, L. and J. C. Miller (1994). Relativistic radiative transfer for spherical flows. *Class. Quantum Grav.* **11**, 1815–1832.

Rezzolla, L. and J. C. Miller (1996). Evaporation of cosmological quark drops and relativistic radiative transfer. *Phys. Rev. D* **53**, 5411–5425.

Richtmyer, R. D. and K. W. Morton (1967). *Difference Methods for Initial Value Problems*. Wiley-Interscience, New York.

Rieth, R. and G. Schäfer (1996). Comment on a relativistic model for coalescing neutron star binaries. pp. 1–6.

Rinne, O. (2008a). Constrained evolution in axisymmetry and the gravitational collapse of prolate Brill waves. *Class. Quantum Grav.* **25**, 135009/1–21.

Rinne, O. (2008b). Explicit solution of the linearized Einstein equations in TT gauge for all multipoles. *Class. Quantum Grav.* **26**, 048003/1–9.

Roache, P. J. (1976). *Computational Fluid Dynamics*. Hermosa, Albuquerque.

Rosswog, S. (2005). Mergers of Neutron Star-Black Hole Binaries with Small Mass Ratios: Nucleosynthesis, Gamma-Ray Bursts, and Electromagnetic Transients. *Astrophys. J.* **634**, 1202–1213.

Rosswog, S., M. Liebendörfer, F.-K. Thielemann, M. B. Davies, W. Benz, and T. Piran (1999). Mass ejection in neutron star mergers. *Astron. ApJ.* **341**, 499–526.

Rosswog, S., F. K. Thielemann, M. B. Davies, W. Benz, and T. Piran (1998). Coalescing Neutron Stars: a Solution to the R-Process Problem? In W. Hillebrandt and E. Muller (Eds.), *Proceedings of the 9th workshop on Nuclear Astrophysics*, pp. 103–106. Max Planck Institut fur Astrophysik, Garching bei Munchen.

Ruffert, M. and H.-T. Janka (1998). Colliding neutron stars. Gravitational waves, neutrino emission, and gamma-ray bursts. *Astro. Astrophys.* **338**, 535–555.

Ruffert, M., H.-T. Janka, and G. Schäfer (1996). Coalescing neutron stars – a step towards physical models. I. Hydrodynamic evolution and gravitational-wave emission. *Astron. ApJ.* **311**, 532–566.

Ruffini, R. and S. Bonazzola (1969). Systems of self-gravitating particles in general relativity and the concept of an equation of state. *Phys. Rev.* **187**, 1767–1783.

Ruiz, M., M. Alcubierre, and D. Nunez (2008b). Regularization of spherical and axisymmetric evolution codes in numerical relativity. *Gen. Rel. Grav.* **40**, 159–182.

Ruiz, M., R. Takahashi, M. Alcubierre, and D. Núñez (2008a). Multipole expansions for energy and momenta carried by gravitational waves. *Gen. Rel. Grav.* **40**, 1705–1729.

Saijo, M., T. W. Baumgarte, and S. L. Shapiro (2003). One-armed Spiral Instability in Differentially Rotating Stars. *Astrophys. J.* **595**, 352–364.

Saijo, M., T. W. Baumgarte, S. L. Shapiro, and M. Shibata (2002). Collapse of a Rotating Supermassive Star to a Supermassive Black Hole: Post-Newtonian Simulations. *Astrophys. J.* **569**, 349–361.

Saijo, M., M. Shibata, T. W. Baumgarte, and S. L. Shapiro (2001). Dynamical Bar Instability in Rotating Stars: Effect of General Relativity. *Astrophys. J.* **548**, 919–931.

Salgado, M., S. Bonazzola, E. Gourgoulhon, and P. Haensel (1994a). High precision rotating netron star models 1: Analysis of neutron star properties. *Astron. Astrophys.* **291**, 155–170.

Salgado, M., S. Bonazzola, E. Gourgoulhon, and P. Haensel (1994b). High precision rotating neutron star models. II. Large sample of neutron star properties. *Astron. Astrophys. Suppl.* **108**, 455–459.

Sandberg, V. D. (1978). Tensor spherical harmonics on S^2 and S^3 as eigenvalue problems. *J. Math. Phys.* **19**, 2441–2446.

Sarbach, O., G. Calabrese, J. Pullin, and M. Tiglio (2002). Hyperbolicity of the BSSN system of Einstein evolution equations. *Phys. Rev. D* **66**, 064002/1–6.

Scheel, M. A., T. W. Baumgarte, G. B. Cook, S. L. Shapiro, and S. A. Teukolsky (1997). Numerical evolution of black holes with a hyperbolic formulation of general relativity. *Phys. Rev. D* **56**, 6320–6335.

Scheel, M. A., T. W. Baumgarte, G. B. Cook, S. L. Shapiro, and S. A. Teukolsky (1998). Treating instabilities in a hyperbolic formulation of Einstein's equations. *Phys. Rev. D* **58**, 044020/1–12.

Scheel, M. A., M. Boyle, T. Chu, L. E. Kidder, K. D. Matthews, and H. P. Pfeiffer (2009). High-accuracy waveforms for binary black hole inspiral, merger, and ringdown. *Phys. Rev. D* **79**, 024003/1–14.

Scheel, M. A., A. L. Erickcek, L. M. Burko, L. E. Kidder, H. P. Pfeiffer, and S. A. Teukolsky (2004). 3D simulations of linearized scalar fields in Kerr spacetime. *Phys. Rev. D* **69**, 104006/1–11.

Scheel, M. A., S. L. Shapiro, and S. A. Teukolsky (1995a). Collapse to black holes in Brans-Dicke theory. I. Horizon boundary conditions for dynamical spacetimes. *Phys. Rev. D* **51**, 4208–4235.

Scheel, M. A., S. L. Shapiro, and S. A. Teukolsky (1995b). Collapse to black holes in Brans-Dicke theory. II. Comparison with general relativity. *Phys. Rev. D* **51**, 4236–4249.

Schinder, P. J., S. A. Bludman, and T. Piran (1988). General-relativistic implicit hydrodynamics in polar-sliced space-time. *Phys. Rev. D* **37**, 2722–2731.

Schnetter, E. (2003). Finding apparent horizons and other 2-surfaces of constant expansion. *Class. Quantum Grav.* **20**, 4719–4737.

Schnetter, E., S. H. Hawley, and I. Hawke (2004). Evolutions in 3d numerical relativity using fixed mesh refinement. *Class. Quantum Grav.* **21**, 1465–1488.

Schnetter, E., B. Krishnan, and F. Beyer (2006). Introduction to dynamical horizons in numerical relativity. *Phys. Rev. D* **74**, 024028/1–20.

Schnittman, J. D. and A. Buonanno (2007). The distribution of recoil velocities from merging black holes. *Astrophys. J. Lett.* **662**, L63–L66.

Schreier, E., R. Levinson, H. Gursky, E. Kellogg, H. Tananbaum, and R. Giacconi (1972). Evidence for the Binary Nature of Centaurus X-3 from UHURU X-Ray Observations. *Astrophys. J. Lett.* **172**, L79–L89.

Schunck, F. E. and E. W. Mielke (2003). TOPICAL REVIEW: General relativistic boson stars. *Class. Quantum Grav.* **20**, R301–R356.

Schutz, B. F. (1980). *Geometrical Methods of Mathematical Physics*. Cambridge University Press, Cambridge.

Schutz, Jr., B. F. (1972). Linear Pulsations and Stability of Differentially Rotating Stellar Models. II. General-Relativistic Analysis. *Astrophys. J. Suppl.* **24**, 343–374.

Schwarzschild, K. (1916). Über das Gravitationsfeld eines Massenpunktes nach der Einsteinschen Theorie. *Sitzber. Deut. Akad. Wiss. Berlin*, 189–196.

Seidel, E. and W.-M. Suen (1990). Dynamical evolution of boson stars: Perturbing the ground state. *Phys. Rev. D* **42**, 384–403.

Seidel, E. and W.-M. Suen (1991). Oscillating soliton stars. *Phys. Rev. Lett.* **66**, 1659–1662.

Seidel, E. and W.-M. Suen (1992). Towards a singularity-proof scheme in numerical relativity. *Phys. Rev. Lett.* **69**, 1845–1848.

Sellwood, J. A. (1987). The art of N-body building. *Ann. Rev. Astro. Ap.* **25**, 151–186.

Sesana, A., M. Volonteri, and F. Haardt (2007). The imprint of massive black hole formation models on the LISA data stream. *Mon. Not. R. Astron. Soc.* **377**, 1711–1716.

Setiawan, S., M. Ruffert, and H.-T. Janka (2006). Three-dimensional simulations of non-stationary accretion by remnant black holes of compact object mergers. *Astron. Astrophys.* **458**, 553–567.

Shapiro, S. L. (1985). Monte Carlo simulations of the $2 + 1$ dimensional Fokker-Planck equation – Spherical star clusters containing massive, central black holes. In J. Goodman and P. Hut (Eds.), *Dynamics of Star Clusters*, Volume 113 of *IAU Symposiumi 113*, pp. 373–412. D. Reidel Publishing, Dordrecht.

Shapiro, S. L. (1989). Thermal radiation from stellar collapse to a black hole. *Phys. Rev. D* **40**, 1858–1867.

Shapiro, S. L. (1996). Radiation from Stellar Collapse to a Black Hole. *Astrophys. J.* **472**, 308–326.

Shapiro, S. L. (1998). Binary-induced gravitational collapse: A trivial example. *Phys. Rev. D* **57**, 908–913.

Shapiro, S. L. (2000). Differential Rotation in Neutron Stars: Magnetic Braking and Viscous Damping. *Astrophys. J.* **544**, 397–408.

Shapiro, S. L. (2004a). Collapse of Uniformly Rotating Stars to Black Holes and the Formation of Disks. *Astrophys. J.* **610**, 913–919.

Shapiro, S. L. (2004b). Formation of Supermassive Black Holes: Simulations in General Relativity. In L. C. Ho (Ed.), *Coevolution of Black Holes and Galaxies*, pp. 103–121.

Shapiro, S. L. (2005). Spin, Accretion, and the Cosmological Growth of Supermassive Black Holes. *Astrophys. J.* **620**, 59–68.

Shapiro, S. L. and M. Shibata (2002). Collapse of a Rotating Supermassive Star to a Supermassive Black Hole: Analytic Determination of the Black Hole Mass and Spin. *Astrophys. J.* **577**, 904–908.

Shapiro, S. L. and S. A. Teukolsky (1979). Gravitational collapse of supermassive stars to black holes – Numerical solution of the Einstein equations. *Astrophys. J. Lett.* **234**, L177–L181.

Shapiro, S. L. and S. A. Teukolsky (1980, January). Gravitational collapse to neutron stars and black holes – Computer generation of spherical spacetimes. *Astrophys. J.* **235**, 199–215.

Shapiro, S. L. and S. A. Teukolsky (1983). *Black Holes, White Dwarfs, and Neutron Stars: the physics of compact objects*. Wiley Interscience, New York.

Shapiro, S. L. and S. A. Teukolsky (1985a). Relativistic Stellar Dynamics on the Computer – Part Two – Physical Applications. *Astrophys. J.* **298**, 58–79.

Shapiro, S. L. and S. A. Teukolsky (1985b). Relativistic stellar dynamics on the computer. I – Motivation and numerical method. *Astrophys. J.* **298**, 34–57.

Shapiro, S. L. and S. A. Teukolsky (1985c). The collapse of dense star clusters to supermassive black holes – The origin of quasars and AGNs. *Astrophys. J. Lett.* **292**, L41–L44.

Shapiro, S. L. and S. A. Teukolsky (1986). Relativistic stellar dynamics on the computer. IV – Collapse of a star cluster to a black hole. *Astrophys. J.* **307**, 575–592.

Shapiro, S. L. and S. A. Teukolsky (1987). Simulations of axisymmetric, Newtonian star clusters – Prelude to 2 + 1 general relativistic computations. *Astrophys. J.* **318**, 542–567.

Shapiro, S. L. and S. A. Teukolsky (1988). Building Black Holes: Supercomputer Cinema. *Science* **241**, 421–425.

Shapiro, S. L. and S. A. Teukolsky (1991a). Black Holes, Naked Singularities and Cosmic Censorship. *American Scientist* **79**, 330–343.

Shapiro, S. L. and S. A. Teukolsky (1991b). Formation of naked singularities – The violation of cosmic censorship. *Phys. Rev. Lett.* **66**, 994–997.

Shapiro, S. L. and S. A. Teukolsky (1992a). Black Holes, Star Clusters, and Naked Singularities: Numerical Solution of Einstein's Equations. *Royal Society of London Philosophical Transactions Series A* **340**, 365–390.

Shapiro, S. L. and S. A. Teukolsky (1992b). Collisions of relativistic clusters and the formation of black holes. *Phys. Rev. D* **45**, 2739–2750.

Shapiro, S. L. and S. A. Teukolsky (1993a). Relativistic Stellar Systems with Rotation. *Astrophys. J.* **419**, 636–647.

Shapiro, S. L. and S. A. Teukolsky (1993b). Relativistic Stellar Systems with Spindle Singularities. *Astrophys. J.* **419**, 622–635.

Shapiro, S. L., S. A. Teukolsky, and J. Winicour (1995). Toroidal black holes and topological censorship. *Phys. Rev. D* **52**, 6982–6987.

Shapiro, S. L. and S. Zane (1998). Bar Mode Instability in Relativistic Rotating Stars: A Post-Newtonian Treatment. *Astrophys. J. Suppl.* **117**, 531–561.

Shen, H., H. Toki, K. Oyamatsu, and K. Sumiyoshi (1998a). Relativistic equation of state of nuclear matter for supernova and neutron star. *Nuclear Physics A* **637**, 435–450.

Shen, H., H. Toki, K. Oyamatsu, and K. Sumiyoshi (1998b). Relativistic Equation of State of Nuclear Matter for Supernova Explosion. *Progress of Theoretical Physics* **100**, 1013–1031.

Shibata, M. (1997). Apparent horizon finder for a special family of spacetimes in 3d numerical relativity. *Phys. Rev. D* **55**, 2002–2013.

Shibata, M. (1998). Relativistic formalism for computation of irrotational binary stars in quasiequilibrium states. *Phys. Rev. D* **58**, 024012/1–5.

Shibata, M. (1999a). Fully general relativistic simulation of coalescing binary neutron stars: Preparatory tests. *Phys. Rev. D 60*(10), 104052/1–25.

Shibata, M. (1999b). Fully general relativistic simulation of merging binary clusters: Spatial gauge condition. *Prog. Theor. Phys.* **101**, 1199–1233.

Shibata, M. (2000). Axisymmetric Simulations of Rotating Stellar Collapse in Full General Relativity —Criteria for Prompt Collapse to Black Holes—. *Progress of Theoretical Physics* **104**, 325–358.

Shibata, M. (2004). Stability of Rigidly Rotating Relativistic Stars with Soft Equations of State against Gravitational Collapse. *Astrophys. J.* **605**, 350–359.

Shibata, M. (2005). Constraining Nuclear Equations of State Using Gravitational Waves from Hypermassive Neutron Stars. *Phys. Rev. Lett.* **94**, 201101/1–4.

Shibata, M., T. W. Baumgarte, and S. L. Shapiro (1998). Stability of coalescing binary stars against gravitational collapse: Hydrodynamical simulations. *Phys. Rev. D* **58**, 023002/1–11.

Shibata, M., T. W. Baumgarte, and S. L. Shapiro (2000a). Stability and collapse of rapidly rotating, supramassive neutron stars: 3D simulations in general relativity. *Phys. Rev. D 61*(4), 044012/1–11.

Shibata, M., T. W. Baumgarte, and S. L. Shapiro (2000b). The Bar-Mode Instability in Differentially Rotating Neutron Stars: Simulations in Full General Relativity. *Astrophys. J.* **542**, 453–463.

Shibata, M., M. D. Duez, Y. T. Liu, S. L. Shapiro, and B. C. Stephens (2006a). Magnetized Hypermassive Neutron-Star Collapse: A Central Engine for Short Gamma-Ray Bursts. *Phys. Rev. Lett.* **96**, 031102/1–4.

Shibata, M., S. Karino, and Y. Eriguchi (2003a). Dynamical bar-mode instability of differentially rotating stars: effects of equations of state and velocity profiles. *Mon. Not. R. Astron. Soc.* **343**, 619–626.

Shibata, M., Y. T. Liu, S. L. Shapiro, and B. C. Stephens (2006b). Magnetorotational collapse of massive stellar cores to neutron stars: Simulations in full general relativity. *Phys. Rev. D* **74**, 104026/1–28.

Shibata, M. and T. Nakamura (1995). Evolution of three-dimensional gravitational waves: Harmonic slicing case. *Phys. Rev. D* **52**, 5428–5444.

Shibata, M. and Y.-I. Sekiguchi (2004). Gravitational waves from axisymmetric rotating stellar core collapse to a neutron star in full general relativity. *Phys. Rev. D* **69**, 084024/1–16.

Shibata, M. and Y.-I. Sekiguchi (2005). Magnetohydrodynamics in full general relativity: Formulation and tests. *Phys. Rev. D* **72**, 044014/1–24.

Shibata, M. and S. L. Shapiro (2002). Collapse of a Rotating Supermassive Star to a Supermassive Black Hole: Fully Relativistic Simulations. *Astrophys. J. Lett.* **572**, L39–L43.

Shibata, M. and K. Taniguchi (2006). Merger of binary neutron stars to a black hole: Disk mass, short gamma-ray bursts, and quasinormal mode ringing. *Phys. Rev. D* **73**, 064027/1–29.

Shibata, M. and K. Taniguchi (2007). Merger of black hole and neutron star in general relativity: Tidal disruption, torus mass, and gravitational waves. *Phys. Rev. D* **77**, 084015/1–22.

Shibata, M., K. Taniguchi, and K. Uryū (2003b). Merger of binary neutron stars of unequal mass in full general relativity. *Phys. Rev. D* **68**, 084020/1–24.

Shibata, M., K. Taniguchi, and K. Uryū (2005). Merger of binary neutron stars with realistic equations of state in full general relativity. *Phys. Rev. D* **71**, 084021/1–26.

Shibata, M. and K. Uryū (2000). Simulation of merging binary neutron stars in full general relativity: $\Gamma=2$ case. *Phys. Rev. D* **61**, 064001/1–18.

Shibata, M. and K. Uryū (2001). Computation of gravitational waves from inspiraling binary neutron stars in quasiequilibrium circular orbits: Formulation and calibration. *Phys. Rev. D* **64**, 104017/1–26.

Shibata, M. and K. Uryū (2002). Gravitational Waves from Merger of Binary Neutron Stars in Fully General Relativistic Simulation. *Progress of Theoretical Physics* **107**, 265–303.

Shibata, M. and K. Uryū (2006). Merger of black hole-neutron star binaries: Nonspinning black hole case. *Phys. Rev. D* **74**, 121503/1–5.

Shibata, M. and K. Uryū (2007). Merger of black hole neutron star binaries in full general relativity. *Class. Quantum Grav.* **24**, S125–S137.

Shibata, M. and K. Uryū (2000). Apparent horizon finder for general three-dimensional spaces. *Phys. Rev. D* **62**, 087501/1–4.

Shibata, M., K. Uryū, and J. L. Friedman (2004). Deriving formulations for numerical computation of binary neutron stars in quasicircular orbits. *Phys. Rev. D* **70**, 044044/1–18.

Shoemaker, D. M., M. F. Huq, and R. A. Matzner (2000). Generic tracking of multiple apparent horizons with level flow. *Phys. Rev. D* **62**, 124005/1–12.

Siegler, S. and H. Riffert (2000). Smoothed Particle Hydrodynamics Simulations of Ultrarelativistic Shocks with Artificial Viscosity. *Astrophys. J.* **531**, 1053–1066.

Sigurdsson, S. (2003). The loss cone: past, present and future. *Class. Quantum Grav.* **20**, S45–S54.

Skoge, M. L. and T. W. Baumgarte (2002). Comparing criteria for circular orbits in general relativity. *Phys. Rev. D* **66**, 107501/1–4.

Sloan, J. H. and L. L. Smarr (1985). General Relativistic Magnetohydrodynamics. In J. M. Centrella, J. M. Leblanc, and R. L. Bowers (Eds.), *Numerical Astrophysics*, pp. 52–68. Jones and Bartlett, Boston.

Smarr, L. and J. W. York, Jr. (1978a). Kinematical conditions in the construction of spacetime. *Phys. Rev. D* **17**, 2529–2551.

Smarr, L. and J. W. York, Jr. (1978b). Radiation gauge in general relativity. *Phys. Rev. D* **17**, 1945–1956.

Smarr, L. L. (1979a). Basic concepts in finite differencing of partial differential equations. In L. L. Smarr (Ed.), *Sources of gravitational radiation*, pp. 139–159. Cambridge University Press, Cambridge.

Smarr, L. L. (1979b). Gauge conditions, radiation formulae, and the two black hole collision. In L. L. Smarr (Ed.), *Sources of gravitational radiation*, pp. 245–274. Cambridge University Press, Cambridge.

Smith, S. C., J. L. Houser, and J. M. Centrella (1996). Simulations of Nonaxisymmetric Instability in a Rotating Star: A Comparison between Eulerian and Smooth Particle Hydrodynamics. *Astrophys. J.* **458**, 236–256.

Smith, T. L., E. Pierpaoli, and M. Kamionkowski (2006). New Cosmic Microwave Background Constraint to Primordial Gravitational Waves. *Phys. Rev. Lett.* **97**, 021301/1–4.

Sopuerta, C. F., N. Yunes, and P. Laguna (2007). Gravitational Recoil Velocities from Eccentric Binary Black Hole Mergers. *Astrophys. J. Lett.* **656**, L9–L12.

Spergel, D. N., R. Bean, O. Doré, M. R. Nolta, C. L. Bennett, J. Dunkley, G. Hinshaw, N. Jarosik, E. Komatsu, L. Page, H. V. Peiris, L. Verde, M. Halpern, R. S. Hill, A. Kogut, M. Limon, S. S. Meyer, N. Odegard, G. S. Tucker, J. L. Weiland, E. Wollack, and E. L. Wright (2007). Wilkinson Microwave Anisotropy Probe (WMAP) Three Year Results: Implications for Cosmology. *Astrophys. J. Suppl.* **170**, 377–408.

Spitzer, L. (1987). *Dynamical evolution of globular clusters*. Princeton University Press, Princeton.

Stairs, I. H. (2004). Pulsars in Binary Systems: Probing Binary Stellar Evolution and General Relativity. *Science* **304**, 547–552.

Stairs, I. H., Z. Arzoumanian, F. Camilo, A. G. Lyne, D. J. Nice, J. H. Taylor, S. E. Thorsett, and A. Wolszczan (1998). Measurement of Relativistic Orbital Decay in the PSR B1534+12 Binary System. *Astrophys. J.* **505**, 352–357.

Stark, R. F. and T. Piran (1985). Gravitational-wave emission from rotating gravitational collapse. *Phys. Rev. Lett.* **55**, 891–894.

Stark, R. F. and T. Piran (1987). A General Relativistic Code for Rotating Axisymmetric Configurations and Gravitational Radiation: Numerical Methods and Tests. *Comput. Phys. Rep.* **5**, 221–264.

Stephens, B. C., S. L. Shapiro, and Y. T. Liu (2008). Collapse of magnetized hypermassive neutron stars in general relativity: Disk evolution and outflows. *Phys. Rev. D* **77**, 044001/1–18.

Stergioulas, N. (2003). Rotating Stars in Relativity. *Living Rev. Relativity* **6**, 1–109.

Stergioulas, N. and J. L. Friedman (1995). Comparing models of rapidly rotating relativistic stars constructed by two numerical methods. *Astrophys. J.* **444**, 306–311.

Stergioulas, N. and J. L. Friedman (1998). Nonaxisymmetric Neutral Modes in Rotating Relativistic Stars. *Astrophys. J.* **492**, 301–322.

Symbalisty, E. and D. N. Schramm (1982). Neutron star collisions and the r-process. *Astrophys. J. Lett.* **22**, 143–145.

Szekeres, G. (1960). On the Singularities of a Riemannian Manifold. *Publ. Mat. Debrecen.* **7**, 285–301.

Tananbaum, H., H. Gursky, E. M. Kellogg, R. Levinson, E. Schreier, and R. Giacconi (1972). Discovery of a Periodic Pulsating Binary X-Ray Source in Hercules from UHURU. *Astrophys. J. Lett.* **174**, L143–L149.

Taniguchi, K., T. W. Baumgarte, J. A. Faber, and S. L. Shapiro (2005). Black hole-neutron star binaries in general relativity: effects of neutron star spin. *Phys. Rev. D* **72**, 044008/1–26.

Taniguchi, K., T. W. Baumgarte, J. A. Faber, and S. L. Shapiro (2006). Quasiequilibrium sequences of black-hole neutron-star binaries in general relativity. *Phys. Rev. D* **74**, 041502/1–5.

Taniguchi, K., T. W. Baumgarte, J. A. Faber, and S. L. Shapiro (2007). Quasiequilibrium black hole-neutron star binaries in general relativity. *Phys. Rev. D* **75**, 084005/1–17.

Taniguchi, K., T. W. Baumgarte, J. A. Faber, and S. L. Shapiro (2008). Relativistic black hole-neutron star binaries in quasiequilibrium: effects of the black hole excision boundary condition. *Phys. Rev. D* **77**, 044003/1–13.

Taniguchi, K. and E. Gourgoulhon (2002). Quasiequilibrium sequences of synchronized and irrotational binary neutron stars in general relativity. III. Identical and different mass stars with gamma=2. *Phys. Rev. D* **66**, 104019/1–14.

Taniguchi, K. and E. Gourgoulhon (2003). Various features of quasiequilibrium sequences of binary neutron stars in general relativity. *Phys. Rev. D* **68**, 124025/1–28.

Taniguchi, K. and T. Nakamura (2000a). Almost Analytic Solutions to Equilibrium Sequences of Irrotational Binary Polytropic Stars for n = 1. *Phys. Rev. Lett.* **84**, 581–585.

Taniguchi, K. and T. Nakamura (2000b). Equilibrium sequences of irrotational binary polytropic stars: The case of double polytropic stars. *Phys. Rev. D* **62**, 044040/1–45.

Tassoul, J.-L. (1978). *Theory of rotating stars*. Princeton University Press, Princeton.

Taub, A. H. (1948). Relativistic Rankine-Hugoniot Equations. *Phys. Rev.* **74**, 328–334.

Taub, A. H. (1959). On circulation in relativistic hydrodynamics. *Archive for Rational Mechanics and Analysis* **3**, 312–324.

Taub, A. H. (1978). Relativistic fluid mechanics. *Annual Review of Fluid Mechanics* **10**, 301–332.

Taylor, J. H. and J. M. Weisberg (1989). Further experimental tests of relativistic gravity using the binary pulsar PSR 1913 + 16. *Astrophys. J.* **345**, 434–450.

Teukolsky, S. A. (1982). Linearized quadrupole waves in general relativity and the motion of test particle. *Phys. Rev. D* **26**, 745–750.

Teukolsky, S. A. (1998). Irrotational binary neutron stars in quasi-equilibrium in general relativity. *Astrophys. J.* **504**, 442–449.

Teukolsky, S. A. (2000). On the stability of the iterated Crank-Nicholson method in numerical relativity. *Phys. Rev. D* **61**, 087501/1–2.

The LIGO Scientific Collaboration: B. Abbott (2007). Searching for a Stochastic Background of Gravitational Waves with the Laser Interferometer Gravitational-Wave Observatory. *Astrophys. J.* **659**, 918–930.

The LIGO Scientific Collaboration: B. Abbott, M. Kramer, and A. G. Lyne (2007). Upper limits on gravitational wave emission from 78 radio pulsars. *Phys. Rev. D* **76**, 042001/1–20.

Thornburg, J. (1987). Coordinates and boundary conditions for the general relativistic initial data problem. *Class. Quantum Grav.* **4**, 1119–1131.

Thornburg, J. (1996). Finding apparent horizons in numerical relativity. *Phys. Rev. D* **54**, 4899–4918.

Thorne, K. S. (1972). Nonspherical Gravitational Collapse–A Short Review. In J. Klauder (Ed.), *Magic Without Magic: John Archibald Wheeler*, pp. 231–258. W.H. Freeman, San Francisco.

Thorne, K. S. (1974). Disk-Accretion onto a Black Hole. II. Evolution of the Hole. *Astrophys. J.* **191**, 507–520.

Thorne, K. S. (1980). Multipole expansions of gravitational radiation. *Reviews of Modern Physics* **52**, 299–340.

Thorne, K. S. (1981). Relativistic radiative transfer – Moment formalisms. *Mon. Not. R. Astron. Soc.* **194**, 439–473.

Thorne, K. S. and A. Campolattaro (1967). Non-Radial Pulsation of General-Relativistic Stellar Models. I. Analytic Analysis for $L \geq 2$. *Astrophys. J.* **149**, 591–612.

Thorne, K. S., R. A. Flammang, and A. N. Zytkow (1981). Stationary spherical accretion into black holes. I – Equations of structure. *Mon. Not. R. Astron. Soc.* **194**, 475–484.

Thorne, K. S. and D. MacDonald (1982). Electrodynamics in Curved Spacetime – 3+1 Formulation. *Mon. Not. R. Astron. Soc.* **198**, 339–343.

Tichy, W. and B. Brügmann (2004). Quasi-equilibrium binary black hole sequences for puncture data derived from helical Killing vector conditions. *Phys. Rev. D* **69**, 024006/1–7.

Tichy, W., B. Brügmann, M. Campanelli, and P. Diener (2003). Binary black hole initial data for numerical general relativity based on post-Newtonian data. *Phys. Rev. D* **67**, 064008/1–13.

Tichy, W., B. Brügmann, and P. Laguna (2003). Gauge conditions for binary black hole puncture data based on an approximate helical Killing vector. *Phys. Rev. D* **68**, 064008/1–11.

Tod, K. P. (1991). Looking for marginally trapped surfaces. *Class. Quantum Grav.* **8**, L115–L118.

Tohline, J. E., R. H. Durisen, and M. McCollough (1985). The linear and nonlinear dynamic stability of rotating $N = 3/2$ polytropes. *Astrophys. J.* **298**, 220–234.

Tohline, J. E. and I. Hachisu (1990). The breakup of self-gravitating rings, tori, and thick accretion disks. *Astrophys. J.* **361**, 394–407.

Tolman, R. C. (1939). Static solutions of Einstein's field equations for spheres of fluid. *Phys. Rev.* **55**, 364–373.

Toro, E. F. (1999). *Riemann Solvers and Numerical Methods for Fluid Dynamics: A Practical Introduction*. Springer-Verlag, Berlin.

Tóth, G. (2000). The div $B = 0$ constraint in shock-capturing magnetohydrodynamics codes. *J. Comp. Phys.* **161**, 605–652.

Uryū, K., M. Shibata, and Y. Eriguchi (2000). Properties of general relativistic, irrotational binary neutron stars in close quasiequilibrium orbits: Polytropic equations of state. *Phys. Rev. D* **62**(10), 104015/1–15.

Uryū, K. and Y. Eriguchi (1999). Newtonian models for black hole-gaseous star close binary systems. *Mon. Not. R. Astron. Soc.* **303**, 329–342.

Uryū, K. and Y. Eriguchi (2000). A new numerical method for constructing quasi-equilibrium sequences of irrotational binary neutron stars in general relativity. *Phys. Rev. D* **61**, 124023/1–19.

Uryū, K., F. Limousin, J. L. Friedman, E. Gourgoulhon, and M. Shibata (2006). Binary neutron stars in a waveless approximation. *Phys. Rev. Lett.* **97**, 171101/1–4.

Usui, F. and Y. Eriguchi (2002). Quasiequilibrium sequences of synchronously rotating binary neutron stars with constant rest masses in general relativity: Another approach without using the conformally flat condition. *Phys. Rev. D* **65**, 064030/1–9.

Usui, F., K. Uryū, and Y. Eriguchi (2000). New numerical scheme to compute three-dimensional configurations of quasiequilibrium compact stars in general relativity: Application to synchronously rotating binary star systems. *Phys. Rev. D* **61**, 024039/1–23.

van der Marel, R. P. (2004). Intermediate-mass Black Holes in the Universe: A Review of Formation Theories and Observational Constraints. In L. C. Ho (Ed.), *Coevolution of Black Holes and Galaxies*, pp. 37–52.

van Leer, B. (1977). Towards the ultimate conservative difference scheme: IV. A new approach to numerical convection. *Journal of Computational Physics* **23**, 276–299.

van Meter, J. R., J. G. Baker, M. Koppitz, and D.-I. Choi (2006). How to move a black hole without excision: gauge conditions for the numerical evolution of a moving puncture. *Phys. Rev. D* **73**, 124011/1–8.

van Riper, K. A. (1979). General relativistic hydrodynamics and the adiabatic collapse of stellar cores. *Astrophys. J.* **232**, 558–571.

Velikhov, E. P. (1959). Stability of an ideally conducting liquid flowing between cylinders rotating in a magnetic field. *Sov. Phys.- JETP* **36**, 1398–1404.

Volonteri, M. (2007). Gravitational Recoil: Signatures on the Massive Black Hole Population. *Astrophys. J. Lett.* **663**, L5–L8.

Volonteri, M. and M. J. Rees (2006). Quasars at z=6: The Survival of the Fittest. *Astrophys. J.* **650**, 669–678.

von Neumann, J. and R. D. Richtmeyer (1950). A method for the numerical calculations of hydrodynamical shocks. *J. Appl. Phys.* **21**, 232–237.

Wald, R. M. (1984). *General Relativity*. The University of Chicago Press, Chicago.

Wald, R. M. and V. Iyer (1991). Trapped surfaces in the Schwarzschild geometry and cosmic censorship. *Phys. Rev. D* **44**, R3719–R3722.

Walsh, D. M. (2007). Non-uniqueness in conformal formulations of the Einstein constraints. *Class. Quantum Grav.* **24**, 1911–1925.

Wasserman, I. and S. L. Shapiro (1983). Masses, radii, and magnetic fields of pulsating X-ray sources – Is the 'standard' model self-consistent. *Astrophys. J.* **265**, 1036–1046.

Weinberg, S. (1972). *Gravitation and Cosmology: Principles and Applications of the General Theory of Relativity*. Wiley, New York.

Wilson, J. R. (1972a). Models of Differentially Rotating Stars. *Astrophys. J.* **176**, 195–204.

Wilson, J. R. (1972b). Numerical Study of Fluid Flow in a Kerr Space. *Astrophys. J.* **173**, 431–438.

Wilson, J. R. (1975). Some magnetic effects in stellar collapse and accretion. *New York Academy Sciences Annals* **262**, 123–132.

Wilson, J. R. (1979). A numerical method for relativistic hydrodynamics. In L. L. Smarr (Ed.), *Sources of Gravitational Radiation*, pp. 423–445. Cambridge University Press, Cambridge.

Wilson, J. R. and G. J. Mathews (1989). *Relativistic Hydrodynamics*, pp. 306–314. Cambridge University Press, Cambridge.

Wilson, J. R. and G. J. Mathews (1995). Instabilities in Close Neutron Star Binaries. *Phys. Rev. Lett.* **75**, 4161–4164.

Wilson, J. R., G. J. Mathews, and P. Marronetti (1996). Relativistic numerical model for close neutron-star binaries. *Phys. Rev. D* **54**, 1317–1331.

Winkler, K.-H. A. and M. L. Norman (1986). Astrophysical Radiation Hydrodynamics. In K.-H. A. Winkler and M. L. Norman (Eds.), *NATO Advanced Research Workshop on Astrophysical Radiation Hydrodynamics*. Kluwer Academic Publishers, Norwell, Massachusetts.

Yamamoto, T., M. Shibata, and K. Taniguchi (2008). Simulating coalescing compact binaries by a new code (SACRA). *Phys. Rev. D* **78**, 064054/1–38.

Yo, H.-J., T. W. Baumgarte, and S. L. Shapiro (2002). Improved numerical stability of stationary black hole evolution calculations. *Phys. Rev. D* **66**, 084026/1–15.

Yo, H.-J., J. N. Cook, S. L. Shapiro, and T. W. Baumgarte (2004). Quasi-equilibrium binary black hole initial data for dynamical evolutions. *Phys. Rev. D* **70**, 084033/1–14.

Yokosawa, M. (1993). Energy and angular momentum transport in magnetohydrodynamical accretion onto a rotating black hole. *Publ. Astron. Soc. Japan* **45**, 207–218.

Yoneda, G. and H. Shinkai (2001). Constraint propagation in the family of ADM systems. *Phys. Rev. D* **63**, 124019/1–9.

Yoneda, G. and H. Shinkai (2002). Advantages of modified ADM formulation: constraint propagation analysis of Baumgarte-Shapiro-Shibata-Nakamura system. *Phys. Rev. D* **66**, 124003/1–10.

York, Jr., J. W. (1989). Intial Data for Collisions of Black Holes and other Gravitational Miscellany. In C. E. Evans, L. S. Finn, and D. W. Hobill (Eds.), *Frontiers in Numerical Relativity*, pp. 89–109. Cambridge University Press, Cambridge.

York, Jr., J. W. (1971). Gravitational degrees of freedom and the initial-value problem. *Phys. Rev. Lett.* **26**, 1656–1658.

York, Jr., J. W. (1979). Kinematics and dynamics of general relativity. In L. L. Smarr (Ed.), *Sources of gravitational radiation*, pp. 83–126. Cambridge University Press, Cambridge.

York, Jr., J. W. (1999). Conformal 'thin-sandwich' data for the initial-value problem of general relativity. *Phys. Rev. Lett.* **82**, 1350–1353.

Yuan, Y.-F., R. Narayan, and M. J. Rees (2004). Constraining Alternate Models of Black Holes: Type I X-Ray Bursts on Accreting Fermion-Fermion and Boson-Fermion Stars. *Astrophys. J.* **606**, 1112–1124.

Zampieri, L., J. C. Miller, and R. Turolla (1996). Time-dependent analysis of spherical accretion on to black holes. *Mon. Not. R. Astron. Soc.* **281**, 1183–1196.

Zel'dovich, Y. B. (1967). Cosmological Constant and Elementary Particles. *Zh. Eksp. & Teor. Fiz. Pis'ma* **6**, 883–884. English translation in *Sov. Phys.-JETP Lett.*, **6**, 316–317.

Zeldovich, Y. B. and I. D. Novikov (1971). *Relativistic astrophysics. Vol.1: Stars and relativity*. University of Chicago Press, Chicago.

Zel'dovich, Y. B. and M. A. Podurets (1965). The Evolution of a System of Gravitationally Interacting Point Masses. *Astron. Zh.* **42**, 963–973. English translation in *Sov. Astron.-A. J.*, **9**, 742–749 (1966).

Zerilli, F. J. (1970). Effective Potential for Even-Parity Regge-Wheeler Gravitational Perturbation Equations. *Phys. Rev. Lett.* **24**, 737–738.

Zhuge, X., J. M. Centrella, and S. L. W. McMillan (1994). Gravitational radiation from coalescing binary neutron stars. *Phys. Rev. D* **50**, 6247–6261.

Zhuge, X., J. M. Centrella, and S. L. W. McMillan (1996). Gravitational radiation from the coalescence of binary neutron stars: Effects due to the equation of state, spin, and mass ratio. *Phys. Rev. D* **54**, 7261–7277.

Index

Printed in the United States
by Baker & Taylor Publisher Services